083760
Aberdeen College Library

PLANTS

Their Use, Management, Cultivation and Biology

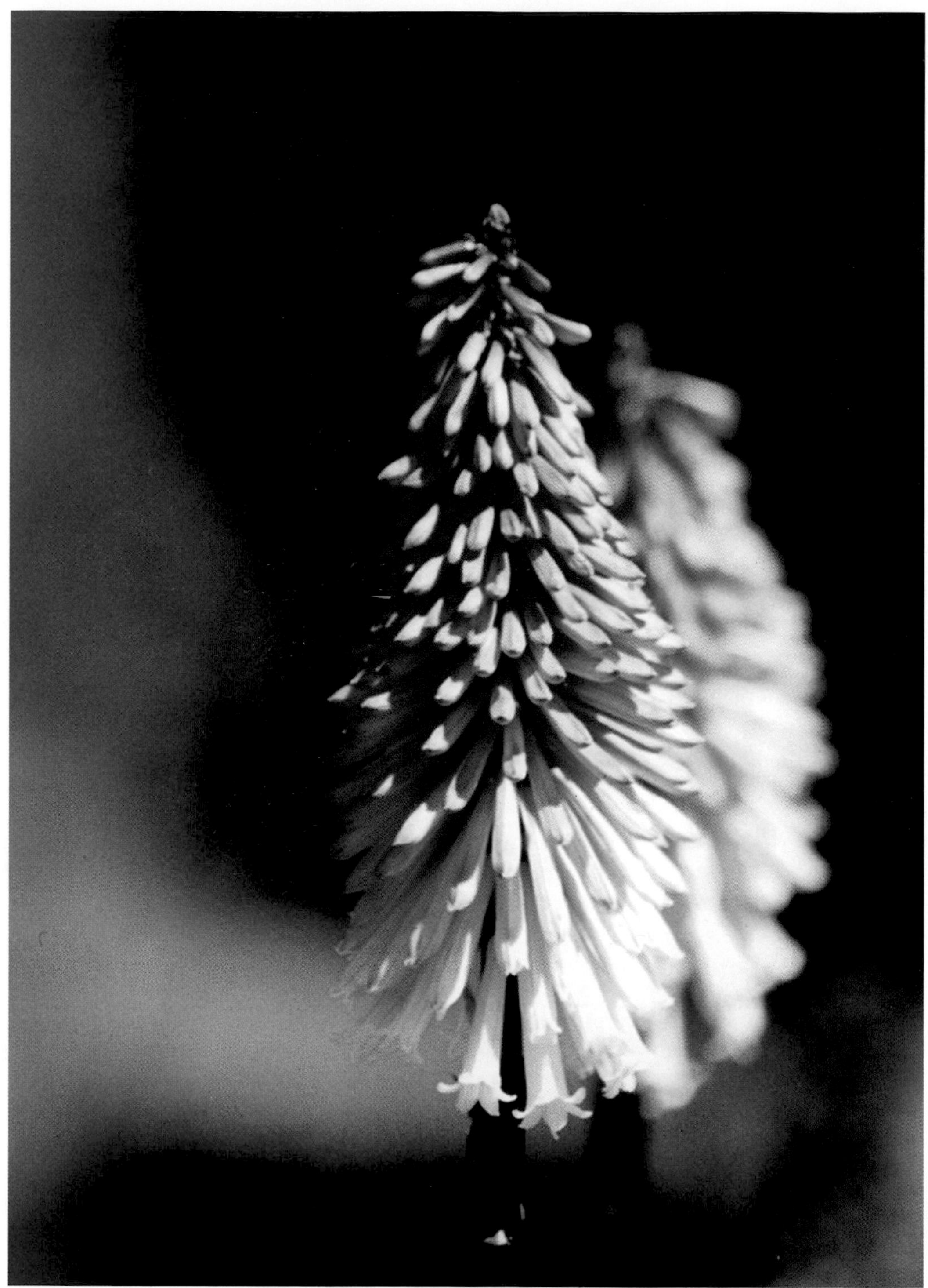

Red Hot Poker (Kniphofia).

PLANTS

Their Use, Management, Cultivation and Biology

A COMPREHENSIVE GUIDE

Bob Watson

THE CROWOOD PRESS

First published in 2008 by
The Crowood Press Ltd
Ramsbury, Marlborough
Wiltshire SN8 2HR

www.crowood.com

British Library Cataloguing-in-Publication Data
A catalogue record for this book is available from the British Library.

ISBN 978 1 84797 028 2

Captions for cover phtoographs
Front cover: A stunning blue *Iris* cultivar. (Bob Watson)
Front flap: Pitchers of an epiphytic *Nepenthes* species in the Dartmouth Community Glasshouse. (Bob Watson)
Rear cover:
Top left: Teasel (*Dipsacus fullonum*). (Ron Mepstead)
Top centre: Fruits of honesty or moonwort (*Lunaria annua*). (Bob Watson)
Top right: Flowers of *Bourgainvillea*. (Bob Watson)
Rear cover below:
Left: *Embothrium coccineum* in the American Garden, Saltwood, Kent. (Bob Watson)
Centre: Group of crocus. (Ron Mepstead)
Right: Developing flowers of bottle brush (*Callistemon*). (Bob Watson)
Lower right: Large-leaved *Rhododendron* species (*Rhododendron macabeanum*). (Bob Watson)
Rear cover drawing:
Left: Cape figwort (*Phygelius capensis*)
Right: Angel's trumpets (*Brugmansia arborea* syn. *Datura arborea*)
Rear flap:
The importance of aspect, orientation and microclimate – a very tropical effect in a country lane, Kingswear, Devon on the River Dart estuary. (Bob Watson)

Typeset by Exeter Premedia Services Private Ltd., Chennai, India

Printed and bound in Malaysia by Times Offset (M) Sdn Bhd

Contents

Acknowledgements and Dedications

Continued thanks to my wife Fay for her forbearance during the production of this book.

Many thanks to Ted Wilson for his inspiration and consistent support throughout, to Jonathan MacDonald, Campbell Logue, Chris Starr and Helen Hall for their support of my previous work and to Crowood for their continued faith in me. Thanks to Shelagh Todd and David Haigh good horticulturists, friends and ex-colleagues. Thanks also to Steve Oliver-Watts, who in the latter part of my career was a tremendous support and inspiration. He may have leapfrogged me, alternating as my colleague then my boss several times, but he has remained an excellent friend and wise counsel.

I am very grateful to Michael Kemp, Penwith District Council, Hayle, Cornwall for his help and plant knowledge. Thanks also to David Molloy, Devon who not only allowed me to photograph freely in his excellent garden, but also put me right on a couple of plants, and prevented the embarrassment of printing incorrect identification (although I am sure there will be others!).

Thanks to Bill and Kay White, Joan Bennison, Dave and Maggie Colverson and Michael Tyler of Kent, John and Joan Hunt, Staffordshire and Klaus and Veronika Frühwirth-Stangl, Kitzbuhel, Austria for allowing me to photograph plants in their garden, and to Mary and John Reid, Kent for permitting me to photograph their magnificent *Embothrium*. Thanks to Tony and Ann Atkin for allowing me to photograph in their garden in Cumbria and to Tricia Cooke, Devon for allowing me to photograph from a vantage point in her garden. Thanks to Ron Mepstead of Kent for supplying his close-up photographs of *Narcissus*, crocus, teasel and clematis and to Kay White for her photograph of shining privet.

Thanks to Ihian and Liz Williamson, Devon and Michael and Annette Emmens, Hampshire for letting me photograph and draw individual plants in their garden. Thanks also to Sarah Taylor, Hampshire who fortuitously opened her garden to the public when I was staying in the area and who very kindly allowed me to take some samples back with me to draw. Thanks to Dartmouth Community Glasshouse and South Hams Operations Team, Dartmouth for their help.

Thanks to Sir John Ropner, at Thorp Perrow Arboretum, Bedale, North Yorkshire, for his faith in me; and thanks to the family Reuthe (particularly Eugene for his tolerance) who in the 1970s introduced me to a very wide range of plants at their nursery at Crown Point.

All diagrams by the author. All photographs by the author except four taken by Ron Mepstead and one by Kay White.

Introduction

This book is an attempt to introduce the diversity of the plant world and the associated diversity of plant management requirements. It is designed to furnish the principles that give a greater understanding of horticultural practice, and is aimed at students of horticulture at National Diploma, Higher National Diploma, RHS Diploma/Master of Horticulture and foundation/first year degree level. However, it should also prove very useful to a wide range of people interested in horticulture and the plant world in general, including informed amateur gardeners, garden and botanic garden managers, other horticultural professionals, and those embarking on horticultural or related courses who have insufficient botanical or biological experience.

The book is intended as a sister volume to *Trees: Their Use, Management, Cultivation and Biology*. Both books hold up on their own, and can be considered separate entities, and may be successfully used as such. Although chronologically this book was written after the book on trees (in film parlance, a prequel) it is an essential precursor to the first volume for those who wish to study plants right across the spectrum. It includes information on soft plants, shrubs, woody climbers, and touches on small trees – and so leads nicely into the concepts of the first volume.

The two books 'collide' therefore at the edge of 'woodiness', and they diverge here also, as this area comprises the final information of the 'plants' book and the commencement of the 'trees' book. The tree book can be used either as an extension to plant knowledge overall, or as a specialist book on trees in its own right and a good lead-in to the principles and practice of arboriculture. Often arborists have weaknesses in their shrub – and other 'horticultural plant' – knowledge, because they tend to concentrate on, and specialize in trees. So, plugging the gaps in these areas often helps with an understanding of all things arboreal. Likewise, horticulturists often have gaps in their knowledge of trees and tree management.

Some of the text is shared by the two books – this is an unfortunate necessity in order that the two volumes may stand alone and not be totally dependent on one another to gain full understanding. There is, for example, a lot of crossover between the two volumes regarding biological text, as it is essential to understand green growth for aesthetic horticulture (and green crops), but equally important to understand the green (primary) growth phases of trees. Similar crossover is also true of some information on flower structure. So, in some instances the two books look at different aspects of the same information, with the different aspects related to the different plant types, not to a radically different biology. However, where it does occur, because of the very wide range of plants used in horticultural and botanical situations, there is a lot of extra information to that found in the tree book – including extra diagrams.

Section One, 'Plant Biology and Principles', commences with an introduction to external features of plants and the recognition of basic plant groups. Being able to recognize these features allows progress within plant identification and nomenclature. Information on flower structure helps further with identification and how floral architecture affects/influences pollination, and subsequent fertilization and fruit formation. The section continues with plant biology, covering plant cells and tissues, plant development, plant physiological processes, and the structure of seeds and its relevance to germination. This book does not attempt to replace good botany books (with the more detailed plant anatomy diagrams that they contain). It does however try to bring the various strands together to aid understanding and to encourage further study. The aim is that this textbook should be progressive, and that the information on plant structure and function is shown to be an important precursor to, and very relevant to, Section Two, 'Plant Management and Practice'.

The basis of plant identification appears early in the book, as work put into accruing knowledge of plant taxonomy (identification, classification and nomenclature) is very important. However, although identification is covered, this book is not intended to be a substitute for good plant identification books, and should not be used as such – in fact in some instances it should be used alongside these to get the best results: the wider the knowledge in this area, the greater the number of options available for any situation. If you have an eye for colourful, bold, dramatic and interesting designs, it is of no use unless you have the necessary repertoire of plants to fulfil the designs imagined. You need to know what species will give the desired effect: for example, it is pointless putting two species together because of their colour if they flower at different times, when the two colours will therefore be separated in time, and the harmonious or contrasting effect will never be fulfilled.

A basic understanding of soils, although really a subject that would sit more comfortably in Section One, is covered under 'practice', only in order to best understand the chapters on site preparation and planting, of which it forms part. The last chapters of the book deal with the management of individual plants and plant collections, and also include the basis of grass (turf) establishment and management. These final chapters are an attempt to help those who already do, or who wish to, manage small or large gardens, including gardens open to the public.

At the back of the book is a glossary of 'terms and concepts'. The glossary is a very important addition because it attempts to highlight important concepts, and furthermore can be used to refer immediately to specific areas of interest or enquiry. There is also a plant genera index (index of generic and common plant names) as well as a main subject index.

The modularization of land-based courses in colleges and universities has created a piecemeal effect to study, and often without the necessary integration of subject matter that was initially promised. On the one hand it offers the undoubted benefits of a flexible entry and learning system, on the other hand it often discourages or prevents progression from the basic to the complex (as large chunks may be left out in the name of early specialization). Furthermore, 'rationalization' also creates a desire to 'dump' certain elements that are expensive and/or difficult to deliver; the effect is to reduce the practical content because it is 'too expensive to deliver', and to reduce the science content 'because it does not directly aid employment'!

As well as loss of continuity, integration and progression, plantsmanship – the knowledge, range and use of diverse plant material – has fallen through the 'gaps' brought about by modularization. This book is therefore a blatant and totally unashamed attempt to bring back the interest in the biology of vascular plants, plant diversity and plantsmanship into horticulture, as true plantsmanship can only be gained with a knowledge of botany and ecology. Perhaps what the profession really needs is a large cohort of up-and-coming 'horti-botanists' (plantspersons) to plug the gaps between all the 'specialisms': the gaps between the out-and-out dendrologists (specializing in exotic woody plants), arboriculturists who tend to specialize in trees only, and horticulturists from specialist sectors – all of whom (and those in associated disciplines) would often benefit from a better overall knowledge of plants in general before they specialize.

'Gardening' has a diverse number of connotations and meanings depending on any individual's present state of knowledge, and whether they are involved professionally or in an amateur capacity. Moreover, there are some 'professionals' who do not have a requisite state of knowledge to fulfil their function well – and conversely, many people of amateur status who do. All, nonetheless, can gain tremendous enjoyment from it.

Gardening can, therefore, be the cultivation of plants at many different levels: small domestic, large-scale domestic, stately home, botanic collection or local authority parks. The list is not exhaustive, and every one of these has its own challenges, and may have wide and varying objectives. They nevertheless share many common areas, and their success depends upon a blend of art and science (aesthetics and biology). Successful cultivation depends upon knowledge of environmental factors and soil conditions, coupled with a knowledge of the best plants for those specific conditions. Superimposed upon this, and only successful if the biological factors are correct, is an eye for aesthetics. The plants themselves, if successfully cultivated, may be sufficiently aesthetic in their own right. However, they are often associated with specific grouping, positioning or design patterns that may enhance their effect.

In order to 'garden' at any reasonable level successfully, it is essential to understand the scientific principles that underpin current practice – the alternative is to inherit plant layouts and learn by rote the timings involved in each maintenance operation, and to carry these out periodically as prescribed. Learning by rote may lead to some success, but is not satisfactory because this level of knowledge does not allow you to make informed management decisions about the successful culture of specific plants, and can lead to disastrous mistakes and losses. Poor underpinning knowledge and lack of understanding of the principles involved severely limits the range of plants that are grown and therefore limits the diversity of aesthetic effects achievable. It also means that the lack of plant/cultural knowledge limits progress and flexibility in design, and therefore the ability to 'develop' and take the collection/garden forwards.

SECTION ONE

Plant Biology and Principles

CHAPTER 1

Longevity and External Features

Introduction to the Main Plant Groups

Plants fall into one of five main groups concerning their longevity: ephemeral, annual, biennial, herbaceous perennial and woody perennial. Their structure necessarily varies accordingly, and only woody perennials have the ability to increase in girth in their second and subsequent years. Ephemerals, annuals, biennials and herbaceous perennials are all actually herbaceous as they have soft 'herby' growth (the derivation of herbaceous). Annuals do not progress into a second season, but herbaceous perennials have a strategy for surviving from one year to the next based on the death and renewal of their aerial growth, and the addition of new underground storage organs.

Ephemeral plants have a fleeting life span and commonly have two life cycles in one season. Each generation dies after setting seed, so procreation of future generations is by seed only. The most commonly quoted examples of ephemerals are the weed plants groundsel (*Senecio vulgaris*) and thale cress (*Arabidopsis thaliana*). Because of their obviously fleeting life span, the term 'ephemeral' is also given to perennial desert plants that flower incredibly quickly (fleetingly) after rain, in that very small 'window' when conditions are suitable.

Annual plants have only one life cycle in a year (they germinate, grow, flower, set seed, and die). Any particular individual is therefore dead within a twelve-month period, and because of growth patterns and their reliance on temperature and light factors it is usually within the confines of the growing season. Annual plants comprise soft, primary growth only, and because they do not live into a second season they have no provision for stem girth increase after year one. So taproots, other storage organs, tough, waxy persistent bud scales and woody stems are absent. Because they cannot overwinter, continuity of the species is by sexual means only – that is, the seed lying dormant in the soil and germinating in the following season. Some annuals are very aesthetic, such as the poached egg plant (*Limnanthes douglasii*) and English marigold (*Calendula officinalis*). However, some are regarded as persistent weeds, including hairy bitter cress (*Cardamine hirsuta*).

Biennial plants have a life cycle spanning two seasons, and typically they produce plenty of soft, photosynthetic vegetation in year one, overwinter as a rosette of large leaves, and then produce flowers, set seed, and die in year two. Year one's vegetative growth produces food at the leaves (via photosynthesis), which is stored for the next season, and in season two the stored food is used to produce flowers.

Edible biennial crops grown for foliage, such as cabbage (*Brassica oleacea*), or roots, such as carrots (*Daucus carota*) and parsnips (*Pastinaca sativa*), are always harvested at the height of their food reserves, to supply the maximum nutrition when eaten. This is during their first season ideally, but for convenience and storage reasons some may be left until the early part of season two. However, harvesting them too late means they are starting to produce flowers and seeds, which severely depletes the nutrient store (the main reason for human consumption) and ultimately they become tough and inedible. High soil nutrition in cultivation and excessively hot weather may encourage some biennial food crops to flower unseasonably in year one – this is known as 'bolting', and renders the crop unharvestable.

Species such as field forget-me-not (*Myosotis arvensis*), which may be annual or biennial, have small leaves and attain little height. They adopt a strategy of many nebulous fine roots (adding up to a large volume of food storage overall), whereas species such as foxglove (*Digitalis purpurea*), which

A cultivar of English marigold (Calendula officinalis).

adhere to the typical large leaf rosette when overwintering, use the strategy of a deep, fleshy taproot for food storage. There is a direct correlation between the relative size of the leaves, the production of foodstuffs (sugars), and the size of the storage organ. Hence foxglove and mullein (*Verbascum bombyciferum*) with their large leaves, and hemlock (*Conium maculatum*) and giant hogweed (*Heracleum mantegazzianum*) are large biennials (with taproots) and very vigorous in comparison to forget-me-nots. Some biennial species have individual plants that are short-lived perennials (for example, some Verbascum species), and some plant species are monocarpic (they die after flowering and setting

Ephemeral-1 year (season)

Annual-1 year (season) **Annual**-1 year (season)

Biennial-year one
(season one)

Vegetative
phase

Seed
dispersal

Flowering (sexual)
phase

Death of
parent plant

Biennial-year two
(season two)

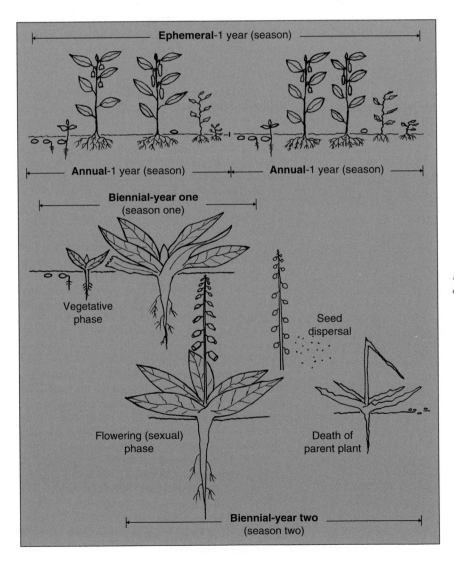

Plant longevity: ephemerals, annuals and biennials.

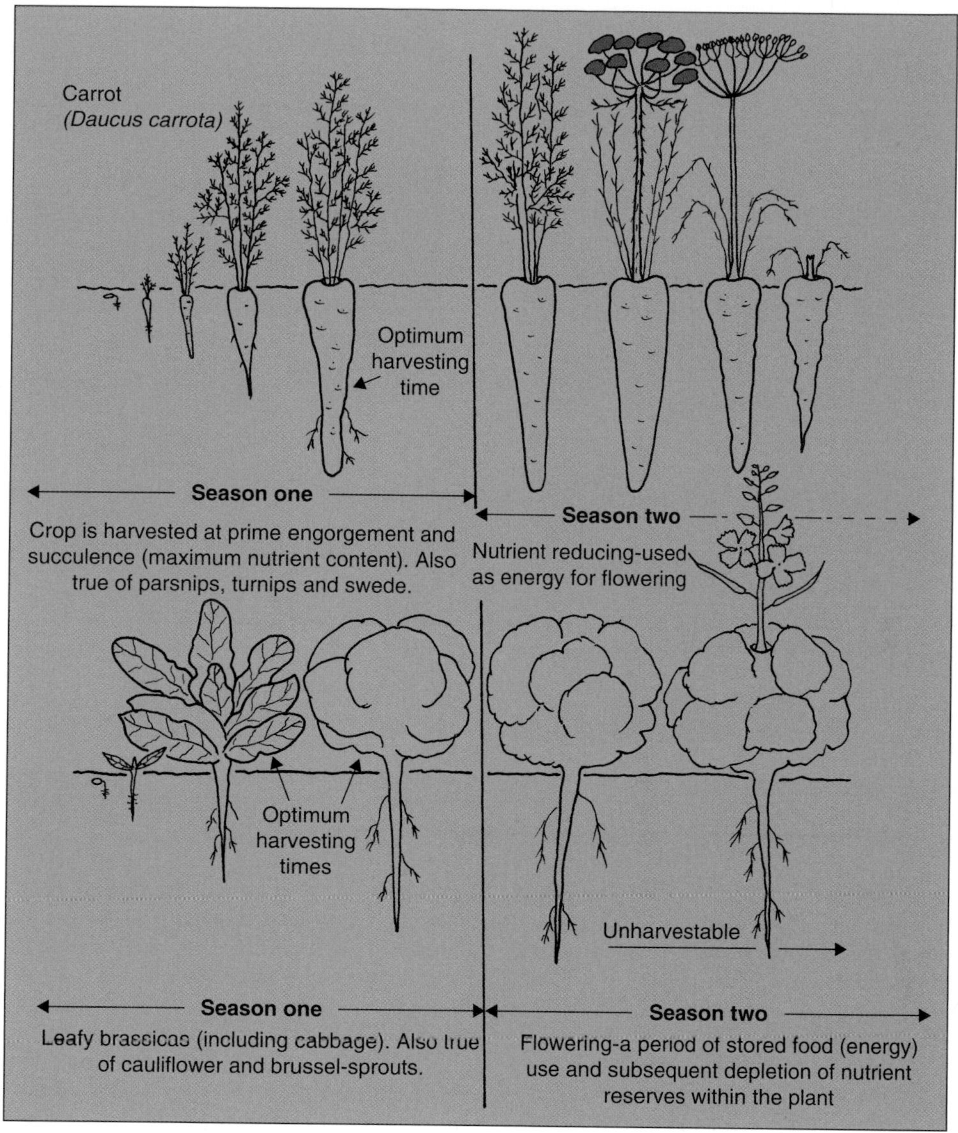

Plant longevity: edible biennials (carrot and cabbage).

seed) such as some species of *Meconopsis*, for instance *Meconopsis paniculata*.

Herbaceous perennials have an indefinite life span typically lasting ten to twenty years, or even longer, depending on species and conditions. They produce new, lush herby growth annually, and attain a maximum height and girth during the growing season, at the end of which that year's growth usually dies back to ground level. The process is repeated annually, with new aerial growth in the second season (and subsequent seasons) fuelled by nutrition derived from underground storage organs. Herbaceous perennials (sometimes called 'hardy' perennials in a garden context) include Michaelmas daisy (*Aster novae-belgae*), day lily (*Hemerocalis spp*), bluebell (*Hyacinthoides non-scriptus*), daffodils (*Narcissus spp*), lilies (*Lilium spp*), cranesbills (*Geranium spp*), bellflowers (*Campanula and Platycodon spp*), wolfbane/monkshood (*Aconitum napellus*) and snowdrop (*Galanthus nivale*).

Foxglove (Digitalis purpurea). Biennial

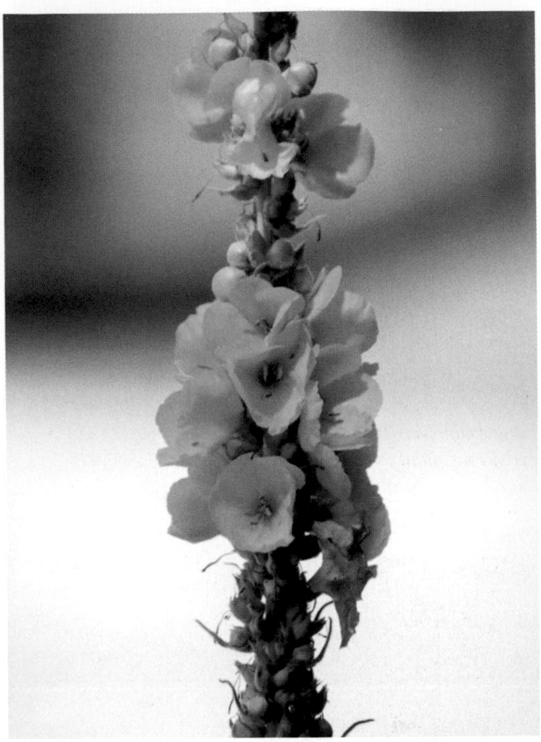

The showy flowers that follow the large rosette of very hairy grey leaves on Verbascum bombyciferum. Biennial

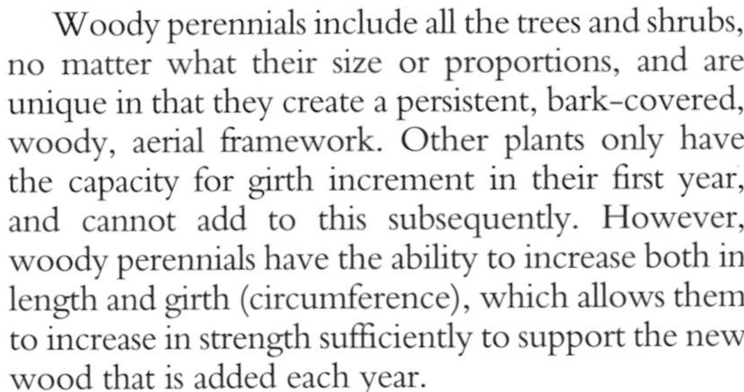

Woody perennials include all the trees and shrubs, no matter what their size or proportions, and are unique in that they create a persistent, bark-covered, woody, aerial framework. Other plants only have the capacity for girth increment in their first year, and cannot add to this subsequently. However, woody perennials have the ability to increase both in length and girth (circumference), which allows them to increase in strength sufficiently to support the new wood that is added each year.

Shrubs are defined as woody perennials, with many persistent stems arising from, or near, ground level; trees are defined as woody perennials with a distinct trunk (or trunks). These are useful definitions, but they still do not cater for the grey area between a very large multi-stemmed shrub and a small bushy tree. Many genera have species of both trees and shrubs within them – for example, the genus *Magnolia* has *Magnolia acuminata* (a large-growing tree), and also *Magnolia stellata* (a bushy, multi-stemmed shrub). Perhaps the major differences between trees and shrubs are height and longevity, as trees do differ from all other plant forms in their longevity and potential height. Shrubs can attain large sizes and survive for long periods of time (commonly twenty to fifty years depending on species). Some tree species may attain great height and great age: 80–150 years is common, 150–250 years is relatively common, and even 250–500 years is not that unusual for some species. In extreme cases trees may attain 100 m or more, and ages of 2,000–5,000 years.

Shrubs go through various phases of growth (germination, developing seedling, semi-mature (sexually mature), mature, over-mature, senescent semi-demise). In some instances in the natural world young, germinated seedlings may develop beneath the framework of the dying parent shrub, as the semi-demise of the crown allows sufficient light to facilitate the process. The phases of tree growth are similar and include the juvenile phases (the excurrent phases, typified by marked apical dominance) of seedlings and developing plants (colloquially called saplings), and then several decurrent phases typified by broadening of the crown – the semi-mature phase (sexual phase) and

Large bellflower (Platycodon grandiflorus mariesii), an herbaceous plant in the Campanulaceae.

the mature phase. The next phases are typified by a recession (lessening of the crown canopy, or stag-headedness) and major splits in the main trunk – over-mature (senescence) and demise (death). Tree seedlings (particularly of the same species) do not often germinate under or near the canopy of the mother tree because the soil is usually too dry and the light factors too low. Also, there are sometimes materials that suppress germination and development produced by the mother tree (allelopathy) to reduce the risk of competition. To get round the problem and ensure procreation without direct competition, some species produce winged seeds, or enlist the aid of birds (or sometimes mammals) to move the seeds away from the influence of the mother tree. Leaves are also now known to add to the allelopathic effect and

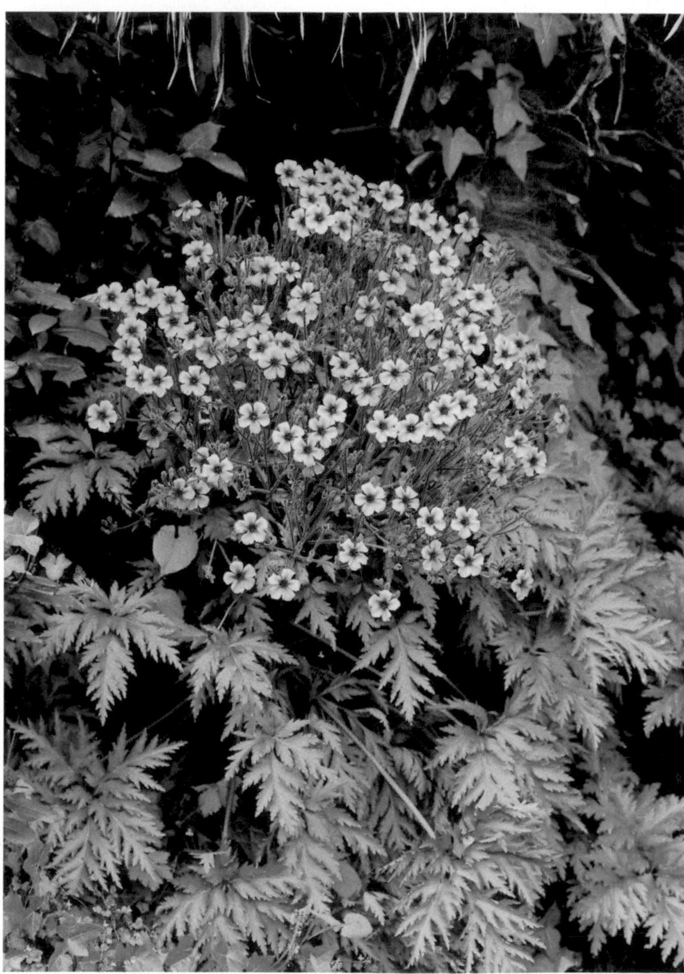

A wonderful crane's-bill (Geranium maderense), an herbaceous plant from Madeira.

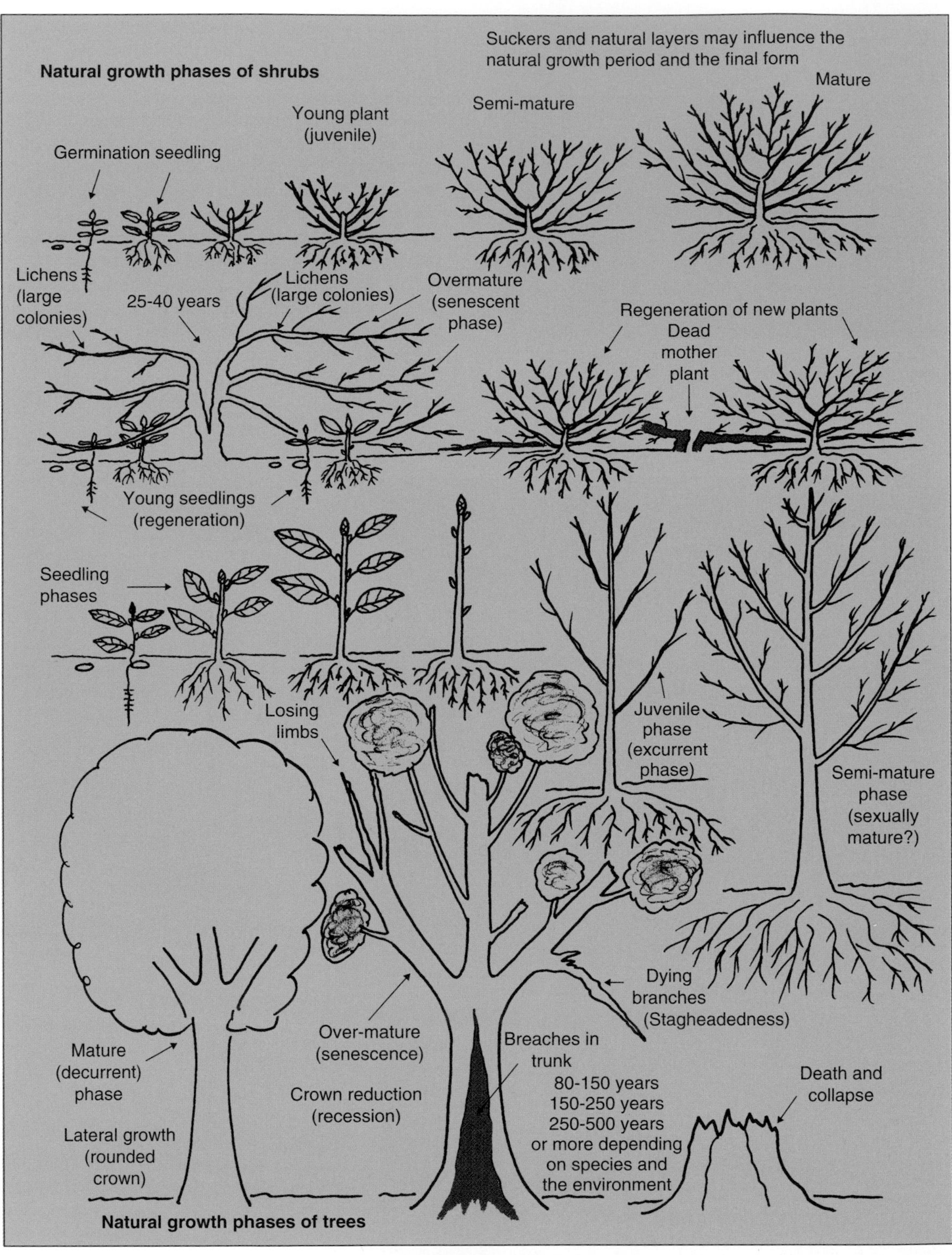

The growth phases of trees and shrubs.

prevent germination because of the build-up of toxic materials that occurs just prior to abscission and are taken to the soil at leaf fall.

The External Features of Plants

The External Features of Green Plants

Green plants comprise a collection of interrelated and interconnected organs, including stems, roots, leaves, buds and flowers. It is the ability of cells to make drastic changes to perform specialized functions, and to form specific congregations, configurations and tissue patterns, that makes the whole system work.

Stems support leaves, facilitate aerial growth, and have provision for continued (and/or lateral) growth in the form of vegetative buds, and may also facilitate the production of flowers at various points along their length. Stems conduct water and dissolved nutrients through the plant, and can act as storage places for starches (condensed sugars) to be used as an energy supply at a later date. We recognize various features of the stem that are useful practically, including nodes, inter-nodes, and leaf axils. A node is the point of attachment where a leaf petiole (leaf stalk) meets a stem, an inter-node is the distance between two nodes, and a leaf axil is the angle formed by the leaf stalk and the stem. The outside of young stems is covered in a soft-celled epidermis (outer skin), below which many cells contain the pigment chlorophyll for photosynthesis and therefore appear green.

Flowers are the sexual reproductive organs of higher plants, and comprise the sexual organs, and various protective (even attractive) outer layers of tissue whilst developing in the bud. They often have highly coloured petals, and the pattern of the petal layers is fairly consistent across most species, even though the petal shapes may vary widely from one species to another. Furthermore, petal numbers (or at least a multiple of the petal numbers) remain constant within a species.

Roots infest soils, spreading outwards into the soil matrix, to harvest moisture and dissolved mineral salts. Roots also provide anchorage and give essential stability to plants within the soil, and act as conducting tissue to transport water, metabolytes and dissolved nutrition from the soil matrix to be delivered to other parts of plants where they are needed. Leaves are the main sugar-producing (photosynthetic) organs of plants, and also comprise the main evaporative surface that creates water movement through the plant (evapo-transpiration), fuelled by energy from the sun.

The Types and Features of Leaves

The main types of leaves recognized in higher plants are cataphylls, cotyledons (seed leaves) and foliage leaves. Cataphylls are forms of scale leaves (including scale leaves at buds, stipules, bracts and bracteoles) that often have a protective role, but may have secondary photosynthetic or even insect attraction roles as well.

Scale leaves include those surrounding and protecting the buds of woody subjects, but also the leaves surrounding the basal shoots of rhizomes beneath the soil. They also include the fleshy scale leaves associated with the internal structure of bulbs, and the papery scale leaves found on the outside of bulbs and corms. Stipules are a leaf formation that may arise as leafy outgrowths at leaf petiole bases, or they may be found associated with subjects such as *Pelargonium zonale* (geranium of horticulture) and tulip tree (*Liriodendron tulipifera*), where they form flat protective flaps each side of the buds, and *Amicia zygomeris* on which they are showy and attractive. Bracts are simple scale leaves that look like, and often take on, the role of highly coloured petals. Handkerchief tree (*Davidia involucrata*) and *Cornus kousa* are good examples of species with showy bracts. Cotyledons are seed leaves found inside seeds whose function is primarily food storage; however, these may take on a photosynthetic role if they appear above soil level after germination.

Foliage Leaves

Although green stems do photosynthesize, foliage leaves are the main photosynthetic organs of the plant, producing large amounts of sugars that can be transported to wherever a high-energy food is needed for growth. Because they contain vascular tissue, and have pore-like apertures called stomata, they are also the main organs of water loss via transpiration. Those having distinct upper and lower surfaces are known as dorsi-ventral leaves (having a dorsal and ventral surface), and are found on all broadleaf species such as dock (*Rumex* species), comfrey (*Symphytum officinale*) and beech (*Fagus sylvatica*), and some linear-leaved species. The more upright leaves of many linear-leaved species have less obvious upper and lower surfaces and are irradiated on both sides. These are known as iso-bilateral leaves – equal (iso) on two sides (bilateral). Conifers have long, thin, tough and heavily waxed leaves (sometimes almost triangular in section, sometimes

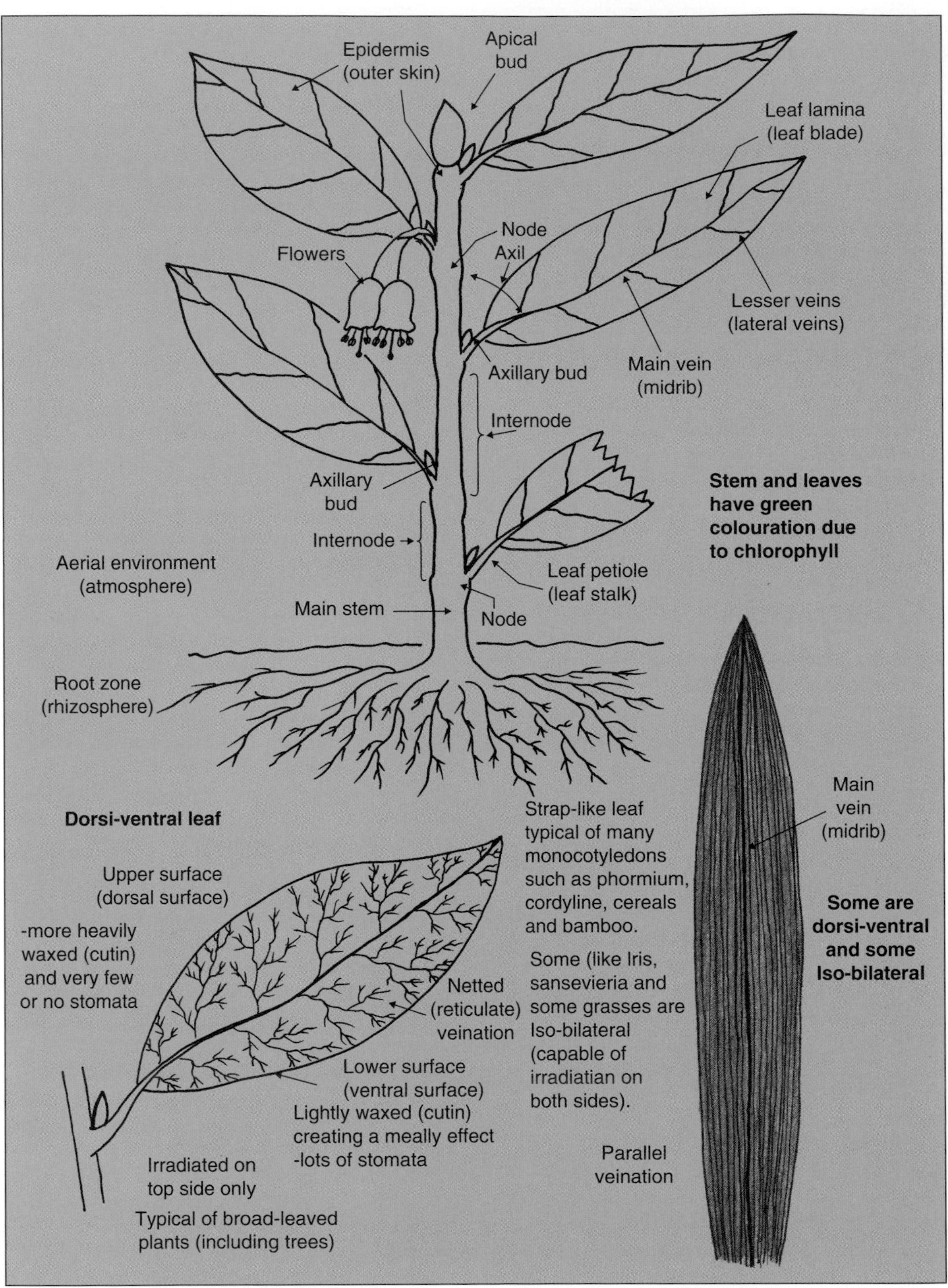

External tissues of the green plant.

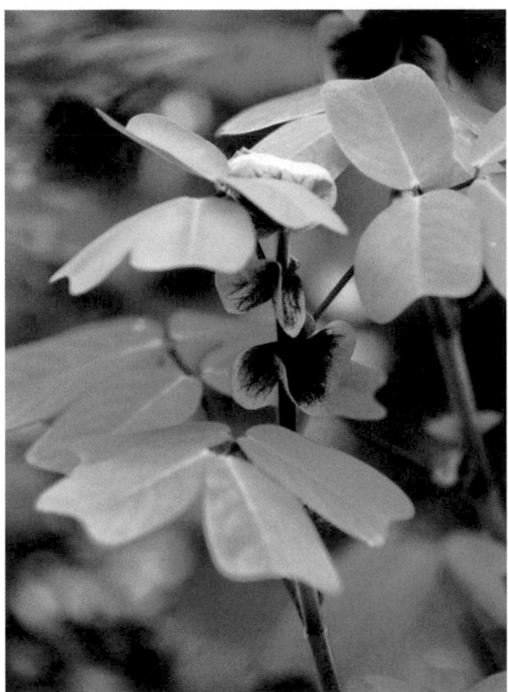

The showy leaf stipules of Amicia zygomeris.

The showy bracts of Cornus kousa chinensis.

flat) known as needles, and in pines are held in bundles of two, three or five by a papery fascicle.

Dicotyledonous subjects (those with two seed leaves in their embryo) have foliage leaves with complex netted (reticulate) venation, whereas monocotyledonous subjects (those with only one seed leaf in their embryo) have leaves with parallel venation. Foliage leaves may be simple, with an entire leaf lamina that either terminates at the base of the plant or is attached to the stem by a leaf stalk (petiole); or they may be compound, where the whole leaf comprises a collection of leaflets attached to the stem by a central rachis (extended petiole). Leaves with a petiole are termed 'petiolate' and may have opposing leaf stipules at the base of the petiole (or rachis), when they are known as 'stipulate' (as found in some roses). Petioles may enclose (wrap around) the stem, in which case they are termed 'sheathing'. Species without petioles are 'sessile' (without leaf stalks). Perfoliate leaves have the stem running through them as if it has punctured the foliage. Where this punctures the centre of the leaf, leaving roughly equal amounts of leaf tissue on each side, it is known as 'connate-perfoliate' (as found in some honeysuckles, some eucalyptus and *Parahebe perfoliata*). Peltate leaves

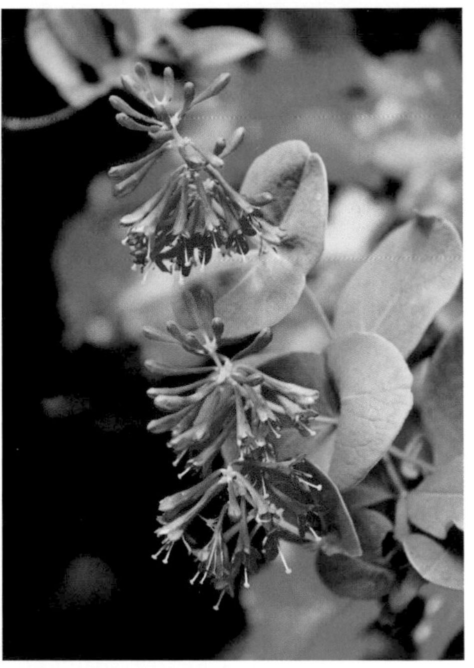

The connate-perfoliate leaves of Lonicera sempervirens.

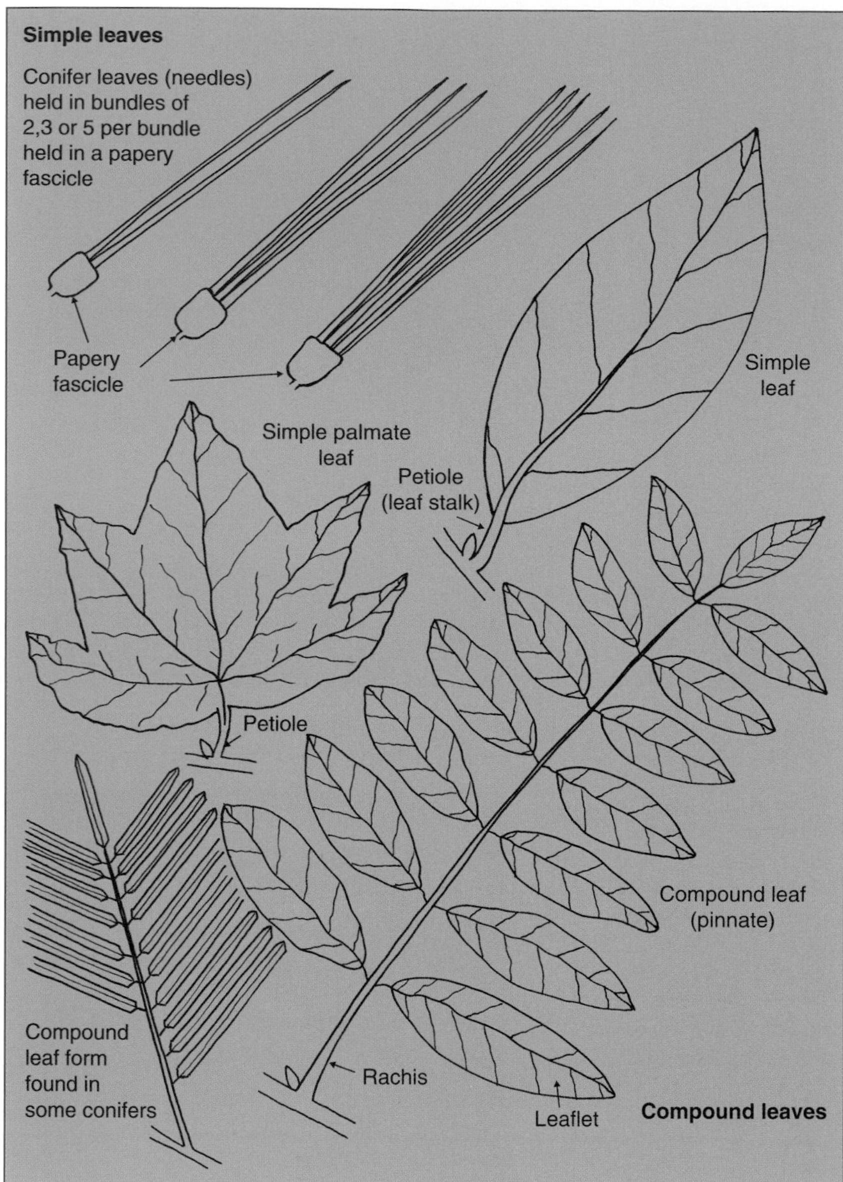

Basic foliage leaf types (simple and compound leaves).

have the petiole attached on their underside near the centre, for example *Pelargonium peltatum*. Simple sword-like leaves arising from one main point, as found in *Iris*, are known as 'equitant'.

Simple leaves have many shapes. Lanceolate leaves are shaped like the pointed head of a lance and have the wider end nearest the petiole; ovate leaves are oval (egg-shaped) with the broadest part nearest the petiole; and, as with all other examples, adding 'ob' to the description inverts the shape – thus oblanceolate and obovate leaves have the narrowest end nearest the petiole and the broadest part farthest from it. Acuminate leaves have a distinctly pointed leaf tip, cordate leaves are heart-shaped, deltoid leaves have almost straight sides forming an equilateral triangle, and rhomboid leaves look somewhat similar but with four sides. Linear leaves are long and narrow with relatively parallel sides, filiform leaves are very thin and cord-like or thin and ribbon-like. Orbicular leaves

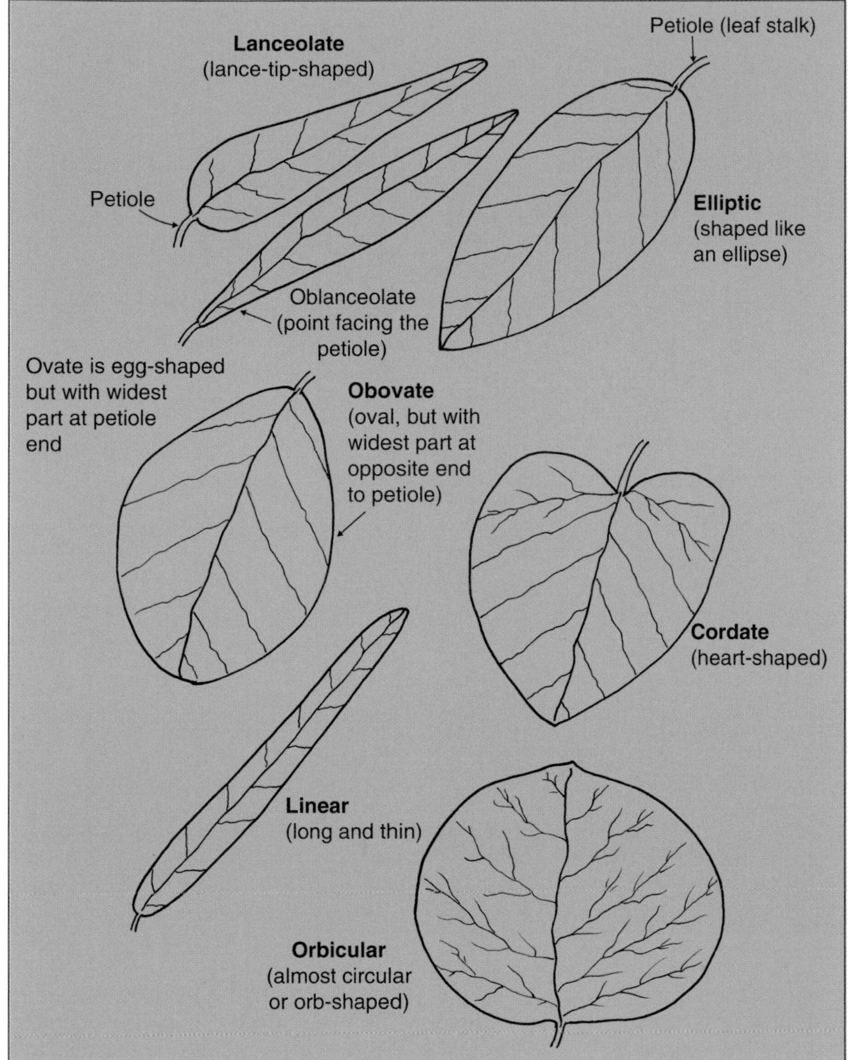

Foliage leaf shapes I (simple leaves).

are very rounded, almost circular in shape, and reniform leaves are kidney-shaped.

Simple palmate leaves have an entire leaf lamina, but have five to seven distinct pointed lobes that radiate round the leaf. *Acer palmatum* (smooth Japanese maple), *Liquidambar styraciflua* (sweet gum) and *Ricinus communis* (castor oil) all have simple palmate leaves. Compound palmate leaves have five to seven individual and free leaflets radiating like fingers attached to the end of the extended petiole (which leads to their alternative name – digitate), for example umbrella plant (*Schefflera*). Palms such as *Trachycarpus fortunei* commence as entire leaves, but split at pre-determined lines to form lots of leaflets. Compound leaves may also be singly pinnate, having a central rachis (extended petiole) with leaflets radiating from each side; or bi-pinnate, having a central rachis with minor branching stems (rachilla), each with leaflets radiating off it. These are all strategies of dividing the leaf into small leaflets yet retaining the overall photosynthetic area and reducing mutual shading to the leaves below.

Trifoliate leaves are simple leaves attached to the petiole that comprise an entire leaf lamina with three distinct yet rounded lobes. Trifoliolate leaves,

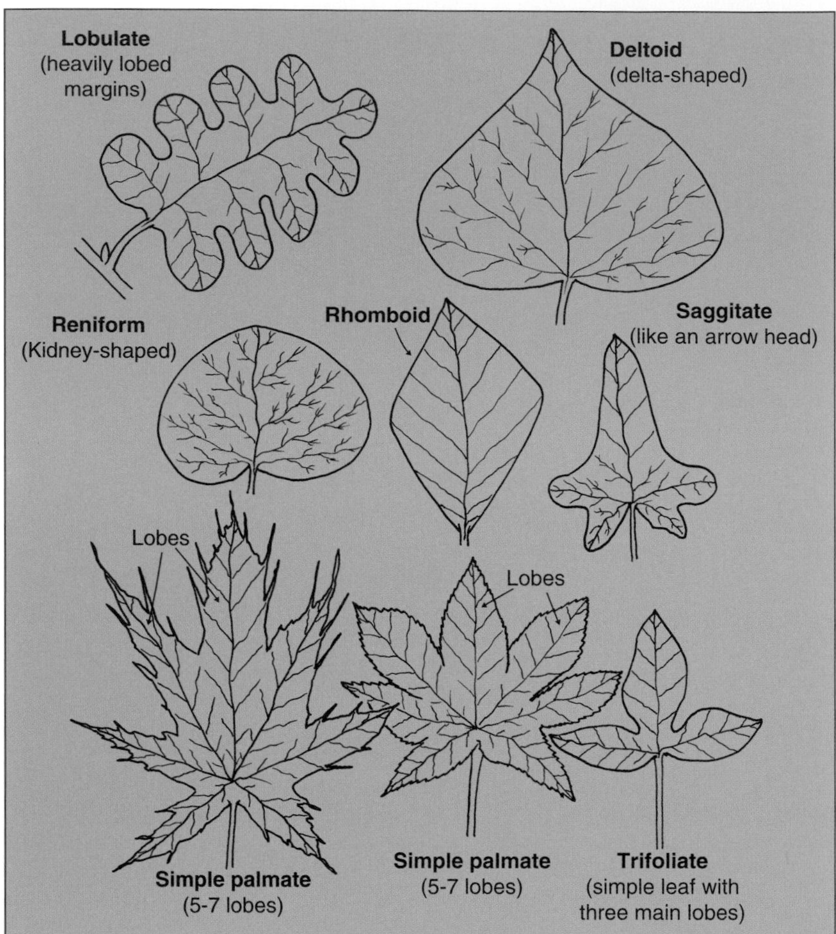

Foliage leaf shapes II (simple leaves).

on the other hand, are a compound form with three distinct and free leaflets attached to the rachis, as found in paper-bark maple (*Acer griseum*) and *Acer henryii*. Leaflets from all types of compound arrangement may be obovate, ovate, elliptic, and so on, and may have specifically shaped leaf bases, tips and margins.

Leaf margins, leaf bases and leaf tips may be very variable and have their own vocabulary to describe them: entire leaf margins, hairy leaf margins, and varying degrees of incision or rounded lobing all exist. Ciliate leaf margins bear a fringe of very fine hairs (cilia); coarse saw-tooth edges to leaf margins are known as 'serrate'; and finely saw-toothed edges are known as 'serrulate'. Coarsely toothed edges are known as 'dentate', and finely toothed edges as 'denticulate'; rounded, undulating lobes appear at lobulate leaf margins, and smaller, finer, less defined lobing is known as 'crenulate'. Leaf margins with cuts into the edges are known as 'incised', and those with deep, open U-shaped incisions with pointed outgrowths have parted leaf margins. Where both leaf margins are the same and distinct in their shape it may define the whole leaf, as in the case of a lobulate leaf – with two equally lobed margins – as found in some oaks, or serrate leaves found in many cherries. Broad, simple leaves where both leaf margins are very deeply incised forming cuts that nearly reach the mid-rib are said to be 'dissected', for example *Acer palmatum* 'Dissectum'.

Leaf tips may be gently rounded (obtuse), pointed (acuminate), have a long pointed bristle (aristate), or may be indented either slightly (retuse), or more markedly (emarginate). Leaf bases may also be obtuse (gently rounded) or cordate (heart-shaped), form ear-like appendages (auriculate), arrowhead shapes (sagittate), be delta-winged

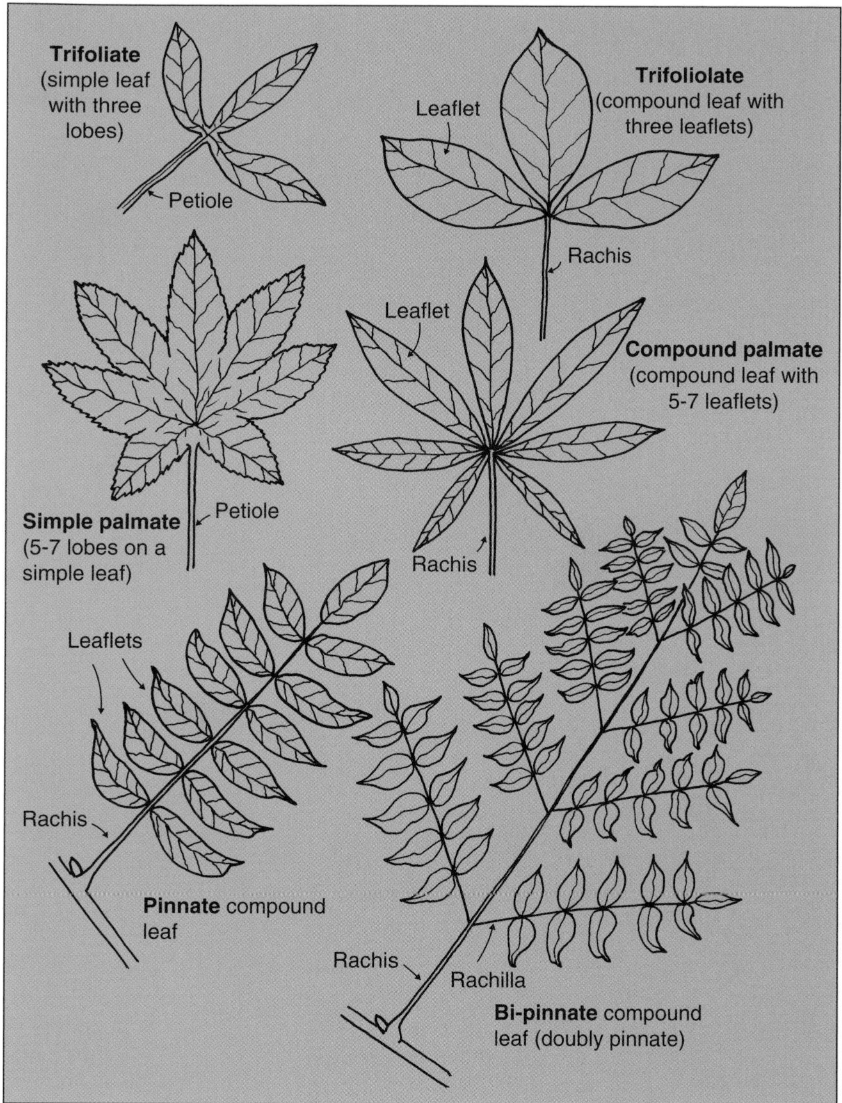

Foliage leaf shapes III (simple and compound leaves).

(hastate), or be flat as if cut straight across (truncate). Leaf bases can be lop-sided (oblique), wedge-shaped (cuneate – like a rabbit's tail), or pointed towards the petiole (attenuate). Sometimes a very distinct leaf base or leaf tip feature will also define the leaf, or indeed the plant (an acuminate leaf having an acuminate leaf tip and an auriculate leaf having ear-like appendages at its base, for example *Rhododendron auriculatum*). Likewise with cordate leaves the base really decides the complete shape of the leaf (heart-shaped), and it is also the same with deltoid leaves.

Although ordinarily leaf size tends to be within a definite range, terrific variation may be witnessed on the same individual plant in some instances. Leaf size differences may be caused by many localized environmental factors, including availability of nutrition, moisture and light, the relative health of the individual plant, and hormonal changes. In some species successive leaves go through a progression of intermediate shapes in their juvenile phases before they attain their optimum shape and ultimate size in their adult phase. Individual plants of common ivy (*Hedera helix*) present saggitate or

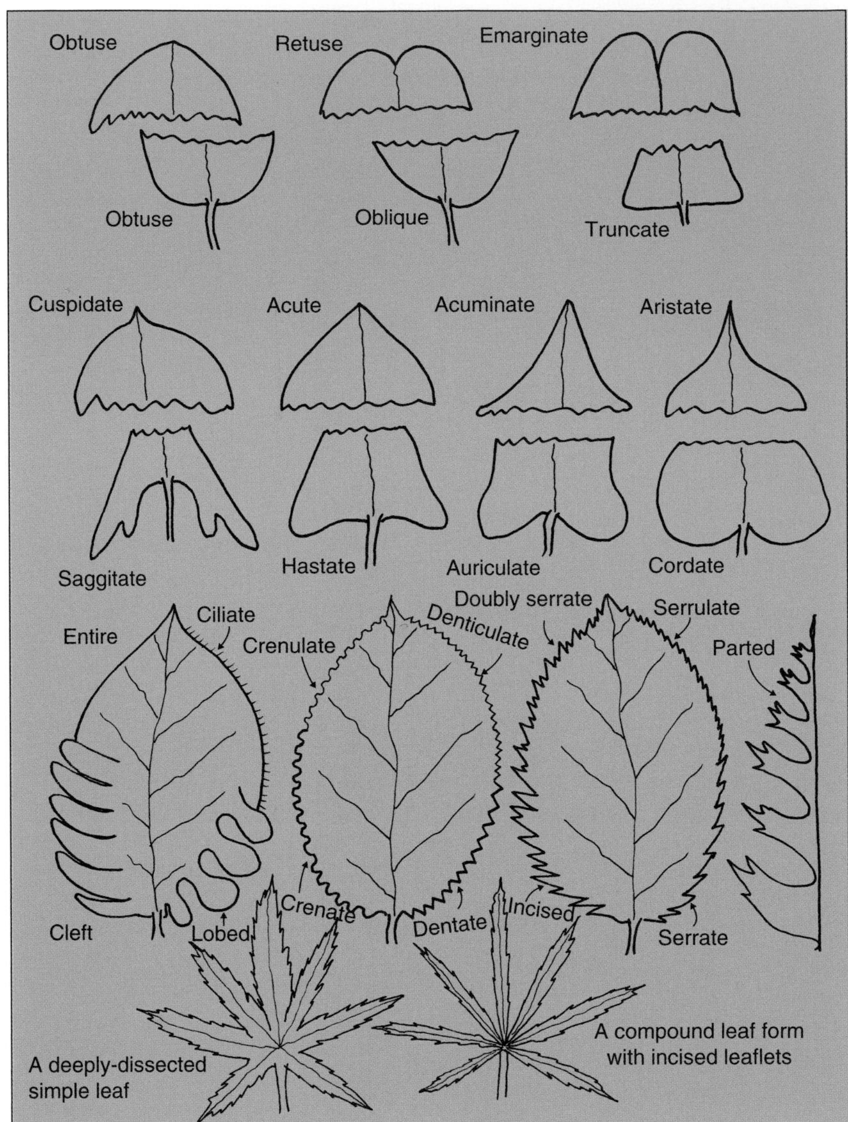

Foliage leaf shapes IV: leaf tips, margins and bases.

pointedly palmate leaf shapes in their juvenile ('climbing') phase – often in poor light because of the shade from the tree canopy – yet they present more widely palmate, even cordate leaves in better light, and distinctly rhomboid leaves during their sexual phase. Many *Eucalyptus* species have rounded juvenile and long linear adult leaves, both of which may appear on the same individual. Tomato (*Lycopersicon esculentum*) presents various complex simple and compound leaf forms depending on local conditions and inherent genetic information. Swiss cheese plant (*Monstera deliciosa)* – named because of the holes in the leaves, not its country of origin (South America, not Switzerland) – shows two distinctly different leaf types on the same plant: heart-shaped (cordate) leaves are produced in good light, and broader leaves with holes and splits (looking like Swiss cheese) in poorer light conditions. The 'split' leaves facilitate some light penetration to other leaves at lower levels.

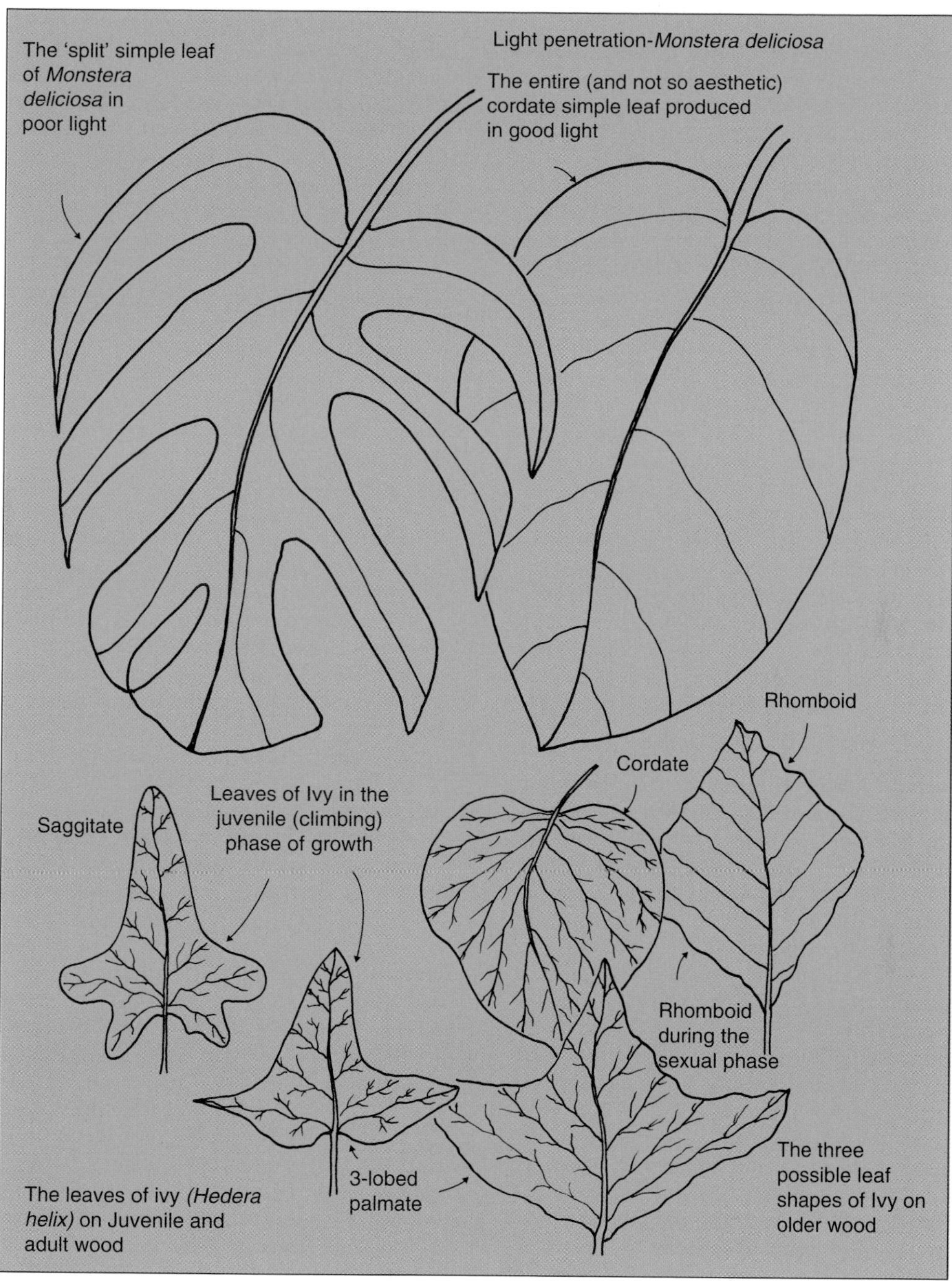

The different leaf types of Monstera deliciosa and common ivy (Hedera helix).

Buds

Buds may comprise undeveloped flowers (flower buds), or provide the potential for the extension of aerial growth (vegetative buds), or a mixture of both (mixed buds). Buds may appear at a shoot tip (apical buds) or at a leaf axil (axillary buds).

Although vegetative and mixed buds usually have leathery, wax-covered bud scales for protection, individual flower buds have a protective outer layer of leaf-like sepals. Vegetative buds at leaf axils provide for lateral growth extension that may be triggered by a hormonal reaction created when a stem apex terminates in a flower bud, or may be induced by the removal of apical bud(s) when pruning. Flower buds often appear in leaf axils, in which case they do not hinder the extension growth at the apex. Flower buds at an apex will often terminate extension growth, and if they do so, they are called terminal buds rather than apical buds. Mixed buds allow for both flowering and continued extension in stem growth, but it is relatively unusual for apical buds to contain mixed buds. However, examples do exist, including *Mahonia* species and their cultivars, which after flowering at their apex, produce thick extension growth that pushes the spent flowers to one side and progresses from the apex.

The External Features of Herbaceous Perennials

Perennial herbaceous species produce soft, herb-like stems (primary growth) during the summer months. Many lose this growth to frost in the winter, but some retain a proportion of old foliage over winter, and renew their foliage sequentially and so retain an evergreen effect. Herbaceous plants can over-winter via various underground (subterranean) storage organs, including bulbs, corms, stem tubers, root tubers, slender and fleshy rhizomes. Other subsurface storage strategies used by herbaceous perennials include some also adopted by biennials, namely either very extensive, fibrous root systems (known as fibrous-rooted crowns, for example *Sedum spectabile*), and the opposite strategy, of thick, fleshy, tap-rooted crowns for example *Rumex*, (dock) and *Rheum* (rhubarb) species.

The stems of herbaceous subjects are renewed on an annual basis, by new systems of primary growth arising as new shoots from the underground storage organs. The ultimate height of the aerial parts of the plant are approximately the same every year, only varying by a few centimetres in height from year to year, depending on the seasonal differences of rainfall, available nutrients, and so on. Increase in size (biomass) of individual plants is attained by subsurface lateral growth by the addition of subterranean organs (for example, the addition of new daughter bulbs or rhizomes). In fibrous and tap-rooted crowns, increase is by the number of annual-growth-derived shoots, never by an increase in aerial stem girth, even if it is persistent over winter. Gradually the old organs become dysfunctional, but the annual addition of new subterranean organs ensures that a good number of young functional organs is maintained.

You cannot change a definition to suit the situation, hence it matters not whether the features of stems are witnessed above or below soil level: if they carry leaves, axillary buds and conducting tissues, they are still stems. Thus many forms of underground organs that exist to aid the overwintering (perennation) of herbaceous plants are, in fact, underground stems, or have underground stems as at least part of their structure. Roots arising from stems (rather than other roots) are adventitious, and adventitious roots arising from various parts of the underground stem are a feature of bulbs, rhizomes and stem tubers. Other organs comprise forms of swollen modified root systems, and others still are complete underground plant systems comprising collections of modified leaves (fleshy and/ or papery scale leaves) and modified stems.

Bulbs are complete underground plant systems, and there are two main types: tunicated and scaly. Tunicated bulbs are complete condensed shoot systems, much of which remains subterranean, with only the foliage leaves and flowers appearing above soil level. They comprise an undeveloped floral apex and foliage leaves surrounded by concentric rings of fleshy scale leaves with a brown (or coloured) papery scale leaf on the outside. Species that have highly coloured outer papery scales gave rise to the name (tunicated) because they look like coloured tunics on the outside of the bulb, and the colour of the 'tunic' is related to the colour of the flowers in some instances. Pink and red hyacinth species, for example, have reddish papery scales; lilac and mauve species have bluish outer papery scales; and white species have silvery outer scale leaves. Tunicated bulbous subjects include common bluebell (*Hyacinthoides non-scriptus*), Spanish bluebell (*Hyacinthoides hispanica*), wild daffodil (*Narcissus pseudonarcissus*) and all other *Narcissus*.

The function of the fleshy scale leaves is to store food, and they are attached to a compact condensed stem at their base (known as the base

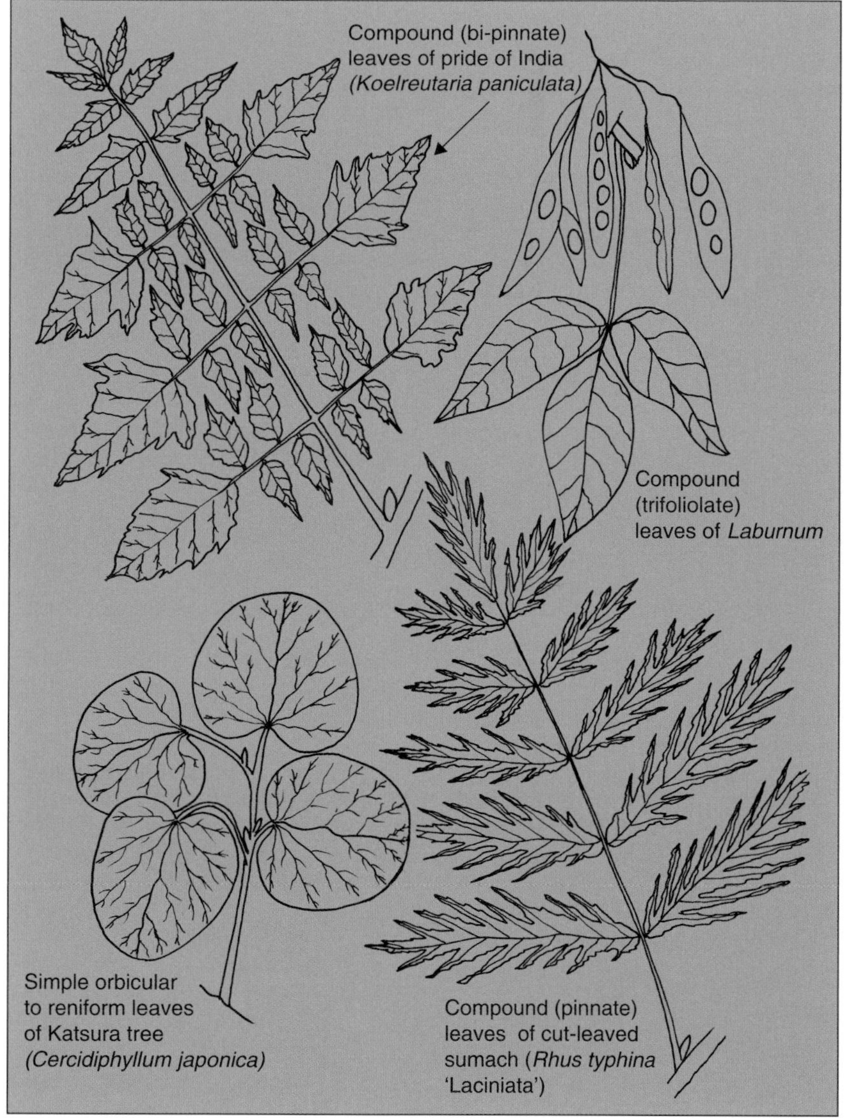

Foliage leaf shapes V (compound and simple leaves).

plate). Fleshy scale leaves (even though they never appear above soil level) form an axillary angle with the condensed stem, and, just as in aerial stems, axillary buds can appear in some of the leaf axils formed. The axillary buds are not destined to be aerial stems, but instead develop into new daughter bulbs; axillary buds on aerial growth and daughter bulbs are very much alike in make-up. Where buds form at the outside edge of the base plate they produce what are known as offsets. These are also destined to become new daughter bulbs; unlike internal axillary buds, they have very little resistance to breaking away and forming a new individual young bulb, so are removed easily from the parent when the bulb is lifted.

Also arising from the condensed stem (base plate), and surrounded by the protective layers of concentric rings of fleshy scale leaves, is a central axis comprising foliage leaves and a developing inflorescence (flower system) – both of which are

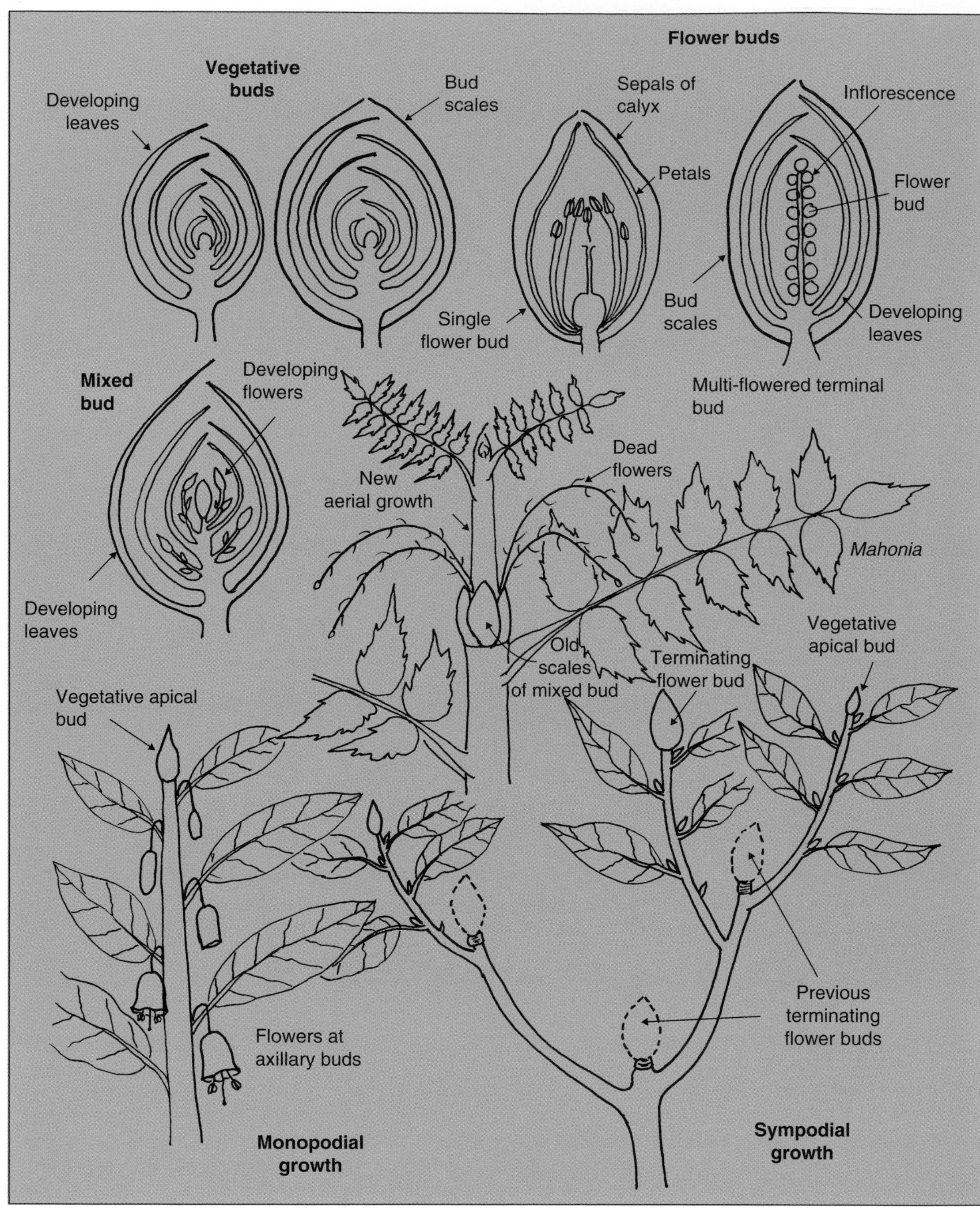

Types of buds.

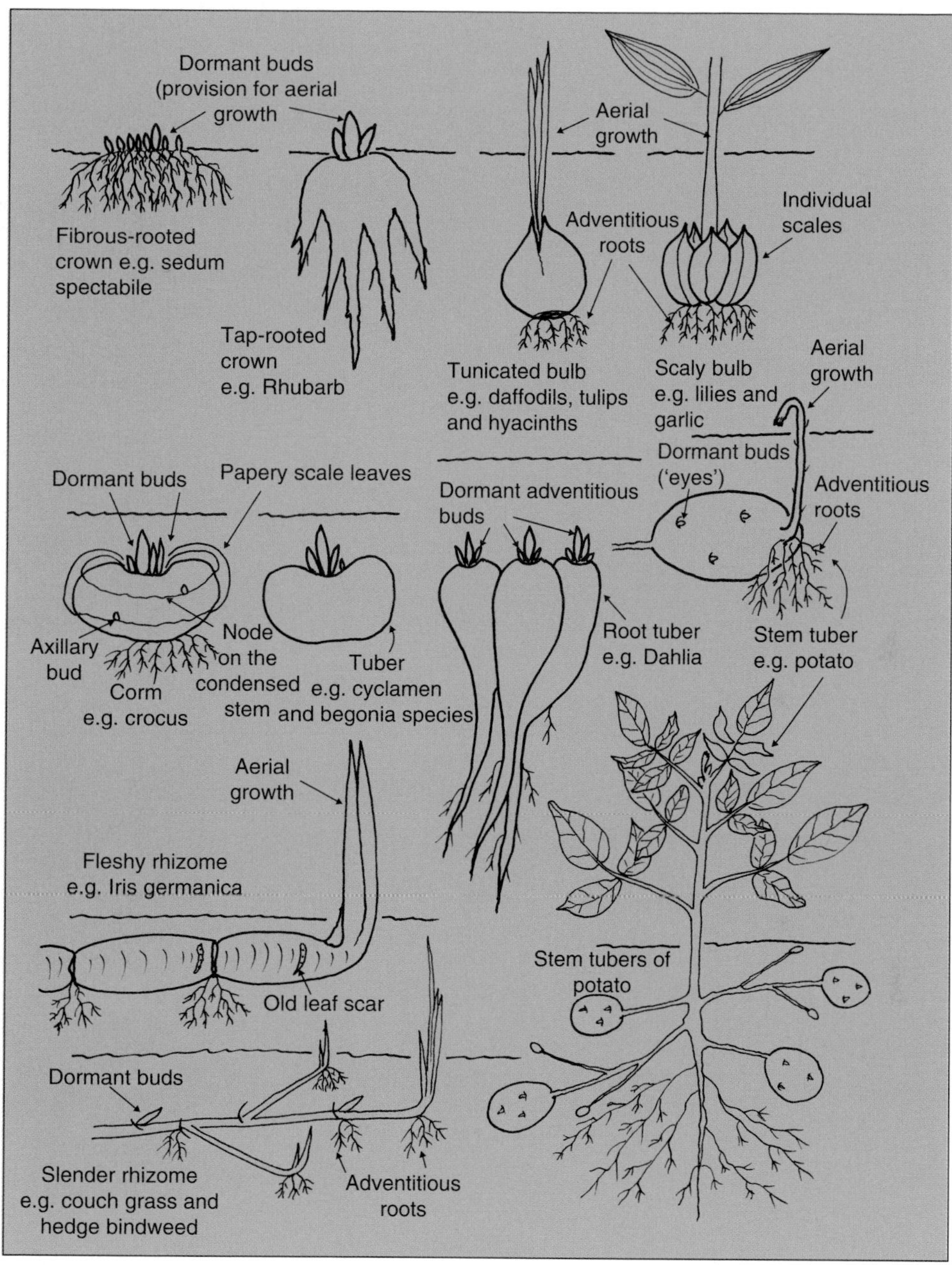

Subterranean organs of herbaceous perennials.

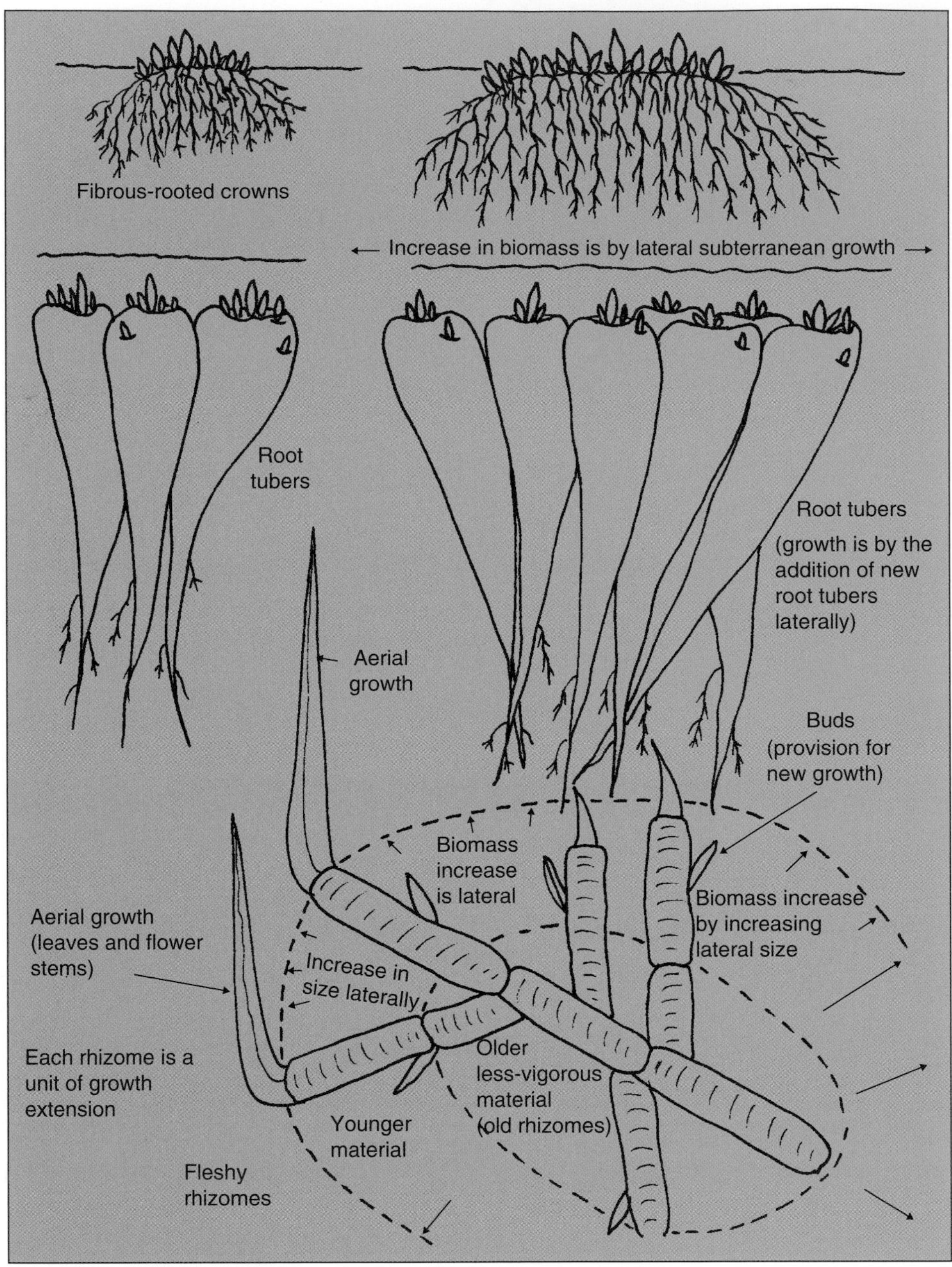

The growth patterns of herbaceous plants.

yellow/green (even in their early phases of development) because they contain chlorophyll. The initial energy supply for the rapid early growth comes from food stored in the fleshy scale leaves; it is later supplemented by sugars produced at the foliage leaves as they appear above the soil surface. Hence the green foliage leaves are very important after flowering to ensure that sufficient sugars are produced by the leaves during the summer, and transported down the phloem (part of the vascular tissue) to the fleshy scale leaves of the underground organ. It is essential therefore to retain the leaves of bulbous subjects after flowering to ensure good energy budgets to facilitate bulb size increases and/or successful flowering in the following season.

Within the scale leaves the soluble sugars are converted to insoluble starches for storage. Obviously oxygen and moisture are essential for the bulb to survive and grow, and it is the coming together of both moisture (which is nearly always present) and favourable temperature that commences enzyme action and initiates growth. Enzyme action causes the digestion of non-soluble starches to soluble sugars, which allows the rapid transportation of sugars throughout the system (for energy). Different enzymes therefore (with a different temperature requirement) may be involved in different species – notably between spring and autumn flowering types.

Simple dissection of common tunicated bulbs such as daffodil or hyacinth shows the details well, and a bulb of onion is ideal for the task because it is easily available (and large in size).

New daughter bulbs and offsets can be removed and grown on in nursery conditions. 'Coring' or 'scooping' may also be used as methods of propagation, or even more intense systems involving dissection of the bulb, such as 'chipping' and 'twin scaling'. Coring involves using an apple corer through the centre of the bulb, and scooping involves scooping out part of the base plate with a sharp knife. The propagation systems depend upon the fact that axillary buds are formed in the angle between the scale leaves and the (albeit condensed) stem (the base plate). Axillary buds are readily produced in this area as a natural process anyway. However, the damage caused by coring and suchlike initiates callus tissue, which could differentiate into a range of tissue types, but because of the location of the developing callus (at a leaf axil), is most likely to form axillary buds that can themselves develop into new daughter bulbs.

Scaly bulbs have some similarities to tunicated bulbs, but they comprise collections of individual fleshy scale leaves radiating round a central axis. Each thick fleshy scale may be naked, or it may have an exterior papery scale and can produce small foliage leaves at its centre. Only when there are sufficient numbers of scales comprising the bulb will it flower, as it is the aggregate effect of the fully engorged scale leaves that ensures sufficient energy. Each scale leaf also has the ability to produce new 'scales', and these can be removed and grown on (multiplied up) as a method of propagation. *Lilium* species and garlic have scaly bulbs – each 'clove' of garlic is a scale leaf. Scaly bulb subjects include giant lily (*Cardiocrinum giganteum*), imperial lily (*Fritillaria imperialis*), snake's head fritillary (*Fritillaria meleagris*) and summer snow flake (*Leucojum aestivum*).

The similarities between aerial axillary buds (condensed shoot systems) and bulbs (condensed subterranean shoot systems) are highlighted by some *Lilium* species (and related cultivars). They produce specific axillary buds in their aerial stem leaf axils known as 'bulbils', which can be removed and 'sown' to produce new scaly bulbs – *Lilium tigrinum* and *Lilium regale* often produce bulbils on their flowering stems. Common onion (*Allium cepa*), crow garlic (*Allium vineale*) and field garlic (*Allium oleraceum*) can all produce aerial bulbils at the apex of their flowering stems, growing within an enveloping, pointed leaf-like spathe.

There are two main types of rhizome, those with vertical and those with horizontal systems – so technically they include all forms of corm and stem tuber as well. Fleshy rhizomes comprise thick, fleshy underground horizontal stems, sometimes shallow systems actually on or near the soil surface. Rhizomes are classified as 'stems' because they have leaves radiating from nodes that form leaf axils, and may have axillary buds within them: so the buds are not adventitious (appearing where they are not expected) because they arise on stems. However, the absorptive roots – often quite rudimentary, and usually arising from nodes – *are* adventitious because they arise from positions on the stems (albeit subsurface stems), and not from an existing root system.

The very large proportions of the fleshy rhizomes of species such as blue flag (*Iris germanica*) and sweet or yellow flag (*Iris pseudoacorus*) and Japanese knotweed (*Reynoutria japonica*) show the details of their underground stems well: they show

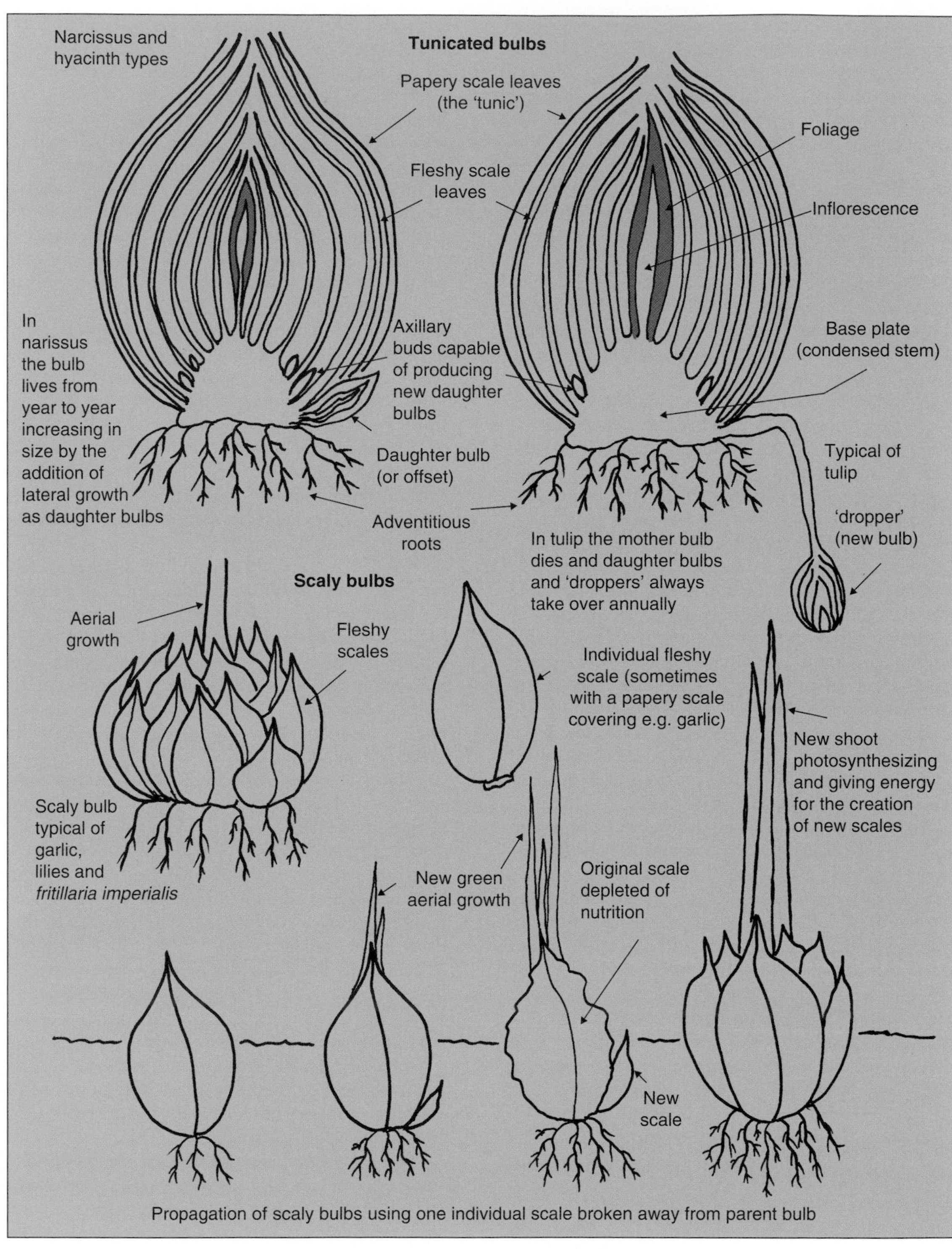

Tunicated and scaly bulbs.

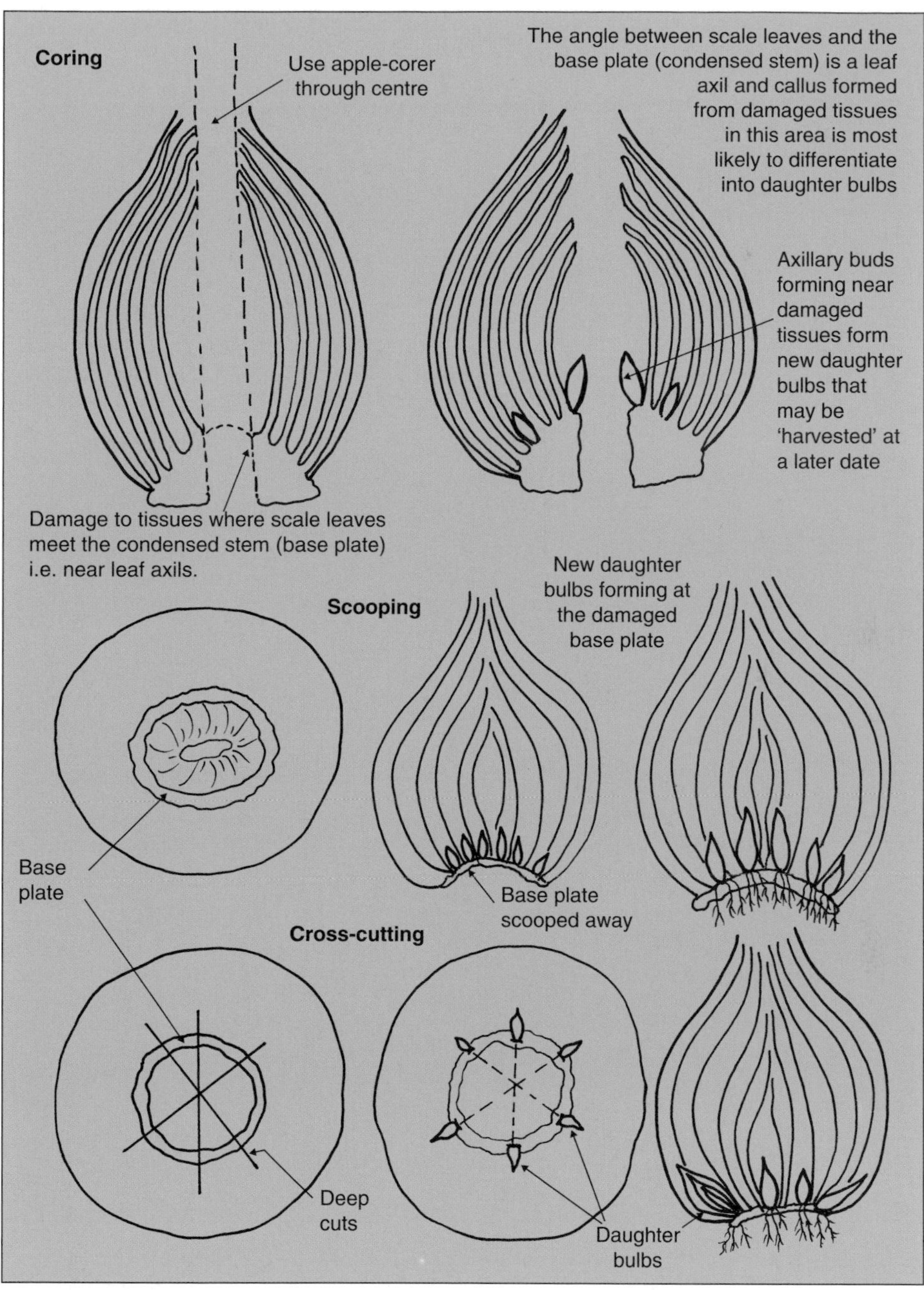

Intensive propagation methods for species and cultivars with tunicated bulbs I (coring, scooping and cross-cutting).

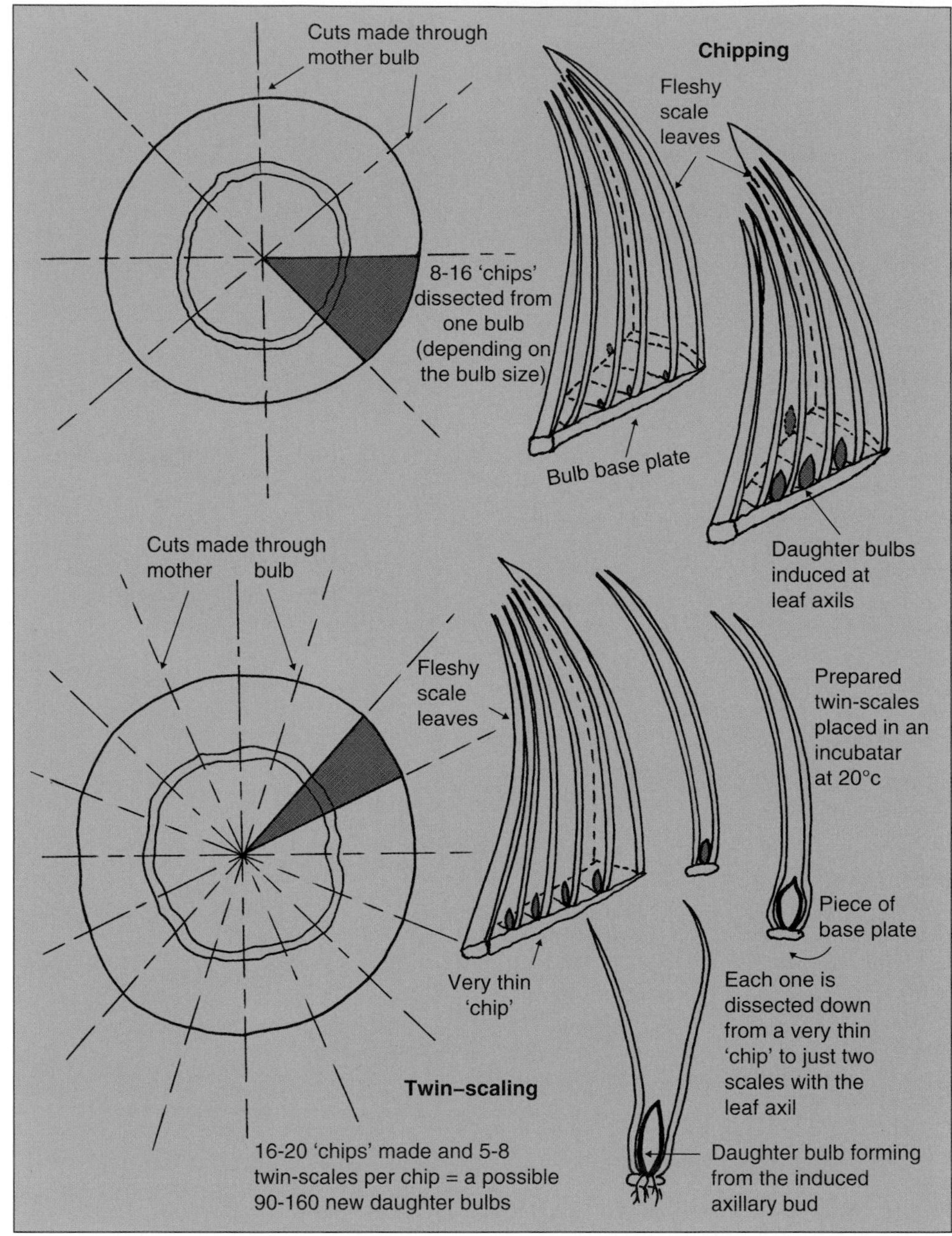

Intensive propagation methods for species and cultivars with tunicated bulbs II (chipping and twin-scaling).

distinct leaf scars at former points of attachment of leaves (nodes), and these also display the smaller scars left by the vascular tissues. Rhizomatous subjects with fleshy rhizomes also include *Clivia miniata,* bracken (*Pteridium perenne*), bamboo and banana.

Slender rhizomes share the external features and have the same storage function as fleshy rhizomes, but because they are thinner than their fleshy counterparts, the features are not quite so easy to discern. There is a correlation between rhizome size and leaf size: large leaves produce lots of sugars that are stored in large, fleshy rhizomes, whereas species with slender rhizomes generally have smaller leaves, but fairly extensive and/or deep rhizome systems that make up a large volume of storage organs but over a greater soil volume. They are often very persistent species because of it, including some herbaceous perennials classified as weeds: hedge bindweed (*Calystegia sepia*), field bindweed (*Convolvulus arvense* – *arvense* meaning 'of the fields'), field or creeping thistle (*Circium arvense* – again, *arvense* meaning 'of the fields'), ground elder (*Aegopodium podograria*) and couch grass (*Agropyron repens*), are all very invasive.

Corms are condensed underground stems (swollen stem bases), but are without the internal fleshy scale leaves associated with bulbs. The scale leaves of corms are found as papery scales surrounding the fleshy condensed stem base only, and they carry axillary buds externally at nodes formed where the papery scale leaves meet the condensed stem. There are usually a low number of nodes on corms, but each associated axillary bud has the ability to swell and form a new daughter corm (cormlet) easily. Individual species of cormous subjects flower at different times: spring (for example *Crocus* and some *Colchicum* species), summer (many *Gladiolus* species) and autumn (many *Colchicum* species). Cormous subjects include angel's fishing rod (*Dierama pulcherrima*) and *Montbretia* (*Crocosmia*).

Some buds at the top of the corm develop into flowering apices, but also at the top of the corm are buds that can form new daughter corms. Some cormous subjects form daughter corms alongside the original (horizontal system), others produce more on top of the original (vertical system) and may produce specialized adventitious contractile roots from the top of the new corm. Contractile roots commence thick and fleshy, but later desiccate and 'contract' to create a 'pulling' force that pulls daughter corms formed at the top of the parent corm into the void created when the previous corm rots away. The system is good, but species such as *Crocosmia* (Montbretia) that produce excessive numbers of new daughter corms annually, and whose old corms are slow to rot away, end up with vertical 'stacks' of corms above soil level. Contractile roots have little or no absorption properties, but fine fibrous roots for absorption are formed as small adventitious groups near nodal regions. On some species new bulbs or corms may be formed on 'droppers', which are swollen nodules on contractile roots. These are commonly found on the corms of *Crocosmia* and *Oxalis latifolia* and the bulbs of tulip.

Stem tubers, as their name suggests, comprise swollen underground stems (vertical rhizome systems) with axillary buds and adventitious roots. The aerial parts of the plant arise from the underground axillary buds, as illustrated by the potato (*Solanum tuberosum*), and also the Jerusalem artichoke (*Helianthus tuberosa*) – this is the main feature that separates Jerusalem from globe artichokes: it is the stem tubers of Jerusalem artichoke, and the inflorescence (single flowers) of globe artichokes that are eaten. Tuberous subjects include *Trillium* species, tuberous begonia (*Begonia hybridatuberosa*), cyclamen (for example *Cyclamen coum, Cyclamen hederifolium,* and *Cyclamen persicum*), dragon lily (*Dracunculus vulgaris*), voodoo lily (*Sauromatum venosum*), arum lily (*Zantedeschia aethiopica*), wild arum (*Arum maculatum*), blue anemone (*Anemone blanda*) and lesser celandine (*Ranunculus ficaria*). Discerning between some forms of corm, stem tuber and fleshy rhizome is not always easy, as they are all forms of fleshy underground stems. However, rhizomes are generally more horizontal, elongate and wide-spreading than tubers (which tend to be more compact and vertical in nature), and rhizomes carry distinct leaf scars along their length.

Root tubers comprise specialized swollen, fleshy roots (that act as food stores) with dormant adventitious buds that are responsible for new aerial growth – the most common example is dahlia. Dormant adventitious buds (shoots) are found near, at, or just above soil level on various modified root systems.

Thick, fleshy, taprooted crowns also have dormant buds at their top, as typified by rhubarb (*Rheum palmatum*), comfrey (*Symphytum officinale*), anchusa (*Anchusa azurea*), dock (*Rumex* species) and dandelion (*Taraxacum officinale*). Species with thick, fleshy taproots often have very large leaves in order to produce lots of sugars and to have high vigour. Subjects with fibrous-rooted crowns (such as *Sedum spectabile*) produce dormant adventitious buds at the base of the old flower stems (or at the top of the root system at soil level) by the end of the season, that facilitate aerial growth for the following year. Food is stored in the myriad fine roots (that add up to a fairly large volume for storage), and they usually have relatively small leaves compared with fleshy-crowned subjects. Without close inspection to discern fine details it is easy to confuse the profuse amounts of thin, slender

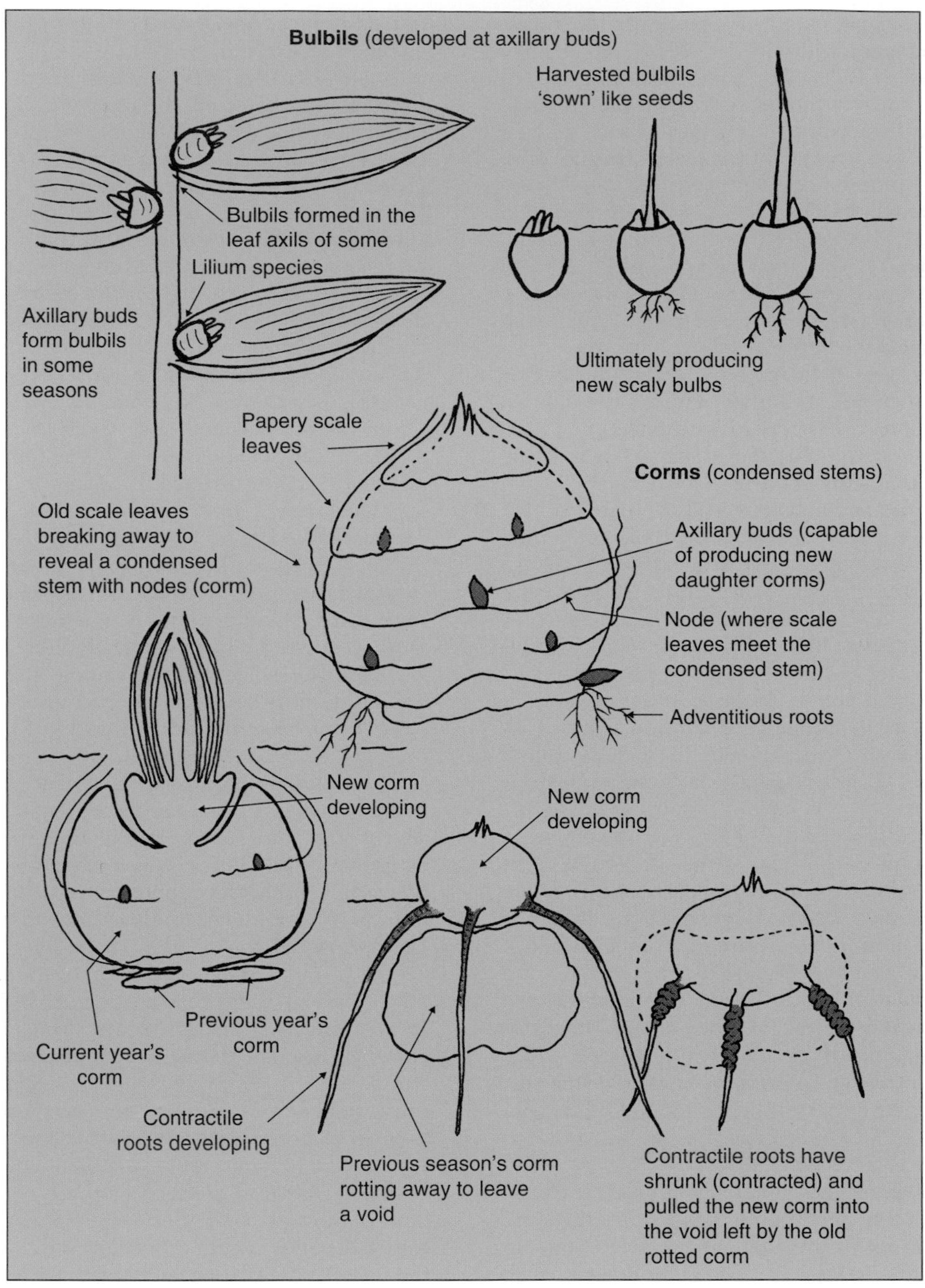
Bulbils (developed at axillary buds)
Harvested bulbils
'sown' like seeds
Bulbils formed in the
leaf axils of some
Lilium species
Axillary buds
form bulbils
in some
seasons
Ultimately producing
new scaly bulbs
Papery scale
leaves
Corms (condensed stems)
Old scale leaves
breaking away to
reveal a condensed
stem with nodes (corm)
Axillary buds (capable
of producing new
daughter corms)
Node (where scale
leaves meet the
condensed stem)
Adventitious roots
New corm
developing
New corm
developing
Previous year's
corm
Current year's
corm
Contractile
roots developing
Previous season's corm
rotting away to leave
a void
Contractile roots have
shrunk (contracted) and
pulled the new corm into
the void left by the old
rotted corm

Bulbils and corms.

rhizomes produced by some species, with fibrous root systems.

Besides the possibility of subterranean stem or root modifications, some herbaceous (and woody) subjects have above-surface (aerial) stem modifications. Stolons are specialized stems that produce adventitious roots along their length (usually, but not solely at nodal regions), and produce new plantlets either from apical buds or axillary buds, or both. The 'runners' of strawberry (*Fragaria*) and spider plant (*Chlorophytum*) are stolons, as are the specialized rooting stems found on blackberry (bramble – *Rubus fruticosus*), and offsets on some *Sempervivum* species that help with their monocarpic nature. When bamboo, banana, *Agave americana* and many *Sempervivum* species flower, the main flowering sections die, so they are said to be monocarpic (mono = once, carpic = referring to the fruit: it fruits once, and then dies). Subjects with stolons include creeping buttercup (*Ranunculus repens*); and the relative ease with which *Cornus stolonifera* roots along its stem led to its specific name.

The stocky 'offsets' or 'toes' of *Agave* and *Sanseveria* are condensed stolons if found above soil level, or rhizomes if found below. However, as they are all technically horizontal rhizome systems, the same organ may be called a runner or a rhizome. 'Creeper' is often erroneously used as a generic name for all types of low-growing rooting stem types, and even some climbers. However, creepers are actually specific in nature, as they are over-ground stolons that root along their length but have no stipules or scale leaves at leaf junctions.

The External Features of Grasses

The leaves of grasses (and cereals) are long and sword-like, and typically divided into two distinct sections: the leaf blade itself (the flat, or partially rolled, expanded part), and a surrounding sheath. The sheath is protective and usually envelopes part of the stem, and may aid stem strength.

At the union where leaf blade and sheath meet is a small piece of delicate tissue called the ligule: its function is thought to be to reduce the invasion of water and soil between the stem and sheath, but this is far from proven. Some species have two ear-like pieces of tissue at the base of the leaf blade called auricles (ear-like appendages). Some species may have both ligules and auricles, whilst other species may have just one or the other.

The External Tissues and Features of Woody Perennial Plants (Arboreals)

Woody plants are unique in their ability to add both length and girth (thickness) annually, and include all the trees and shrubs (by definition they have persistent woody stems and overwinter as free-standing woody structures, and do not die down on an annual basis). It is the capacity for continued girth increment that builds up tissue layers and makes them strong enough to cope with the additional weight added annually, and creates a large, self-supporting, aerial structure, and a stable subsurface network of roots.

Deciduous woody perennials lose their leaves in the autumn, and overwinter as a woody framework with scale-leaf-protected dormant buds as provision for the following season's growth. Evergreen woody perennials do lose their leaves, but it is a sequential process, and they retain at least the current season's whorl and the previous season's whorl of leaves. In exceptional circumstances – in ideal conditions of good light, sufficient moisture, and mild temperatures – three (or rarely, four) whorls of leaves may be retained. However, two to three is the norm, as anything but ideal conditions brings on premature leaf senescence.

Flowers on woody plants may emanate from the current season's growth, but more commonly come from older growth. Those that form apical flower buds on the current season's growth will terminate vegetative extension at that point, producing mainly multi-branched (sympodial) growth patterns. Other species flower from leaf axils or on lateral 'spurs' that do not terminate apical growth patterns. These tend to make more extension growths at the apex, and although having some branching, and are therefore not truly monopodial (having growth in only one direction), they can be more upright in habit than other species. It is usually a mixture of both sympodial and monopodial growth that forms the structural pattern of any one species. Trees during their juvenile phases are not sexually mature, so flowering does not occur; they are therefore mainly monopodial in their growth patterns, and tend to be tall and slender – a strategy that aids their competition for light with other trees.

Addition to the structure in length is brought about by the soft green growth at their stem tips, and is known as 'primary growth'. Addition to the structure in girth annually is brought about by tissues that are added to the primary tissues, and

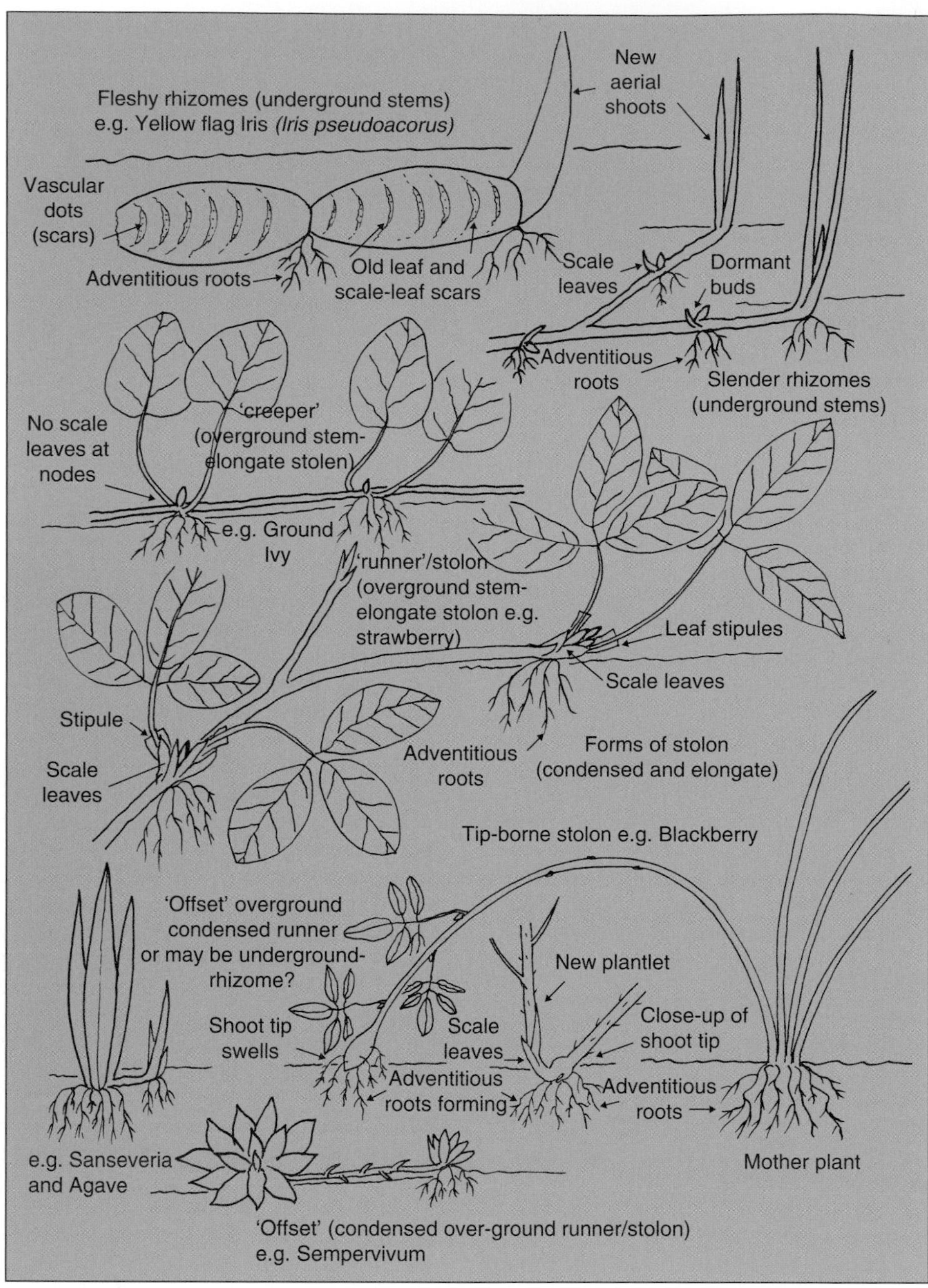

Natural vegetative methods of propagation.

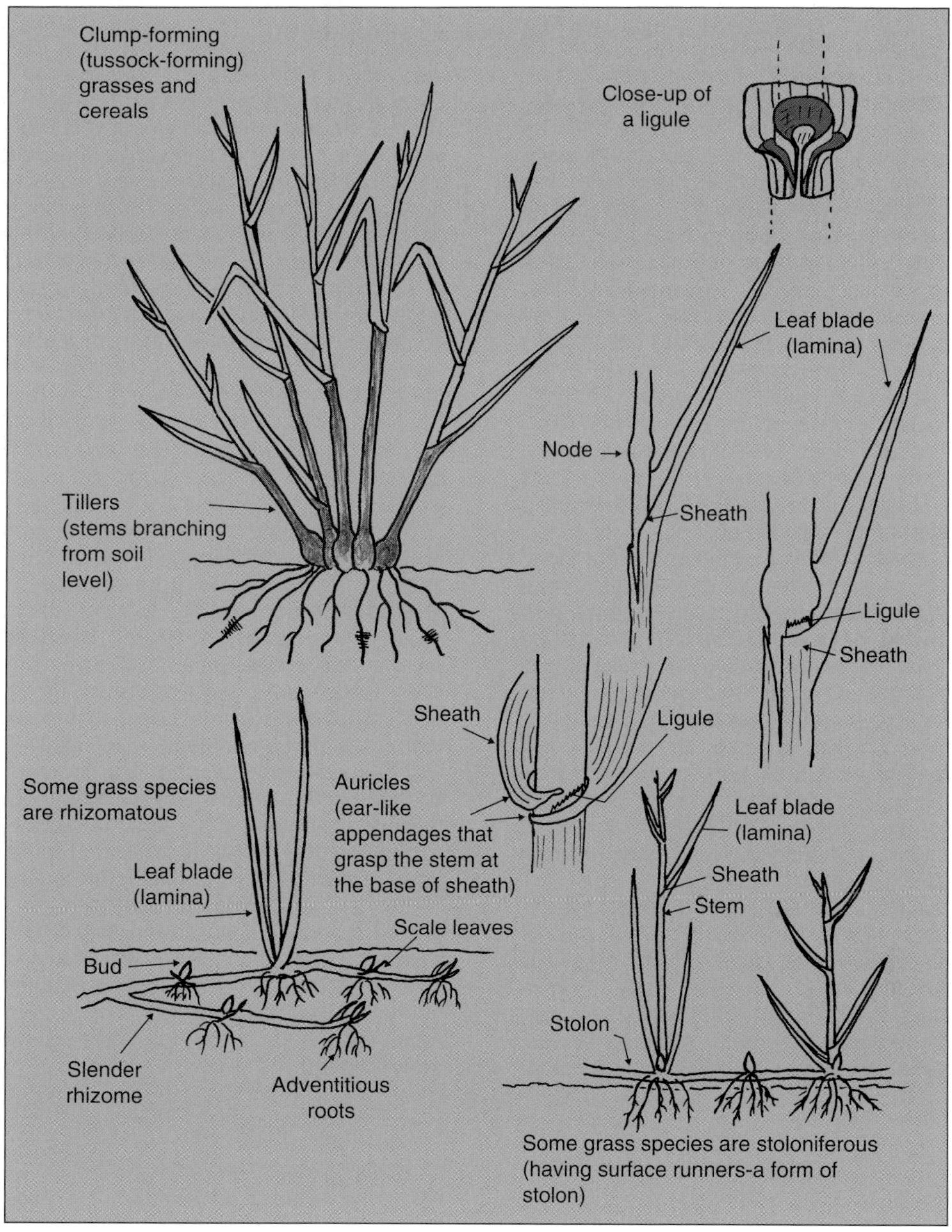

The external features of grasses.

is known as 'secondary growth' ('secondary thickening'). The tissues found in green stems (and the early seedling stages) of trees, because they have only primary growth patterns, are similar in anatomy to green plants in general, whereas the secondary tissues are unique to woody plants.

As the season progresses, and just before the close of growth in the first year, the internal tissues of the soft extension growth go through a strengthening process known as lignification to prepare the wood for the winter; outwardly they become covered with bark.

PLANT STRATEGIES: GRASSES

The growing points of grasses responsible for the production of new leaves and stems remain low in the growing system, and may even be near or under the soil level. Thus the growing points of rhizomatous grass species are below the soil surface; on stoloniferous types they lie on the soil surface; and in tussock-forming types they are at the base of the plant (not that far from the soil surface). However, the other feature of grasses that ultimately affects their habit is whether lateral branching occurs very low down and breaks through the side of the sheath, or grows inside the sheath some way up the plant before breaking away. This varies from species to species. Some species (bents) have abrupt knee-like bends in their stems (known as geniculate growth) to aid their stem strength and low-growing ability.

The growing points of grasses are not found in the tips of the leaves, so if grass leaves are damaged, the low growing points produce new replacements because they remain undamaged by the heights of cut involved in mowing and grazing. If this were not the case, every time a lawn was mown, all the growing tips of the grass plants would be removed and the plants would soon become exhausted and die. Rhizomatous and stoloniferous grasses happily regenerate from dormant adventitious buds found along their length, and tussock-forming types from their base.

If the tops of leaves are cut off with scissors, because there is no provision for repair or extension growth, the leaves do not magically become re-pointed: instead they remain with their straight cut tops until they die, and are replaced by new individual leaves. The same exercise carried out on other monocotyledonous plants, even those with very large leaves, will have the same effect: thus a straight cut removing the tip of the leaf of mother-in-law's tongue (*Sansevieria trifasciata*) will remain until the individual leaf dies and is replaced. The removal of the very sharp and dangerous spines on the leaf tips of *Agave americana*, so that the leaves remain as aesthetic features until they die and are replaced, hangs on this. Death and replacement of old senescent leaves, rather than renewal of old existing leaf tissues, is of course the norm. It is the absence of growth points in the leaves of both herbaceous and woody plants that leads to natural senescence (and in some cases abscission), so is not at all unusual in the plant world.

Regular grass mowing encourages lateral growth known as 'tillering'. Grasses have rudimentary stems that carry several leaves, and all grasses produce stems to hold their flowers (their inflorescence) at height, because they are wind-pollinated (anemophilous). The relatively large scale of tussock-forming species and cereals shows the basis of the external features well. Grass stems are initially formed at the basal meristem, and they become aerial because they contain intercallary meristems (secondary meristematic regions). These form at positions up the stem (at nodes) rather than forming at the apices, and allow for extension in height by producing tissues for new aerial growth that is pushed ahead of them, not left behind by them, as found in the stem growth of other plants. Hence, when leaf tips are removed by mowing, although they do not re-point, and remain with a flat cut tip, the leaves continue to grow in length.

The cork of bark forms a very strong, protective outer layer covered with suberin (a tough, waterproof wax) and punctuated by lenticels. Lenticels are pores that facilitate the gaseous exchange of oxygen, carbon dioxide and water vapour; they are easily recognized as slightly raised dots, or more extensive areas of raised bark, of a different colour to the surrounding bark. Close inspection of common elder (*Sambucus nigra*) shows very discernible lenticels (as raised light brown areas), and also green (chlorophyll) just below the bark surface. Antarctic beech (*Nothofagus antarctica*) has such distinctly white lenticels that it could arguably be grown for this feature alone, and all forms of peeling bark birch have prominent, easily seen lenticels.

Other recognizable features of woody stems include leaf scars (left after leaf fall), and girdle scars that are created when the bud scales surrounding the apical buds drop off. Leaf scars are typified by various sized crescent- or urn-shaped scars (depending on the shape of the petiole of the species), with very distinct 'vascular dots/vascular plugs'. If at leaf fall (abscission) the vascular traces

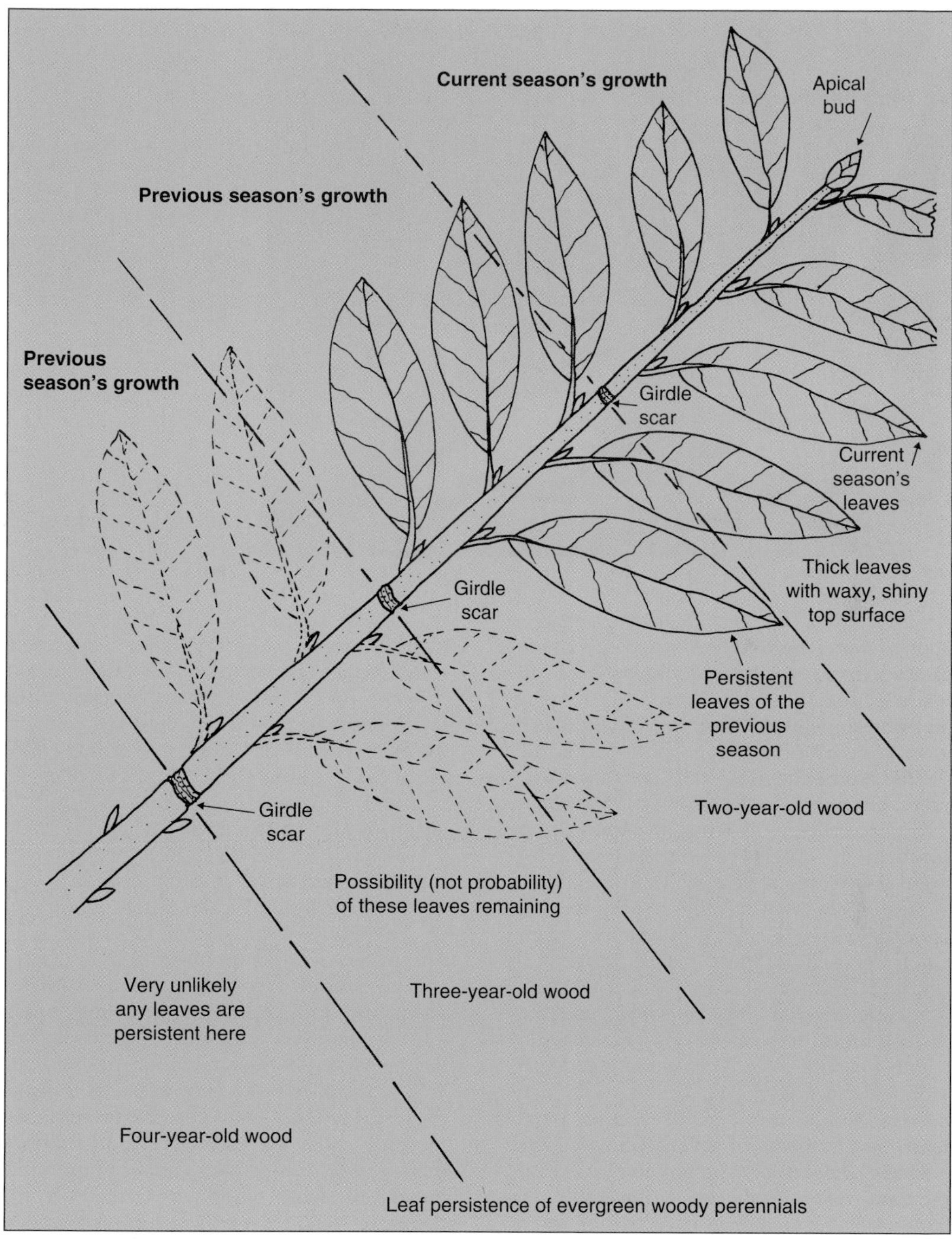

External features of woody stems I (evergreen woody perennials).

Lenticels (and chlorophyll below the bark) of common elder (Sambucus nigra).

that supplied water from the roots to the leaves were just left open, they would allow the loss of vast quantities of water from the system. To prevent this from happening, several weeks before leaf fall a corky layer (abscission layer) is formed in the tissues of the leaf petiole (leaf stalk) across the vascular tracts. The waterproof cork cuts off the water supply to the leaves, and instigates leaf fall. A deciduous stem in winter, or two to three year (or older) stem at any time of year, will show leaf scars and the associated vascular plugs. Some species, because of the relatively large proportions of their leaves and leaf petioles (leaf stalks), show bud scales, girdle scars and leaf scars particularly well: horse chestnut (*Aesculus hippocastanum*), angelica tree (*Aralia sinensis*), tree of heaven (*Ailanthus altissima*) and foxglove tree (*Paulownia tomentosa*) show wonderful leaf scars.

Leaf scars bear a number of vascular plugs related to the number of main veins in the leaves. Horse chestnut, which can have five or sometimes seven leaflets to each distinctly compound leaf, will therefore have five or seven 'vascular plugs' at each leaf scar. Because after the leaf fall of deciduous subjects there are no petioles, and therefore no axils (the angle between the stem and the petiole), buds can no longer technically be called axillary at this stage, therefore they are known as lateral buds – that is, responsible for side growth.

Even thick, leathery, suberin-covered bud scales (which are forms of modified leaves) need a water supply in order to remain turgid and avoid desiccation. Hence they are connected to the normal vascular system, and the corky dots are actually minute vascular traces that are now 'plugged' with cork to prevent them from 'bleeding' moisture and nutrients when the bud scales fall. Rapidly expanding green extension growth causes the bud scales to drop away in spring, and the scars left by bud scales represent a microcosm of the main foliage system involving leaf scars. On species of small proportions they may require a hand lens to view them. Bud scales may be green, pale pink, salmon pink or red, and can be quite aesthetic on some species, particularly if they hang persistently for some time after bud burst.

The distance between girdle scars is decided by the green growth phase (primary growth), as this is the only tissue that can increase in length. From then on the stems only increase in girth, and the distance between girdle scars remains fixed; they therefore offer a perfect chronological record of growth patterns produced by a tree or shrub influenced by its previous specific environmental conditions.

The root systems of woody species comprise three main elements: the thick, bark-covered persistent roots that increase in girth annually and act as very substantial water-conducting tissue; the

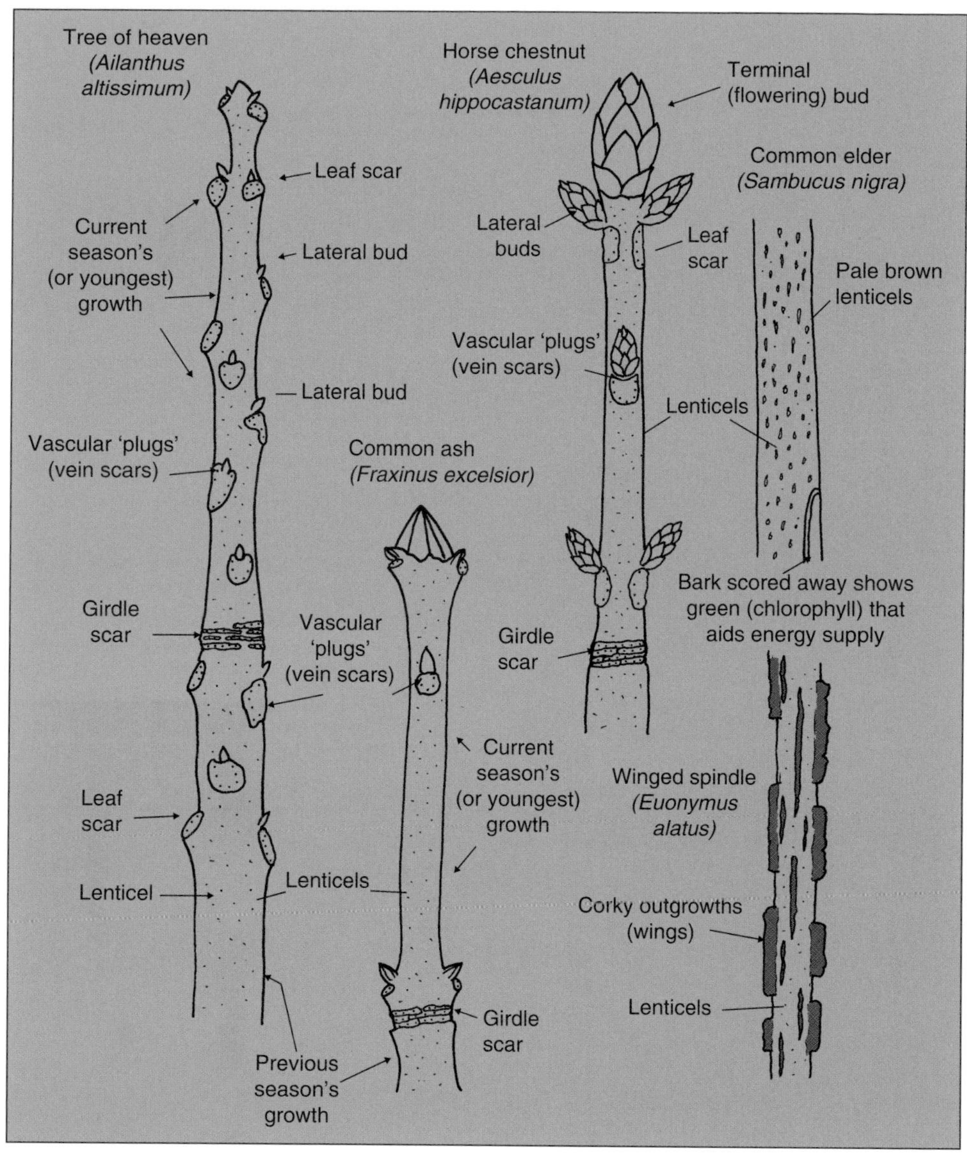

The external features of woody stems II (deciduous woody perennials in winter).

bifurcating, ramifying lateral roots; and the very fine, moisture-absorbing roots (rootlets) with associated fine root hairs. Externally the stem-to-root interface can be seen if root flare formed at the junction is near the soil surface, but it can be far less obvious on small shrubs than on trees. It is now known that the root plates of healthy trees and shrubs are much wider than the crown spread, and although there is evidence of many fibrous, water-absorbing roots at the drip-line level, many main roots (and their associated finer roots) extend way past the drip-line/perimeter of the crown. Hence, stability is given by shallow but very wide root plates.

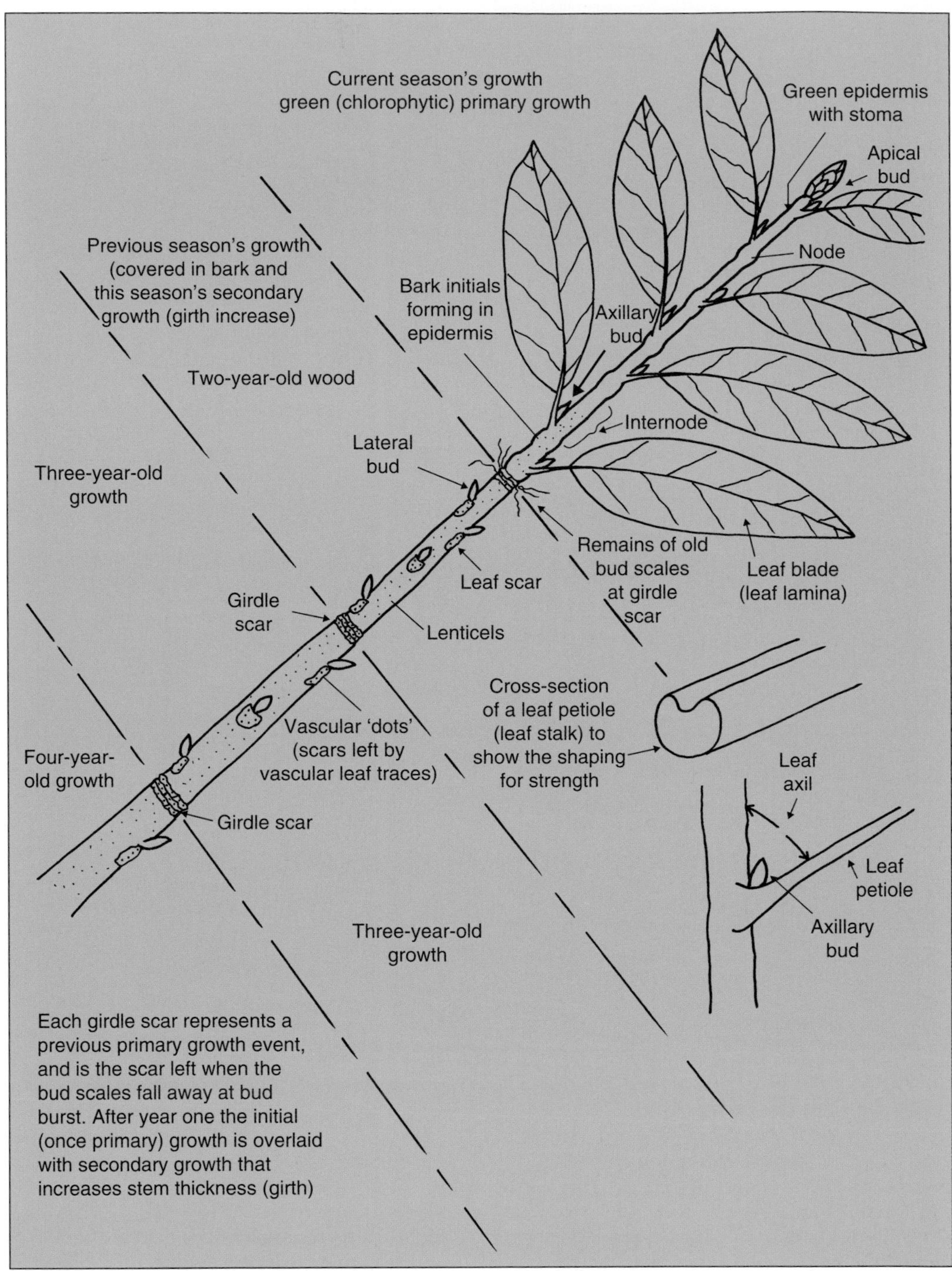

The external features of woody stems III (deciduous woody perennials in summer).

Arborescent Plants

Arborescent plants are not technically arboreal (having all of the features of woody plants) but attain tree-like proportions or have some tree-like features. Arborescent plants include palms (often associated with lots of thick coarse fibres on their stem), and a few other monocotyledonous plants in the Liliaceae and Agavaceae families, such as *Yucca*, *Cordyline*, *Dracaena* and *Agave*. An arborescent species of the bird of paradise plant (*Strelitzia gigantea* syn. *Strelitzia nicolai*) also attains tree-like heights. Palms do not increase in girth annually (which is considered the main discerning feature of woody plants). However, some Cordylines and Dracaenas do increase in girth annually, albeit slowly, and have a rudimentary bark – but they do not have the same secondary growth patterns as true arboreal plants. The dragon tree (*Dracaena* draco) is slow growing, but forms a round-headed tree of large proportions over time. A very large specimen at Icod de los Vinos (Tenerife) is thought to be 600–1,000 years old, and does have a very thick, bark-covered trunk – illustrating just how large some arborescent plants can be. Some Pteridophytes (ferns), tree ferns (such as *Dicksonia antarctica*), and some distant relatives of the conifers, the Cycads also have arborescent forms (for example *Cycas revoluta*) that attain tree-like proportions.

Annual gains in height are facilitated by a central growth bud that develops sets of either large frond-like leaves (true palms, ferns and Cycads) or whorls of lancelolate leaves (*Cordyline*, *Dracaena* and *Pandanus*). The rate of growth varies with the species, but they gain height relatively slowly (by only a set of leaf bases per year). In true palms, because there is little or no annual increase in girth, the thickness of stem laid down in any particular growing season remains with the plant for the rest of its life. Hence, some palm trees show restricted stem girth in places, indicating that the conditions involved in development in that year were poor when compared with others.

Bamboos are also arborescent monocots, but they have a different growth pattern to those described above, because new stems arise from rhizomes (underground stems). Stems increase in height rapidly, because they have the provision of many grass-like plants (intercalary meristems) that allow growth to occur, not just from the stem apex, but from several points up the stem. Their rate of growth in length is very fast and they do thicken in girth slowly, which makes them quite strong and flexible, but they also gain mutual support by the fact that the developing stems arising from the rhizomes are very close together.

Banana also attains tree-like proportions, but is in fact an herbaceous plant with the stem ('trunk') actually formed by tightly rolled, green, leaf-like material. Banana does not therefore have any arboreal features such as bark and heavily lignified tissues – all tissues are primary, just like any other herbaceous subjects of smaller proportions. Beneath the soil it has very large, fleshy rhizomes like bamboo. Because of the very large areas of green photosynthetic material, the stems grow rapidly and become quite large. Banana stems die after fruiting (monocarpic). The death of fruiting stems (coupled with the management systems and incredible vigour, proportions and lateral growth of the rhizomes) creates 'creep', whereby over several years the centres of the plants move large distances from their original planting positions.

Sub-shrubs

Sub-shrubs are shrub-like plants that do not fully lignify at the end of every season, but instead lignify fully at their thickened stem bases over time. They naturally grow in warmer climates and have soft, lush new growth that if necessary can be sacrificed in unseasonal conditions, years, or one-off seasons, or if grown in colder-than-normal climates, and can regenerate readily from the remaining lignified (or partially lignified) older growth. Sub-shrubs may be nurtured in sheltered niches in poorer climates, and then pruned when frost damage cuts them back. This system is on the borderline of herbaceous when plants are grown in colder climates than they are used to, but sub-shrubs do have some secondary growth and therefore girth increment at their bases (and they are sometimes considered to be arborescent because of this).

Sub-shrubs often survive in harsher climates but not without sustainable damage (they survive, rather than thrive). Nevertheless, the range of low temperature conditions in which sub-shrubs can still survive is relatively limited, so they are usually treated as for Mediterranean plants by horticulturists (even if they actually come from elsewhere). Examples include *Coraria terminalis xanthocarpa, Mimulus aurantiacus, Phlomis fruticosa, Abutilon vitaefolium album, Brugmansia* (*Datura*), *Salvia fulgens* and *Salvia concolour.*

What may be a hardy shrub in one climate may be a sub-shrub in another, slightly harsher climate. *Paulownia tomentosa,* in mild climates a true woody

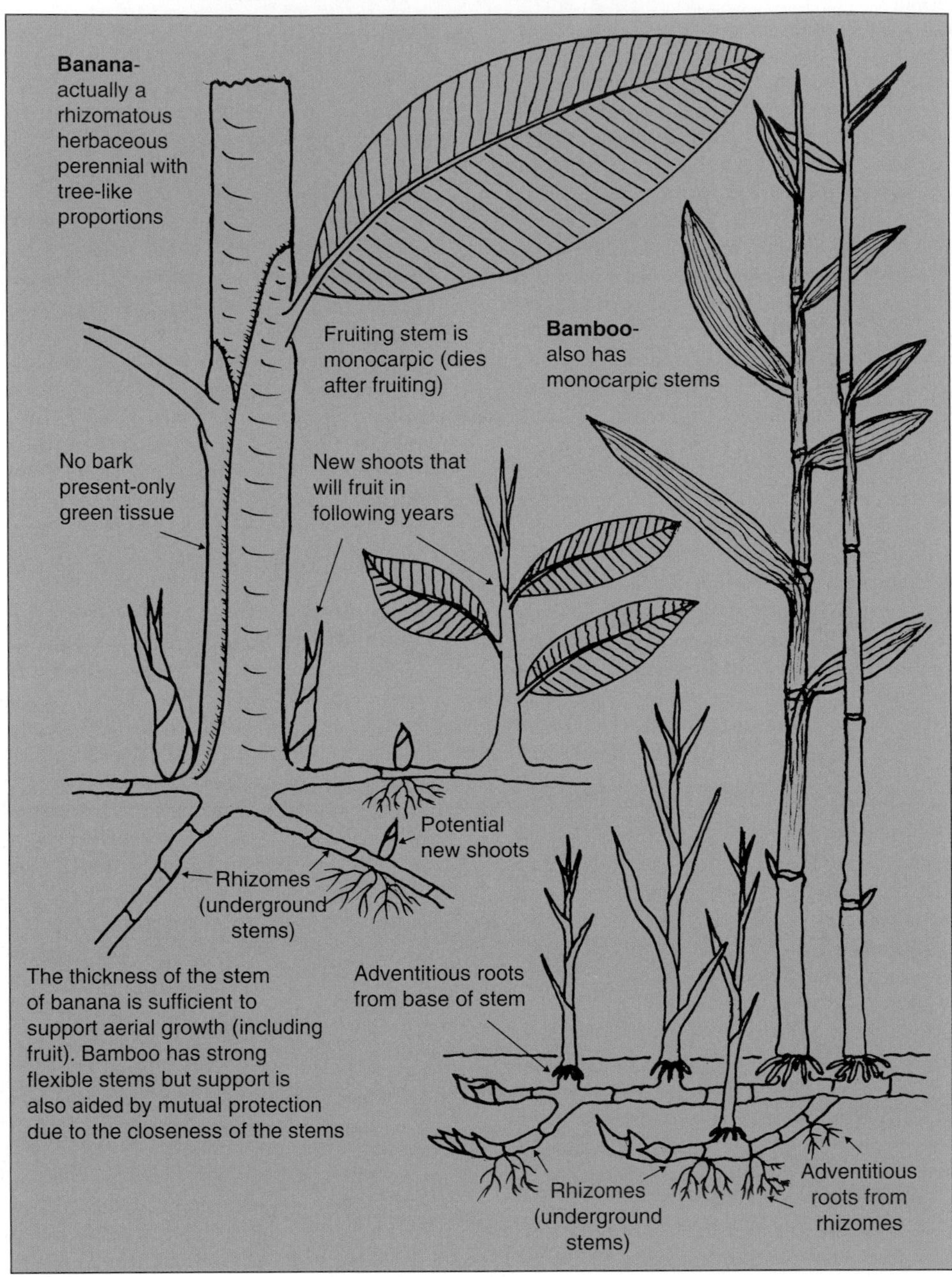

Arborescent plants I (monocots): banana and bamboo.

arboreal of tree-like proportions, in harsher, colder climates is an arborescent sub-shrub (albeit large) with soft herbaceous-type new growth. *Fremontadendron californicum* is an excellent sub-shrub with large yellow flowers. However, it produces very vigorous growth and seems to lack the trigger for good girth increment, which makes it rather top-heavy; it therefore does not make a stable free-standing plant in most climates. It prefers the shelter of a wall for success anyway, so tying it in

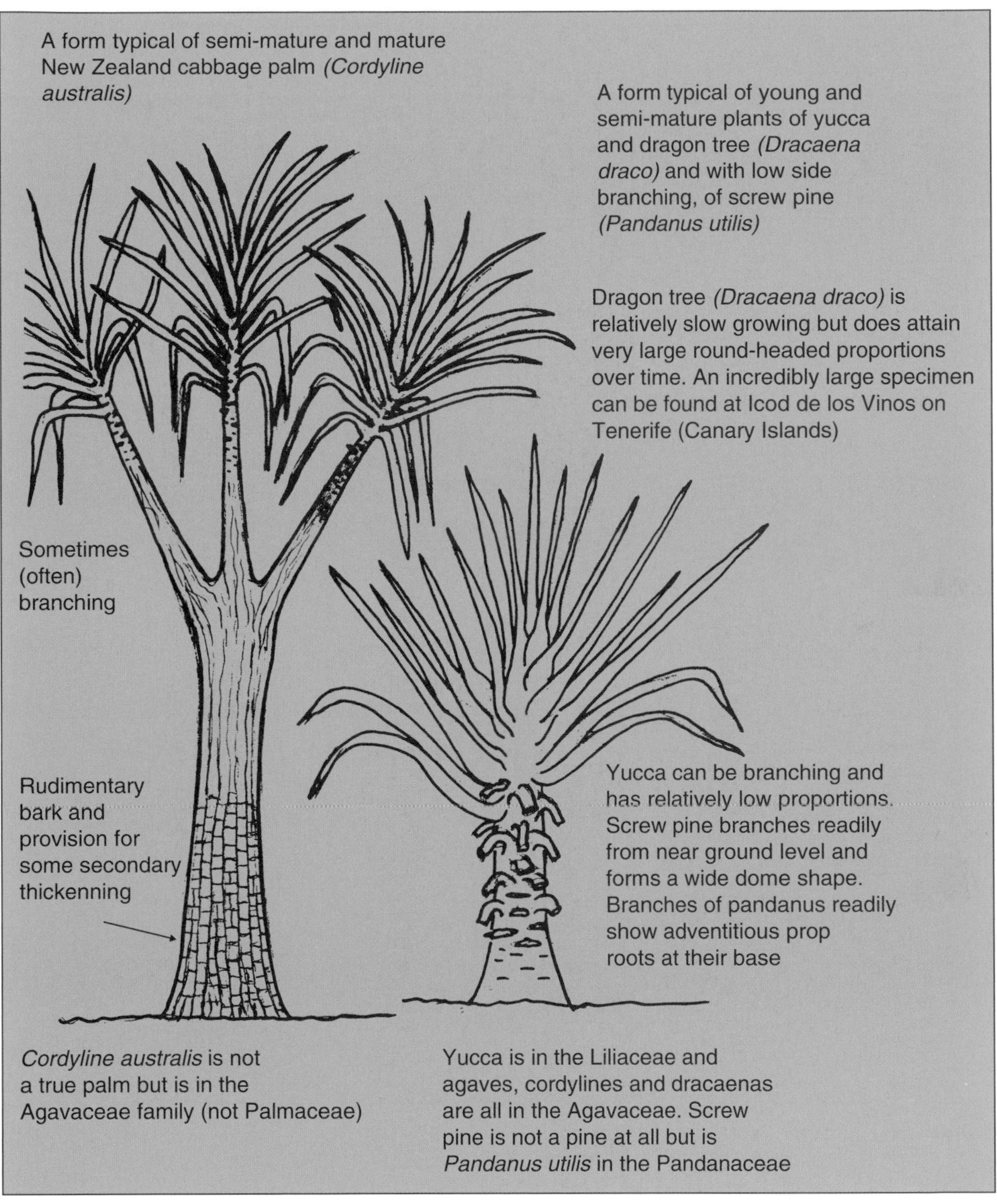

A form typical of semi-mature and mature New Zealand cabbage palm *(Cordyline australis)*
A form typical of young and semi-mature plants of yucca and dragon tree *(Dracaena draco)* and with low side branching, of screw pine *(Pandanus utilis)*
Dragon tree *(Dracaena draco)* is relatively slow growing but does attain very large round-headed proportions over time. An incredibly large specimen can be found at Icod de los Vinos on Tenerife (Canary Islands)
Sometimes (often) branching
Rudimentary bark and provision for some secondary thickenning
Yucca can be branching and has relatively low proportions. Screw pine branches readily from near ground level and forms a wide dome shape. Branches of pandanus readily show adventitious prop roots at their base
Cordyline australis is not a true palm but is in the Agavaceae family (not Palmaceae)
Yucca is in the Liliaceae and agaves, cordylines and dracaenas are all in the Agavaceae. Screw pine is not a pine at all but is *Pandanus utilis* in the Pandanaceae

Arborescent plants II (monocots): Cordyline, Yucca and Dracaena.

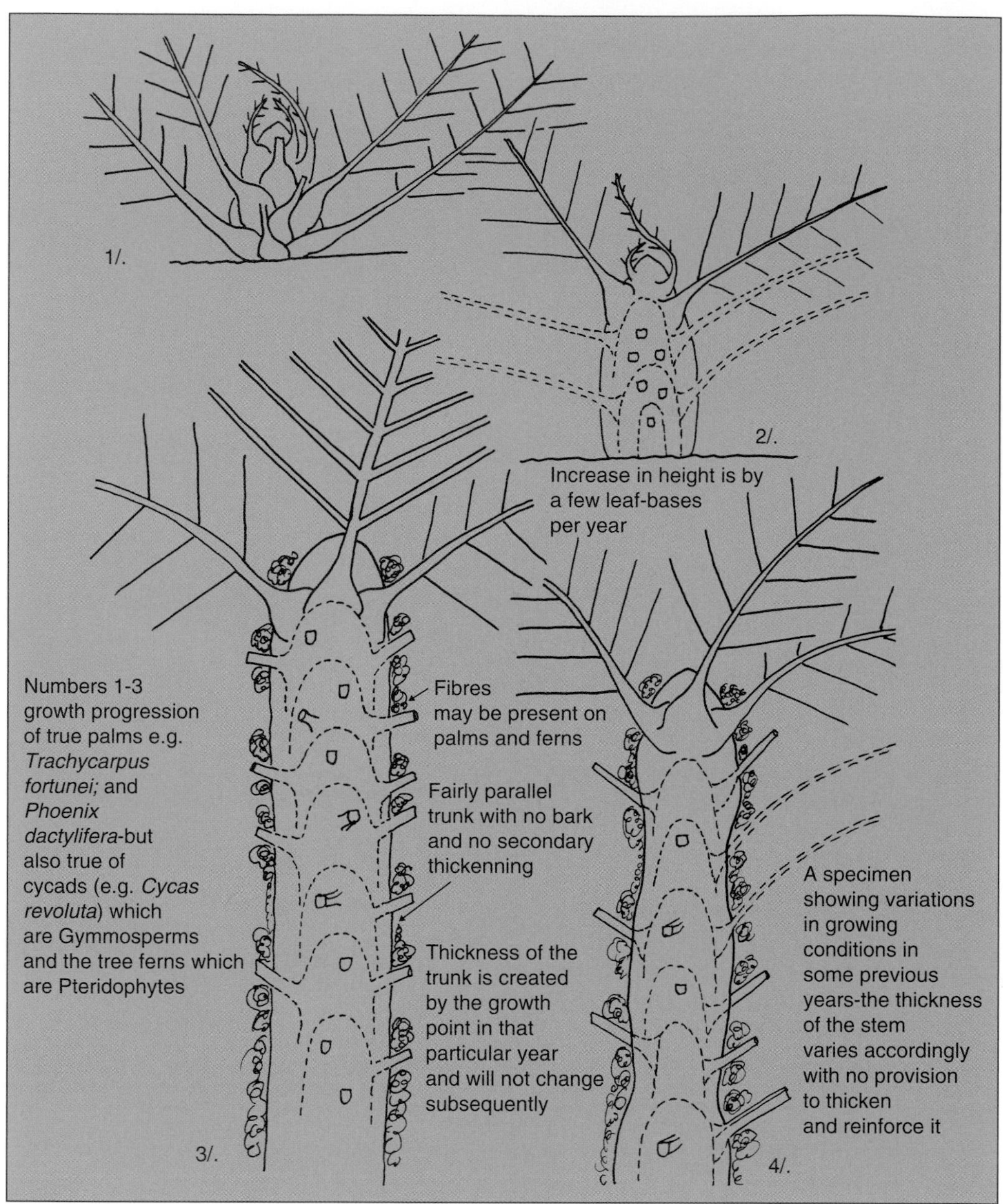
1/.
2/.
Increase in height is by
a few leaf-bases
per year
3/.
4/.
Numbers 1-3
growth progression
of true palms e.g.
Trachycarpus
fortunei; and
Phoenix
dactylifera-but
also true of
cycads (e.g. *Cycas*
revoluta) which
are Gymmosperms
and the tree ferns which
are Pteridophytes
Fibres
may be present on
palms and ferns
Fairly parallel
trunk with no bark
and no secondary
thickenning
Thickness of the
trunk is created
by the growth
point in that
particular year
and will not change
subsequently
A specimen
showing variations
in growing
conditions in
some previous
years-the thickness
of the stem
varies accordingly
with no provision
to thicken
and reinforce it

Arborescent plants III (monocots): true palms.

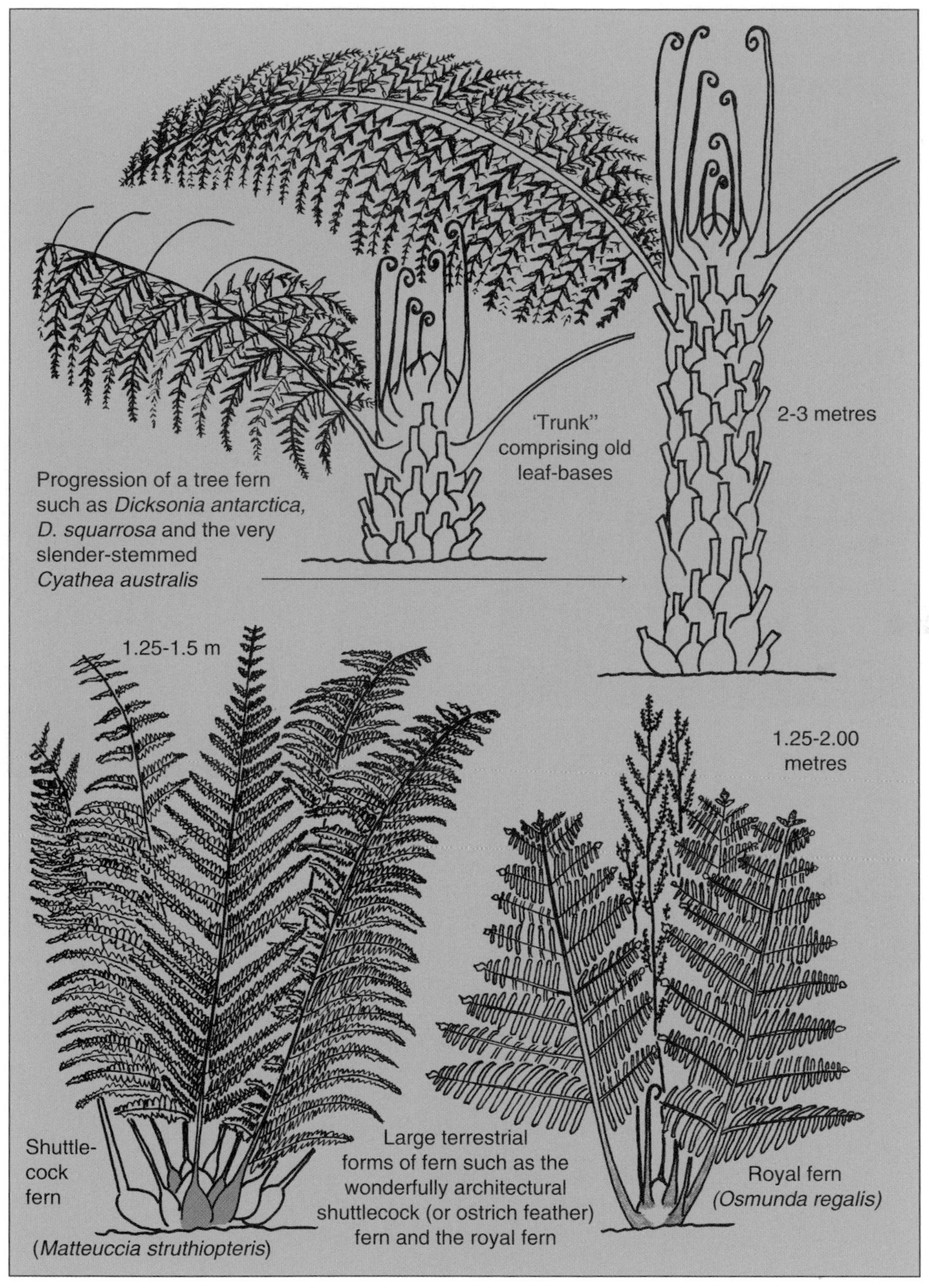

Arborescent plants IV: tree ferns (Pteridophytes).

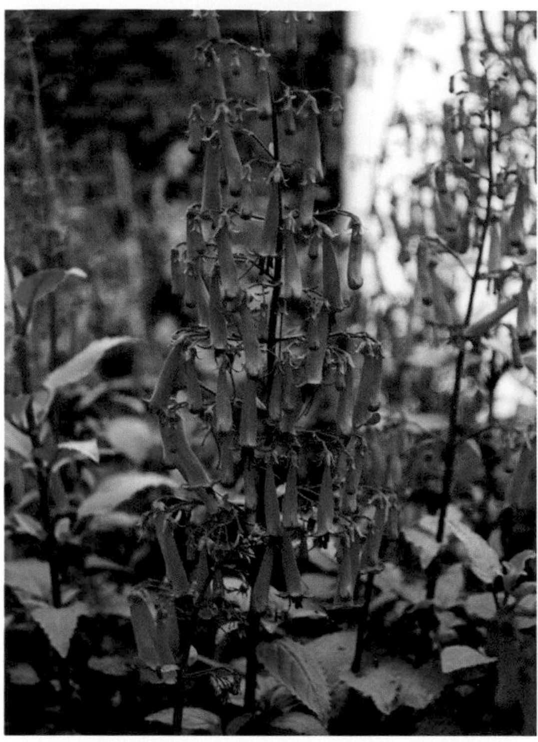

Flower of cape figwort (Phygelius capensis).

Abutilon vitaefolium album.

to wires and/or against a wall is the best way to manage it.

Where does sub-shrubiness end and arborescence begin? The question highlights the fact that there is actually a gradation between all the plant groupings, as not all want to remain specifically within the categories to which they have been arbitrarily assigned. Some individual annuals may be biennial and some individual biennials may be short-lived perennials. Sub-shrubs like figworts can be pruned back and retained in an almost herbaceous habit (if it were not for their woody base). Large sub-shrubs merge into shrubiness, and large shrubs merge into trees.

CHAPTER 2

Taxonomy: The Identification, Classification and Nomenclature of Plants

Taxonomy is the systematic study of plants involving their identification (recognition), classification (arranging into a hierarchy) and nomenclature (naming), and is based upon their common features (similarities) as well as a comparison of their differences (dissimilarities).

The Conventions of Nomenclature and the Binomial System

Plant names are registered and acknowledged by an international body responsible for nomenclature, and are governed by strict conventions. Common names of plants, although widely used, will vary from region to region within a country, and also between countries, and are therefore very unreliable. In order to rectify this, scientific names are used, and Latin was chosen as the main language of nomenclature (naming) because it is universal, and can be understood by academics and botanists anywhere in the world. However, Arabic and Greek derivatives may also be present in plant nomenclature (and modern languages in certain instances). For this reason, we do not refer to 'Latin names', but rather to 'scientific or botanic names'.

There is a universal convention in the way scientific names are written down: the genus (generic name) commences with a capital letter, and the species (specific epithet) is always in lower case letters. Where the generic name and specific epithet are used alone, it forms the complete specific name and is known as the Binomial System of Nomenclature (bi = two and nomial = naming). The Binomial system is used on a day-to-day basis by horticulturists, arboriculturists, foresters and ecologists whenever they discuss plants. The binomial system is used for convenience, but the two names alone – for example, *Primula denticulata* – are not always sufficient to classify a particular plant fully, so a third word (an extra epithet), such as the variety or cultivar, may also be necessary. The binomial system actually comprises a small part of a much larger classification hierarchy (originally devised by Carl Gustav Linne, latinized to Linnaeus).

If, in a book, article or research paper, a genus has been written once, as, for example, *Rhododendron*, then reference to the same genus again can be shortened to *R.* Hence the convention is that *Rhododendron sanguineum, R. yakusimanum, R. luteum* are all understood to be in the genus *Rhododendron*. Likewise *Magnolia wilsonii, M. stellata, M. acuminata* and *M. campbellii* are all understood to be species of *Magnolia*.

In scientific journals and documents the scientific name may be followed by the 'authority', which is the full name, or the first initial and abbreviated name of the person (or persons) attributed with authenticating, or originally naming the plant. Linnaeus appears as an authority so many times that his name is abbreviated to a capital 'L': for example *Acer platanoides* L. (Norway maple). Many plant names have synonyms – different names for the same plant. They are perfectly legitimate scientific names used previously for a species and now fallen out of use because of reclassification. Sometimes the original authority reclassifies a plant, in which case the same authority's initials will follow both the new classification and the old classification (the synonym). However, reclassification is more likely to be carried out by a different authority (usually after a review of the original), so a different name or initial will follow the new scientific name. Examples include *Campsis grandiflora* (Thunb.) K. Schum. syn. *Campsis chinenesis* (Lam.) Voss., *Mrytus ugni* Mol. syn. *Eugenia ugni* (Mol.) Hook., and *Myrtus luma* Mol. syn. *Myrtus apiculata* Hort. Initials in brackets denote a name given by an authority, even originally as a varietal form, but later reclassified as a species by the authority whose initials appear after it.

Although generic and specific names are the most regularly used, the full classification hierarchy involves many divisions, such as kingdoms, sub-kingdoms phyla, sub-phyla, classes, sub-classes, orders, families, genera, species, sub-species. The important conventions include the use of capital letters, lower-case letters, particular word endings, inverted commas, and a set pattern to denote order, family, genus, species, and so on.

Orders are recognized because they always commence with a capital letter and have a consistent word ending: they end in '-ales'. Examples include Ericales (the order containing the Ericaceae family and the genera *Erica* and *Rhododendron*), and Liliales (the order containing the Liliaceae family and the genus *Lilium*).

Families comprise groups of genera with definite similarities and affinities. Genera within a family also have dissimilarities, which are equally as important when identifying and classifying them. Family names always start with a capital (upper case) letter, and end in -aceae or -ae, for example Rosaceae, which is a very large family whose members include roses, but also potentillas, geums, photinias, chaenomeles, mountain ash, cherries, plums, apples, pears and cotoneasters.

Genera (the plural of genus) comprise groups of species (or sometimes, a single species) that have greater resemblance to one another than just their family affinities. The convention for writing generic names is that they must always commence with a capital (upper case) letter and in scientific publications are printed in italic, *Rhododendron, Lilium, Primula, Cotoneaster, Hemerocalis, Aster* and *Begonia* being examples. The generic name may be derived from an original classical Latin or Greek name and may be helpful as it can describe features of the plant when translated. *Rhododendron*, for example, is Greek for 'rose-red tree' (*rhodo* = 'rose red' and *dendron* = 'tree'), named because the first rhododendrons that were classified were very large tree-type arboreals that bore red flowers. The whole genus now bears this name whatever the flower colour of the different species. There are many genera of plants within the orchid group, and

Phormium tenax fruits.

the name 'orchid' is derived from the Greek for testicle, as the pseudobulbs of many orchid species resemble testicles in outline. *Hyacinthoides*, now the preferred generic name of bluebell, perfectly describes the flower type (*oides* = 'looks like' – meaning that it 'looks like a hyacinth'). *Streptocarpus* describes the spirally twisted fruit (*strepto* = 'spiralled' and *carpus* = 'fruit'). New Zealand flax (*Phormium tenax*) also has spiralled fruits, but was not named because of it.

Generic names may be named after their discoverer, or as a form of recognition for a plant hunter; they include *Davidia involucrata* (the handkerchief tree) named after Armand David, a French plant hunter; *Tradescantia* named after John Tradescant; *Fuchsia* after Leonhart Fuchs; and *Buddleia* after the Reverend Adam Buddle. The same generic name can only be used once, therefore incorporating a person's name into a generic name can only ever be once for the same person. If a genus has only one representative species within it, it is known as a 'monotypic genus', for example *Trochodendron araliodes* – there are no species other than *aralioides*.

Species show the greatest similarities and mutual resemblance, and species is the smallest main unit of classification. Specific epithets are often very descriptive; they are written in lower case letters, and in italic in scientific publications. They may indicate a specific feature of the plant, the country of origin, or, again, commonly the name of the discoverer (or one of his/her associates or peers).

Unlike generic names, specific epithets, bearing the name of botanists and plant hunters, can be used more than once – obviously not within a specific name for the same plant, but attached to another genus, thus describing a completely different plant, yet acknowledging the same person. Some specific names are very helpful, and when translated give a description of the plant.

There are quite a few specific epithets that describe leaf shape or leaf features of some sort. The suffix '*folia*' obviously refers to leaves, but specific names that include '*phyllus*' '*phyllum*' or '*phylla*' within them also refer to leaves: *Cotoneaster glaucophyllus*, for example, refers to the glaucous (blue-grey/silvery blue) coloration on the leaves, as does *Rhododendron glaucophyllum* (on the underside). The specific epithet *glauco-album* (whitish-blue) is a perfect description of the underside of the leaves of *Vaccineum glauco-album*. *Syringa microphylla* is a dwarf lilac with small leaves (*micro* = small and *phylla* = leaf). *Sophora microphylla, Azara microphylla, Columnea microphylla* and *Cotoneaster microphyllus* all accurately describe the small leaves of these species, whereas *Magnolia macrophylla* is a *Magnolia* species with very large leaves (*macro* = large and *phylla* = leaf). *Parvifolia* also refers to small leaves, and examples include *Viburnum parvifolium*, *Ulmus parvifolius*, *Vaccineum parvifolium* and *Rosa parvifolia* (Burgundian rose).

The scientific name *Hydrangea quercifolia* describes perfectly a *Hydrangea* species with oak-shaped leaves (*quercus* = oak and *folia* = leaves). Even though the leaf shape is variable – because not all oak species have rounded-lobed leaves, as some have more sharply pointed lobes – the description still fits well.

Pyrus salicifolia translates (with its generic name) as willow-leaved pear (willow = *Salix*). *Ficus lyrata* describes the lyre-shaped leaves of the species, and *Cyclamen hederifolia* is a cyclamen with leaves like ivy (*hederifolia*: *Hedera* = ivy). *Ligustrum lucidum* (shining privet) and *Geranium lucidum* (shining geranium) are both described in their scientific and common names (*lucidum* = shining) and make reference to their shiny leaves. *Lavandula dentata* (French lavender) and *Ceanothus dentatus* fairly accurately describe the fine-toothed margins to the leaves, and *Primula denticulata* the finer toothing on the leaf margin of these species. *Olearia macrodonta* accurately describes the large toothing on its leaf

Examples of Specific Epithets

alba (*um*) = white e.g. *Morus alba* (white mulberry)
albo-sinensis = white Chinese e.g. *Betula albo-sinensis* (white Chinese birch)
alnifolia = leaves like alder (*Alnus*) e.g. *Clethra alnifolia*
barbata (*um*) = bearded (stiff, bristly hairs) e.g. *Rhododendron barbatum*
campanulata (*um*) (*us*) = bell-shaped flowers e.g. *Rhododendron campanulatum*
campestre (*is*) = of fields or plains e.g. *Acer campestre* (field maple)
cardinale (*is*) = crimson/cardinal red e.g. *Lobelia cardinalis, Delphinium cardinale* and *Gladiolus cardinalis* (all with red flowers)
coccinea (*eum*) = scarlet e.g. *Musa coccinea* (scarlet banana – with yellow flowers sheathed in scarlet bracts)
darwinii = after Darwin e.g. *Berberis darwinii*
floribunda (*um*) (*us*) = very free-flowering e.g. *Malus floribunda*
foetid (*us*)/*foetidissima* = smelling (foul smell) e.g. *Iris foetidissima* (stinking iris) and *Helleborus foetidus* (stinking hellebore)
fuchsioides = flowers like a fuchsia e.g. *Begonia fuchsioides*
pseudo = false e.g. sweet flag (*Iris pseudoacorus*) translates as 'false rush'
rivulare (*is*) = next to water e.g. water avens (*Geum rivulare*)
rosmarinifolia (*us*) = having foliage like rosemary e.g. *Grevillea rosemarinifolia*
semperflorens = always flowering e.g. *Begonia semperflorens*
somnifera (*um*) = pertaining to sleep e.g. *Papaver somniferum* (opium poppy)
vulgare (*is*) = common e.g. *Calluna vulgaris* (common heather, or ling) and *Senecio vulgaris* (common groundsel)
wilsonii = after Ernest 'Chinese' Wilson e.g. *Magnolia wilsonii*

Flowers of Shining Privet (Ligustrum lucidum). (Photo: Kay White)

margins, and *Olearia ilicifolia* alludes to the sharp, holly-like (*Ilex*-like) toothing on the leaf margins.

Ligustrum ovalifolium describes its oval leaves well, and *Kiringeshoma palmata* (a beautiful yellow-flowered herbaceous plant), *Rheum palmatum* (rhubarb) and *Acer palmatum* all describe their simple palmate leaf shape. Three species of New Zealand mint (*Prostanthera)* are separated by their leaf shape as reflected in their specific names: *Prostanthera ovalifolia* (with oval leaves)*, Prostanthera rotundifolia* (with rounded leaves) and *Prostanthera cuneata* (triangular leaves). *Peperomia rotundifolia* is also easily recognized by its rounded leaves.

Schefflera digitata (umbrella plant) describes its digitate (compound palmate) leaves, a leaf shape that it shares with horse chestnut, but a feature that is not mentioned in the scientific name of that species. However, the specific name of *Quercus dentata* (a wonderful oak species with impressively large leaves) could be criticized, as the leaf margins are in fact heavily lobed rather than toothed – perhaps *Quercus lobulata* would have been more accurate? – whereas *Zelkova serrata* (a relative of elm) and *Azara serrata* are fairly accurate in their leaf margin description, although some might consider *serrulata* (more finely serrated) to be even more so. You can see, therefore, that the degree of serration or lobing is subjective, and will vary from person to person and with what the original is compared.

Cordifolia and *cordata* describe a heart-shaped leaf. However, they tend to be used to describe generally broad leaf types, and some broad-leaved subjects that do not have a specifically heart shape may carry the specific name *cordata. Tilia cordata* is perfectly descriptive of the leaf shape; however, the common name centres on the comparative leaf size (small-leaved lime), not its heart shape. *Diascia cordata* (a low-growing herbaceous subject) does have heart-shaped leaves, albeit small ones, as does *Philodendron cordatum* (a tender, glasshouse evergreen climber) – but does *Macleaya cordata*, or are they just large and broad? *Bergenia cordifolia* has broad leaves with two equal rounded basal lobes, which give rise to the common name 'elephant's ears' – but is the tip of the leaf acute enough to create a heart shape? Again, it is a matter of opinion: one person's cordate could be another person's deltoid, and because there is no sensible chronological order to plant discovery and their naming, and therefore no order to the comparisons made, apparent anomalies of various types will arise.

The difference in degree or obviousness of any feature observed by individuals has to be accepted from the outset. *Elaeagnus macrophylla*, for example, does have large leaves when compared with other *Elaeagnuses*, but not necessarily so when compared with other plants in general; and *Senecio grandifolius* has larger leaves compared with other *Senecios* (25–45 cm long). However, neither has leaves that are as large as some individuals of *Acer macrophylla* or *Magnolia macrophylla*. Furthermore, there is nothing to say that either *Acer macrophylla* or *Magnolia macrophylla* necessarily have the largest leaves within their respective genera. So, although the specific names of these plants do tell us these species have very large leaves, other species within the genera, perhaps discovered at a later time, could have even larger leaves. Also, the authority may have decided on describing another feature in the specific name instead – often because the more descriptive name has already been used for another species. *Magnolia obovata* (syn. *Magnolia hypoleuca*) has very large leaves that in some individual specimens may be larger than those of *Magnolia macrophylla*. However, the fact that the large leaves are broader at the end furthest from the leaf petiole (obovate) was chosen as the main feature for the specific name. *Cotinus obovatus*, with its obovate leaves, is discernible from *Cotinus coggygria* with its ovate leaves. However, no mention of leaf shape is made in the scientific name of the latter, yet *Forsythia ovata, Eryngium agavifolium* (leaves like an agave) and *Drimys lanceolata* accurately reflect the leaf shape of these species.

Ranunculus aconitifolia makes excellent reference to its aconite-looking leaves (for example, *Aconitum napellus* – monkshood), and *Sagittaria sagittifolia* (a water plant) uses its arrow-headed leaf shape as a descriptive feature in both its generic and specific names.

The specific epithet *acuminata* is used to accurately describe the pointed leaf tips of *Clethra acuminata, Cotoneaster acuminata* and *Monstera acuminata. Mahonia aquifolia* describes the sharp points on the leaves, as does *Ilex aquifolium* (common holly). *Sophora tetraptera* describes the four (*tetra*) wings (*ptera*) of its fruits, and *Maclura pomifera* also describes its fruit in its specific name – that is, having a fruit that looks like an apple (*pome*), though a rather wrinkled apple in this case.

The specific epithet *ciliata(um)* refers to a fine fringe of hairs, and is an accurate description of both *Bergenia ciliata* and *Rhododendron ciliatum*. Nevertheless, the use of the epithet *ciliata* does not necessarily infer that it is always the leaves that are fringed with hairs. *Menziesii ciliicalyx* refers to the fringe of hairs on the sepals of the calyx – and thankfully the specific name accurately describes both the hair type and its location on the plant. The use of *hirsuta* (also describing hairs) is not exclusive to leaves, as in hairy bitter cress (*Cardamine hirsuta)*, because its use in *Penstemon hirsuta* refers to the hairs on the flowers. *Pelargonium tomentosum* (peppermint geranium) is named because of the very fine woolly hairs on the leaf surface (*tomentosum*), and not because of the very pungent smell that it emits. Furthermore, the very fine hairs might better have been described as 'pubescent' – and not only that, pubescence can be the factor giving softness (mollifying) and sometimes described in the specific name by the epithet *mollis*. *Alchemilla mollis* (lady's mantle) and *Geranium molle* (dove's foot crane's-bill) both describe their softness (created by pubescent hairs).

Geranium sylvaticum (wood crane's-bill) and *Geranium dissectum* (cut-leaved crane's-bill) have their common names translated perfectly in their botanical names (*sylvatica* = of the woods, and *dissectum* = cut-leaved). *Geranium pratense,* like *Geranium sylvaticum*, describes its chosen habitat (as *pratense* means 'of the meadows'), and furthermore this is reflected in its common name (meadow crane's-bill). *Tragopogon pratense* (goat's-beard) also describes its habitat (rather than its quite stunning heads of dandelion-like fruits).

Specific epithets don't appear to discern between trifoliate (simple) and trifoliolate (compound) leaves, hence any reference to leaves comprising either three lobes or three leaflets tends to be written as *trifoliata* even when sometimes *trifoliolata* (with the extra 'ol') would be more accurate. Perhaps the extra 'ol' may have not been used, or has been lost over time in proof reading? *Poncirus trifoliata, Ptelea trifoliata and Akebia trifoliata* (with three leaflets on a compound trifoliolate leaf) are good examples – *Akebia quinata* describes a similar species with five leaflets to the compound leaf.

Ribes sanguineum (flowering currant) and *Rhododendron sanguineum* describe their blood-red flowers, likewise with *Geranium sanguineum* (bloody crane's-bill). The specific epithet *cyanus* or *cyaneum* refers to a blue coloration and appears in the scientific name *Tillandsia cyanea* to describe the purple-blue bracts of this species, and in *Allium cyaneum* to describe the blue flowers of this ornamental onion. However, many of the other ornamental onions may also have blue flowers (including chives), but it could not be used again within the same genus – so other features are chosen instead. *Stewartia pseudocamellia* (meaning false camellia) refers to the similarity of the flowers to the camellia, *Magnolia stellata* describes its star-shaped flowers, and *Lonicera fragrantissima* and *Viburnum fragrans* describe their very fragrant flowers. *Phacelia campanularia* describes its bell-shaped (campanulate) flowers, whereas a related species, *Phacelia tenacetifolia,* describes the similarity in foliage to tansy.

The specific epithets *flava(um), lutea(um)* and *aurea(um)* refer to a yellow or gold coloration, and *xantho* and *aurant* refer to some part of the plant being orange or gold. However, the degree of yellowness, orangeness or goldness varies, and they are therefore often used interchangeably. Good examples include *Xanthoria parietina*, where the prefix is adopted in the generic name to describe the wonderful orange coloration of this common lichen, and *Mimulus luteus*, *Gentiana lutea* and *Erigeron aureus* describing the yellow flowers of the species. *Rhododendron xanthocodon* and *Begonia xanthina* have orange/yellow flowers, and *Mimulus aurantiacus* and *Erigeron auranticus* both describe their orange flowers. *Crepis aurea* is a yellow-flowered species in the dandelion (and lettuce) group and *Romulea flava, Allium flavum, Crataegus flava, Sarracenia flava* and *Crocus flava* all have yellow flowers. *Linum flavum* is the name for the yellow (or orange) flax, *Glaucium flavum* is perfect to describe the yellow-flowered, glaucous-green horned poppy, and *Colchicum luteum* describes the yellow flowers of this crocus-like species. *Rhododendron luteum, Digitalis luteum* and *Paeonia lutea* also describe their yellow flowers. Yet *luteum* does not appear in the specific names of other yellow-flowered plants in different genera (for example *Erythronium americanum* has yellow flowers) because other features have been chosen instead – in this case the country of origin rather than flower colour.

The specific epithet *officinale(is)* means the plant has been traditionally used for medicinal (and/or culinary) purposes, and refers to the fact that it would have been sold commercially for the purpose (*officinalis* = of the shops, of the office). *Mandrago officinalis* (mandrake) has long been associated with both herbal medicine and witchcraft, *Borago officinalis* (borage), *Rosmarinus officinalis* (rosemary) and *Hyssopus officinalis* have traditional culinary and herbal medicine usage. *Styrax officinalis* and *Jasminum officinalis* must also have useful medicinal properties, and *Rosa gallica* var. *officinalis* is called the apothecary's rose because of its history of medicinal use.

Giant yellow gentian (Gentiana lutea).

Crassifolia describes plants with thick leaves, and names a complete family of succulents with this feature (Crassulaceae), and within the family, *Crassula* describes a genus of very thick-leaved species including *Crassula arborescens*. The epithet *arborescens* describes an almost woody habit, and should be reserved for plants that do not have all the features of trees but attempt to attain their aerial proportions. The epithet *arborea* refers to tree-like features, and is generally reserved for larger growing species within a genus that normally contains smaller, shrubby-type plants. However, the two are often interchanged (arguably incorrectly) and so may cause some confusion – they are therefore both best viewed as meaning 'tree-like' when looking at plant names. *Aloe arborescens* and *Crassula arborescens* perfectly describe their arborescent nature, as they are monocotyledons with the provision to attain relatively large proportions. However, bladder senna (*Colutea arborescens*), *Heliotropium arborescens* and *Artemesia arborescens* are dicotyledonous sub-shrubs that actually have some truly woody features – like the provision for secondary thickening at their base – but so is *Brugmansia (Datura) arborea*, with the epithet *arborea* not *arborescens*. Both types may share the ability to have a bark covering over some of the thicker stems, and it may be this that groups them with the common epithet, but there is, however, a difference between the two types botanically, and *suffruticosa(um)(us)* would be more accurate for sub-shrubs, as it means 'woody at the base'. The term 'tree peony' is really only delineating between herbaceous and shrubby forms of peony, and *suffriticosa* within the scientific name is far more accurate than the vernacular term 'tree peony'. The epithet *fruticosa* means 'shrubby' (for example *Potentilla fruticosa* to describe the shrubby, rather than herbaceous forms of potentilla). *Rhododendron arboreum, Amelanchier arborea, Ceanothus arboreus* and *Oxydendrum arboreum* all accurately describe these tree-like individual species, and their larger proportions and bark covering gives them the right to be placed under the epithet.

Squamata, squamatous and squamosus make reference to the plant having scales, and is used to accurately describe the profusion of brown scales on the fungus dryad's saddle (*Polyporus squamosus*). It could have been used to describe the very sharp triangular scales associated with juniper species, as they are important identifying features when comparing junipers with other conifers (that have much softer foliage). However, the scientific name *Juniperus squamatus* actually refers to the flaky scales of bark that are produced on mature plants, and not, in this case, the sharp scales that they possess.

The suffix '*oides*' always infers that the subject looks like another species. *Schizophragma hydrangeoides* resembles the woody climbing species *Hydrangea petiolaris*, and its flowers resemble those of hydrangea anyway. Oddly, *H. petiolaris* was named for its long leaf stalks (petioles), and not its climbing habit. *Tillandsia usneoides* resembles those grey, fruticose (shrubby) hanging lichens in the genus *Usnea*. *Sisyrinchium graminoides* makes reference to the resemblance of its small sword-like leaves to plants in the Graminaceae (Poaceae) grass family. *Hebe cupressoides* describes a hebe with whip-cord foliage looking like *Cupressus*, and *Nothofagus betuloides* describes a tree that resembles a birch (*Betula*) in its leaf shape. *Shortia soldanelloides* has flowers that are not unlike those in the genus *Soldanella*, and *Sedum sempervivoides*, although being a *Sedum*, actually has features not unlike a *Sempervivum*. The true *Sempervivum*, *Sempervivum arachnoideum*, describes the spider-like webbing that appears all over the plant.

Senecio maritima, *Alyssum maritimum* and *Armeria maritima* (thrift or sea pink) describe their preferred location – maritime conditions. *Cicerbita alpina* (alpine sowthistle), *Aquilegia alpina* (alpine columbine), *Ribes alpinum* (mountain currant) and *Alchemilla alpina* (alpine lady's mantle) all describe their natural habitat. *Dicksonia antarctica* and *Nothofagus antarctica* both describe their geographic location, as do *Chaenomeles japonica* and *Mahonia japonica*. Three species of nettle tree or hackberries (*Celtis*) are separated by their different geographic locations, a fact that is reflected in their specific names: thus *Celtis australis* (southern nettle tree) comes from southern Europe and North Africa, *Celtis occidentalis* (western nettle tree) is from North America, and *Celtis sinensis* (Chinese nettle tree) from China, Korea and Japan. The epithet *cambrica* refers to western Britain (Wales in particular), and *orientale(is)* is frequently used to describe plants from the East (the Orient): *Papaver orientale* (oriental poppy) and *Meconopsis cambrica* (Welsh poppy) are good examples. Notice that both species have 'poppy' as a common name, yet they are in different genera – but the same family (Papaveraceae).

The epithet *praecox* describes the earliness of flowering, as witnessed by *Stachyrus praecox, Thymus praecox, Weigela praecox, Rhododendron praecox* and *Chimonanthus praecox*. Unfortunately the changes in climate that we are now experiencing alter the flowering times of many plants, and for several consecutive years many of those that naturally flower in the early spring are encouraged to routinely flower in the autumn instead.

Of the epithets that describe the habits of plants, both *prostrata* and *procumbens* describe the fact that they lie flat to the ground and have a creeping habit. The epithet *horizontalis* obviously also describes a similar habit, but both *repens* and *reptans* infer a very low creeping habit (with the ability or propensity to root along the stem). *Ceanothus prostratus, Gaultheria procumbens* (partridge berry), *Loiselaria procumbens, Juniperus horizontalis, Ajuga reptans, Rhododendron repens* and *Mahonia repens* are good examples.

The specific name *nutans* means nodding, and describes beautifully the nodding action of the 'flower heads' (collection of bracts) of *Billbergia nutans*. *Cernus* also means nodding.

Specific epithets may therefore be meaningful, even descriptive, but they can only give very small snippets of information, and they have been chosen and prioritized by someone who may not have had the same idea of the most prominent features as you.

Subspecies denote a distinct variation from the species, and are usually applied to similar forms but from different geographical locations – so the variations may be induced by the environmental differences of the different locations. Subspecies epithets are written in lower case, and subspecies or subsp. is acceptable. Examples include *Cyclamen coum* subspecies *coum* and *Cyclamen coum* subspecies *caucasicum* found in two different locations in the wild.

Varieties (Varietas) indicate a variation of the species that occurs in the wild, and is commonly a minor variation in plant structure or morphology such as variant leaf or flower shapes. Varietal names are written in lower case, and because it appears as a third name it is sufficient to indicate that it is a varietal epithet, and nothing need be added. However, there are two other acceptable alternatives: using *Betula utilis jaquemontii* (a selected variety of Himalayan birch with stark white stems) as an example, it may also be written as *Betula utilis* var. *jaquemontii* or *Betula utilis* variety

The nodding flowers of Billbergia nutans.

The nodding flowers of Enkianthus cernus var. *rubens.*

jaquemontii. Other examples include *Punica granatum nanum* (dwarf pomegranate), *Cardiocrinum giganteum* var. *yunnanense* (a variety of giant lily from the Yunnan in China), and *Enkianthus cernus* var. *rubens* (a form with nodding red flowers). *Daphne mezereum* var. *album* (a white-flowered form of *D. mezereum*), *Clematis heracleifolia* var. *davidiana* (a varietal form of herbaceous clematis), and *Cyclamen hederifolia* var. *album* (a white-flowered form of the hardy *Cyclamen hederifolia*) are also good examples. *Embothrium coccineum lanceolatum* (Chilean fire bush) describes both its lanceolate (spear-shaped) leaves and scarlet flowers. *Coriaria terminalis xanthocarpa* describes the position of the flowers (and therefore the fruits) at the terminal buds, and the varietal name describes the orange/yellow fruits of this plant. The variety *Rhododendron glaucohyllum* var. *tubiforme* describes both the glaucous underside to its leaves, and the fact that instead of having open flat petals like the type, it has petals that form a tubular corolla.

Some types are put together in a loose group known as a '*forma*'. *Forma* denotes that variations are minor (not definitive differences at varietal level), and that the group procreates sexually as a loosely related form. Variations can include showing a colour difference from the straight species. Hence many plants formerly considered to be varieties are now more comfortably classified as forms – for example, the white variation of drumstick primula is classified as *Primula denticulata* forma *alba* by many authorities (it may also be written as *Primula denticulata* f. *alba*). Other herbaceous examples include *Dicentra spectabilis* forma *alba* (the white form of bleeding heart), and *Epilobium angustifolium* forma *alba* (the white form of rosebay willow-herb). Woody examples include *Clianthus puniceus* forma *alba* (the white form of lobster claw), *Philadelphus delavayii* forma *melanocalyx* and *Cytisus scoparius* forma *andreanus* (common broom). The divisions between sub-species, variety and forma are not always as distinct as they might be, and many authorities disagree about the classification of plants at this level.

Cultivars have no wild equivalent: they are not variations of a species, but may be hybrids of unknown origin with the influence of two, or more than two, parents. Alternatively they may be found as the result of a genetic mutation (as a 'sport' – a colour break, or chimera), or propagated vegetatively from juvenile material. In all cases they are selected and kept in cultivation by human intervention, and cannot exist independently without human beings. Cultivar names always start with a capital letter, they are often (but not always) in modern language, and are placed in inverted commas; for example, *Camellia* 'Donation' (a large pink semi-double-flowered cultivar) and *Callistemon citrinus* 'Splendens' (a cultivar of Australian bottlebrush with large red flowers).

There are many commonly recurring cultivar names, and they often describe habit: for example 'Pendula' for weeping types, 'Fastigiata' for upright (fastigiate) types, 'Pyramidalis' for pyramidal forms and 'Columnaris' for columnar forms. Cultivar names can often include a description of colour: thus *Daphne odora* 'Aureo-marginata' (having a yellow/gold leaf margin), *Fritillaria imperialis* 'Lutea' (yellow-flowered emperor fritillary) – even *Choisia ternata* 'Sundance' hints at its yellow leaves. 'Erythrophylla'

Callistemon citrinus 'Splendens'; also showing the extension growth occurring past the flower.

means red leaves (or the underside of leaves, as in *Begonia* 'Erythrophylla'; 'Erythrocarpa' describes red fruits, as in the vivid red double-winged fruits of *Acer psuedoplatanus* 'Erythrocarpa'; and 'Xanthocarpa' or 'Fructo-lutea' describe yellow fruits, as does 'Flava' and 'Bacciflava'. Examples include *Sorbus aucuparia* 'Xanthocarpa', describing the orange/yellow fruits (it may also be seen written as *Sorbus aucuparia* 'Fructo-lutea'); *Ilex aquifolium* 'Bacciflava' (yellow-berried holly); and *Pyracantha rogersiana* 'Flava' (yellow-berried firethorn). 'Purpureum' means purple (usually, but not always, referring to the leaves), and 'Atropurpureum' means particularly deep purple, as in *Acer palmatum* 'Atropurpureum' and *Berberis thunbergii* 'Atropurpureum'.

'Aconitifolia (um)' describes cultivars with aconite-type leaves (meaning deeply cut, like *Aconitum*) as in *Acer japonicum* 'Aconitifolium'; and 'Flore Plena' indicates double flowers, as in *Cardamine pratense* 'Flore Plena' (the double-flowered cuckoo flower). *Caltha palustris* 'Flore Plena' is a double-flowered cultivar of kingcup, *Convallaria majalis* 'Flore Plena' is a double-flowered cultivar of lily of the valley, and *Ranunculus acris* 'Flore Plena' is a double-flowered cultivar of common buttercup. However, the absence of 'Flora Plena' does not mean that the cultivar is not double-flowered: *Anemone nemerosa* 'Vestal', for instance, is a double-flowered form of wood anemone that was named for its 'whiteness/purity' rather than its 'doubleness'. Some authorities are not keen on two or more words forming a cultivar name, and there is a move to reduce this to one only. Hence, what was previously *Prunus avium* 'Flore Plena' (double-flowered wild cherry) is now often seen as *Prunus avium* 'Plena', still indicating that the cultivar has double flowers.

Nevertheless, the use of more than one cultivar name is very commonly featured in ornamental conifers, and is probably because many new cultivars of conifers are created by taking cuttings from shoots in their juvenile form. The effect of this is to produce a new cultivar with very feathery foliage and very different in appearance to the parent plant. Because the juvenile foliage forms can only be carried on by human intervention (as it will never become sexually mature and produce seeds – and if it did, it would produce the common type, not the new form – then it has to have a cultivar name. If either juvenile foliage on the original mother plant or the new cultivar has a genetic breakdown, another potentially new and different cultivar can be created, by taking cuttings from the coloured (or different) material. This new cultivar has now to be classified, and in these cases a second cultivar name is added to describe the different appearance of the new one. *Chamaecyparis lawsoniana* 'Minima' (a dwarf form of Lawson's cypress with green foliage, and *Chamaecyparis lawsoniana* 'Minima Aurea' (a yellow form of the dwarf cultivar) is a good example, but there are many more in the conifer world – particularly cultivars of Lawson's cypress. *Chamaecyparis lawsoniana* 'Ellwoodii' (with grey-blue feathery foliage) and *C. l.* 'Ellwoodii Aurea' (with yellow foliage), *C. l.* 'Columnaris' (with columnar habit and green foliage) and *C. l.* 'Columnaris Glauca' (columnar habit and blue-grey foliage) are good examples.

Hybrids and Cultivars

Cultivars include hybrids that are known to have occurred naturally in cultivated areas, and hybrids produced on purpose by human intervention. For natural hybridization to occur, the plants involved must be in the same locality. There is no control over the process in the wild – other than whether plants are sexually compatible, promiscuous (cross easily) or apomictic (do not mix: that is, do not cross-fertilize) – nor any control of what happens to the resulting progeny. Natural hybrids in the wild, because of all the influencing factors and high mortality rate, are low in number. Natural hybrids also occur in plants now displayed alongside one another that would normally be separated in space by their geographic isolation, and have now been brought together in a botanic collection. This creates unique situations, and puts plants close enough together to hybridize when they would normally be hundreds, or thousands, of miles apart. This brings some unique new plants that may be of both botanic and commercial interest.

When hybrids are formed by two different species in the same genus, they are known as interspecific (or species) hybrids. The convention for showing this is to place a multiplication sign [×] between the generic name and the specific name, for example *Camellia* × *williamsii, Berberis* × *lologensis, Begonia* × *hybridatuberosa, Eucryphia* × *nymansensis* and *Arbutus* × *andrachnoides*. Sometimes hybrids present the phenomenon known as 'hybrid vigour', where the resultant hybrid is more vigorous than either of the two parents involved. Cultivars that occur after the original hybrid cross has been developed, are written with the parent cross first (when this is known), followed by the cultivar name: for example, *Camellia* × *williamsii* 'Donation', *Eucryphia* × *nymansensis* 'Nymansay', *Crocosmia* × *crocosmifolia* 'Citronella'.

When hybrids are formed by parents in two different genera the resulting plant is called an inter-generic, bi-generic, or genus hybrid. The convention used to indicate this is to place a capital X in front of the generic name: the most common example normally quoted is X *Cupressocyparis leylandii*. The parents of this tree are *Cupressus macrocarpa* and *Chamaecyparis nootkatensis* (you will note that the generic name *Cupressocyparis* is itself a hybrid of the two generic names *Cupressus* and *Chamaecyparis*). X *Gaulnettya (*a hybrid between *Gaultheria* and *Pernettya*, and X *Fatshedera lizei* (a hybrid between a cultivar of *Fatsia japonica* and a cultivar of *Hedera helix*), and X *Mahoberberis aquisargentii* (a hybrid between *Mahonia* and *Berberis*) are also common examples of inter-generic hybrids.

Cultivars may occur by genetic mutation, essentially by genetic freaks involving colour breaks known as chimeras, including flecked colours in flowers and leaf variegations. Variegated cultivars of plants become variegated by a genetic breakdown of the tissues, which reduces the amount of chlorophyll in the leaves. Variegations tend to be green and cream, green and white or green and silver – often, but not always, in a broad margin surrounding the leaf edge. The epidermal cells of leaves are transparent, and the normal leaf colour is given by the other pigments, (mainly chlorophyll), showing through them. When genetic malfunctions create lower chlorophyll content, the cells that remain have little or no pigment, and transparent cells reflect light in countless directions; because of this they present to the human eye as white, cream or silver. The same phenomenon occurs in white flowers, and when leaf-mining insects eat the leaf mesophyll (containing the pigments), the separated transparent epidermis showing as silver or white.

Because variegation is a product of genetic breakdown, it is just as likely that the unstable features that created the breakdown can also themselves malfunction. The result is that all, or some parts of, the variegated plant revert back to tissue with a full complement of chlorophyll once more. The reverted green tissues, although welcome from a food manufacture viewpoint, are unwelcome if the plant was purchased for its (arguably aesthetic) variegated effect, so are removed during management.

In general, and illustrated by most examples, the lower chlorophyll content of variegated plants makes them less vigorous than their all-green counterparts. In the wild, this disparity of vigour causes the variegated material to be shaded out by the more vigorous green growth, and so the material affected by the genetic breakdown is short-lived, and probably never seen. Where such events, which are relatively common, happen in botanic gardens and conditions of cultivation, they are seen before they succumb to shading out. If they are considered to be of interest or commercial importance they may be propagated asexually (vegetatively) by budding, grafting, or cuttings, as they cannot be produced sexually (by seed). This is another instance therefore, along with all other forms of cultivar, where human intervention ('cultivation') is needed to perpetuate the clone.

Other genetic variations that can be selected to perpetuate cultivars include purple leaf types, such as *Fagus sylvatica* 'Rohanii', which is a stable cultivar with purple-coloured dissected leaves. There are clonal weeping forms of many common trees that are perpetuated as cultivars; these include *Cercidiphyllum japonicum* 'Pendulum' (weeping katsura), and *Fagus sylvatica* 'Pendula' (weeping beech). Fastigiate (upright) forms are also selected from commonly occurring genetic abnormalities, and most common tree species have representative fastigiate forms. Examples include fastigiate purple beech (*Fagus sylvatica* 'Dawyck Purple' and fastigiate hornbeam (*Carpinus betulus* 'Fastigiata'). However, *fastigiata* is also used as a specific epithet on rare occasions to indicate a species (rather than a cultivar) with an upright form, for example *Cassiope fastigiata*.

Understanding that scientific names cannot possibly tell you everything you need to know about a plant, and that the 'authority/author' may not have given prominence to the same features for 'essential' inclusion in the plant name as others might have done, helps in some ways. But because of this, certain classifications appear to be anomalous, until you look further. Red clover (*Trifolium pratense*), for instance, refers to the colour of the flower in the common name, but there is no reference to red (*rubra*) in the scientific name. Instead, the author used the specific epithet *pratense*, referring to the plant's common habitat – in fields, meadows and pasture. Furthermore, red clover is a different species to white clover, and not just a colour variation (variety) of white clover, or vice versa. White clover therefore is not *Trifolium pratense album*, but is *Trifolium repens*, a different species to red clover, and has a specific epithet that refers to its low-growing, creeping and rooting nature. Even though it inhabits very similar natural terrain to red clover, there is no reference to its habitat (meadows), or to its white flower colour (*alba*) in the scientific name. Obviously only one *Trifolium* species can be named *Trifolium pratense* anyway, as the specific epithet can only be used once within the same genus. Nevertheless, the 'authority' may give 'priority' over a specific feature and choose to use it within the name, but it may not seem so obvious to others at a later date, or be as descriptive as it might be.

Euonymus alatus is a good name for a shrubby *Euonymus* with 'wings' of cork (corky outgrowths of bark) on the stems, because *alatus* means 'with wings'. However, it does seem odd to call the non-winged version *Euonymus alatus aptera* (*aptera* meaning 'without wings'), as this translates as 'winged *Euonymus* without wings'. The smooth version might have been better named something else, such as *Euonymus decidua*, allowing the winged type to be named as a variety of the smooth

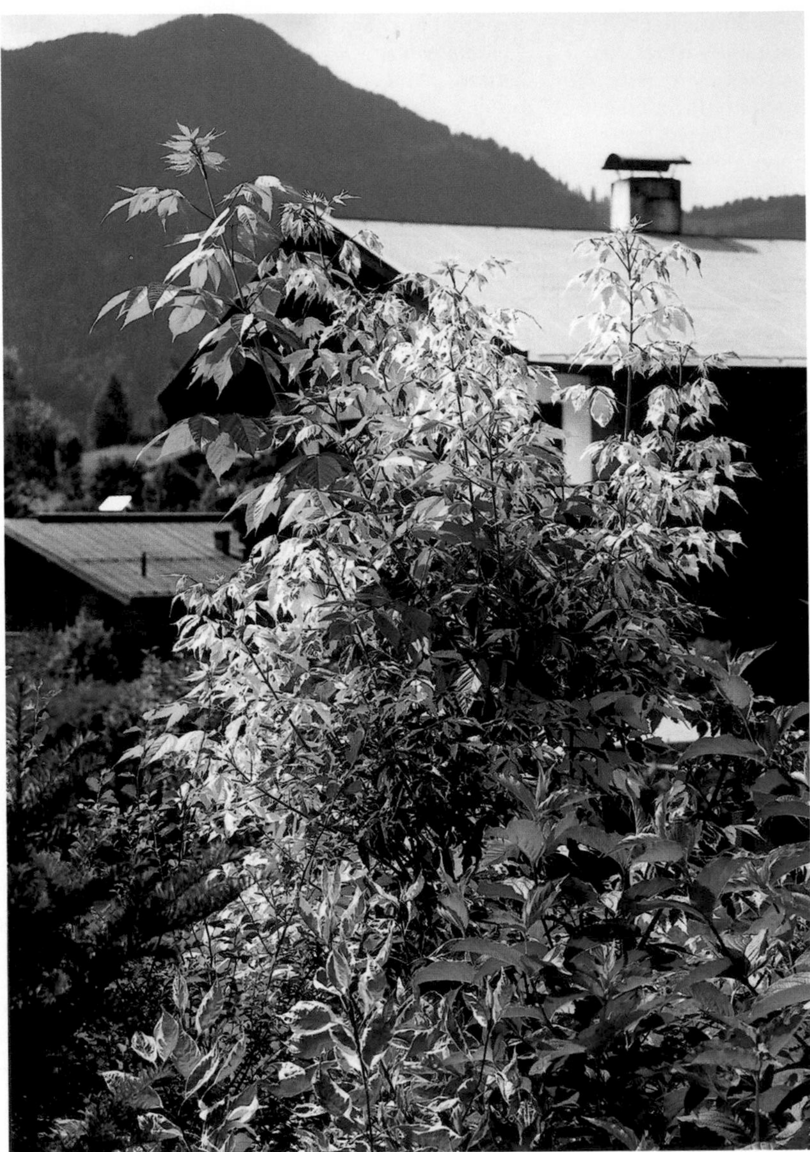

Variegated form of box elder (Acer negundo 'Variegata'*), showing reversion to green.*

type: *Euonymus decidua alatus*. However, the nomenclature as it stands would suggest that the winged species was discovered first, and the smooth (non-winged) varietal form afterwards, thus creating this seemingly nonsensical classification. What if the smooth (non-winged) variety had been found first: would it have been called *Euonymus glabra* (*glabra* means 'smooth'), or *Euonymus apterus*? Why would it? Because, until its discovery at a later date, there would be no other species or variety to compare it with for its smoothness or 'wingedness', so its name would not include either. Furthermore, if either *Euonymus glabra* or *Euonymus apterus* had been used for the original species, the name would still have been anomalous on discovery of the winged variety. It would either be *Euonymus glabrus alatus* (smooth with wings) or *Euonymus apterus alatus* (non-winged with wings), so neither would have helped avoid the anomaly. Unfortunately, once initiated, the anomalies continue. Trying to remedy these anomalies at a later date has always proved to be difficult, and has often foundered in the past.

A build-up of plant knowledge is therefore essential to aid your classification skills. Certain elements become easier with use and familiarity, but you cannot second-guess the system, and ultimately you will need to increase your plant repertoire and begin to know what parts of the plant a specific name refers to. There is often a strong resistance to changes in nomenclature, no matter how sensible they appear to be. In recent years *Acer japonicum* 'Aureum', the golden/yellow-leaved cultivar of the slow-growing Japanese maple, is now considered to be actually a cultivar of *Acer shirazawanum*, not *Acer japonicum* – you will see it classified in both ways in different reference books. These anomalies should not be viewed as anything but anomalies, because in the main, the system works extremely well.

The Main Classification Hierarchy

Although the binomial system serves all those interested in plants on a day-to-day basis, the information is actually taken from a much larger hierarchy. At the highest level of the hierarchy (and therefore farthest away from species level), all living organisms are divided into two main groups according to the way their genetic material exists in their cells: the prokaryotes and the eukaryotes. Prokaryotes (including bacteria and blue-green algae) have cells that do not contain paired chromosomes or a clearly defined nucleus sac (nuclear envelope) to separate the DNA from the cytoplasm. Eukaryotes, on the other hand, include all those that have a distinct nuclear envelope and DNA with paired chromosomes (both plants and animals). At the next level down the eukaryotes are split into five main kingdoms: the animals, monera, protista, fungae, and the plants (Plantae), all of which have distinct features.

Each kingdom is subdivided into phyla, and phyla are divided into sub-phyla, and then into classes (sub-classes in some instances), then orders, families, sub-families genera and species. If we work our way upwards through the hierarchy, 'species' denotes specific, detailed information that will categorically identify the plant, while 'genera' denotes more general similarities, broader than specific information, and within larger groupings with similar likenesses that puts them into family groups. By the time we get to the level of kingdom, this denotes the fundamental difference between animals, plants, bacteria, algae and fungi, the main deciding factors being whether they are motile/mobile, have blood or not, whether they can photosynthesize, and so on.

The plant kingdom (Plantae) is divided into two sub-kingdoms: the thallophyta and the embryophyta. The embryophyta includes all plants that reproduce from a multi-cellular embryo that is still attached to the parent plant, so includes a large amount of plants of horticultural interest. The embryophyta is further divided into two main phyla: the bryophyta and the tracheophyta. The bryophyta include the mosses and liverworts – all those plants without well developed vascular tissues. Those plants in the tracheophyta, on the other hand, have well developed vascular tissues to aid water (and dissolved mineral salt) transport, and can attain heights and volume far above the bryophytes. All developed vascular plants are tracheophytes, and are further divided into two sub-phyla: the seed-bearing plants (spermatophyta) and the non-seed-bearing plants (pteridophyta).

Pteridophyta

Pteridophytes include the ferns and horsetails, and because they have well developed vascular systems, it allows them to attain large sizes; however, they bear spores rather than seeds. The sub-phylum pteridophyta is further divided to include the class Filicales (ferns) and the class Lycopodiales (club mosses). Club mosses are technically not mosses (bryophytes) at all, but are actually small-leaved ferns. They have inherited this vernacular term because ostensibly they look like mosses with branched stems, and they also share the same habitats as mosses.

There is a wide diversity of shapes, sizes and forms in the fern world (class Filicales). Terrestrial forms naturally inhabit rocks, woodland, rotten logs and heath, whereas epiphytic forms grow on the outside tissues of the trunks and branches of tree species. Some species, such as the common polypody (*Polypodia vulgaris*), can exist successfully as either terrestrials or epiphytes if the correct environmental conditions prevail (usually high rainfall/high humidity). New, unfurling fronds, with their characteristic crooked shape, are known as 'crosiers'. A few species have long, simple leaves (for example hart's tongue fern), or long and simple with lobed/indented leaf-margins (as the common polypody and the hard fern), and some have very unusual complex leaf shapes (for example the maidenhair fern, *Adiantis*). However, a predominant number of species have large compound (pinnate) or doubly compound (bipinnate) fronds with many small leaflets (pinnae), which may themselves have small lobes called 'pinnules'.

Most species have large photosynthetic fronds to facilitate a sufficient energy supply to support their large aerial size. Species with very large leaves thrive in moist conditions and are usually shade or semi-shade dwellers, as this environment reduces their potential for water loss. Smaller-leaved ferns tolerate lower moisture levels and do not have quite such an urgent need for the organic litter and moist soils of woodlands as their larger-leaved counterparts. But because of their developed root system and large area of photosynthetic (and moisture-losing) tissue, they do need deeper soils than mosses or liverworts to succeed. Many species have fibrous-rooted crowns, and some species have underground storage systems such as rhizomes (e.g. bracken, *et al*) which can also act as a vegetative method of reproduction. However, spores are the predominant system of reproduction in most species.

Spores are produced asexually (vegetatively) in small sacs (called *sori*) borne on the underside of pinnae. When spores fall on to a moist substrate they develop into prothalli (small, heart-shaped, chlorophytic areas of fundamental tissue that has not differentiated into the normal organs of stem, root and leaf). The asexual prothallus is a prerequisite of reproduction, and from the prothallus develops the male and female (sexual) reproductive structures. Fertilization of the female structure occurs when the male cells migrate to them, which can only happen in moist conditions. Because ferns are spore-bearing, the developing plant body is known as a sporophyte, not a seedling.

The Spermatophytes

The spermatophytes, as their names suggests, bear seeds, but the way the seeds are borne on the plant leads to the next main division (classes): the gymnosperms and the angiosperms.

The Gymnosperms

The gymnosperms reproduce by seeds that are not enclosed in an ovary, but are borne on the surface of the scale of a cone. They have a fully differentiated vascular system, and dividing cells that form leaves, stem and roots. In fertilization, only the egg cell is fertilized, unlike the angiosperms where a second fertilization creates the endosperm. Three sub-classes of the gymnosperms (the cone or strobili bearers) are distinguished: the cycads (Cycadopsida), the Gnetopsida, and, most importantly from a forestry, arboriculture and horticulture viewpoint, the conifers (Coniferopsida). Within the sub-class Coniferopsida there are three orders that contain living (that is, non-extinct) ancient species: Ginkgoales, Coniferales and Taxales. The order Ginkgoales has one representative genus and one species within that genus, (monotypic) *Ginkgo biloba* (maidenhair tree). The order Coniferales includes all of the main conifer genera including firs (*Abies*), cedars (*Cedrus*), pines (*Pinus*), cypresses (*Cupressus*), false cypresses (*Chamaecyparis*), *Sequoia* and *Sequoiadendron*. The order Taxales is represented by *Taxus* (the yews), *Nothotaxus*, and *Torreya* (nutmeg).

The Angiosperms

Angiosperms include the main seed-bearing plants used in agriculture and horticulture. They produce their seeds enclosed in an ovary that may consist of one or several, free or fused carpels. Reproduction (seed production) of angiosperms involves a double fertilization, producing both a zygote (from fused male and female gametes) and an endosperm (food store). Angiosperms are the most advanced of the vascular plants and are sub-divided into two sub-classes: monocotyledons (those with one seed leaf) and dicotyledons (those with two seed leaves). Within the two sub-classes the plants are divided into orders, and families on the basis of the arrangement of the floral parts and developing fruits.

To summarize, most vascular plants (tracheophytes) are in the spermatophyta and are therefore either angiosperms (those having seeds in enclosed ovaries), or gymnosperms (those having naked seeds – cones). Ferns such as bracken are not in the spermatophyte, because even though they have developed vascular systems (tracheophytes) they bear spores not seeds, so are neither angiosperms or gymnosperms – they are instead in the pteridophyta.

Plant Identification

Identification commences with general outward appearances, gradually refining the information with more detail. Knowing that a species is deciduous rather than evergreen, for example, is not sufficient to enable recognizing the species, but it does rule out the species that it cannot be. The main plant groupings based on longevity (annuals,

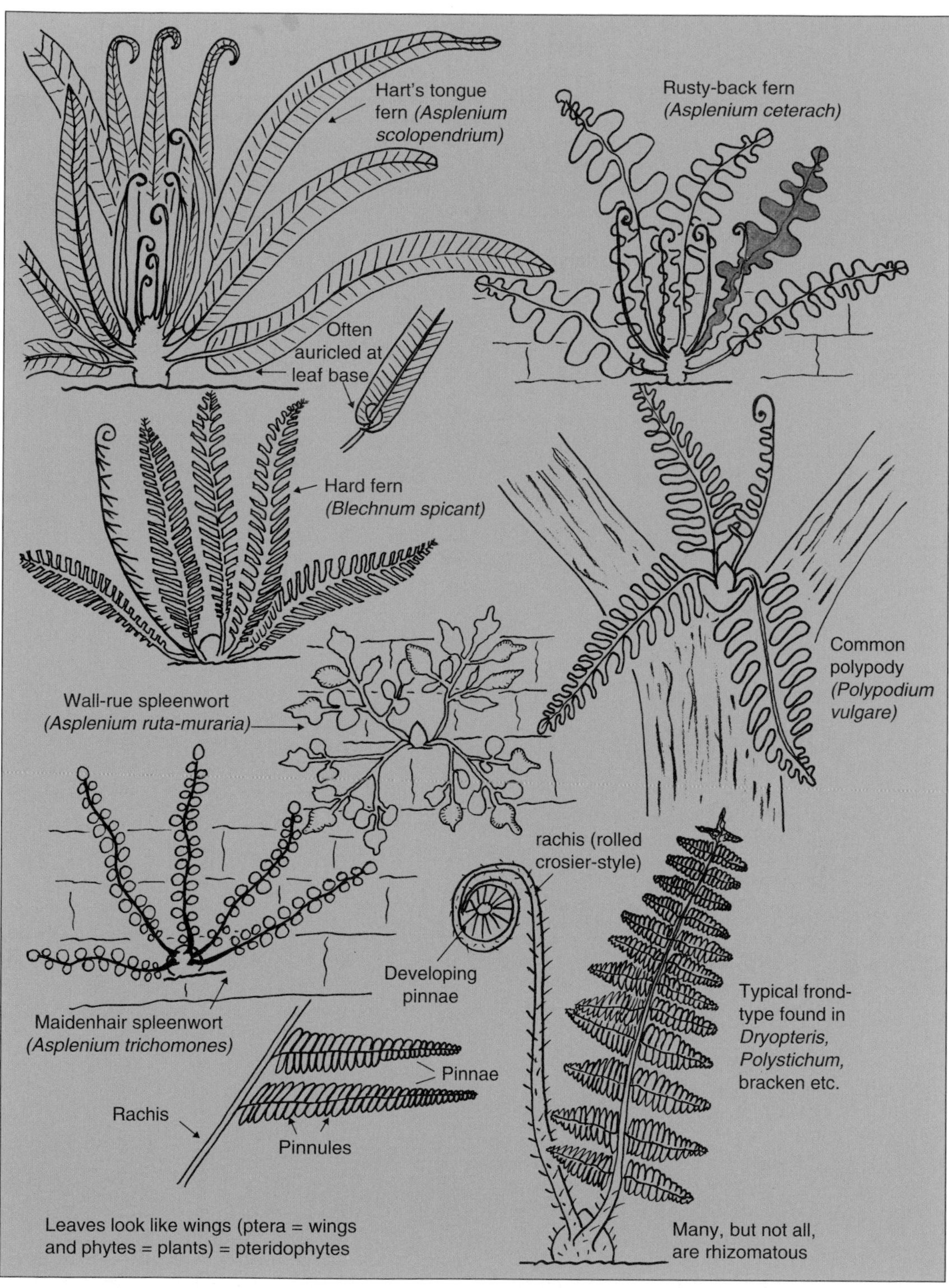

Ferns (Pteridophytes): non-flowering vascular plants – terrestrial and epiphytic types.

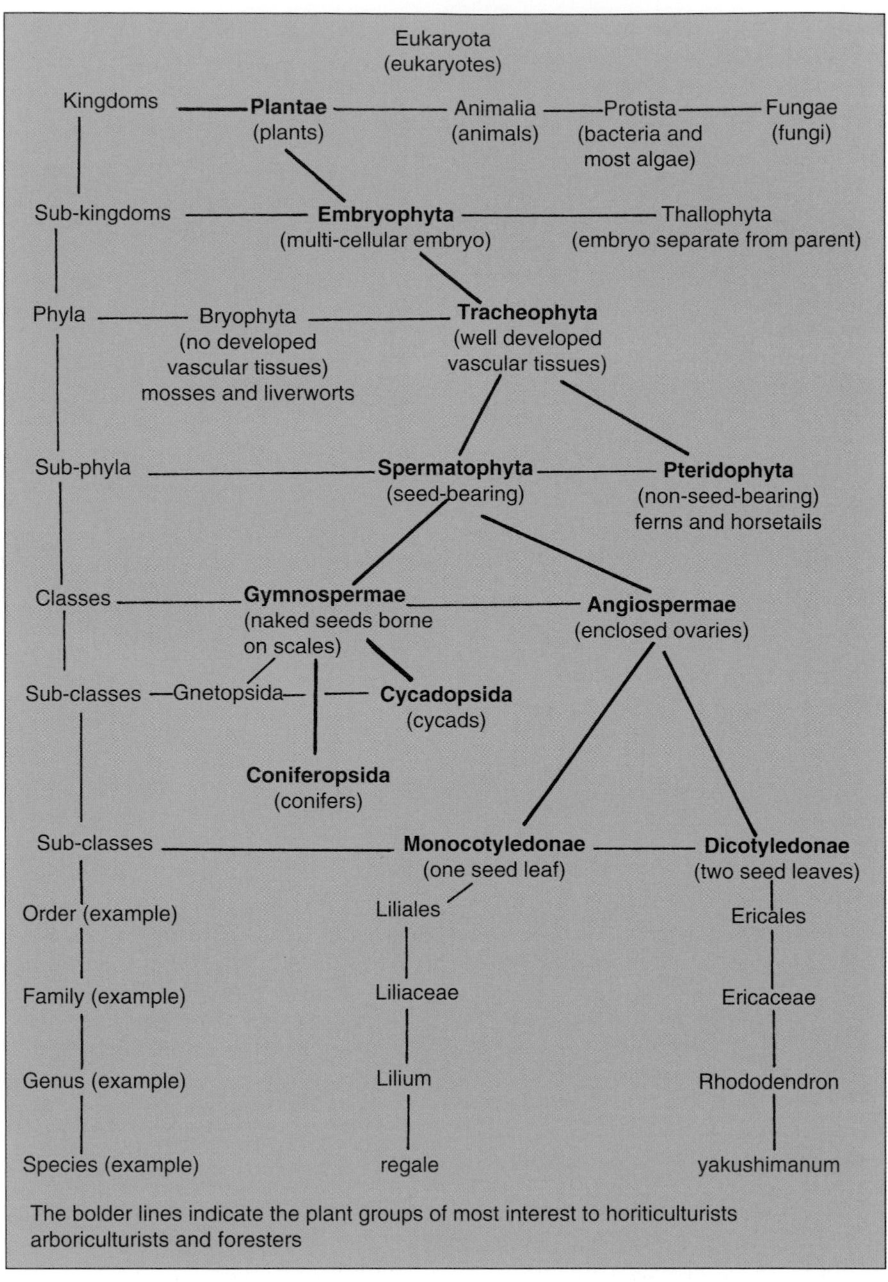

The taxonomic hierarchy.

biennials or perennials), types of storage organs and woodiness are very general, and they vary considerably within families and across genera; so flowers, fruits and anatomical features are used as the main indicators of similarities or differences. In the absence of flowers or fruits, leaves, stems, bark and buds may all help with identification, and it may also be necessary to consider minute details of parts of the plant to finally decide on its place in the system (classification). Because plant recognition and identification is a comparison exercise, specific features of an individual species are compared against known, already documented species.

Monocotyledons and Dicotyledons

One of the major divisions (sub-classes) of flowering plants with enclosed ovaries (the angiosperms) involves the difference between those having one (mono) seed leaf (cotyledon) in their seed (monocotyledonous plants), and those having two (di) seed leaves within their seed (dicotyledonous plants). This might seem to be of little importance, but in fact the differences between the two types are very marked, and are fundamental to the later development of the plants. The external features that can be discerned between the two can aid identification at sub-class and species level. The major division separates the grasses, cereals, palms and bamboos from the broad-leaved plants (including broad-leaved weeds and trees).

The monocotyledons are the smallest group of the two sub-classes and have strap-like, or sword-like, linear leaves (or leaflets) with parallel leaf veins, and petals and sepals in threes or multiples of three. Banana has very large, broad leaves, but the veins run parallel off the main vein, and floral parts are in threes. So for the same reasons, both the banana and the bird of paradise plant (*Strelitzia reginae*) are monocotyledonous. Monocots comprise mostly small herbaceous perennial species with subterranean organs, such as bulbs, corms and rhizomes – examples include lilies, *Narcissus, Crocus, Tricytris, Hedychium* and all the grasses. Nevertheless there are a few larger-growing (arborescent) monocotyledons, including palms, bamboo, *Cordyline, Agave, Yucca* and *Draceana*.

Dicotyledonous plants (two seed leaves in their embryo) are the largest group of the two sub-classes. Dicotyledons have broad leaves (often with serrated or dentate edges, or divided into lobes) and the leaves have netted (reticulate) veins (a complex system of venation with main veins, lateral veins and sub-lateral veins). Petals and sepals are in whorls of fours or fives (or multiples of four or five). Dicotyledons comprise a wide range of plant types, from annuals, biennials, herbaceous perennials and many woody plants (arboreals); examples include wallflower, potato, beans, cabbage, broad-leaved weeds (including dock and thistle), and trees and shrubs such as hazel, sycamore, beech, roses, apples and plums.

Types and numbers, and the freedom or fusion of floral parts feature heavily in the identification process, as does the structure and outward appearance (morphology) of resulting fruits. Trilliums are monocotyledonous, so a plant cannot be a *Trillium* if it has four or five petals – but it could be a double *Trillium* if it has six, because three petals (or multiples of three) denote that it is a monocotyledonous species.

Geranium robertianum (herb Robert) has five petals, *Pelargonium zonale* (the geranium of horticulture) has five petals, *Pentaglottis sempervirens* (green alkanet) has five petals. Furthermore they all have fibrous root systems (not bulbs, corms, or rhizomes) and, like potato (*Solanum tuberosum)* and lesser celandine (*Ranunculus* ficaria) that have tubers, they bear broad leaves with netted (reticulate) venation, so they are all dicotyledonous. All snowflakes, including *Leucojum vernum* (spring snowflake), *Iris pseudoacorus* (sweet flag or yellow iris), *Iris foetidissimum* (stinking iris) and all other irises have petals (actually perianth segments) and some other parts in multiples of three – a common feature of monocotyledons. Furthermore, all of the examples have long, linear leaves and parallel leaf veins, and are bulbous, cormous or rhizomatous, so are all monocotyledonous.

Thus, recognizing whether a plant is a monocot or a dicot may not tell us what the plant is, but it can sometimes tell us what a particular species cannot be. Flowers and the resulting fruits are the main identifying features, because although other features may aid identification, they may also be very misleading. Leaves are notoriously fickle when it comes to identification, not least because they vary in size and shape on the same species in different environments. Plants within the same family may have wide-ranging differences, so favoured environmental conditions may give some clues to the identity of a species, but only within a small, specific and known range of possibilities. Skunk cabbage (*Lysichiton americanus*) is a herbaceous species favouring moist soil conditions in the temperate climate of North America, whereas Swiss cheese plant (*Monstera delicious)* is a woody climber (a liana) from South America. They have vastly different environmental needs for success, are obviously in very different genera, but both species are in the Araceae family (characterized by having flowers comprising a spathe and spadix). The fruit of *Monstera deliciosa* is not only edible (with the consistency of banana) but is also very tasty (with a taste akin to pineapple), as indicated by the specific epithet *deliciosa*. But care must be taken, as many plants in the Araceae family are very poisonous.

Other Identifying Features

Buds are an important identifying feature of woody plants (perhaps more so than in other plant forms) and may be very useful to add to the evidence that leads towards a positive identification. *Sorbus aucuparia* (rowan/mountain ash) has dark purple bud scales covered in, and masked by, grey pubescent hairs; these may be easily rubbed off to reveal the shiny, purple to purple-black bud scales below. *Sorbus Americana*, on the other hand, has pointed buds with red bud scales.

Deciduous subjects have thin, membranous leaves, which they lose in the autumn, and they overwinter via a network of dormant buds (scale-covered embryo shoots). Evergreen species produce thicker, more persistent leaves, with more layers of tissue than their deciduous counterparts, and they retain more than one whorl of leaves – the current season's and the previous season's as a minimum. The particularly thick, waxy cuticle on the upper epidermis of evergreens reduces water loss, helps protect the leaves to aid longevity, and produces a notably shiny effect.

Fragrance and pungence are often underestimated in plant identification: there are distinct smells associated with many plants, especially when tissues are crushed. However, there are some plants with really distinct smells even when the foliage or other parts are just gently brushed, including *Fritillaria imperialis* whose foliage and scaly bulbs smell very strongly, and the foliage of *Euodia hupehensis*. Unfortunately, pungent natural oils can also be misleading so their use for identification has to be tempered with other botanic information. *Alliaria petiolata* (Jack-by-the-hedge, hedge garlic or garlic mustard), for instance, has oils that smell of garlic. However, a fairly cursory look shows the plant not to be typical of garlic or onion; in fact, the netted venation, broad leaves and four-petalled flowers prove that it is not even a monocot. Closer inspection proves it to be a *Crucifer* (cabbage and cress group), and this is confirmed by the long pod-like fruits (*siliquas*) that follow the flowers. All that smells of garlic is not necessarily garlic! Likewise peppermint- and lemon-scented pelargoniums are neither peppermint nor lemon.

Plant Families

Plants are usually listed alphabetically by genus. However, studies of the similarities and dissimilarities within family groups can be a very useful exercise to aid the understanding of plant classification, plant recognition, and sometimes their individual cultural needs. Becoming adept in recognizing family traits and resemblance (and even better, generic features) is a very useful tool, as it enables you to start the reference process that will ultimately lead to successful identification. Anything that gives a starting point for reference must be good. Recognizing particular fruit shapes and types from families gives immediate help, and recognizing the fruit shapes of those in the different genera within the same family gives even more help in identifying a plant – particularly if other identifying features are also used to add more 'evidence' for your decision.

The range of plant types within a family can be wide, and, unlike species, traits at familial level are diverse. Members of families cut across the longevity groupings and may have representatives from annuals to woody perennials, but subjects within the same family (or even the same genus) do not necessarily have the same underground organs. Obviously this is absolutely specific at species level, but at generic level there may be wide variation. Species within the genus *Iris*, for instance, may vary widely: *Iris pseudoacorus, Iris germanica* and *Iris siberica* all have fleshy rhizomes in order to overwinter, whereas *Iris reticulata* and *Iris histrioides* are both bulbous. *Crocus* and *Crocosmia* are in the Iridaceae family and have corms; *Colchicum* is in the Liliaceae and also has corms, although most species in the Liliaceae have scaly bulbs.

Genera and species of succulent types from arid regions, bulbous forms and woody forms may all appear in the same family by virtue of their flower and fruit structure. It is common for some members of the Crassulaceae (a family of mostly succulent plants) to have small plantlets develop on their leaf margins. Furthermore, some genera within the family present plantlets more than others, for example *Bryophyllum* and *Kalanchoe*. However, some members of the genus *Polystichum* also feature plantlets (bulbils) on their leaves. *Kalanchoe* species are succulents with leaves modified to cope with arid and intense sunlight conditions. *Polystichum setiferum* is a fern with large membranous fronds that inhabits the forest floor in poor light conditions and high humidity. So the propensity to produce plantlets on leaves is no indication of family or generic similarities, but could be one of the deciding factors at species level.

Initially it is best to make comparisons of those species with the most obvious likenesses or dissimilarities. However, some will not be quite so obvious, and will take much closer scrutiny in order to understand their relationships. Gradually it becomes easier to know what to look for, and just what the identifying features are, and you may even disagree with the purported similarities and the reasons for allocating a plant to a specific group, family, genus or species. It is by no means cut and dried in all instances, and a constant review and update of research leads to changes in classification all the time. Just in the last two decades *Stransvaesia* has been included in the genus *Photinia* (still both in the Rosaceae). *Cimicifuga* has now been included in the genus *Actaea* (still both in the Ranunculaceae) and *Senecio greyii* became *Brachyglottis* (reserving *Senecio* for other plants within the group, and remaining in the same family – now Asteraceae). The familiar family Compositae now becomes Asteraceae (named after the genus Aster), and the equally familiar family Labiatae (named because of the two lips – labia – of the flowers) now becomes Lamiaceae (named after the genus *Lamium* within it). Both garlic with its scaly bulb, and onion with its tunicated bulb, are no longer in the Liliaceae but are now placed in the Alliaceae (named after the genus *Allium*) along with *Agapanthus* and all those related plants with the typical ball-shaped flowerhead. Grasses belong to the Graminaceae family but some place them in the Poaceae, and bamboos, although having many similarities to grasses in general, obviously also have many differences, but are also in the Graminaceae/Poaceae. The Papilionaceae family has been variously called, or sub-divided into, the Fabiaceae (after beans – *Fabia*), Papilionatae and, more recently, Leguminosae (after their pod-like fruit – legume).

Often families are divided into sub-sections, and in some cases these sub-sections are taken up on an everyday basis. The term 'brassica', for example, is commonly used by gardeners to refer to cabbage-like plants. Its derivation is as a tribe within the family Cruciferae – however some now place all *Crucifers* in the Brassicaceae. Changes rarely come thick and fast, so the effect is one of relative stability.

Rhododendron species are not only classified in genera and species, but are also held in series and sub-series that denote certain affinities. Obviously, all rhododendrons are in the same family (Ericaceae) and the same genus (*Rhododendron*); however, their other similarities across the specific differences are recognized by grouping them into 'series': *Rhododendron barbatum* and *R. strigillosum,* for example, are different species but share the barbate (coarsely hairy) leaf petiole that puts them in the same series (series *Barbatum*). However, their foliage types and flowers are distinct enough one from another to put them in different sub-series. So, *Rhododendron barbatum* is in both the *Barbatum* series and sub-series, whereas *R. strigillosum* is in the *Barbatum* series and the sub-series *Maculifera.* Deciduous and evergreen azaleas are actually rhododendrons in the *Azalea* series, which has led horticulturists to use the common general name 'azalea' for this large group. Deciduous rhododendrons, such as *Rhododendron atlanticum, R. viscosum, R. viscosum glaucum, R. occidentale, R. schlipenbachii and R. kaempferi* all belong to the *Azalea* series. So the plant generally known as *Azalea lutea* (the wonderfully scented, yellow-flowered deciduous shrub) is actually botanically *Rhododendron luteum* – in a similar way to *Pelargonium zonale* being known as geranium. The very large-leaved species *Rhododendron rex, R. hodgsonii* and *R. falconeri* are all in the *Falconera* sub-series because of the similarities they share.

It may seem a strange way to tackle recognition, identification, classification and nomenclature, but the rest of this chapter is dedicated to plants categorized into families in alphabetical order. Within the families some examples of representative genera, and species commonly used in horticultural practice, are discussed. It is always best to gradually add to your repertoire by remembering a few at a time until they stick, initially remembering only those that are of particular interest or importance to you, and those that you favour. It helps to break down lists into smaller bite-size chunks, and by repeating those that you have come across before, they are liable to remain in the memory, especially when the same names keep cropping up under different criteria.

Because some of the relationships will be better understood after the information on flower structure in the next chapter, just briefly study the common family groups by quickly scanning the lists. If you have not tried the exercise before, it may be beneficial to pick on a family, some of whose representatives you already know, and then consider their similarities and wide-ranging dissimilarities. Even if you already have a good plant repertoire you might still come across both obvious and surprising relationships.

You will notice that word endings of generic and specific names are matched wherever possible: for example *Mimulus aurantiacus, Acanthus spinosus.* However, this is not always possible, and sometimes just the last letter is matched – for example,

Aphelandra squarrosa.

Developing flower of Acanthus spinosus.

Acanthus mollis – or there is no match at all – for example *Acer pseudoplatanus*. The meanings and explanations of some specific epithets (and some generic names) have already been discussed, but others are highlighted and discussed within the family lists.

A Few Important Plant Families

Acanthaceae

This family comprises mostly clump-forming herbaceous perennials with very large leaves (the pattern of acanthus leaves was used as decoration on columns and pillars by the Romans). Examples are *Acanthus mollis* (soft bear's breeches), and *Acanthus spinosus* (spiny bear's breeches).

Aphelandra squarrosa (zebra plant) is a tender perennial.

Aceraceae

This is the maple family and comprises deciduous woody perennials. It includes *Acer pseudoplatanus* (sycamore); *Acer platanoides* (Norway maple); *Acer campestre* (field maple); *Acer davidii* (striated, striped or snake-barked maple); *Acer palmatum* (smooth Japanese maple); and *Dipteronia sinensis*.

Agavaceae

Within the family the consistent feature is the formation of whorls or rosettes of highly fibrous leaves typical of many monocotyledonous species. Succulent types have rosettes of thick, fleshy, pointed leaves that sometimes terminate in a spine, and have some water-storing capacity. Many of the genera within the family produce large plants, including some *Agave* species (whence the family name comes). *Agave Americana* has flower spikes

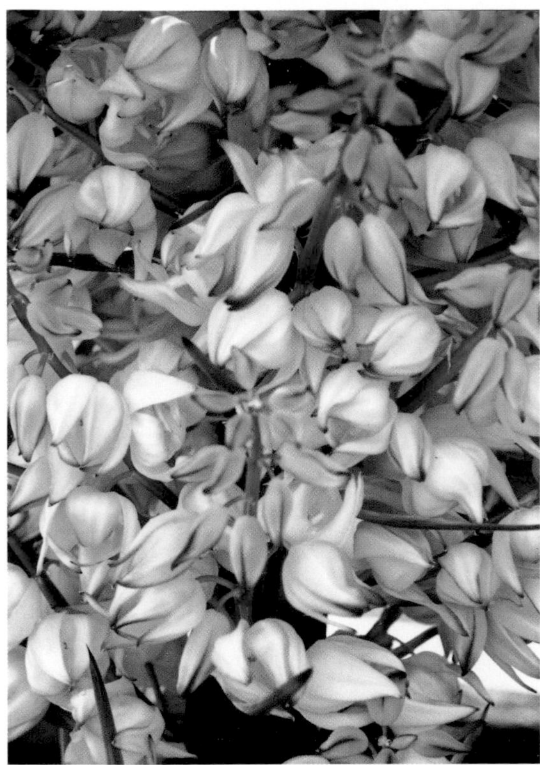

Flower of Yucca species. (Agavaceae)

that attain tree-like proportions. It is difficult to decide on the age of sexual maturity (although young plants do not flower), but both sufficient age and specific environmental conditions together seem to be the trigger for flowering. This species is unfortunately monocarpic, so the main mother plant dies directly after flowering. However, young offsets from the base of the mother plant remain alive and can develop to form new main plants. Flowers bear some similarities to lilies, and some species now in the Agavaceae were formerly in the Liliaceae.

Other species include dragon tree (*Dracaena draco**), *Beschorneria yuccoides**, cabbage palm (*Cordyline australis**), *Yucca filamentosa**, *Sansevieria trifasciata* (bow-string plant).

Alliaceae

Garlic (*Allium sativum*), onion (*Allium cepa*), chives (*Allium schoenoprasum*), leek (*Allium porrum*), *Allium campanulatum*, *Agapanthus africanus*.

Amaryllidaceae

A family of mostly bulbous subjects including *Amaryllis, Galanthus nivalis* (common snowdrop), *Clivea miniata, Narcissus pseudonarcissus* (wild daffodil) and *Nerine bowdenii*.

Apiaceae (Umbeliferae)

Formerly the family Umbeliferae; it includes ground elder (*Aegopodium podogaria*), *Angelica archangelica* (syn. *Angelica arvensis*), *Astrantia major,* celery, hemlock (*Conium maculatum*), hogweed (*Heracleum sphondylium*), giant hogweed (*Heracleum mantagazzianum*), cow parsley (*Anthriscus sylvestris*), carrot (*Daucus carota*), alexanders (*Smyrnium olustratum*), common fennel (*Foeniculum vulgare*), parsnip (*Pastinaca sativa*) and parsley (*Petroselinum crispum* syn. *Petroselinum sativum*). Many are biennials or short-lived perennials.

Asphodelaceae

Asphodlines, asphodels and red hot pokers.

Asteraceae (Compositae)

Formerly the Compositae family; daisy-like/aster-like flowers. The Asteraceae family incorporates many Lactifers including dandelion, lettuce and the sowthistles. It also includes sunflower, thistle, chrysanthemum, dahlia, chamomile and daisies, and many shrubby woody perennials (for example *Olearia* x *haastii, Olearia macrodonta, Brachyglottis* (syn. *Senecio*), *Ozothamnus ledifolius*) and the woody climber *Mutisia decurrens*.

Annuals

Lettuce (*Lactuca sativa*), common sowthistle (*Sonchus oleraceus*), groundsel (*Senecio vulgaris*), annual tickseed (*Coreopsis*), pot marigold (*Calendula officinalis*), cornflower (*Centaurea cyanus*), sunflower (*Helianthus annuus***)*, annual rudbeckia, French marigold (*Tagetes patula*), African marigold (*Tagetes erecta*).

Biennials

Cichorium intybus (chicory).

**Dracaena draco* is a reference to the red exudates that looks like 'dragon's blood'; *Beschorneria yuccoides* is a reference to the leaves looking like *Yucca* ('oides' = 'looks like', and *australis* means 'southern' – a reference to its native habitat New Zealand); and *filamentosa* is a reference to the filamentous fibres that break away from the leaf edges of this species. *Helianthus* is a direct translation of 'sunflower' in Greek (Helios = sun, and anthus = flower) – and *annuus* obviously refers to its annual life cycle.

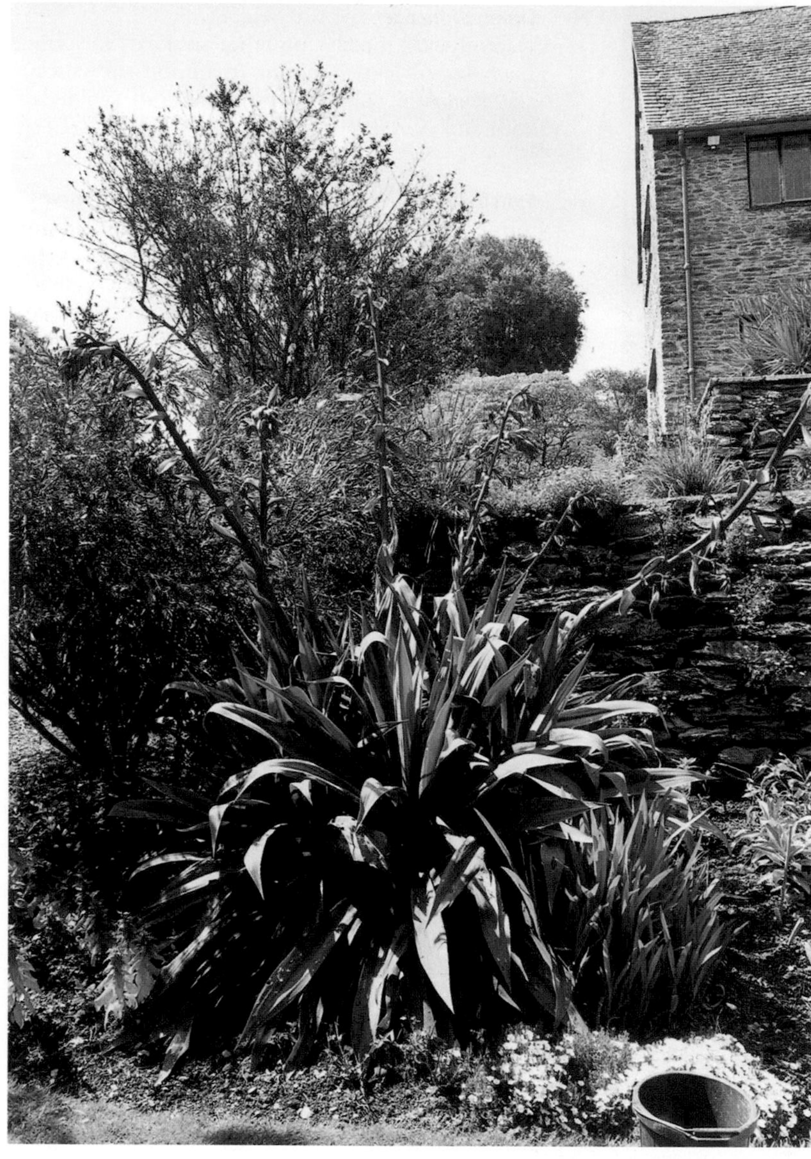

Beschorneria yuccoides. (Agavaceae)

Herbaceous Perennials
Dandelion (*Taraxacum officinale*), *Bellis perennis* (common daisy), dahlia species, thistle species, chrysanthemum species, artichoke, perennial sowthistles, mayweed, alpine sowthistle (*Cicerbita alpina* syn. *Lactuca alpina*), *Ligularia prezwalskia, Ligularia dentata (*syn. *Senecio clivorum*), *Osteospermum* (syn. *Dimorphotheca*), *Achillea millefolium,* edelweiss (*Leontopodium alpinum*), *Artemesia, Aster* (Michaelmas daisies), *Tanacetum, Hieraceum* (hawkweed), *Doronicum, Helichrysum, Coreopsis, Gazania, Rudbeckia, Echinops.*

Boraginaceae
Borage, forget-me-not, comfrey.

Annuals
Borago officinalis (borage).

Developing flower of a typical Allium species. (Alliaceae)

Biennials and Short-lived Perennials
Echium pininana, Myosotis (forget-me-nots).

Herbaceous Perennials
Anchusa, Pulmonaria★ (lungwort), *Symphytum* (comfrey).

Caprifoliaceae
Sambucus nigra★ (common elder), *Linnaea borealis*★ (twinflower), honeysuckles such as *Lonicera periclymenum, Leycesteria formosa* (Himalayan honeysuckle), *Abelia, Symphoricarpus albus* (snowberry), viburnums (including *Viburnum farreri* syn. *Viburnum fragrans*) and *Weigela.*

Caryophyllaceae
Carnations, pinks and phlox.

Annuals
Corn cockle (*Agrostemma coeli-rosa*), *Saponaria* (soapwort).

★*pulmonary* refers to the lungs (herbal medicine).
★*nigra* = black, a reference to the dark purple/black berries; *borealis* is a reference to northern forests (boreal forests), the native habitat of the species.

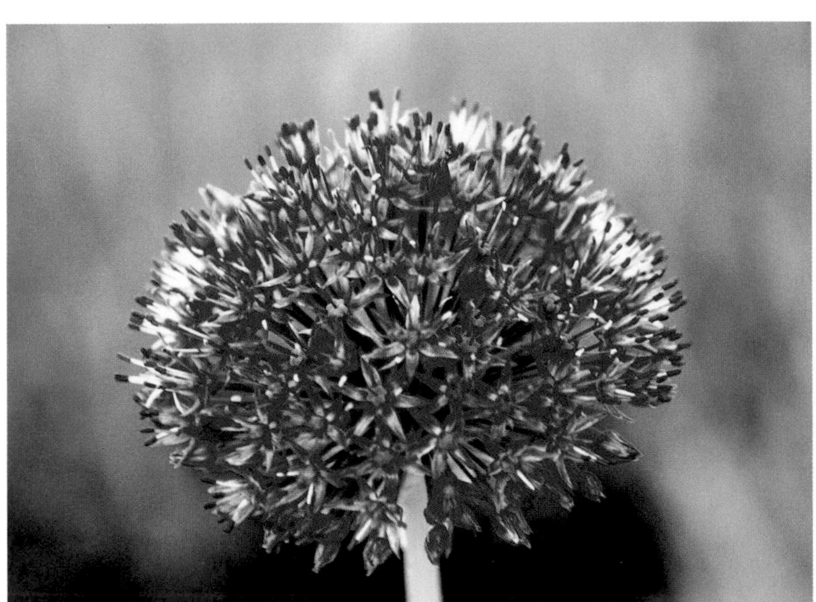

Close-up of Allium cyanus. (Alliaceae)

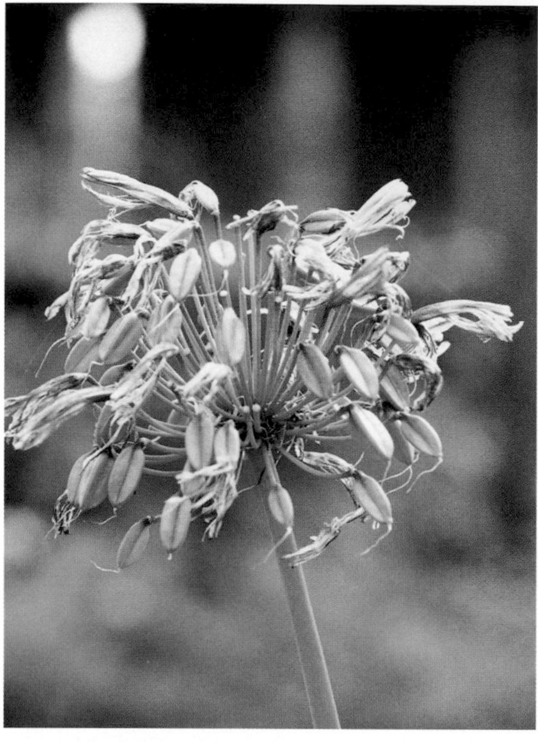

The deteriorating flower and developing fruit of Agapanthus africanus. (Alliaceae)

Herbaceous Perennials
Carnations, phlox, red campion (*Silene dioica*), ragged robin (*Lychnis flos-jovis*), *Dianthus* (pinks), *Lychnis, Gypsophila.*

Cruciferae (Brassicaceae)
This includes the Brassica tribe – the cabbage group.

Annuals
Rape, mustard, hairy bitter cress (*Cardamine hirsuta*), *Lobularia maritima* (syn. *Alyssum maritimum*), moonwort (*Lunaria annua*).

Biennials
Brassicas such as cabbage (*Brassica oleracea capitata*) and kale (*B.o. acephala*) that are harvested as leafy crops. Cauliflower (*B.o. botrytis*) grown for its white inflorescence, and Brussels sprouts (*B.o. gemmifera*) grown for the production of the large, rounded, leafy axillary buds. Swede (*Brassica rutabaga*), turnip (*B. rapa*), radish (*Raphanus sativus*) and kohl-rabi (*Brassica oleracea caulorapa*), all grown for their fleshy swollen roots.

Herbaceous Perennials
Aubretia, Allysum, Aurinia saxitilis (syn. *Alyssum saxitilis*), *Draba aizoides* (yellow whitlow grass), *Iberis,* stocks (*Mathiola*), wallflower (*Chierianthus*), *Cardamine pratense* (cuckoo flower).

Clivea miniata. (Amaryllidaceae)

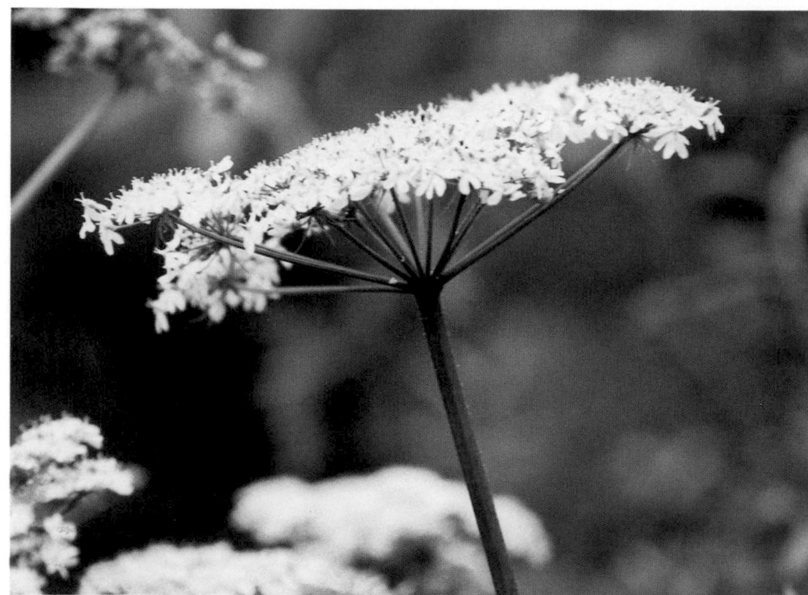

Hogweed (Heracleum sphondylium), with typical umbel of flowers (like the spokes of an umbrella). Apiaceae (Umbeliferae)

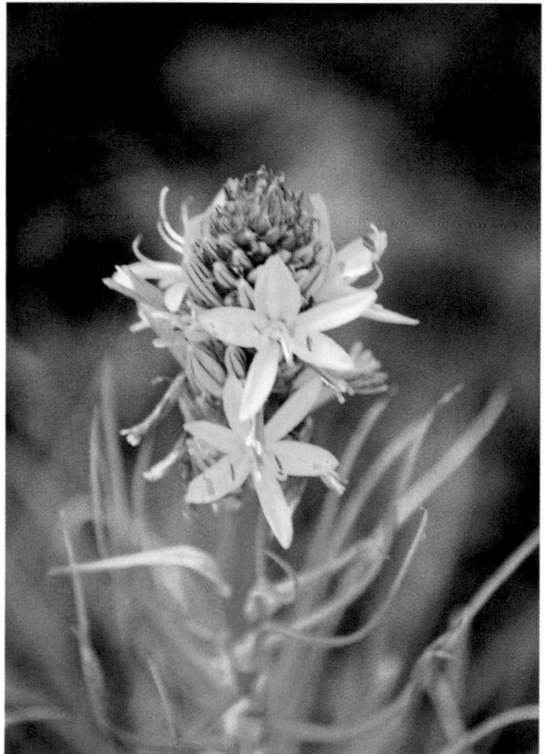

Flower of Asphodeline lutea. (Asphodelaceae)

Ericaceae

The Ericaceae family includes the genera *Rhododendron*, *Pieris*, *Vaccinium*, *Gaultheria*, *Pernettya*, *Cassiope* and heathers (*Daboecia*, *Erica* and *Calluna*). Some examples of these are as follows:

Rhododendrons: *Rhododendron arboreum*, *Rhododendron auriculatum*★, and *Rhododendron yakusimanum*.

Strawberry trees: *Arbutus unedo*.

Heathers: *Erica carnea*, *Erica tetralix*, *Calluna vulgaris and Daboecia cantabrica. Cassiope lycopodioides*★, *Cassiope fastigiata*, *Vaccineum glauco-album*, *Vaccineum myrtillus*, *Arctostaphyllus*, *Enkianthus campanulatus*★, *Gaultheria*, *Zenobia*, *Oxydendrum arboreum*, *Menziesia ciliicalyx*, *Phyllodoce*, *Pieris floribunda*★, *Pieris formosa var forrestii*★, *Kalmia angustifolia*, *Kalmia angustifolia forma rubra*★, *kalmia latifolia*★, *Ledum groenlandicum*★.

★*auriculatum* refers to the ear-like appendages (auricles) at the base of the leaves of this subject. *lycopodioides* is a reference to the fact that this species of *Cassiope* looks like a low-growing tussock of club moss in the genus *Lycopodium*. *campanulatus* is a reference to the bell-like flowers, *floribunda* is a reference to the free-flowering nature of the species (abundant flowers), and var. *forrestii* describes the plant as a variation from the species, and named after the plant hunter George Forrest. *Kalmia angustifolia forma rubra* tells us it is the red-/pink-flowered form, and *Kalmia latifolia* tells us it is the broader-leaved species (*lati* = wide and *folia* = foliage). *groenlandicum* is a reference to its native habitat in the tundra of Greenland.

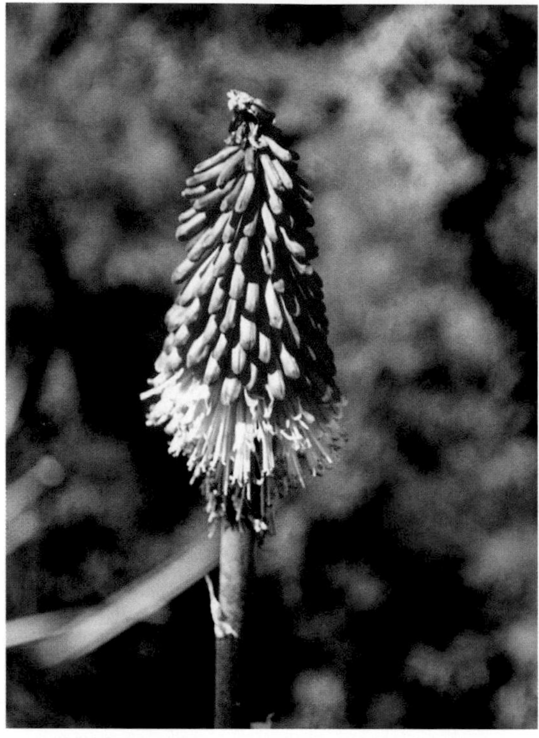

Flower of Kniphofia species (red hot poker). (Asphodelaceae)

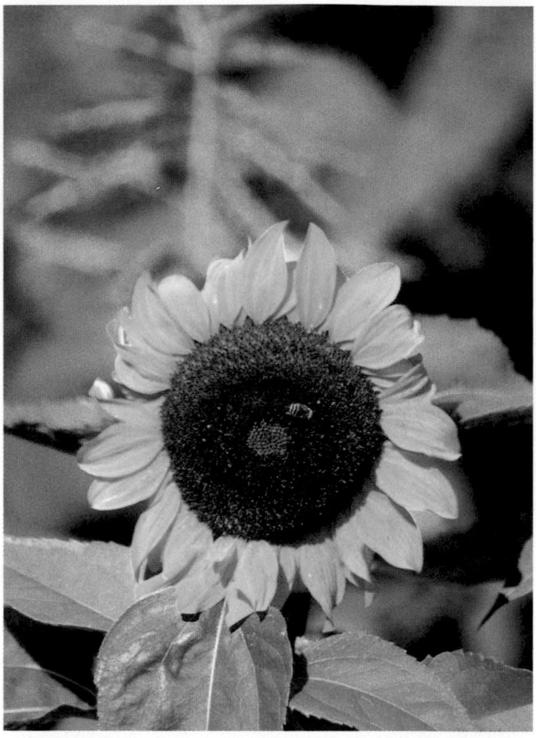

Sunflower (Helianthus annuus). Asteraceae (Compositae)

Geraniaceae

This family comprises the clump-forming herbaceous perennial *Geranium* species, and semi-succulent perennial sub-shrubs in the genus *Pelargonium*. Both types present the typical 'crane's-bill/stork's-bill shaped' fruits after fertilization.

Herbaceous Perennials

Geranium psilostemon syn. *Geranium armenum*, *Geranium pratense*, *Geranium maderense*.

Succulent and Semi-succulent Types

Pelargonium zonale types (zonal pelargoniums – the geraniums of horticulture, whose leaf has dark zones), *Pelargonium regale* types (regal pelargoniums), *Pelargonium peltatum* types (ivy-leaved, trailing geraniums/pelargoniums – whose leaves are peltate), *Pelargonium tomentosum* (the peppermint geranium of horticulture).

Hydrangeaceae

Herbaceous Perennial

Kirengeshoma palmata.

Woody Shrubs

Deutzia, hydrangeas – for example *Hydrangea hortensis, Hydrangea aspera subsp. aspera* (syn. *Hydrangea villosa), Hydrangea aspera subsp. sargentianum, Hydrangea paniculata* and *Philadelphus.*

Woody Climbers

Hydrangea anomola subsp. petiolaris syn. *Hydrangea petiolaris, Shizophragma hydrangoides* (Japanese hydrangea vine).

Iridaceae

Rhizomatous iris: *Iris pallida* (bearded iris types), *Iris siberica, Iris ensata* syn. *Iris kaempferi, Iris germanica.*

The large-growing biennial Echium pininana. (Boraginaceae)

Bulbous iris: *Iris latifolia, Iris reticulata, Iris bakeriana, Iris danfordiae, Iris historioides.*

Cormous species: *Crocosmia, Crocus, Dierama pulcherrima, Freesia, Gladiolus, Schizostylis coccinea**, *Watsonia meriana.*

Fibrous-rooted types: *Sisyrinchium*

Lamiaceae (Labiatae)

Formerly Labiatae, this family includes deadnettles and mints; they are distinguished by their two-lipped flowers (*labia* = lips). White dead nettle (*Lamium album*), *Ajuga reptans, Monarda didyma* (bergamot), *Coleus, Lavendula, Mentha, Nepeta, Phlomis fruticosa**, *Teucrium fruticans** (shrubby germander), *Perovskia atriplicifolia, Salvia splendens, Mentha* × *piperita* (peppermint), *Mentha sauveolens* (apple mint), *Nepeta* (catmint), *Prostanthera rotundifolia* (mint bush), *Prunella* (self-heal), *Rosmarinus officinalis* (rosemary), *Stachys byzantina* syn. *Stachys lanata, (lamb's ear) Thymus praecox* (thyme).

**Schizostyllus* indicates a split style, and *coccinea* describes the red/pink colour of the flowers.

**fruticosa* and *fruticans* are references to the shrubby form of these species, as some others in this genus are herbaceous (*fruticose* = shrubby).

Honeysuckle (Lonicera periclymenum). Caprifoliaceae

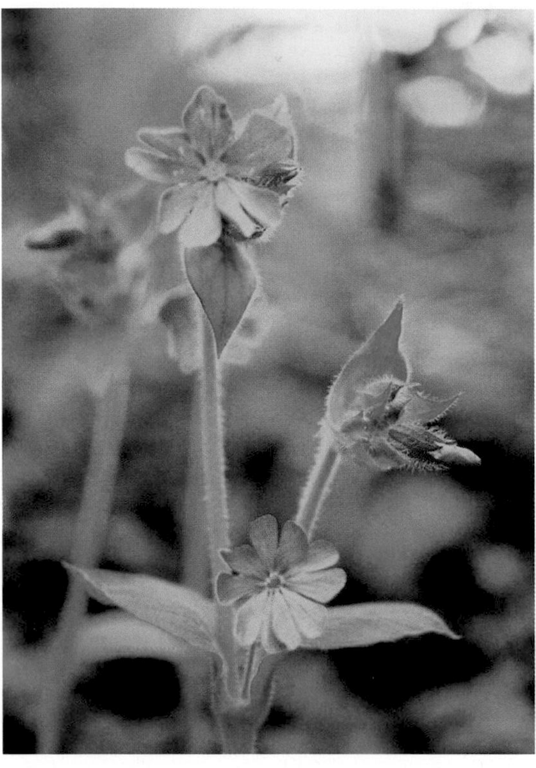

Red campion (Silene dioica). Caryophyllaceae

Lactifers

Lactifer is not a family classification, but comprises a group of very disparate plants within the angiospermae (across families and genera) that share the common feature of having specialized cells or ducts that excrete a white milky fluid known as lactate (latex). Examples include *Lactuca* (lettuce group), sowthistles and some maples. Within the same family some genera (and some species within a genus) may be lactifers and others not.

Latex exuded by Lactifers contains sugars, proteins, oils and mineral salts. Its function is not known, but it is thought to be involved in wound healing, and perhaps like tannins, terpenes and phenols, a resistance to pathogens. Latex exudes freely from damaged tissues and usually congeals easily. However, it may cause excessive 'bleeding' from prepared stem cuttings of woody perennials such as *Ficus elastica* and its cultivars, in which case the cut ends are dipped in water to speed up the congealing process. *Ficus elastica* and its many cultivars are commonly known as 'rubber plants' because they exude latex, but commercially rubber is tapped from *Hevea brazilliensis*, another Lactifer.

For Leguminosae see Papilionaceae.

Moonwort (Lunaria annua); most other crucifers have white or yellow flowers, including cabbage. Cruciferae (Brassicaceae)

Cut-leaved geranium (Geranium dissectum), showing the crane's-bill-shaped central structure to the flowers. (Geraniaceae)

Lauraceae

This is the laurel family, however there is only one true laurel, *Laurus nobilis* (bay or sweet laurel). The other so-called common laurel is in fact a prunus (*Prunus laurocerasus*). The name *Laurus nobilis* was chosen as it was used by the Romans in wreaths to honour their noble warriors.

Liliaceae

The lily family: the Liliaceae comprises clump-forming herbaceous perennials (many of which are bulbous) and includes lilies and bluebells. It formerly included onions and garlic (now in Alliaceae), and also a large group of succulents (aloes). Within this grouping there are rosette-forming, ground-hugging perennials and arborescent (tree-like) perennials, including a number of red hot pokers (*Kniphofia*). The disparity of forms leads some authorities to put the aloes in their own family (Aloaceae), and some include *Kniphofia* in the Aloaceae, while others consider them to be in the Asphodelaceae. The same criteria led to a separation of the arborescent forms such as *Cordyline* and *Yucca* from the Liliaceae into the Agavaceae.

Bulbous types: *Lilium regale* (regal lily), *Lilium lancifolium* syn. L. *tigrinum* (tiger lily), *Lilium longiflorum* (Bermuda lily, Easter lily), *Lilium pyrenaicum* (yellow turkscap lily), giant lily (*Cardiocrinum giganteum**), *Tulipa*, *Chionodoxa*, *Colchicum*, bluebell (*Hyacinthoides** *non-scriptus* syn. *Endymion non-scriptus*), *Ornithogalum* (star of Bethlehem), *Scilla scilloides*.

Tuberous types: *Erythronium americanum*, *Erythronium grandiflora*.

Magnoliaceae

This family incorporates magnolias and tulip trees. Magnolias include *Magnolia stellata*, *Magnolia wilsonii*, *Magnolia sinensis*, *Magnolia acuminata* (cucumber tree); *Magnolia grandiflora* and *Liriodendron tulipifera* (tulip tree).

Malvaceae

This family incorporates mallows and hibiscus, including *Malva palustris* (common mallow), hollyhock (*Alcea rosea* syn. *Althea rosea*), *Malva moschata*, *Sidalcea*.

Woody perennials: *Hibiscus syriacus*, *Hibiscus rosa-sinensis**, *Abutilon vitaefolium*, *Lavatera*, *Hoheria lyallii*.

**giganteum* is a reference to the very large proportions of this species. *Hyacinthoides* = 'looks like a hyacinth'. *rosa-sinensis* is a direct translation of red Chinese hibiscus.

Hydrangea hortensis. *(Hydrangeaceae)*

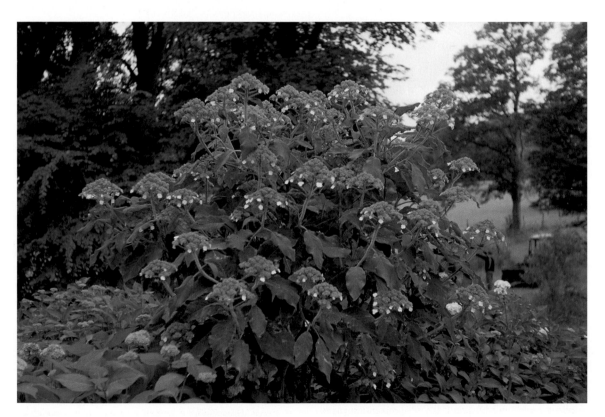

Hydrangea aspera subsp. sargentianum. (Hydrangeaceae)

Jerusalem sage (Phlomis friticosus). Lamiaceae (Labiatae)

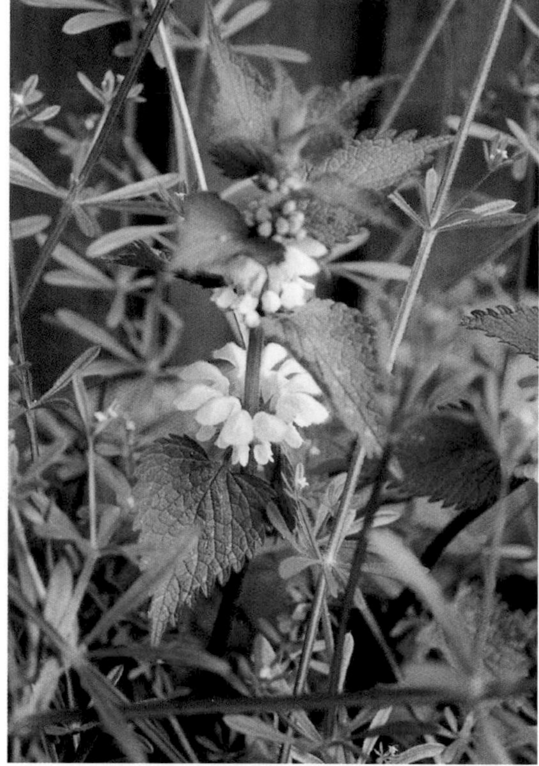

Common white deadnettle (Lamium album). Lamiaceae (Labiatae)

Myrtaceae

The gum trees and myrtles: *Callistemon citrinus, Feijoa sellowiana, Eucalyptus gunnii* (cider gum); *Eucalyptus pauciflora subspecies niphophila* (snow gum), *Myrtus luma, Myrtus communis.*

Oleaceae

This family incorporates olives and ashes.

Trees: *Fraxinus* (ash), *Ligustrum lucidum* (shining privet), *Olea europaea* (olive).

Shrubs: *Forsythias, Osmanthus delavayi, Syringa vulgaris* (common lilac).

Climbers: *Jasminum nudiflorum* Jasminum polyanthum*.*

Papaveraceae

Poppies and meconopsis: *Papaver nudicaule* (Iceland poppy), *papaver orientale* (oriental poppy), *Papaver somniferum* (opium poppy), *Papaver rhoeas* (field or corn poppy), *Meconopsis betonicifolia* (blue poppy), *Meconopsis cambrica* (Welsh poppy), *Romneya coulteri* (Californian poppy), *Corydalis, Dicentra spectabilis* (bleeding heart).

Papilionaceae (Leguminosae)

Formerly Leguminosae; the pea and bean family can include sub-family Mimosaceae. It includes garden peas, sweet peas, broad and haricot beans, lupins, clover, robinea, Acacia and laburnum – all those with a simple-podded fruit (legume).

Annuals

Broad bean, runner bean, garden pea (*Pisum sativum*), mangetout, sweet pea (*Lathyrus odoratus)*, alfalfa.

**nudiflorum* is a reference to flowering on bare (leafless) wood; *polyanthum* describes it as 'free-flowering' (*poly* = many, and *anthum* = flowers).

Tulips – note that the outside tissues of the flowers are green. (Liliaceae)

Herbaceous Perennials
Lupinus (lupin), *Trifolium repens* (white clover), *Trifolium pratense* (red clover), *Lathyrus latifolius* (everlasting or perennial pea), *Mimosa pudica* (sensitive plant).

Woody Perennials
*Acacia dealbata, Albizia, Bauhinia, Caesalpina gilliesii, Cassia corymbosa**, *Cassia didymobotria, Erythrina crista-galli* (coral tree), *Clianthus puniceus, Indigofera gerardiana, Lupinus, Cytisus, Ulex, Spartium junceum**, *Wisteria sinensis, Cercis siliquastrum* (Judas tree); *Cladrastis lutea** (yellow wood); *Gleditsia triacanthos** (honey locust); *Laburnum*; *Piptanthus nepalensis* (Nepalese laburnum); *Robinia pseudoacacia** (false locust or false acacia), *Colutea arborescens, Genista, Sophora microphylla, Sophora tetraptera.*

Primulaceae
Primulas and primroses: *Primula auricula* (auriculas), other primulas including *Primula japonica, Primula*

**corymbosa* refers to the inflorescence (arrangement of the flowers) – large corymbs of yellow flowers. *junceum* refers to the fact that the species looks like (but is not) a rush (*Juncus* species). *lutea* = yellow and describes the colour of the wood. *Triacanthos* = three thorns (*Tri* = three and *acanthos* = thorns), describing the fact that sets of three thorns appear all over the stems. *pseudoacacia* = false acacia.

Flower of the large-leaved magnolia with two-lobed leaf (Magnolia officinalis f. biloba). Magnoliaceae

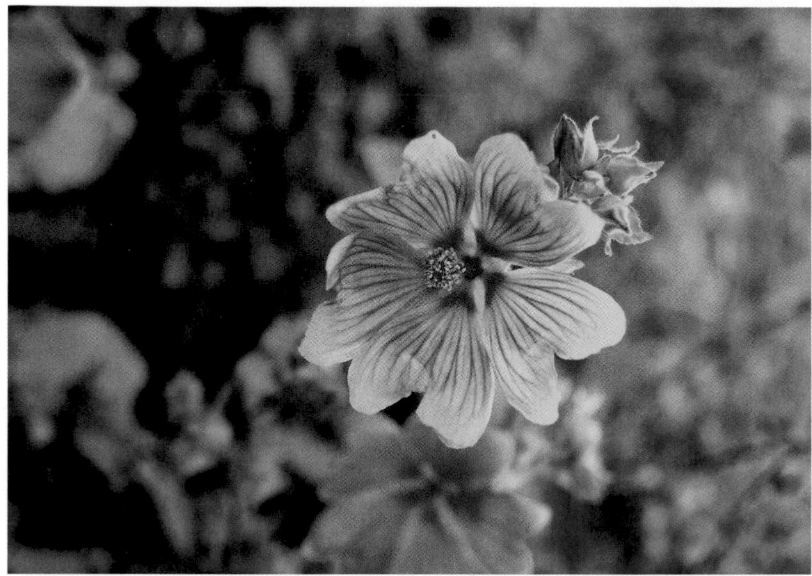

A cultivar of tree mallow (Lavatera). Malvaceae

denticulata (drumstick primula), *Primula florindae* (Himalayan, or giant cowslip). *Cyclamen persicum, Cyclamen hederifolia (*syn. *Cyclamen Neapolitan), Soldanella alpina* (alpine snowbell).

Proteaceae

Banksia, Protea, Embothrium coccineum, Grevillea, Telopea truncata (New Zealand waratah).

Ranunculaceae

Buttercups, and so on. Mostly herbaceous perennials (including herbaceous forms of clematis), but also includes a large group of self-supporting climbers in the genus clematis.

Herbaceous Perennials

Clematis integrifolia, Ranunculus ficaria (lesser celandine), *Ranunculus repens* (creeping buttercup), *Ranunculus acris* (meadow buttercup), *Anemone japonica* (*Anemone* × *hybrida*), *Anemone nemerosa* (wood anemone), *Actaea pachypoda* (syn. *Actaea alba*) (white baneberry), *Actaea rubra* (red baneberry), *Pulsatilla, Aquilegia, Caltha palustris* (kingcup), *Delphinium, Helleborus foetidus** (stinking hellibore), *Helleborus orientalis* (Lenten rose), *Helleborus niger* (Christmas rose), *Nigella damascena* (love-in-the-mist), *Trollius* (globeflower), *Aconitum napellus* (monkshood).

Woody climbers

Clematis alpina, Clematis tangutica, Clematis armandii, Clematis orientalis.

Rosaceae

The rose family. The Rosaceae comprises many woody perennials, but is also represented by many herbaceous perennials.

Woody Perennials (Shrubby)

The roses include *Rosa arvense** (field rose), *Rosa canina** (dog rose), *Rosa moyesii, Rosa omiensis pteracantha** (wing-thorned rose). *Chaenomeles, Cydonia* (quince), *Photinia, Potentilla, Eriobotrya japonica* (loquat), *Pyracantha**, *Amelanchier* (snowy mespilus or June snowberry); *Cotoneaster*; *Prunus laurocerasus** (cherry laurel), *Prunus lusitanica** (Portugese laurel).

**foetidus/foetidid* = stinking

**arvense* = 'of the fields', *canina* = 'pertaining to dog'. *pteracantha* is a direct translation of 'winged thorn' (*ptera* = wing, and *acantha* = thorn). *Pyracantha* is a direct translation of its common name 'firethorn' (*Pyra* = fire and *acantha* = thorn). *laurocerasus* is a direct translation of the common name 'cherry laurel' (*lauro* = laurel and *cerasus* = cherry). *lusitanica* refers to Portugal.

Forsythia × intermedia. (Oleaceae)

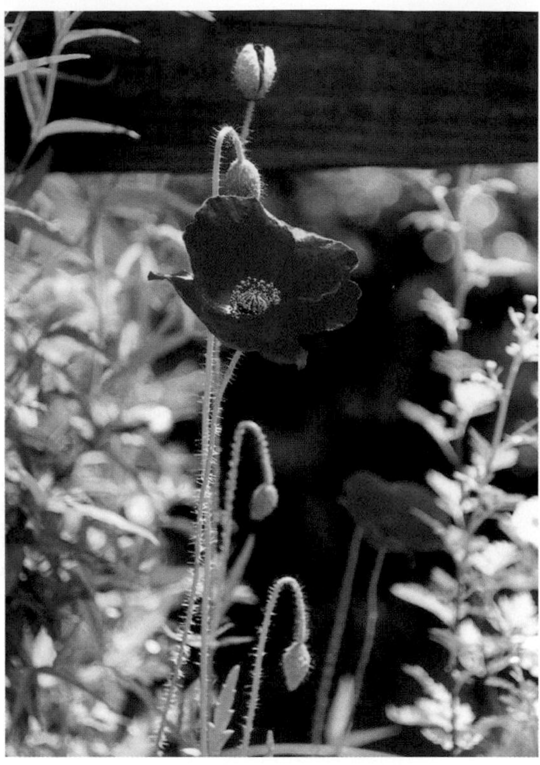

Corn or field poppy (Papaver rhoeas). Papaveraceae

Woody Perennials (Tree Forms)
Crataegus laciniata (oriental or cut-leaved hawthorn); *Crataegus monogyna* (common hawthorn); *Malus sylvestris* (crabapple); *Malus trilobata**, *Mespilus germanica* (medlar), *Prunus avium** (wild cherry); *Prunus padus* (bird cherry); *Pyrus* (pear); *Sorbus* (rowans and whitebeams).

Herbaceous Perennials
*Acaena, Alchemilla mollis** (lady's tresses). *Geum, Fragaria, Potentilla, Filipendula.*

Scrophulariaceae
Annual: *Nemesia strumonium.*

Biennials: *Digitalis purpurea* (foxglove), *Verbascum bombyciferum* (mullein), *Verbascum lychnitis* (white mullein)

Herbaceous Perennials and Sub-shrubs
Verbascum densiflorum, Scrophularia (figwort), *Digitalis grandiflora, Veronica, Antirrhinum* (snapdragon), *Cymbalaria muralis** (ivy-leaved toadflax), *Diascia rigescens, Mimulus aurantiacus* (syn. *Mimulus glutinosus*), *Veronica perfoliata** (syn. *Parahebe perfoliata*), *Penstemon, Isoplexis canariensis.*

**trilobata* refers to the three (tri) lobes of the leaf. *avium* refers to birds, but does not lead to the common name 'bird cherry', as this is reserved for *Prunus padus*. The reference here is because of the importance of birds in the germination of the seeds – germination is good after the fruits pass through the gizzard of birds. *Mollis* = Soft (hairs). *muralis* is a reference to the natural habitat of this plant (growing on walls), *perfoliata* refers to the leaf form where the stem appears to pierce a rounded leaf,

Everlasting pea (Lathyrus latifolius). Papilionaceae (Leguminosae)

Laburnum. Papilionaceae (Leguminosae)

Embothrium coccineum. (Proteaceae)

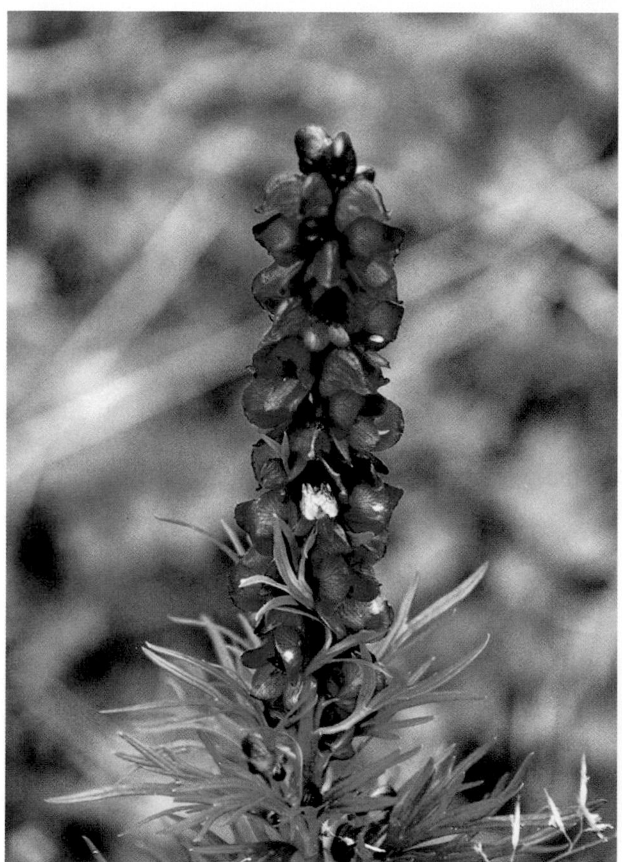

Monkshood (Aconitum napellus). Note: This plant is very poisonous. It is difficult to relate the structure of the flowers in the photograph to common buttercup; however, it is the central structure of pointed seed pods (follicles) that puts them in the same family, Ranunculaceae.

Flower of Rosa rugosa. (Rosaceae)

Flower of Geum cultivar 'Mrs Bradshaw'. (Rosaceae)

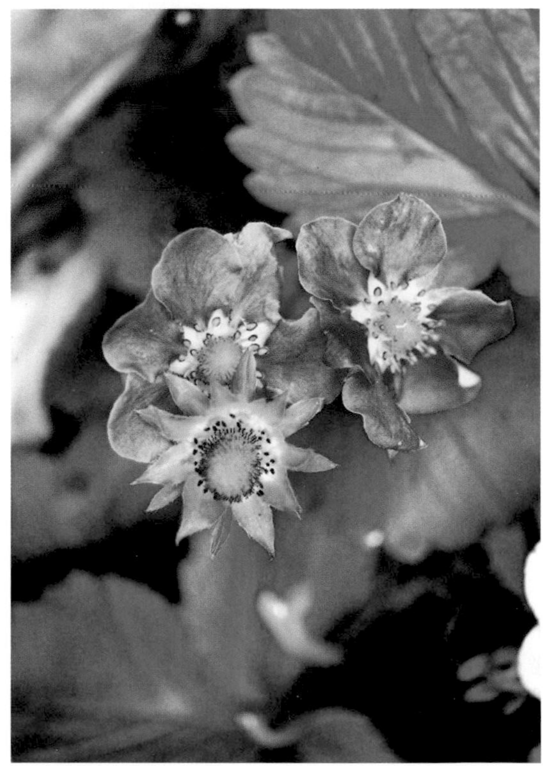

Flower of Fragaria (ornamental strawberry) cultivar. (Rosaceae)

Flower of the foxglove tree (Paulownia tomentosa). (Scrophulariaceae)

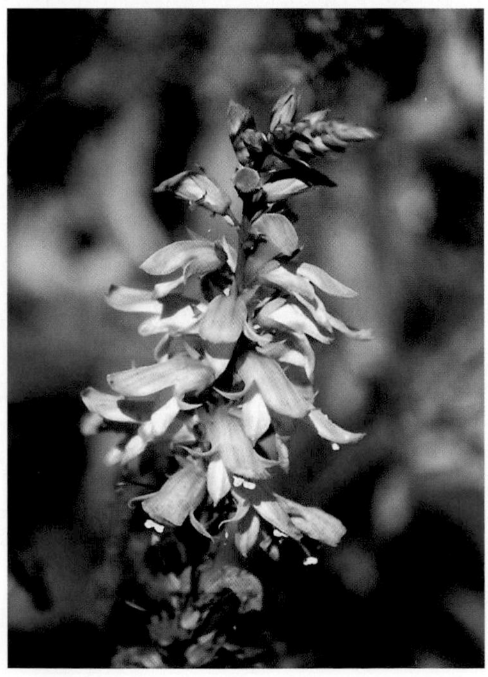

Isoplexis canariensis a shrubby but tender relative of Foxglove. (Scrophulariaceae)

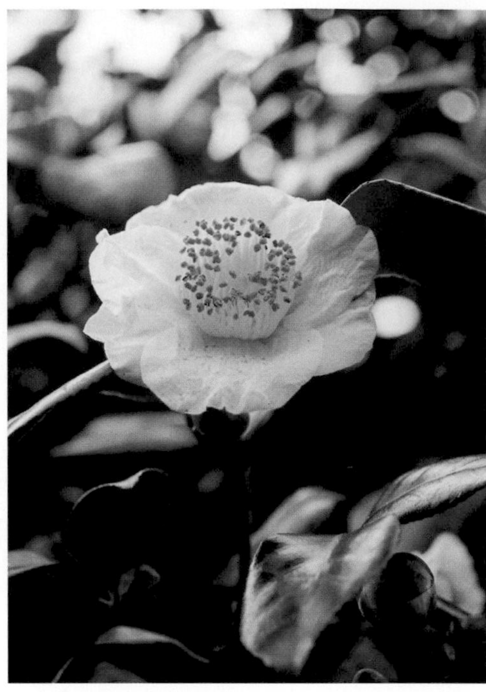

Flower of Camellia cultivar, showing the large numbers of stamen. (Theaceae)

Woody Perennials
Paulownia tomentosa (foxglove tree), hebes including *Hebe cupressoides**, *Hebe macrantha* and *Hebe salicifolia**.

Styraceae

Snowdrop trees: *Halesia carolina, Halesia monticola, Styrax japonica* (Japanese snowbell), *Styrax obassia* (fragrant snowbell).

Theaceae

*Camellia sinensis** (tea), *Camellia japonica, Stewartia pseudocamellia**, *Stewartia sinensis**.

**cupressoides* refers to the stems looking like cupressus (conifer), and *salicifolia* like salix (willow) leaves – salixifolia, *sinensis* = from China, *pseudocamellia* = false camellia (looks like camellia).

CHAPTER 3

Flower Structure and Fruit Development

Flower Structure and Floral Architecture

Flowers are the most important identifying characteristic of plants, and are the organs of sexual reproduction unique to the higher plant group spermatophyta (the seed-bearers); they are not found on ferns, fungi, lichens, algae, mosses, liverworts or bacteria.

Plants classified in the subdivision of the spermatophyta called the angiosperms bear seeds in enclosed ovaries, and this large group comprises all the grasses, the main flowering plants, and all broad-leaved trees and shrubs; it therefore includes both the monocotyledonous and dicotyledonous groups. Others bear seeds protected only by rudimentary scales in structures forming cones: these are the so-called naked seed-bearers (the gymnosperms), and include the conifers and cycads.

The floral architecture of angiosperms comprises parts considered to be modified leaves arranged in whorls (or sometimes spirals). Complete flowers have four whorls of modified leaves, and they are always in the same order. Listed from the outside whorl inwards they commence with the calyx, next the corolla, then the androecium (the male organs), and right in the centre (affording maximum protection) the gynoecium (the female organs). No matter how complex the floral pattern and petal shapes appear to be, most will adhere to the four whorls in the same order. However, in some species one or more of these whorls may be missing altogether and are said to have incomplete flowers, and may have no calyx, or a calyx but no petals, or their calyx and corolla fused together.

The calyx is the outside whorl and is made up of individual tough, leathery, wax-covered sepals that give excellent protection to flower buds. The number of individual segments (sepals) usually mirrors the number of petals in the flower (or the petals are a multiple of the sepal number). The sepals of the calyx may be green, or they can be more highly coloured (maroon, purple or red); they may shrivel soon after fruit formation, or in some instances will remain very persistent. Some *Rhododendron* species in the *thomsonii* series have calices that form a colourful feature for some considerable time after the petals have withered; *Philadelphus delavayii forma melanocalyx*★ is a form with particularly dark purple calices (★*melano* = dark).

The corolla comprises individual petals that in most instances are highly coloured in order to attract insects. Insect attraction is a very important function of the corolla, as pollination, fertilization and eventual seed production (sexual reproduction) is the main function of flowers. However, during flower bud development the sexual organs of angiosperm flowers are not only protected by the tough leathery calyx, but also by the soft, folded petals of the corolla internal to it.

Flowers may be single sexed and carry only female or only male parts; these are known as 'imperfect' flowers, and usually have less significant corollas (petals) so are often wind pollinated. All the conifers, but only some of the angiosperm species, have monosexed flowers. Individual flowers that carry both sexual organs (stamen and carpels) are hermaphrodite (bisexual) and are known as 'perfect' flowers, often comprising highly evolved, petal-bearing flowers, with nectar-producing glands, which attract insects in order to achieve a more efficient pollen transport, and a more accurate pollen placement.

However, some flowers may have relatively insignificant petals and may instead have highly coloured bracts (other forms of modified leaves that look like petals). Bracts are found below floral

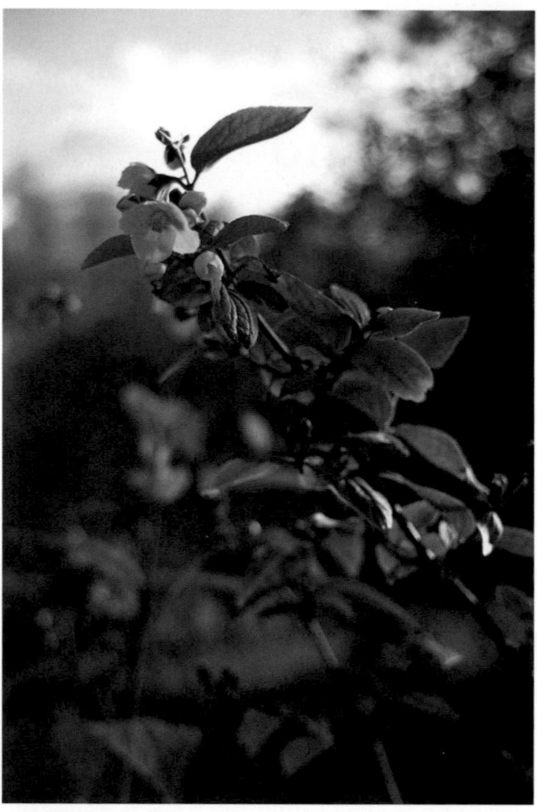

Philadelphus delavayii forma *melanocalyx*.

structures (they subtend the flowers), not where the whorl of petals is normally situated, yet they often take on the insect-attraction role normally performed by petals. Bracts are found in species such as *Davidia involucrata* (the handkerchief tree), *Cornus kousa, Cornus florida, Bilbergia nutans, Euphorbia pulcherrima* (poinsettia), *Bougainvillea glabra* and *Leycesteria formosa* (Himalayan honeysuckle). Looking at the modified leaf characteristics of bracts acting as petals makes it easier to visualize the whorls of sepals and petals making up the flower also as modified leaves – as do the green flowers of *Helleborus foetidus* and *H. argutifolius* (syn. *H. corsicus*), where the sepals function as petals and really do look like green leaves. The transition from leaf-like appendages to male and female organs is not so easy to visualize. However, studying the sexual organs of non-flowering plants such as ferns, their complex shapes, and the way they form from a green leaf-like thallus, helps with this concept.

The Male and Female Parts of the Flower

The andreocium is the collective name for the male parts of the flower, and comprises collections of individual stamen, with each stamen made up of an anther and filament. The anther is the pollen-bearing organ and comprises two large lobes, and a stalk-like filament for support, with each lobe divided into two elongated pollen sacs. Pollen is released as the sac splits longitudinally when the anther is ripe.

Geranium maderense, showing the hairy calices of the flowers covering the petals, and where the petals have fallen the calyx remains persistent.

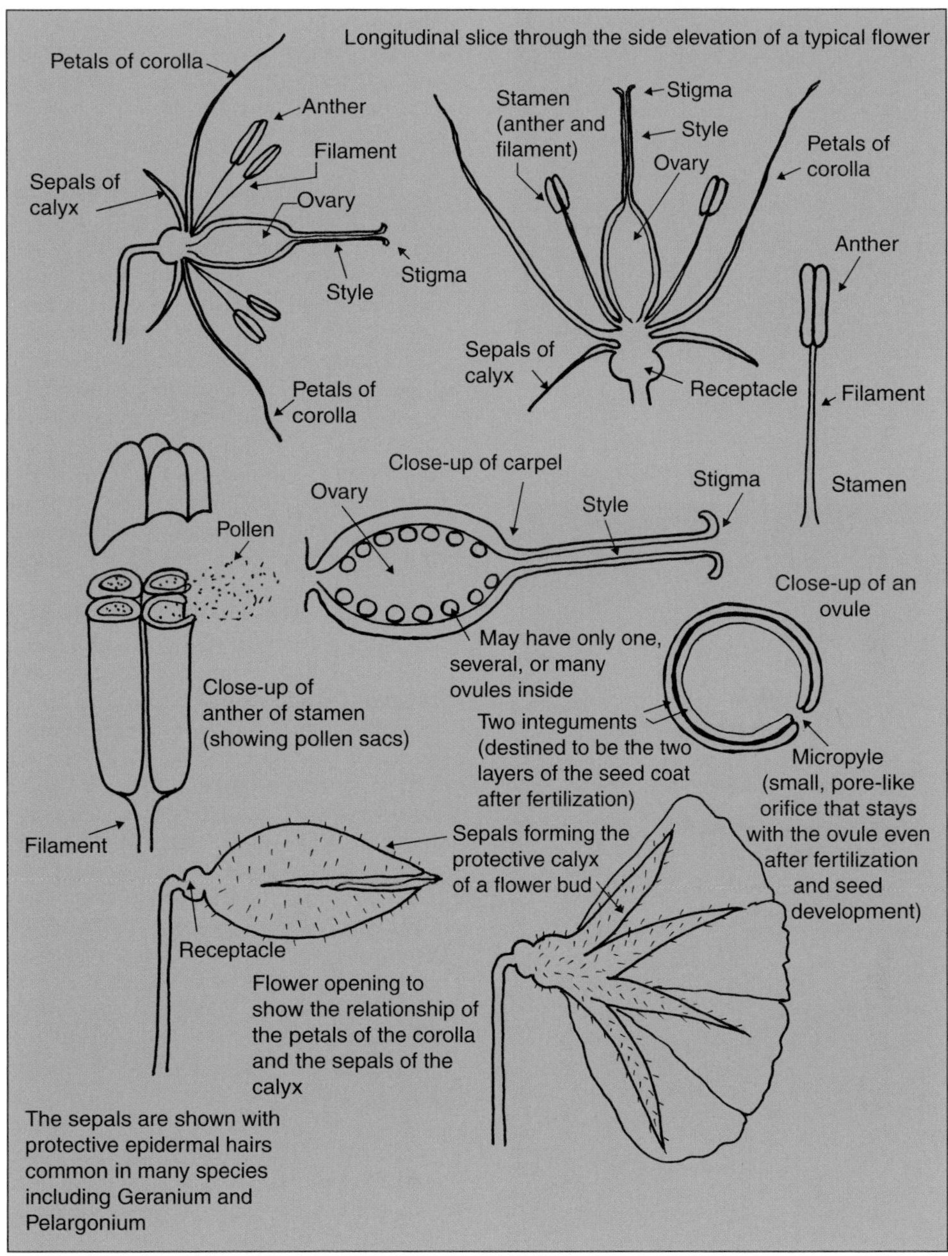

The basic flower structure of angiosperms I: perfect flowers – the four whorls of modified leaves.

The gynoecium (gynaecium) is the collective name for the female parts of the flower, and forms the innermost whorl of modified leaves. The gynoecium comprises a central structure called the pistil, and the pistil may comprise only one carpel, two carpels, or many carpels, either fused together, or free from one another. Each carpel comprises three main parts: a stigma, a style and an ovary. The stigma forms a receptive area for pollen, because in suitable conditions it exudes a sticky

The red bracts of poinsettia (Euphorbia pulcherrima).

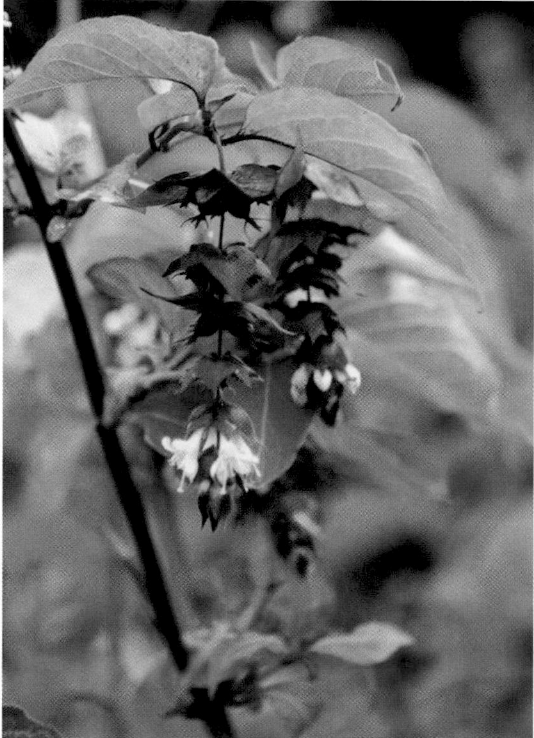

The red bracts of Himalayan honeysuckle (Leycesteria formosa).

liquid (stigmatic fluid) that has adhesive properties, and will, therefore, physically stick pollen grains in place. However, the exudate also has chemical properties that are hormone-based, and can create either a compatible or a hostile reception for pollen grains. The stigma forms an orifice to the style (a slender tube lined with soft tissue), and acts as the regulatory entrance (by both chemical means and orifice size) to the ovary; ovules (destined to be seeds after fertilization) are found within the ovary.

All the floral parts are attached to the swollen end of the flower stalk known as the receptacle. Floral parts would desiccate if they had no moisture supply, so in the same way that very small-scale leaves of buds have vascular traces (veins) to service them with moisture, likewise the whorls of floral parts also have vascular traces to each one. The stalk-like filament of a stamen, for example, usually has a single vascular trace running through it: the trace runs through the flower stalk, pierces the receptacle and enters the filament. The same is true of petals, but they have a network of minor veins coming off the main veins, not as developed, but nevertheless not unlike the system of veins found in foliage leaves.

A fairly cursory inspection of flowers with the naked eye (or better still, with a hand lens) will soon show that, although the stigma and stamen of different species can be described as diverse, they do nevertheless adhere to basic, fairly consistent and easily recognized shapes. The corolla of flowers can be quite simple, or quite complex in their nature. Petals may be free, or may be fused together, they may radiate simply from the centre as in *Malus, Prunus, Philadelphus, Geranium* and *Cistus*, or may be fused into a corolla tube as in foxglove (*Digitalis purpurea*), *Rhododendron cinnabarinum, Gloxinia, Incarvillea* and *Streptocarpus*.

Corolla shapes are classified by their floral symmetry, where regular flowers are actinomorphic and have approximately equidistant petals of a similar shape radiating from the centre of the flower. Even though there may be an uneven number of petals in the corolla, any line drawn to pass through the centre of the flower will divide it into two equal halves, and two equal numbers of petals (including parts of petals). Irregular flowers are termed zygomorphic, and have petals that are dissimilar in shape and form, are not equidistant from one another, and do not radiate equally from a central point. There is, therefore, only one line that will divide the flower into two equal halves.

All floral parts may be free or fused together, and the amount of fusion may vary between species – it may be that the entire length of some floral parts are fused together, or sometimes only at their bases. The sepals of calices may be fused

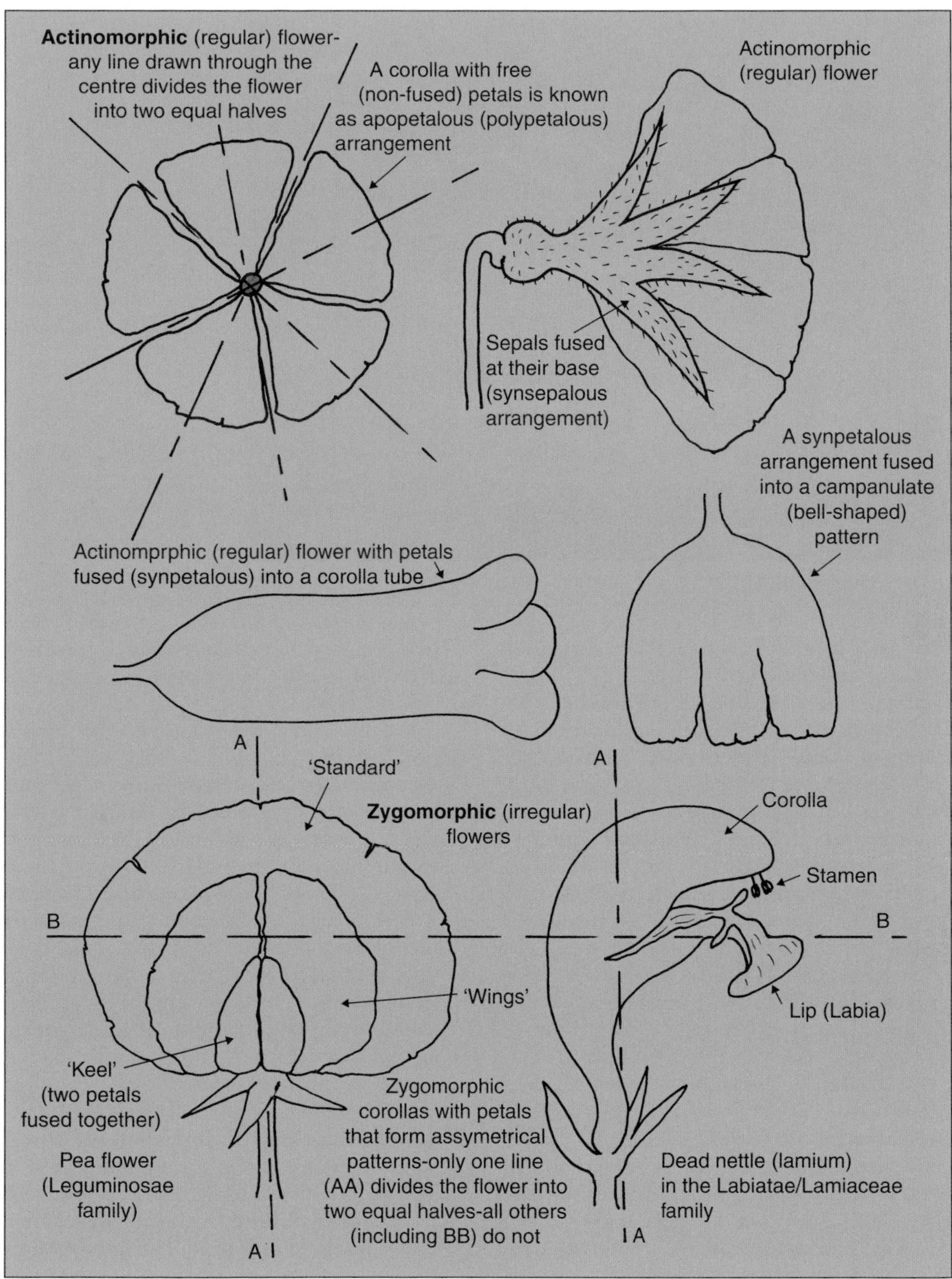

Basic flower structure of angiosperms II: floral symmetry – regular and irregular flowers.

Green leaf-like, petal-like sepals of Helleborus foetidus. Also showing the central structure of follicles typical of Ranunculaceae.

together to form a tube-like or bell-like structure such as is found in *Correa backhousiana*.

Corolla tubes may be long and cylindrical, or they may be bell-shaped (as found in *Campanula*, whose name describes the bell-shaped flower). They may even form very complex-looking structures with lips and pouches – as found in *Salvia* (in the Lamiaceae family) and many orchids. Often, but not always, there are distinct lobes at the mouth of the corolla that give an indication of the number of fused petals involved in the tube. There is also terminology to describe the freedom or fusing of the petals that form the corolla, and the sepals that form the calyx. Arrangements with free (non-fused) petals are known as apopetalous, whereas those with part, or all, of their petals fused together are known as synpetalous. Likewise, free (non-fused) sepals are known as aposepalous arrangements, and those with fused parts (or all) of their sepals are synsepalous. Some plants have two layers of highly coloured petals forming the corolla – a feature known as 'hose-in-hose', as exhibited by Christmas cactus (*Schlumbergera bridgesii*) and lobster cactus (*Schlumbergera truncata* syn. *Zygocactus truncata*).

The perianth is the collective name for the calyx and the corolla combined, and some species have calices and corollas that are fused together into one unit, so that the component parts (petals and sepals) are indistinguishable from one another. In this case the flower is said to have a fused perianth, and the coloured, fused, petal-like structures are known as perianth segments (sometimes called tepals).

Stamen may have anthers that are fixed and non-moving (adnate), or they may have hinged (versatile) anthers. Stamen may be individually free, or fused together at their base (sometimes forming a ring of basal tissue), or they may be fused to the sides of the corolla – whether the corolla is formed by individual petals or by a corolla tube. But no matter what their guise, they are easily recognized, because, although some species have only two or three stamen, they are more commonly found in large numbers compared with other floral parts. Stamen, unlike sepals and petals, do not normally reflect the number (or a multiple of the number) of other floral parts. Stamen are also recognized by the relatively large anthers that at some time in their existence (when conditions are correct) will spew out pollen grains. Members of the Rosaceae family such as *Rosa, Malus* and *Chaenomeles* all have an abundance of free stamen, with the gold-coloured anthers typifying many species. *Camellia* species (and cultivars) and *Stewartias* (in the same family) have masses of stamen, and are excellent to show the profusion of easily recognized stamen that appear 'free' but are actually fused into a crown-like ring at their base.

By far the best plants for stamen recognition are the many species and cultivars of large-flowered lilies. The very large proportions of species such as

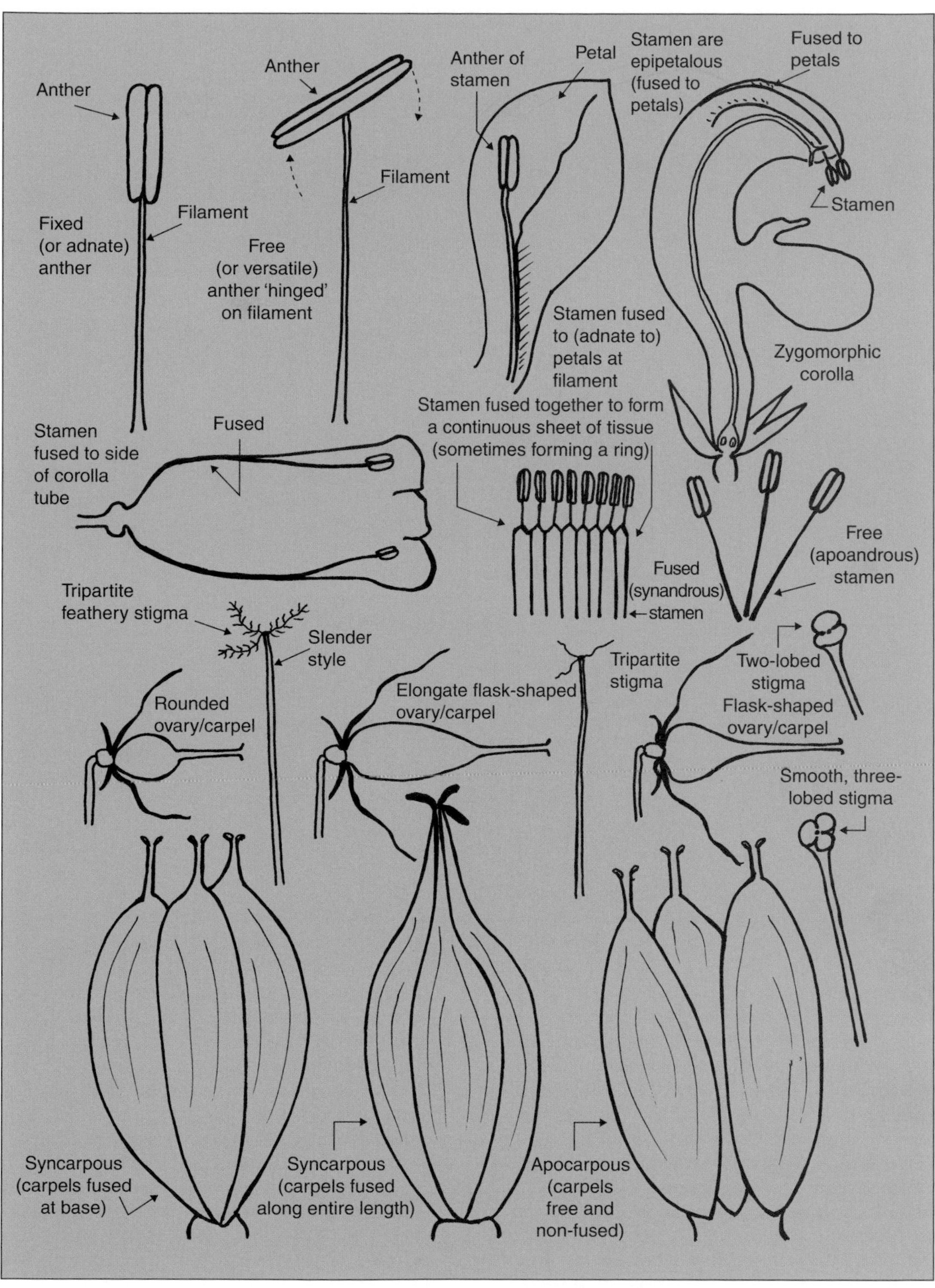

Basic flower structure of angiosperms III: variety in male and female organs – fused and free floral parts.

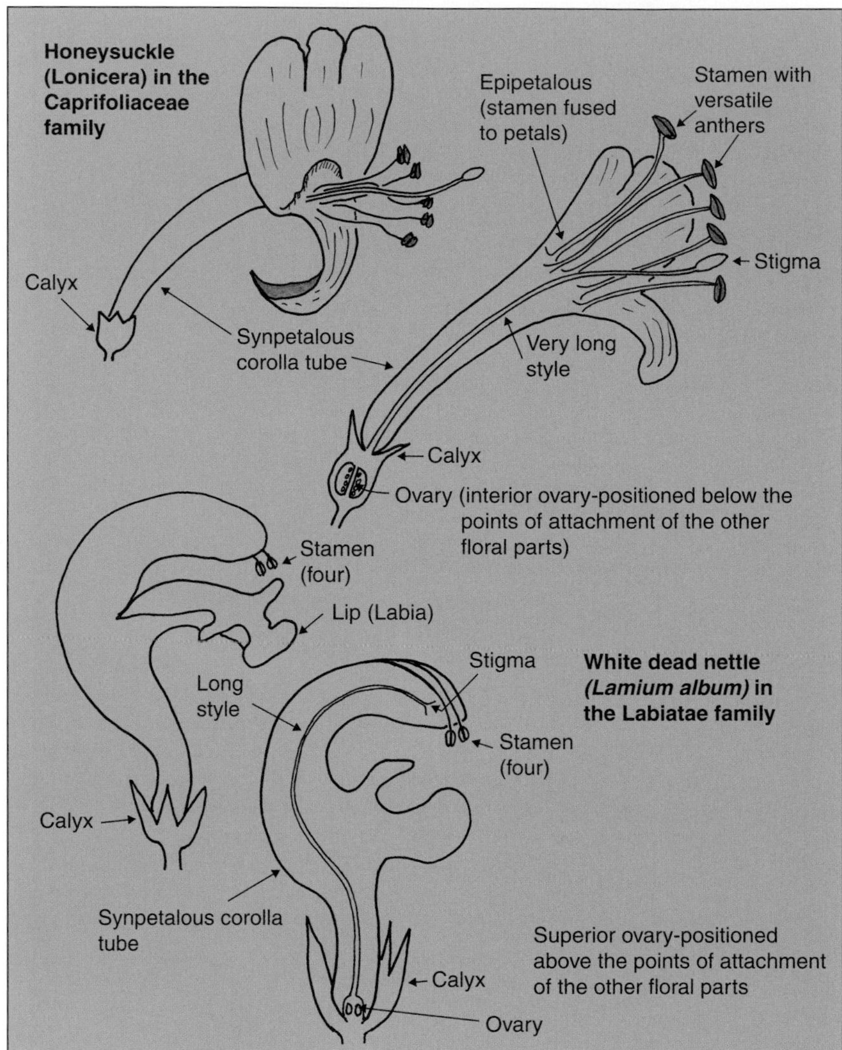

Basic flower structure of angiosperms IV: zygomorphic (irregular) flowers.

Lilium regale makes them excellent subjects for easy dissection and recognition of the floral parts. They usually have very large and obvious versatile stamen with the anther hinged on the filament (although some types do have fixed (adnate) stamen) producing profuse amounts of pollen that stains heavily.

The pistil is the main component of the gynoecium and may only comprise one carpel (in which case it is termed a simple pistil), but where more than one carpel exists (a compound pistil) these may be fused together along their vertical length (including the styles). In this case the stigma may appear to be one organ, or more commonly, it may reveal that it actually comprises two or three (or more) carpels because of the number of distinct lobes at its end. If only part of the carpel lengths are fused the individual styles are left free.

Lilies also boast distinct stigmas attached to a very large thick style, and the three-lobed stigma-end denotes that the species has a pistil comprising three carpels fused together. Stigmas may be smooth and rounded, or feathery at their ends. The flower buds of lilies comprise three outer green segments fused along their length, taking on the role of the calyx (sepals) and enclosing three petals internally. As they develop, the six petal-like parts (perianth segments) at a cursory

The corolla tube of Incarvillea delavayi, an herbaceous plant.

The corolla tube of Rhododendron cinnabarinum, hybrid.

glance are indistinguishable from one another, but in the early stages the outer whorl of three sepals are protective and green initially, then change to the same colour as the petals. On the other hand the inner whorl is more like true petals from the outset, but go through a colour change themselves. Fully developed perianth parts are only distinguishable by their position in the pattern, with the outer whorl of three being sepals, and the inner whorl of three being petals. Fully developed perianth parts removed from their positions (and therefore removing the evidence of which whorl they came from) are difficult, if not impossible to distinguish from one another, even with close inspection. At some time pollen exudes from the anthers and the stigmas become shiny and sticky with stigmatic fluid – usually, but not always, after the initial pollen-shedding process from the anthers.

Within each carpel, the hollow void that forms the ovary (and carries the ovules) presents various possible patterns, depending on the species. Carpels may only have one ovary void (locule) as found in the simple pod-like fruits (legumes) of pea and bean, or they may comprise several locules, each with ovules inside. These different design patterns can vary widely within genera but are species specific, and because each ovule is attached to the part of the carpel wall known as the placenta, ovule attachment patterns are known as the placentation. A short stalk known as the funicle attaches the ovule, and it is through this that the rudimentary vascular strand forms to service the ovule with moisture.

Particular patterns of petals should not fool you into thinking that the floral architecture is more complex than it actually is. There are examples of variations from the normal four whorls of modified leaves (including incomplete flowers), but most complete flowers show the same patterns and layers.

Variations of Floral Architecture

Primulaceae Family

Primula vulgaris (common primrose) has regular flowers with petals radiating equally around the centre. *Cyclamen* is in the same family (Primulaceae), but the petal pattern of the corolla looks totally different because some of the petals are bent backwards and do not radiate equally around a central point (irregular flowers). However, the most basic dissection will soon reveal that the flower does in fact comprise the typical four whorls, with the persistent sepals of the calyx on the outside, the stamen, and a long central style with stigma attached.

Papilionaceae Family

Flowers of species in the Papilionaceae (formerly Leguminosae) family appear complex because the corolla comprises petals with strange shapes, for which there is specific terminology. The large

The flower of Lilium showing the outer sepals and inner petals.

The flowers of Lilium showing distinct three-lobed stigma and versatile anthers covered in pollen.

Flower of Lilium clearly showing the stigma and versatile anthers of the stamen.

petal that stands at the 'back' of the pattern is the 'standard' (or *vexillum*), the petals internal to this are the 'wings' (or *ala*) and are often fused together, and the two fused petals that enclose the pistil are known as the 'keel' (or *carina*). The stigma only has one orifice into one style, as there is only one carpel in the pistil (and therefore only one ovary), because the pod-like fruits of legumes are simple fruits. The stamens are fused together into a collar-like structure at their base.

Genus Fuschia

The flowers of *Fuchsia* have a calyx that comprises four very thick fleshy sepals that are usually highly coloured. The thickness of the sepals leads you to believe that they are in fact tepals (sepals and petals fused together), but closer inspection shows that all four of the expected whorls of modified leaves are present. The calyx, because of the strong colour of the sepals, is as prominent as, or more prominent than, the four petals of the corolla. Maroon, deep purple or red sepals of the calyx often accompany white or cream petals of the corolla. There is therefore a contrast in colours between petals and sepals that greatly adds to the aesthetic appeal of fuchsias. *Fuchsia* 'Thalia' is particularly useful for illustrating floral architecture, as the sepals of the calyx are coral pink, and the petals of the corolla are also pink (with only a minor difference in shade) and are much smaller and more discreet than other *Fuchsia* types. Removing and carefully dissecting a flower of *Fuchsia* 'Thalia' or F. 'Koralle' when it is still in bud, recognizing that there are indeed four whorls making up the flower, and identifying the parts, is a good exercise to do.

Fuchsia – showing in this case that the 'skirt' (petals) is a different colour to the sepals.

Genus Narcissus

The flowers of the daffodil (*Narcissus*), in the Amaryllidaceae family, have no discernible calyx (that is, no noticeable sepals) as they have a tubular perianth (formed by the calyx and corolla fused together into one tissue). Furthermore, because the fused perianth tube also has the stamen fused to it (thus having three, not two, tissues fused into one), it has specific terminology. When three whorls (calyx and corolla forming the perianth tube and the androecium) are all fused at the base, it is known as a 'hypanthium'. The hypanthium of *Narcissus* terminates at its widest section (at the mouth) with perianth segments (known as tepals) that radiate round in a circular fashion, corolla- or petal-style. Extending further still from the end of the hypanthium (including the perianth tube) is a 'skirt' (not unlike the 'skirt' of fuchsias) that looks like true petals (a corolla) protruding from it: this

Flower of Narcissus: in this instance the corona is a different colour to the rest of the perianth tube (hypanthium). (Photo: Ron Mepstead)

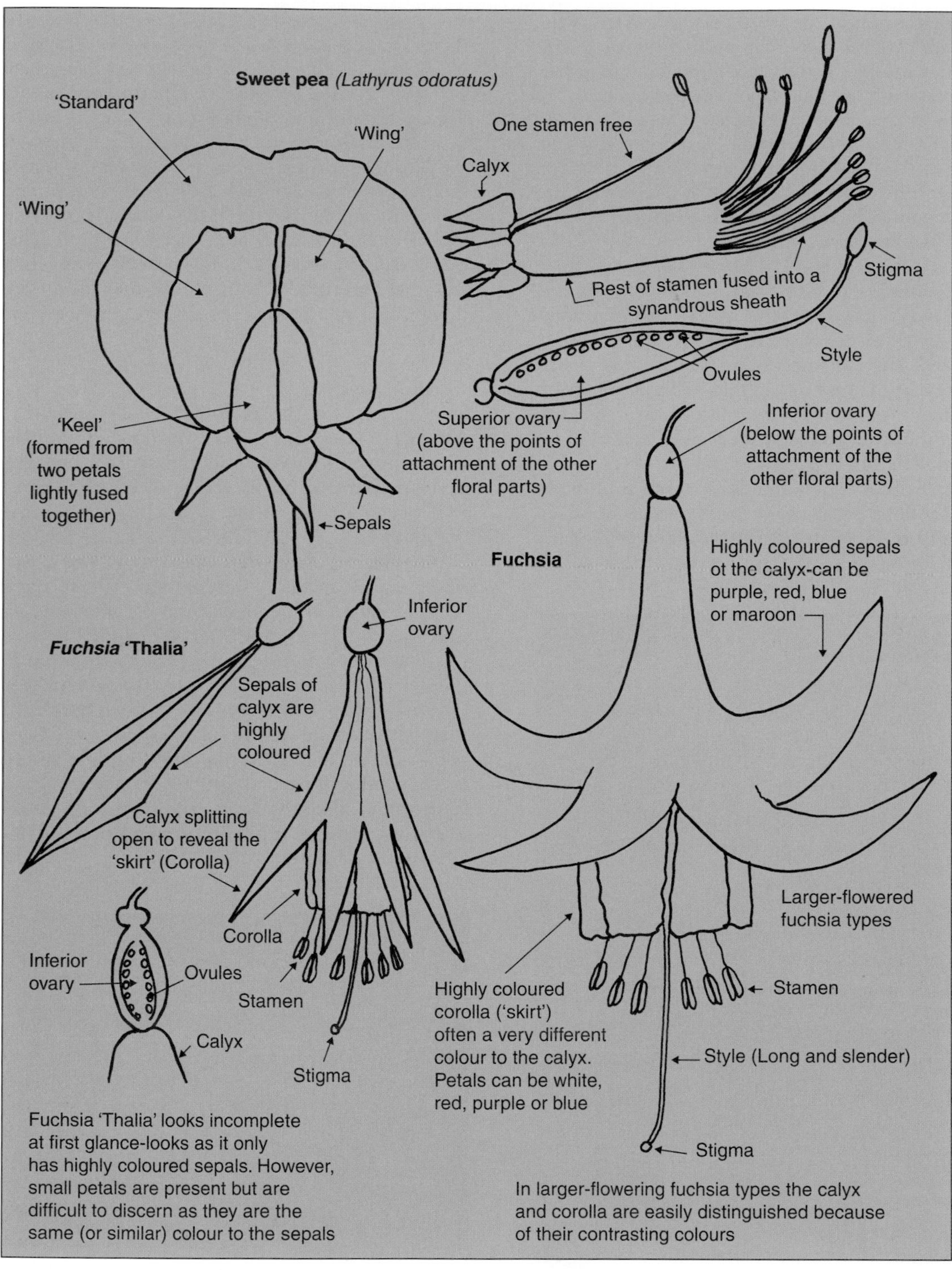

Basic flower structure of angiosperms V: sweet pea and Fuchsia.

is known as the 'corona'. In some *Narcissus* species and cultivars the skirt-like corona is the same colour as the rest of the hypanthium tube, but in other instances it is different in colour.

The stigma of *Narcissus* is central, and the ovary is inferior (borne below the points of attachment of the other floral parts) and on a short stalk known as the scape. The leaf-like structure at the first joint below the flower is a spathe (another specialized leaf form, this time with an enclosing nature). In some species modifications of the spathe form the surround for the floral apex – for example arums.

Liliaceae Family

The outside tissues of the flower bud of tulips (members of the Liliaceae) are green (and very leaf-like) initially, and because they are a different colour to the inside tissues, they are considered to represent the calyx. But as the flowers open, both the outside and the inside of the sepal-like parts become suffused with the flower colour. So the inner whorl of three colourful petal-like structures are considered to be petals, but the three outer ones are considered to be either sepals or fused perianth segments (tepals), which take on the function of the sepals of the calyx initially and then later the function of the petals.

Liliaceous flowers have six stamen, carpels that are fused together, and a stigma usually having three divisions, with a style to each one of the three fused ovaries and a superior ovary.

Genus Iris

Iris gives its name to the family Iridaceae: it has an inferior ovary and a very complex-looking set of perianth segments with an unusual and definitive structure. There are three perianth segments at the centre of the flower that stand erect and are known as standards. The three outer perianth segments are known as 'falls' because they drop away from the usually more erect standards. The 'haft' of the fall is folded at a midrib-like line along its length (for strength?), and the wide, lobe-like tip of the fall (the blade) drops downwards at 90 degrees from the haft on most species. The stamens usually lie within the fold of the haft.

Genus Clematis

In *Clematis*, what are considered to be the sepals are four in number and the most prominent feature, so they take on the function of both the calyx (sepals) and the petals, as they are brightly coloured and fulfil the insect attraction role. They are classified as tepals by some. A close inspection of species such as *Clematis tangutica, C. orientalis, C. montana* and *C. alpina* shows just four very thick and fleshy petal-like coloured sepals as the outside whorl, whereas some hybrid clematis have up to ten sepals. In some types the petals have evolved to form strange appendages (known as staminodes – stamen with colourful petal-like outgrowths) situated inside the whorl of petal-like sepals of the calyx. Others have thick, coloured hair-like stamen filaments (corona filaments).

A similar situation is witnessed in the common passion flower (*Passiflora caerulea*), but there are ten segments to the outer whorls, and the outermost five (perhaps representing sepals) are definitely exterior to an inner whorl of five (representing petals?). However, because a very notable whorl of highly coloured, hair-like modified petals (corona filaments) exist internal to this, perhaps all the outer whorls should be regarded as sepals? The outside tissues of the sepals/petals are green and very leaf-like, with a protective role. The three stigma-tipped styles of *Passiflora* are almost horizontal, but curve downwards over the five very large stamens.

Orchids

No matter how complex the flowers of orchids may look, they actually comprise the same basic whorls of modified leaves as other angiosperms. Some authorities consider that these petal-like structures that make up the perianth consist of two layers of tissue indistinguishable one from another, and they therefore call all of them perianth segments. Most consider the outer tissues to be sepals and the inner ones to be petals. Orchid flowers are irregular (zygomorphic) and have their floral parts in threes (three sepals to the calyx, and three petals to the corolla – or six perianth segments?). However, one sepal (the dorsal sepal) is more prominent than the others, and stands sentinel at the rear of the flower structure. Furthermore, the lowest petal of the three (known as the 'labellum') is more prominent than the other two, and can be highly coloured and quite complex in shape – like lips (which is why it is called the labellum: labia = lip).

The shape of the labellum may have the appearance of the arms and legs of a human being (for example the twayblade – *Listera ovata* – and the aptly named 'man' orchid, *Aceras anthropophorum*), or an inflated bladder-like appearance as found in lady's slipper orchid (*Cypropidium calceolus*), or the pouch-like appearance of *Paphiopedilum*

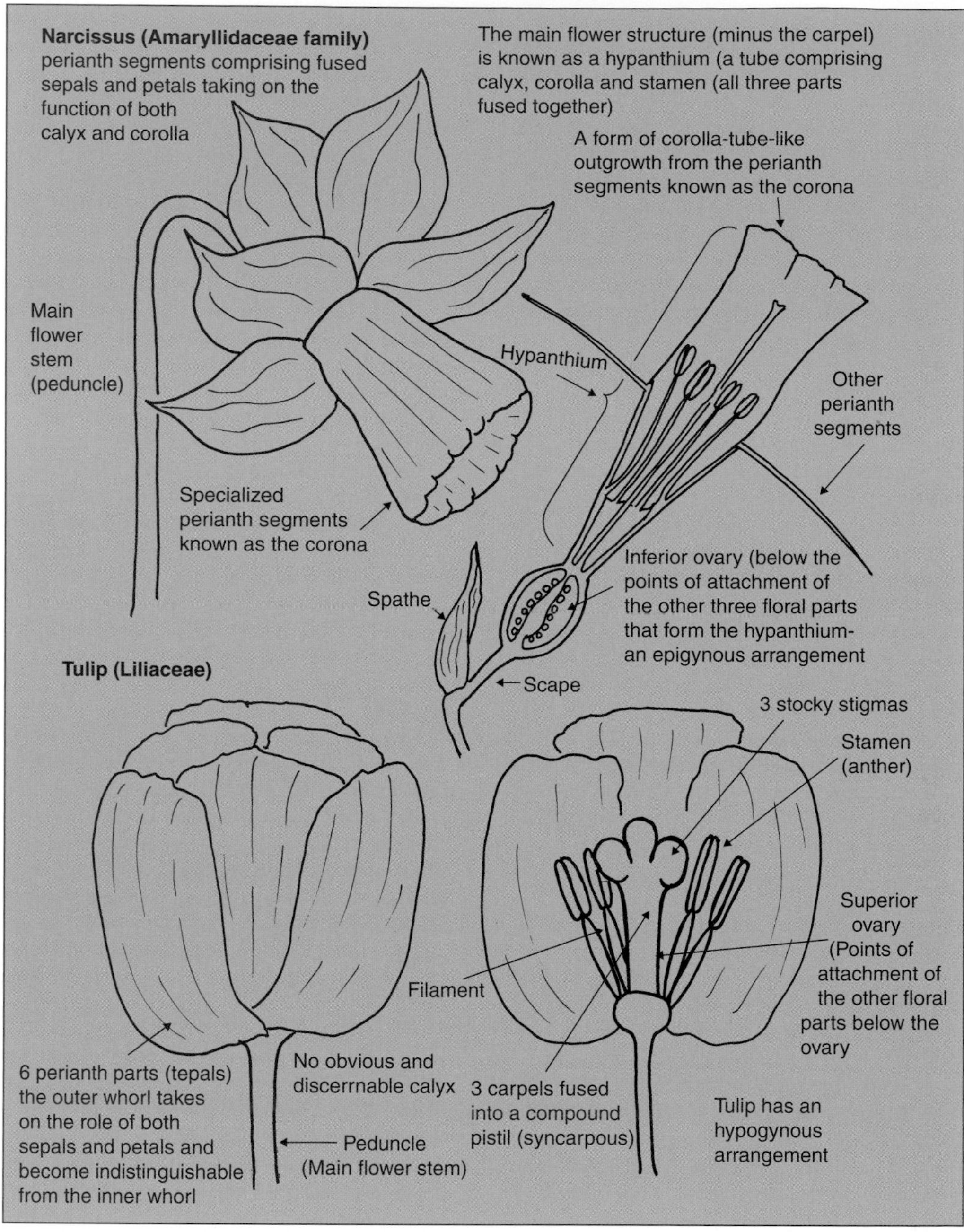
Narcissus (Amaryllidaceae family)
perianth segments comprising fused sepals and petals taking on the function of both calyx and corolla
The main flower structure (minus the carpel) is known as a hypanthium (a tube comprising calyx, corolla and stamen (all three parts fused together)
A form of corolla-tube-like outgrowth from the perianth segments known as the corona
Main flower stem (peduncle)
Hypanthium
Other perianth segments
Specialized perianth segments known as the corona
Inferior ovary (below the points of attachment of the other three floral parts that form the hypanthium- an epigynous arrangement
Spathe
Scape
Tulip (Liliaceae)
3 stocky stigmas
Stamen (anther)
Superior ovary (Points of attachment of the other floral parts below the ovary
Filament
6 perianth parts (tepals) the outer whorl takes on the role of both sepals and petals and become indistinguishable from the inner whorl
No obvious and discerrnable calyx
Peduncle (Main flower stem)
3 carpels fused into a compound pistil (syncarpous)
Tulip has an hypogynous arrangement

Basic flower structure of angiosperms VI – Narcissus and tulip.

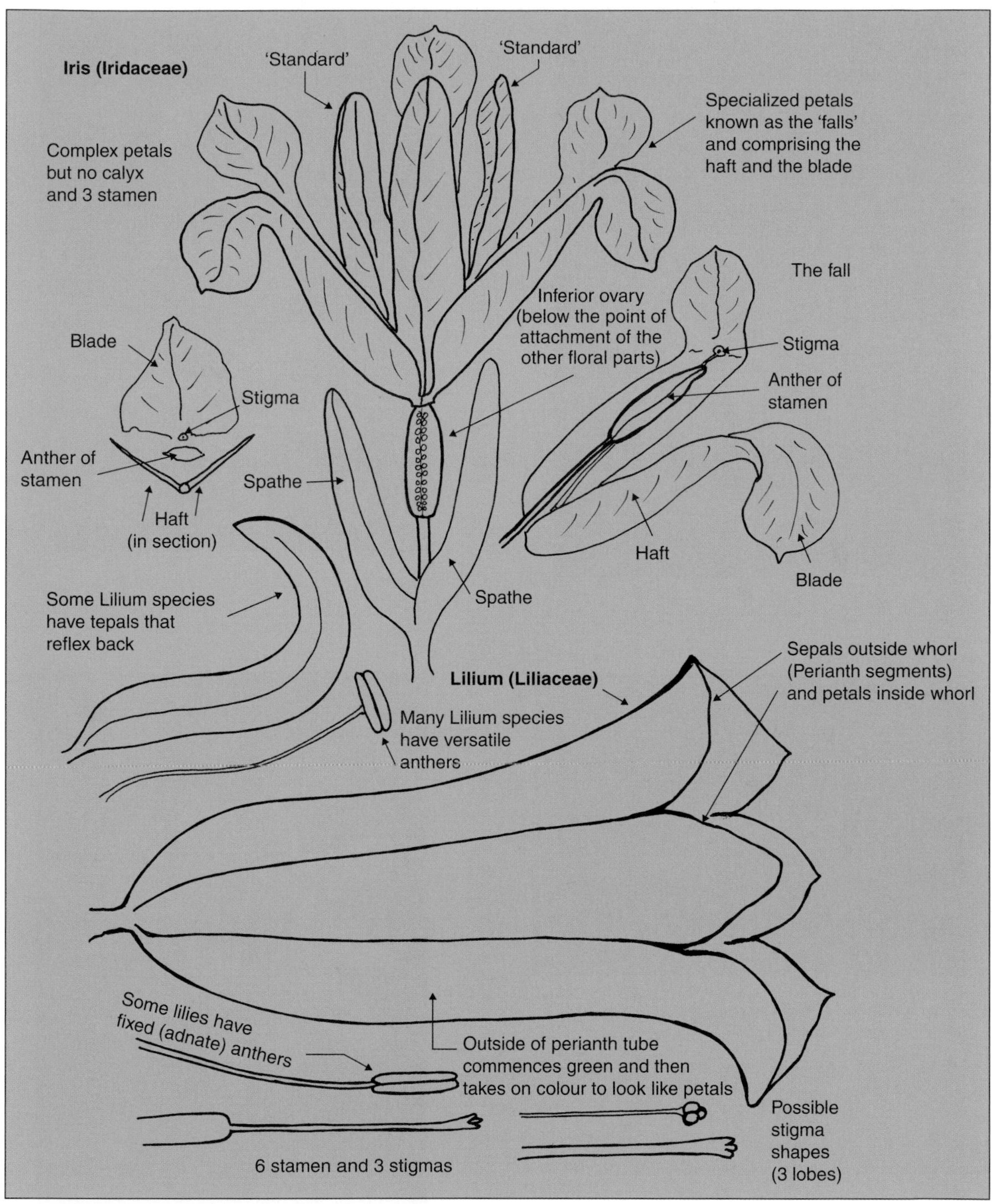

Basic flower structure of angiosperms VII: Iris and Lilium.

Flower of Iris.

Flower bud of passiflora showing very leaf-like outer layers.

Close-up of a hybrid clematis showing the thick, hair-like stamen filaments. (Photo: Ron Mepstead)

Flower structure of the passion flower (Passiflora caerulea) showing the white sepals, coloured hair-like corona filaments (remnants of petals?), stigmas on three curved styles, and five large anthers of stamen facing downwards.

species. The pouch-like shape is not that far removed from the pitcher-like shape of the modified leaves of *Nepenthes* species (a group of insectivorous plants). In orchids it is the insect-attracting flowers (themselves comprising whorls of modified leaves) that bear this shape, whereas in the carnivorous *Nepenthes* species it is the modified foliage leaves that bear the shape, and attract insects

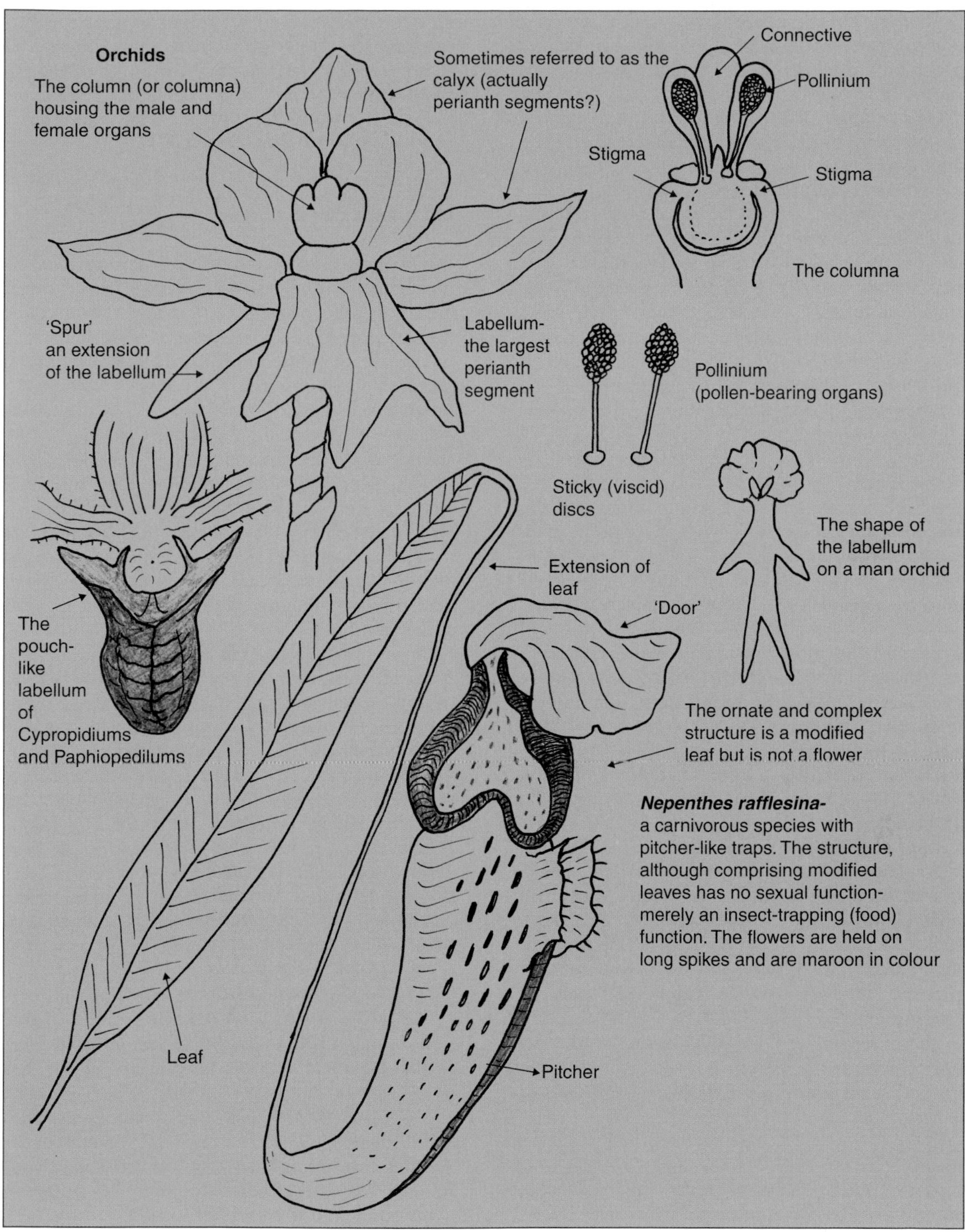

Basic flower structure of angiosperms: the flowers of orchids versus the modified leaves of pitcher plants (Nepenthes).

in order to trap them; the flowers are borne on a different part of the plant. Perhaps the idea of floral parts considered to be modified leaves is not so difficult to envisage after all!

The columna (or column) forms the central part of the flower, and the sexual organs (the stamen and the carpels) lie within it. Often the stamens are fused together, and quite often the stamen and carpels are fused into the columna as a raised and prominent central structure. The male organs (pollinidia) have sticky discs at their base, which actually adhere to visiting insects to aid pollination. Orchid species often have very distinct honey guides (nectar guides) comprising dark lines, blotches or lines of dots. Nectar guides notify insects of the whereabouts of the nectaries that exude nectar, and in the process aid pollination.

Araceae Family

Members of the Araceae family – including the arum lily (*Zantedeschia aethiopica*), dragon lily (*Dracunculus vulgaris* syn. *Arum dracunculus*), mouse plant (*Arisarum proboscidium*) and wild arum or cuckoopint (*Arum maculatum*) – have incomplete flowers. They do not have all four whorls of modified leaves, and the most prominent whorl that does exist (the very leaf-like spathe) is a defining feature of the family and substitutes for (and functions as) the corolla. The spathe is usually green or white, or more highly coloured with brown or purple blotches – the spathes of both *Arum dracunculus* and voodoo lily (*Sauromatum venosum*) are heavily blotched, and even the green, very leaf-like spathe of *Arum maculatum* sometimes has purple blotches on the outside. The spathe is like a single, open-fronted, tube-like tepal, as both the outside and inside tissues of the fleshy spathe can be recognized, but they are fused together as one, and no calyx exists. The flower has a central, rod-like structure called the spadix, which itself may be heavily pigmented. The spadix is a spike and carries all the remaining floral parts, having congregations of small sessile staminate (male) flowers at the top, and small sessile pistillate (female) flowers at the base.

Hibiscus

Members of the *Hibiscus* group (the Malvaceae family) have complete, perfect flowers with sepals, and coloured, open, urn-shaped or bell-shaped corollas and an erect central structure holding both the male and female organs. The central structure ostensibly looks like a spadix in some ways, but in *Hibiscus* actually comprises the styles fused together, to which the stamen are attached; but although the stigmas appear on the top, the long thread-like styles connect to the ovary well below the stamen. Common mallow (*Malva palustris*) and *Hibiscus rosa-sinensis* show this very well.

Inflorescence

Inflorescence is a description of the way in which flowers are arranged on the plant. Flowers can be borne on stalks bearing a single flower or a collection of flowers: they may be at a stem apex, in which case they often terminate extension growth, and are known as terminal flowers; or they may be situated in leaf axils (in which case they are termed axillary) – some species may have examples of both arrangements on the same plant. The inflorescence of the palm *Trachycarpus fortunei* is so intimately close to the leaf bases that the inflorescence is termed interfoliar. The main flower stalk is known as the rachis, and lesser stalks are known as peduncles; the small stalks that actually carry the flowers (sometimes branching off the peduncle) are known as pedicels.

If one single flower is borne on a single stem (pedicel) the inflorescence is termed solitary: for example camellias, tulips and most large trumpet daffodils. However, *Narcissus cyclamineus* and *Narcissus triandrus* may have one or two flowers per stem, *Narcissus* Tête-à-Tête' has two flower heads per stem, and some *Narcissus jonquilla* types can have five or more flowers per stem. The solitary flowers of the primrose set it aside from the other *Primula* species with their more complex inflorescences. Included in the more complex forms of inflorescence are racemose and cymose arrangements, where there are several flowers on a single (or branched) stalk.

The arrangement of flowers in their inflorescence is also relevant to the chronology of their development and opening. In racemose arrangements the most mature flowers are at the base of the raceme, and the youngest flowers are at the tip, as found in *Mahonia japonica* and *Prunus padus* (bird cherry). Racemes are not branched and are therefore termed monopodial (growing mainly in one direction). Racemes may be simple or compound, and the simplest form of raceme is the spike, which is a collection of flowers attached directly to the main stalk (rachis) without lesser flower stalks (pedicels). When pedicels off a peduncle do not exist, or if a main stalk (rachis) does not

Orchid flower showing the columna and labellum.

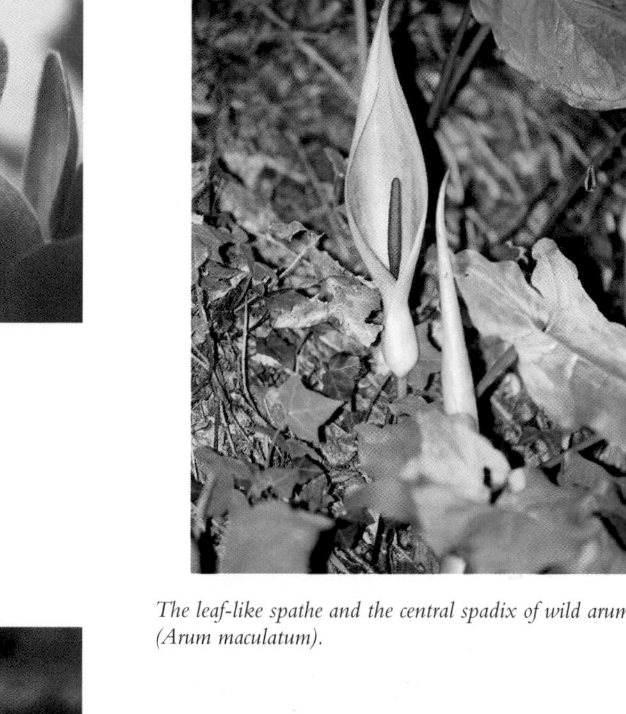

The leaf-like spathe and the central spadix of wild arum (Arum maculatum).

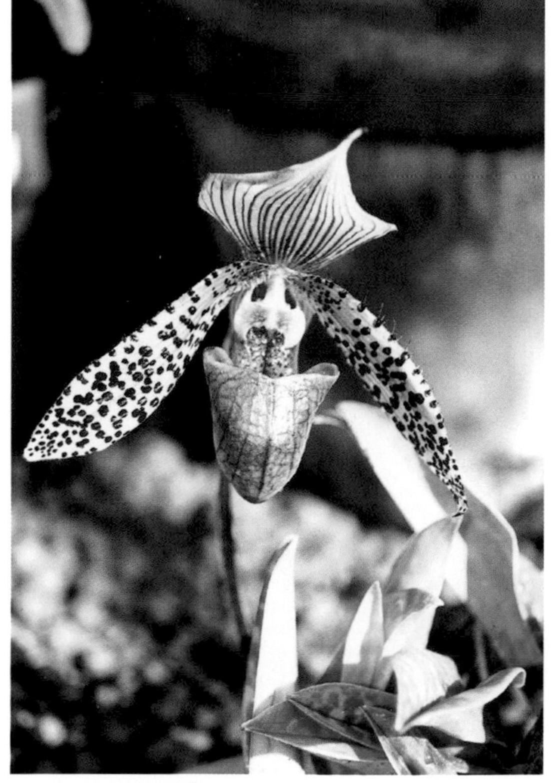

The pouch-like labellum of the orchid Paphiopedilum.

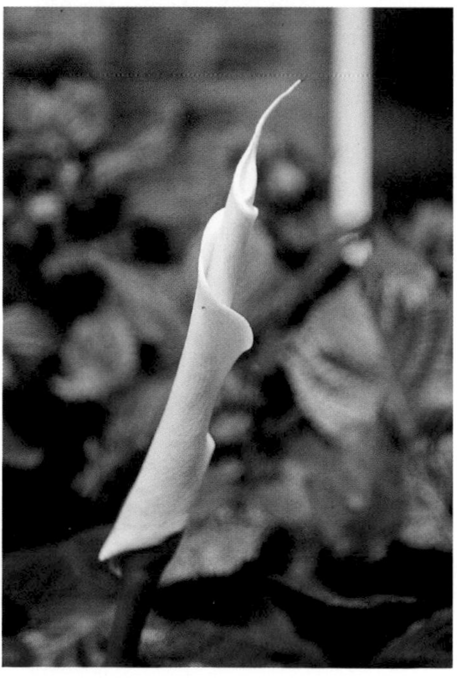

Spathe of the arum lily (Zandtedeschia aethiopica) unfurling.

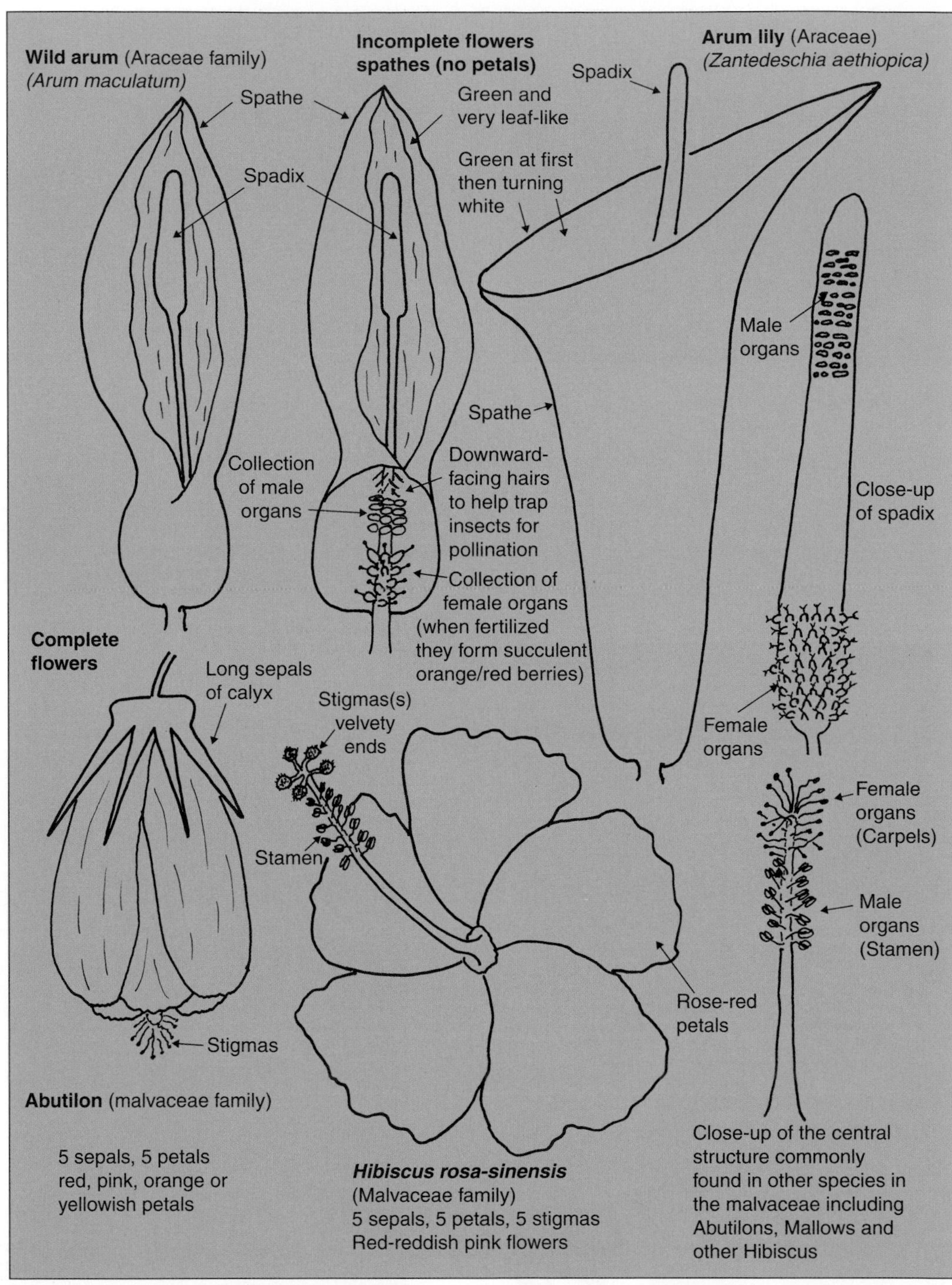

The basic flower structure of angiosperms: incomplete flowers versus complete flowers – hibiscus and arums.

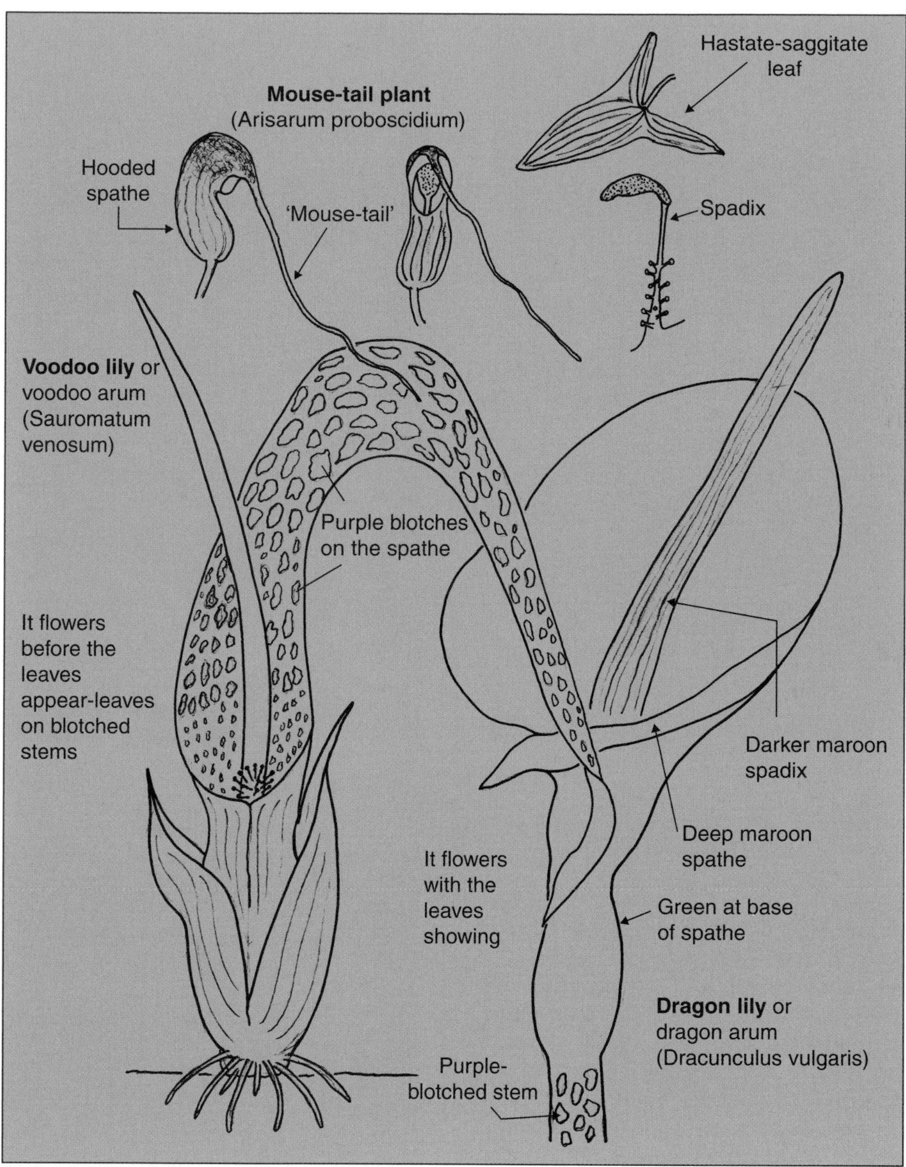

Basic flower structure of angiosperms: comparing arums.

exist, the flower is said to be sessile: so flower spikes are sessile. Or if, as is commonly exhibited, the flowers bear very short stalks, they are often still termed a flower spike, albeit incorrectly – for example the flowers of foxglove and hyacinth.

Compound racemes are collections of simple racemes on the same main stem, for example a panicle as found on *Hydrangea paniculata* (as described in its scientific name) and the larger flowered *Hydrangea paniculata* 'Grandiflora'. A corymb is a specialized form of simple raceme, and an umbel (which gave its name to the Umbelliferae, now Apiaceae, family) is a specialized form of corymb that may be simple or compound. An umbel basically consists of flowers set on stalks (pedicels) of approximately equal length, like the spokes of an umbrella. Polyanthus flowers are

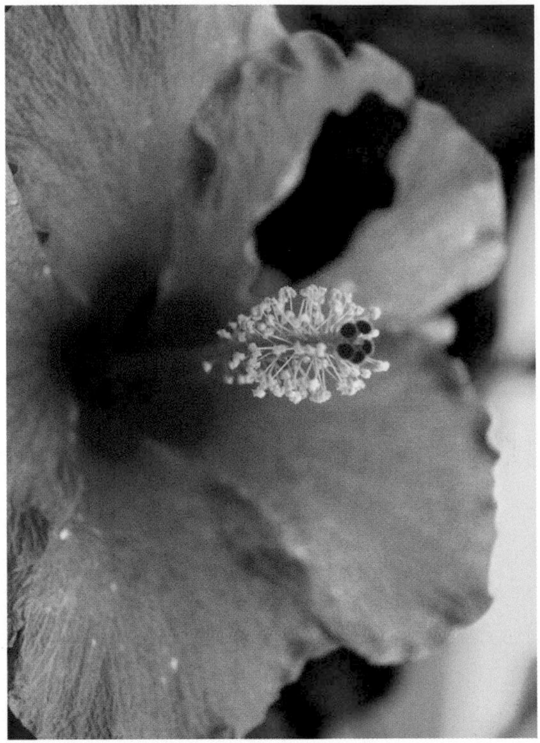

Flower of the cultivar of Chinese hibiscus (Hibiscus rosa-sinensis) showing the shape of the stigma – five lobes and the stamen behind.

Candelabra primulas.

Primula viallii.

borne on an umbel, which distinguishes them from the solitary flowers of primrose.

In cymose arrangements the youngest flowers are found at the base of the cyme, and the oldest flowers at the tip, or in the centre of flat-topped structures, as found on cluster-flowered rose species and the amur maple (*Acer tartaricum* subspecies *ginnala*). Cymes may be simple or compound. Monochasial cymes are simple cymes with their growth on one side only; dichasial cymes are compound (double) cymes with growth on two sides. But all are branched in some way, and are said to have sympodial growth.

Members of the Asteraceae (formerly Compositae) family have an unusual floral design and include all the daisy-like flowers. Their flower head is really a collection of small individual flowers (sometimes known as florets). So, what appears to be a single flower actually comprises many flowers, all of which are held on a specialized form of condensed spike called a capitulum. The capitulum is rather pin-cushion-like and holds many individual florets in a collective, composite cluster that originally gave the name to this important family. Included in the Asteraceae are aster, dahlia, dandelion (*Taraxacum officinale*), oxeye daisy (*Leucanthemum vulgare*), groundsel (*Senecio vulgaris*), chrysanthemum and sunflower (*Helianthemun annuum*).

There are two main types of floret (small flowers) that make up composite heads: ray florets

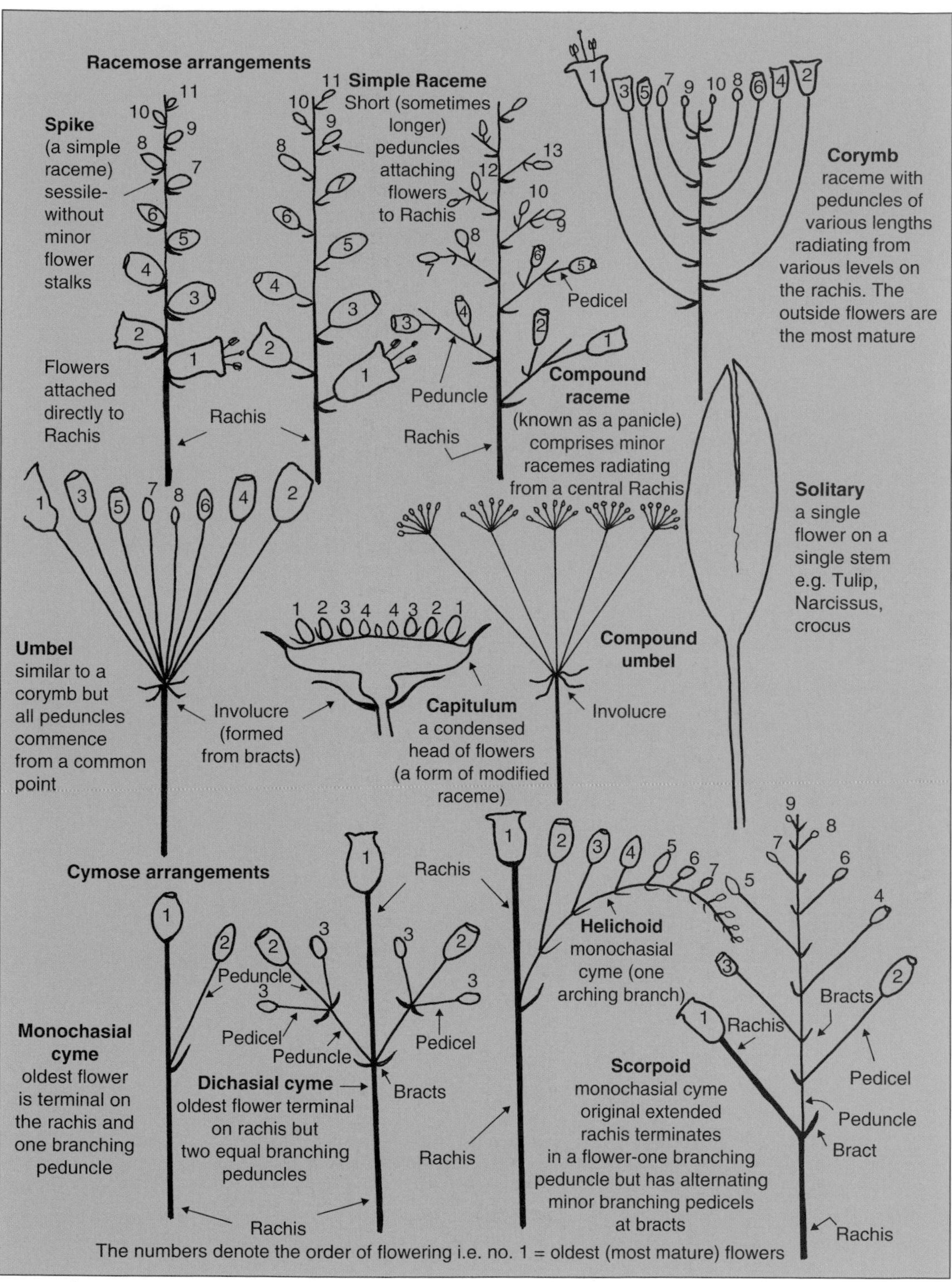

Inflorescence: the arrangement of flowers on stems – cymose and racemose arrangements.

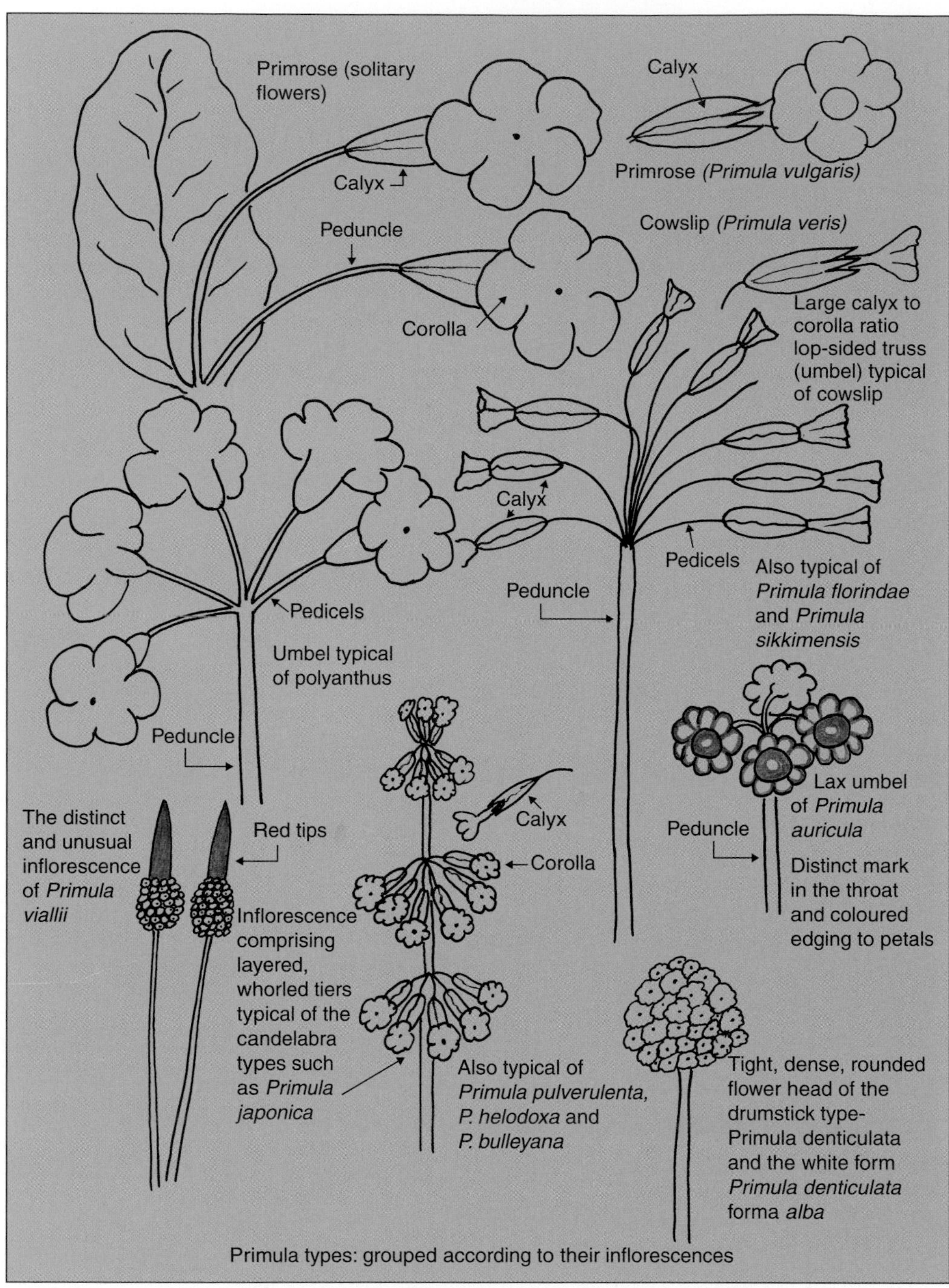
Primrose (solitary flowers)
Calyx
Calyx
Primrose (Primula vulgaris)
Peduncle
Cowslip (Primula veris)
Corolla
Large calyx to corolla ratio lop-sided truss (umbel) typical of cowslip
Calyx
Pedicels
Also typical of Primula florindae and Primula sikkimensis
Peduncle
Pedicels
Umbel typical of polyanthus
Peduncle
Lax umbel of Primula auricula
Peduncle
Distinct mark in the throat and coloured edging to petals
The distinct and unusual inflorescence of Primula viallii
Red tips
Calyx
Corolla
Inflorescence comprising layered, whorled tiers typical of the candelabra types such as Primula japonica
Also typical of Primula pulverulenta, P. helodoxa and P. bulleyana
Tight, dense, rounded flower head of the drumstick type- Primula denticulata and the white form Primula denticulata forma alba
Primula types: grouped according to their inflorescences

Inflorescence: Primula types (the distinguishing features of primulas).

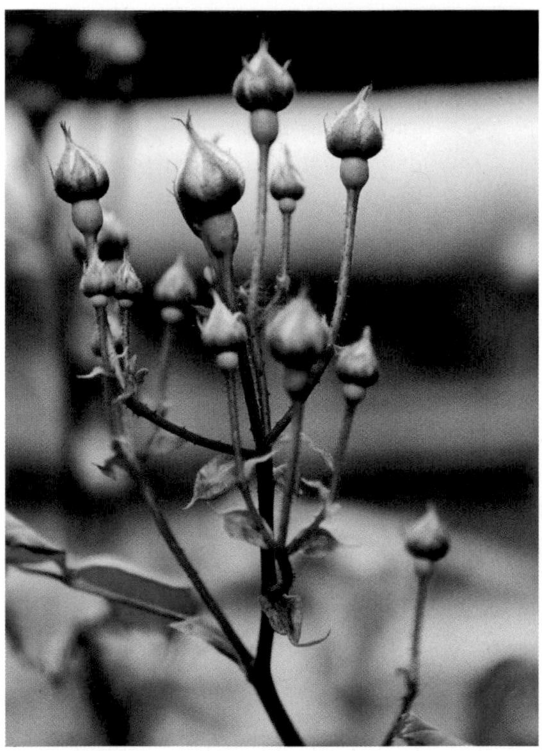

The dichasial cyme of the rose.

(or strap florets), and disc florets. Florets may be perfect or imperfect flowers, and both (or only one) type may be present, depending on the species. Ray florets are those small flowers that radiate around the composite head, each one terminating to look strap-like or petal-like (creating the typical daisy look); they are usually pistillate (female), but may be sterile. Disc florets usually carry both sexual organs (so are perfect flowers), and a corolla tube with very short (almost indiscernible) petals, and they are commonly presented in the centre of the flower structure to form the cushion-like effect (known as the cone) found in the centre of 'single' daisy-like flowers. Hence, seeds (actually nut-like fruits-achenes) can be carried in some of the individual flowers making up the flower head, but not in others.

Although some composites carry fairly developed sepals at each disc floret, many are known for the production of hair-like appendages (where the sepals should be) that make up the calyx. These much reduced structures become very conspicuous in species such as dandelion (*Taraxacum officinale*), creeping thistle (*Circium arvense*) and goat's beard (*Tragopogon pratensis*), where they form a feathery parachute-like carrier (known as the 'pappus') for each individual developed fruit. In some species the achene is attached to the pappus by a long stem and is known as 'beaked', and in other species this is absent ('non-beaked'). Developing florets on the capitulum head (essentially forming a terminal bud) are enclosed in dense overlapping scale-like leaves (bracts), collectively forming what is known as the 'involucre'. These structures cannot be classed as sepals (that enclose a conventional flower bud), because they actually enclose a collection of flowers (developing florets), not a single flower, so they necessarily have to have a different terminology. Collections of bracts are always referred to as involucres, and lead to the botanic name of handkerchief tree (*Davidia involucrata).*

Dandelion (Dan-de-lion, from the French *dents*, teeth, and *de lion*, of the lion because of the very dentate leaves of some forms) has prominent ray florets, and less obvious disc florets, although fairly low-level observation will soon find them in the centre. The single daisy ('day's eye' – an observation of the flowers' tendency to 'open' in good sunlight and 'close' in dull light, known as 'photonasty') has very prominent ray and disk florets. The sunflower (*Helianthus*, from *helios* = sun, *anthus* = flower) was so named, not because the massive flower head looks like the radiating sun (which it does), but because the flower follows the path of the sun during the day. Because sunflowers have a large proportion of their surface area covered with fertile disc florets, a large part of the floral head can be covered in the characteristically large seeds in late summer. Rayless mayweed (*Chamomilla suaveolens*) has disc florets only (no ray florets, so giving rise to its common name). A synonymous common name, pineapple mayweed, comes from a description of the shape of this yellow-green floral structure (and/or perhaps describing the pungent smell?).

The pattern of disc florets is distinct in other types. Single-flowered daisy types, *Chrysanthemum* and *Dahlia*, all have relatively low central cushions with disc florets, whereas both *Rudbeckia* and *Echinacea* have much larger and more pronounced central cone-like central structures made up of disc florets, this leading to their common name of 'cone' flowers. *Echinops* comprise a thistle-like globe of disc florets only, and some species comprise ray florets only, for example *Crepis vesicaria* (beaked hawk's-beard).

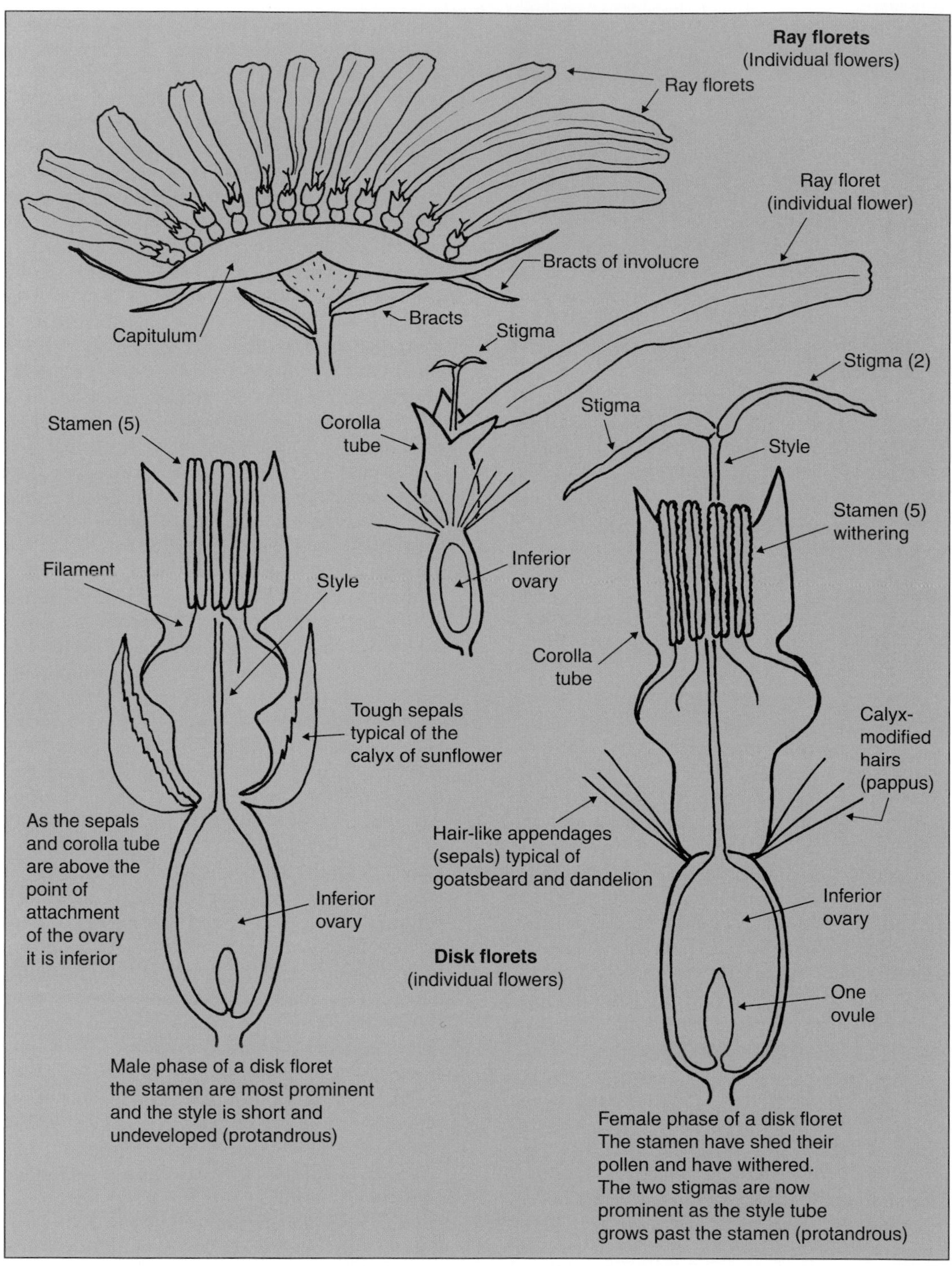

The structure (and inflorescence) of composite flowers in the Asteraceae.

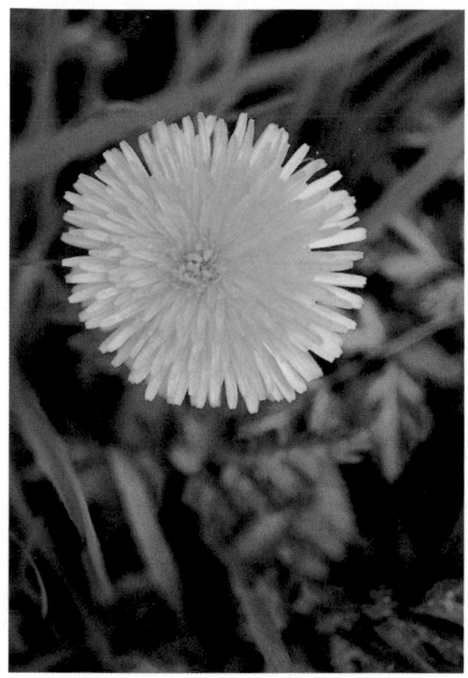

The flower of the dandelion (Taraxacum officinale).

Flower of the field thistle (Circium arvense).

Alpine sowthistle (Cicerbita alpina).

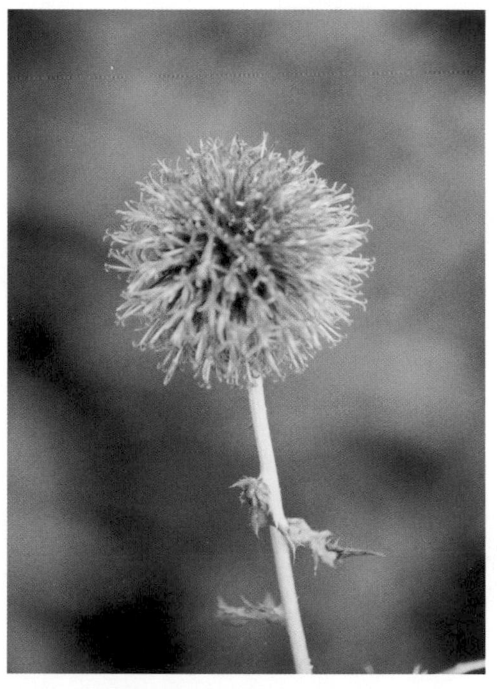

The flower of Echinops.

Chrysanthemum maximum.

Close-up of a flower in the Asteraceae with composite head showing ray and disk florets.

The Flowers of Grasses

The inflorescence of common grasses, cereals and bamboo comprise either an upright spike as typified by wheat (*Triticum*), barley (*Hordeum*), crested dog's tail (*Cynosurus cristatus*), sheep's fescue (*Festuca ovina*), perennial ryegrass (*Lolium perenne*), timothy (*Phleum pratense*) and couch grass (*Agropyron repens*); or a more open, loose panicle as found in the bents, for example brown top bent (*Agrostis tenuis*), common oat (*Avena sativa*), annual meadow grass (*Poa annua*) and bamboo.

Although the scale differs between species (*Stipa gigantea* is a good species to look at for easy-to-see detail), florets (individual flowers) that make up the inflorescence are relatively small. Florets have no showy, highly coloured or overtly conspicuous parts (including petals) as they are anemophilous (wind pollinated), which is aided by either the upright or semi-pendulous nature of the inflorescence. Florets are usually hermaphrodite (bisexual), and are held together in small groups known as spikelets, often comprising three flowers, and attached to the spikelet by short stalks called rachilla. In upright spikes, the spikelets are held close to the main flowering stalk to give the dense, upright, ear-like fruit collection typical of cereal crops such as wheat, barley and rye. Those species with panicles have the spikelets attached to the main flower stalk by slender, arching sub-divisions of the flower stalk, as in wild oat (*Avena fatua*).

The large, globe, thistle-type composite heads of globe artichoke that separate it from Jerusalem artichoke, which has yellow sunflower-type heads.

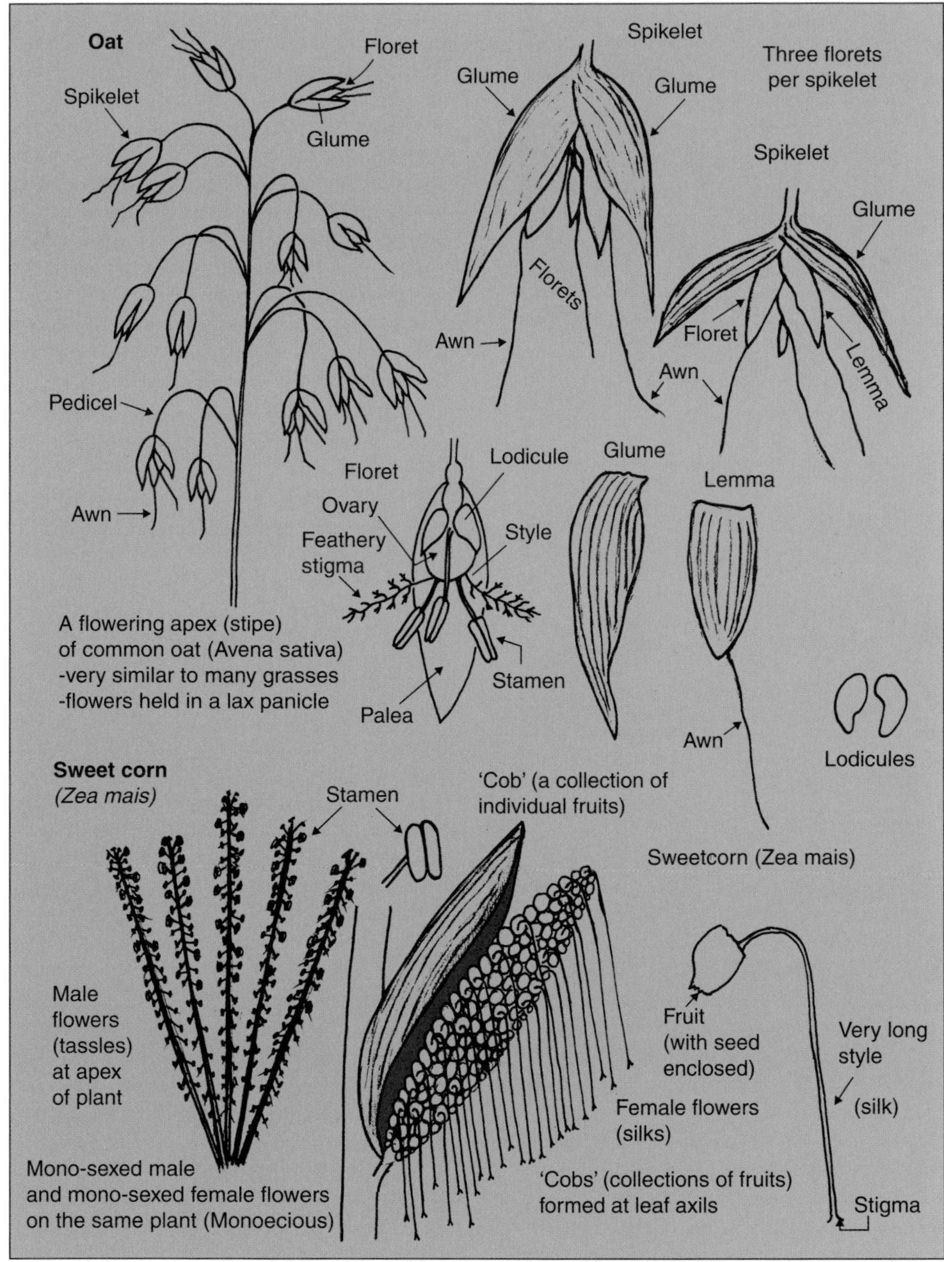

The flowers of grasses and cereals.

External to each spikelet (the small collection of florets) are two tough bracts called glumes, which may almost completely envelop the developed spikelet (as in wheat), or may be more open, allowing the developed florets of the spikelets to show. In either case, the glumes are obviously protective in nature and carry out the role performed by sepals (or because they protect a collection of florets, involucre bracts) in other angiosperm flowers. Internal to the glumes, and enclosing each

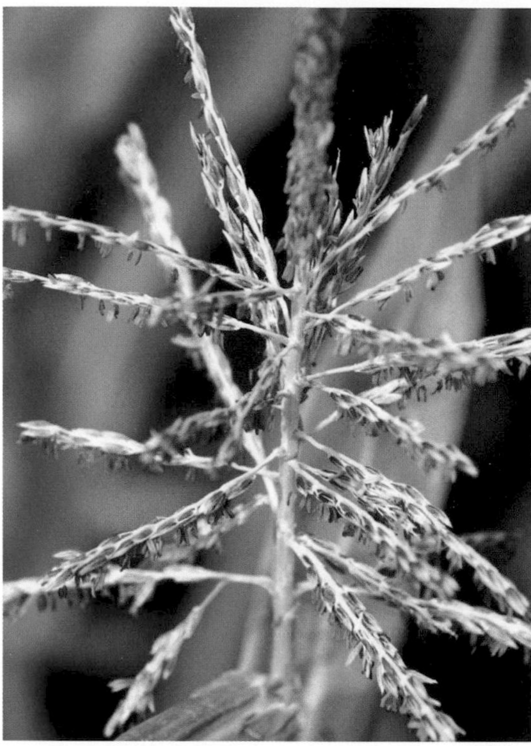

The male flowers ('tassels') of sweetcorn.

individual floret for extra protection (therefore more accurately akin to sepals), are two bract-like (even, valve-like) sheaths known as the 'lemma' and the 'palea'. The lemma often has a very long, hair-like appendage known as the awn attached to its tip – typified by the very bristly nature of the inflorescence of barley (including wild barley, *Hordeum murinum*). The palea is often overlapped by the larger lemma to form a very protective outer shield enclosing the all-important sexual organs of each individual floret. The sexual organs of perfect-flowered species usually comprise two fused carpels and three stamens. Typically, the stamen have elongate anthers, and initially have short filaments that may lengthen prior to pollen dispersal and protrude through the gap left between the lemma and the palea (because the lemma and palea separate slightly as the florets mature). The elongate anthers hang free of the floret to facilitate wind pollination.

Some species in the grass and cereal group have imperfect flowers, the most common example being *Zea mais* (sweetcorn), which carries monosexed male flowers and monosexed female flowers on the same individual plant (a monoecious arrangement). Hence all individual plants have the capacity to produce 'cobs' of corn, providing that pollination and fertilization is successful. Male flowers form in a panicle at the apex of plants, and appear as great clusters of stalks (collectively known as 'tassels') with masses of pivoted anthers attached. Female flowers form lower down (in leaf axils) and appear in large congregations, with each individual flower presenting exceptionally long styles hanging like silken threads, at first green and later turning red, and colloquially known as the 'silks'. Each small fruit has its own hanging style, and the aggregate fruit (known as the cob) comprises hundreds of individual fruits (caryopses), held together in the typical torpedo shape associated with sweetcorn. The developing aggregate fruit (cob) is protected by two green, leaf-like (glume-like), enveloping bracts known as the husk.

Pollination

When the anther sacs of stamen are fully developed they split open and shed pollen. If the ambient temperature and moisture content is correct the pollen will 'flow', and pollination (the transfer of pollen from the anther of a stamen to a stigma) can take place. If all other conditions are correct, the result of such transference is usually fertilization (the fusion of the male and female elements of the plant).

Types of Pollination

Pollen is produced within the anthers and is dependent on vectors such as wind /air currents, water, mammals and insects to carry out the process of pollination. Self-pollination occurs when the pollen from the anthers of a flower alights on to the stigma of the same flower, and it also occurs when pollen from the stamen of a flower alights on the stigma of another flower on the same plant. Cross-pollination occurs when pollen from the stamen of one flower alights on to the stigma of another flower on a different plant.

Cross-pollination is 'preferred' because it is beneficial to have a degree (or the possibility of a degree) of variation in the sexual process, and the slight variations induced can be the difference between the success or otherwise of a whole race of plants. Cross-pollination (and therefore cross-fertilization) increases the chances of these minor

The flowers of bamboo with similar structure to other grass flowers in the Graminaceae (Poaceae) family.

Male and female flowers on the same plant of sweetcorn (monoecious arrangement).

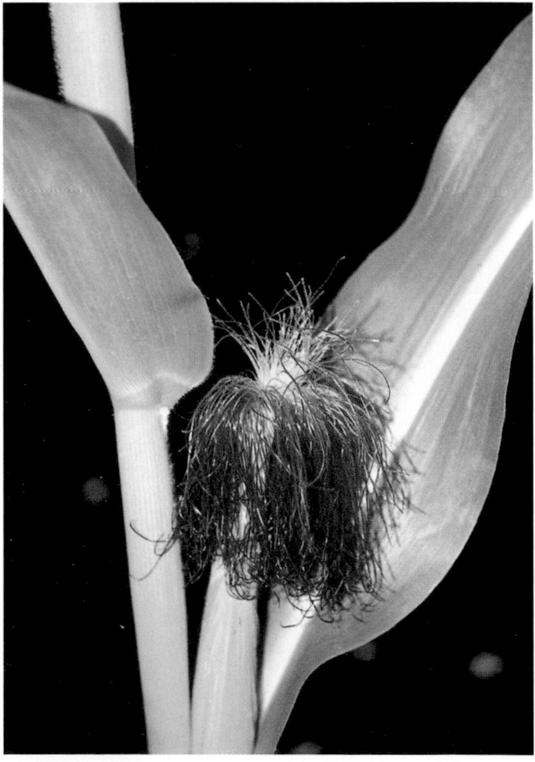

The female flowers of sweetcorn (Zea mais) showing the prominent red styles and stigmas ('silks').

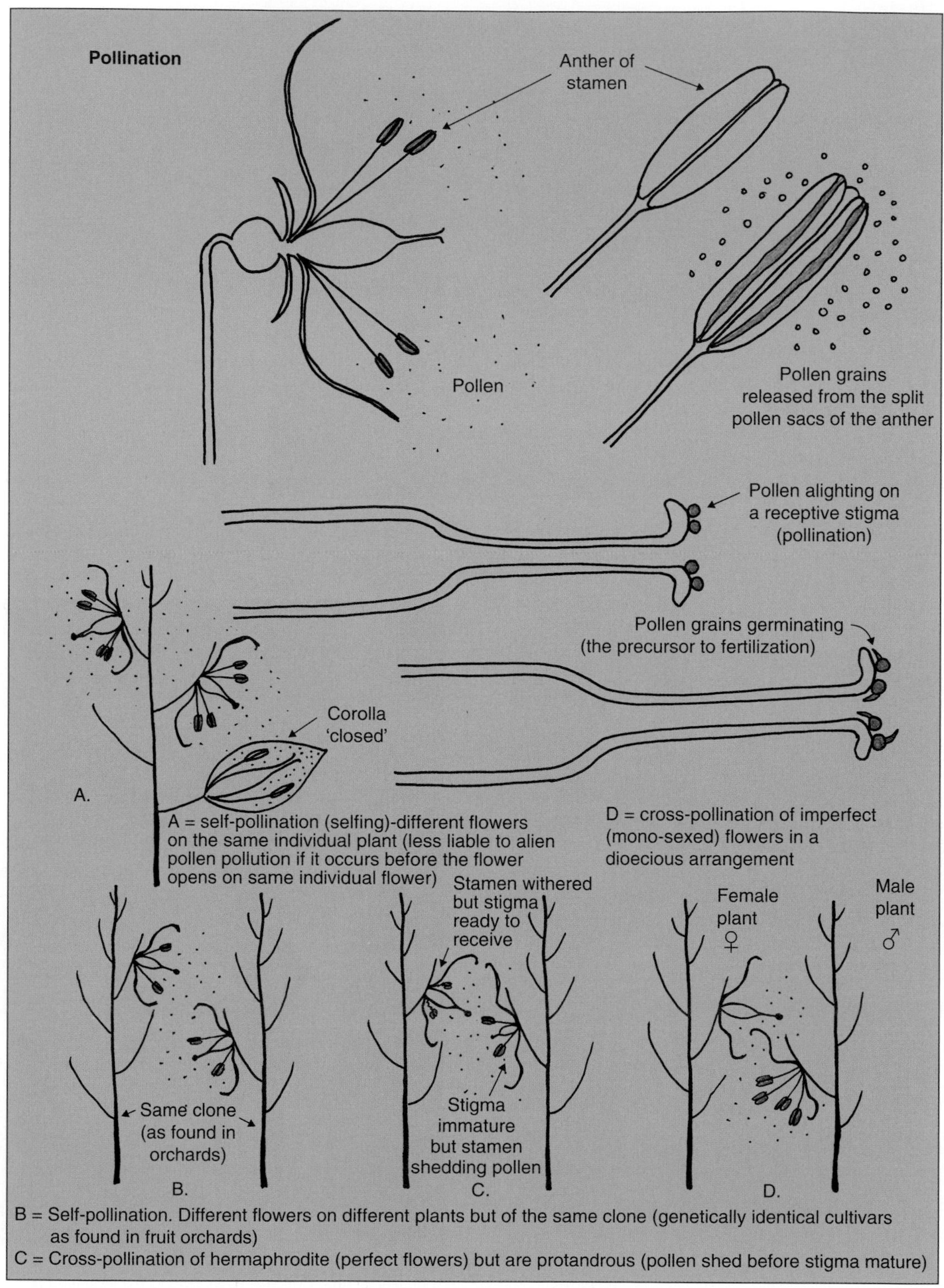
Pollination
Anther of stamen
Pollen
Pollen grains released from the split pollen sacs of the anther
Pollen alighting on a receptive stigma (pollination)
Pollen grains germinating (the precursor to fertilization)
Corolla 'closed'
A.
A = self-pollination (selfing)-different flowers on the same individual plant (less liable to alien pollen pollution if it occurs before the flower opens on same individual flower)
D = cross-pollination of imperfect (mono-sexed) flowers in a dioecious arrangement
Stamen withered but stigma ready to receive
Female plant ♀
Male plant ♂
Same clone (as found in orchards)
Stigma immature but stamen shedding pollen
B.
C.
D.
B = Self-pollination. Different flowers on different plants but of the same clone (genetically identical cultivars as found in fruit orchards)
C = Cross-pollination of hermaphrodite (perfect flowers) but are protandrous (pollen shed before stigma mature)

Pollination, the transference of pollen from an anther to a stigma: types of pollination.

variations – variations that could instil extra hardiness, for example, or any feature that might encourage success in a specific environment (or a drastically changing one!).
Commercial fruit tree orchards comprise trees that are produced by vegetative means, and because all trees of the same cultivar derive initially from the same mother plant, they share the same genetic material, and each one is therefore the same clone. Pollen transferring from the anthers to the stigma of the same flower, or of different flowers on the same individual plant, or flowers from different plants of the same clone (that is, genetically identical individuals), is all classed as self-pollination. Hence the pollen from the cultivar Cox's orange pippin (a clone) alighting on the stigma of any other flower of Cox's orange pippin (on the same or different individual plants) is self-pollination.

Pollen transferred from the stamen to the stigma of flowers on different individual plants (where they are not clones), even from mono-sexed flowers (dioecious plants), is classed as cross-pollination. Cross-pollination via alien pollen from other cultivars would be a major problem in orchards if seeds, rather than fruits, were the desired outcome, because any progeny raised from such seeds would be very variable and would not produce young plants of the desired cultivar. It is for this reason that commercial cultivars are produced by asexual (vegetative) systems such as budding and grafting. However, because the desired outcome of an orchard is good quality fruit, good cross-pollination (leading to cross-fertilization) is essential for the production of good fruit crops in commercial top fruit orchards (and soft fruit plantations). In apple crops the presence of pollen often leads to initial apple development, but poor or incomplete fertilization often ultimately leads to premature fruit drop of partially developed fruit (known as June drop).

Sufficient amounts of pollen, temperature for pollen flow, moisture/humidity for pollination, and water at critical times for the fruit development process, all influence fruit production. Imported beehives are commonly used to aid fruit pollination on a commercial scale. Planting selected trees of pollinating cultivars that cross over in flowering time with the existing cultivars can be systematically mixed in with fruit-tree stands, or even single branches of pollinating cultivars, formed by budding on to the main branch-work of fruiting types, can be added throughout the orchard matrix. The very profuse-flowering *Malus* 'Golden Hornet' is commonly used for this purpose.

Fertilization

The male gametes are carried in the pollen grains (in the sperm nucleus), and the female gametes are held in the ovules (in the egg nucleus), and both via meiosis (reduction division) have a haploid (half) number of chromosomes. Fertilization is the fusion of a male and female element (gametes) to form a zygote. Zygotes carry the features of both male and female elements, and at fertilization unite the two half chromosome numbers to reinstate the full number for the species.

In angiosperms, ovules are held in the ovary (a hollow cavity in the carpel). Ovules carry an egg nucleus and an endosperm nucleus; the latter provides the generative power for the production of the food store for the developing seed after fertilization. When pollen grains alight on to a receptive stigma, if all other compatibility factors (including correct temperature) are favourable, they germinate. As they germinate they form long pollen tubes (germ tubes) that enter via the aperture at the stigma and progress down the style towards the ovary. Pollen tubes contain two nuclei: a sperm nucleus and a generative nucleus. It is thought that the generative nucleus is present to give the energy for the germination process.

When the germ tube (pollen tube) reaches the ovary it heads towards an ovule, which it then enters via the micropyle (the small pore piercing the casing formed by the integuments that later become the two layers of the testa, or seed coat). On entry, the germ-tube tip disintegrates and releases the sperm nucleus. At this point the sperm nucleus and the egg nucleus fuse to form a zygote (fertilization). Each ovule has to be visited by an individual germinating pollen grain in order to be fertilized. Hence fruits containing only one seed will only need one successful pollen grain, whilst those with 100 seeds will need 100 viable and compatible pollen grains.

Placentation

Fruit capsules of angiosperms can be of many shapes and sizes, and the various patterns that they take on are specific (and definitive) to the species. The way that ovules are arranged within the fruit – that is, the positions of attachment of the ovules – is known as the 'placentation', as the tissue where the ovules attach to the carpellary wall is known as the placenta. Small pieces of stalk-like tissue (known as the 'funicle', or 'funiculus') attach ovules (and later fertilized seeds) to the placenta. Various patterns and designs of placentation

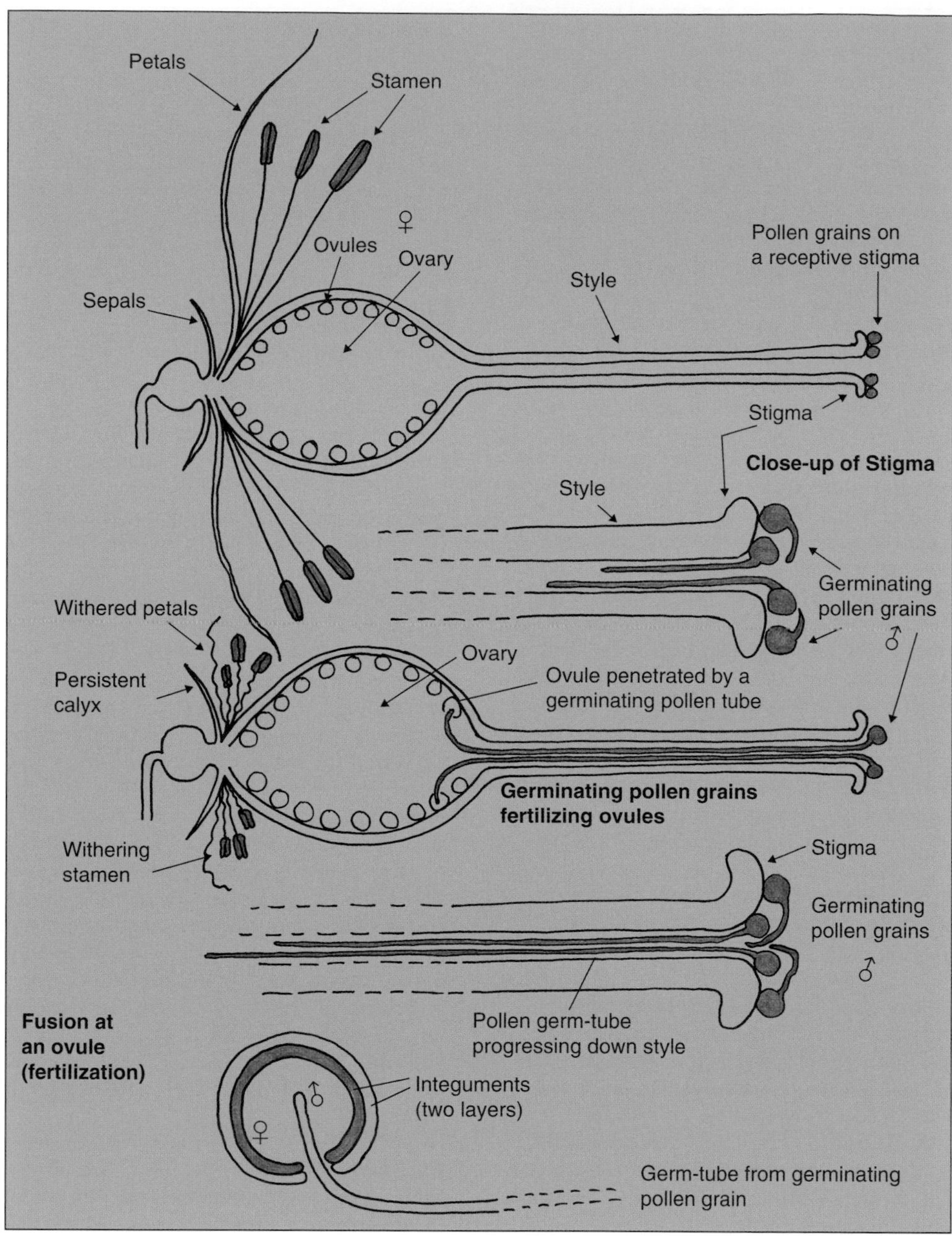

Fertilization.

Plant Strategies

Flower Structure and Pollination

Floral architecture has an important bearing on the pollination process. Monosexed flowers may have small, insignificant petals, such as the white petals of both male and female holly (*Ilex*), or they may have showy corollas, as found in both the male and female flowers of red campion (*Silene dioica*). However, many monosexed flowers consist of collections of rudimentary scales enclosing either several stamen or several pistils (comprising carpels), and are known as catkins.

Imperfect single-sexed male catkins comprise only stamen (known as staminate flowers), and female catkins only pistils (known as pistillate flowers). Single-sexed flowers are often shaped to facilitate wind pollination, and are known as anemophilous (wind-friendly). Male catkins nearly always hang down and are relatively elongate specifically to facilitate wind movement. When their anther sacs are fully developed they split open and shed pollen, and the wind carries it to the female flower. The system is a bit 'hit and miss' in that sufficient pollen needs to arrive at the female flower: there is a large element of luck involved. Hence, male flowers produce very large amounts of pollen so there is a chance of at least some arriving at the correct place.

Obviously the process is influenced by a particular wind direction at the time of pollen release, and the period of time for which it is released. Furthermore, the female flowers need to be ready to receive pollen at the time of its release. Both the temperature and the atmospheric humidity affect the production and 'flow' of pollen. Warming temperatures (and relatively high humidity) encourage pollen flow, and the pollen is very visible at this stage, as a whitish to yellow material on the stamen. Common hazel (*Corylus avellana*) is wind-pollinated, and if branches are shaken on warm days in early spring, the male catkins shed profuse amounts of yellow pollen visible to the human eye.

Insect Pollination

Species of plants with insect-pollinated flowers have evolved towards a more efficient system of cross-pollination because of the feeding habits of insects, and are known as entomophilous (insect-friendly). Insect-pollinated flowers have mostly highly coloured petals and are usually perfect (hermaphrodite) flowers with both male and female organs on the same flower. They may have other mechanisms to encourage insect visitations, including nectaries in the base of the petals containing nectar (a high-energy sugary solution), and the highly visible nectar guides on the corolla.

Many types of insect are involved in pollination, including flies and beetles; these insect species tend to have non-puncturing probosci (tongue-like appendages). Lepidoptera (the moths and butterflies) have sucking mouthparts in their adult phases consisting of a long, rolled-up proboscis that when unfurled can be used to access fairly deep flower corollas. Hence, different floral architecture will accommodate various insect tongue lengths. Actinomorphic flowers with open radiating petals and short distances/depths to the nectaries may be successfully visited by a wide range of insects – even those with shorter probosci. Flies and bees have slightly shorter probosci than some other insects, and although they can access the nectaries of some difficult zygomorphic (irregular) flowers, they cannot access nectar as deep as moths and butterflies can.

Strange zygomorphic flower shapes, including lips and bladders, may hide pollen-producing organs (stamen) and protect them from damaging rain to prevent a wash-out when pollen is ready to flow, and some actually act as landing platforms for insects when they visit. Only specific sizes of insect may breach the floral outgrowths and reach the protected pollen, but sometimes insects that are otherwise too large have to resort to actually chewing through the floral epidermis and cutting a hole in the side of the corolla tube in order to access the nectar. This is a relatively common sight on flowers with highly zygomorphic and complex corolla tubes. Such drastic actions (and in fact all insect-foraging processes) are often encouraged/motivated by a fragrance that is associated with previously successful nectar and/or pollen harvesting. The fragrance of flowers is a powerful insect attractant, with greater fragrance in the evening or day, attracting either nocturnal or diurnal insects.

Nectaries themselves are not dissimilar to hydathodes involved in the process of guttation. Spurs, pockets and other corolla outgrowths often form nectar reservoirs that fill from the secretions from the nectaries;

for example the green-petalled flowers of *Helleborus foetidissima* (stinking hellebore) have petaloid, corolla-tube-like nectaries at their base. Although flower colour may be a strong element of attraction, for insect pollination to work effectively the production and harvesting of nectar is important. Some insects actually feed on pollen, and most collect pollen on their body hairs when visiting the nectaries. Mimicry exists as an insect attractant in a few species, where the floral architecture mimics female insects, and male insects pollinate the flower at (what turns out to be simulated) copulation. e.g. Bee orchid.

Chlorophyll is the main leaf pigmentation, and gives the green coloration. However, carotenoides (carotenes and xanthophylls) at various levels also influence leaf pigmentation – in fact xanthophylls (yellow and golds) are actually chloroplast pigments. Carotenes also feature in flower pigmentation and are responsible for stable yellow and orange colorations in floral tissues, where they are produced in specialized plastids. The darker flower colours (blues, purples, pinks and reds) are produced not by coloured plastids, but by pigmentation in the cell sap (the fluid in the vacuoles of live cells). These pigments are known as anthocyanins* and they act (or rather react) just like litmus solution (or other indicator/reagent solutions) that alter colour according to the prevailing pH. This reaction accounts for the apparent anomaly of *Hydrangea* 'Blue Wave' being blue on acid soils and yet decidedly pink on alkaline soils. The litmus-like reaction also accounts for the gradation of flower-bud colour (from pink when tightly closed, to blue when open) of *Myosostis* (forget-me-not), and other common members of the Boraginaceae family (including *Anchusa* and *Echium*). White flowers have very little pigment, and their appearance as white comes from the fact that, although their cells are actually transparent or opaque, the different orientations of the cells within the different tissues reflects light at different angles. The human eye sees this 'scatter-gun' reflection effect as white light.

The unusual floral architecture of the arum lily group, with their enveloping outer spathe that encloses a central spadix (which is actually a spike holding the sessile male and female flowers), is geared towards insect pollination. The spathes of plants in the Araceae often have brown or dark blotches that may look bizarre, but probably make the spathe look like dung or animal faeces of some sort, and many produce a strong, foetid odour not unlike carrion or rotting meat, both of which help to attract insects. Voodoo lily (*Sauramatum venosum*) and dragon lily (*Arum dracunculus*) are very good examples of those with this effect. Arums actually trap insects, as insects fall into the bulbous base of the spathe, and the effect is for the rapid movements of the panicking insects to aid the pollination of the female flowers situated at the base of the spadix.

The simple pitfall mechanism involved is obviously very similar to that found in insectivorous/carnivorous species of plants, but in this case it is pollination, not nutrition, that is the ultimate aim. The container-like structure of the bulbous part of the spathe traps the insects after they fall into it. The insects are unable to leave the trap both because of the downward-pointing hairs and because the main part of the spathe above the bulbous section is coated with slippery, oil-like secretions on its internal layer. Furthermore, the structure of the inside epidermis of the spathe has specialized cells that also decrease the frictional hold for insects. Insects, if they are still alive, are released after their ordeal because the inside tissue of the spathe becomes more tractable as it becomes flaccid following the pollination process. Whether it is the confusion and disorientation of the insects, or the physical restriction at the neck of the bulbous section, or a mixture of both that prevents insects from flying out, is difficult to say.

Mimulus aurantiaca (syn. *M. guttata)* has a motile stigma (the lobes of the stigma 'close' when touched by humans – and by insects), and *Sparmannia africana* has motile stamen (easily seen when you touch the stamen of a young, fully formed flower). This will almost definitely have a bearing on the pollination mechanism in some way.

Sexuality and Compatibility

Species with single-sexed flowers can have different sexuality depending on how the flowers

---

**antho* = flower, and *cyanin* = dark blue. Note the generic name of blue cornflower = *Cyanus*.

Bee orchid (Ophrys apifera).

Sparmania Africana.

Mimulus aurantiacus (in the left-hand flower the stigma lobes are open, and in the right-hand flower they are closed).

are arranged on the plant. Monoecious arrangements have single-sexed male and single-sexed female flowers on the same individual plant; examples include sweetcorn (*Zea mais*), hazel (*Corylus avellana*), common walnut (*Juglans regia*) and alders (*Alnus* species). Alders, because of their profusion of very distinct male and female flowers, are good examples to look at to illustrate this concept.

Dioecious arrangements have single-sexed male flowers on one individual plant, and single-sexed female flowers on a different individual plant; they include all willows (*Salix* species) and poplars (*Populus* species). Red campion (*Silene dioica*) gets its specific name because of its dioecious arrangement. Male flowers of common holly (*Ilex aquifolium*) will flower, but will never fruit, because they do not carry female organs; conversely, female holly can have fruits, but needs the presence of pollen from male plants in order to do so. However, *Ilex aquifolium* 'Pyramidalis' is a hermaphrodite that fruits well without a second pollinating plant. Most *Skimmia* species are dioecious, but there are also some hermaphrodite forms including *Skimmia* 'Foremanii', and also some forms that are grown for their flower, rather than berries: for example, *Skimmia* 'Rubella'. *Viburnum davidii* and *Hippophae rhamnoides* (sea buckthorn) are dioecious so need both male and female forms to produce purple and gold berries respectively. Where species present three different flower types within a race (male, female and bisexual) they are known as trioecious species.

Protandry and Protogyny

In the natural world, cross-pollination and any resulting fertilization is 'preferred' to self-pollination because it can create small, yet sometimes important, differences between individual plants. The dioecious strategy obviously facilitates cross-pollination. However, in insect-pollinated species with perfect flowers, if homogamy exists (male and female flowers maturing simultaneously), then self-pollination (and self-fertilization) can occur and other mechanisms are therefore needed to encourage, or ensure, cross- rather than self-pollination. The maturity of the male and female organs being separated in time (dichogamy) is one such strategy, with time delays of one to two weeks being quite common. Many species display protandry, where the male organs (stamen) develop fully before the female organs. Protogyny is where the female organs develop before the male organs (the stigma ready to receive pollen before the stamen is ready to release it), and is not so common.

Protandrous and protogynous strategies separate the pollen from the stigma on the same individual flower and/or plant, but allow the transference of pollen to another, more developed individual flower on another plant within wind-carrying or insect-carrying distance. Thus the pollen from a ripe stamen on plant A will distribute its pollen to arrive on a receptive stigma of the more advanced plant B. Plant B would have already shed its pollen (maybe to a further developed plant C, where the stigma was already receptive – or perhaps to waste). The process works because of the differing maturity and development of individual plants in favoured locations or particularly good environments. Whether they are perfect flowers or imperfect flowers in monoecious (or even sometimes dioecious) arrangements, the sexual organs are usually separated in time in their maturation/development in order to facilitate the benefits of cross-pollination.

Compatibility of Pollen and Stigma

The compatibility of pollen is essential for successful pollination and ultimate fertilization. Even in homogamous species, if the pollen from the stamen is not compatible with the stigma on the same flower they cannot self-pollinate successfully – another mechanism to ensure cross-pollination with all of its benefits. Compatibility is a complex thing and many factors can affect it, including the physical shape of the exine (the outer skin of pollen grains) and the receptiveness of the stigma (the adhesive sticky exudate is both present and not toxic to the pollen). The genetic closeness (or otherwise) of the species – their 'bloodline' – is also important, as successful cross-pollination (then cross-fertilization) of plants that are not in the same species (or very closely related species) is rare. So the lack of compatibility of pollen because of a hormonal (chemical) secretion (and protandry or protogyny) prevents the self-pollination of many species with perfect flowers, and also some monoecious species.

Dimorphism

Dimorphism (producing two different flower types on the same species: *di* = two and *morph* = outward shape) commonly occurs in *Primula* flowers, including common primrose (*Primula vulgaris*), where it involves two different style lengths – known as heterostyly (*hetero* = different). Some individual plants have short styles (placing the stigma low in the corolla) and are known as 'thrum-eyed types'; others have flowers that bear long styles (placing the stigma at the entrance, or throat, of the corolla) and are known as 'pin-eyed types'. These differences occur on different individual plants but of the same species and within the same colonies. The different locations of the style encourages cross-pollination by insect visitation, because the body of an insect will pick up pollen from the stamen of one flower with a short style (thrum-eyed), but because of the short style will not touch the stigma of that flower. When visiting another flower on another individual plant with the thrum-eyed feature, it touches the stigma as it delves for nectar, thus pollinating the flower. Other species may be even more complex and show trimorphism (heterostyly involving not two, but three different style lengths).

Failsafe Mechanisms

Viola species produce two types of flowers: open types that are pollinated ordinarily; and closed types that bear no resemblance to what we normally recognize as a flower, but more resembling pouch-like buds. The flowers within the closed, unopened buds self-pollinate and act as a separate mechanism over and above the normal open-pollinated flowers, producing races of self-pollinated and races of open-pollinated progeny from the same plant. This phenomenon

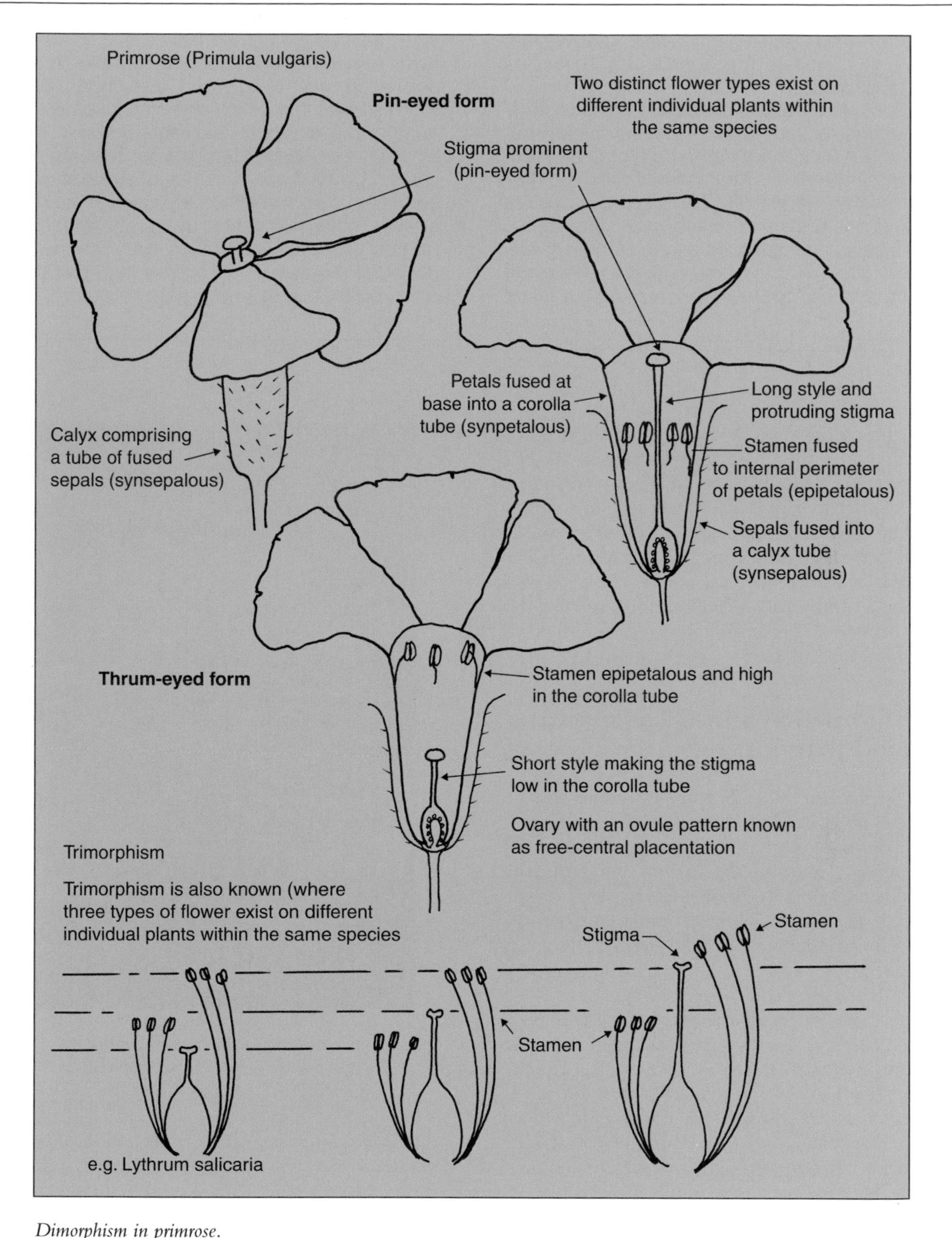
Primrose (Primula vulgaris)
Pin-eyed form
Two distinct flower types exist on different individual plants within the same species
Stigma prominent (pin-eyed form)
Petals fused at base into a corolla tube (synpetalous)
Long style and protruding stigma
Calyx comprising a tube of fused sepals (synsepalous)
Stamen fused to internal perimeter of petals (epipetalous)
Sepals fused into a calyx tube (synsepalous)
Thrum-eyed form
Stamen epipetalous and high in the corolla tube
Short style making the stigma low in the corolla tube
Ovary with an ovule pattern known as free-central placentation
Trimorphism
Trimorphism is also known (where three types of flower exist on different individual plants within the same species
Stigma
Stamen
Stamen
e.g. Lythrum salicaria

Dimorphism in primrose.

is known as 'cleistogamy', and species that have this fail-safe mechanism are known as cleistogamous.

Some crane's-bills (*Geranium* species) have a mechanism that favours cross-pollination, but at the last minute will self-pollinate, rather than not pollinate at all. After normal cross-pollination the carpels, with their fertilized eggs inside, start to desiccate and go brown. As they do so, the tensions set up in the drying carpellary wall cause each ovary to bow outwards whilst still being attached to the other carpels by a fused end at the stigma. Dehiscence occurs as the ovary eventually splits open. If, however, cross-pollination has not occurred, there is a change in the tissues that encourages the fused stigma ends of the carpel to desiccate first and split, bowing the tissue backwards towards the corolla. Under these circumstances the stigmas bend back and touch the last grains of the flower's own pollen left behind as the stamen wither. Hence, self-pollination takes place at the last possible moment, but only if the preferred cross-pollination has not already occurred.

exist. The ovules may be attached to the wall of the ovary, in which case it is known as parietal placentation. Ovules may be attached to tissues at the centre of the fruit that separates the carpel into several compartments (locules), a pattern known as axial placentation. Entire carpels comprising only one void (one locule), with ovaries attached to a specialized central tissue that does not section the carpel into several compartments (locules), have free central placentation.

Fruit Formation and Development in Angiosperms

After fertilization the zygote develops into a complete plant embryo enclosed within a seed. Accompanying the changes in the individual seeds is a massive change in the carpel containing the fertilized ovules, as its tissues start to swell drastically and increase in internal volume to accommodate the swelling seed(s). The patterns of locules (placentation) are relevant when discussing carpels, and the developing and developed fruit phases.

Fruits that are formed as swollen carpels with enclosed seeds are termed 'true fruits'. However, there are also fruits known as 'false' fruits, or pseudocarps, that form by organs other than the carpel swelling at fertilization. The most common of these is the production of a fruit from a swollen receptacle – the swollen end of the flower stalk holding the floral parts as in apple and pear.

The term 'fruit' bears no reference to whether it is edible or not – it is a botanical term for the structures as described. Similarly succulence or nuttiness is not a guarantee of edibility. However, many fruits are edible to a wide range of animals (including humans), insects and birds. Many are edible, in that they are not poisonous and can be eaten, but whether you would want to or not is another matter. Hence edibility, palatability and taste are different things – and many, of course, are very poisonous so are not 'edible' at all.

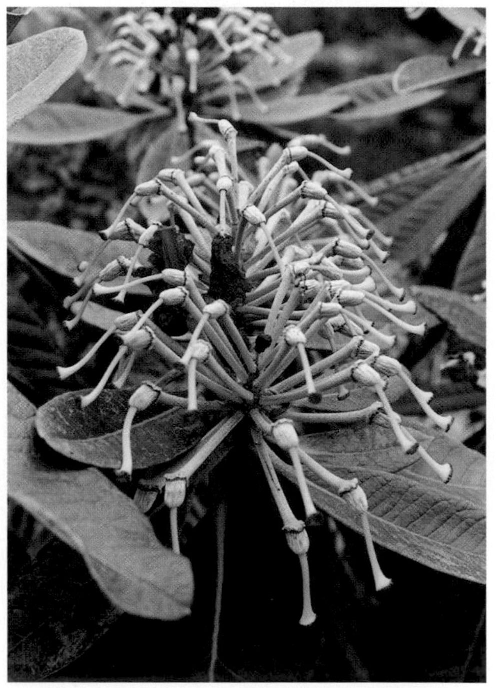

Close-up of the stigma and swollen carpels after fertilization of large-leaved Rhododendron species.

The development of fruits is greatly influenced by the floral architecture.

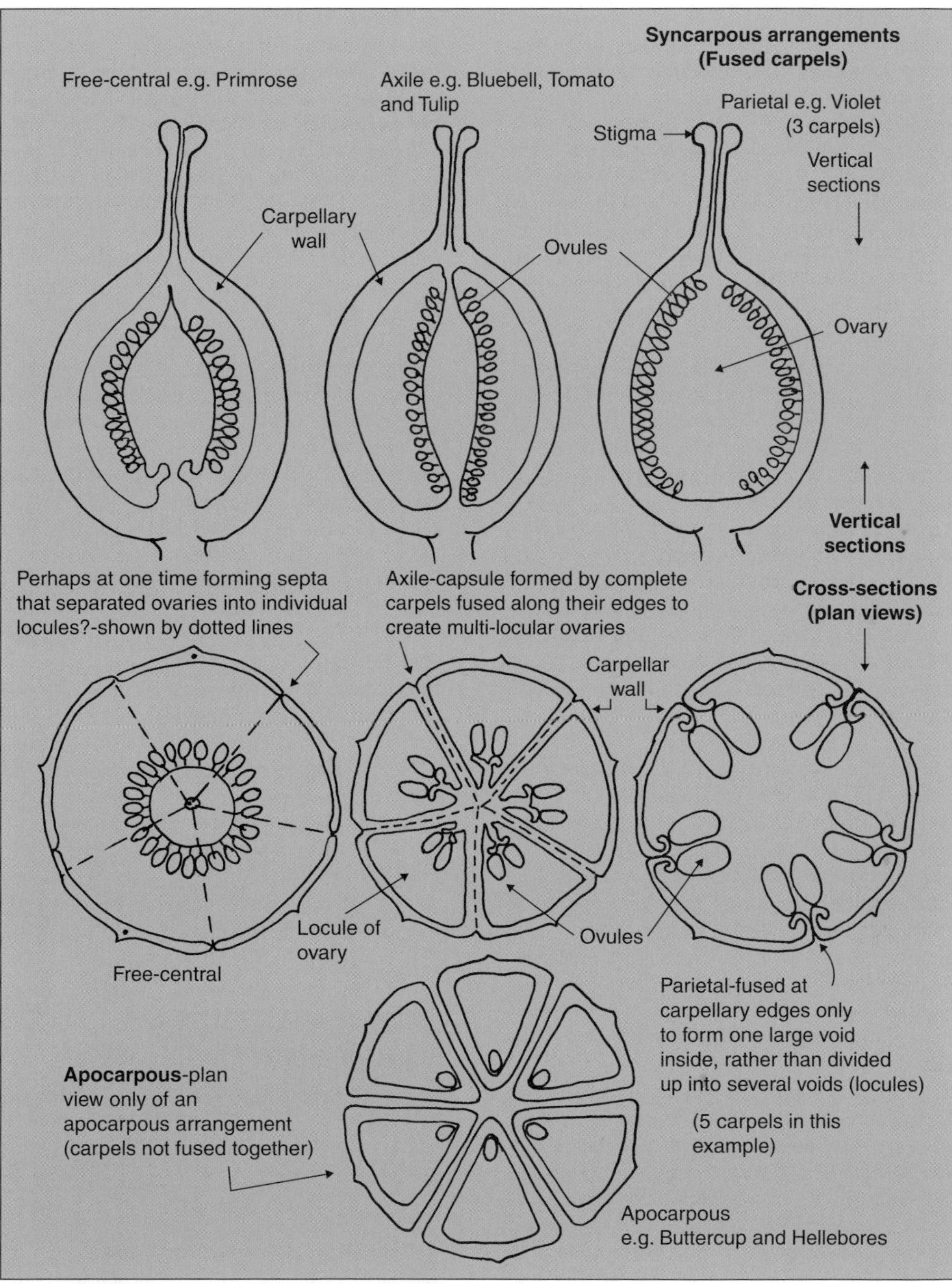

Placentation patterns (ovule arrangements within ovaries).

The position of the calyx, corolla and stamen on the receptacle, the shape of the receptacle, the shape of the carpel initially (and after fertilization), and the position of the ovary, all affect the type of fruit formed. Where the other floral parts arise from a position below the point of attachment of the gynoecium it is known as a hypogynous arrangement (hypo = below). Flowers with hypogynous arrangements have convex-shaped receptacles borne below the gynoecium. This structure produces a superior ovary – *tulipa* has hypogyny and a superior ovary. Perigynous arrangements have a cup-shaped receptacle that surrounds (but does not completely enclose) the gynoecium (peri = around, and gynous = of the gynoecium). However, because the ovary is attached to the end of the receptacle and therefore above the point of attachment of the other floral parts, the flower is also said to have a superior ovary.

Where the point of attachment of the other floral parts (calyx, corolla and stamen) is in front of/above the gynoecium, fruits develop from an inferior ovary (for example *Fuchsia, Iris* and *Narcissus*). If, during development, they leave any persistent parts (the remains of the style or the calyx), these are at the end (tip) of the fruit, and not near the pedicel (stalk).

False fruits of *Malus* (apples) and *Pyrus* (pears), known as pomes, have concave receptacles that completely enclose the ovary, which means that the point of attachment of the other floral parts is around the rim of the receptacle. Because the other floral parts are attached above the ovary, this is known as an epigynous arrangement (epi = above and gynous = of the ovary), and is said to have an inferior ovary. As the pome develops, the swelling receptacle completely encloses the carpel, and retains some of the floral parts (notably the persistent calyx) at the front end of the false fruit.

Both perigynous and epigynous arrangements are represented in the Rosaceae family. Even species with very discernible sepals (of the calyx) and clearly recognizable, radiating petal structures (such as cherry and plum), may have a hypanthium. Cherry, for example, has obvious sepal segments, but also has fused tissue (comprising petals and stamen) that is itself fused to the calyx at the base (making up the three fused tissues of the hypanthium). Typical plum and cherry fruits are formed with the carpel being inside a concave receptacle that surrounds but does not fully enclose the carpel (a perigynous, not an epigynous arrangement). Because they are true fruits, and it is the carpel that swells (not the receptacle), the fruit develops with the swollen carpel growing past the point of attachment of the other floral parts. There is therefore no persistent calyx attached to the tip of the developed fruit. However, in the early stages of development, the stigma and style are clearly seen attached to the tip.

Bearing a hypanthium in a perigynous arrangement creates a superior ovary. Bearing a hypanthium in an epigynous arrangement is indicative of species with an inferior ovary, for example *Narcissus*. Ovaries in this situation are 'fixed' to (adnate to and continuous with) the hypanthium.

After fertilization, fruits will develop to be either fleshy and succulent, as found in plum, cherry, peach, tomato, loquat and nectarine; or as dry fruits that desiccate to hard outer tissues. Various shapes arise out of fruit development, and each one is specific to (and characteristic of) the species, for example nuts of hazel, pods of species in the Leguminosae (Paplionaceae) family, and the winged fruits of sycamore.

Dry Fruits

Dry fruits can be dehiscent or indehiscent. Dehiscent fruits actually split open and disperse (or even dispel/propel) their seeds; the term 'dehiscence' is used to describe the release of materials from split tissue edges, including pollen leaving the longitudinal opening of a ripe anther, and seeds leaving the splits and cracks of a ripe dry fruit. Dehiscence in all its forms is a function of the inequality of the drying of tissues (of varying thickness) creating tensions – thick tissues dry slowly, whereas thin

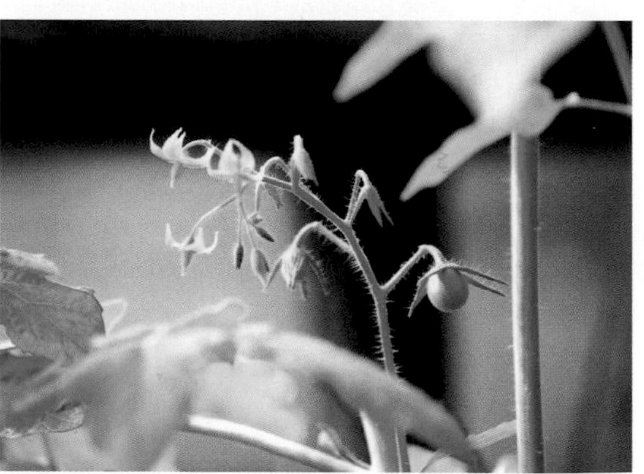

Fruit formation of the tomato.

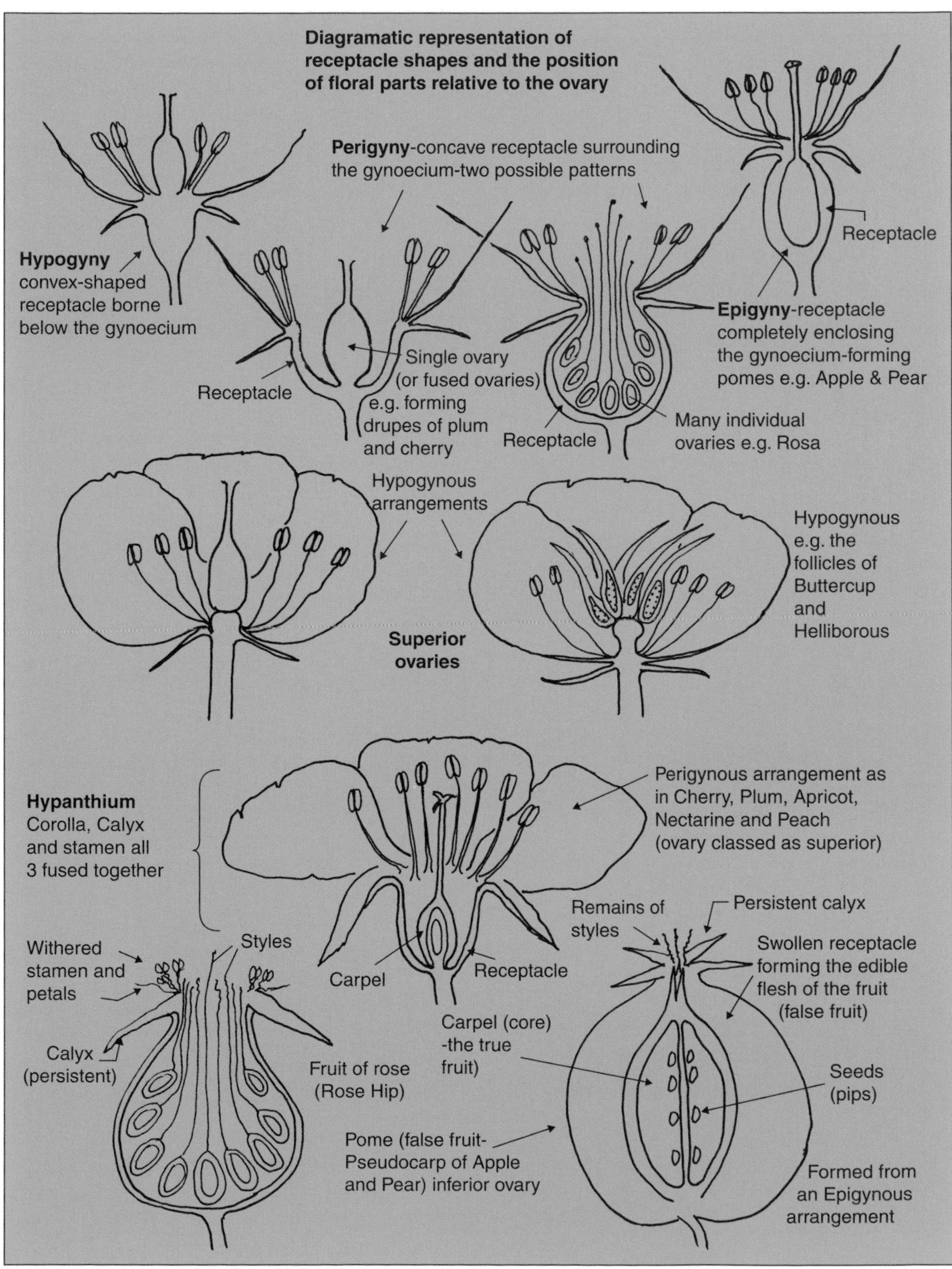

Fruit formation I: hypogyny, perigyny and epigyny.

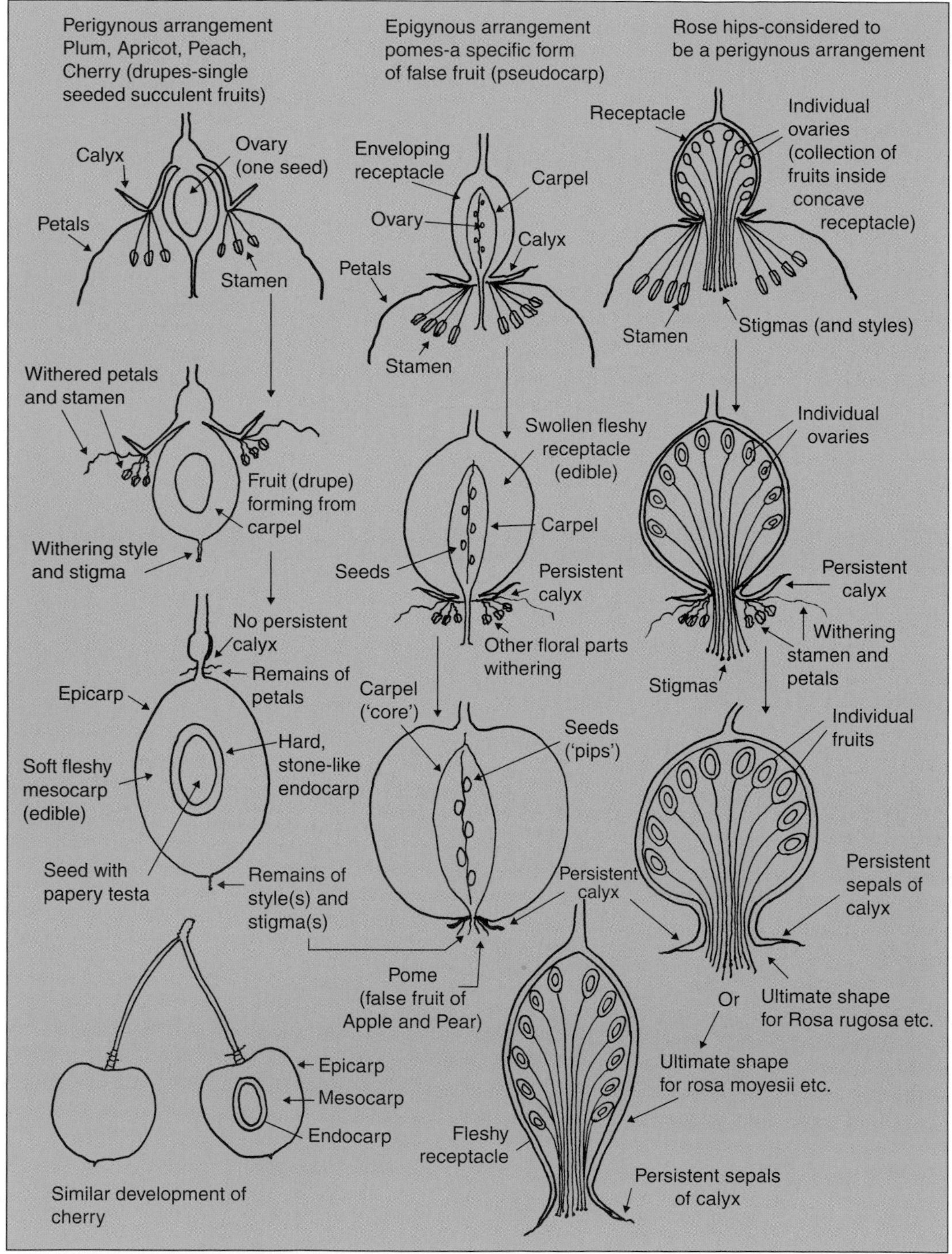

Fruit formation II: Rosaceae.

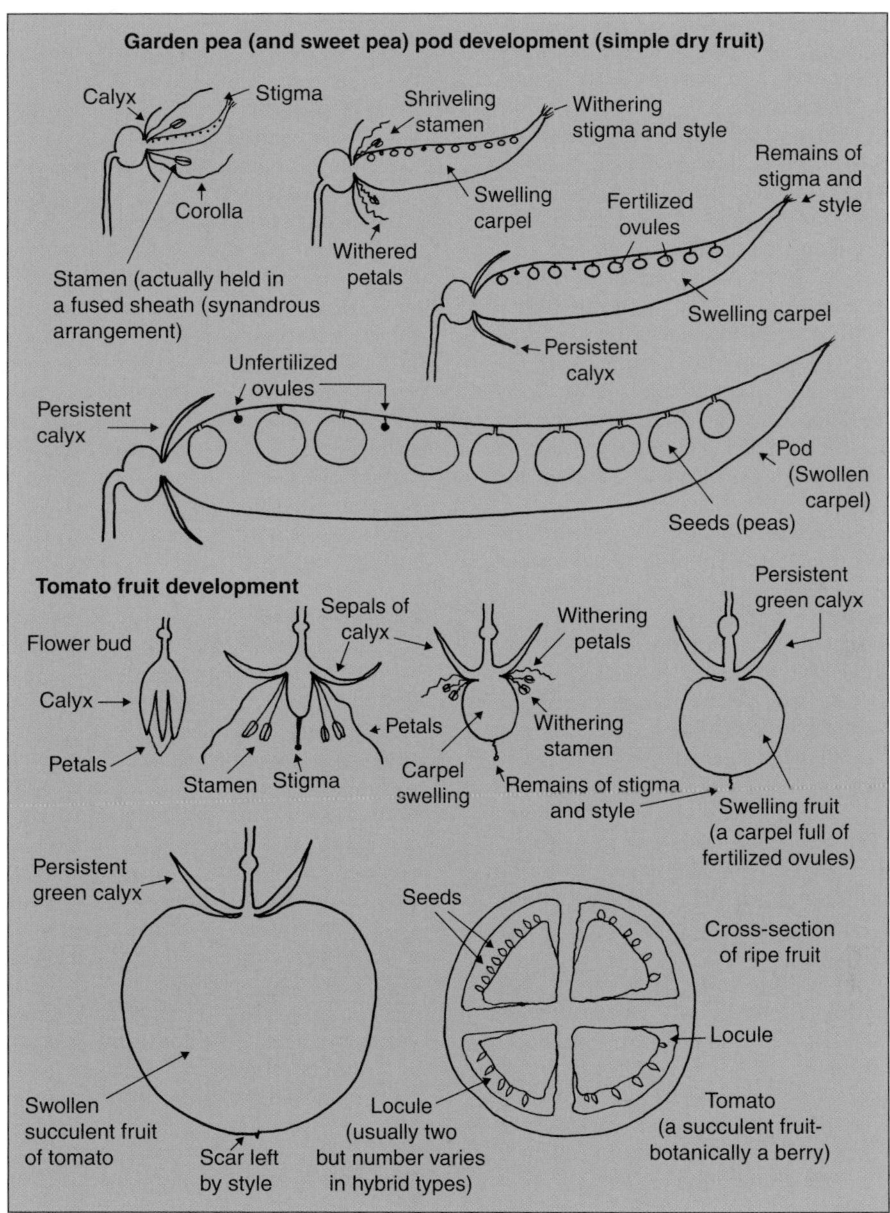

Fruit formation III: dry and succulent fruits (pea and tomato).

tissue layers dry quickly, which sets up powerful stresses. Very propulsive systems are still driven by the differential in drying of the carpellary tissues, but to a much greater degree.

The success of hairy bitter cress (*Cardamine hirsutum*) and Himalyan balsam (*Impatiens balsamifera*) as invasive weeds is dependent on a very efficient explosive method of dispersal. Himalayan balsam has capsule-like fruits that do not desiccate to a hard brown tissue, but remain green and with some moisture content. However, the differential drying of the different thicknesses of tissues 'fires' the seeds long distances, and this, coupled with the plants close proximity to water (that acts as a very effective vector), accounts for the relative success and invasiveness of the species along the banks of rivers and canals – even though it is a native of a different country.

Dry, indehiscent fruits, on the other hand, do not split open at their carpellary edges, and instead rely upon other methods of dispersal, such as wings and wind currents.

Simple fruits consist of one single carpel, and are typified by the pods derived from a single ovary in a single carpel: this is the main identifying feature of the Leguminosae (Papilionaceae) family. The carpel of these 'pods' usually splits along both sutures at dehiscence – for example pea, bean, *Robinea* and *Laburnum*. However, in these instances the seeds are not propelled with any great force, but are merely released from the drying, twisting carpel (fruit pod). Sometimes the pod can be thin, membranous and papery as in bladder senna (*Colutea arborescens*).

Compound fruits have more than one carpel. Siliquas are dry fruits formed from a superior ovary, an arrangement that helps classify the Cruciferae (Brassicaceae) family (for example cabbage, mustard, cress and wallflower). Siliquas comprise a long, linear pod-like fruit formed by two carpels fused together, and divided into two longitudinal halves, separated by a thin membrane called the false septum (or replum). Similar structures, but more rounded or different in shape, are known as siliculas, as found in *Lunaria annua* (honesty) and *Capsella bursa-pastoris* (shepherd's purse).

Capsules are dry fruits with several ovary voids (locules) created by several carpels fused together, some of which allow the seeds to disperse via mechanisms such as the censer (or pepperpot) system. Field poppy (*Papaver rhoeas*), oriental poppy (*Papaver orientale*) and opium poppy (*Papaver somniferum*) are good examples of this system – apertures (pores) created in the capsule by differential desiccation allow the small seeds to be thrown out as the flower stalk moves in the wind – these are known as poricidal capsules. Capsules comprising valve-like compartments, such as in the horse chestnut (*Aesculus hippocastanum*), are known as valvate capsules.

Achenes comprise a hard-shelled nut with one seed inside, and they exist as the hard nutty seeds on the outside of soft aggregate fruits such as strawberry, and as individual fruits associated with the follicles of *Paeonia, Ranunculus* and *Magnolia*. Individual follicles comprise a single unfused carpel with a pointed hair-like tip, and which split down one side of the carpel only. They can have many seeds inside, but may also be associated with single achenes attached to the main fruit by hair-like appendages (as in *Magnolia*). Achenes can be associated in a grouped fruit cluster forming an aggregate fruit ('etaerio'), with hair-like appendages protruding from each one (as in *Clematis tangutica*).

The fruits of plants in the Asteraceae (Compositae) family comprise a single-seeded nut (achene), and many have a very efficient wind-dispersal mechanism that develops from hair-like appendages attached to the achene, resembling a parasol (or parachute) and known as a pappus. This is responsible for the very efficient dispersal of dandelion (*Taraxacum officinale*), goat's-beard (*Tragapogon pratensis*), groundsel (*Senecio vulgaris*) and many thistle species. Achenes also exist as ash fruits (keys) with single wings, and as the split fruits (schizocarps) of maples (*Acers*), where the whole structure comprises two achenes lightly fused together at a suture line and having two wings (samaras). Some authorities separate achenes, as described above, and true nuts, which include acorns and hazelnuts (without wings).

Succulent Fruits

Succulent fruits rely upon the intervention of mammals or birds for their main method of dispersal: these eat the fleshy part, and then transport the seed, depositing it elsewhere. Succulent fruits can occur individually as pomes (for example apples), drupes (for example plums), or as berries; or as aggregate fruits (etaerios), with either a collection of attached drupes (for example blackberry and raspberry), attached achenes (as in strawberry), or achenes enclosed within a soft fruit, as in rose.

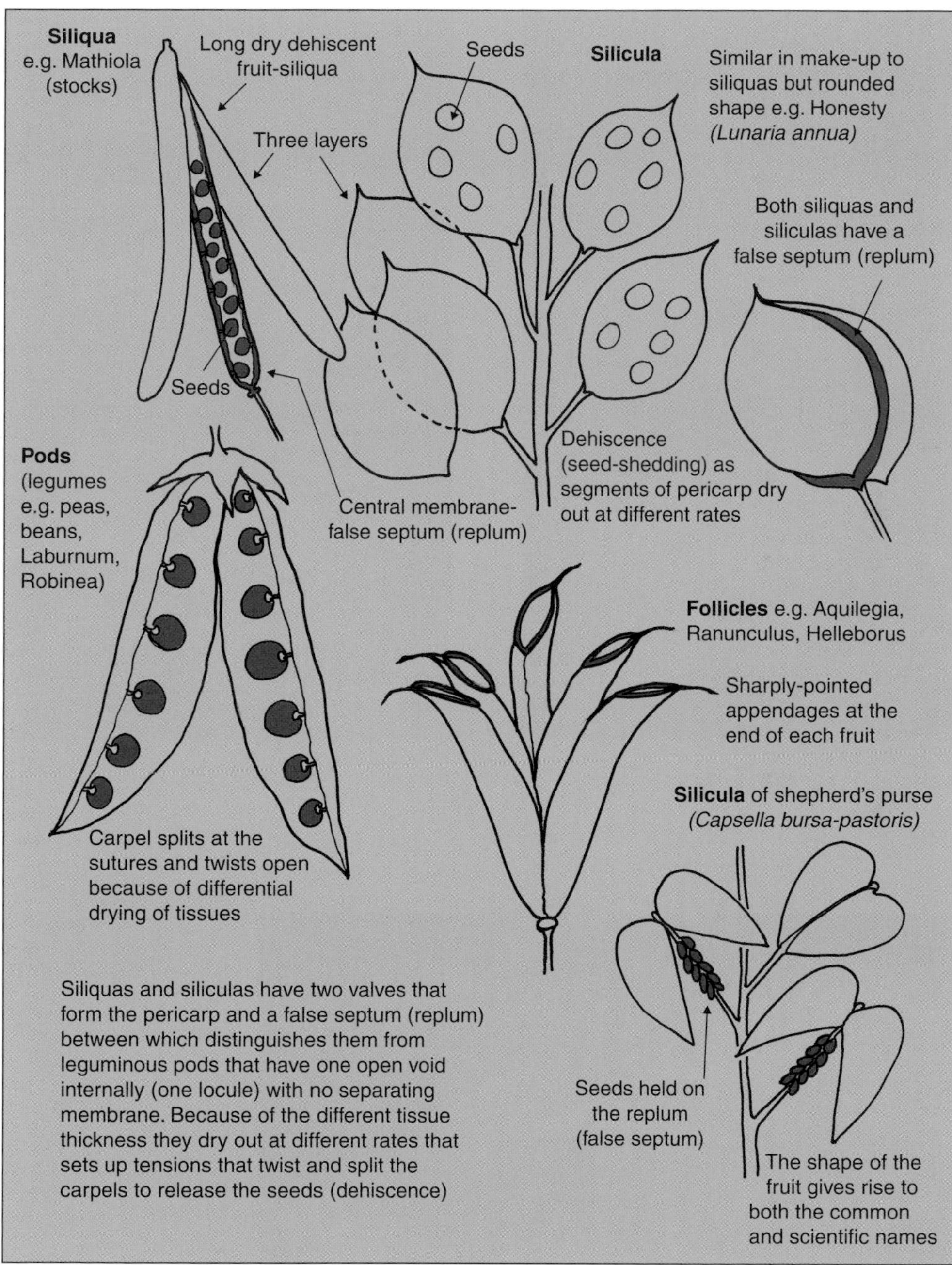
Siliqua
e.g. Mathiola (stocks)
Long dry dehiscent fruit-siliqua
Three layers
Seeds
Seeds
Silicula
Similar in make-up to siliquas but rounded shape e.g. Honesty *(Lunaria annua)*
Both siliquas and siliculas have a false septum (replum)
Pods
(legumes e.g. peas, beans, Laburnum, Robinea)
Central membrane-false septum (replum)
Dehiscence (seed-shedding) as segments of pericarp dry out at different rates
Follicles e.g. Aquilegia, Ranunculus, Helleborus
Sharply-pointed appendages at the end of each fruit
Silicula of shepherd's purse *(Capsella bursa-pastoris)*
Carpel splits at the sutures and twists open because of differential drying of tissues
Siliquas and siliculas have two valves that form the pericarp and a false septum (replum) between which distinguishes them from leguminous pods that have one open void internally (one locule) with no separating membrane. Because of the different tissue thickness they dry out at different rates that sets up tensions that twist and split the carpels to release the seeds (dehiscence)
Seeds held on the replum (false septum)
The shape of the fruit gives rise to both the common and scientific names

Fruit formation IV: dry dehiscent (seed-shedding) fruits.

Flowers and developing siliculas of moonwort (Lunaria annua).

Pod-like fruits (siliquas) of wallflower.

Developed fruits (siliculas) of moonwort.

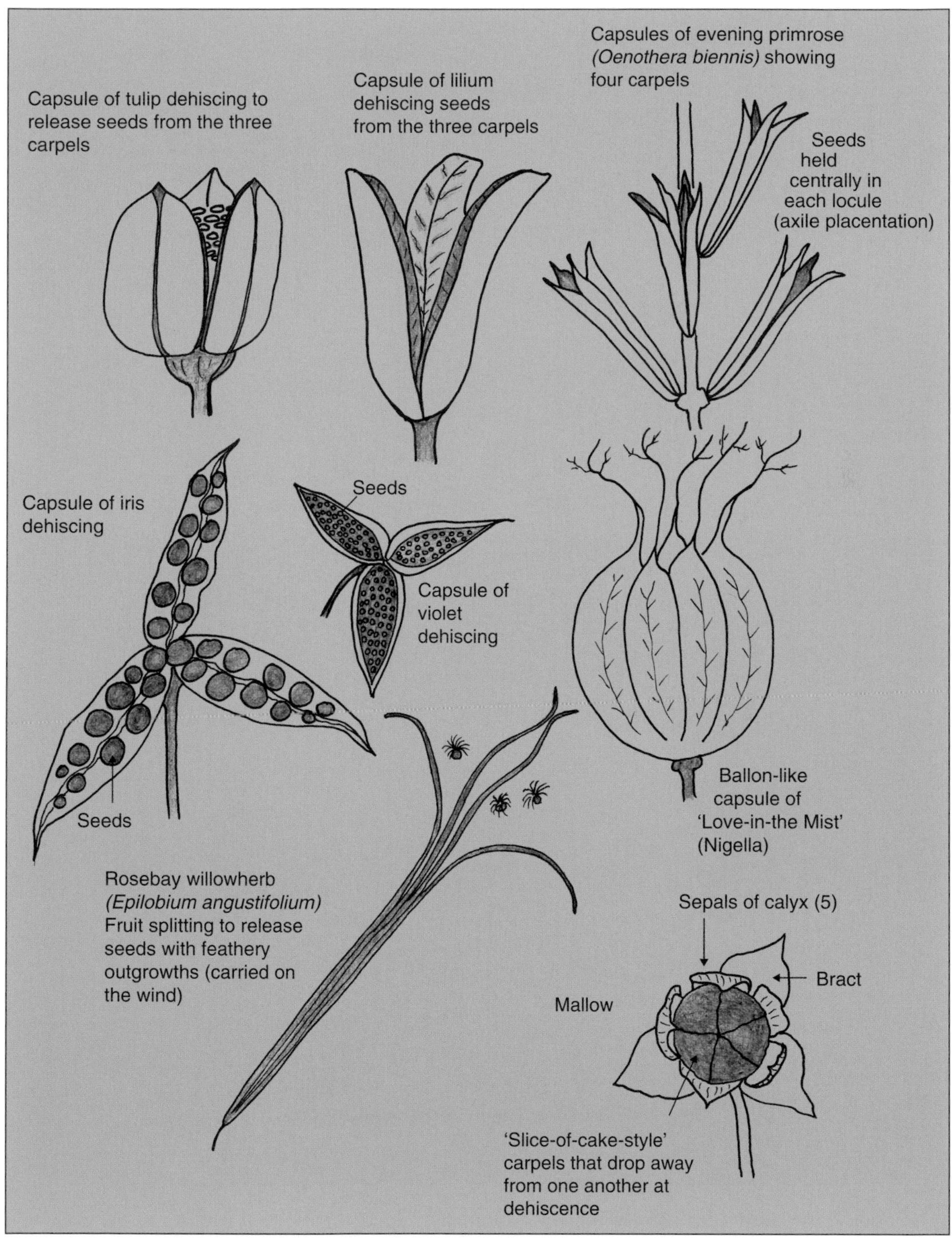
Capsule of tulip dehiscing to release seeds from the three carpels
Capsule of lilium dehiscing seeds from the three carpels
Capsules of evening primrose (Oenothera biennis) showing four carpels
Seeds held centrally in each locule (axile placentation)
Capsule of iris dehiscing
Seeds
Seeds
Capsule of violet dehiscing
Ballon-like capsule of 'Love-in-the Mist' (Nigella)
Rosebay willowherb (Epilobium angustifolium) Fruit splitting to release seeds with feathery outgrowths (carried on the wind)
Sepals of calyx (5)
Bract
Mallow
'Slice-of-cake-style' carpels that drop away from one another at dehiscence

Fruit formation V: fruit capsules.

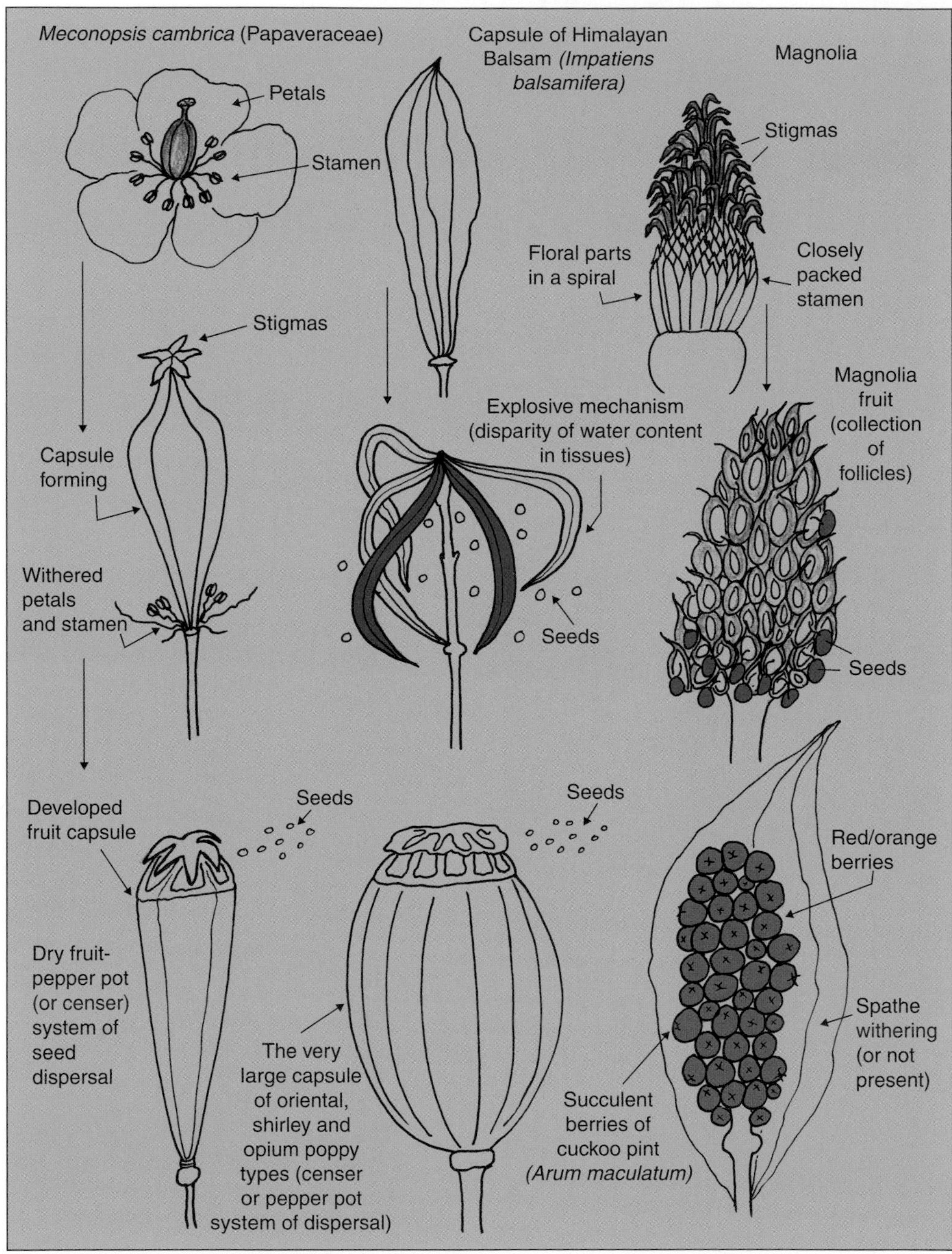

Fruit formation VI: dry capsules, follicles and succulent berries.

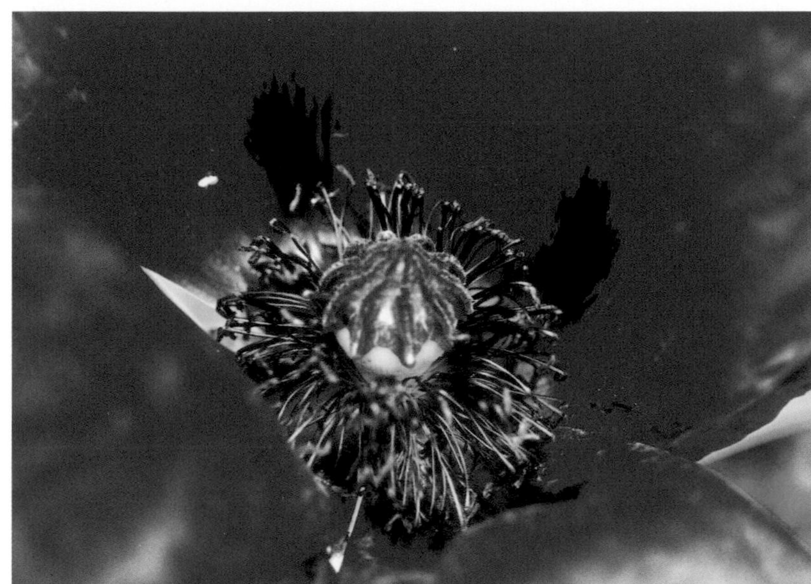

Seed capsule of the oriental poppy (Papaver orientalis) forming after fertilization.

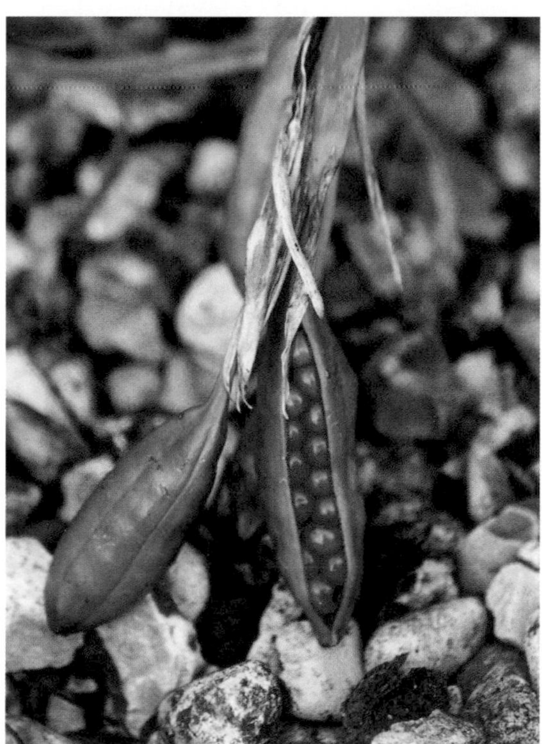

Fruit of Iris foetidissima dehiscing.

Aggregate fruits comprising individual achenes with hair-like appendages on Clematis tangutica.

Fruits (follicles) of Paeonia.

Fruit (follicles) of Magnolia.

Seeds with soft, red arils protruding from the follicles of Magnolia.

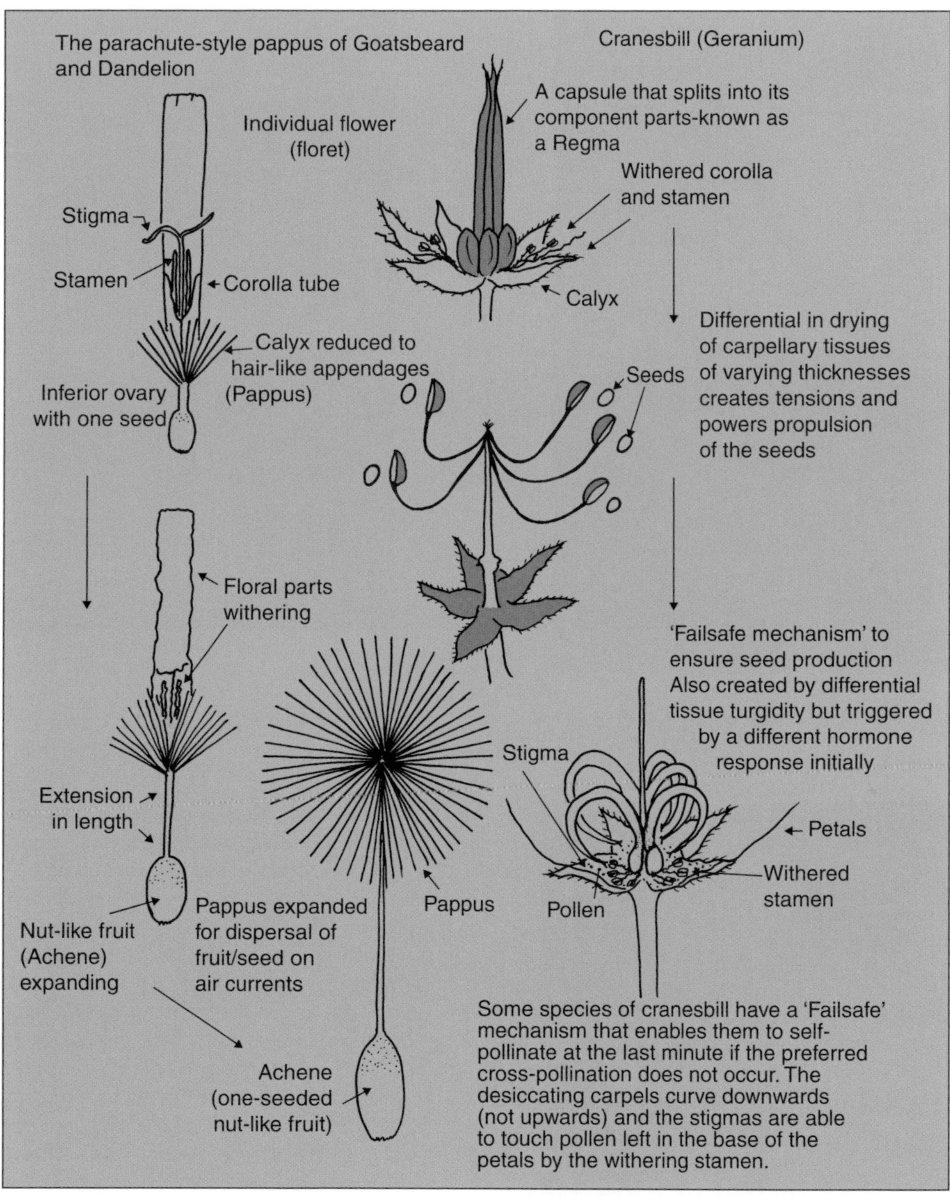
The parachute-style pappus of Goatsbeard and Dandelion
Individual flower (floret)
Stigma
Stamen
Corolla tube
Calyx reduced to hair-like appendages (Pappus)
Inferior ovary with one seed
Floral parts withering
Extension in length
Nut-like fruit (Achene) expanding
Pappus expanded for dispersal of fruit/seed on air currents
Pappus
Achene (one-seeded nut-like fruit)
Cranesbill (Geranium)
A capsule that splits into its component parts-known as a Regma
Withered corolla and stamen
Calyx
Differential in drying of carpellary tissues of varying thicknesses creates tensions and powers propulsion of the seeds
Seeds
'Failsafe mechanism' to ensure seed production Also created by differential tissue turgidity but triggered by a different hormone response initially
Stigma
Petals
Withered stamen
Pollen
Some species of cranesbill have a 'Failsafe' mechanism that enables them to self-pollinate at the last minute if the preferred cross-pollination does not occur. The desiccating carpels curve downwards (not upwards) and the stigmas are able to touch pollen left in the base of the petals by the withering stamen.

Fruit formation VII: dry fruits propelled by wind and explosion.

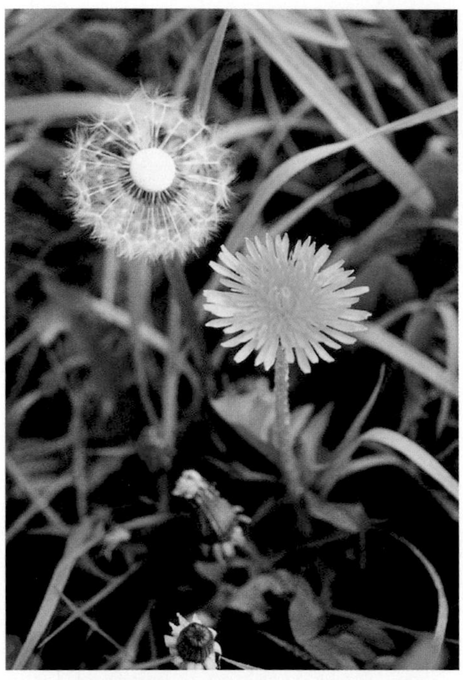

Pappus of the dandelion (Taraxacum officinale).

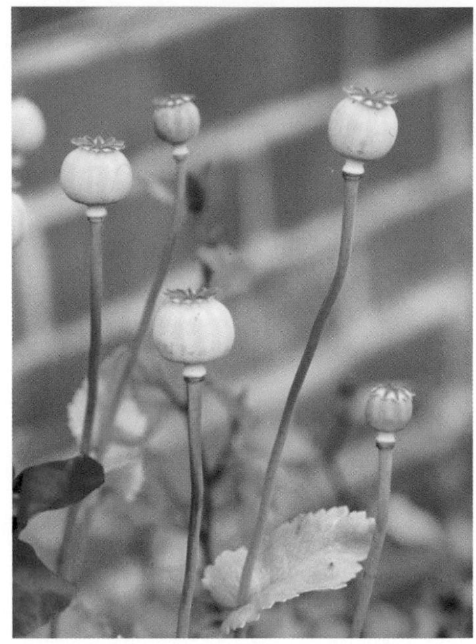

Seed capsules of the poppy (Papaver).

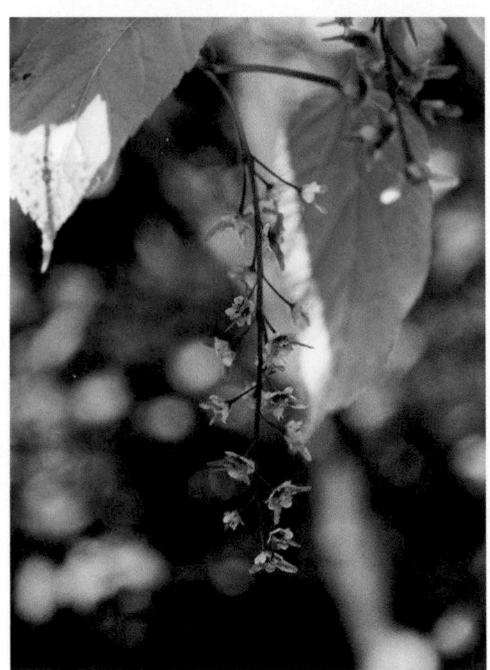

The two-winged fruits of Acer capillipes developing.

Rose fruits (hips) comprise a fleshy exterior enclosing a collection of individual achenes. In apple (a pseudocarp known as a pome*), the fleshy edible part is formed by the swollen receptacle, and the true fruit (enveloped by this tissue) is the 'core' which comprises the tough tissues of the carpel (pericarp) that enclose the seeds ('pips').

Berries are distinguished from drupes in that each berry comprises a soft, succulent fruit resulting from a compound ovary (containing more than one seed); good examples are *Passiflora Sorbus, Pyracantha*, *Cotoneaster*, and some, but not all species of *Crataegus*. Berry-like fruits containing only one seed are actually drupes, for example plum, cherry, peach and common hawthorn (*Crataegus monogyna*) – monogyna describing that there is only one seed in the gynoecium.

Drupes comprise a tough, waxy, outer skin (known as the epicarp), which surrounds the fleshy (usually edible) middle part of the fruit (mesocarp), which in turn surrounds the hard, stone-like endocarp (inside wall of the fruit). Inside the endocarp

*Pomology is the study of apple and pear-like fruits.

The fruits of Rosa rugosa.

(the 'stone' of plums, peaches, apricots and cherries) is the seed (the 'kernel') covered in a papery 'testa' (seed coat). What may not be quite so obvious is that a walnut, although seemingly bearing the features (and the common name) of a nut, botanically is actually a drupe. It is classified as a drupe because each fruit comprises one papery-scale-covered seed (the edible part) inside a hard shell, but enclosed in a pulpy/fleshy casing. So under this classification, coconut (*Cocos nucifera*) is also a drupe. The fruits of *Quercus* (oak) and *Corylus* (hazel) comprise single-seeded fruits without a soft, pulpy, fleshy casing, so are nuts in the true sense of the word.

Aggregate fruits (etaerios) comprise either a collection of dry achenes (for example *Clematis*), or a collection of soft, succulent, single-seeded fruits (drupes or druplets) held on hard, swollen inedible tissue (usually the receptacle), and include raspberry and blackberry. Each druplet is formed from an individual fertilized carpel, and the carpels (with associated styles and stigmas) are easily seen in the developed flower and development phases of the fruit. Mulberry fruits look similar in structure but are syncarps (multiple fruits with united carpels).

The fruit of strawberry is a false aggregate fruit comprising a swollen receptacle (not a swollen carpel), holding an aggregation of small, nut-like fruits (achenes). Each achene is formed from an individual fertilized carpel. The less edible, nutty pieces of strawberry, in contrast to the soft edible pulp of the swollen receptacle, are witness to achenes being present in the structure.

The succulent fruits of tomato and fuchsia are actually berries by botanical definition. Fuchsia develops from an inferior ovary whereas tomato develops from a superior ovary, as witnessed by the persistent calyx being behind (at the back of) the fruit. Tomato, persimmon and the false aggregate fruit strawberry have persistent green calices (the plural of calyx) that are always left intact when picking and marketing the fruit. It is the collection of soft druplets that are the edible parts of raspberry and blackberry, not the swollen receptacle, so these are harvested minus the cone-shaped receptacle ('the plug') and the attached calyx.

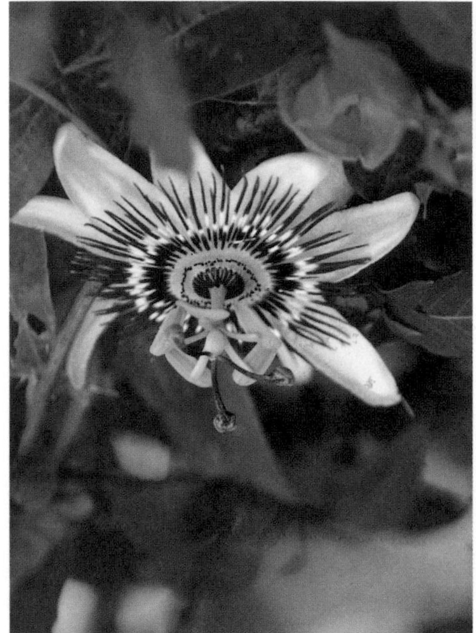

Flower of Passiflora caerulea showing the sexual organs after fertilization, and the commencement of fruit development above the downward-curving stamen and below the downward-curving styles with stigmas at their ends.

The developing fruit of Passiflora caerulea with the three-pronged stigma on the end.

Fruits (berries) of Mahonia hanging in a raceme.

The colourful, developed succulent fruits (berries) of Passiflora caerulea.

Berberis darwinii after fertilization, with the fruits (berries) beginning to develop.

The white fruits (berries) of Cornus alba that give the plant its name. (alba = white)

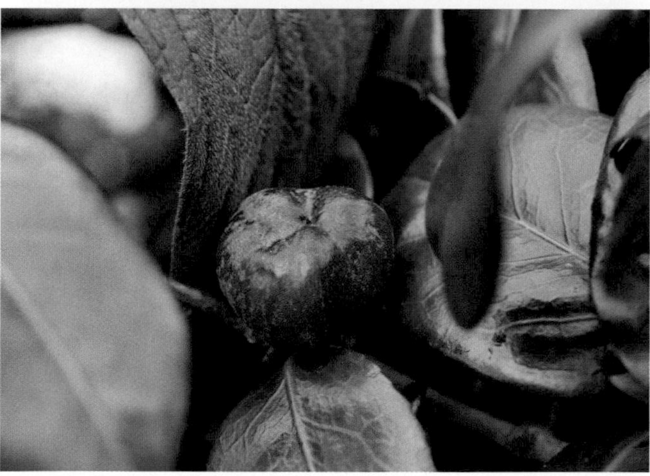

The fruit of Camellia.

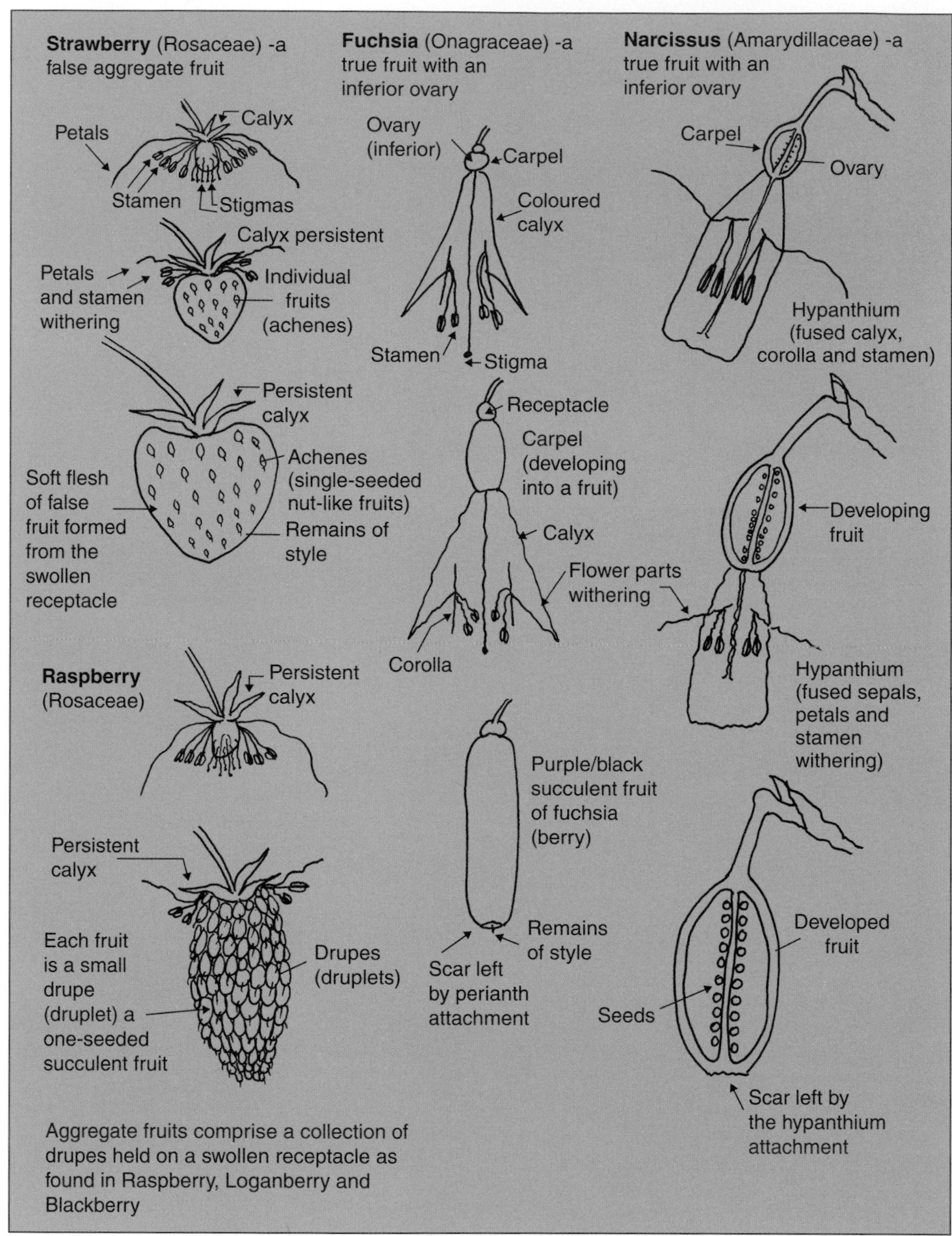

Fruit formation and development of the strawberry and raspberry, and Fuchsia and Narcissus.

Fruit development (aggregate fruit) of the blackberry.

Parthenocarpic Fruits

Fruits formed without fertilization are known as parthenocarpic, and sometimes they produce seeds, and sometimes they do not (but they swell up as if doing so). Banana, seedless orange, seedless grapes and cucumber are examples. Sometimes it is pollen that, although not actually fertilizing, triggers a hormonal response that causes the fruit to swell; but some parthenocarpic fruits are formed without this trigger.

CHAPTER 4

Seed Structure and Germination

Seeds

Seeds contain plant embryos that develop after the fusion of the male and female gametes at fertilization (the sexual process) of plants. Embryos have the potential for creating a complete, new, fully developed plant in the right conditions, and they carry genetic material from both male and female parents (or male and female parts of the plant).

After dehiscence (seed shedding), or after the rotting of succulent fruits, the seed(s) drop to the soil, and it is the soil that normally gives the necessary environment for the germination process. But because in the wild it is a lottery just where seeds fall (or are placed by mammals or birds), they may not always end up in a good germinating environment; hence, even very large amounts of seed production may not ultimately bring about very many plants. Furthermore, methods of holding back germination (until good environmental conditions are likely?) are inbuilt in some plant species.

Fecundity (Fertility), Viability and Mortality

Fecundity is a measure of fertility – the relative ease or otherwise with which a flower will become fertilized and an embryo formed. Obviously, therefore, poor fecundity reduces the 'successful' outcome immediately.

Viability is the ability of seeds to lie in a dormant (non-active) state for long periods of time, yet remain alive (viable), so that at a later stage (whenever the correct conditions permit) the viable seed may germinate. External signs such as relatively plump, succulent-looking seeds may not actually indicate a live embryo within. In some instances tests to prove viability may be necessary to prevent a wasted seed-sowing operation on any particular batch. Methods include cutting into samples (destroying the seeds) to check for good tissues inside, and flotation in water – good seeds 'sink' and 'empty' seeds float. The period of time a seed may remain viable is known as its longevity.

Mortality rate describes the number of seeds that may die at any time after seed production by the plant. Hence seeds may die before, during or after germination. Losses (death) may be due to unusually low air and soil temperatures, drought, being eaten by mammals or birds, being disturbed at any time during the process, or insufficient oxygen. The presence of toxic materials in the soil atmosphere will also kill seeds (granular forms of toxic herbicides rely upon this to kill dormant and germinating 'weed' species).

Hence in the natural world, even if fecundity and viability are reasonably high, other factors may render the mortality rate high eventually, and the final number of plants that are 'successful' to any developmental stage may be very low. The controlled, good germinating environments and husbandry/management techniques used by horticulturists greatly reduce mortality rates, and therefore (providing they commence with viable seed initially) significantly increase the 'success' rate.

Seed Structure

The basic seed comprises an embryo attached to seed leaves (or a seed leaf) called cotyledons, enclosed by a hard, tough, often leathery outer seed coat called the testa. Impressed into the testa is a distinct scar (the hilum) that denotes the original point of attachment of the seed (via the funicle or funiculus) to the carpellary wall of the fruit. There is also a small orifice or pore called the micropyle that was initially important for penetration of the germinating pollen tube during fertilization. The micropyle continues to be important in the developed seed in facilitating water uptake as a precursor to germination. An external inspection of relatively large seeds such as runner bean or broad bean clearly shows their important external features. The same main features exist on most seeds, but the size of the seed makes a difference to the relative ease (or otherwise) of their recognition.

The cotyledons are considered to be part of the embryo. Each embryo comprises an embryo shoot called the plumule, an embryo root called the radicle, and an axis joining the points together. The top part of the axis above the point of attachment of the cotyledons is called the epicotyl, and the lower part of the axis below the point of attachment of the cotyledons is called the hypocotyl.

Seeds having two seed leaves are classed as dicotyledonous, and those with only one seed leaf are classed as monocotyledonous, a classification that divides the angiosperm world into two main groups. Some seeds have more than two seed leaves and are known as polycotyledonous (having many cotyledons), typified by conifers (gymnosperms, not angiosperms).

In some species the seed leaves are thick and fleshy and engorged with starch, and in these instances the cotyledons act as the food supply to the attached embryo when it starts to develop at germination. Seeds with food stored in their cotyledons are called non-endospermic seeds (without endosperms); oak, broad bean, pea and walnut are all non-endospermic (they all have fleshy, starch-storing cotyledons).

Other species have a food supply that is within the seed/fruit casing but is external to the embryo (not unlike a nutrient sac or reservoir), called an endosperm; these seeds are termed endospermic. Many monocotyledonous plants have endospermic seeds, including palms, cordylines, all grasses, and cereals (such as sweetcorn – *Zea mais*), whereas the majority of dicotyledonous plants have non-endospermic seeds. However, examples of dicotyledonous plants with endospermic seeds do exist, and in these cases the cotyledons are very thin and membranous, and their function is to absorb nutrition from the endosperm and transport it (via the vascular tissue) to the embryo; castor oil plant (*Ricinus communis*, from whence the deadly poison ricin is extracted) is an example of a dicotyledonous endospermic species. The seeds of conifers do not form a true endosperm in a second nuclear fusion at fertilization; however, they do produce a nutritive tissue that is sometimes referred to as the endosperm.

Large seeds such as broad beans and peas that have been soaked in water overnight dissect very easily. The engorged tissues readily show the important external features, and careful removal of the testa with a sharp knife or scalpel will reveal the internal features (the use of a basic x 10 hand lens helps even further). Because of the relative abundance and ease of harvesting sycamore seeds (fruits) and acorns in the autumn, several individuals can be soaked and dissected so that you may become familiar with the structure. Dissecting through a single separated nut (achene) of sycamore (*Acer pseudoplatanus*) reveals that the enclosed seed has an embryo with two cotyledons (seed leaves). The seeds of sycamore (and all other maples) are non-endospermic and have two thin cotyledons to store food, and although they are rolled up tightly 'Swiss roll style' they are obvious, because even at this early stage of development they already show a pale green coloration due to the presence of chlorophyll. The presence of chlorophyll heralds the fact that the cotyledons will appear above soil level after germination, and photosynthesize (epigeal germination).

Dissection of an acorn of oak (*Quercus*) reveals two fat, fleshy, cream-coloured, starch-rich cotyledons (with the consistency of cheese) firmly enclosing a small embryo that leaves its impression in the tissues. Even though on occasions three seed leaves are produced, oak remains classified as dicotyledonous, because two seed leaves are the norm. The seed leaves of oak, pea and bean have no chlorophyll present because they will not appear above soil level to photosynthesize, and they leave their cotyledons (seed leaves) below soil level, even after the depletion of their nutrition by the embryo (hypogeal germination).

Large amounts of stored nutrition are needed for the germination process, to give the necessary energy for mass cell division, and rapid growth and development of the embryo. Food when stored in cotyledons (non-endospermic seeds) is usually starch based, whereas endospermic seeds often have oil-based or fat-based food stores.

Germination

The major requirements for germination are moisture (soil moisture), oxygen (in the soil atmosphere), favourable temperature, and in a few instances light. In all circumstances the first three requirements are essential for the germination process. If one (or more) of these requirements is not met, the seed remains in an enforced dormancy known as quiescence, and will not germinate;

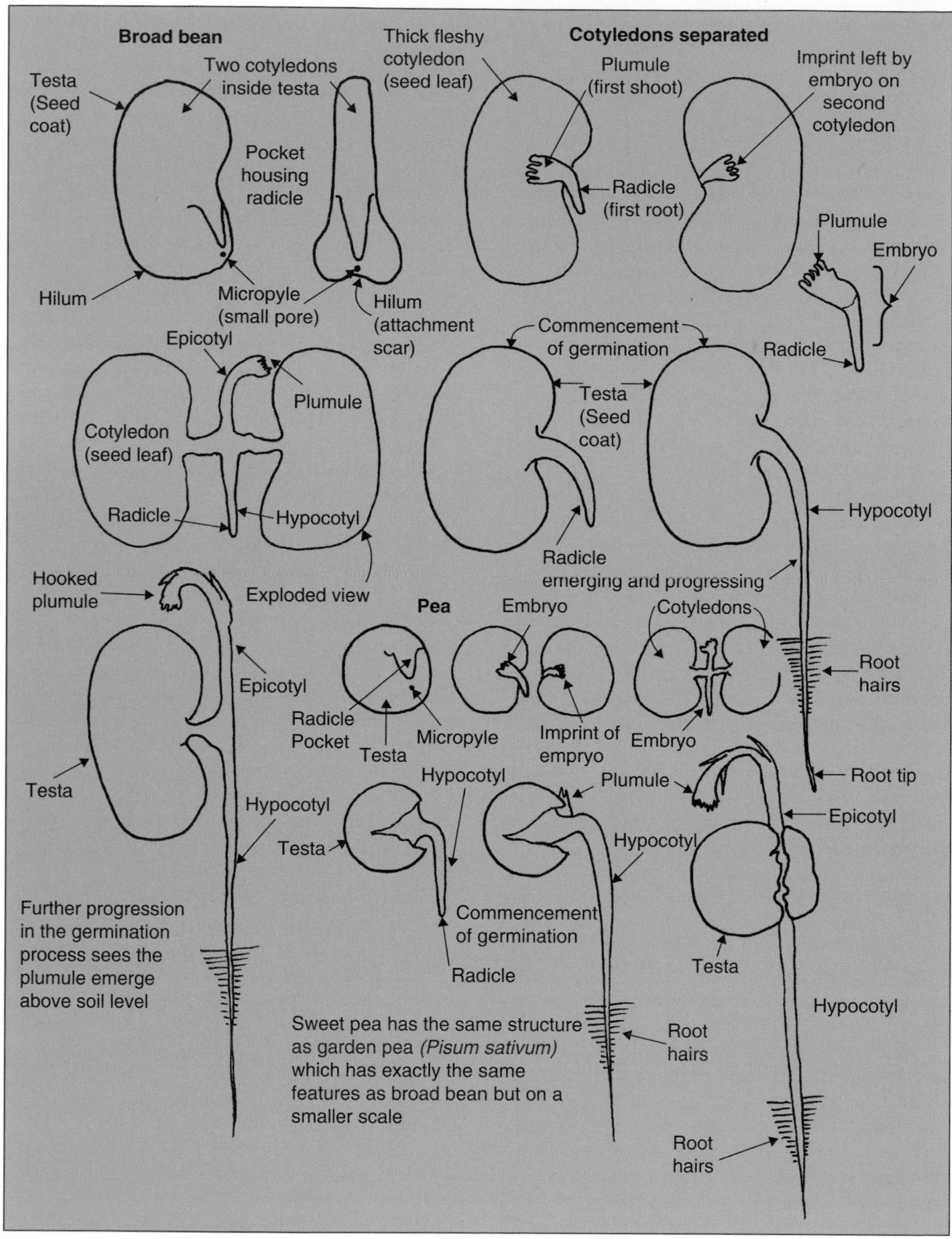

The external features of seeds: broad bean and pea.

or even worse, it will commence germination, use up its finite energy store, and then die because of the incorrect sub-surface and/or aerial environment.

The condition of dormancy is the normal resting situation for seeds until all the major requirements for germination are met, and this condition allows the seed to remain alive (viable), and yet resting, for long periods of time. However, in many instances dormancy may continue after these conditions are met. This can cause disappointment for individual plant enthusiasts raising small numbers of plants, and major difficulties (including loss of cost effectiveness) for large-scale commercial plant producers, as large areas of seedbed are tied up (and need to be kept irrigated and weed free) with no end result (or financial recompense).

Dormancy is therefore any condition that prevents germination of the seed after the major requirements of moisture, oxygen and favourable temperature have been met. The reasons for dormancy include having immature (not fully developed) embryos within the seed, and food substances not being fully broken down within the embryo. Other reasons include the testa (seed coat) being too tough for the radicle to penetrate, the tough and waxy testa being totally impermeable to water, and/or the presence of hormonal inhibitors. Some species have a low temperature requirement (normally associated with winter – vernalization) for a prescribed period of time, or a short, sharp, hot temperature (as in a bush fire) which, if not fulfilled, prevents germination.

The Germination Process

A seed in a dormant state has a water content of approximately 10 per cent of its own weight, which is important in combating drought and extreme temperature fluctuations. Before the germination process can commence, the seed must imbibe more water and increase its moisture content. The testa comprises two distinct tissue layers that are derived from the two original layers (integuments) of the ovule. The outside layer is tough and waxy, having a protective role, whereas the inner layer is soft (sometimes with velvety hairs) and has an absorptive role. Water is taken in through the micropyle (the small pore), and initially engorges cells in the radicle (primary root), priming it for rapid extension growth. The cells of the interior layer of the testa also absorb this water rapidly, and as more water is imbibed through the micropyle, it engorges more of the soft cells of the inner layer of the testa, putting the outer tissues of the testa under stress. Eventually the outer layer cracks, causing splits and fissures that greatly increase the water uptake of the embryo, and ultimately facilitate the progress of the developing radicle.

The progression of the radicle through the soil leads to a taproot formation as it gets larger. Plants such as carrot, parsnip and dandelion live their limited life with this original swollen radicle (with limited lateral roots) as their main underground structure and food store. Woody species, however, tend to have taproots that only persist for a limited period of their long lives, and fairly soon in the development of their seedlings, lateral roots create a more complex and sometimes fibrous root system. There are gradations of this complexity, but in general, the persistence of taproots into mature phases (long thought to be common in mature trees) is unusual, rather than the norm (Cutler and Glasson). Many (but not all) monocotyledonous species abort the original taproot fairly early on in their development in favour of adventitious support roots (prop roots) arising from the base of the stem.

Water not only engorges cells, but also commences enzyme action, which only operates in a very specific regime (as enzymes must have both moisture and a correct temperature before they are activated). Enzymes are proteins, and they need co-enzymes such as magnesium ions to activate them – you can therefore see the importance of the presence of magnesium in leaf tissues for nutrition. Furthermore, changes brought about by the first enzyme reaction create a substrate necessary for the second reaction. Hence, sequential orders of enzyme action occur, and the very nature of these step-wise reactions accounts for the action or non-action of enzymic pathways, depending on the environment they find themselves in.

Basically, enzymes will only work on very specific substrates, and without their presence the process cannot occur, and without the specific order of substrate, production cannot progress to conclusion. This is as true of germination as it is of the growth and development of bulbous subjects, and in part (along with temperature, moisture and day length) accounts for seasonality in plants. The presence of a range of very specific enzymes, able to be activated in a specific order, explains the different temperature requirements of seeds from different species, and the reason for some bulbous species growing then flowering in the spring, whilst others commence growth then flower in the autumn.

Enzymes digest stored starches that are not soluble in water, and convert them to soluble sugars that can be transported in water solution to any part of the embryo needing a high energy supply. The speed at which enzymes can break down the stored foods and convert them to soluble sugars depends on what the stored food is. Some species have stored starch, and some species have proteins or oils as a food store. Proteins take longer than starches to break down, and oils take longer still.

The energy stored as starch, oils or fats within the seed is solely responsible for the growth of the embryo up to this point. However, once the green cotyledons or the embryo shoot breaks the soil surface and becomes illuminated, then photosynthesis produces sugars to top up the energy supply for continued growth and development. This highlights the importance of the correct depth at seed sowing, as seeds sown too deeply will utilize their finite stored energy from within the seed before they break the soil surface where photosynthesis can replenish it.

The soluble sugars are used to maintain existing tissue by providing the large amounts of energy demanded by the respiration needs of both the shoot-tip and root-tip meristems in order for cells to divide and mature rapidly. The first part of the embryo to receive this energy supply is the radicle (primary root), which is situated, usually in a shaped pocket, very close to the micropyle (water-imbibing pore). Because of its close proximity to the micropyle its tissues are the first to be able to imbibe water.

It is the rapid cell division, cell maturation and cell enlargement (caused because the cells are engorged with water) that makes the radicle the first part of the embryo to elongate and break through the testa, no matter what the species. The ongoing and developing radicle anchors the seed in the soil, and also begins to absorb even more moisture to trigger further enzyme action within the embryo. Eventually, as more sugars are released from stored starches, either the hypocotyl or the epicotyl carrying the plumule (primary shoot) grows towards the light.

At this stage in the development of the germinating embryo, it depends on which point of the axis (the epicotyl or hypocotyl) grows more quickly, as to whether the seed leaves (cotyledons) remain below the soil level or appear above soil level. If the epicotyl (the upper part of the axis, or rudimentary stem) develops more quickly than the hypocotyl (the lower part of the rudimentary stem), then the plumule appears above the soil surface. However, the seed leaves (cotyledons) remain below the soil surface, so have hypogeal germination (hypo = below and geal = germination). In these instances the cotyledons are a food source (and facilitate food transport) only, and have no photosynthetic role. It is the true leaves that develop from the plumule that emerges above soil level, which take on this role. This process is consistent within species, and can even help to identify certain species. Oak, broad bean and pea species have hypogeal germination, and all show true foliage leaves (with the normal characteristics of the species) or intermediate leaves (not the cotyledons) above the soil surface at germination.

If the hypocotyl (the lower part of the rudimentary stem below the point of attachment of the cotyledons) develops more quickly than the epicotyl, the cotyledons are pushed above soil level and have epigeal germination (epi = above), where the cotyledons take on a photosynthetic role. Thus a more quickly extending epicotyl results in hypogeal germination, and a more quickly extending hypocotyl results in epigeal germination.

Epigeal germination is characterized by the cotyledons enclosing the plumule as a protective measure (and also the crooked hypocotyl that acts to further protect the tip growth as it progresses through the soil). Thus Seed leaves, not true foliage leaves, appear above the soil level in the first instance. Some epigeal subjects retain the protective, closed cotyledons (and the surrounding testa) above the soil surface for a short time after germination and development. In most instances, however, the testa is removed by friction during the journey through the soil.

Both sycamore (*Acer pseudoplatanus*) and tomato (*Lycopersicon esculentum*) have epigeal germination and produce two distinct green lanceolate seed leaves that appear above the soil surface. The true foliage leaves are then produced with subsequent growth from the initial apical bud. Sycamore has palmate foliage leaves, and tomato may take on one of many possible shapes, from intermediate leaves (that are heavily lobed, simple leaves) to various truly compound types (pinnate forms). Beech has two rounded, thick and fleshy seed leaves that appear above soil level, and in the early stages are held tightly to the epicotyl (developing upper stem) near the apical bud of the seedling.

Castor oil (*Ricinus communis*) has two rounded, penny-like green seed leaves, and like all epigeal germinators, relies upon the friction of the soil to help remove the testa as the faster developing and crooked hypocotyl drags the seed through it. As it does so, the remains of the endosperm (a white

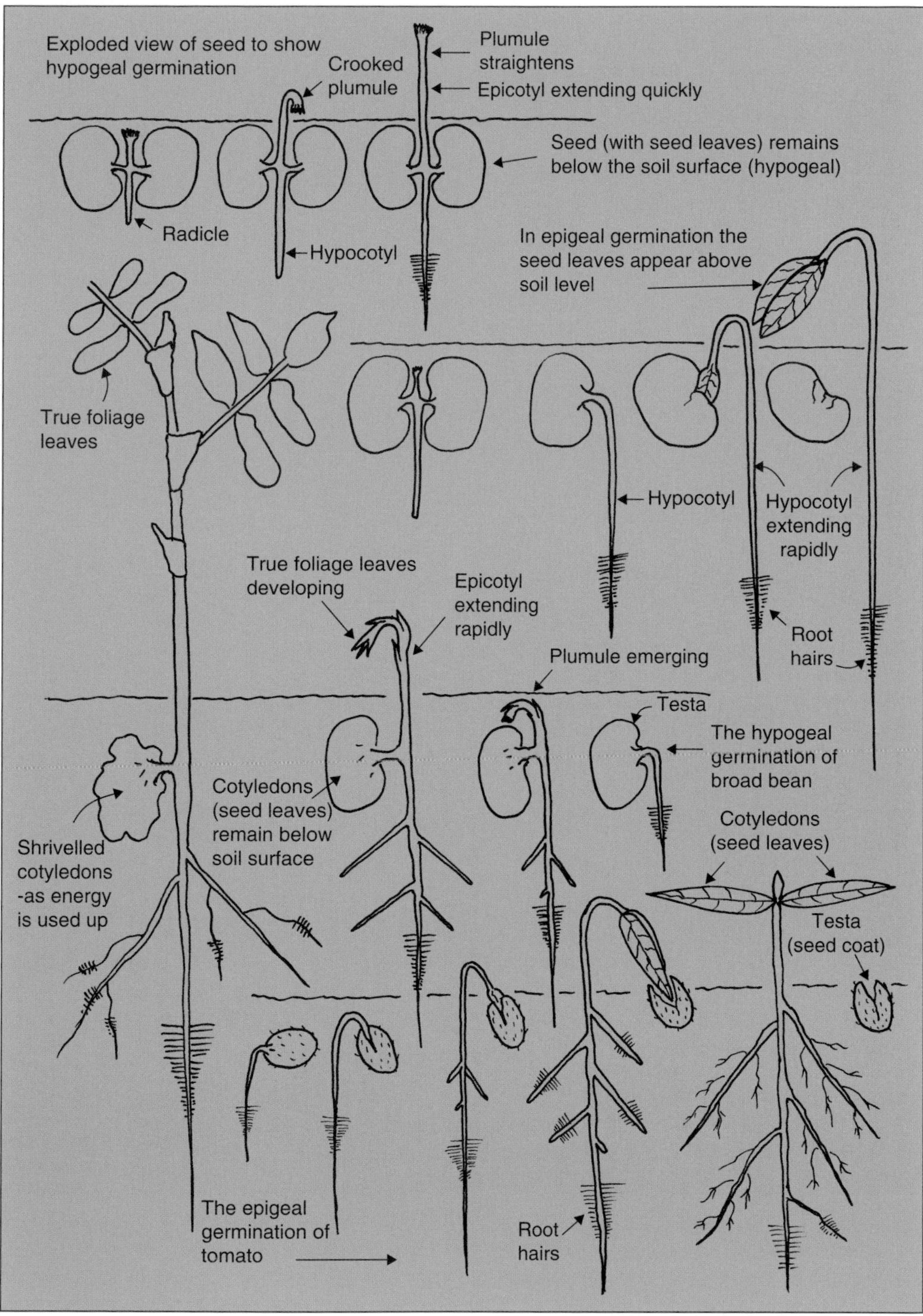

Hypogeal and epigeal germination.

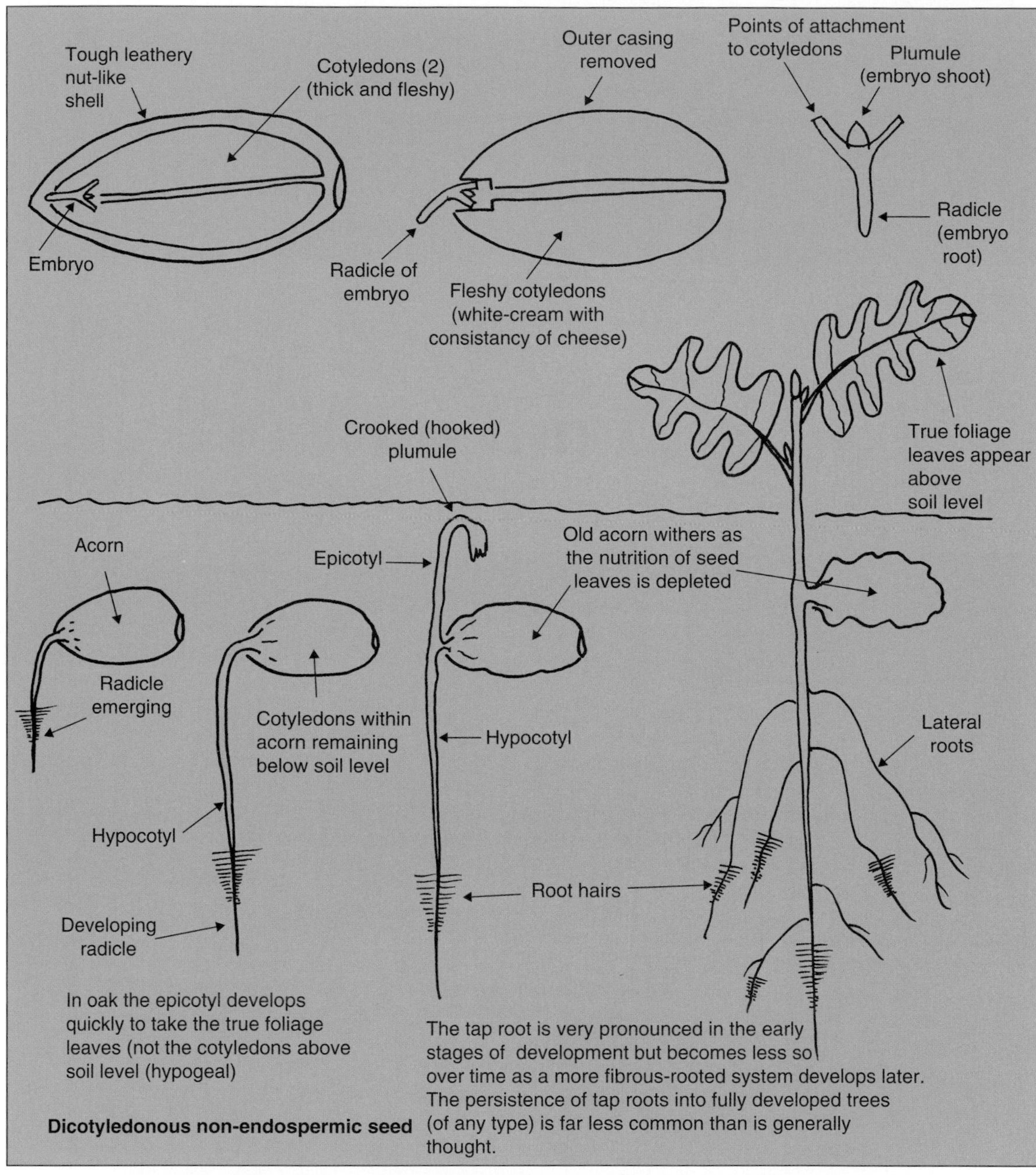

The seed structure and hypogeal germination of oak (Quercus).

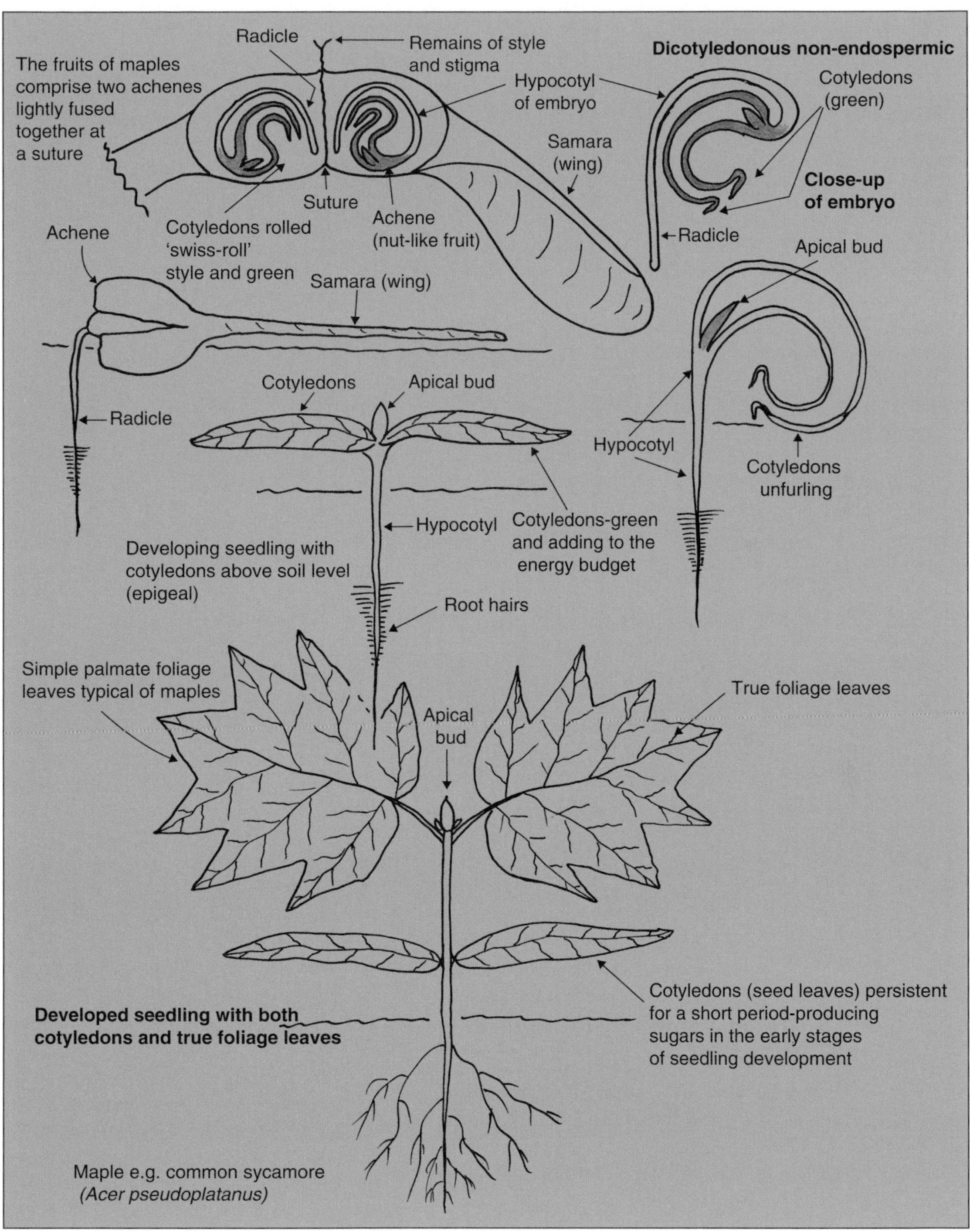

The seed structure and epigeal germination of maples (Acer).

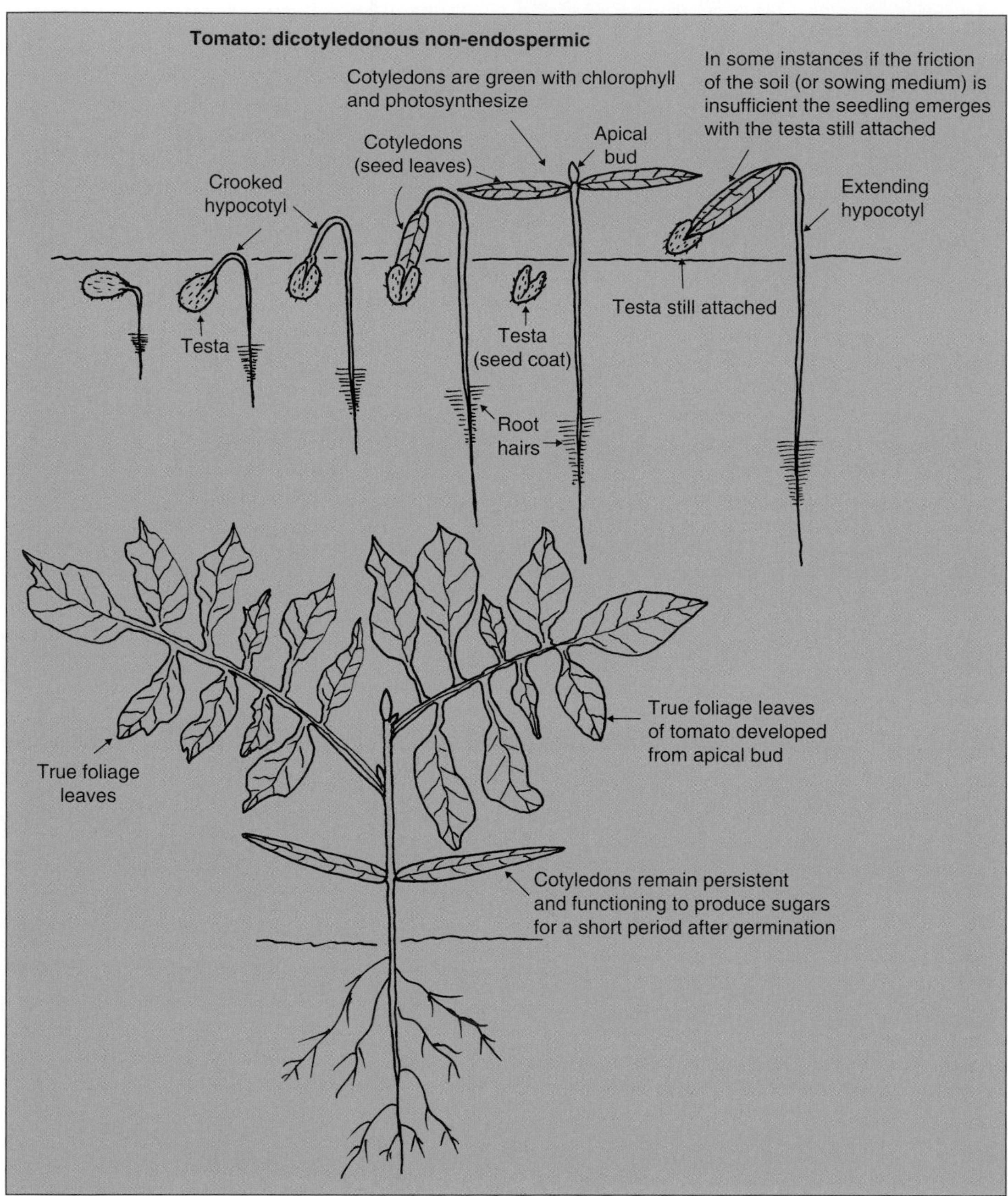

The epigeal germination and development of the tomato.

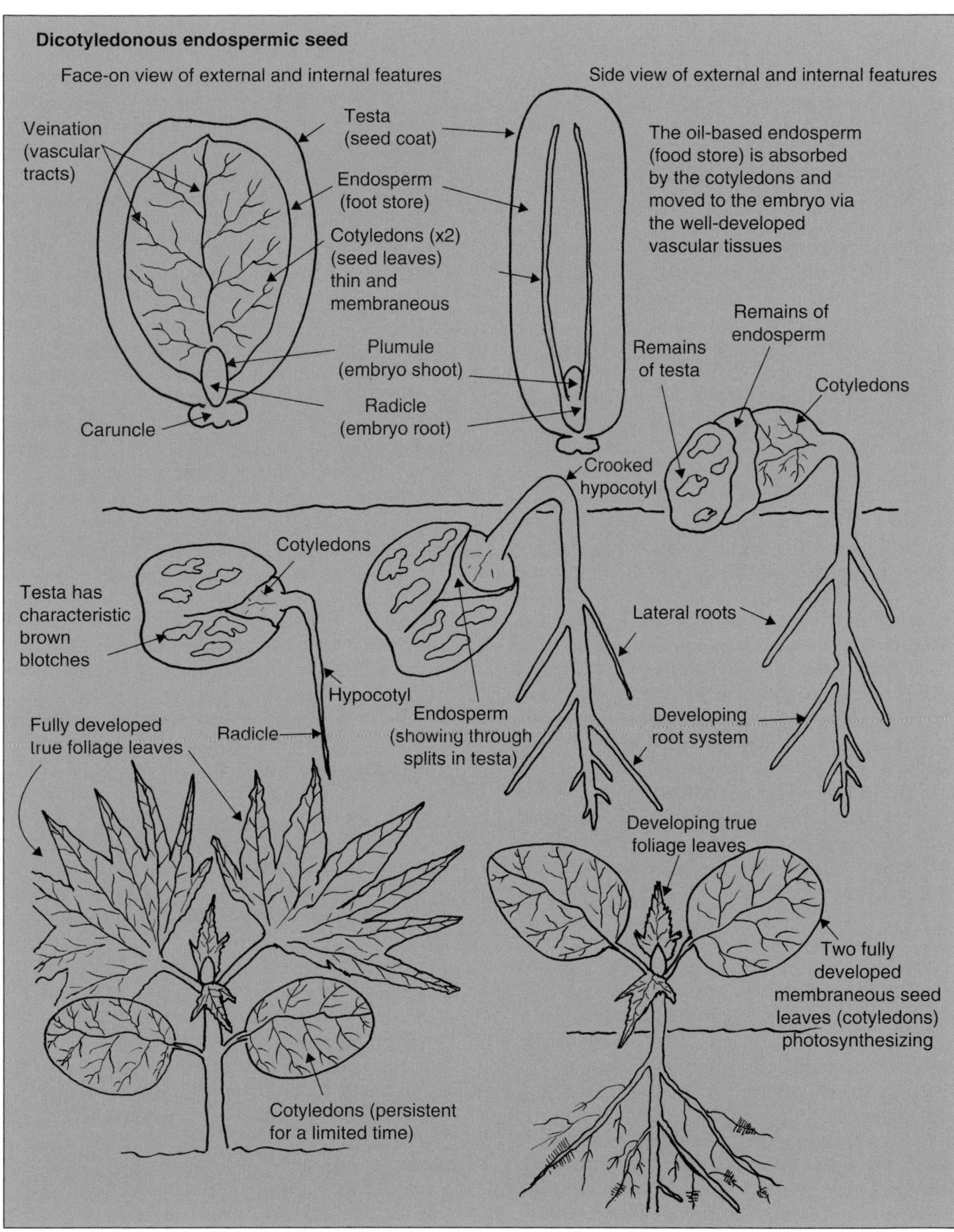
Dicotyledonous endospermic seed
Face-on view of external and internal features
Side view of external and internal features
Veination (vascular tracts)
Testa (seed coat)
Endosperm (foot store)
Cotyledons (x2) (seed leaves) thin and membraneous
Plumule (embryo shoot)
Radicle (embryo root)
Caruncle
The oil-based endosperm (food store) is absorbed by the cotyledons and moved to the embryo via the well-developed vascular tissues
Remains of endosperm
Remains of testa
Cotyledons
Crooked hypocotyl
Cotyledons
Testa has characteristic brown blotches
Lateral roots
Hypocotyl
Endosperm (showing through splits in testa)
Developing root system
Fully developed true foliage leaves
Radicle
Developing true foliage leaves
Two fully developed membraneous seed leaves (cotyledons) photosynthesizing
Cotyledons (persistent for a limited time)

The seed structure and epigeal germination of the castor oil plant (Ricinus communis).

material) can be seen through the splits in the testa. The true foliage leaves are simple palmate. Runner bean (*Phaseolus vulgaris*) has two very large, rounded, simply shaped foliage leaves appearing above the soil surface: these are the most notable, because although the cotyledons do appear above the soil surface (epigeal germination), they are often hidden by the developing true foliage leaves.

Sweetcorn is a monocotyledonous, endospermic species. The 'seeds' of sweetcorn are actually complete fruits, each one comprising a single seed attached to, but external from the endosperm (food store), and all enveloped by the pericarp. The single, shield-shaped cotyledon is known as the scutellum, and it remains below soil level (hypogeal germination). Along with most other monocotyledonous plants in the grass and cereal group, sweetcorn has an apical sheath known as the coleoptile (= covering of shoot), and the true leaves are rolled longitudinally within it. The leaves that show above soil level are true foliage leaves, and the first of these splits the coleoptile open during liberation. It also has a root sheath, known as the coleorhiza (= covering of root), and similar descriptions could be used for wheat (most other cereals) and grasses. Lateral roots develop quickly during the germination and development processes, and are mostly (but not always) adventitious, arising from anywhere on the basic stem area.

Onion (and many ornamental plants in the onion group, *Allium*) are endospermic and have epigeal germination. So even though they are monocotyledonous, the single cotyledon appears above soil level, and again the seed coat (testa) is removed by the movement of the seed through the soil (by friction), this time as the crooked cotyledon (not the hypocotyl) develops. Later in the development of the seedling the plumule breaks away (is liberated) from the base of the enclosing cotyledon to reveal the first true foliage leaves. Roots are adventitious as they arise from the swollen stem base (developing bulb).

Temperature

Different species have different temperature requirements for their successful germination. Fluctuations can be very wide between species, and can vary from temperatures just above freezing to as high as 28°C. Most common plants germinate within the 15–20°C range. Hence it is not necessarily a high temperature that is required, but a suitable temperature for the enzymes within the seed. For example, attempting to germinate lettuce in temperatures of 15–20°C will not be successful, but in temperatures between 5° and 12°C approximately they germinate well. Likewise, some hardy woody subjects that you would assume need low germinating temperatures, actually prefer slightly higher temperatures for success – for example *Arbutus unedo* (Killarney strawberry tree) and *Camellia*.

Some species need a low (not usually sub-zero) temperature as a pre-germination condition to break dormancy. Other species, naturally growing in areas that regularly have bush fires, need a hot temperature as a pre-germination trigger – for example, some protea and eucalyptus species from South Africa and Australasia. Temperature will also affect the absorption of water – for instance, lower temperatures cause slower water absorption. Slow or rapid water absorption has repercussions for all the processes associated with water movement through the embryo, and mitosis (cell division) increases in warm favourable conditions.

Oxygen

Oxygen, found in the soil atmosphere, is essential for the respiration of tissues within the germinating embryo, as the rapid growth experienced at germination involves high respiration rates.

The three major requirements for germination are found in well managed soils, natural woodland soils, and potting/seed-sowing media (composts), highlighting the importance of well drained soils so that moisture is available, yet not in excess so it precludes air. Some formulations of soil substitutes/seed-sowing media lack the friction of natural soils, so when epigeal subjects pass through the medium at germination, poor resistance results in the testa remaining on the end of the germinating seed, and still enclosing the cotyledons, after it breaks the surface. Tomato sometimes needs help by removing the testa by hand. However, in general terms, the reliably uniform nature of seed-sowing media, and the uniform germination environment provided, leads to more successful germination than in natural conditions.

Light

Some species are light sensitive. Some *Begonia* species, for example, germinate much more successfully if sown on the surface of their sowing medium. This is related to both the shallow depth of sowing needed for such a fine seed, and its light requirement.

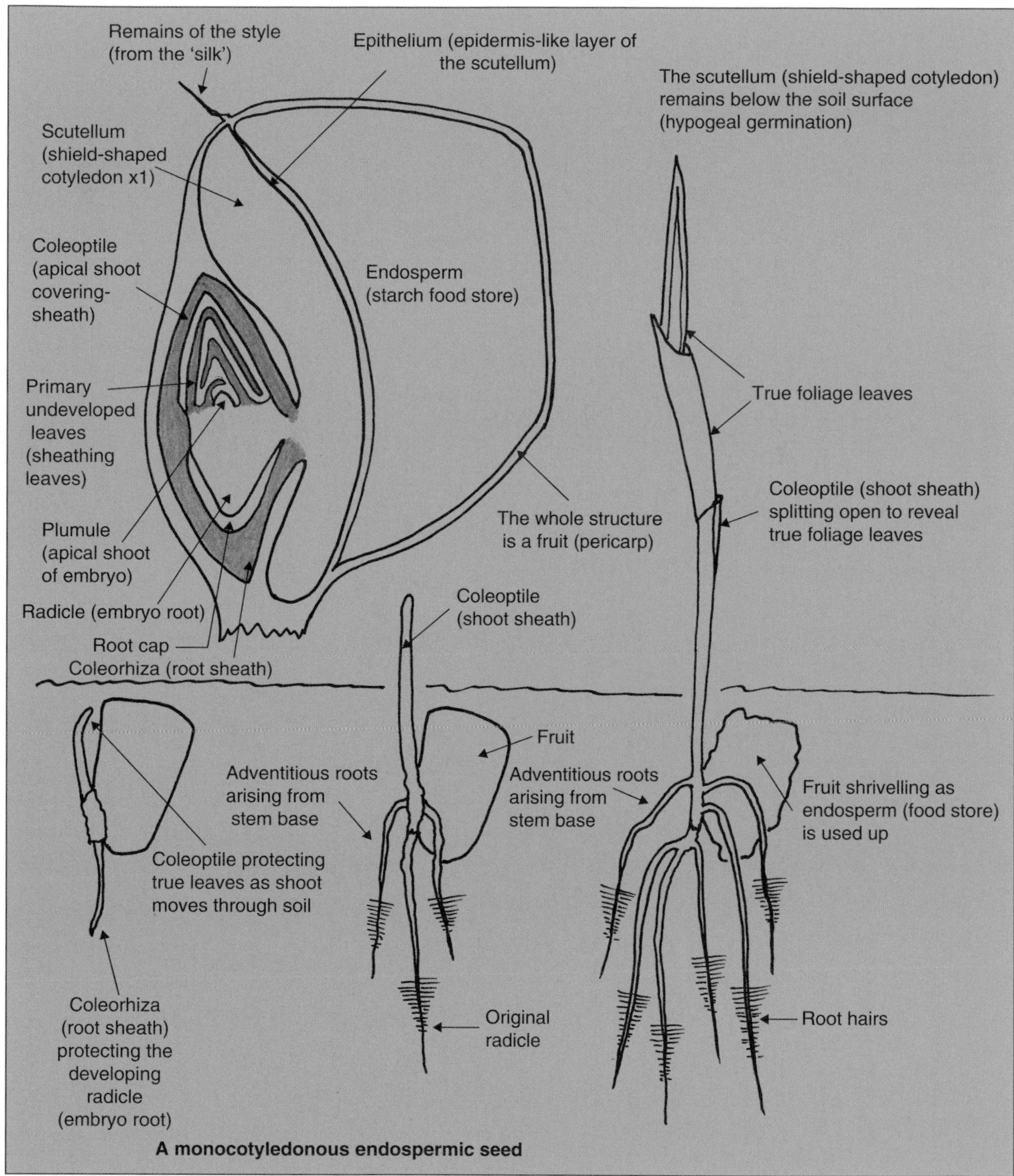
Remains of the style
(from the 'silk')
Epithelium (epidermis-like layer of the scutellum)
The scutellum (shield-shaped cotyledon) remains below the soil surface (hypogeal germination)
Scutellum (shield-shaped cotyledon x1)
Coleoptile (apical shoot covering-sheath)
Endosperm (starch food store)
Primary undeveloped leaves (sheathing leaves)
True foliage leaves
Coleoptile (shoot sheath) splitting open to reveal true foliage leaves
The whole structure is a fruit (pericarp)
Plumule (apical shoot of embryo)
Radicle (embryo root)
Coleoptile (shoot sheath)
Root cap
Coleorhiza (root sheath)
Fruit
Adventitious roots arising from stem base
Adventitious roots arising from stem base
Fruit shrivelling as endosperm (food store) is used up
Coleoptile protecting true leaves as shoot moves through soil
Original radicle
Root hairs
Coleorhiza (root sheath) protecting the developing radicle (embryo root)
A monocotyledonous endospermic seed

The seed/fruit structure and hypogeal germination of sweetcorn (Zea mais).

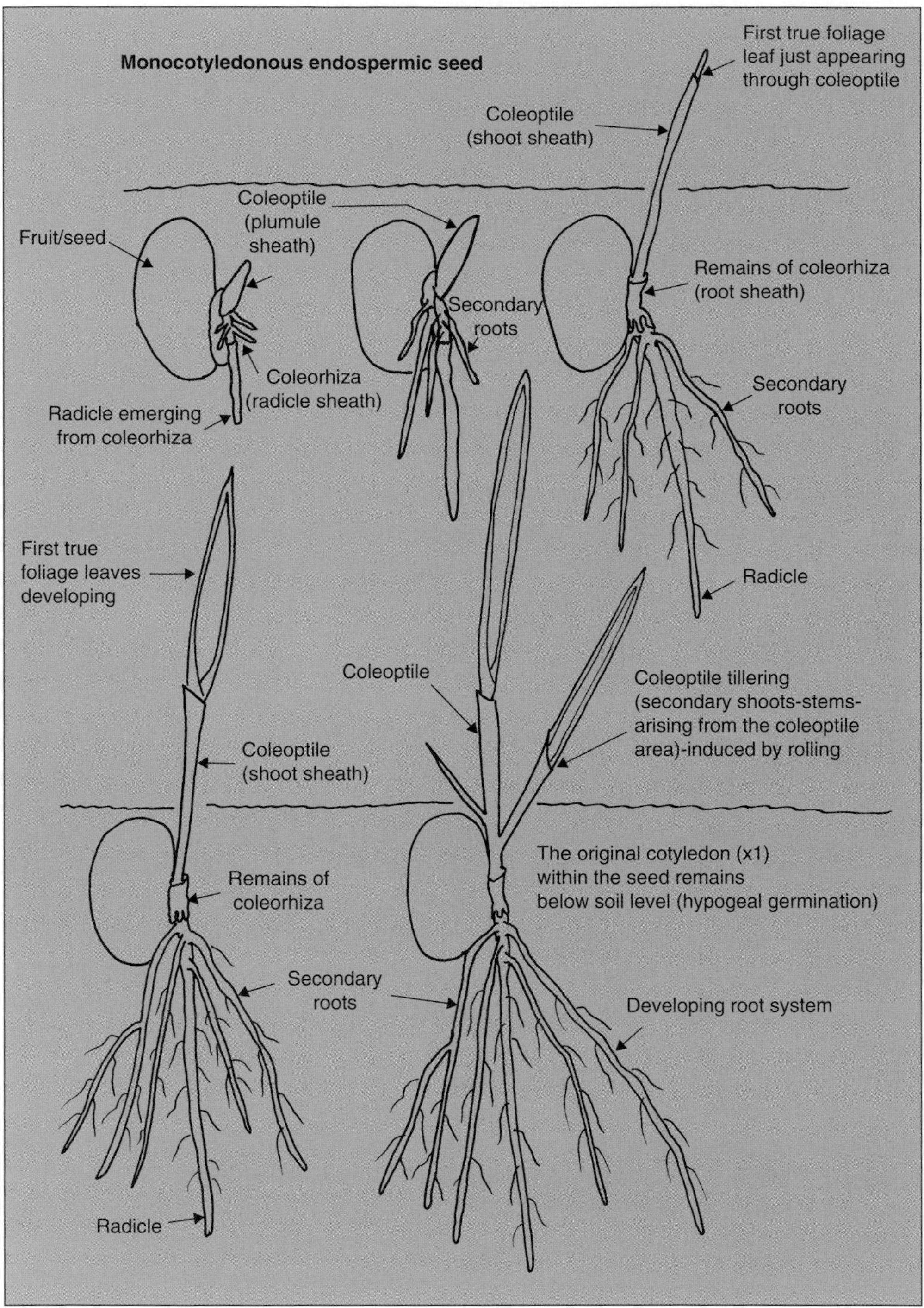

The hypogeal germination of cereals and grasses (for example, wheat).

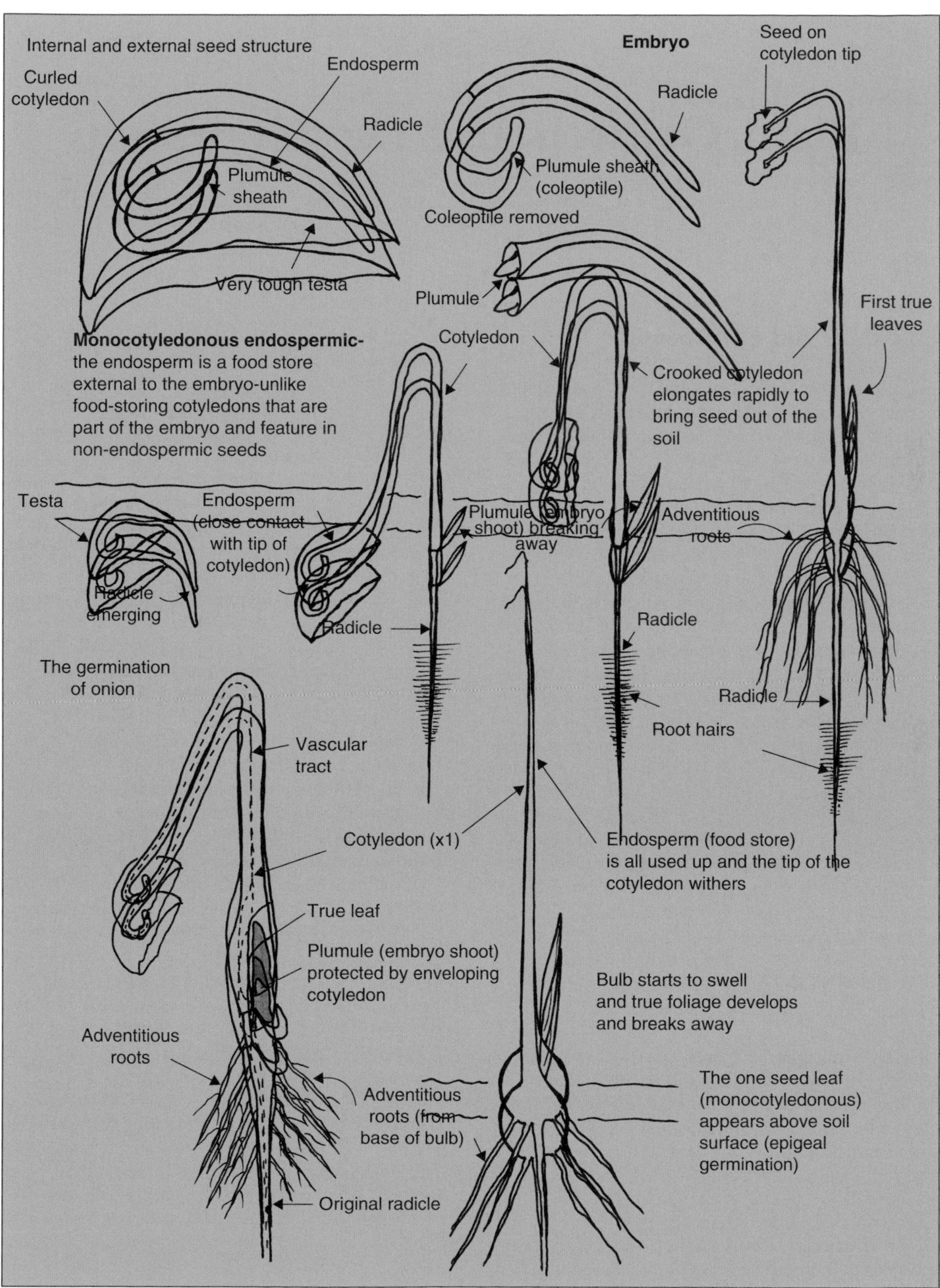

Internal and external seed structure
Curled cotyledon
Endosperm
Radicle
Plumule sheath
Very tough testa
Monocotyledonous endospermic- the endosperm is a food store external to the embryo-unlike food-storing cotyledons that are part of the embryo and feature in non-endospermic seeds
Embryo
Radicle
Plumule sheath (coleoptile)
Coleoptile removed
Plumule
Seed on cotyledon tip
Cotyledon
Crooked cotyledon elongates rapidly to bring seed out of the soil
First true leaves
Testa
Endosperm (close contact with tip of cotyledon)
Radicle emerging
Radicle
Plumule (embryo shoot) breaking away
Adventitious roots
Radicle
Root hairs
Radicle
The germination of onion
Vascular tract
Cotyledon (x1)
Endosperm (food store) is all used up and the tip of the cotyledon withers
True leaf
Plumule (embryo shoot) protected by enveloping cotyledon
Bulb starts to swell and true foliage develops and breaks away
Adventitious roots
Adventitious roots (from base of bulb)
The one seed leaf (monocotyledonous) appears above soil surface (epigeal germination)
Original radicle

The epigeal germination and development of onion (and other alliums, including many ornamental forms).

CHAPTER 5

Plant Growth and Development

Plant Cells and Cell Division

Plants comprise collections of cells of many different types, and held together in specific groups. Cells are initially soft-bodied (soft-walled) and their entire internal volume is full of viscous (treacle-like) cytoplasm, but many (though not all) cells go through drastic changes over time. Cells are produced in the first instance by rapid division at specialized areas known as meristems (meristematic areas), and the basis of all plant-growth patterns for stems, leaves and roots is laid down at the stem-tip and root-tip meristems. Groups of specialized cells are collectively known as tissues, and congregations of specific tissues, with specific functions, form specialized organs of the plant. The patterns of tissues involved in plant organs are fairly consistent, and so we consistently recognize stems, roots, leaves and flowers.

Plant cells can divide, produce various chemical/hormonal substances, use energy, produce energy, and accumulate/store various substances to be used in the future for metabolic processes. The whole of the living part of the cell (including the cytoplasm) is called the protoplasm. Cell walls are cemented together by a pectin-based substance at the middle lamella, and although the basic plant cell wall (secreted by the protoplasm) comprises cellulose, other materials can be deposited on its inside.

Cells in meristematic areas at root and shoot tips are regularly multiplied, and produce identical daughter cells. The system is continuous because some cells retain their malleability in order to divide again and again. Specialized meristems (cell 'factories') that produce sheets of brick-like daughter cells are called 'cambia' (singular 'cambium'), and along with the amorphous mass of shoot-tip and root-tip meristems, from whence they originated, are responsible for all new tissue patterns, and ultimately all plant growth.

All tissues commence initially as soft, pliable cells known as parenchyma cells, and once produced, take on a specific function depending on their place in the tissue or organ pattern. Freshly divided cells can then go through a three-fold process of maturation that involves many changes (elongation, vacuolation and differentiation), but new cells are continually produced by the meristems. During the specialization process (known as cell differentiation), cells change (sometimes drastically) to be perfectly adapted for their chosen function. As they differentiate they may become strengthened with extra cellulose, still remaining fairly soft, but losing the ability to divide (collenchyma or collenchymatous cells). Or, if they are in positions where particular strength is needed, they can add varying degrees of the wax lignin to their cell walls (lignification) and form various types of sclerenchyma (sclenchymatous) cells. Hence cells initially produced by division at meristems will differentiate during their maturation process, and the degree of differentiation varies depending on the ultimate role of the tissues.

The plant body is built up of somatic cells (plant body cells) that have differentiated to perform specific functions, and have congregated into tissues (groups of cells with similar functions). Seeds (plant embryos) comprise cells that have previously gone through a sexual process (sexual cells), but even though they have a different starting point, their subsequent divisions and processes are very similar to somatic cells (plant body cells).

It is both the addition of, and elongation of, specific cells that is responsible for plant growth in length. Cell elongation creates an increase in cell volume, which puts the cytoplasm under stress, forming open areas (or one large area) known as 'vacuoles' (vacuolation). Vacuoles contain aqueous solutions of dissolved sugars, nutrient salts and other metabolites (including water-soluble pigments) known collectively as 'cell sap'.

After vacuolation, the cytoplasm, instead of being an amorphous mass filling the increasing cell volume, forms a thin membranous layer on the inside of the cell wall and is known as the 'cytoplasmic membrane'. The cytoplasmic membrane is itself enveloped by two membranes: the tonoplast, which separates the cytoplasm from the vacuole; and the plasmalemma, which separates the cytoplasm from the cell wall. The tonoplast, therefore, forms a sac surrounding the vacuole, which is full of cell sap, and the two membranes have some control over the passage of substances from one cell to another. The tonoplast and plasmalemma cannot be seen under a light microscope.

Minute, pore-like gaps in the cell wall are called plasmodesmata, and they facilitate the movement of cytoplasm from one cell to another, a process known as 'cytoplasmic streaming'. If it were not for this ability of materials to move from one cell to another, individual cells would have no relationship with one another, and would merely lie alongside one another with no interaction, which would not support life at any level. Even the ability of cells to pass water from one to the other through the semi-permeable cell wall (by osmosis) is essential for living cells to become turgid (engorged with water) and react to their environment. Although strength is attained through lignified xylem cells (and sometimes fibres), the attitude (uprightness) of green plants (and green parts of plants) is upheld by cell turgor. Hence, reductions in cell turgor at times of water scarcity leads to flaccidity (wilting). Woody parts of plants are retained in their attitude because they comprise lots of lignified tissues (including fibres and a large volume of enveloping xylem tissues added to on an annual basis).

Because of the difficulties of visualizing minute cells and structures at cellular level, even with good access to photomicrographs and prepared microscope slides, there will always be conceptual problems. One of the main conceptual problems is picturing pores, pits and inclusions within cells that are minute themselves. Much of the information has to be taken on trust in the first instance, and then once we believe that the anatomy exists as explained, evidence presents itself for its existence. External features give 'clues' to, and sometimes verify the existence of, internal structures. Thus the cork-covered vascular traces seen in leaf scars do not show us the structure of vascular bundles, but they do, at least, indicate that the anatomy as described is correct. And the stress lines, cracks and sloughing of bark are evidence to the annual increase in girth created by the addition of secondary tissues internally.

Organelles

Organelles are cell inclusions that are suspended in the cytoplasm, and include the nuclei, mitochondria, endoplasmic-reticulum, dictyosomes, microtubules, microbodies and plastids (chloroplasts, leucoplasts, amyloplasts and chromoplasts). Organelles carry out metabolic and physiological processes at a biochemical level. The cell cytoplasm acts as a medium to support the organelles, and allows them to pass metabolytes from cell to cell by diffusion. Cunningly, however, water, cell sap, and more viscous fluids such as the cytoplasm, can move from cell to cell, but important cell inclusions (organelles), notably the nucleus, cannot pass through the very fine pores of the plasmodesmata (the pores in the cell wall). The important organelles, therefore, remain within their parent cell, thus preventing the genetic and metabolic chaos that would otherwise ensue.

Mitochondria are responsible for the processes of respiration. It is thought that endoplasmic reticula are involved in protein synthesis and the interchange of cytoplasm from cell to cell via the plasmodesmata. Golgi Apparatus is a collective term for the Dictyosomes in a cell, and it is thought that Dictyosomes synthesize the precursors of cell wall components. Microbodies accumulate enzymes of different types that have very important roles within the plant, amongst other things the responsibility for the digestion of starch (which is insoluble in water) and converting it back to sugars (that are soluble).

Plastids are organelles involved in pigmentation, and include chloroplasts (green pigments) that make up the chlorophyll found in leaves and green stems, and are important as a catalyst in photosynthesis (producing sugars); also leucoplasts (colourless pigments), and chromoplasts (yellow and red pigments). Chromoplasts include carotenoides (yellow and red pigments – carotenes and xanthophylls). It is thought that chromoplasts trap radiant light energy, at wavelengths corresponding to their colour spectrum, and pass it on to chlorophyll. Amyloplasts store starch.

Nuclei carry genetic information, and comprise a sac-like envelope called the 'nuclear envelope' containing a viscous liquid known as the 'nucleoplasm', and a dense body (or several dense bodies) known as a 'nucleolus' (nucleoli). Genetic information is stored (as DNA) on chromosomes, and is disseminated from the nucleus – information necessary for cell division (and all other functions of the cell).

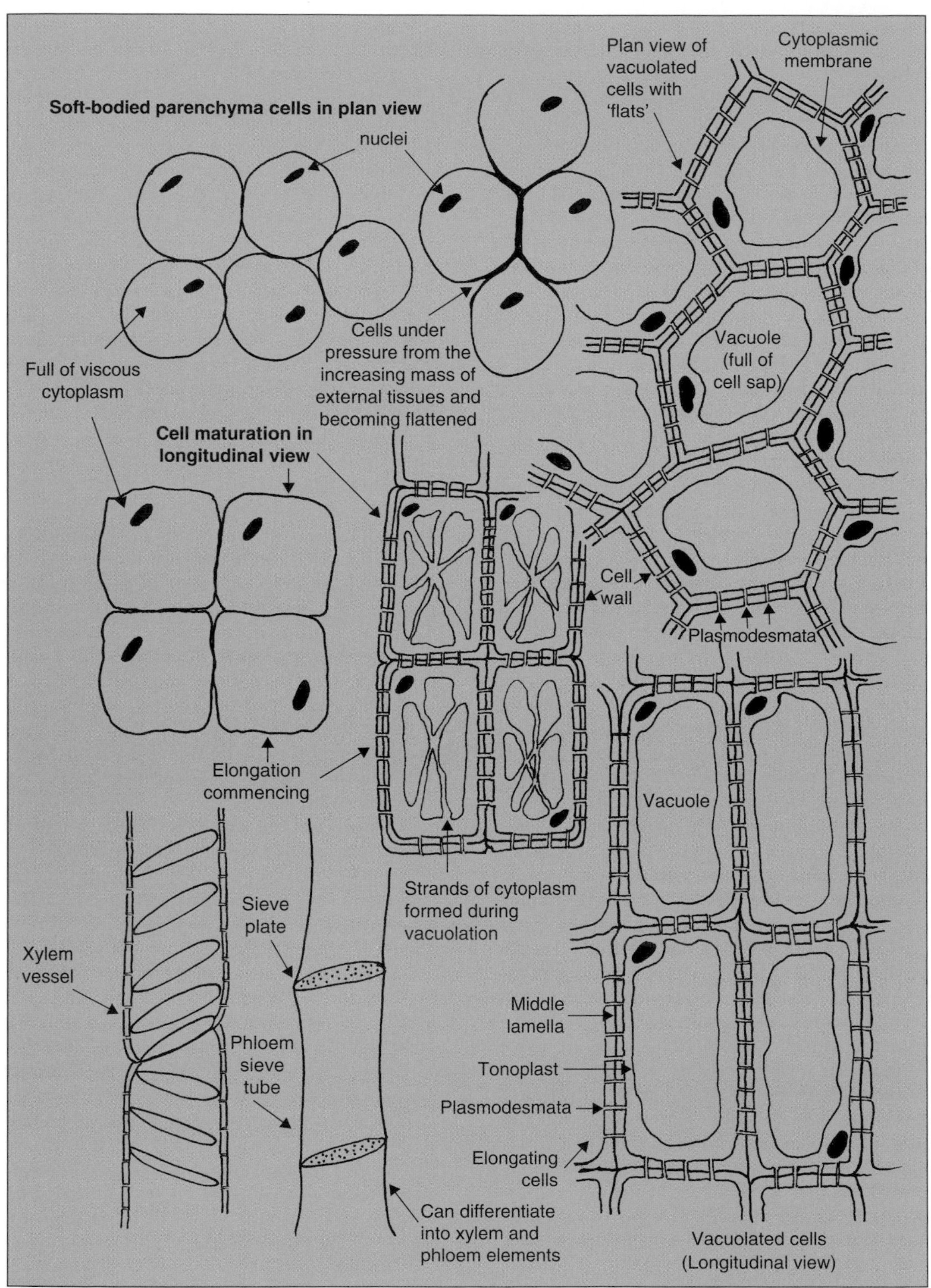

Cells and cell maturation.

Cell Division

The ability of cells to divide is absolutely fundamental and essential for the life and growth of plants. There are two separate and totally independent cellular systems, and both involve the division of parenchyamtous cells in meristematic regions. However, one system is responsible for the production of cells for the main plant body, called somatic (body) cells, and the other is responsible for the production of sexual cells and is only found in the male and female sexual organs. Division of somatic cells is known as mitosis (mitotic division), and division of sexual cells is known as meiosis (or reduction division).

Cell nuclei contain long, helical threads of DNA with different genes along their length in particular arrangements. Prior to cell division each DNA strand spiralizes with protein – these are known as chromosomes, because the thickened spirals, when stained, can be seen under a light microscpope as coloured bodies (chromo = coloured, and soma = body). Chromosomes with their genes attached are present in homologous pairs (pairs with the same structure and having the same gene number). Genes determine the characteristics of plants, and the gene make-up (genetic potential) of the plant is known as the 'genotype'.

The homologous pairs of chromosomes are attached to sphere-like areas known as centromeres to form chromatids, with one chromosome pair coming from the male side, and one pair from the female side.

Somatic (plant body) cells all have a set number of chromosomes known as the diploid number, and represented as 2n pairs of chromosomes. The asexual division of somatic cells produces daughter cells with exactly the same number of chromosomes as the mother plant. As somatic cells go through mitosis (mitotic division), each chromosome duplicates itself, and the DNA helix, which has also divided in two longitudinally, replicates itself to form a complete helix again after division, thus reinstating the genetic status quo. Mitosis occurs in all main meristematic areas of the plant, and is responsible for the production of all new cells that make up the plant body.

Mitosis

In mitotic cell division, the chromatids (spiralized pairs of homologous chromosomes) in the nucleus polarize within the cell, so that half of each chromatid (not half the number of chromatids) goes north and half of each chromatid goes south, bursting out of the nuclear envelope as they do so. The soft, cellulose cell wall of the parenchymatous somatic cells invaginates, and forms a cell plate, which eventually forms a new cell wall between the two polarized sets of divided chromatids. A new nuclear envelope encloses the separated chromatids, and each set of halved chromatids replicates to form entire chromatids with the original diploid (2n) number of chromosomes in each daughter cell.

Mitotic divisions of somatic cells comprise four dynamic, yet distinguishable phases, which, in chronological order, are prophase, metaphase, anaphase and telophase. The resting phase between divisions is known as the interphase. The phases can be readily seen under a light microscope, even though some details of the various structures involved are sub-microscopic. Root squashes, produced by macerating the tissues at a root-tip meristem, because they are relatively easy to prepare, are often used to gain cells showing the various phases of mitosis. Mitosis results in two daughter cells with identical genetic information in each cell, and typical of parenchymatous cells they can divide again, or differentiate to perform their specific function, depending where they are in the tissue configuration of the plant.

Polarity of Cell Division

During the division of the nucleus a mitotic spindle forms from spindle fibres at the equator, and the polarity of these influences the polarity of dividing cells. Cells may either be totally disorganized/disoriented (as in the production of callus), to be oriented at a later date during differentiation; or they may be anticlinal or periclinal. Anticlinal divisions are the most common in root-tip and shoot-tip meristems, and occur when the spindle fibres are parallel with the root-shoot axis, so are mainly responsible for growth extension in length. Periclinal divisions occur when the spindle fibres are at right angles to the root-shoot axis. These are more common in cambia, so along with radial divisions, are responsible for the addition of cells on both sides of the cambium, and the increase in girth that this creates, both in primary growth and later in the secondary growth of woody plants.

Meiosis

Sexual cells (in pollen and mother cells of ovules) go through a process known as meiosis when they divide. In the first phase of meiotic division each chromatid does not split into two halves as happens in mitosis, but instead, the number of

Diagrammatic representation of mitosis

Mitosis occurs at all meristems (including cambia) and produces somatic (plant body) cells. All the cells are genitically identical with $2n$ pairs of chromosomes.

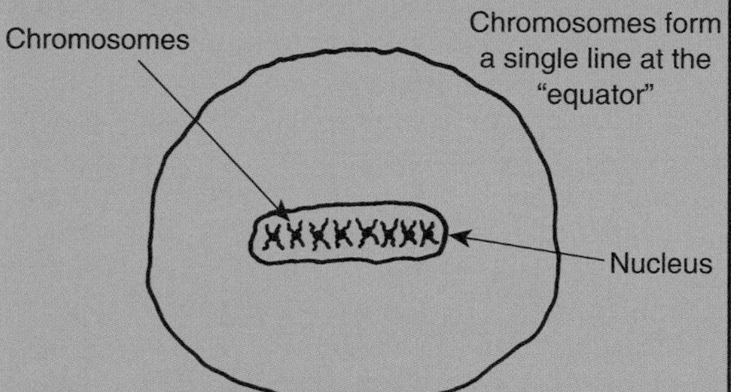

Chromosomes split in two with half of each going "North" and half going "south".

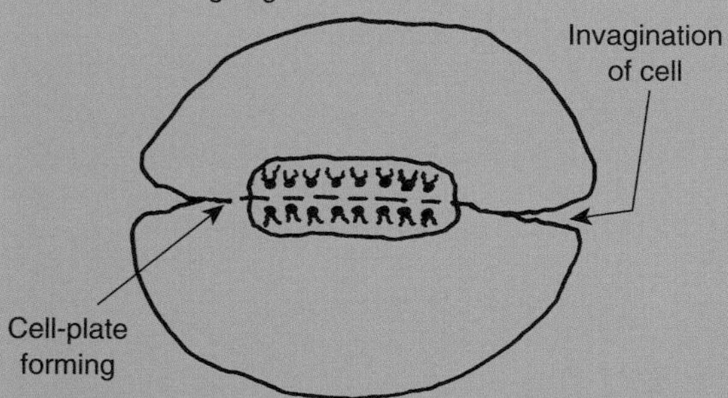

DNA replicates on each half chromosome reinstating both sets of chromosomes with correct (normal) chromosome number—diploid number.

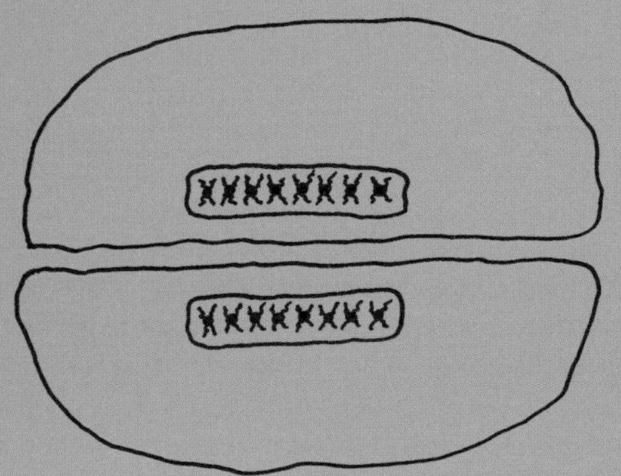

Two identical daughter cells formed with $2n$ chromosomes.

Diagrammatic representation of meiosis

Meiosis occurs in cells at the sexual organs-pollen (at anthers) and eggs (within the ovary). Meiosis involves two divisions. At the first, chromosomes form a double line at the "equator" and half go "south" and half migrate "north".

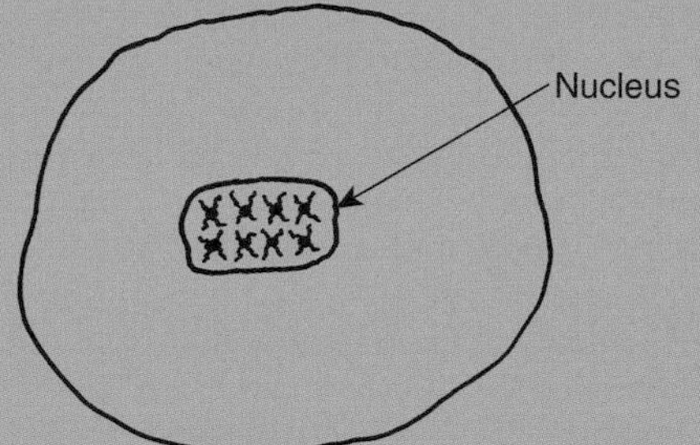

As it is half the number of chromosomes that migrate (not half of each chromosome) the chromosome number is halved within each cell.

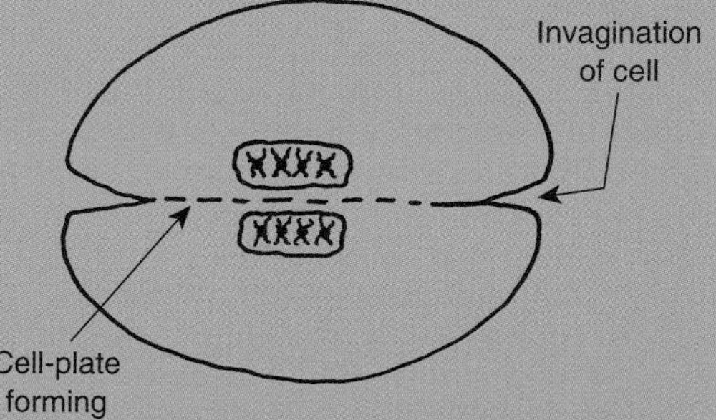

The second division is mitotic, producing four identical cells (each with half the chromosome number (n)).

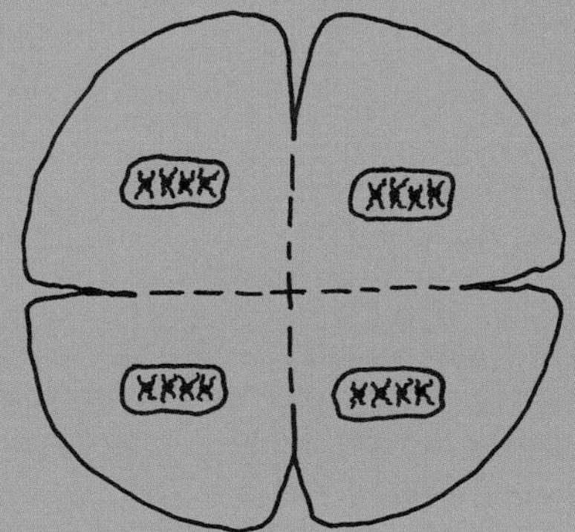

At the fusion of sexual cells (fertilization) male haploid and female haploid cells unite to reinstate the normal chromosome number in the embryo.

Cell division: mitosis and meiosis.

chromatids is divided in two – half going north and half going south. Thus two dissimilar daughter cells are created with half the normal number of pairs of chromosomes for the species (known as the haploid number and represented as n).

In the second phase a mitotic-style division occurs, which replicates the existing haploid daughter cells. Four daughter cells now exist, with half the chromosome number (haploid). This process occurs both in the male sexual parts (the pollen grains), and in the female sexual parts (the ovules).

At fertilization the male gametes (haploid cells in the pollen with half the chromosome number n) and the female gametes (haploid cells in the ovule also with only half the chromosome number n) fuse to form a zygote. Zygotes have cells with the normal chromosome count (diploid number) typical of the species. So at fertilization, the fusion of the male and female cells reinstates the diploid number (2n), and carries genetic information from both the male and female elements of the plant. The diploid cells produced will now go through mitosis to form identical somatic daughter cells, and create an embryo within the seed from cells carrying the correct chromosome number for the species. If the sexual cells did not commence with half the number of chromosomes, they would double the number of chromosomes typical for the species at each fusion, resulting in chaotic growth.

Genetics and Genetic Inheritance

We now know that specific characteristics of plants are carried on chromosomes as genes that act as an encoded pattern and determine heredity. Furthermore, we know that within the nucleus of plant cells is a helical structure (DNA) that has the ability to split longitudinally at cell nuclear division and divide into two equal halves. Moreover, after dividing in two, and separating to take one half into one cell and the other half into the other divided cell, each half of the helix has the ability to replicate itself, thus re-instating the missing complimentary half to form complete genetic information within each cell once more. This process is the basis of mitotic cell division, whereas the process of meiosis halves the chromosome number in both the pollen and the egg nucleus, and it is only reinstated to the full count at fertilization.

On the one hand, passing on genetic information at mitosis ensures the replication of species with constant features. On the other hand, provision for variation is also very important so that individual plants can be flexible and adaptable in the face of environmental/climatic changes (locally and regionally). The constancy of identical somatic (body) cells via mitosis is the norm. Although these cells can then differentiate to form vastly different cells able to perform specific functions, and environmental factors can affect the final outcome, tissue types, organ types, and complete plants do remain fairly constant in their form within species.

Provision for variation is by the sexual process (via meiosis), and minor variations occur all the time through this process. Plants with extreme variations that render the plant unsuitable to its present environment are culled naturally, and the unsuitable genetic information is therefore not passed on to the next generation. Minor variations that render the plant even more suited to its present or changing environment are upheld by its relative success and its ability to pass on the combination of genes that led to that success, and communities of that plant will thrive. The normal division of the sexual cells (meiosis) and gamete fusion (fertilization) is often sufficient to facilitate any changes and adaptations necessary for success. However, included in meiosis is the possibility of not only the normal recombination of genes from the male and female elements, but also the possibility of gene reassortment and other gene combinations.

The extra provisions for change are brought about by the movement of genes between chromatids and from one chromosome to another. The tails of the chromatids of homologous pairs of chromosomes are very intimate, and touch one another at various points along their length (known as a chiasma – plural = chiasmata). At chiasmata it is relatively common for portions of the tails of chromatids to break off and then fuse to the tails of other intimate chromatids, forming chromatids with different gene combinations. This process is known as 'crossing over', and it can cause various combinations and reassortments of genes along the chromatid tails, resulting in different outcomes for the resulting plants. Furthermore, the possibilities of crossing over and gene recombination complicate the already uncertain outcome of hybridization even further.

The pioneer of genetics, Gregor Mendel, carried out some very rigorous experiments to determine heredity. Without Mendel's initial work our level of knowledge and understanding of how genetic inheritance works would be very different. The use of microscopy, and further experimentation based on his work and initial concepts, has helped us to understand more.

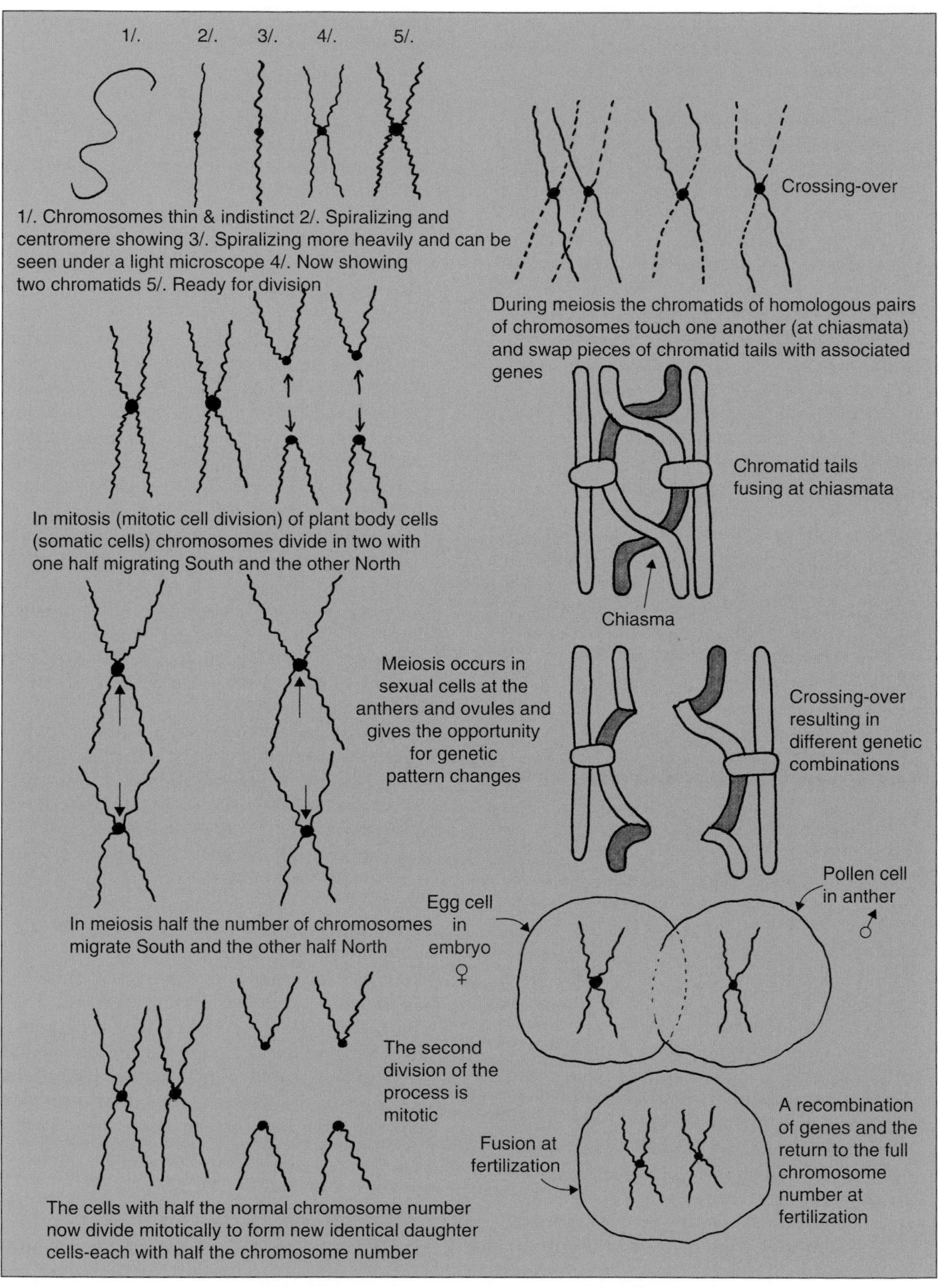
1/.
2/.
3/.
4/.
5/.
1/. Chromosomes thin & indistinct 2/. Spiralizing and centromere showing 3/. Spiralizing more heavily and can be seen under a light microscope 4/. Now showing two chromatids 5/. Ready for division
Crossing-over
During meiosis the chromatids of homologous pairs of chromosomes touch one another (at chiasmata) and swap pieces of chromatid tails with associated genes
In mitosis (mitotic cell division) of plant body cells (somatic cells) chromosomes divide in two with one half migrating South and the other North
Chromatid tails fusing at chiasmata
Chiasma
Meiosis occurs in sexual cells at the anthers and ovules and gives the opportunity for genetic pattern changes
Crossing-over resulting in different genetic combinations
In meiosis half the number of chromosomes migrate South and the other half North
Egg cell in embryo ♀
Pollen cell in anther ♂
The second division of the process is mitotic
Fusion at fertilization
A recombination of genes and the return to the full chromosome number at fertilization
The cells with half the normal chromosome number now divide mitotically to form new identical daughter cells-each with half the chromosome number

Crossing over and recombination.

The way that genetic information of plants is carried from generation to generation is via the zygote (the seed) created by the fusion of male and female gametes at fertilization. Most plants cross-pollinate (and therefore cross-fertilize) for success. Yet some species self-pollinate (and therefore self-fertilize) as the norm. Garden pea (*Pisum sativum*) is one such plant that has specific forms that successfully self-pollinate, as pollen is normally ready and dehisces before the corolla opens, avoiding pollution by alien pollen (and therefore breeds true – apomictic). This was one of the reasons that Mendel chose garden pea as a research tool for his experiments. The other reason is that types of garden pea have specific characteristics that are so defined that they can be traced through the process. Flower colour, ultimate height and certain seed types were chosen, and the appearance (or not) of these traits was tracked through the system from one generation to another.

Hybridization may be carried out by emasculating (removing the stamen) of the pollen donor (before the corolla has opened) and transferring the pollen to the stigma of the egg donor (by removal of the corolla before it opens naturally). But a muslin bag has to be placed over the pollinated egg donor flowers to ensure there is no interference with the fertilization process by alien pollen from air currents or visiting insects.

When strains of tall plants were crossed (hybridized) with strains of dwarf plants within the species, races of intermediate plants were not the result: there was no dilution or merging of the two characteristics. Instead, all the plants resulting from seeds harvested from the cross (the first filial generation, or F1) were in fact tall. Furthermore, it mattered not which type (tall or dwarf) was the pollen donor (male) and which the egg donor (female): the result was the same. The plants bred true for tallness – tallness is therefore considered to be a dominant trait, and dwarfness a recessive trait dominated by the 'tall' genes.

In Mendel's experiment, all the tall progeny resulting from the first cross were allowed to self-fertilize as normal, and the seeds harvested and sown – and this time the resultant plants were not uniform. In fact there were both tall and dwarf types, but with more tall than dwarf plants at a ratio of three tall to one dwarf. Even if, instead of letting the tall plants from the first cross self-pollinate naturally, they had been hybridized (by emasculation and bagging), then the plants resulting from the harvested seeds (now the second filial generation, or F2) would have been the same. The dominant characteristics to recessive characteristics would come out at a 3:1 ratio. Hence, crosses between smooth-seeded and wrinkled-seeded types result in a 3:1 ratio of smooth to wrinkled in the F2 generation (wrinkled is recessive), and F2 crosses between purple- and white-flowered types result in 3:1 purple (as purple is dominant).

The characteristics, both dominant and recessive, are carried over by genes, and the process is best illustrated by using symbols. If T is used to represent 'tall' (dominant), and t to represent 'dwarf' (recessive), then the F1 cross gives T × t = Tt, which always shows as the tall (dominant) phenotype. Because genes are held in pairs, then TT or tt are both possible, as TT is found in self-pollinated tall plants, and tt in self-pollinated dwarf types. However, at fertilization between two plants only half the chromosome number (and associated genes) are found within each gamete, and only when the two gametes come together at fertilization to form the zygote is the number restored. Hence, crossing tall and dwarf plants gives Tt i.e. all tall. TT, Tt, tT and tt are pairs of genes known as alleles, and the male gamete carries one of the alleles and the female gamete the other. Furthermore, TT and tt are known as homozygous alleles (two alike in the zygote), and Tt or tT are known as heterozygous alleles (two dissimilar in the zygote). In the F2 cross the following combinations of alleles are possible: TT, Tt , tT or tt, and probably in equal amounts. Hence the first three types appear as tall phenotypes, and the last one (tt) is the only type with recessive genes only, so appears as dwarf (3:1, tall : dwarf ratio).

Therefore seeds offered for sale (including commercially) as progeny from self-pollinated types will come true to type (for their main traits), and results will be as expected. Seeds resulting from F1 hybrids will remain consistent to their dominant main features, as variations do not exist until the second generation. F1 seeds of common half-hardy annuals and glasshouse perennials – including French marigolds, African marigolds, zonal pelargonium and trailing pelargoniums – are commonly offered for sale by seed producers. In these cases seeds have to be produced each time by repeating the initial F1 cross of the two parents. So seed to produce stock plants of the original parents has to be held and grown each year in order to recreate the cross and harvest seeds from the egg donor plant on an annual basis. There is therefore an ongoing cost involved in the production of F1 seeds, and they can be quite expensive to buy.

You will also notice that when comparing like for like (species for species), F2 seeds are usually cheaper than F1: this is because they can be self-pollinated, thus relieving the need to produce both parents each year. F2 seeds will, of course, have some variations within the progeny. However, some types will give acceptable levels of variation – sometimes acceptable because of the lower cost.

Plant Breeding

In many instances the pollen from the stamen of a flower is not compatible with the stigma on the same flower and they cannot self-pollinate successfully. However, some species with hermaphrodite flowers can self-pollinate, and inducing this on a true species can be used to produce progeny that comes 'true to-type' and will have all of the features of the mother plant (apomixis – not mixing, not cross-fertilizing). Taking pollen from a male mono-sexed flower and placing it on to the stigma of a female mono-sexed flower on the same plant of a monoecious species will have the same effect (as long as alien pollen is precluded). Hence, 'selfing' is sometimes an option to produce young plants of true species that will have all of the features expected of that species.

Specialist hybridizers often set out to 'create' plants with specific aesthetic and/or commercial features, sometimes with great difficulty, and sometimes without success at all. Their methods revolve around ensuring the environment for successful pollination and fertilization between specially chosen subjects. Sometimes plant breeders trying to 'create' plants within their chosen and favoured genus have a fairly good knowledge of the potential gene pool – but even so, predicting the outcome is very difficult. In many cases the process may exaggerate or modify existing known features of the plant group, and on some occasions genetic information, that until now has remained latent (recessive), is brought back to prominence during the process. Genetic information may be released to show outwardly for the first time by the effect of different combinations of the gene pool that are achieved during the sexual process, and can produce novel plant material that may or may not have commercial importance.

Producing new hybrids involves two parent plants, and it must be evident that hybridizing a species with a plant that is already a hybrid, or two plants that are both themselves hybrids, leads to very complex possible gene combinations. Some rose hybrids, for example, may have the influence of many parents, and the traits of some may remain latent (recessive), or may be unlocked by a double recessive gene. The creation of new, innovative hybrids of woody perennial plants is particularly complicated because of the relatively long time-scale some take before becoming sexually mature and producing flowers for fertilization. The process of hybridization to create new plants, and certainly any attempt at prediction of the likely outcome, is further complicated by the inter-change of genes that can occur via reassortment, recombination and crossover at chiasmata.

Natural pollination via visiting insects or wind-borne pollen is always possible at critical points in the plant breeding process. In the natural world this may or may not be a benefit. In the world of induced hybridization (or induced self-pollination) this is 'pollution' that will almost certainly be detrimental, and will not achieve the result intended. To avoid promiscuity (the problems of alien pollen pollution), botanic garden staff and plant breeders 'bag' the chosen flowers with muslin bags. Muslin is used as it prevents the ingress of both pollen grains and insects but still allows a flow of air. The timing for 'bagging' is critical, and in the case of induced 'self-pollinating', involves tying the bag securely around the stem of the chosen flower before the flower begins to open to ensure no alien pollen affects the outcome. Those that 'self' naturally will do so within the bag. Those that need 'help' may need pollen from other flowers on the same plant, in which case pollen is harvested before the flowers open (or after bagging) by removing the corolla. Harvested pollen is then placed on the stigma of other bagged flowers, or just before the petals open naturally, and the pollen transfer is carried out swiftly and the flower quickly bagged (or re-bagged).

In the case of attempts to produce new hybrids, both the flower chosen to be the pollen donor (the male donor) and the flower chosen to be the pollen recipient (the female recipient) are 'bagged' before they open. Pollination of flowers is carried out by first emasculating the pollen donor – removing the stamen when pollen begins to flow (or removing the pollen with a brush or feather), and placing the pollen on to the receptive stigma of the female recipient. The female recipient flower (the egg donor) is then re-bagged to prevent the ingress of alien pollen. Bagged flowers are then monitored; once the fruit starts to form they can have the bag removed, as pollen (from any source) is no longer a threat to the result.

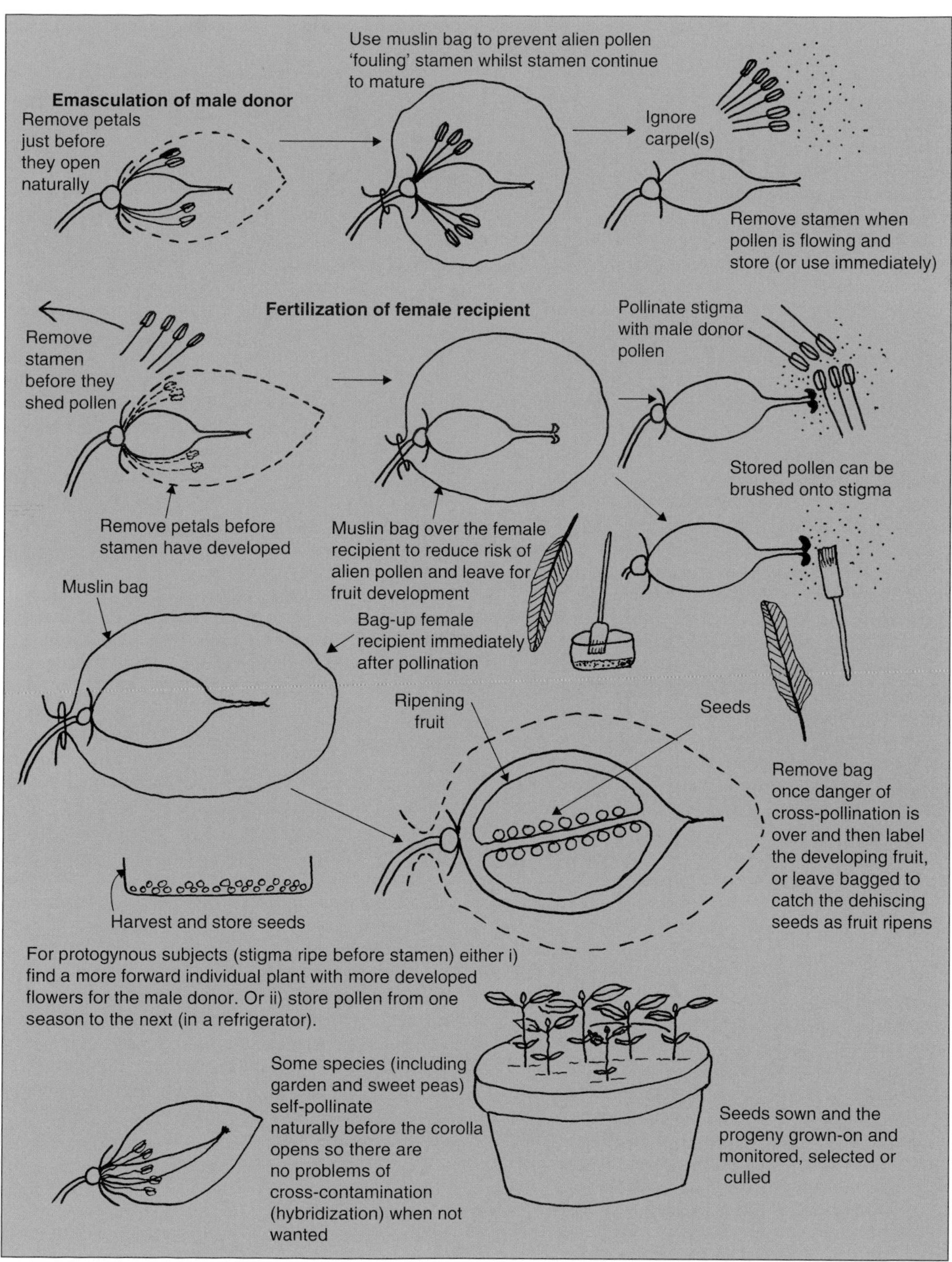

Use muslin bag to prevent alien pollen 'fouling' stamen whilst stamen continue to mature
Emasculation of male donor
Remove petals just before they open naturally
Ignore carpel(s)
Remove stamen when pollen is flowing and store (or use immediately)
Fertilization of female recipient
Remove stamen before they shed pollen
Pollinate stigma with male donor pollen
Stored pollen can be brushed onto stigma
Remove petals before stamen have developed
Muslin bag over the female recipient to reduce risk of alien pollen and leave for fruit development
Muslin bag
Bag-up female recipient immediately after pollination
Ripening fruit
Seeds
Remove bag once danger of cross-pollination is over and then label the developing fruit, or leave bagged to catch the dehiscing seeds as fruit ripens
Harvest and store seeds
For protogynous subjects (stigma ripe before stamen) either i) find a more forward individual plant with more developed flowers for the male donor. Or ii) store pollen from one season to the next (in a refrigerator).
Some species (including garden and sweet peas) self-pollinate naturally before the corolla opens so there are no problems of cross-contamination (hybridization) when not wanted
Seeds sown and the progeny grown-on and monitored, selected or culled

Plant breeding: emasculation, pollination and bagging.

Ostensibly, the process is as simple as taking the pollen from the stamen of a male donor, and placing it on the stigma of the female organ of a female recipient. In practice it is not quite so easy, as in some species the pollen from the flower of the donor plant is not compatible with the stigma of the intended recipient. Lack of compatibility between any two intended species is a major problem for plant breeders. Non-compatibility caused by inherent differences, including the exine shape (outer skin shape) of the pollen grains, and chemical (hormonal) rejection of pollen, usually means that hybridization cannot be carried out. However, non-compatibility due to the disparity of maturity times of sexual organs is solvable.

Disparity in maturity of sexual organs is typified by ripe anthers releasing pollen, but the unripe stigma not exuding the sticky exudate that is essential as a precursor to successful pollination. However, although less common, it can also be true the other way round, where the stigma matures before the stamen. In either case the problem can be rectified by storing pollen until the female recipient is ready to receive, which is obviously easier in the case of pollen being ready before the stigma (protandry), and not so easy if the stigma is ready before the pollen (protogyny). In the latter case, the best method is to take pollen from a more advanced individual plant, directly to the mature stigma of the less advanced recipient. If pollen viability is retained throughout by correct storage, and the stigma is ready to receive, then successful pollination, and fertilization, should be the result.

The difficulties do not cease at this point, as any progeny resulting from the process (germinated seedlings) cannot be screened for aesthetics, fruiting or commercial potential – or any other parameters – with any objective accuracy because of their immaturity. However, it is often the case that decisions have to be made at this juncture because of the financial implications of growing thousands of seedlings on to maturity. Many potentially commercial plants must have been culled over the years during this process! Plants are monitored and assessed for shape, habit, flower and/or fruit, and their relative unusualness compared with existing types. Screening of perennial plants is a long process and involves growing them on to sexual maturity, which may be within only two years for herbaceous perennials, but may be five, ten, fifteen or twenty years for some woody perennials – so may be a very expensive business.

Although the terminology has fallen out of use, the resultant progeny of any cross are known as 'grexes'. Thus, unnamed grexes can be numbered until their relative good or bad points are determined, and only then can each individual plant be given a cultivar name. But as with any system it can be abused, so holding several grexes, even several generations of grexes, from the same cross and releasing them as a definitive cultivar, is not acceptable. This is because each grex is technically unique, and although it may bear very similar characteristics to another grex of the same parents, because it is a sexual process it will not be the same clone. Thus very accurate records (stud books), and a process of registering and only releasing the correct plant under any cultivar name, is essential in order to uphold the integrity of the system. *Eucryphia* × *nymansensis* had several grexes labelled Nyeman's 'A', Nyeman's 'B' and so on, but the clone selected – the grex Nyeman's 'A' – to produce as a cultivar was named *Eucryphia* × *nymansensis* 'Nymansay'.

Plant breeding is carried out by professionals on a large scale, but also by many skilled amateurs on a small scale, with some excellent results and with relatively little financial input. However, the need to recoup the outlay of commercial plant-breeding ventures has led to very aggressive marketing systems and even copyrighting new plants. It has also led to the increasing use of micropropagation techniques to quickly bulk up stock in readiness for sale on the back of strong marketing campaigns. Once a new perennial plant cultivar is selected it has to be propagated asexually from then on to retain its clonal status and avoid the unacceptable variations associated with seed production. Conventional methods of propagation are relatively slow when compared to the tens of thousands of plants that may be produced in a year using micropropagation techniques. However, although an increasing number of plants are propagated in this way, even now, not all species and cultivars lend themselves to this method, and some still have to be propagated by more traditional systems.

---

*For more information on plant breeding, see *Breeding New Plants and Flowers* by Charles W. Welch (Crowood Press)

Types of Cells and Tissues

Plants comprise cells organized into tissues (groups of similar cells), and the tissues, in turn, are organized into groups (of dissimilar types) to form organs (stems, roots, leaves and flowers).

The ultimate size and role of cells is predetermined by the genetic information within them. However, the final outcome, size and function of cells are all influenced by the position of the cells within the tissue configuration, the available space, and the stresses (tension and compression) set up within the tissues and their immediate environment (including the chemical/hormonal environment). Furthermore, gravity, water content (turgidity or flaccidity) and light (day length and light intensity, for those that are illuminated) may all influence the outcome. Even those cells that are not influenced themselves by light may have chemical triggers passed to them from those cells that are. So there is a very complex set of possible parameters that may ultimately affect the final cell design, function and type. Nevertheless, the cell types and tissue patterns involved in specific organs are fairly consistent.

There are three main types of somatic (plant body) cells, each of which may be classified further, depending on their specific function in the plant: cells with soft cellulose cell walls, cytoplasm and nuclei (parenchyma cells), cells with extra cellulose in their walls (collenchyma cells), and those that are heavily lignified with a thick layer of the wax lignin on their inside wall (sclerenchyma cells).

There are three very important waxes involved in plant structure: lignin, cutin and suberin. They have various roles in the plant, mainly strengthening, protection and waterproofing. Cutin mostly appears on the outside of the tissues of leaves and stems. Suberin is present on the outside of bud scales, the leathery coating (testa) of seeds, and on the cork of bark, and, like cutin, has a protective and waterproofing role. All three waxes are very difficult to break down by bacteria and other organisms. Many cells go through phases of cell wall thickening as part of their differentiation, and lignin is the usual product deposited as a secondary material on the inside of cell walls to give increased structural strength. Unfortunately, the greater rigidity (less flexibility) created by waxes being deposited inside cells prevents cell division – so once the drastic changes associated with cell differentiation occur, cell division is no longer possible. However, some cells do remain soft and pliable throughout their life, and remain meristematic (they retain the ability to divide).

The varying degrees of lignification can be reflected in the ultimate function of the cells: for example, non-lignified parenchyma cells in meristematic areas, collenchyma cells in some stems and petioles for a degree of extra strength, and specialized sclerenchyma cells for both the transport of fluids and mechanical strength. Sclerenchymatous tissues are heavily lignified, and although the classification covers any group of heavily lignified cells, they may also have a specific name depending on their function: for example fibres, xylem vessels (a form of conducting tissue), and sclereids (that form the hard shell of nuts).

Tissues are designed to perform specific functions, including division, protection and transporting fluids (vascular tissues). Groups of parenchyma cells are known as parenchymatous tissues, and these have a specific name depending on their position in the plant; this could include the cortex (packing tissue), or the epidermis (outer skin) of a stem, root or leaf, all of which are classed as protective tissues. Cambia, the guard cells on leaf epidermi, and the mesophytic tissues of leaves are also parenchymatous. Cambia function by the rapid division of brick-like parenchymatous cells, and because most types of parenchyma tissue remains meristematic, it can replace cells by cell division.

The tissues of green plants comprise primary growth patterns, and many plants complete their life cycle only ever having primary growth. However, woody perennials produce and retain primary growth patterns at their shoot tips that increase stem length, but also add secondary tissue patterns (secondary growth), the most recent of which overlays both the original primary growth in its original position and any previous secondary growth. The effect of adding secondary tissues on an annual basis is to increase stem girth.

Vascular Tissues

In the early phases of stem development the vascular tissues (veins) are arranged in bundles, and in plants with a short life span this is their final pattern also. Woody plants commence with bundles, and as the plant develops, new tissues form within the bundles and new bundles are added, all of which eventually coalesce, and the vascular tissues ultimately form a complete ring. Sometimes cells are named ahead of their ultimate predetermined role: for example, xylem parenchyma (destined to be xylem when it differentiates). Vascular bundles comprise groups of phloem and xylem cells (and

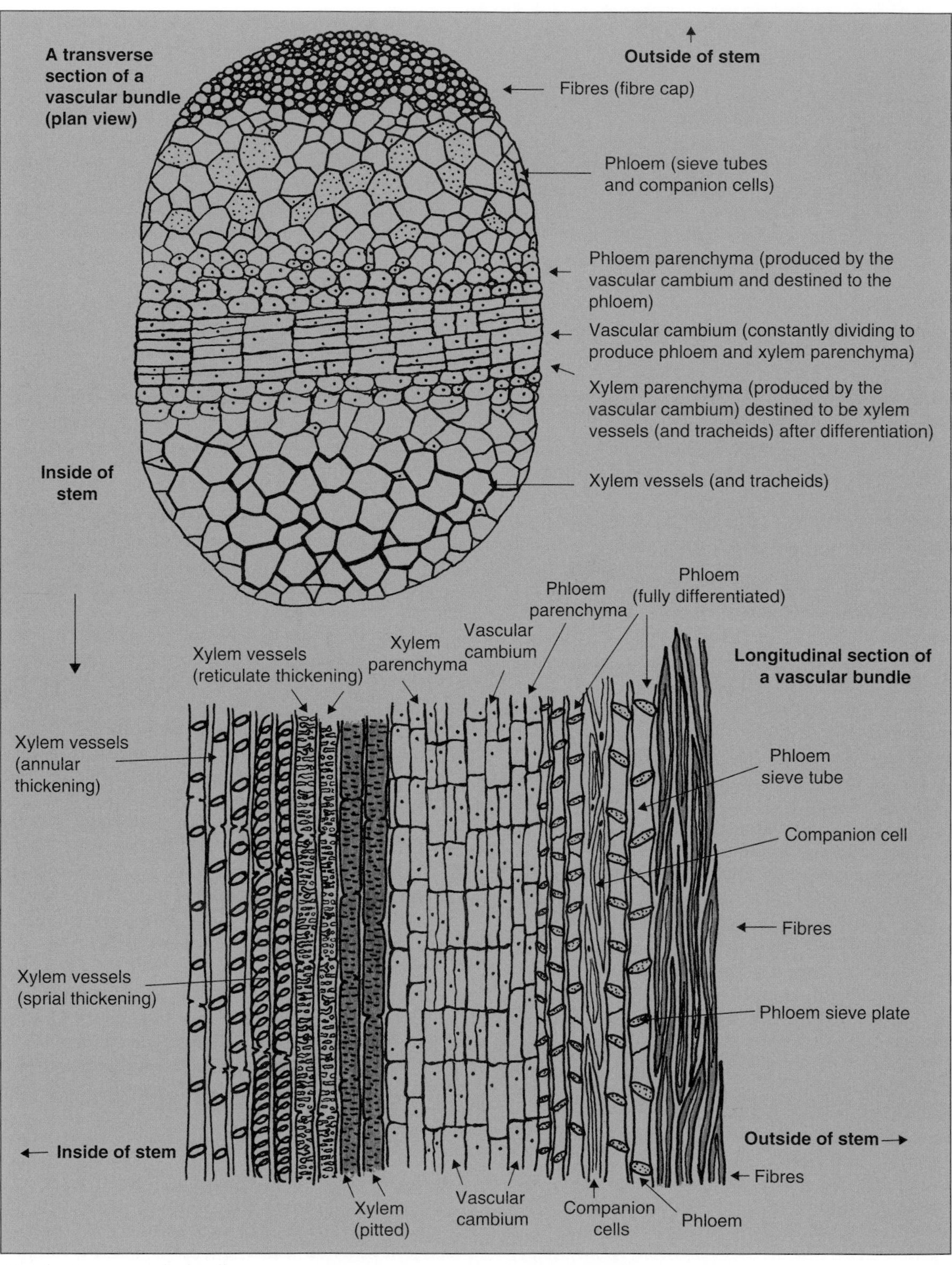
A transverse section of a vascular bundle (plan view)
Outside of stem
Fibres (fibre cap)
Phloem (sieve tubes and companion cells)
Phloem parenchyma (produced by the vascular cambium and destined to the phloem)
Vascular cambium (constantly dividing to produce phloem and xylem parenchyma)
Xylem parenchyma (produced by the vascular cambium) destined to be xylem vessels (and tracheids) after differentiation)
Inside of stem
Xylem vessels (and tracheids)
Phloem (fully differentiated)
Phloem parenchyma
Vascular cambium
Xylem parenchyma
Xylem vessels (reticulate thickening)
Longitudinal section of a vascular bundle
Xylem vessels (annular thickening)
Phloem sieve tube
Companion cell
Fibres
Xylem vessels (sprial thickening)
Phloem sieve plate
Inside of stem
Outside of stem
Fibres
Xylem (pitted)
Vascular cambium
Companion cells
Phloem

Vascular tissues: vascular bundles.

associated non-conducting fibres) sandwiching a vascular cambium. The vascular cambium is a very active meristematic region responsible for producing new xylem parenchyma on the inside, and new phloem parenchyma on the outside – tissues that are destined to become fully functioning xylem and phloem after differentiation.

Xylem vessels facilitate the transport of water (originating from the soil) to all parts of the plant in order to keep cells turgid (engorged with water), and also transport dissolved nutrient salts and other metabolytes suspended in the water essential for healthy plant growth. Xylem vessels form the veins of plants and are open, fluid-conducting, cylinder-like structures (with no end walls, as they disintegrate); they have heavy lignification (with various patterns of lignified thickening – for example annular, spiral, reticulate and pitted) that aids structural strength. Tracheids (sometimes found in other plants mixed with xylem vessels, but most commonly found as the main conducting tissues of conifers) have a similar function and structure to xylem vessels, but have sloping end walls where the cells meet, and set up a hydraulic pressure not found in xylem vessels. Xylem elements are often considered analogous to a plumbing system, and in some ways (regarding conducting water) this is true. However, xylem vessels do not comprise sealed copper-pipe-like tubes, but are instead more akin to brandy snaps in their structure, offering fairly easy lateral leakage from the system. Because of the water containment and transport of xylem elements, but also due to the possibility of sideways movement out of the vessels, a drainage system in reverse is probably a more accurate analogy.

Phloem tissue comprises tube-like cells known as 'sieve tubes', with only partially disintegrated end cell walls (known as 'sieve plates'). Sieve tubes are responsible for the 'downward' movement of sugars (produced by photosynthesis). 'Downward' is not an accurate term, as sometimes the phloem tissues are horizontal, or may even be oriented upwards; however, it is the description usually used, because it describes the overall downward progression of sugars from the leaves towards the roots. Sieve plates slow the downward movement of sugars so that the energy they supply can be used at various sites on the 'downward' journey. Unused sugars are stored by condensing to starches, which are insoluble, and therefore cannot move through the system. Storage of starches occurs in cortical tissues of the stem, and in particular in the cortical tissues of the roots. Starch acts as an energy source for the future, and can be released by enzyme action digesting the starch back to soluble sugars that may be transported in solution throughout the plant to wherever they are needed – particularly the meristematic areas, as they are high-energy users. Integrated with the sieve-tube elements are live parenchymatous cells known as 'companion cells'.

Phloem tissue is differentiated but not lignified; instead, sieve-tube elements may have extra cellulose thickening, and the associated companion cells are parenchymatous anyway. The soft, slippery, slimy effect that is experienced on a freshly cut log after debarking is because the freshly exposed phloem tissue cells are cellulose-based and full of sugars. The lack of lignification of phloem tissues is sometimes compensated for by being closely associated with long fibres (in close bundles) for added mechanical strength yet continued flexibility. Fibres are relatively long when compared with other cell types, and are overlapping sclerenchymatous cells that have thick, heavily lignified cell walls surrounding a hollow centre (lumen). Fibres have strengthening, not conductive roles, and are often found in stems, leaves and leaf petioles accompanying the other tissues, and may be associated with both phloem and xylem tissues for added strength. Xylem vessels and fibres (both sclerenchymatous tissues) are not unlike fibreglass rods (or reinforcing rods) running longitudinally through the tissues, giving much greater strength to plant organs, but allowing some flexibility. The presence of fibres allows wilting stems to 'bow' but not 'crease' or 'cripple', which would otherwise cause irreversible soft tissue damage.

Collenchymatous cells, with their extra layer of cellulose, can also help support leaf petioles. Rods of collenchymatous tissues found in the corners of the square stems of sunflowers allow them, with primary tissues only (including lignified primary xylem tissues), to attain such great heights in one season, without falling over.

The Tissues of Stems, Roots and Leaves

Stem Tissues

Green shoot tips produce primary growth. At a green shoot apex the once amorphous mass forming the apical meristem begins to divide, at first showing hardly any difference to the surrounding tissues, but later showing very different and distinct tissue types as they differentiate over time.

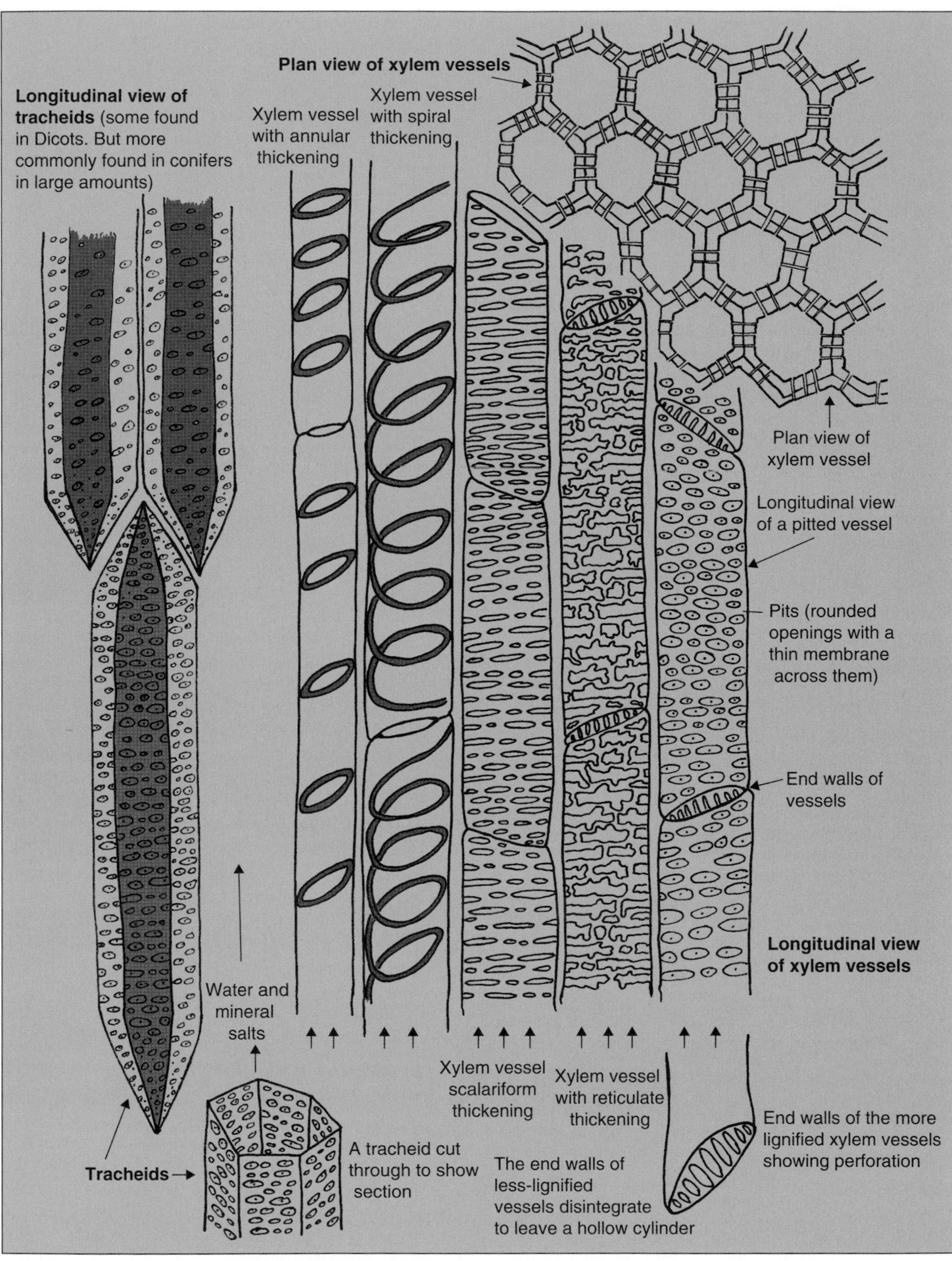
Plan view of xylem vessels
Longitudinal view of tracheids (some found in Dicots. But more commonly found in conifers in large amounts)
Xylem vessel with annular thickening
Xylem vessel with spiral thickening
Plan view of xylem vessel
Longitudinal view of a pitted vessel
Pits (rounded openings with a thin membrane across them)
End walls of vessels
Longitudinal view of xylem vessels
Water and mineral salts
Xylem vessel scalariform thickening
Xylem vessel with reticulate thickening
End walls of the more lignified xylem vessels showing perforation
Tracheids
A tracheid cut through to show section
The end walls of less-lignified vessels disintegrate to leave a hollow cylinder

Vascular tissues: xylem vessels and tracheids.

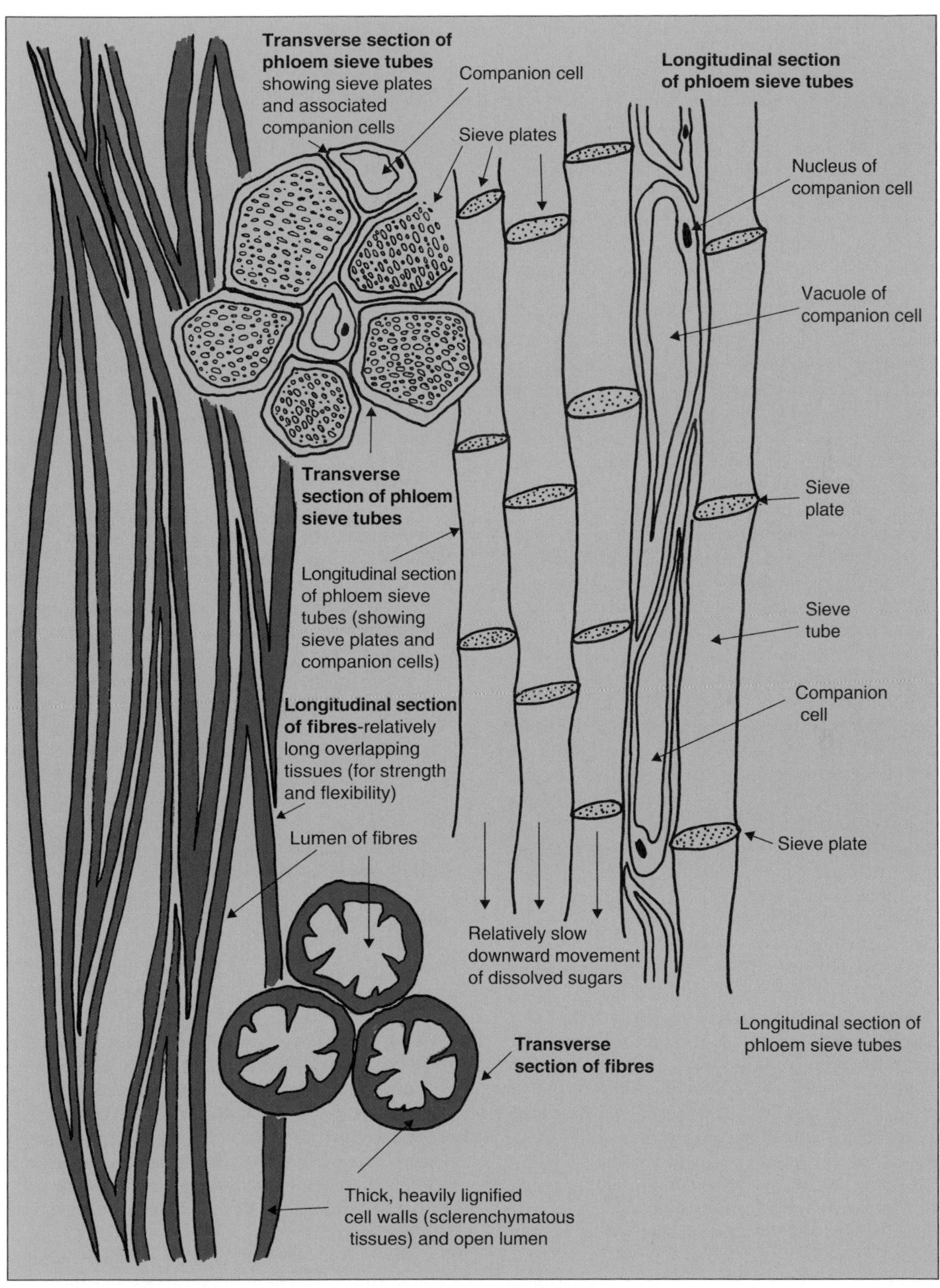

Vascular tissues: phloem and fibres (non-transporting strengthening tissues).

The tissues nearest to the stem apex divide and differentiate to form the tunica (on the outside) and the corpus (main plant body) on the inside of this. Within the corpus (responsible for producing all the main plant body tissue patterns), procambial cells are produced.

The procambium is meristematic and produces cells destined to be vascular strands (but not yet obviously so, as they are very difficult to discern in the early phases of development). These soon-to-be vascular tracts are therefore 'potential' at this stage, but are destined to be fully functional as vascular tissue by virtue of their position in the stem tissue configuration, and to become 'actual' as the cells differentiate. Prior to them actually functioning, water is moved to engorge cells via diffusion and osmosis – which functions perfectly well over the very small distances involved. Efficient vascular vessels for water transport are only required as the volume and length of tissue groups increases and water is needed more speedily to prevent flaccidity and desiccation. Therefore, tissues that commenced as meristematic parenchyma in the apical meristem ultimately form procambium that in turn differentiates to vascular tissues with its own vascular cambium for further tissue production. You can see now why the terminology procambium is used: pro = coming before.

The corpus not only produces the tunica on its outside (destined to be the epidermis of the stem) but also produces undeveloped primordial leaves. The primordial leaves are left behind to develop fully as the stem increases in length. Furthermore the increasing stem length telescopes out the distance between the leaves: that is, inter-nodes lengthen during the increase in stem length, thus the distance between leaves increases to a predetermined maximum for the particular species under its specific set of environmental conditions. Both space and time allow the leaves left behind by the process to develop fully.

Green plant stems comprise an external protective skin (epidermis) surrounding a ground tissue (the cortex - a form of packing tissue), and with strands of vascular tissue (vascular bundles) running through the cortex. Epidermal cells are transparent, and the main green effect comes from large amounts of chlorophyll in the outer rows of cortex. The pattern of vascular bundles in stems differs in monocotyledonous and dicotyledonous plants, monocotyledonous plants having their vascular bundles (veins) dispersed randomly throughout the cortical tissue, and dicotyledonous plants having their vascular bundles arranged in a distinct ring within the cortical tissue (when viewed in the transverse section of the stem). The difference seems small but is, in fact, of fundamental importance.

In monocotyledonous plants the vascular bundles are said to be 'closed', as the vascular cambium is not long-lived and it dies out after a relatively short period of time. Thus, once the initial vascular cambium cells have developed they no longer have the ability to add to existing tissue, and they do not increase stem diameter in following years. Although tree-like (arborescent) monocots do exist, they are unusual rather than the norm. The few examples of arborescent monocotyledons that show significant increase in stem girth (for example *Cordyline australis* and *Dracaena draco*) do so by adding more vascular bundles initiated within the ground tissue (cortex), not by the addition of new (secondary) xylem and phloem from the initial vascular cambium.

Annual plants complete their life cycle in the primary growth condition. Biennial and herbaceous plants also survive with primary growth only on their aerial parts. However, their continuance from one year to the next may involve small amounts of secondary tissues for strength in their underground organs.

In dicotyledonous woody plants the vascular bundles are said to be 'open', as the vascular cambium is long-lived, and adds secondary xylem and phloem tissue to the existing bundles, as well as new vascular bundles added to the original ring. This facilitates the commencement of secondary growth, and the increase in stem girth annually (secondary thickening) allows true woody plants (arboreals) to attain great size.

Plant Roots

In the same way that shoot-tip meristems are important in stem production, plant roots develop from the tissues at root-tip meristems – both are apical meristems as they are positioned at growth apices, but there is a fundamental difference between them. Some outer tissues of stems have chlorophyll (as they will be illuminated), whereas subterranean root tissues would be wasting their important energy budget by producing chlorophyll when they will not be illuminated. However, not all root systems are devoid of chlorophyll, as some specialized forms of aerial root may have chlorophyll present.

Root anchorage gives plant support and stability in the soil. The absorption of soil moisture is facilitated by fine fibrous rootlets with associated

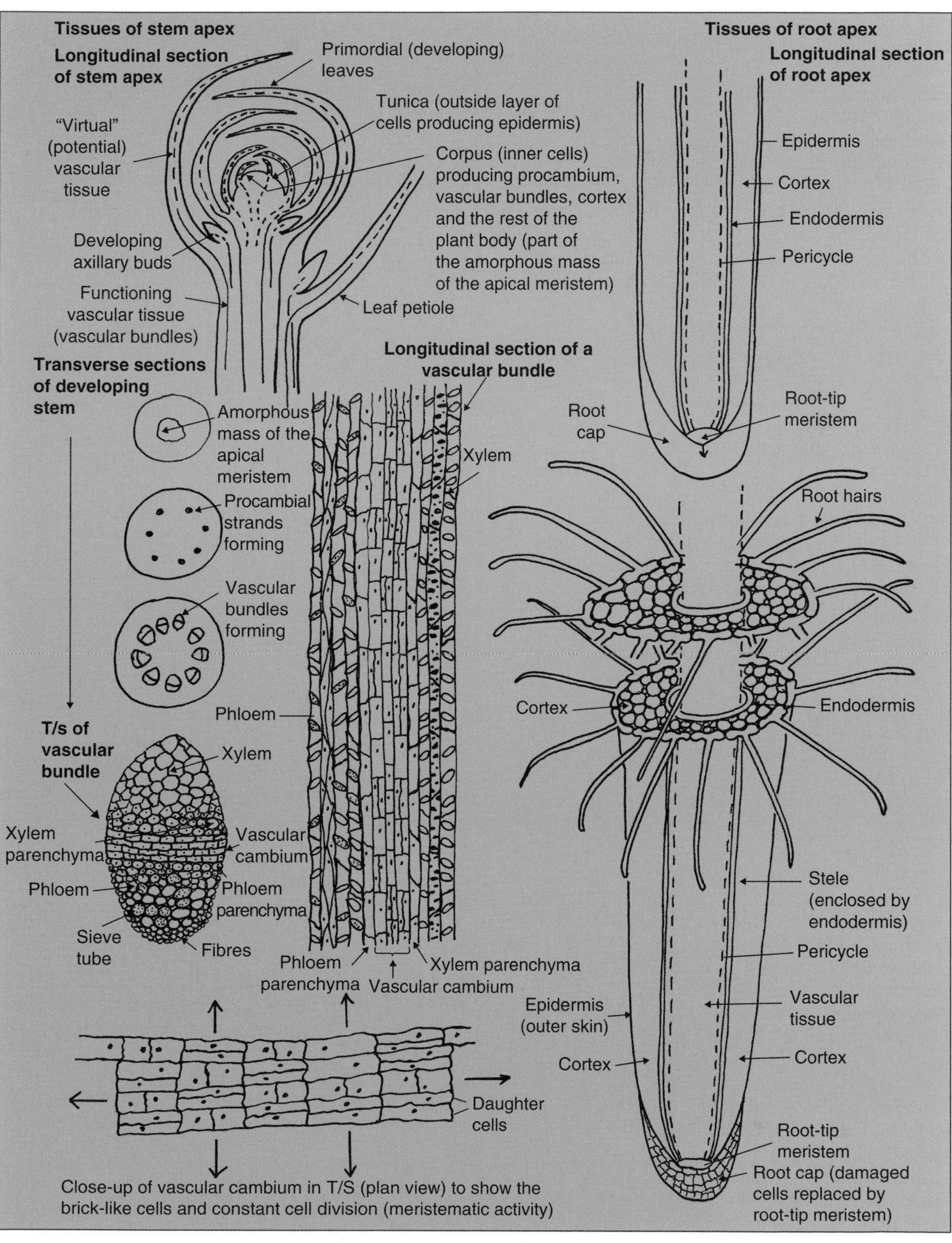

Comparing tissues of young stem and root apices.

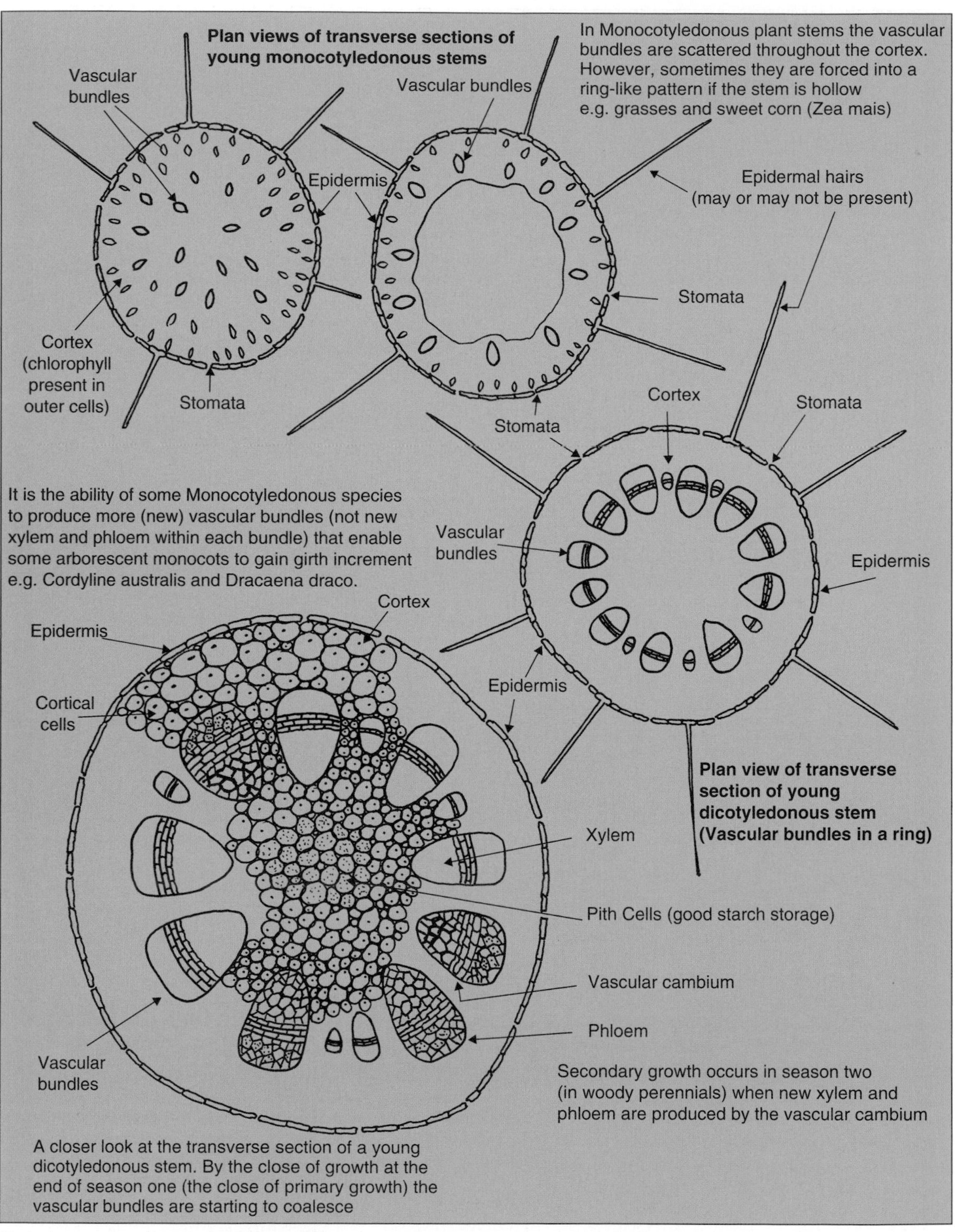

Plan views of transverse sections of young monocotyledonous stems
In Monocotyledonous plant stems the vascular bundles are scattered throughout the cortex. However, sometimes they are forced into a ring-like pattern if the stem is hollow e.g. grasses and sweet corn (Zea mais)
Vascular bundles
Vascular bundles
Epidermis
Epidermal hairs (may or may not be present)
Stomata
Cortex (chlorophyll present in outer cells)
Stomata
Stomata
Cortex
Stomata
It is the ability of some Monocotyledonous species to produce more (new) vascular bundles (not new xylem and phloem within each bundle) that enable some arborescent monocots to gain girth increment e.g. Cordyline australis and Dracaena draco.
Vascular bundles
Epidermis
Cortex
Epidermis
Cortical cells
Epidermis
Plan view of transverse section of young dicotyledonous stem (Vascular bundles in a ring)
Xylem
Pith Cells (good starch storage)
Vascular cambium
Phloem
Vascular bundles
Secondary growth occurs in season two (in woody perennials) when new xylem and phloem are produced by the vascular cambium
A closer look at the transverse section of a young dicotyledonous stem. By the close of growth at the end of season one (the close of primary growth) the vascular bundles are starting to coalesce

Basic tissue patterns in stems (primary growth).

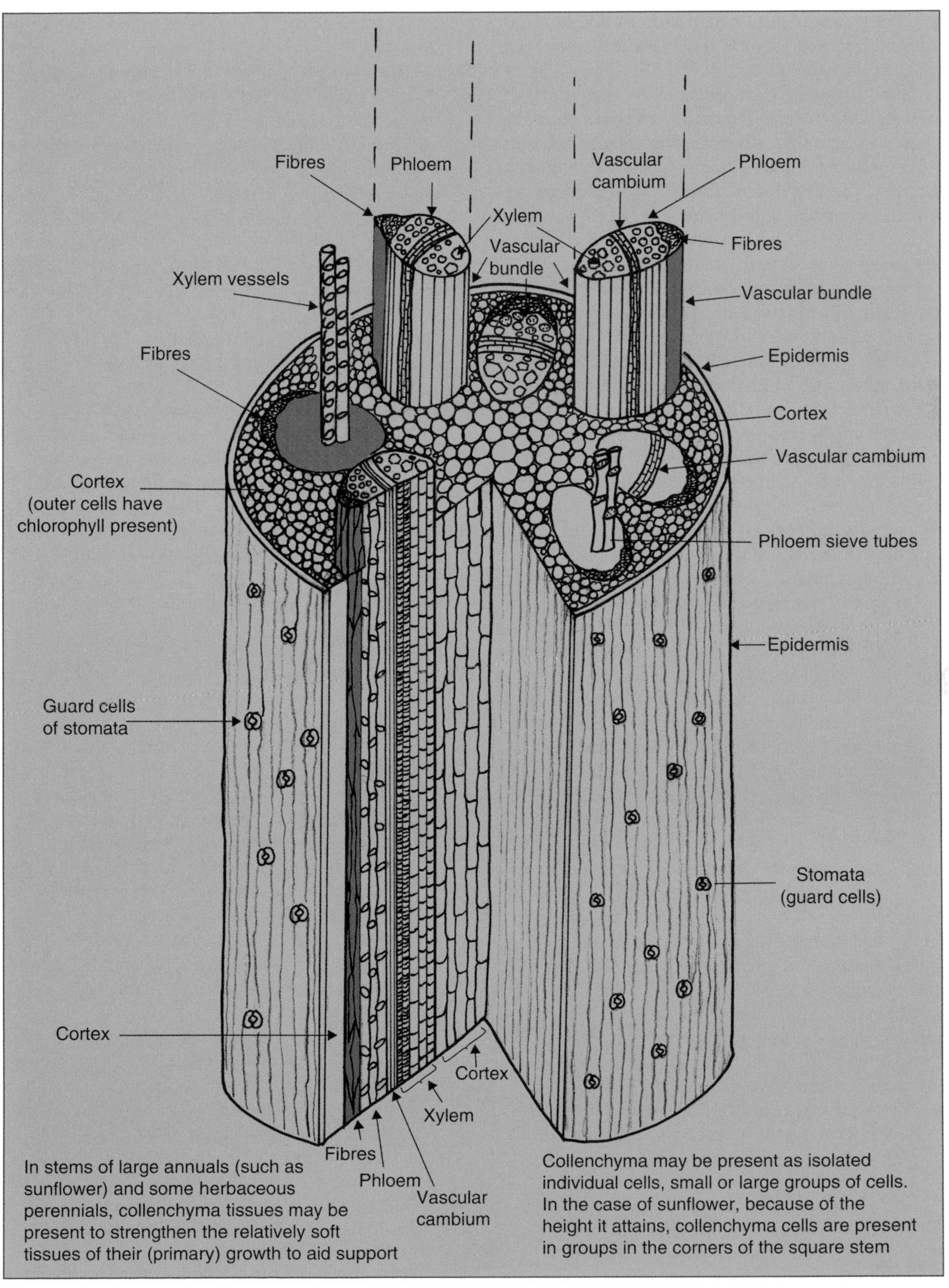

Longitudinal view of young dicotyledonous stem showing the vascular bundles.

root hairs, and the conduction of water and dissolved nutrient salts from the soil is through the xylem of root systems.

Broadly speaking, in soft annuals and biennials there are two main types of root system, namely taproot systems and fibrous root systems, and in herbaceous perennials, fibrous roots, taproots and root tubers (root modifications). Root systems of woody perennials are more complex as they include both soft absorptive tissues and bark-covered non-absorptive roots.

Taproot systems are derived from the original root (primary root or radicle) of a germinating seed. The root develops downwards and will usually have lesser roots, known as lateral roots or secondary roots, branching off the main taproot, and sometimes minor roots branching off the lateral roots (termed sub-lateral or tertiary roots), for example, carrot (*Daucus carrota*).

Fibrous root systems are formed when the original radicle, or part of it, naturally dies, is damaged as it hits an obstacle in the soil, or is pruned, or killed. The demise of the original root encourages a complex lateral system of dense fibrous roots to form. Fibrous-rooted begonia (*Begonia semperflorens*) owes its common name to the fact that it bears fibrous roots rather than the tubers or rhizomes of other *Begonia* species. Unfortunately, the common name was not translated to the scientific name – ever-flowering (*semperflorens*) was chosen instead. Some herbaceous perennial species such as *Sedum spectabile* have fibrous root systems, as do petunias and busy lizzies (*Impatiens*).

Many woody species commence with a taproot system just after germination. However, these rarely persist, and often a more branching, or fibrous, system results. Even oak, which commences with a very strong taproot in the early stages of its development after germination, usually forms a more laterally branched root system fairly quickly afterwards. Woody plants such as *Rhododendron* and *Camellia* have very fibrous root systems.

Anatomy of the Root

The tissues of a young root comprise a central core of vascular tissues known as the 'stele', surrounded by cortex, and an outer epidermis. The root has distinct recognizable zones, namely root cap, root tip, the area of elongation, and the area of differentiation that houses the piliferous (root hair) region and the tissues above. The function of the root cap is to protect the delicate root tip as the root progresses through the soil, and is sacrificed and replaced regularly in order to protect the important root-tip meristem efficiently. Both growth in root-tip length and the replacement of damaged root-cap cells are facilitated by the same root-tip meristem.

The stele differentiates from an original procambium strand within the cortex at the very tip of the root apex. Young, first xylem (protoxylem) and first phloem (protophloem) differentiate at the root tip from the procambium, with alternating xylem and phloem producing the typical star-shaped configurations. The xylem forms the 'arms' and centre of the 'star', and the phloem forms in the 'bays', and between the xylem and phloem a vascular cambium initiates. The tissue configuration of the stele is distinct and comprises set numbers of alternating groups of xylem and phloem, and the different configurations have specific names, depending on how many arms there are to the xylem 'star' at the centre: thus two arms to the star = diarch, three arms = triarch, four arms = tetrarch, five arms = pentarch, and so on. During the early stages of development of dicotyledonous plants they have up to seven arms on the xylem star, whereas monocotyledonous plants have eight or more (known as polyarch systems). So not only do the vascular patterns of stems and leaves differ between monocots and dicots, but also the vascular patterns of their roots.

Different parts of the roots on the same plant can have different tissue configurations: hence small roots may be diarch, and other larger or more vigorous roots may be tetrarch or pentarch. However, the radicles of seeds at germination usually have a constant number of xylem arms to the star.

Each root hair is a single specialized epidermal cell that absorbs soil water, which is then transported by osmotic pressure across the cortex of the root towards the vascular tissue (at which point transpiration pull takes over as the main power of water movement through the plant). If the delicate root hairs were at the extreme root tip, the advancement of the root through the soil would destroy them before they could carry out their function, so the area of elongation separates the root tip from the root hair region (piliferous area). Root hairs are actually short-lived anyway, and are constantly replaced by new root hairs forming lower down the root. Their position at the area of differentiation, rather than the elongation zone of the root, gives them a period of time to carry out their absorptive function prior to being destroyed.

The cortex (or ground tissue) of the root comprises soft, parenchymatous cells whose function is

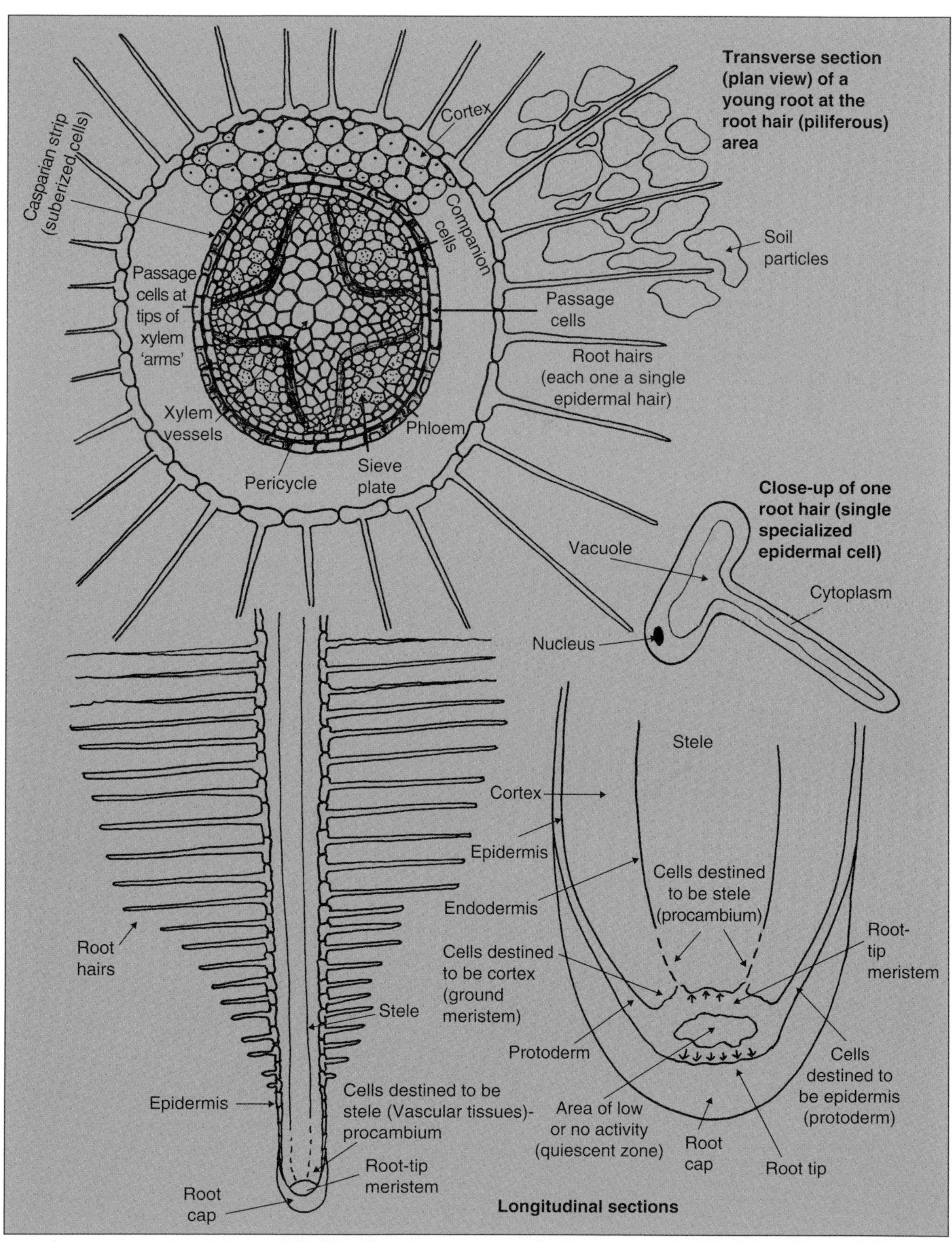

Tissues of young root at root tip and root hair (piliferous) region.

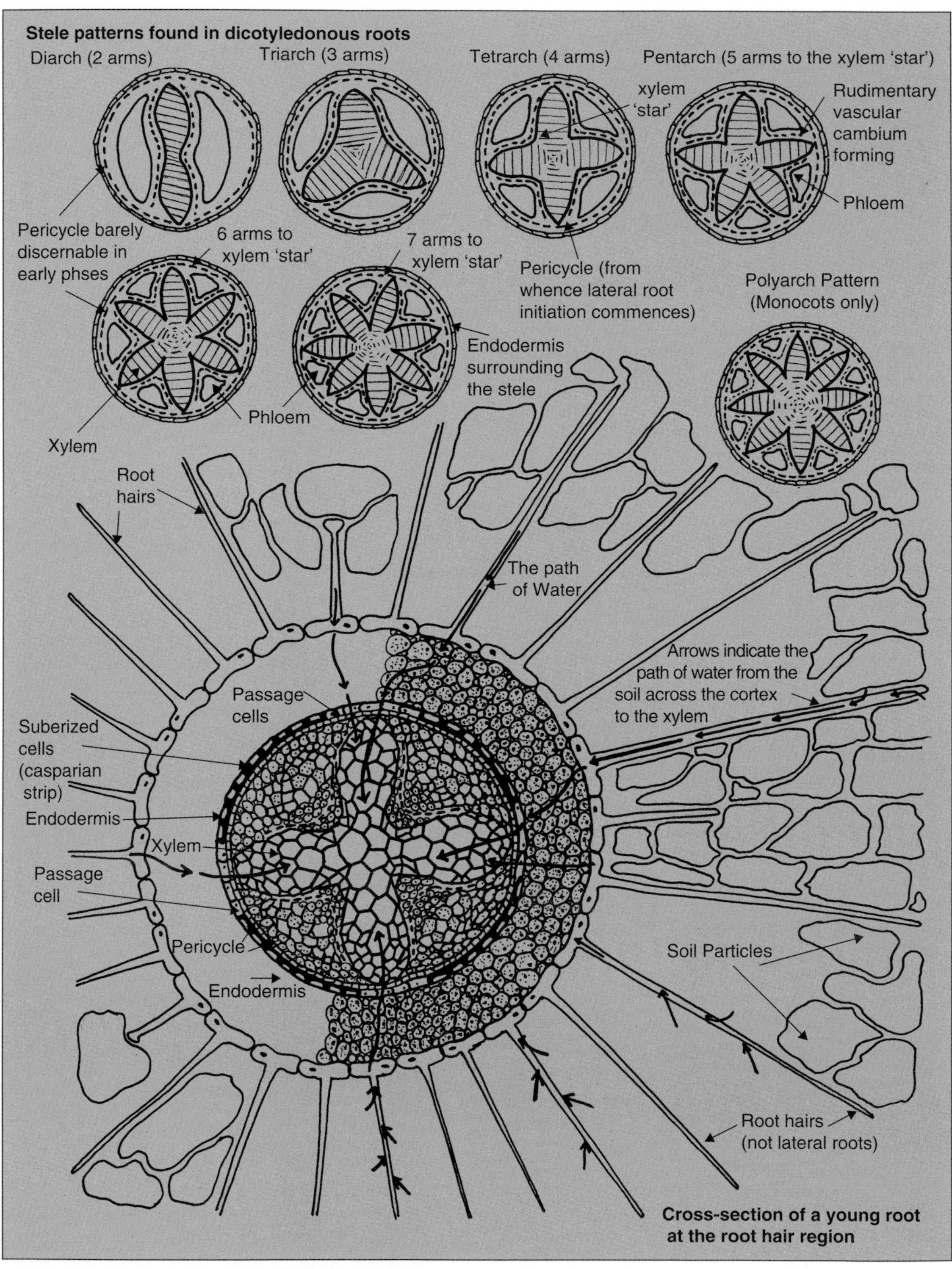

Possible stele patterns and the root hair region (piliferous area).

to contain the other tissues, including the stele (and afford them some protection), but also to store starches. Some species develop pith in the centre, comprising parenchymatous cells that are very efficient at starch storage.

Effectively enclosing the tissues of the stele is a circle of specialized cells (formed from interior cortical cells) known as the endodermis (the inside skin). Some of the cells of the endodermis are waterproofed with strips of suberin wax (known as 'casparian strips'). The function of these is probably to stop water (coming from the fine roots and moving laterally across the cortex) arriving at the phloem tissue where it would be of no use. Instead, it channels the water to the ends of the arms of the xylem tissue through unsuberized 'passage cells' where it can be taken upwards by transpiration pull. Internal to the endodermis is a ring of specialized cells known as the 'pericycle'. It is from this tissue that new lateral roots are initiated, and as they progress, the developing lateral roots push through the tissues of the endodermis and cortex, ultimately breaching the epidermis.

The tissues of a young stem and root emulate one another in some ways. However, the main differences are of fundamental importance. Apart from notable exceptions (mainly those specialized for extra support), young roots are absorptive and will therefore always have epidermal hairs somewhere along their length. Stems, on the other hand, may or may not bear epidermal hairs, and where they do appear they are either moisture-retentive (not absorptive as individuals, but collectively acting like a sponge) or secretary – as in stinging nettles. The vascular tissues of young roots are enclosed in a central stele, whereas the vascular tissues of a young stem comprise separate vascular strands (veins or vascular bundles). The root tip has an apical meristem that produces both cells that extend growth and increase root length, and cells that form the protective, sacrificial root cap. The meristem at a shoot apex produces cells to extend stem length and produce new developing leaves upon that stem.

Root Modifications

Specific root adaptations exist, including root tubers and taproots that comprise swollen, fleshy roots modified to be efficient food stores.

Adventitious roots are roots that arise from stems, rather than from other roots, or other parts of the root system. They arise, therefore, from tissues where they are not expected. In woody plants lateral branches arise from internal (endogenous) tissues, and epicormic shoots arise from external (exogenous) tissues (the vascular cambium of the sap wood). Likewise, lateral roots normally arise from endogenous (internal) tissues originating at the pericycle. However, adventitious roots from stems (which have no pericycle) and lateral roots from older roots, also with no pericycle (because it has already been used in the production of the periderm), arise from other tissues – usually the vascular cambium near the phloem.

Examples of adventitious roots on woody plants include aerial roots that push their way through the bark. Their function is initially to gain extra moisture from the soil, but as they grow older and increase in girth, they may take on a more supportive role, in which circumstances they are known as 'prop roots'. The roots at the base plate (condensed stem) of bulbs, and the moisture-absorbing, support roots arising from the stems of woody climbers such as common ivy (*Hedera helix*) and climbing hydrangea (*Hydrangea petiolaris*) are common examples of adventitious roots.

The aerial roots of epiphytic orchids are specialized to perform different functions. Some grow into the crevices of the host and anchor the plant to it: they are specifically designed for support, and the absorptive function of these roots is low. Other roots are adapted to both anchor and absorb moisture, and they have specialized absorbent root tips for the task. Roots that radiate outwards into the surrounding environment a little way from the host, even upwards rather than downwards, are known as 'nest' roots because they form basket-like 'nests' that fill with plant (and insect?) debris, whose nutrients can be absorbed by the orchid. Some epiphytic orchids have a white spongy sheath to their roots (known as the 'velamen') that soaks up and holds water to aid their survival. Many aerial roots contain pigments at their tips (mainly chlorophyll), so the root tips are stimulated towards the light (positively phototropic). Other stimulants also affect the roots of orchids, as most root types may change to support roots at any time to hold the plant firmly in place. Orchids, like many other plants (including herbaceous and woody perennials), commonly have mycorrhizal associations with their roots.

Mycorrhiza

Mycorrhiza is a generic term for a group of beneficial fungi that infect ('colonize') plant roots. The symbiotic (mutually beneficial) relationship that they form with the roots of many herbaceous and woody perennial plants is often essential for continuing plant health. The fungi take sugars from the root tissues of their host plant and in return

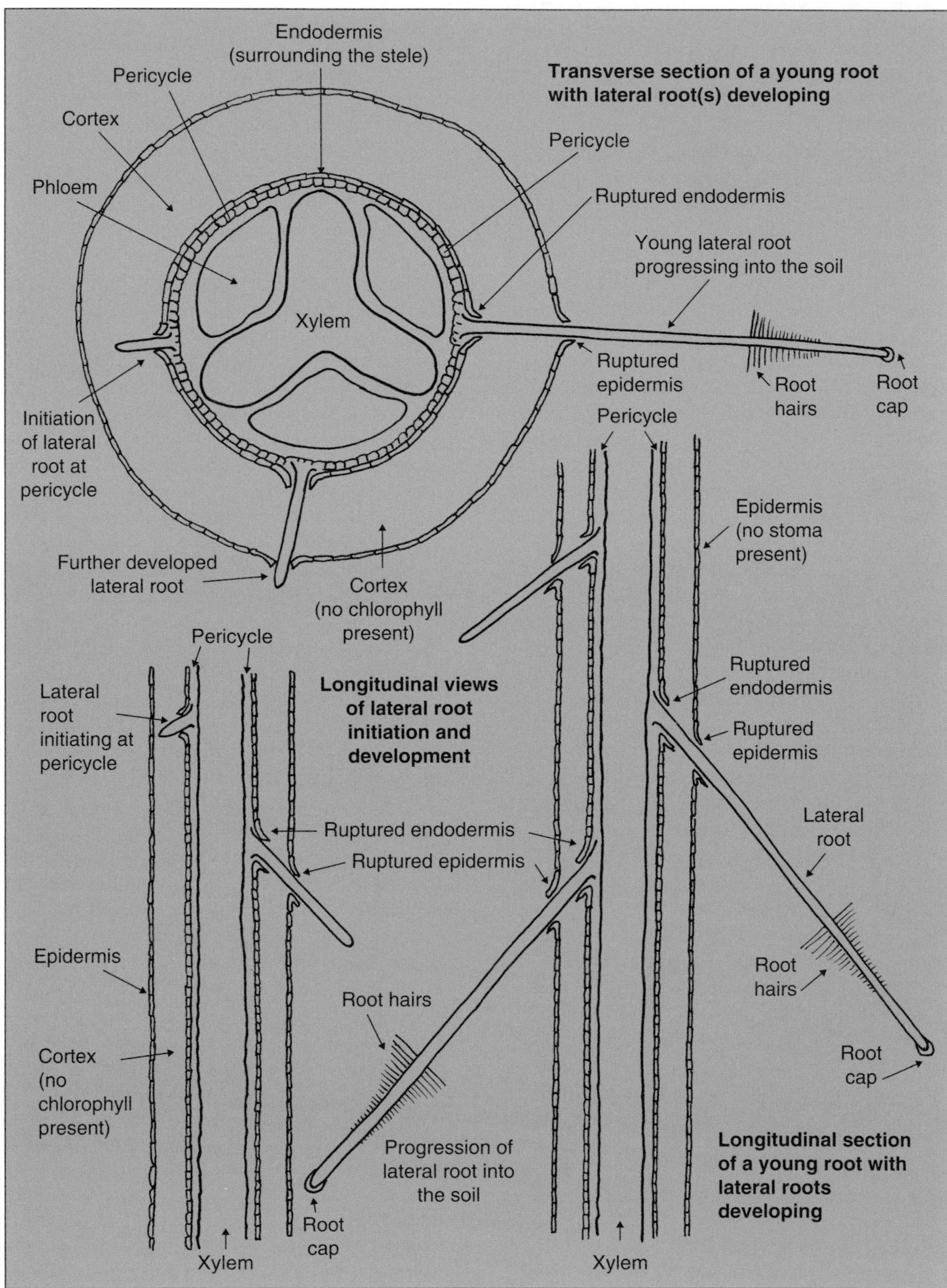

Tissues of a young dicot root showing lateral root development.

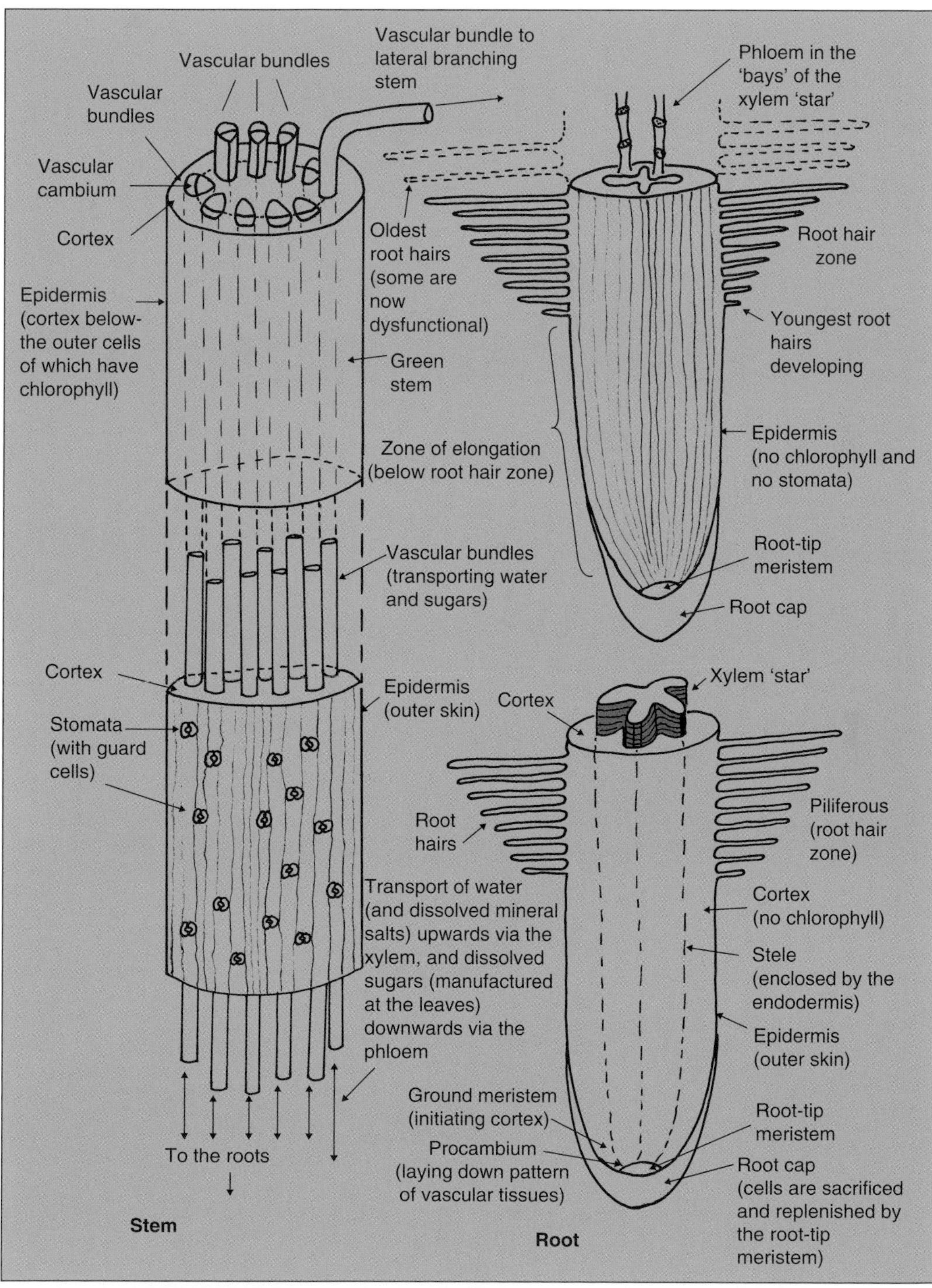

Young green stem versus young root apex (longitudinal views).

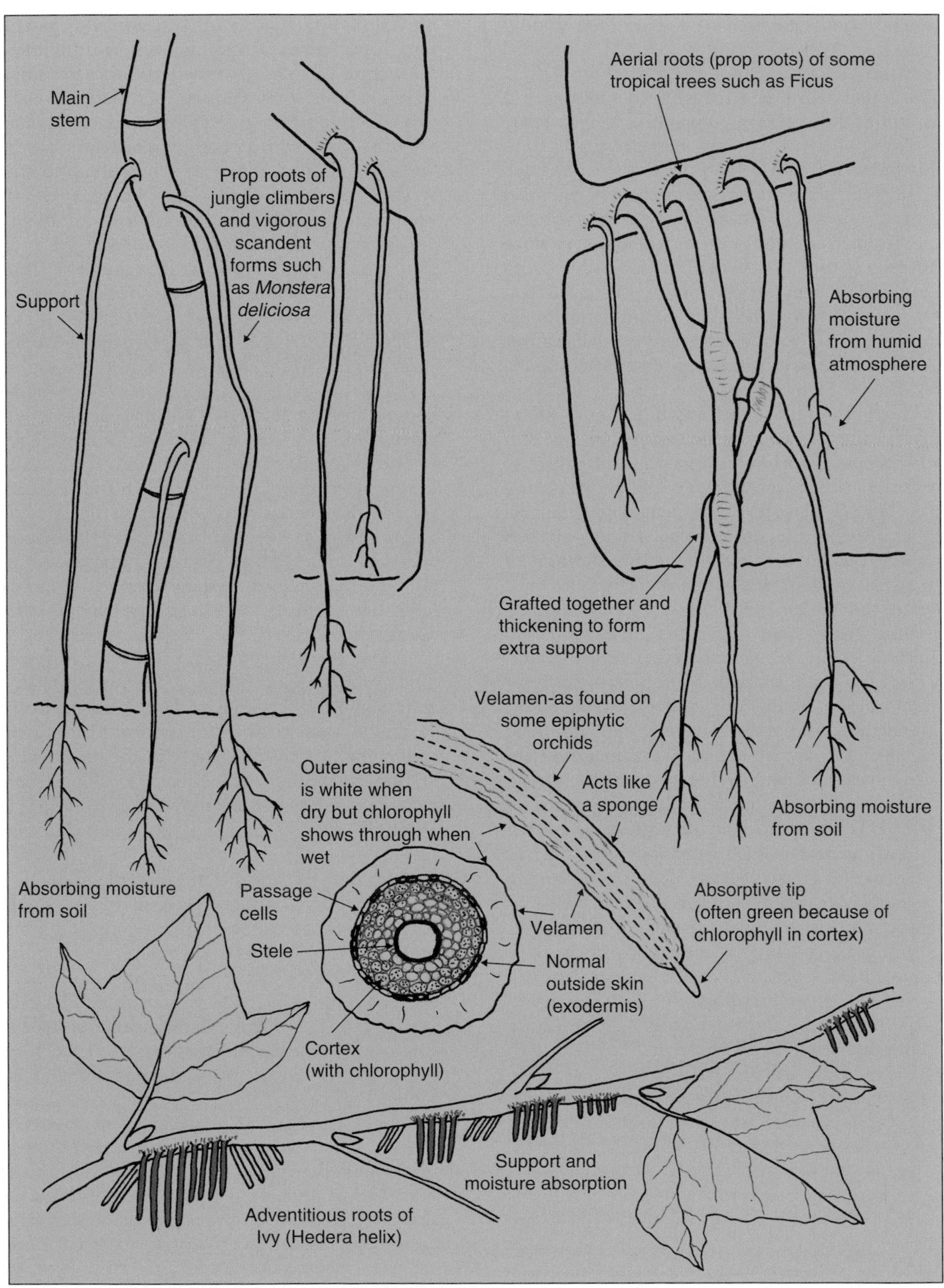
Main stem
Prop roots of jungle climbers and vigorous scandent forms such as *Monstera deliciosa*
Support
Absorbing moisture from soil
Aerial roots (prop roots) of some tropical trees such as Ficus
Absorbing moisture from humid atmosphere
Grafted together and thickening to form extra support
Absorbing moisture from soil
Velamen-as found on some epiphytic orchids
Outer casing is white when dry but chlorophyll shows through when wet
Acts like a sponge
Passage cells
Stele
Velamen
Normal outside skin (exodermis)
Cortex (with chlorophyll)
Absorptive tip (often green because of chlorophyll in cortex)
Support and moisture absorption
Adventitious roots of Ivy (Hedera helix)

Adventitious roots (including prop roots) and root modifications.

they increase the absorptive area of the root system and process various nutrients found in the soil matrix (notably phosphorous). Phosphorous is naturally abundant but found as poorly soluble salts, and mycorrhiza render normally insoluble forms of phosphorous soluble (and therefore available) to the host plant.

The importance of mycorrhiza is often underestimated, as they colonize a vast number of trees, herbaceous and shrubby species, and their absence leads to a relatively quick degeneration of the host plant. Some plant species can be host to a few different mycorrhiza, but many mycorrhiza are specific to one particular host.

The two most common divisions of the mycorrhizal world are ectomycorrhiza (those with much of their tissues external to the plant roots that they colonize) and endomycorrhiza (those with most of their tissues inside the cortex of the colonized host). Many, but not all, woody plant species host ectomycorrhiza, and most herbaceous subjects host endomycorrhiza. Ectomycorrhiza = ectophytic mycorrhiza (living outside the plant) – also known as sheathing mycorrhiza. Initial infection may be by spores within the soil, hyphae (mycelial strands) entering the young absorptive roots and rootlets, or both. Mycorrhiza tend to colonize the very fine roots rather than the thicker, absorptive roots.

The internal hyphae of ectomycorrhiza form a close relationship with the cortical tissues of plant roots, and form a distinct pattern known as the 'Hartig net'. However, their main distinction is external to this, as they form antler-like (coral-like) outgrowths of hyphae (vegetative strands of mycelium) on the outside of the young roots that they infest. They also form small segments that 'bud' or 'bead off' from one another, and they are often covered in fine hairs to give a cotton-bud effect. Ectomycorrhiza are easily identified from harvested hand specimens of fine rootlets by these stocky, often hairy, outgrowths, many becoming much branched. Although the main lateral roots are typified by a tendency to grow downwards or horizontally affected by several stimuli, other smaller roots often grow upwards towards the soil surface (and may even infest leaf litter). Mycorrhiza are often abundant in these particularly well aerated fine roots that lie just below the soil surface.

The corraloid outgrowths are much thicker than the fine sub-lateral rootlets of the plant, and considerably thicker than the root hairs (as these are only one cell thick). Furthermore, the hyphal outgrowths of ectomycorrhiza are not equidistant on the root like root hairs, and they are branched not straight (like root hairs), sometimes protruding centimetres out into the soil; and nor do they have the tissue patterns (epidermis, cortex, endodermis, pericycle and stele) associated with young roots. Root hairs are parallel-sided and uniform, whereas mycorrhiza outgrowths are neither. The tips of the hyphal outgrowths have fine membranes that absorb water, nitrogenous salts, and phosphates in particular, but the other parts of the corraloid outgrowths may also absorb water and nutrient salts, so they have the major benefit of greatly increasing the absorptive area of the root system. Often, root hairs are missing from infected/colonized areas as they are surplus to requirements. There is evidence to suggest that even in winter when plant activity is very low, mycorrhiza (still feeding off stored plant carbohydrates) can release phosphorous to the plant in readiness for growth in the following season.

Many orchids and woodland and ericaceous plants (from heathland habitats) have very strong mycorrhizal associations that are essential for their wellbeing. The degeneration of many woodland and heathland plant species when planted into 'landscaped' soil that has been stacked prior to use, bears witness to the essential nature of mycorrhizal associations. Stored, heaped soils are anaerobic, or only partially aerobic, throughout large proportions of their profile, and most soil organisms die under these conditions (mycorrhiza are particularly susceptible), so the soil denatures relatively quickly over time. Along with ensuring the correct pH of soils, the intentional (by inoculation) or unintentional (by leaf litter and other organic material) reinfection of these soils by mycorrhiza often decides the death or otherwise of the plant species planted in them – not necessarily their establishment, but certainly their future development. So important is the association of mycorrhiza to heathland plants such as ling (*Calluna vulgaris*) and whortleberries (*Vaccineum* species) that the seeds (and internal embryos) are often infected prior to dispersal from the mother plant. Many conifer seedlings will not develop successfully without the presence of mycorrhiza, and in extreme cases some species of orchid, and some ericaceous plants, will not even germinate without the presence of specific mycorrhiza. Some authorities consider that because ectomycorrhiza can protrude into the soil several centimetres, they become associated with the hyphae of neighbouring mycorrhizal colonies on the same plant species, and offer a system of chemical messaging from one plant to another via their intimate connections.

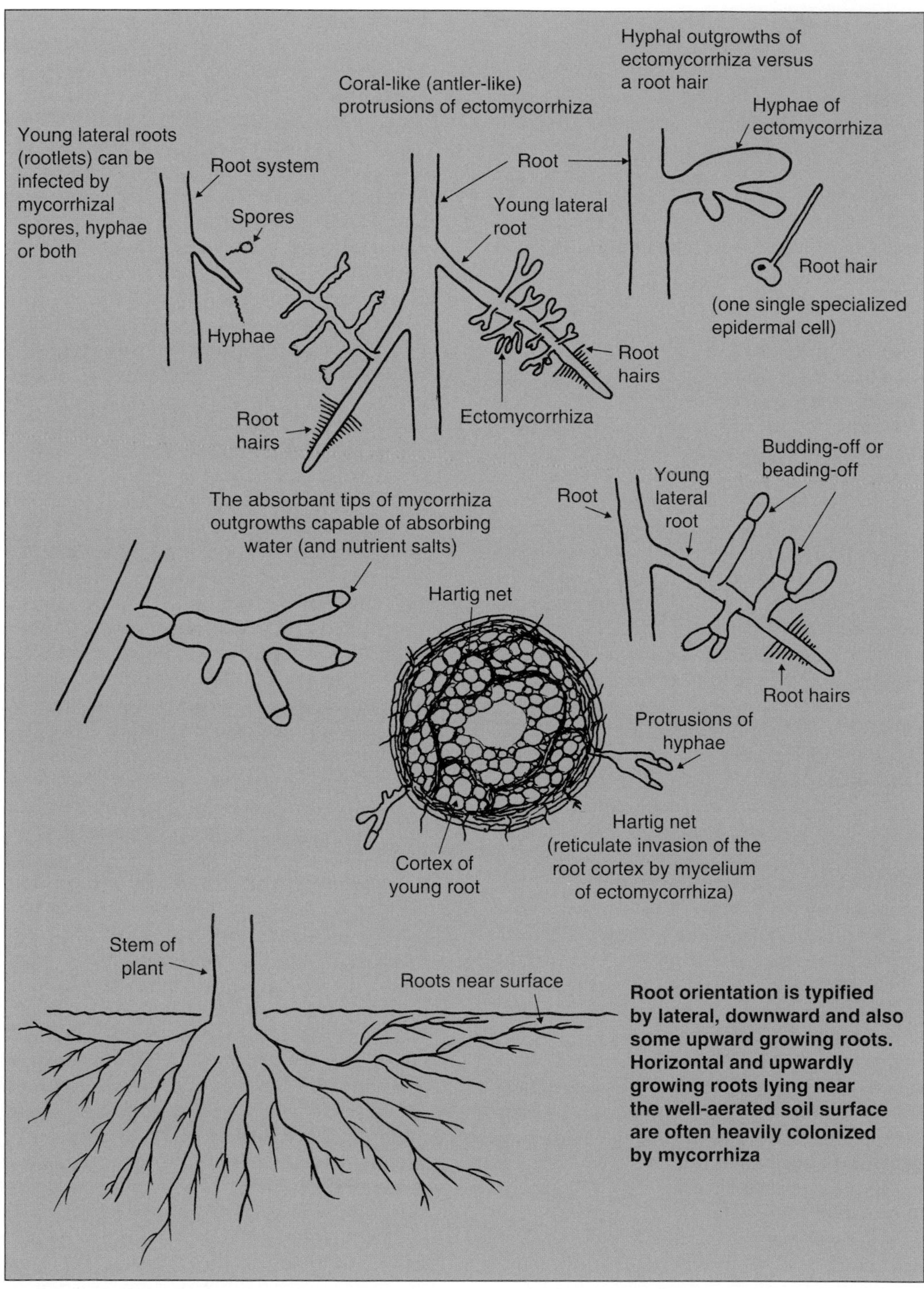

Mycorrhiza colonizing root systems.

Plant Hormones

The control of cell, tissue and subsequent organ development comes under the influence of their genetic make-up and also various environmental factors, including light and moisture. However, the main control of tissue development is by chemical triggers (hormones). Different hormones have different effects, and different concentrations of the same hormone may have varying degrees of influence. Furthermore, some hormones create a specific growth rate because of their relative influences and the balance of their concentrations. Some hormones encourage (promote) growth whilst others inhibit growth. Inhibitors induce bud dormancy, and can cause seed dormancy by inhibiting the growth of plant embryos at germination – for example, abscisic acid. They may also reduce/control the effect of growth promoters when both types of hormone are present – an effect responsible for the rapid regrowth of shoots after pruning. The hormone groups auxins, cytokinins and gibberellins are common in plant tissue.

Auxins

Auxins are produced at the buds at shoot tips in relatively large amounts, and at lower order buds as well, but less so than at the apical buds. Auxins inhibit the processes of both leaf and fruit abscission (leaf fall and fruit drop), and naturally occurring auxins in apical buds inhibit the devlopment of lateral buds. Hence when an apical bud is removed by pruning, the lateral buds develop very quickly afterwards. Auxins in toxic amounts are also used as selective herbicides, where they disrupt the normal plant growth patterns and ultimately kill weed plants. Auxins as growth promoters are responsible for cell division at meristems, for initiating fruits, for cell growth (cell elongation), initiating callus at wounds, and for initiating root development (including from callus on stem cuttings). Their role in rapid cell elongation is also responsible for the orientation of stems and roots caused by positive and negative phototropism (where organs grow towards or away from a stimulus).

Auxins initiate cambial activity at the cut surface of stems prepared as cuttings and induce callus production (disorganized parenchyma cells), and its subsequent differentiation to any tissue type necessary facilitates rooting. Some of the callus tissue will heal the wound, but because of the position of the developing callus, with poor light and additional auxins taken to the cut surface by gravity, some tissue will develop into roots (rather than new aerial shoots, which require more light). The natural build-up of auxins at nodal regions (which encourages root development) makes the preparation of nodal stem cuttings the favoured choice. However, some forms of inter-nodal cutting may be used successfully, though even these tend to root more readily at the node. The progression of newly developed roots on cuttings is via the new root-tip meristem. Callus production at graft unions spewing from the vascular cambium exposed at the cut surfaces of both the scion and the understock is also initiated by auxins, and as elsewhere, the callus can differentiate into any other tissue patterns. In this instance, as long as the correct and favourable conditions are maintained, its position in the system encourages the initiation and differentiation of common vascular tissues – shared by scion and understock – so creating one functioning plant out of two pieces of different tissue.

Cytokinins

Cytokinins can work with auxins to promote/increase cell division and cell development – the incredible growth rate in one season of very vigorous annuals such as the sunflower is due to the cytokinins present. Some plant species can produce roots from the cut surface of a leaf petiole (leaf stalk), or indeed from any damaged tissue at a midrib or secondary vein. Severing the main veins (midribs) releases the vascular cambium at the main vascular bundle of the leaf. Whilst it is true that exposed vascular cambium usually produces callus (soft, disorganized parenchyma cells), and that callus has the ability to differentiate into any other tissue type, it is not always the case that this occurs every time in natural conditions. Callus in full light will usually differentiate into shoots, and callus in subdued or low light usually into roots. But in order for the vascular cambium exposed at a severed leaf midrib (or lesser vein) and inserted into a rooting medium to differentiate into both root and shoot tissue, the conditions have to be correct.

Auxins are responsible for roots forming on the cut petioles of leaf cuttings, and cytokinins are responsible for the initiation of aerial growth buds that develop to create a complete new plant. But not all (in fact, not many) species show this ability freely under normal propagation conditions, and many have never been induced to produce complete plantlets (including aerial shoots) from the cut surface at all. However, some species do have this ability, and are successfully propagated by leaf cuttings. The ability has to include tissues that differentiate into the organs of aerial growth

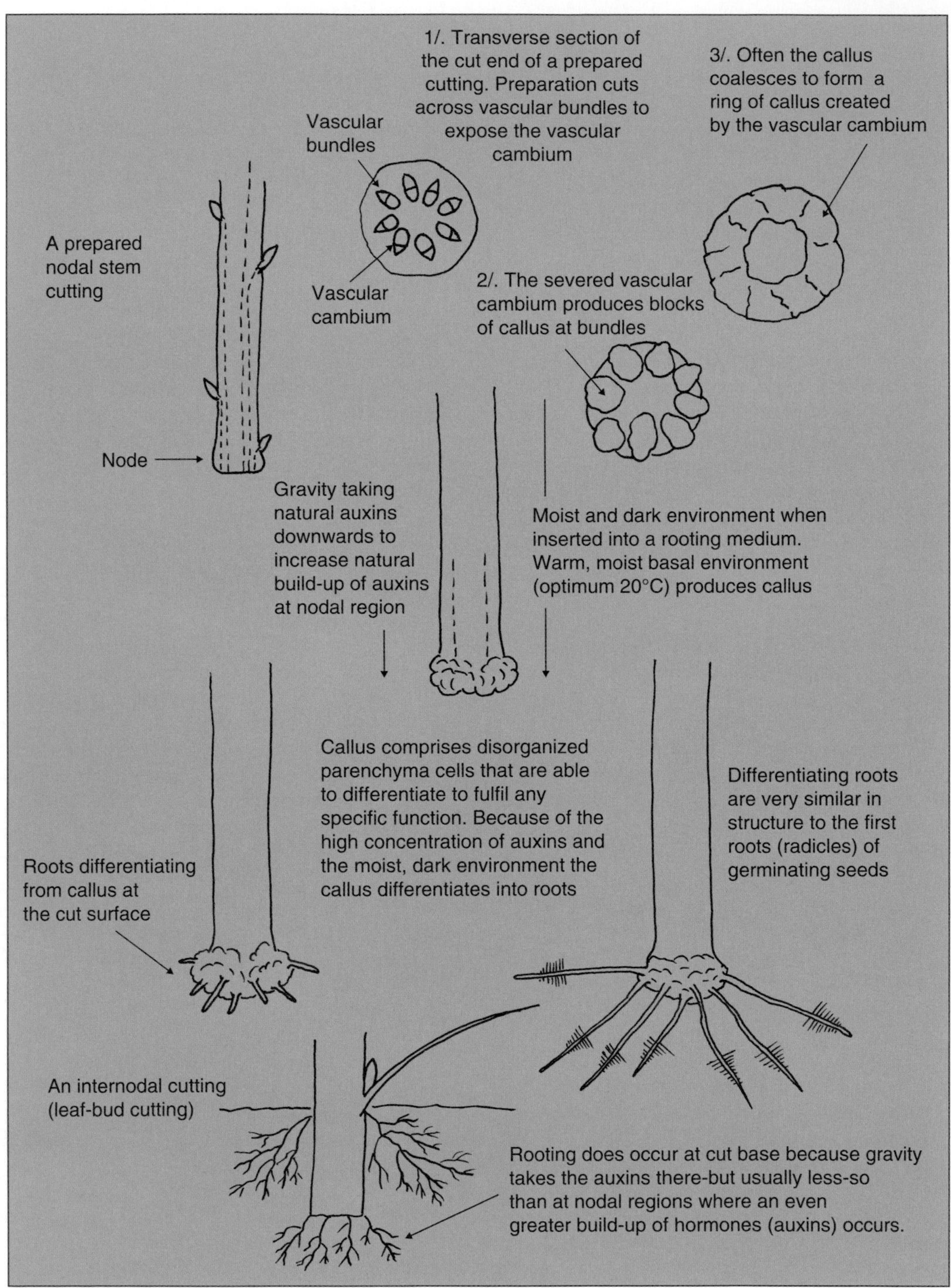
1/. Transverse section of the cut end of a prepared cutting. Preparation cuts across vascular bundles to expose the vascular cambium
3/. Often the callus coalesces to form a ring of callus created by the vascular cambium
Vascular bundles
A prepared nodal stem cutting
Vascular cambium
2/. The severed vascular cambium produces blocks of callus at bundles
Node
Gravity taking natural auxins downwards to increase natural build-up of auxins at nodal region
Moist and dark environment when inserted into a rooting medium. Warm, moist basal environment (optimum 20°C) produces callus
Callus comprises disorganized parenchyma cells that are able to differentiate to fulfil any specific function. Because of the high concentration of auxins and the moist, dark environment the callus differentiates into roots
Differentiating roots are very similar in structure to the first roots (radicles) of germinating seeds
Roots differentiating from callus at the cut surface
An internodal cutting (leaf-bud cutting)
Rooting does occur at cut base because gravity takes the auxins there-but usually less-so than at nodal regions where an even greater build-up of hormones (auxins) occurs.

Production of roots on a nodal stem cutting.

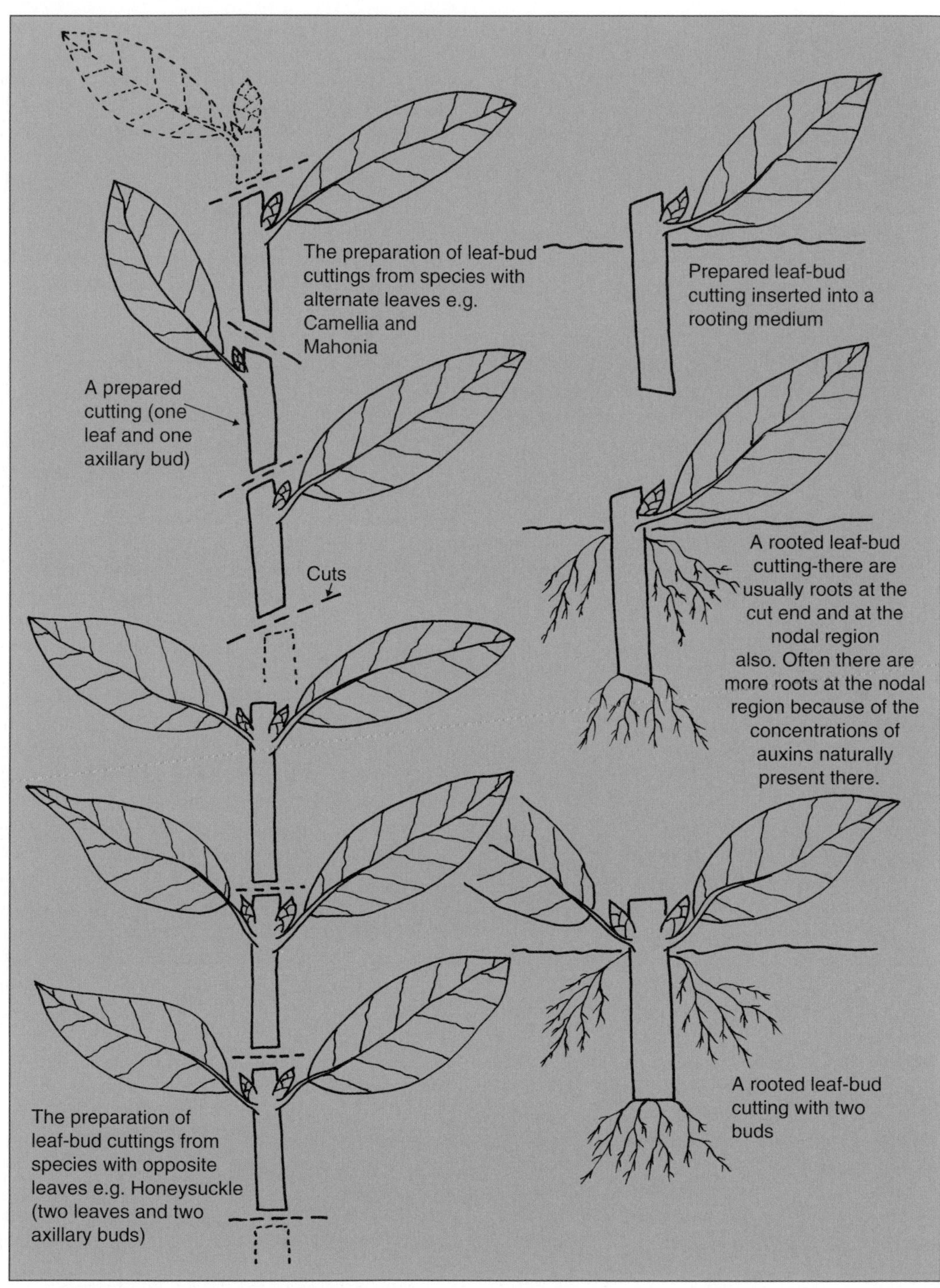

The preparation of leaf-bud cuttings from species with alternate leaves e.g. Camellia and Mahonia
Prepared leaf-bud cutting inserted into a rooting medium
A prepared cutting (one leaf and one axillary bud)
Cuts
A rooted leaf-bud cutting-there are usually roots at the cut end and at the nodal region also. Often there are more roots at the nodal region because of the concentrations of auxins naturally present there.
A rooted leaf-bud cutting with two buds
The preparation of leaf-bud cuttings from species with opposite leaves e.g. Honeysuckle (two leaves and two axillary buds)

Leaf-bud cuttings (a form of inter-nodal cutting).

as well as root systems, because a root system alone on a leaf with no provision for continued growth and regeneration, and a very limited life expectancy, is a futile exercise.

The production of plants from leaf cuttings involves specific soft, leafy subjects. Methods include leaf-petiole cuttings (for example *Peperomia, Saintpaulia)*, whole leaf-lamina cuttings (*Begonia hybridatuberosa*), leaf lamina squares and strips (*Begonia masonorum, Begonia rex*), lacerated full leaf laminas (*Begonia* x *hybridatuberosa* and *Begonia rex*) and split leaf laminas (*Streptocarpus capensis*). After rooting, the initial leaf degenerates (as the dividing and differentiating tissues use up the available nutrition from it), and nutrition from then on is provided by the developing aerial shoots from the new plantlet that is produced.

Gibberellins

Gibberellins present in seeds and buds can override the influence of inhibitors and initiate germination and bud development. They may also override the need for cold periods (vernalization) in seeds and developed plants of some species, and encourage germination (or flowering – for example in carrot). Giberrellins are implicated in the initiation of flowers and fruits, and the transformation of leaves from juvenile to adult, as found in some *Eucalyptus* species and in ivy (*Hedera* species).

Leaf Tissues

Leaf initials (leaf primordia) are formed by the multiplying shoot-tip (apical) meristem, and along with their undeveloped petioles, are left behind to finish development as the stem length gradually increases and 'telescopes' out, ultimately forming the typical leaf stalk (petiole) and the leaf lamina (leaf blade). Leaf-lamina development to its ultimate shape and size is continued by marginal meristems, which are short-lived and only present during the development stages. Leaves develop fully and then ultimately die, because after the leaf primordia are produced they gradually develop to be remote from the apical meristem and have no meristem of their own in the leaf tip. In areas where there is no longer a servicing meristem, replacement of cells is not an option, so the replacement of leaves is as complete new ones, produced from new sites, and replaced in their entirety in new parts of the plant (and at dormant buds in deciduous trees).

Procambium strands in the stem apex develop into the main veins, and branching lateral veins develop from the vascular cambium of the main vein. The main vascular bundles that run to the leaf are known as leaf traces, and the number of leaf traces varies from species to species, but for subjects with small, simple leaves it is usually only one that is continuous with the main central vein (midrib). However, in very large, simple leaves there may be more than one prominent vein (often one midrib and two slightly smaller, nevertheless important, main veins). Hence very large-leaved species, such as rhubarb, have three leaf traces, not one. Species with trifoliate (three distinctly lobed) leaves will usually have three leaf traces, whereas compound leaves with three individual leaflets (trifoliolate) or more leaflets (pinnate and bipinnate) have a main leaf trace to the rachis (extended leaf stalk), and leaflets are serviced from lateral bundles branching off the main leaf trace. On deciduous woody subjects the number of leaf traces can easily be seen after leaf fall as vascular dots within the leaf scars. Vascular leaf traces are specialized for leaves, and are slightly different in structure to other vascular bundles.

Foliage leaves are the main photosynthetic organs of the plant; they contain chlorophyll, which is essential for the photosynthetic (sugar-producing) process, and is responsible for their green coloration. The outside tissue of leaves, like green stems, comprises a soft-celled epidermis with a protective waxy outer layer called the cuticle. The cuticle and the epidermis are transparent, but cells below this layer have chlorophyll. In fact, chlorophyll is named after its role in leaves, as chlorophyll means 'green leaf' (chloro = green, and phyll = leaf). Leaves also act as an area of evaporation, driving the transpiration stream that moves water through the plant to cool it, to carry dissolved nutrient salts, and to engorge soft cells.

Most plants (including most trees and shrubs) have mesophytic foliage leaves, comprising an upper and lower epidermis (epi = outside, and dermis = skin) sandwiching middle tissue called the mesophyll (meso = middle, and phyll = leaf). The mesophyll comprises an open, spongy tissue called the spongy mesophyll, and a denser regimented tissue called the palisade mesophyll. The palisade mesophyll tissues are heavily infested with chloroplasts and orientated vertically in a soldier-like row to ensure maximum illumination. Vascular tissue (venation) penetrates the mesophyll layers.

Mesophytic leaves are either dorsiventral or isobilateral. Dorsiventral leaves have a definite top and bottom, and are illuminated mainly on their top side – as in deciduous trees, broad-leaved weeds and suchlike. Isobilateral (being relatively

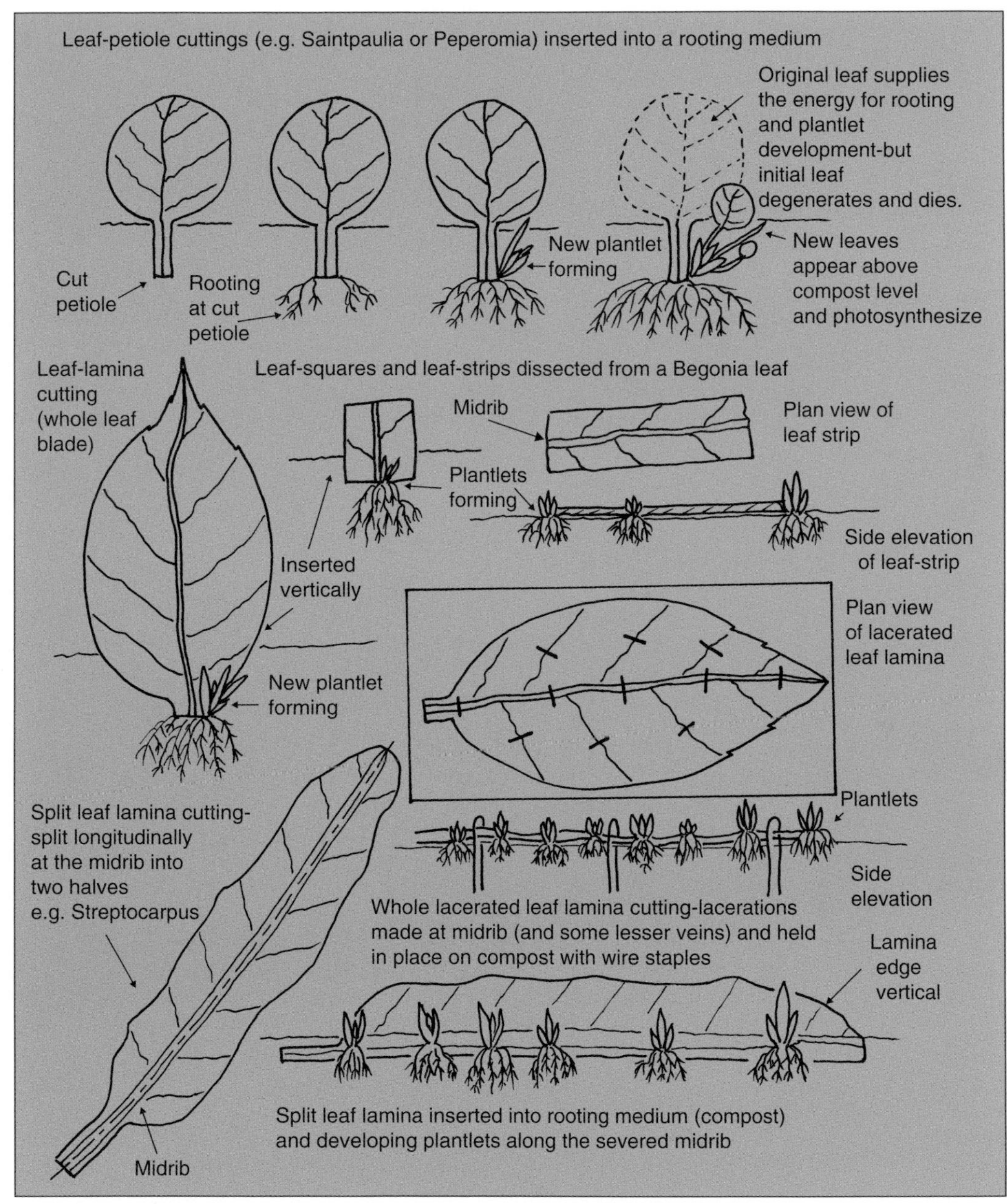

Leaf cuttings.

equal on both sides) leaves have no leaf stalk (sessile), are held vertically or nearly vertically, and are illuminated on both sides; they are common in many monocotyledonous types such as *Iris*, some grasses, *Sanseveria*, *Agave*. However, not all monocotyledonous subjects have isobilateral leaves, even though they may be sword-like, because if the leaves (or part of the leaves) are held horizontally, or near horizontal, they may be dorsiventral. *Lilium* species, for example, have

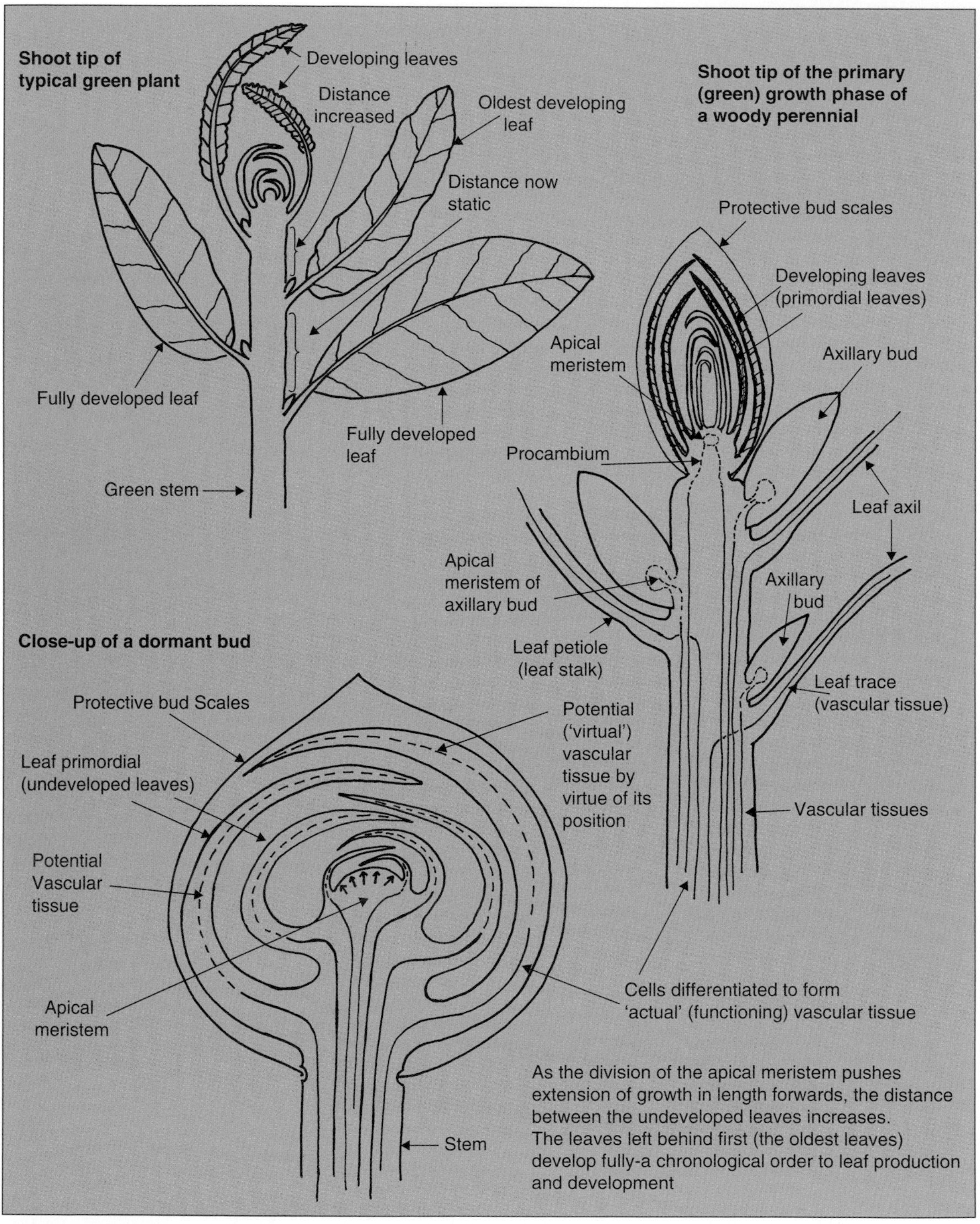

Shoot tips, and extension in stem length (primary growth) and leaf development.

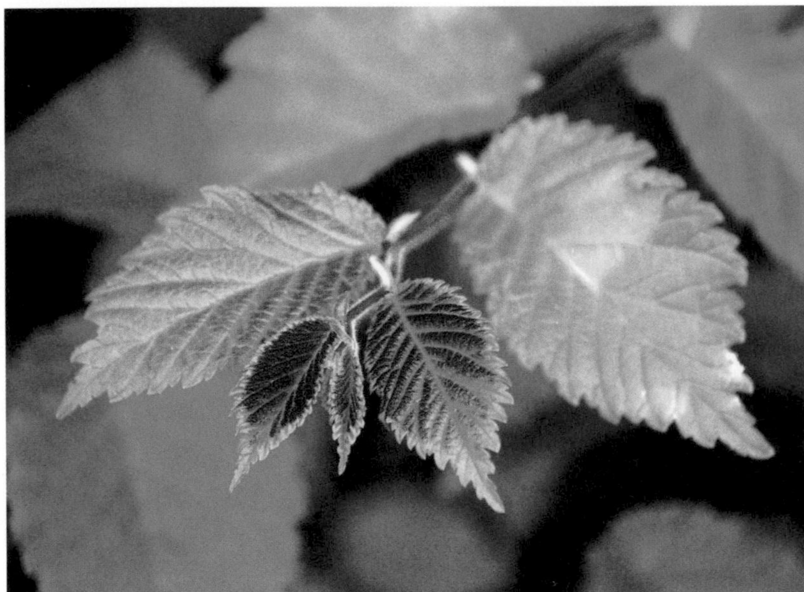

The simple leaves of wych elm (Ulmus glabra) in their development phase, showing the developing venation, and that pigments other than chlorophyll are present within them.

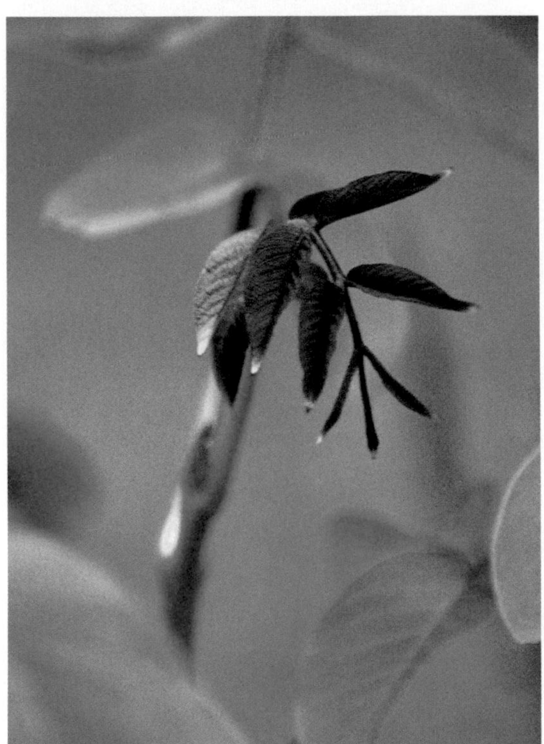

The compound leaves of common walnut (Juglans regia) in their development phase, showing the development of the rachis and leaflets.

dorsiventral leaves, with a typical palisade layer just below the upper epidermis and a spongy mesophyll with many inter-cellular spaces. Isobilateral leaves are typified by their vertical attitude, and may have a 'v'-shaped cross-section, and instead of a specialized palisade and spongy mesophyll, they have loosely packed parenchymatous cells just below both epidermi, that are infested with many chloroplasts for photosynthesis and functions for both tissue types.

The leaf petiole (leaf stalk) of dorsiventral leaves attaches the leaf lamina to the stem, and there is a distinct raised area of the lamina known as the midrib. Running through the cortex of the leaf petiole and midrib is the vascular tissue that branches out from the midrib to a system of lesser veins within the leaf lamina. The upper epidermis has a waxy, waterproof cuticle to reduce water loss and protect the leaf. The lower epidermis is punctuated by a large number of microscopic pores called 'stomata'. Associated with each stoma are two kidney-shaped guard cells that regulate gaseous exchange (including water-vapour movement in and out of the leaf) and a cavity in the spongy mesophyll that facilitates air and water exchange, and forms an integral relationship with the open spongy tissue.

The size of the stomatal opening is altered depending on the turgidity or flaccidity of the two guard cells surrounding it. The two guard cells are

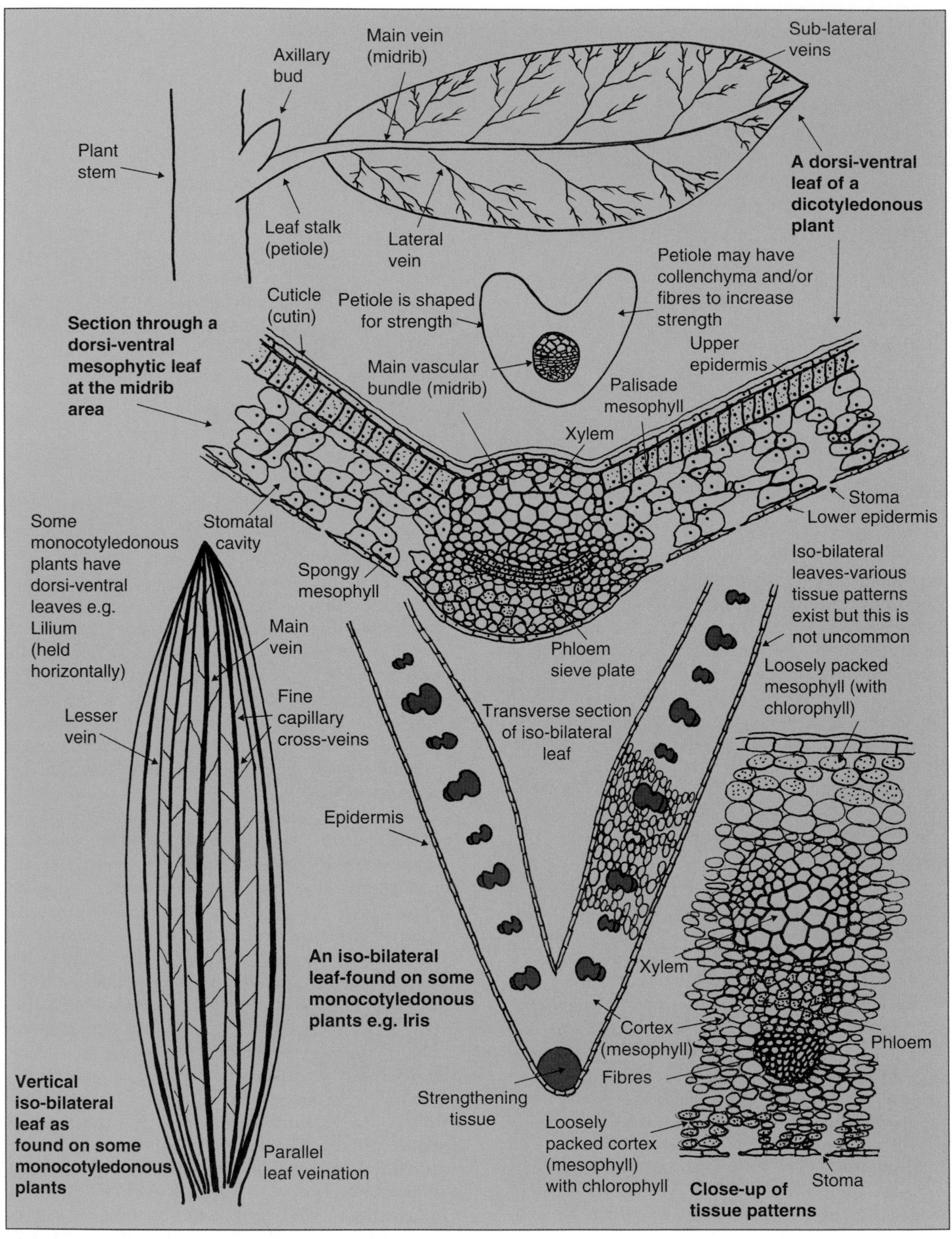

Dorsiventral and isobilateral foliage leaves.

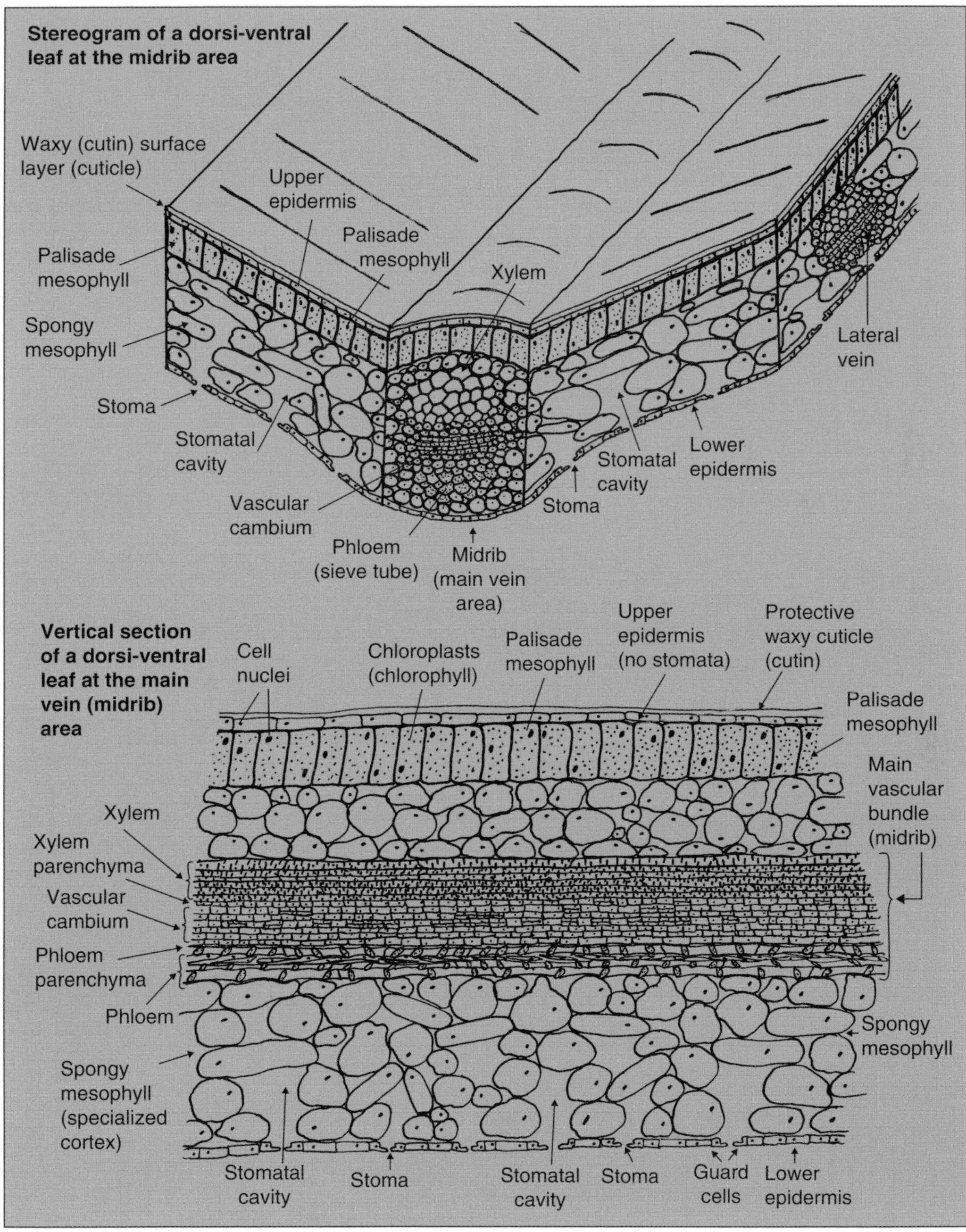

The tissues of a mesophytic dorsiventral leaf.

thickened on their inner cell walls with thickening that is rather 'spring-like' in its action, and holds the two cells 'closed' (as the default position) when the cells are flaccid. Turgor pressure from cells engorged with water will 'open' the cells (or rather the stomatal aperture) against the spring-like pressure created by the thickened cell walls. Hence in times of low water content, the stomatal pore is effectively closed, thus preventing any further loss of water from the leaf tissue; but in times of high water content turgor pressure will open the stoma and allow water to leave the leaf tissue, and also allow carbon dioxide to enter (which is important for the photosynthetic process). Thus in times of very low water availability the guard cells close the stomata to reduce moisture loss – though unfortunately this also reduces, or stops, photosynthesis.

Besides available water via the roots, the rate of water loss via evaporation affects the guard cells. Hence good soil moisture and highly humid conditions will retain the open stomata and allow continued photosynthesis. Guard cells are also affected by light, as during the day photosynthesis within the guard cells produces sugars so that cell sap within them becomes highly concentrated. The concentration gradient set up between the guard cells and the surrounding epidermal cells causes water to move into the guard cells by osmosis, and thus the cells are engorged and the stomata 'open'. At night the production of sugars ceases, and hence the cell sap concentration reduces and water will leave the guard cells, thus 'closing' the stomatal pores. Water leaving (evaporating/volatizing from) the surfaces of the leaves (mainly from the underside via the stoma) creates the driving force for transpiration (the movement of water through the plant) powered by the sun.

Summary and Review of the Concepts of Primary Growth

Primary growth involves an increase in stem and root length, and the development of stems and roots to the end of season one. Meristems found at all plant growth tips (roots and shoots) lay down the new cells, which are destined to form the typical patterns associated with the tissues and organs of plants. Growth (increase in length) manifests itself not only by the addition of new cells, but also by cell elongation during the cell maturation process, a process that greatly increases the obvious visual effect of growth in length.

There are no meristems at the apices of leaves. Instead, the leaf tissue patterns are laid down by the stem apical meristem, and the tissues are left to develop into fully formed leaves as the stem increases in length and progresses ahead of the leaf primordia it lays down. Short-lived marginal leaf meristems aid development to their full size and structure, and cells destined to be veins differentiate to form fully functioning vascular tissue.

Prior to the full development of leaves (and at undeveloped stem apices), water is moved relatively efficiently by diffusion and osmosis. This would not be sufficient for movement over the much larger distances involved in developed organs to satisfy their instant moisture requirements, and fully differentiated conducting vessels (xylem) of the vascular tissues are necessary to prevent large areas of tissues from desiccating.

The degree of change, or how far down the line of differentiation particular cells go, depends on the ultimate function of those cells, and this is decided by their place in the tissue configuration, and the environmental and hormonal triggers they are subjected to. Differentiation may be minor, leaving changed, but nevertheless still soft tissues, in which case the cells may remain alive, with nuclei; in some instances they are still able to divide, and in others they are not. Or, differentiation may involve drastic changes, such as laying down waxes such as lignin (lignification), or more rarely, suberin, inside the cell wall, setting up the cells and tissues for their specialized functions within the plant.

The vascular cambium on division and differentiation produces all the cells of vascular tissue (the veins). The epidermal cells of green stems and new roots are soft, parenchymatous cells, whose differentiation does not include the addition of the wax lignin to the inside of the cell walls. Xylem cells, tracheids and fibres, on the other hand, differentiate to include large amounts of lignin, which gives mechanical support and strength but renders the cells 'dead' as they no longer have either cytoplasm or nuclei.

In the green plant the whole of the tissue patterns for aerial growth are laid down in the apical meristem, and their subsequent development gives rise to extension in length (elongation), and their subsequent differentiation gives rise to all the specialized cells needed for successful existence. Provision for new green aerial growth is also potentially given by meristems in the axillary buds (because they share the same tissue patterns as apical buds, including meristems). This growth type, which is primarily involved in extension in length, with only a limited predetermined girth produced in its first year, is primary growth.

PLANT STRATEGIES

Leaf Modifications

Floating water plants have specialized open tissues (aerenchyma) within the leaf petiole and leaf lamina that help buoyancy; a good example is the leaves of water lilies (*Nymphaea* species) floating on the surface whilst their rhizomes are bedded into the silts and muds of the bottom of the pond or lake.

It is quite common for the leaves of some monocotyledonous plants to have very thin-walled bubble-shaped (bulliform) cells. These induce the opening, folding or rolling of leaves, depending on the various turgor (water) pressures present in the leaves at any one time, and are found in the epidermal layers of, for example, sweetcorn (*Zea mais*).

Leaf modifications also include xerophytic adaptations to the leaves of species in very dry climates, the twining petiole of some *Clematis* species, and the leaf tendrils of some pea species – leaf extensions that aid plant support. Spines on the leaf margins (barrier leaves) of species such as common holly (*Ilex aquifolium*), *Desfontainea spinosa* and sea holly (*Eryngium maritimum*) are good examples of leaf modifications, as are the very sharp spines on the tips of the leaves of *Agave americana*.

The leaves of some *Bryophyllum* and *Kalanchoe* species (succulents) have small plantlets (propagules) on their leaf margins – for example *Kalenchoe daigremotiana*. Some fern species (notably the *Polystichum divisilobum* group) and *Begonia Sutherlandii* also show this phenomenon.

Some species of plant are insectivorous (carnivorous), and have adapted to gain all their nutrition from trapped insects because they live in highly acidic soils of very low nutrient status.

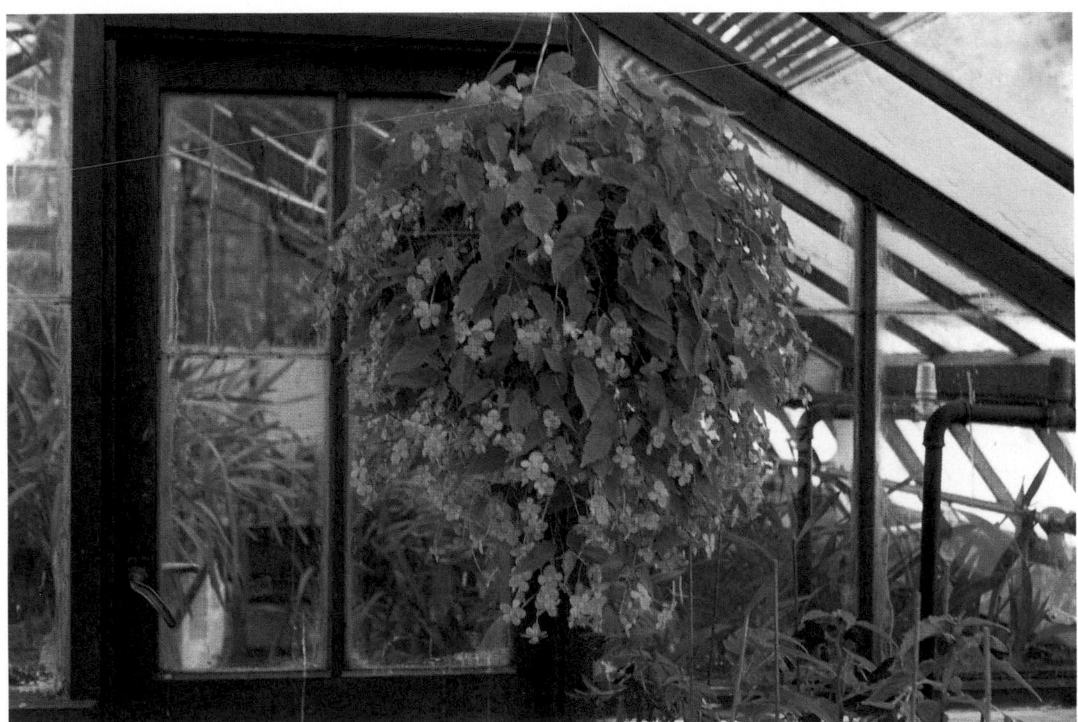

Begonia sutherlandii.

In all instances the 'trapping' organ is formed by a modified leaf, either formed into 'pitchers' as rolled leaves, or with closing valves.

Motile Parts

Besides those found in insectivorous plants, other plant movements are also known. The 'closing/folding' effect of the leaves of the sensitive plant (*Mimosa pudica*) is triggered by touch, and is thought to be facilitated by differential turgor pressure in the rachis (the extended petiole) and rachellae (smaller branching leaf stalks) of the leaf. Although touch is the trigger in these instances, it is the discharging of the electrolytic forces within the organ (and the associated rapid loss of cell turgor) that causes the movement, rather than a stimulus towards, or an irritation away from it.

Root-tip meristems lay down all the basic patterns for root growth, and they are responsible for root extension (growth in length) as the young cells elongate; and the differentiation of parenchymatous cells laid down by the meristem creates all the specialized tissues found in roots. Specific cells differentiate into xylem and phloem, and others into soft packing tissues (cortex), and external cells into epidermis.

The basic tissue patterns in stems and roots are different in the early stages of development. At the close of primary growth in dicotyledonous subjects the vascular bundles in the root take on the formation of a 'star', and in the stem they form a ring of individual bundles but beginning to coalesce. Ultimately, after secondary growth, both form identical concentric rings of tissues.

Dicotyledonous Woody Stems

Woody plants add new primary growth as extension in length every year, and they also add secondary tissues externally to the primary tissue configuration on an annual basis, to form a distinct woody framework. It is the capacity for girth increment, formed by secondary growth, that sets woody plants (arboreals) apart from all other plant groups.

In woody plants nearing the end of year one, the epidermal cells of the once green stem become peppered with bark initials (periderm) that displace the soft tissues and ultimately coalesce to form a uniform covering. Furthermore, within the stem the vascular bundles are no longer arranged as individual bundles, but have coalesced until the phloem, xylem, and vascular cambium tissues form continuous rings in plan view (cylinders in three dimensions). The cambium layer between the xylem and phloem remains meristemetic, and produces new xylem cells on its inside, and new phloem cells on its outside. The addition of these cells gradually increases the stem girth and puts the older, most external phloem tissue under extreme pressure so it cracks and sloughs off. Furthermore, the epidermis, some of which has already been replaced with bark tissue, also comes under stress and cracks away.

Apical meristems comprise an amorphous mass of tissues able to divide in all directions to increase the volume of meristematic tissue. Cambia, on the other hand, although they are initiated from early shoot apex and root-tip meristems in procambial areas, tend to specialize in the directions of their cell division. Cambia form brick-like cells in circles, rings and cylinders of tissue by having rapid and continued cell division within the circumference of the 'ring' (and either side of the circumference). Hence the addition of cells by meristems, which would extend organ length, continues in primary tissues but is not the norm in secondary tissues – only an increase in stem or root girth occurs. Furthermore, it is essential that the increase in circumference of the cambium is quick enough to facilitate the ever-increasing thickness of tissues created by divisions inside and outside the cambial ring. In the tissue configuration of young dicotyledonous stems, divisions of the intrafasicular cambium (vascular cambium within the bundles) creates thicker tissues by adding both extra xylem and phloem to the bundles, but this is to a finite level in year one. The ability of meristematic tissues between the vascular bundles (the interfasicular cambium) to produce many more new bundles eventually leads to the bundles coalescing into a complete circle of tissue when seen in plan view, and a cylinder of tissue when seen in three dimensions.

Meristematic activity of the vascular cambium produces new xylem parenchyma on its inside, and new phloem parenchyma on its outside. Phloem parenchyma differentiates into phloem tissue (sieve tubes). Xylem parenchyma is destined by virtue of its position to be xylem, and differentiates over a

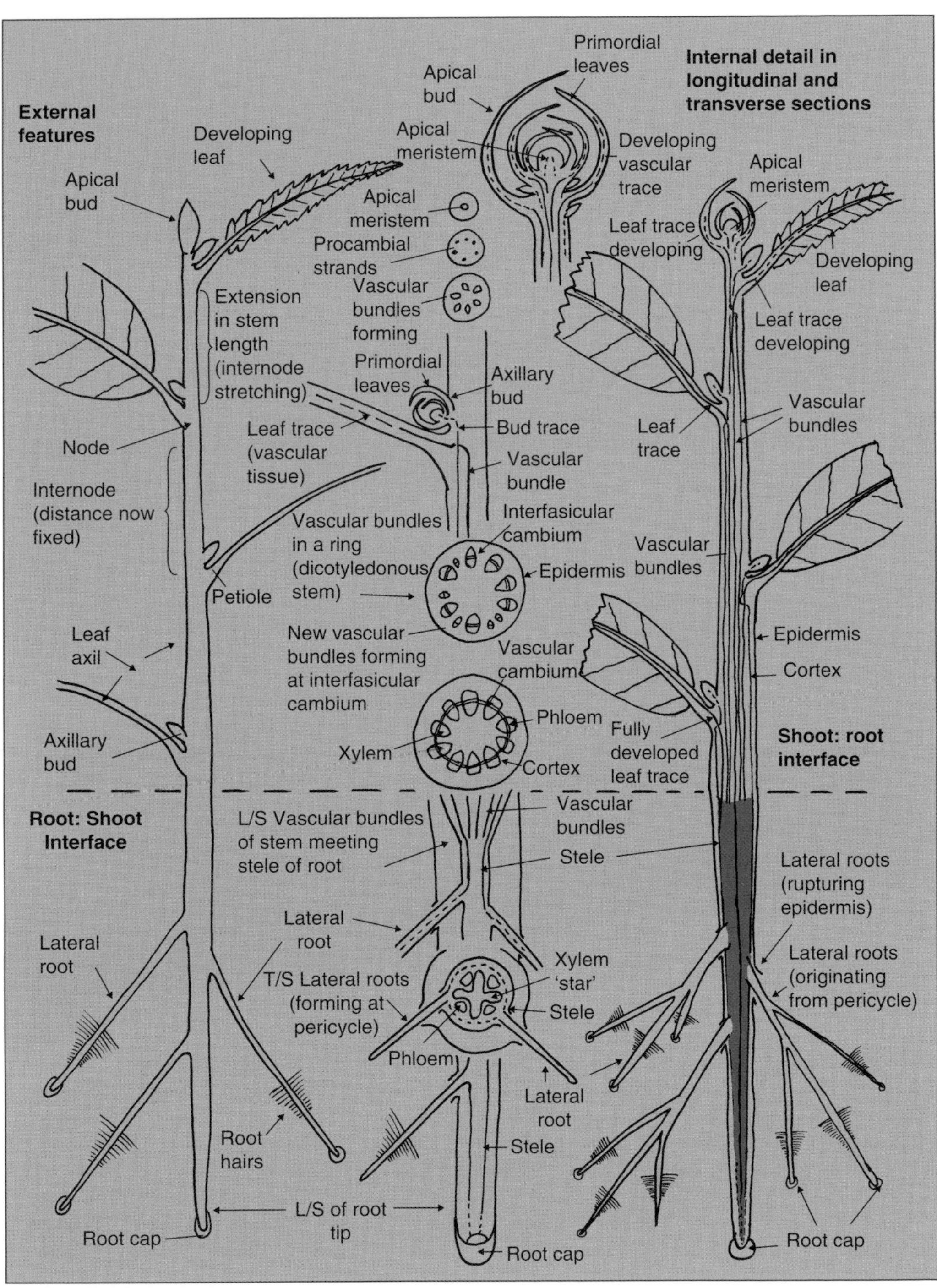

The external and internal features of the roots and stems of a green plant compared.

relatively short period of time to become tube-like cells that facilitate the upward movement of water and dissolved nutrient salts from the roots to the leaves and buds.

It is the addition of xylem tissues internal to the vascular cambium and phloem tissues external to it, plus the radial additions, that cause the large increase in stem circumference. When the vascular cambium puts on xylem parenchyma internally, there is nowhere for it to go, as each season's additions compress the trapped internal tissues. The only way that new xylem can be laid down internally is by the ring diameter of the vascular cambium increasing, which is facilitated by lots of periclinal divisions (parallel to the root/shoot axis), and also radial divisions (increases within the circumference). As the stem circumference increases rapidly it puts terrific pressure on the tissues outside the vascular cambium ring (the phloem, and the rest of the cells that make up the bark). It is these tissues that are regularly sloughed off and replaced. Phloem therefore always comprises new, well functioning cells.

Bark Production

Bark is produced by the cork cambium, which is meristematic, but because the cells it produces are not brick-like (as in other cambia), it is considered by some not to be a cambium at all, and the term phellogen (to describe the generation of phellum – cork) is used instead.

On young stems approaching the end of their first season, the change from soft epidermis with waxy cuticle to suberized bark may occur gradually by the invasion of bark initials that shed soft epidermal cells when they come under pressure. Or, as is common in many species, a periderm forms below the epidermis, initiated in the cortex, which sloughs/displaces the green epidermis as it develops, and the stoma present on the stem are displaced and sloughed off with the outer tissues. The result is a cork-covered stem, punctuated with fixed-aperture lenticels, rather than the stoma with regulatory guard cells. Lenticels are small pores comprising breaches in the bark full of softer, lightly packed cells, creating apertures that facilitate gaseous exchange by diffusion. The drastic external changes to primary growth are hormonally induced, and triggered by decreasing day length coupled with decreasing temperatures, and are obviously a precursor to the onset of winter – that is, setting the plant up for the vagaries of the winter climate. Soft, green, primary growth would not overwinter well in temperate or colder climates.

The periderm is formed before the other tissues external to it fracture, and 'lies in wait' to lay down new material in the spaces left by the fracture. The periderm is associated with older crushed phloem, some secondary phloem, and crushed phloem fibres, because it permeates these tissues. Everything outside the periderm in a developing root or stem is dead or dying and sloughs away, taking the old fused phloem with it. The periderm is so called because it forms the outermost skin (peri = outside, and derm = skin), and its three main tissues are the phellogen, phelloderm and phellum. The phellogen (cork cambium) is the central tissue and is meristematic – a cell-generating tissue that produces phelloderm (cork cortex) on its inside, and phellum (cork) on its outside. Bark therefore comprises four tissues: the phellum, phellogen, phelloderm and old phloem.

Cork cambium (phellogen) produces cells radially to keep up with the increasing stem girth, so that as much tissue as possible remains covered with protective bark at all times, but it also divides rapidly periclinally to produce new cells on each side of it. The tissues laid down internally to the phellogen are known as phelloderm (cork cortex), and have a packing tissue role where it interfaces with the phloem below it. The external tissues (phellum or cork) crack under the stress of the increasing stem girth, and are constantly sloughed off externally (or may be retained for long periods before sloughing – a natural and continuous process).

The growth of woody stems is therefore accomplished by the increase of two concentric rings of tissue (vascular cambium and phellogen). The increase in stem girth comes by the addition of secondary xylem and phloem tissues produced by the cylinder of vascular cambium, and the phellogen ring has to increase drastically to accommodate this. The vascular cambium also has to divide rapidly in diameter in order to accommodate the new xylem cells it is producing internally.

Xylem cells are dead. They have in fact 'differentiated themselves to death', in that they have changed so drastically from their initial soft, malleable condition that they now have heavily lignified cell walls, and no cytoplasm in which to suspend nuclei. But although xylem cells are dead, they still function at this stage, acting as a vessel for water transport because of their tube-like configuration, and they also have a very important role in giving structural strength to stems and roots. The xylem has good functioning tissue for some way inside the ring of vascular cambium,

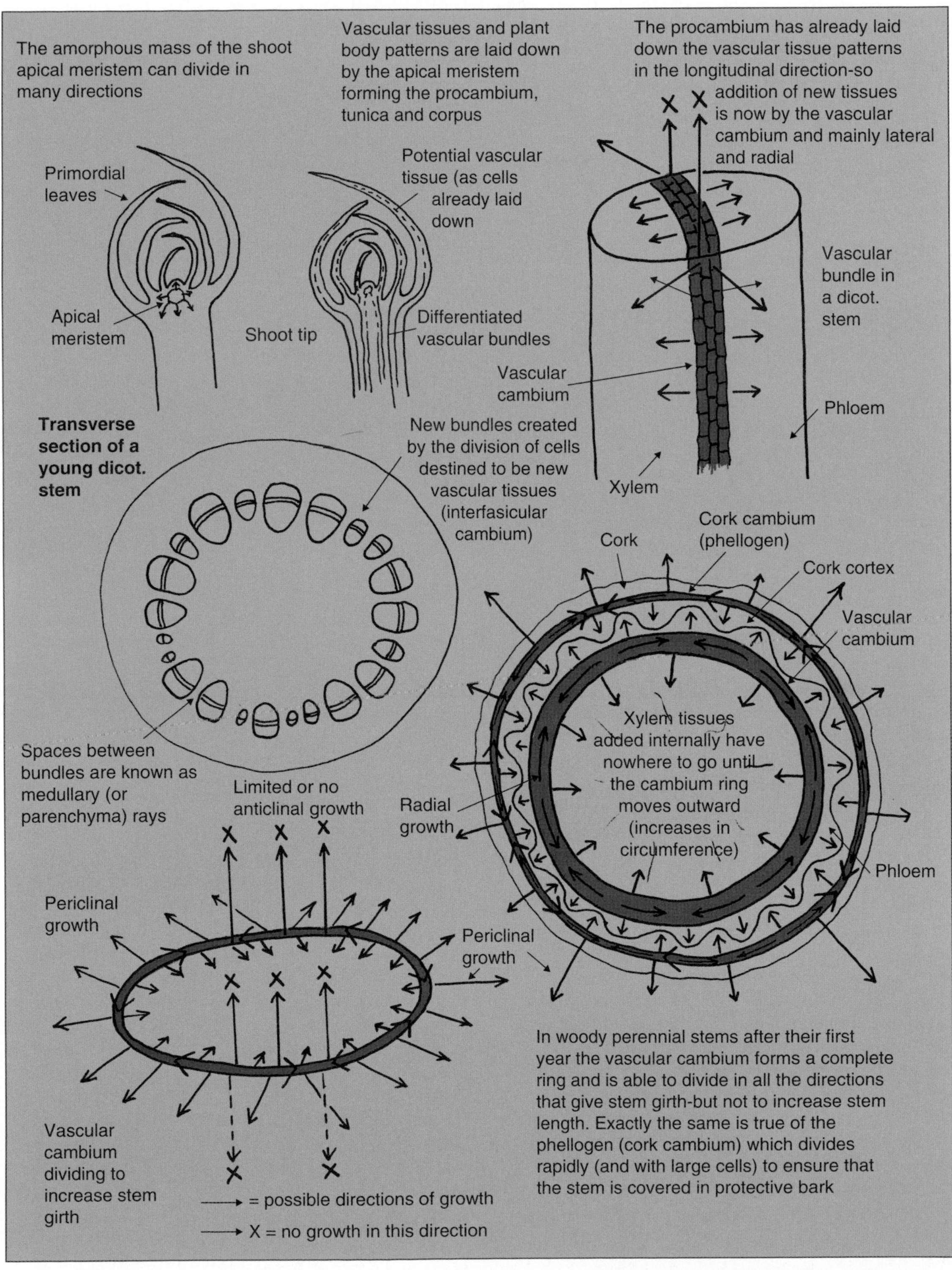

The amorphous mass of the shoot apical meristem can divide in many directions
Vascular tissues and plant body patterns are laid down by the apical meristem forming the procambium, tunica and corpus
The procambium has already laid down the vascular tissue patterns in the longitudinal direction-so addition of new tissues is now by the vascular cambium and mainly lateral and radial
Primordial leaves
Apical meristem
Shoot tip
Potential vascular tissue (as cells already laid down
Differentiated vascular bundles
Vascular bundle in a dicot. stem
Vascular cambium
Phloem
Xylem
Transverse section of a young dicot. stem
New bundles created by the division of cells destined to be new vascular tissues (interfasicular cambium)
Spaces between bundles are known as medullary (or parenchyma) rays
Cork
Cork cambium (phellogen)
Cork cortex
Vascular cambium
Xylem tissues added internally have nowhere to go until the cambium ring moves outward (increases in circumference)
Radial growth
Phloem
Limited or no anticlinal growth
Periclinal growth
Periclinal growth
Vascular cambium dividing to increase stem girth
= possible directions of growth
X = no growth in this direction
In woody perennial stems after their first year the vascular cambium forms a complete ring and is able to divide in all the directions that give stem girth-but not to increase stem length. Exactly the same is true of the phellogen (cork cambium) which divides rapidly (and with large cells) to ensure that the stem is covered in protective bark

The addition of cells to meristems and cambia.

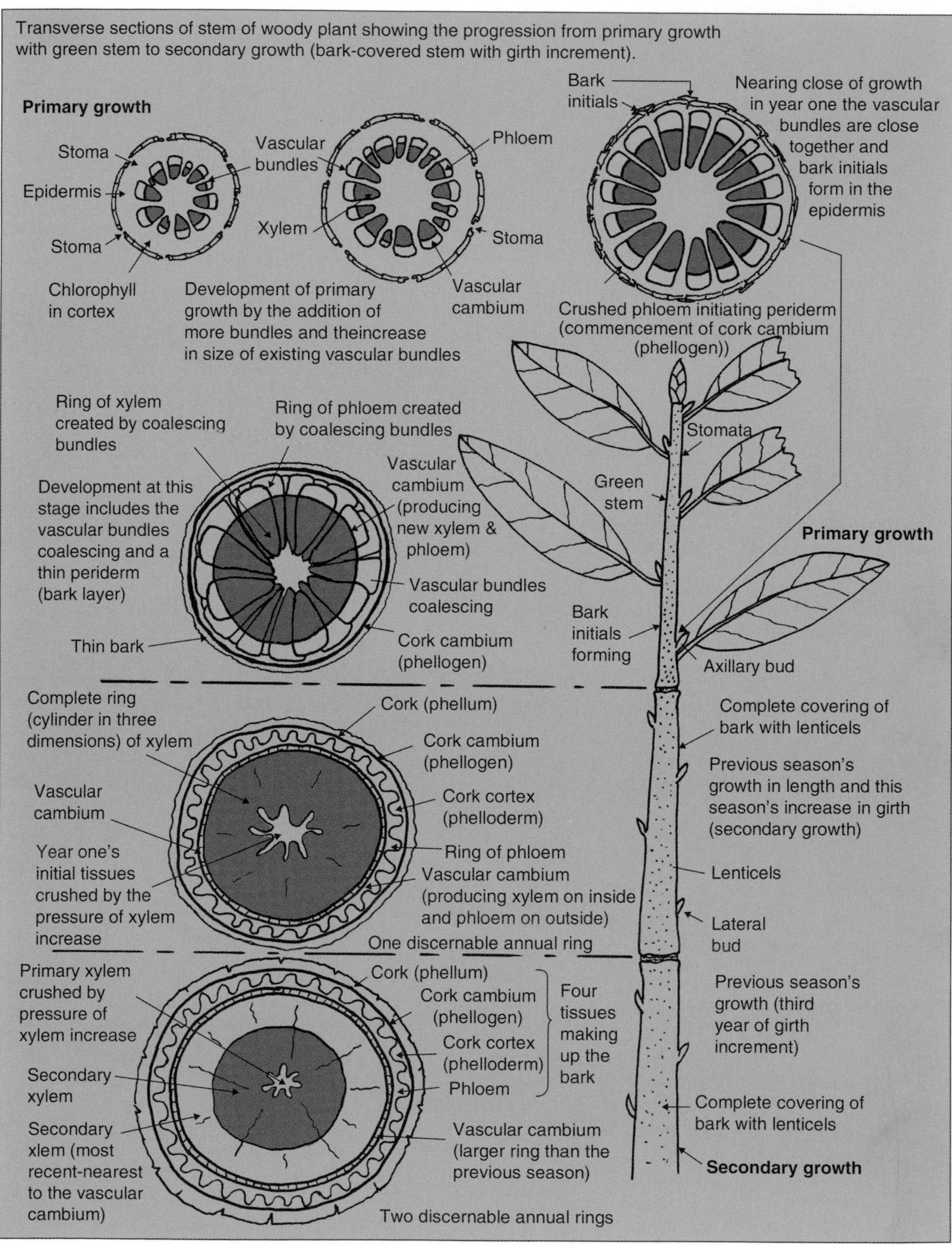
Transverse sections of stem of woody plant showing the progression from primary growth with green stem to secondary growth (bark-covered stem with girth increment).
Primary growth
Stoma
Epidermis
Stoma
Chlorophyll in cortex
Vascular bundles
Phloem
Xylem
Stoma
Vascular cambium
Development of primary growth by the addition of more bundles and theincrease in size of existing vascular bundles
Bark initials
Nearing close of growth in year one the vascular bundles are close together and bark initials form in the epidermis
Crushed phloem initiating periderm (commencement of cork cambium (phellogen))
Ring of xylem created by coalescing bundles
Ring of phloem created by coalescing bundles
Development at this stage includes the vascular bundles coalescing and a thin periderm (bark layer)
Vascular cambium (producing new xylem & phloem)
Vascular bundles coalescing
Thin bark
Cork cambium (phellogen)
Stomata
Green stem
Primary growth
Bark initials forming
Axillary bud
Complete ring (cylinder in three dimensions) of xylem
Cork (phellum)
Cork cambium (phellogen)
Vascular cambium
Cork cortex (phelloderm)
Year one's initial tissues crushed by the pressure of xylem increase
Ring of phloem
Vascular cambium (producing xylem on inside and phloem on outside)
One discernable annual ring
Complete covering of bark with lenticels
Previous season's growth in length and this season's increase in girth (secondary growth)
Lenticels
Lateral bud
Primary xylem crushed by pressure of xylem increase
Cork (phellum)
Cork cambium (phellogen)
Cork cortex (phelloderm)
Phloem
Four tissues making up the bark
Secondary xylem
Secondary xlem (most recent-nearest to the vascular cambium)
Vascular cambium (larger ring than the previous season)
Two discernable annual rings
Previous season's growth (third year of girth increment)
Complete covering of bark with lenticels
Secondary growth

Secondary thickening (girth increment) of woody stems.

but further in, the oldest xylem tissues become dysfunctional (later to form heartwood). The very first xylem cells (primary xylem) become crushed by the terrific pressure caused by the addition of secondary tissues. Ultimately the primary xylem is no longer discernible as a separate and distinct tissue pattern – unlike the rings of secondary xylem tissues (wood) added annually from then on, which create and leave distinct annual rings. Annual rings hold an historical record of the growth patterns within any year, which remain with the tree or shrub all its life.

Tree and shrub growth can be best envisaged as woody growth of one year being overlain by the next season's growth – having a 'core' of original growth overlain with a 'skin' of newer growth annually. All growth/tissues once laid down remain where they are, and new tissues are superimposed upon them on an annual basis, as described by the core:skin hypothesis. The effect of this is to increase girth, to create taper (a strong shape), and to build up tissues from which new extension growth can be produced.

The Roots of Dicotyledonous Woody Species

Inside the endodermis of the root is a single layer of parenchyma cells known as the pericycle, the cells of which become meristematic in woody perennials and ultimately become the site for the production of a periderm (outside skin). The periderm includes a phellogen (cork cambium) that produces corky tissues externally on old roots, as ultimately roots develop a corky outer bark layer, like stems. The pericycle is also the primary site for the initiation of lateral roots. The cambium of the root is produced between the tissues of the xylem and the phloem, when the cells in that area become meristematic and make radial and periclinal divisions. Where this tissue touches the pericycle it becomes continuous with it, eventually forming a ring of root cambium.

The root cambium produces new (secondary) xylem on its inside and new (secondary) phloem on its outside, the effect of which is to compress the initial (primary) xylem internally, and push the initial (primary) and secondary phloem outwards. Furthermore, the increase in xylem tissues forces the cambial tissues into a circle, not unlike the tissue configuration of a stem.

Lateral root initiation commences at the meristematic pericycle, when some cells in this area mass to form a root apex and gradually organize themselves into new root tissues. As they develop they force their way through first the endodermis, and then the other outer tissue layers (the cortex first and then the epidermis). Hence lateral roots, just like lateral branches, arise from internal tissues as the norm (endogenous). However, there are variations from the norm, as the same meristematic tissues at the pericycle are also responsible for the production of the periderm that will ultimately cover old roots externally. If the pericycle is used up forming the periderm (as it is in older, cork-covered roots), then lateral root production occurs instead from areas of the vascular cambium near the phloem. Furthermore, just as stems are able to form specialized epicormic stems originating from external tissues (exogenous) and not creating the normal strength inherent in branch unions, likewise root systems can produce new roots derived from external tissues rather than the internal pericycle, such as the aerial roots of orchids; and, as already discussed, stems can also produce forms of adventitious roots from their external tissues.

Summary and Review of Woody Growth

Understanding that new growth in girth overlays old existing growth, that extension in length occurs in primary (green) growth only, and that it remains where it is laid down; and that increase in girth is achieved by two concentric rings of meristematic tissues (the vascular cambium and the phellogen), is to understand woody plant growth. Green stems are responsible for extension in length. Leaves and green stems are punctuated with pores called stomata (whose aperture is regulated by guard cells), whereas bark-covered stems are punctuated by permanently open, non-regulatory lenticels. Thus both stomata and lenticels facilitate the interaction of woody plants with their environment – the aerial environment only in the case of stomata (leaves and green stems), and both aerial and sub-surface aerobic environment in the case of lenticels (cork-covered stems and roots).

Healthy bark has a moist interface with the phloem (considered to be an integral part of the bark tissues) on its inside, and the phloem has a moist interface with vascular cambium internal to it. The external cork cells form plates or layers of tissue that ultimately slough off the outside – not unlike reptilian skin. As the old bark sloughs away, the new replacement bark produced by the phellogen (cork cambium) fills the gaps. The new bark is produced to replace old discarded bark, fill the existing spaces between bark plates, and form a complete sheet under the older bark tissues, thus ensuring integrity throughout.

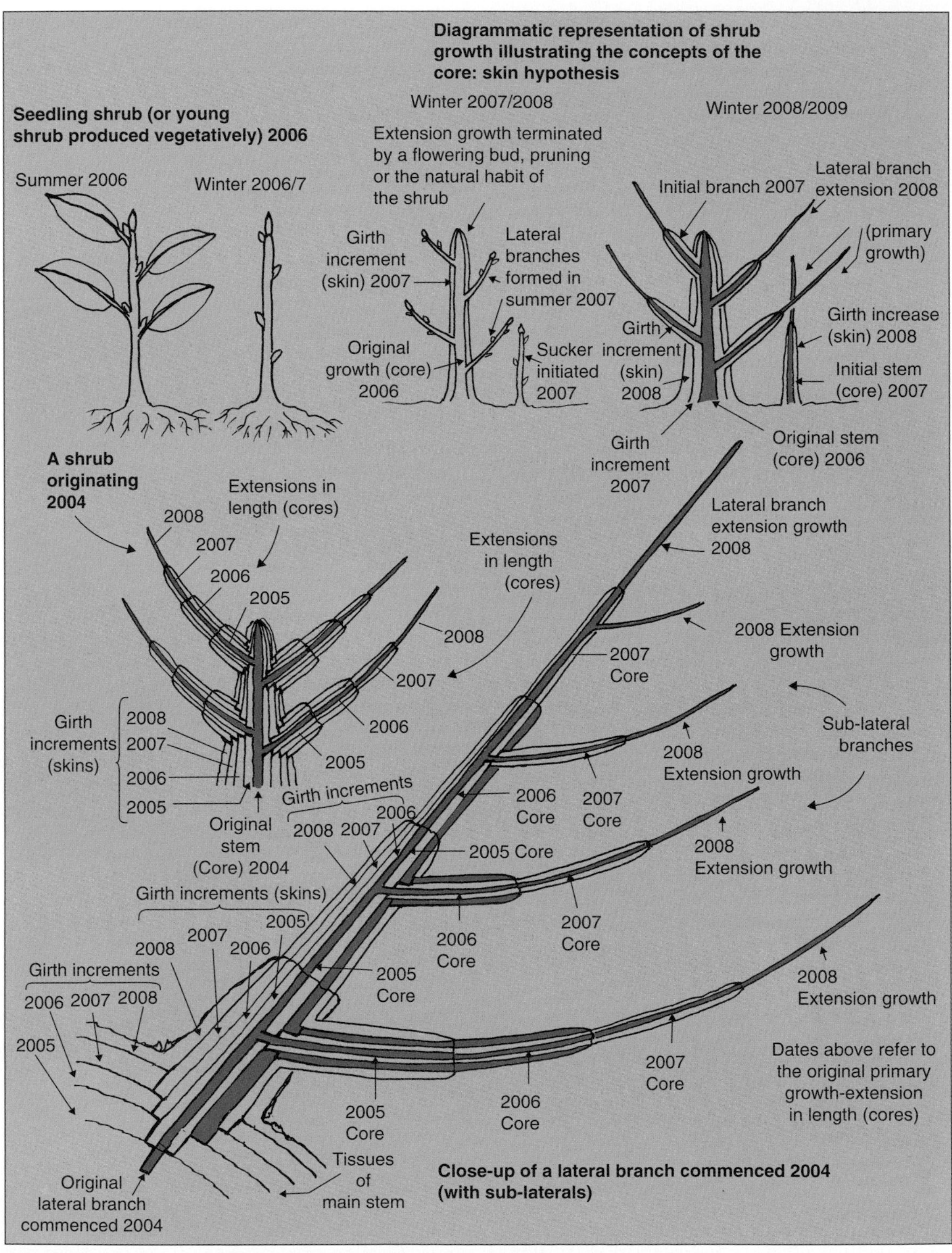
Diagrammatic representation of shrub growth illustrating the concepts of the core: skin hypothesis
Seedling shrub (or young shrub produced vegetatively) 2006
Summer 2006
Winter 2006/7
Winter 2007/2008
Extension growth terminated by a flowering bud, pruning or the natural habit of the shrub
Girth increment (skin) 2007
Lateral branches formed in summer 2007
Original growth (core) 2006
Sucker initiated 2007
Winter 2008/2009
Initial branch 2007
Lateral branch extension 2008
(primary growth)
Girth increment (skin) 2008
Girth increase (skin) 2008
Initial stem (core) 2007
Girth increment 2007
Original stem (core) 2006
A shrub originating 2004
Extensions in length (cores)
2008
2007
2006
2005
Girth increments (skins)
Original stem (Core) 2004
Lateral branch extension growth 2008
2007 Core
2008 Extension growth
Sub-lateral branches
2006 Core
2007 Core
2005 Core
Girth increments
2008 2007 2006
Girth increments (skins)
2008 2007 2006 2005
Girth increments
2006 2007 2008
2005
2005 Core
2006 Core
2007 Core
2008 Extension growth
Dates above refer to the original primary growth-extension in length (cores)
Tissues of main stem
Original lateral branch commenced 2004
Close-up of a lateral branch commenced 2004 (with sub-laterals)

The core: skin hypothesis as illustrated in shrubs.

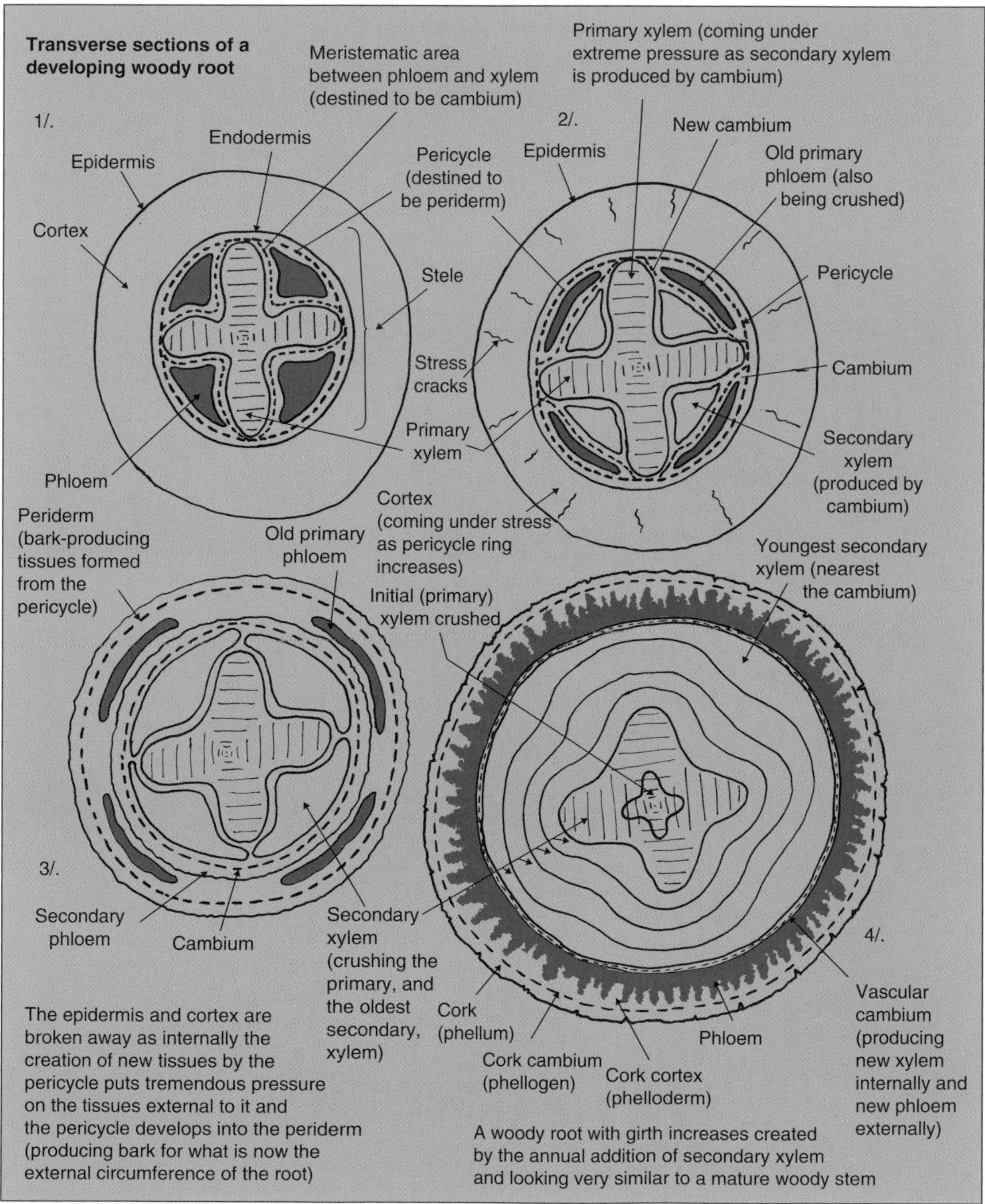
Transverse sections of a developing woody root
Meristematic area between phloem and xylem (destined to be cambium)
Primary xylem (coming under extreme pressure as secondary xylem is produced by cambium)
1/.
Endodermis
Epidermis
Pericycle (destined to be periderm)
Cortex
Stele
Stress cracks
Primary xylem
Phloem
2/.
New cambium
Epidermis
Old primary phloem (also being crushed)
Pericycle
Cambium
Secondary xylem (produced by cambium)
Cortex (coming under stress as pericycle ring increases)
Periderm (bark-producing tissues formed from the pericycle)
Old primary phloem
Initial (primary) xylem crushed
Youngest secondary xylem (nearest the cambium)
3/.
Secondary phloem
Cambium
Secondary xylem (crushing the primary, and the oldest secondary, xylem)
Cork (phellum)
Cork cambium (phellogen)
Cork cortex (phelloderm)
Phloem
4/.
Vascular cambium (producing new xylem internally and new phloem externally)
The epidermis and cortex are broken away as internally the creation of new tissues by the pericycle puts tremendous pressure on the tissues external to it and the pericycle develops into the periderm (producing bark for what is now the external circumference of the root)
A woody root with girth increases created by the annual addition of secondary xylem and looking very similar to a mature woody stem

Tissue patterns of the root and central stele showing the root development of a woody dicotyledonous plant (trees and shrubs).

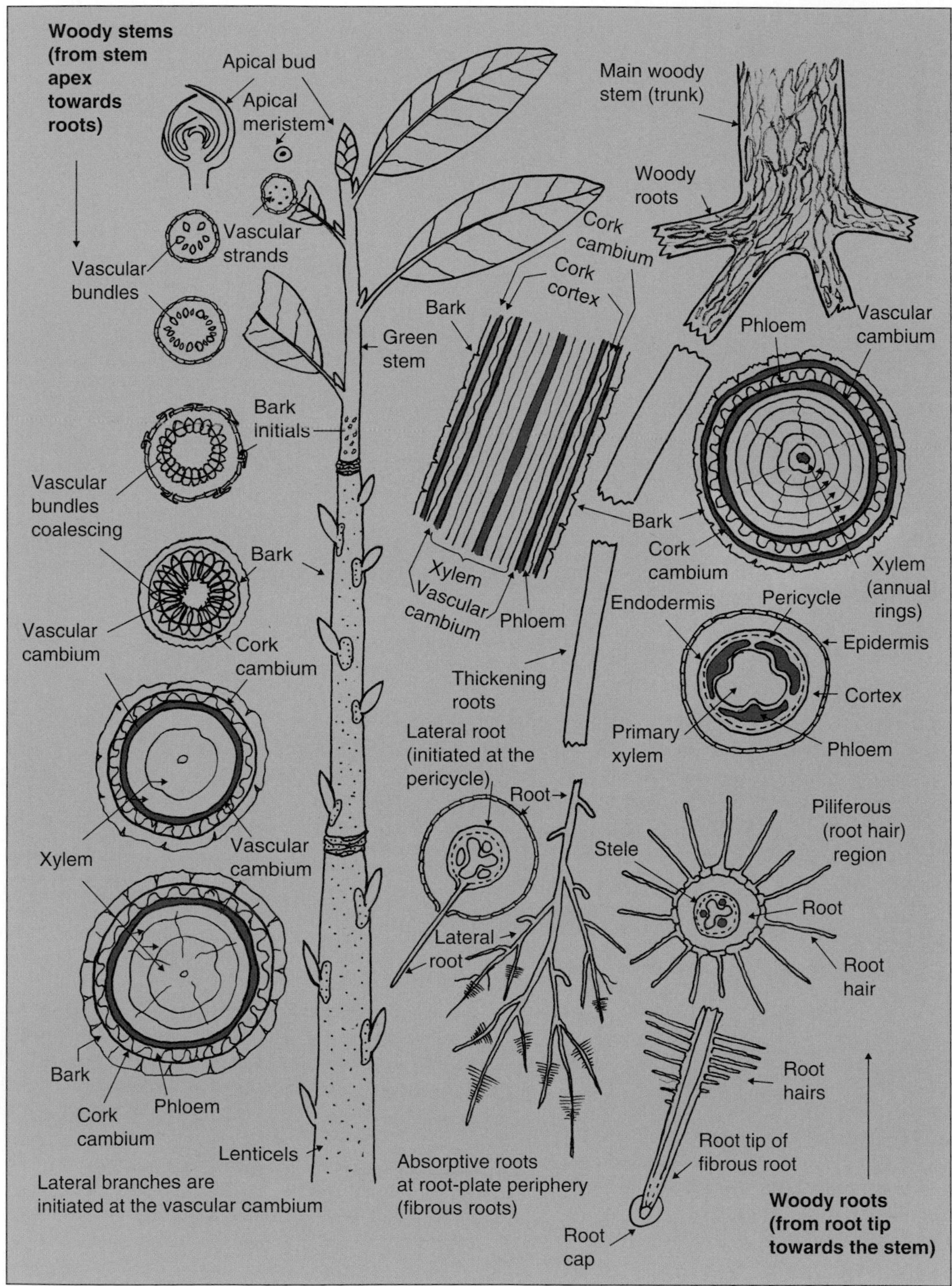

Comparing woody stem and woody root development.

CHAPTER 6

The Physiological Processes of Plants

When discussing plant physiology it is useful to remember that energy cannot be destroyed, but can only change its state (be converted to a different form). Energy can be released as heat or light, as chemical energy, or stored in solids (such as plants) to be released at a later date by dissipation into the atmosphere (where it may be reused by temporarily being incorporated into complex compounds). The energy (from the sun) trapped in green plants as sugars and starches can be used by converting it into different products, but can never be destroyed, even though the individual plants may be.

The main physiological processes of plants are photosynthesis, respiration and transpiration. Photosynthesis involves the production of sugars, whereas respiration involves releasing energy from the sugars produced by photosynthesis, and transpiration concerns the movement of water through the plant from the soil matrix to the atmosphere.

Photosynthesis

Green plants obtain their nutrition by both photosynthesis and by absorption of minerals from the soil. Photosynthesis, however, accounts for approximately 95 per cent of the dry matter of the plant body, whilst mineral absorption only accounts for about 5 per cent. Nevertheless, the presence of certain minerals is essential for protein synthesis and the energy for plant growth – particularly those minerals involved in the chlorohyll molecule.

Photosynthesis occurs in leaves and green stems, and is an anabolic (energy-producing) process producing carbohydrates (glucose/soluble sugars) using carbon dioxide and water as the main ingredients, and with chlorophyll as a catalyst in the presence of light (photo = light, and synthesis = manufacture).

Mesophytic leaves are perfectly set up for photosynthesis. Movement of moisture and CO_2 in and out of the leaf tissues is facilitated by stomata (regulated by the guard cells), and the spongy mesophyll allows easy movement of gases through the internal tissues, including oxygen, carbon dioxide and water vapour. The palisade layer has the densest concentration of chloroplasts, and because the epidermis is transparent, and leaves can orient themselves to face the sun by tropic movements (positive phototropism), the soldier-like, upright orientation of the palisade cells allows maximum solar radiation. This commences the process of photosynthesis by 'activating' the chlorophyll.

A temperature range that supports plant life is essential for photosynthesis. Water is essential but only needed in minute amounts, and is supplied by atmospheric moisture and cell sap. Air containing carbon dioxide and water vapour is taken in at the stomata and collects in the large pore spaces (stomatal cavities) and intercellular spaces within the spongy mesophyll. Chlorophyll activated by solar energy reacts with water and splits water into hydrogen and oxygen (photolysis): $2H_2O = 2H_2 + O_2$. Oxygen is released via the stomata into the atmosphere, which is very important for respiration. The hydrogen reacts with carbon dioxide to ultimately form carbohydrates (glucoses – soluble sugars) that are transported through the plant via the phloem and may be stored as starches (condensed sugars) in stems and roots. At a later date enzyme action can convert starches back to soluble sugars so that they may be used during respiration to provide energy where it is needed.

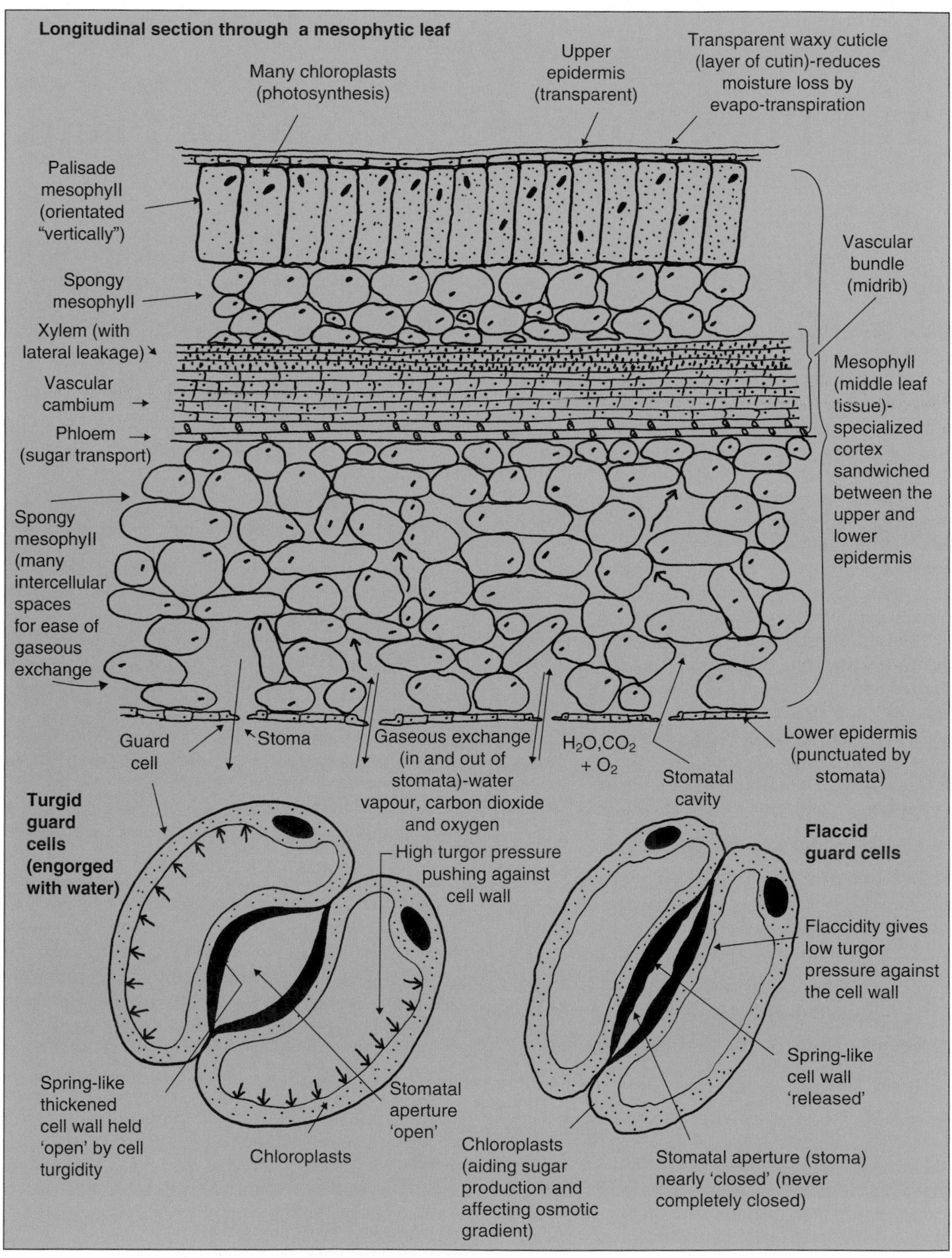
Longitudinal section through a mesophytic leaf
Many chloroplasts (photosynthesis)
Upper epidermis (transparent)
Transparent waxy cuticle (layer of cutin)-reduces moisture loss by evapo-transpiration
Palisade mesophyll (orientated "vertically")
Spongy mesophyll
Xylem (with lateral leakage)
Vascular cambium
Phloem (sugar transport)
Vascular bundle (midrib)
Mesophyll (middle leaf tissue)-specialized cortex sandwiched between the upper and lower epidermis
Spongy mesophyll (many intercellular spaces for ease of gaseous exchange
Guard cell
Stoma
Gaseous exchange (in and out of stomata)-water vapour, carbon dioxide and oxygen
H_2O, CO_2 + O_2
Stomatal cavity
Lower epidermis (punctuated by stomata)
Turgid guard cells (engorged with water)
High turgor pressure pushing against cell wall
Spring-like thickened cell wall held 'open' by cell turgidity
Chloroplasts
Stomatal aperture 'open'
Flaccid guard cells
Flaccidity gives low turgor pressure against the cell wall
Spring-like cell wall 'released'
Chloroplasts (aiding sugar production and affecting osmotic gradient)
Stomatal aperture (stoma) nearly 'closed' (never completely closed)

Leaf tissues and their relevance to physiological processes.

A simplified reaction for photosynthesis may be shown thus:

$6CO_2$	$+ 6H_2O$ +	Light and	$= \star\ C_6H_{12}O_6 + 6O_2$
Carbon	+ Water +	Chlorophyll	⋆ Glucose + Oxygen
dioxide		(Energy)	given off

Photosynthesis releases oxygen, converts inorganic substances into stable organic substances with high potential chemical energy (it converts CO_2 and H_2O into carbohydrates), and 'fixes' carbon in the plant in the form of carbohydrates (the carbon store or carbon sink). The carbohydrates stored in plants (in very large amounts in trees) form the basis of the carbon cycle.

Light intensity and duration, carbon dioxide concentrations and temperature, all affect the rate of photosynthesis, which is limited by the lowest factor and/or a limiting maximum. If temperature and carbon dioxide concentrations are kept constant, the rate of photosynthesis increases with an increase in light intensity. However, the rate then levels off (light saturation). If the carbon dioxide concentration is increased, photosynthesis increases, then levels out – the extra CO_2 allows the use of extra available light. If the temperature is continuously increased, the rate of photosynthesis increases until the CO_2 is the limiting factor – because the stomatal apertures reduce in the increasing temperature.

Under controlled glasshouse conditions some growers enrich the atmosphere with extra CO_2 by burning propane gas, which brings concentrations up from the normal 380 ppm of our atmosphere to nearer 1,000 ppm. If other factors are not limiting, then this increase in CO_2 will greatly improve photosynthesis – particularly if additional artificial light sources are used in naturally poor light conditions.

Light

Both light intensity and quality are important. High light intensities are good for photosynthesis, but are unfortunately usually associated with high temperatures that close the guard cells, and therefore reduce the CO_2 content, which limits or stops the process. Solar radiation must give the correct wavelengths. Red light is important to photosynthesis, as is blue. Yellow and green light are also important for photosynthesis, but far less so than red or blue.

Leaves absorb (and reflect some) light within the part of the spectrum visible to the human eye. Light that is absorbed by leaves gives up its energy to the system, and reflected light from pigments found in plant tissues of both leaves and flowers allow us to see the specific colours associated with particular plant organs.

Light comprises quanta of energy ('chunks' or particles) called photons. However, the energy of a photon is not the same for all types of light, but is inversely proportional to the wavelength – the longer the wavelength, the lower the energy. The wavelengths involved are electromagnetic in the visible light spectrum, and range between 400–700 nanometers – the colours of the rainbow. Photons can give up their energy under certain conditions – light can be absorbed by pigments (including chlorophyll) that act as catalysts for photosynthesis by absorbing energy from its photons.

The main photosynthetic pigments are chlorophyll 'a', chlorophyll 'b' and the carotenoids. Chlorophyll is found in the chloroplasts of cells in green stems and leaves – the tissues of primary growth. Chlorophyll 'a' has a formula of C55 H72 O5 N4 Mg, and chlorophyll 'b' has a formula of C55 H70 O6 N4 Mg. Note the importance of hydrogen and oxygen (found in water), and nitrogen and magnesium (both found as soluble salts in soil and carried by water) in the formulae for the chlorophyll molecules. The most important of the carotenoid pigments are the carotenes (responsible for reds, oranges, purples and maroons) and the xanthophylls (responsible for golds and yellows).

Photosynthesis has two main stages: the light stage (photochemical stage), and the dark stage (chemical stage). During the light stage, both chlorophyll b and the carotenoids absorb light in the violet, blue and blue-green part of the spectrum (400–500 nm approx), whereas chlorophyll a absorbs light from the yellow and orange part of the spectrum (625–700 nm approx.). The green and yellow-green part of the spectrum (525–625 nm approx.) is largely reflected, rather than absorbed. Hence we see the green of leaves very easily with the human eye.

C3 and CAM Plants

Most plants photosynthesize via what is known as the 'C3 pathway'. One of the compounds formed in the process in light splits into two

---

⋆$C_6H_{12}O_6$ is a general formula for glucose.

molecules of a three-carbon compound PGA (phosphoglyceric acid), and this three-carbon sugar gives rise to the name 'C3 plants' for those photosynthesizing in this manner. Most annual, herbaceous and woody perennials photosynthesize during the day via the C3 pathway. However, plants such as succulents and cacti that live in very hot and arid conditions have evolved to close their stomata during daylight hours in an attempt to reduce moisture loss. Plants of this type open their stomata at night (diurnal closure and nocturnal opening). This is very helpful for water conservation, but unfortunately the closed stomata do not allow the egress of carbon dioxide during the day when light factors for photosynthesis are good. Plants of this type therefore adopt a different system for their photosynthesis – a system known as 'crassulaic acid metabolism' (CAM).

Respiration in Plants

Respiration is a catabolic (energy-using/releasing) process, and involves the breakdown (oxidation) of sugars within individual cells, with the associated release of energy stored in these substances. Sugars produced during photosynthesis are the most common substrates oxidized in plant respiration. However, sugars can be converted to related carbohydrates such as sucrose, starch, fats and proteins (with the essential inclusion of N, P and S from the soil), which are all organic compounds and act as plant food – as a substrate for respiration. Enzymes are responsible for the oxidation of food substrates, and specific enzymes are compartmented within the cytoplasm and cell organelles, for example chloroplasts and mitochondria. Therefore there are specific reactions within particular compartmented areas of an individual cell.

The energy released by respiration is used in a prioritized way: it gives the energy for the production of cell protoplasm (assimilation), cytoplasmic streaming, the movement of minerals across cells (active transport), synthesis of proteins, and the maintenance of existing cells and tissues. It is only the remainder that can be used for the addition of biomass (plant growth) created by cell division/cell replacement, cell growth and development, and areas of repair/wound healing. New biomass is therefore not added until the maintenance of existing tissues is carried out, and the sexual process may take up to 60 per cent of available energy at any one time. Hence a healthy energy budget must be maintained in order to maintain and replace old tissues, and add new tissues to increase the size of the plant and for reproduction. Energy is also needed for both transpiration and photosynthesis, but in both cases this energy is gained from the sun.

Carbon dioxide gas is released as a by-product during respiration, and along with respiring animals (including ourselves), and decomposing organic matter, accounts for much of the 0.03–0.04 per cent of carbon dioxide found in our atmosphere.

Plants respire aerobically (in the presence of air) and may be represented (albeit rather simplistically) by the following reaction:

$C_6H_{12}O_6 + 6\,O_2$	=	$6CO_2$	+ $6H_2O$	+ Energy *
Glucose + Oxygen	=	Carbon dioxide	+ Water	+ 673 Calories

Some species of lower organisms (a limited number of species of bacteria and fungi) can respire without oxygen, which allows them to function under very poor conditions. This ability enables organisms to partially break down organic matter to peat under waterlogged conditions – a process that is very slow. Anaerobic respiration releases only small amounts of energy.

Respiration rates are notably high in areas of rapid cell division: shoot tips, root tips, and germinating seeds. Respiration is affected by ambient air temperatures, and in general terms as temperature increases, so does the respiration rate. However, this is limited ultimately because very high temperatures will inhibit enzyme action, which is essential for the process to continue. Conversely, respiration rates are reduced in lower ambient temperatures – which accounts for the lower metabolic rates of plants in winter.

Photosynthesis releases large amounts of oxgen into our atmosphere, and uses some of the carbon dioxide derived from respiration. Respiration needs oxygen and releases carbon dioxide, and the point at which the rate of carbon dioxide production from respiration is compensated exactly by the rate of carbon dioxide uptake during photosynthesis is known as the 'compensation point'. Below the compensation point the rate of photosynthesis is less than the rate of respiration, and therefore carbon dioxide is evolved. Above the compensation

*At a biochemical level, photosynthesis and respiration are far more complex than the simplified formulas would have us believe.

point the rate of photosynthesis exceeds the rate of respiration, and therefore oxygen is evolved.

Water Movement in Plants

Water moves into plants by absorption at the roots, and moves through them by diffusion, osmosis, and transpiration. Absorption occurs through membranes such as cell walls, and accounts for the initial passage of water from the soil into root hair cells. Diffusion is the movement of molecules of water powered by concentration gradients within cells, and involves the relatively slow movement of water from cell to cell and through inter-cellular spaces. Diffusion is very useful in keeping cells hydrated when water moves over very short distances in thin tissue layers and in areas of new growth before vascular tracts have differentiated.

Osmosis is a form of diffusion, where the movement of water is relatively rapid through a semi-permeable membrane (the cell wall) from a less concentrated solution towards a more concentrated solution. Water will therefore move from the soil (normally a low concentrated solution) towards cell sap (usually more highly concentrated). Water also moves across tissues (like the cortex of the root) powered by osmosis.

Transpiration

Transpiration is a process involving the upward movement of water through plants via the xylem (vascular tissue). Water travels from the root hair region (piliferous layer) up the plant to the leaves, and once at the leaves it evaporates from their large surface area via the stomatal apertures. The evaporation is fuelled by the sun's energy, and the term 'evapo-transpiraton' is often used to describe water loss that occurs through this system.

Because of the strong affinity of water molecules for one another, water is held in both the fine and the medium pore spaces between soil particles, against the force of gravity, and the soil acts as a reservoir of moisture for plants. Although osmotic pressure cannot release moisture from the very fine pore spaces (micropores), it is strong enough to take up water from the medium-sized pores (mesopores).

Each fine root hair is an individual epidermal cell that differentiates to specialize in water uptake and invades the soil particles. Initially water is absorbed through the cell wall, and once entering the root hair, is influenced by the osmotic gradient that it encounters (created by adjacent cells of the cortex). As water enters a cell, it dilutes the cell sap, making it less concentrated than its adjacent partner, and water will therefore move into the more concentrated adjacent cell. In this way water moves laterally across the cells of the cortex until it reaches the endodermis. At this point water cannot pass through the suberized casparian strip, but can pass through the 'passage' cells, which effectively channel water towards the tube-like cells of the xylem. Once water enters the xylem vessels (or tracheids) via the pits and pores in their lateral walls, the power of transpiration pull (fuelled by the evapo-transpiration at the leaf surfaces) takes over, and it moves swiftly into the transpiration stream.

Water may naturally be in one of three states: solid (as ice), liquid (as water) or gaseous (as water vapour). There is no chemical difference between the three states, only a physical difference brought about by the amount of energy present in the water. For water to change in state from ice to liquid requires an energy input (usually in the form of heat), and for it to change in state from liquid to vapour requires even more energy input. As energy is gradually given to the system the state does not change immediately. Instead, ice remains as ice until sufficient energy is given to take the ice up to the next level (liquid), and in order for liquid water to receive enough energy to form vapour (steam), it requires considerable energy input. Consider an electric kettle full of liquid water receiving energy from the mains supply: until a sufficient amount of energy has been supplied (over several minutes) that excites the water molecules and takes the water to the next level of energy (in this case vapour), it remains as a liquid. This illustrates the quantum leaps of energy needed to change the state of water.

Transpiration (the upward movement of water through the plant) is fuelled by solar energy (at leaf surfaces in particular). As sheets of water molecules are energized/excited by the sun's rays, they volatize into the atmosphere and leave the leaf surface as a gas (water vapour). The molecules of moisture evaporating from leaf surfaces are so closely allied to the rank of molecules next to them in the xylem that they pull them up to replace the deficit (cohesion). The adhesive properties of water molecules make sure that the water stays firmly against the side of the xylem vessel walls, and as long as the molecular chain

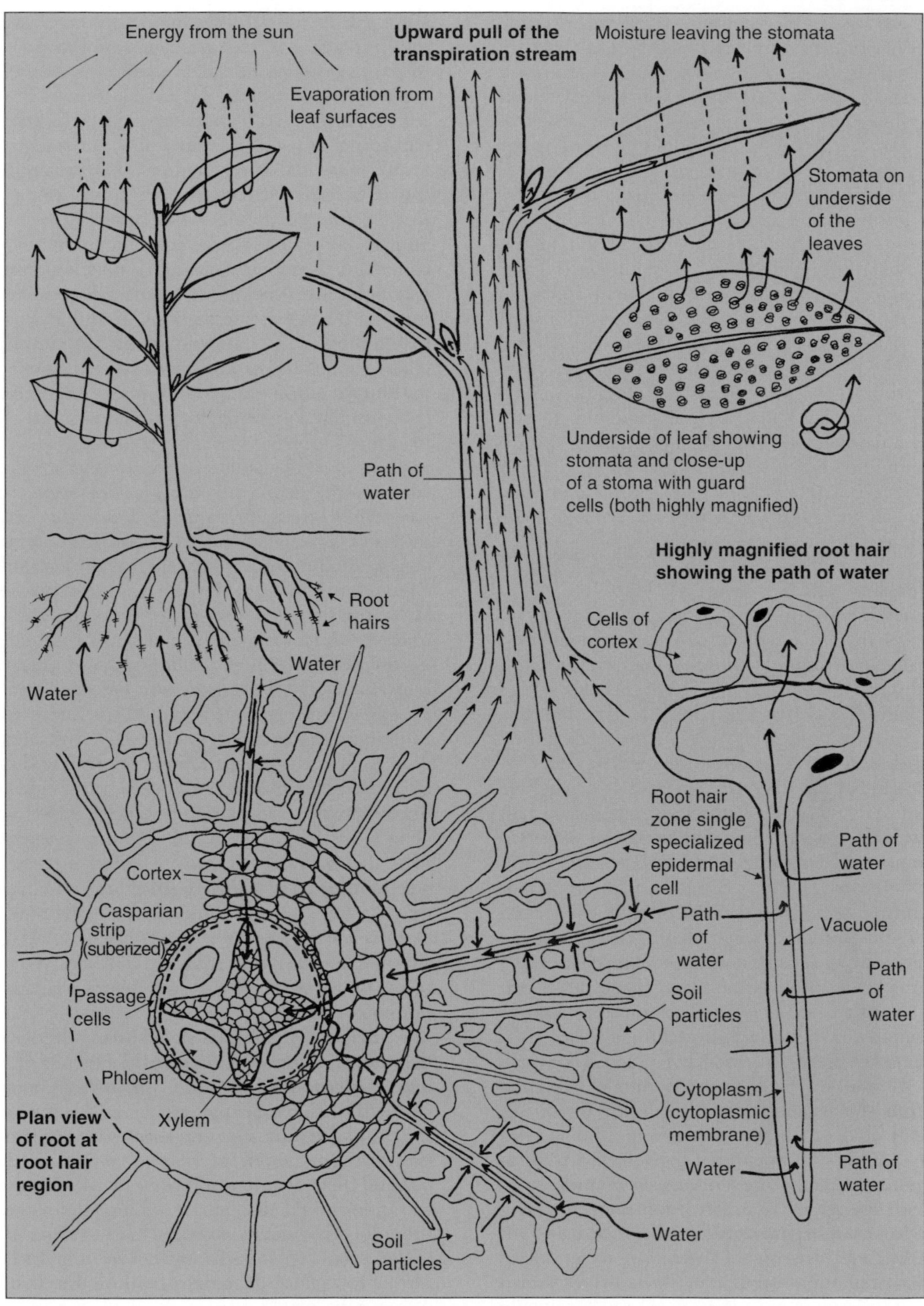

Transpiration (and transpiration pull).

is unbroken, the transpiration stream will remain entire. Transpiration pull is therefore directly attributable to the surface tension, adhesive and cohesive properties of the water molecules, and the theory describing this process is known as the cohesion-tension theory. Other theories of transpiration cannot account for the incredible distances that water will travel upwards in trees (over 100m (328ft) in coastal redwoods).

An added help (albeit a small one) to the upward movement of water in the plant is root pressure, believed to be caused by osmotic pressure within the cells and tissues of the roots. However, this can only ever be responsible for a rise in the column of water of a few feet above ground level, and never more, hence it does not explain the movement of water through large arboreal subjects.

Experiments to prove the cohesion-tension theory, carried out by Strasburger in the 1890s, involved erecting scaffold around a 20m high oak tree to support it. The tree was cut through to create a suspended canopy remote from, but still in line with, the root system. A large reservoir tank (rather like a giant vase) was then inserted below the severed trunk and on top of the root plate, and filled with water.

Initially the tank was monitored to see how water levels fluctuate in different conditions, and as expected, water levels dropped rapidly in warm conditions. The water obviously left the tree via evaporation at the leaves and in general terms the leaves remain turgid as the vascular system keeps water flowing to them. To ensure that the water level is not dropping via evaporation directly from the water surface itself, more recent repeats of the experiment include a cover being placed over the tank. Furthermore, a blue dye (fushin) was placed in the water to track the path of the water. Hence, if the water is actually moving upwards from the tank to the canopy, leaves removed from the system should, and did, present the blue dye.

In order to determine that live tissues are not involved, and are not responsible for some sort of 'pumping mechanism' to fuel the process, all living cells were killed by adding a strong phytotoxin to the reservoir water (picric acid). It was found that the process continued even after all cells within the tree were dead – proof that the process functions via the dead tube-like xylem vessels purely by molecular affinity (cohesion), and no form of living cells are required.

The effect of the cohesion and adhesion of water molecules is to have a continuous stream (transpiration stream) of water moving through the plant. Water travelling through the plant delivers dissolved nutrients and metabolytes to their area of use, activates enzymes, cools the plant in hot weather, and gives strength and turgidity to soft cells by engorgement.

Transpiration is a continuous process whose rate will differ affected by certain environmental and atmospheric conditions and the age of the plant. The rate of transpiration is faster in young plants (or young parts of plants) than in old plants (or older parts of plants). Obviously it can only be the soft green parts (primary growth) of the plant that are affected because they are the only parts with stomata.

The process is regulated by the availability of water at root level, by the health of the plant, by the leaf surface area, and by the guard cells surrounding the stomatal pores. Temperature will influence the transpiration rate dramatically, primarily because high temperatures will volatize water molecules readily and therefore evapotranspiration rates will increase. In general terms there is a direct correlation: as the temperature rises, the transpiration rate increases accordingly.

Guard cells are very important in regulating the transpiration rate, as most water loss occurs through stomata, and not via ordinary leaf epidermal cells (as these are covered with a thick, waxy cuticle to prevent water loss). Stomata and associated guard cells are found on the underside of dorsiventral leaves, but are also found on any green tissue including stems. In high temperatures, or when water is scarce at root level, the guard cells become flaccid and therefore reduce (rather than completely close) the stomatal opening, which reduces the flow of transpiration to guard against excessive moisture loss. However, the reduced flow of transpiration that does continue will act as a cooling agent for the plant. In cooler air temperatures and in conditions when water is readily available at root level, guard cells become engorged and turgid, and in this state they open the stomatal pore wider and increase potential transpiration. However, the lower temperatures actually reduce the transpiration rate, so the cells remain fully engorged, and the plant remains erect.

Tracheids, the main conducting tissues of conifers, are very heavily lignified, so much so that they only leave very small pits in the cell walls. They differ in their structure to xylem vessels as they are tapered where the cell ends meet. This is thought to help prevent embolisms (air in the transpiration stream), because conifers

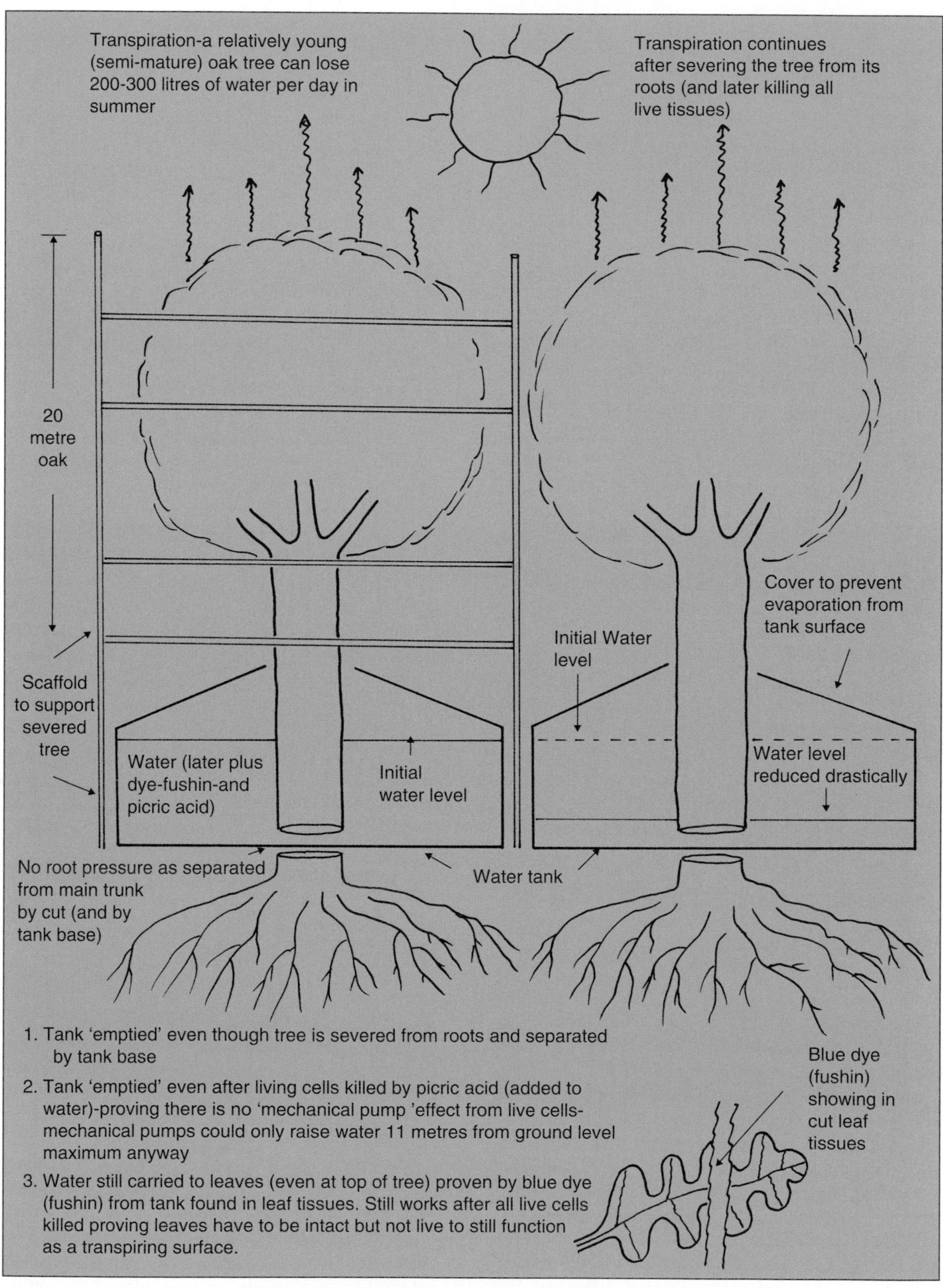

The concepts of Strasburger's experiment.

are found at altitude and regularly encounter very low temperatures in these conditions.

In general terms, the better the light conditions the faster the transpiration rate, as opening of the stomata depends on good light conditions. Good light is necessary for guard cells to photosynthesize and produce sugars. In doing so, the change in cell sap concentration created by the extra sugars in solution, sets up an osmotic gradient that engorges the guard cells with water and holds the stomatal aperture open. Furthermore, if light falls directly on the leaf, the energy transmitted to the leaf will cause a rise in temperature, and with it, a rise in transpiration rate.

The faster that air moves over the leaf lamina the faster the transpiration rate, as rapid air movement prevents the congestion of moist, saturated air near the stomata and moves the moisture away from them quickly. Hence, as the humidity in the atmosphere increases, so the transpiration rate decreases – practical operations such as 'damping down' under glass, mist propagation units and polyethylene covers over rooting cuttings all take advantage of this to reduce the chances of flaccidity (wilting). In fact, the whole idea of 'nursery' conditions for plants concerns reducing mortality when their physiology needs support. It's as true of seeds sown in ideal conditions as it is when traumatic damage is caused to tissue on prepared cuttings and on cuttings and layered plants with poor initial root ratios – too poor to support the amount of transpiring top growth. It is also true of freshly lifted plants with damaged roots and poor root:shoot ratios. All these need shelter from drying winds, shaded foliage (but enough light for photosynthesis if for a protracted period), extra moisture added, and available moisture retained to ensure success. Offering these conditions in the nursery retains cell turgidity and allows the normal physiological processes to function whilst root-to-shoot ratios return to a self-supporting level.

If soil at the root zone is cold, water becomes slightly more viscous, and therefore uptake by the plant, osmosis and transpiration is slower than in warmer conditions. In very dry soil conditions the lack of soil moisture means that the surrounding soil will have a greater concentration of solutes than the cell sap of the root hairs of the plant, therefore osmosis will be slow. If this dry condition persists, ex-osmosis (or reverse osmosis) will occur – water will leave plant tissues rather than enter them, and cells become plasmolysed (the cell cytoplasm collapses away from the cell wall). Plasmolysis causes irreversible damage, and the plant wilts (not temporary wilting, but a permanent condition). Reverse, or ex-osmosis, makes the guard cells close, which slows down the transpiration rate to a minimum; but it is in vain as tissues will actually die, and typically manifest initially as a brown 'scorch' at leaf margins and soft green tissue apices. The same effect is encountered if excess concentrations of mineral salts (as artificial fertilizers) are applied to the soil around plant roots.

Technically osmosis cannot be classed as being 'in reverse' or 'ex', as the direction of the water quoted in the definition of the process is relevant to the concentration gradient, not the plant. However, because the welfare of the plant is regarded as the main aim, water leaving the plant is considered to be the reverse of what is required.

The vascular tissues (vascular bundles) that run from the roots through the stems branch off at leaf traces (serving the midrib of the leaf), ensuring moisture both for cell turgidity and photosynthesis via the lateral leakage of the xylem and the network of lateral and sub-lateral veins. There are also vascular junctions at every nodal region serving undeveloped buds or developed lateral shoots, and at lateral branch unions on woody plants.

Specific vascular tissues of leaf traces, bud traces and the vascular tracts associated with branching stems, although continuous with the vascular tissues above them, form leaving a 'gap' in the xylem bundle or cylinder. Thus whilst serving the lateral growth, it restricts the progression of fluids vertically at that point. If the vascular tissue were completely continuous from the root, through all side branches and leaves to branches and leaves above them, it would increase the chances of any infections being passed on to higher parts of the plant. The 'gap' in the xylem cylinder limits this possibility, although it does not eradicate it completely, both because of the (albeit interrupted) continuance of tissues, and the relative lateral 'leakiness' of the xylem system.

Lenticels in the bark are, in effect, the woody stem's equivalent of stomata, and allow gaseous exchange including oxygen and water vapour through soft parenchymatous cells packed into the lenticel pore. Oxygen for respiration diffuses through the parenchyma cells and can arrive relatively quickly at internal cells. Unlike stomata, lenticels are not regulated by guard cells, and water vapour can leave woody stems by diffusing through the packing cells and causing desiccation if the water deficit is not rectified by water from the roots. However, water loss via diffusion at lenticels is much slower than via evaporative leaf surfaces. Deciduous subjects during the summer lose water via both stomata and lenticels, but during winter (due to leaf fall) they only lose water via the lenticels, and as temperatures are lower (and rainfall higher) at this time of year, both metabolism and water loss are very low anyway.

Guttation

Guttation is the loss of water in a liquid form (rather than as a vapour) from the leaf, and is known in several hundred species of plants, including tomato and *Fuchsia*. Busy lizzy (*Impatiens glandulifera*) and Himalayan balsam (*Impatiens balsamifera*) are good examples of species that guttate freely from the glands (sometimes red) apparent on their leaf margins. Liquid exudes and congregates in a specialized structure at the glands called a 'hydathode'. The liquid exuded contains high levels of dissolved salts that tend to leave a crystalline crust on the leaf surface when it evaporates away. Root pressure is most probably responsible for the guttation process, and it occurs mainly in warm, humid environmental conditions, especially at night when there is plenty of moisture in the soil but the transpiration rate is very slow.

Photoperiodism

Light intensity, quality and wavelength all have a bearing on the efficiency of photosynthesis. However, leaf abscission and plant seasonality are influenced instead by day-length (and temperature). Day-length describes the period of time a plant is illuminated – or by definition the converse: how long a plant is in darkness. The day-length response of plants determines the vegetative growth, their flower-bud initiation and their flowering times.

One of the most common plants used to illustrate this phenomenon is the chrysanthemum. Perennial chrysanthemums naturally initiate flower buds in shortening days, and commonly flower in the short days (long nights) of late summer/early autumn, when days are shorter than fourteen hours. Armed with this information, photoperiodically responsive plants such as chrysanthemum can be induced to flower at any time of the year (which leads to the terminology 'all year round' – AYR – chrysanthemums of commercial cut flower production). In naturally short days their growth in stem length is induced by the use of artificial lighting (which is used to extend day-length by giving an illuminated night break). Following this, once stems are at the required length, flower-bud initiation is then induced by turning off the lights (and thus returing the plants to their natural short day conditions).

Conversely, in naturally long days, stem length increases ordinarily in the available day-length, and once stems are at the correct length, blackout shading is used to reduce natural day-length and commence flower-bud initiation. After the bud initiation process is triggered, final flower development continues normally (without shading) in whatever the natural day-length. The number of hours of extra illumination or extra shade, and the number of weeks involved to complete the process, varies between species and cultivars – but only by relatively small amounts. Tables of the relative photoperiodic responses (and the necessary subsequent treatment for each type) are readily available so that actual flowering times can be calculated for any set of circumstances.

Poinsettia (*Euphorbia pulcherrima*) and tobacco are also short-day plants (which accounts for the production techniques of poinsettia under glass). Conversely, some plants only flower in long days (over ten hours of daylight) – summer in temperate regions – such as the petunia. Many plants are not affected by day-length and are known as day-neutral plants, such as the begonia, that can flower at any time of year.

Photoperiodic responses are triggered by various concentrations of the pigment phytochrome, a photoreceptor within the plant found in the leaves. The perceived responses of the plant occur because there are two distinct forms of phytochrome: P660 and P730. P660 absorbs red light at wavelengths around 660 nanometres, whereas P730 absorbs red light at the far end of the spectrum (wavelengths that peak at 730 nanometres). The two phytochrome forms are inter-convertible (can convert/revert to one another). Thus daylight with different periods of irradiation of specific wavelengths will create more of one type over the other. Long-day plants flower when a greater concentration of P730 has built up, and short-day plants when more P660 has built up.

PLANT STRATEGIES

Adaptations for Survival in Exposed and Arid Environments

Plants have adopted various strategies that resist water loss, and various modifications of growth that aid moisture retention. Excessive water loss can be because of extreme heat, such as that experienced by desert plants, or because of exposure and severe winds, and temperature fluctuations as experienced by conifers and alpines at high altitudes, and coastal and sand-dune plants. Mechanisms to facilitate the suppression of moisture loss (xerophytic adaptations) include succulence, extensive root systems, small leaves, and particularly thick, waxy, waterproof cuticles.

Resistance to moisture loss can also be achieved by extra lignified tissues produced several layers deep, with a very waxy outer layer – this is often accompanied by modified leaves or stems typified by cacti spines and cladodes. Butcher's broom (*Ruscus aculeatus*) has very tough, leaf-like flattened stems with sharply pointed (acuminate) tips that take over the role of photosynthesis in the absence of leaves. These structures (cladodes) are considered to be leaf-shaped stems because they arise from the axil formed by the main stem and the cataphylls (scale leaves), and carry their flowers (and the resulting green – later turning red – berries), suggesting that although they look like leaves, they function as flower-carrying stems.

Plants losing excess moisture also lose valuable photosynthesis time, because, in an attempt to regulate the loss of moisture, the stoma close in response. Unfortunately, closed stoma not only mean the reduction of moisture loss, but also the shut-down, or at best the reduction, of photosynthesis. Hence in conditions of ideal light, and potentially good photosynthesis, if there is moisture stress, there may be no sugar production at all.

Cacti have evolved to have a swollen, water-retaining structure, which looks like a collection of large green leaves, but is in fact a stem system comprising cladodes (or phylloclades) that takes over the main photosynthetic (sugar-producing) role. In order to retain moisture, the leaves have been reduced to hard spines with no sugar-producing remit, but have become a spiny barrier to protect against browsing. Because this leaves cacti open to moisture loss from the large green stems, they often store large amounts of water and close the stomata on their stems during the day. Diurnal rather than nocturnal closure of the stoma means there have to be different metabolic pathways for these species in order to produce sugars at night (they are known as CAM plants – crassulaic acid metabolism).

Succulent plants have leaves, but take on many of the above modifications. They store extra water in their stems and/or their leaves (which gives them their characteristic swollen appearance), some have diurnal closing stomata (CAM plants), and often have spines substituting for some of their leaves. The succulent *Senecio articulatus* (syn. *Kleinia articulata*) has typical swollen stems and small leaves. Desert plants in general often have shallow and wide root systems to maximize on light rain, or conversely, very deep taproots (such as *Welwitschia mirabilis*), or extensive roots, as some *Acacia* species, to gain moisture from lower levels. Another common modification is a well developed ability to reproduce by vegetative means, as water is essential for germination. Propagation is sometimes possible by leaf cuttings (some *Sedum* and *Crassula* species) or specialized plantlets can develop at leaf margins (common in many *Kalenchoe* and *Bryophyllum* species).

The phyllodes of some Australasian *Acacia* species are considered to be leaf modifications because of their origins witnessed in immature leaves. The word 'phyllode' includes the prefix 'phyll', meaning leaf, because of this. Young stems of the *Acacia* species involved carry immature pinnate leaves with a discernible rachis and leaflets, whereas older stems carry mature leaves where the rachis/petiole and leaflet laminas are indistinguishable from one another (phyllodes); or the rachis is large and flattened in a leaf-like form. Examples include *Acacia verticillata* (prickly Moses) and *Acacia longifolia* (Sydney golden wattle).

The spines on *Berberis* species are considered to be reduced and modified leaves. The true foliage leaves are borne on a short stem that arises from the axil (the angle of which is formed by the main stem and the three spines radiating from it at that point – the node). Common broom (*Cytisus scoparius*) has very small trifoliate leaves to reduce moisture loss.

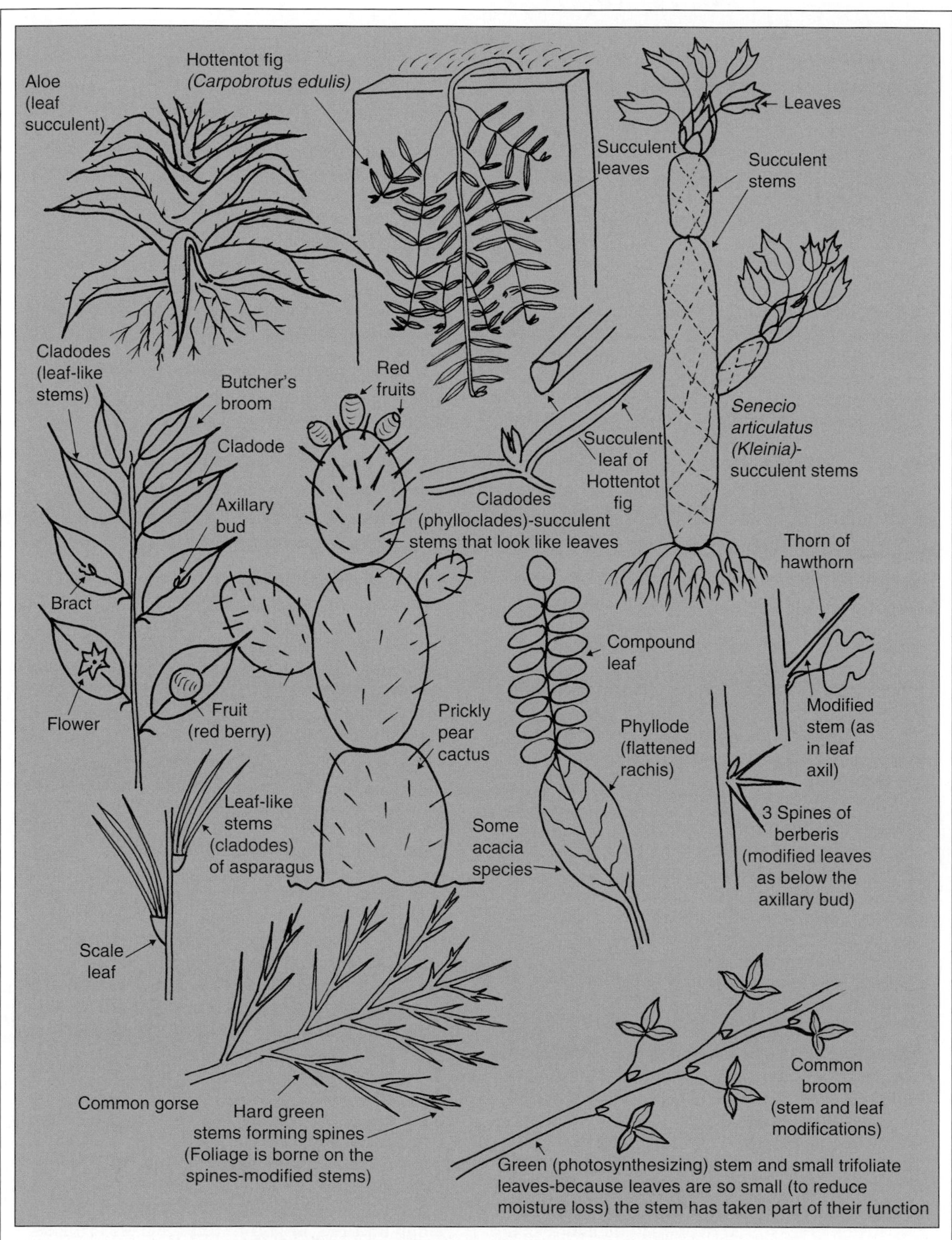

Plant strategies: leaf and stem modifications to reduce moisture loss.

The small leaves and stature of alpine rose (Rhododendron ferrugineum).

The very small encrusted leaves and ground-hugging nature of Sedum spathulifolium.

However, to counteract the associated loss of photosynthetic area of the leaves, the stems have a high chlorophyll content and they are furrowed/fluted (corrugated) to increase the surface area. Gorse (*Ulex europaeus*) has also adopted a strategy of having a green, photosynthetic stem (some of which has actually evolved to spines), but unlike cacti does not store large amounts of water, and it has small leaves. The extreme modification of leaf spines offers an effective physical barrier against harmful grazers and browsers.

Plant Size Modifications

Plant evolution, the size of whole plants, and parts of plants such as leaves, are all greatly influenced by environmental conditions. Some plants, naturally growing at high altitude (alpine

Stachys byzantina (syn. S. lanata).

conditions) or tundra, have evolved to become less wind resistant by having low stature, often rounded (aerodynamic) shapes, and very small leaves – often hairy or mealy, waxy or encrusted. Examples include many *Saxifraga*, *Androsace, Dionysia* and *Draba* species, including *Draba aizoides*; *Cassiope lycopodioides*, *Andromeda polifolia*, *Phyllodoce caerulea* and *Cyathodes colensoi* are also good examples. *Sorbus reducta* is low in stature and has small compound leaves (broken up into very small leaflets); it is also deciduous, with wax-encrusted bud scales – another mechanism to reduce moisture loss in exposed environments. Most *Lewisia* and *Echiveria* species, on the other hand, have taken the succulence route regarding adaptations to add to their low stature. They are therefore exceptionally useful to grow in cracks and crevices in walls and other dry sites – with success not only gauged by thriving healthy growth, but also very successful flowering in these conditions. Succulent species endemic to coastal cliffs or mountainous regions, as well as low stature, may also have a floury-looking, waxy encrustation over their inflated leaves/stems to aid moisture retention, for example *Sedum spathulifolium*.

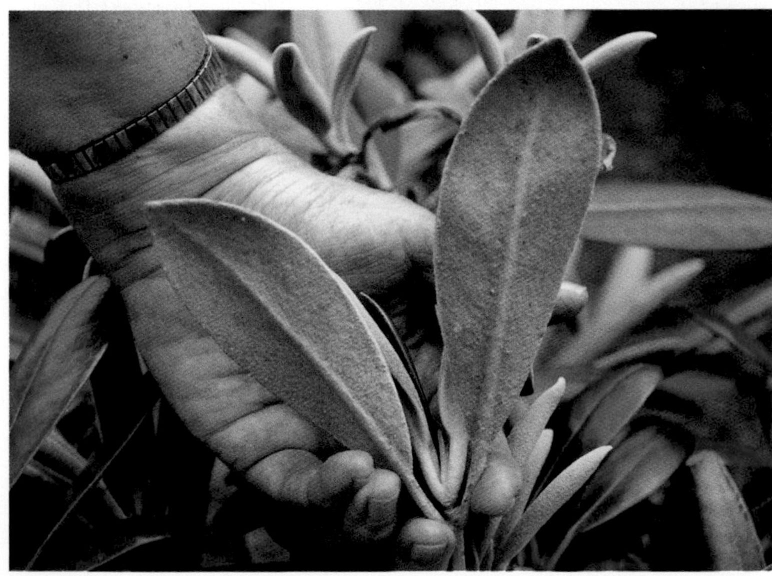

Indumentum on a Rhododendron yakushimanum hybrid.

Very large leaf of Senecio grandifolius with indumentum to reduce water loss.

Hairs

Hairs also help retain moisture, and this adaptation is not always reserved for small-growing alpine plants. *Pulsatilla* species are only relatively low-growing, as they are naturally found amongst grasses and other plants that give some form of protection anyway, and although they have relatively large leaves they do have very hairy leaves, calices and petals for their protection. Protective hairs may be in relatively small numbers around the leaf margins, or in larger numbers on the top surface of the leaf and the young stem. They may comprise masses of thick, felt-like hairs on the underside of the leaves (the main stoma-bearing surface). Some arboreal (woody) species have specialized, dense, grey, white, brown or rust-brown felty hairs on the underside of their leaves to retain moisture: this is known as 'indumentum'. Woolly, felty hairs on the upper surface of leaves and on young stems are termed 'tomentosum', often incorrectly used interchangeably with the terminology 'indumentum'. The hairs of indumentum and tomentosum reduce water losses from stem and leaf stoma by acting as a sponge and increasing the relative humidity around the stomatal pores; they are adopted by both evergreen and deciduous subjects.

Many *Rhododendron* species have white, cream, brown or rust-coloured indumentum, and on some it can be very aesthetic. *Rhododendron campanulatum, R. fulvum, R. falconeri, R. ficto-lacteum, R. bureavii, R. mallotum, R. yakushimanum, R. roxianum Rhododendron mallotum, R. metternichii, R. smirnovii, R. macabeanum,* and specific forms of *R. arboreum* (such as *R. a. cinnamomeum)*, are examples that have good indumentum. Other tree-like *Rhododendron* species may have what is known as 'plastered indumentum', where the length of the hairs is minimal or non-existent and just looks like a coloration on the lower leaf surface. Examples include *Rhododendron arboreum*, *R. sino-grande, R. calophytum*, and *R. praestans.* Some species and cultivars of evergreen *Magnolia* (including *Magnolia grandiflora)* and shrubs such as *Senecio grandifolius, Brachyglottis greyii, Brachyglottis monroei, Olearia* × *hastii, Olearia ilicifolium* and *Olearia macrodonta* are good examples of the adoption of indumentum as a moisture-retaining strategy. Deciduous examples include *Sorbus aria* (whitebeam), *Viburnum lantana,* and *Populus alba* (white poplar). Other examples of species with dense matted hairs on their leaves (and sometimes stems) include sub-shrubs such Jerusalem sage (*Phlomis fruticosa*) and herbaceous types such as lamb's ear (*Stachys byzantina*).

Leaf Size

Moisture loss from leaves is related not only to environmental conditions, but also to the surface area of the leaf and leaf structure. Species such as *Paulownia tomentosa** and banana, because of their large evaporative leaf surface area, lose very large amounts of water via their leaves. The correlation stretches further, because large-leaved species, due to their high moisture-loss potential, can only exist in deeper, moister soils and in more sheltered environments with higher rainfall than their smaller-leaved counterparts. Large-leaved species cannot thrive in exposed environments, so they do not appear in them: the problems of large leaves being severely damaged in exposed locations alone would be enough to prevent their success in this type of environment. Large-leaved species have very good sugar production because of their high surface area of chlorophytic tissue, and species tend therefore to be comparatively thick-stemmed and vigorous. However, there is a trade-off, as large leaves tend to mutually shade one another, thus reducing some of the potential benefits.

Leaf orientation and arrangement (phyllotaxy) can help by offsetting leaves from one another to reduce mutual shading, and gain extra irradiation. Temperate species such as horse chestnut, and tropical species such as *Schefflera*, adopt a particular compound palmate or digitate (like fingers) leaf shape in order to reduce mutual shading and therefore increase illuminated surfaces; this means they keep their overall leaf surface area, but it presents as many smaller connected leaflets, the 'splits' in the leaves allowing some light penetration to the ranks of lower leaves – compound leaves of all types (for example, *Acacia dealbata* and *Gleditsia triacanthos*) facilitate this strategy to some degree. You may remember that Swiss cheese plant (*Monstera deliciosa)* presents large, lacerated, simple leaves in poor light conditions, thus reducing the mutual shading of lower leaves, but in good light conditions where mutual shading is not a problem, it presents large cordate leaves with entire leaf laminas.

Conifers have evolved to have large areas of photosynthetic tissue, but made up of many small, needle-like leaves, rather than a lesser number of large ones. Small leaves are less likely to become physically damaged, and each individual leaf loses less moisture than larger leaves. Furthermore, each leaf is very aerodynamic in shape and offers little wind resistance, and is covered in a thick waxy cuticle all round to prevent desiccation. The stomata on conifers are set in small hollows (depressed) in the leaves, and trap still, moist air, thus reducing stomatal moisture losses. On a microscopic scale, the depression at the stomatal opening gives a greater distance for water to diffuse, and this slows down transpiration.

Thick, waxy cuticles are typical on most woody evergreens. The extra layers of cutin (the wax found on the upper epidermis of leaves) reduce moisture loss both by acting as a waterproof layer and by helping to reflect excessive sunlight – thus keeping the leaf cooler and reducing transpiration. Common holly (*Ilex aquifolium*) has leaves with very waxy cuticles that also have sunken (depressed) stoma, like many conifers.

Other leaf modifications include some species that can cigar-roll their leaves so as to trap still, moist air near the stomata, and again lengthen the diffusion path of water – some *Rhododendron* species do this, as does ling (or common heather) *Calluna vulgaris,* lavender and *Ammophila* (marram grass).

Many species from warmer climates or with warm summers have xerophytic adaptations that include concentrated, viscous, oils in the leaves. Resins, terpenes and oils all aid water retention by helping to reduce transpiration losses, and are found in Western red cedar, resinous pine species, eucalyptus, lavender, rosemary, thyme and citrus species.

Besides some deciduous species having adaptations such as indumentum to reduce moisture loss in summer, they have also adapted so as not to lose moisture in winter conditions, and to overwinter successfully by their deciduous habit and wax-coated dormant buds.

Tropisms (Plant Movements)

The direction of growth of leaves, shoots and roots of plants are affected by various stimuli. Plant movements away from, or towards stimuli

---

*tomentosa (having tomentosum to reduce moisture loss).

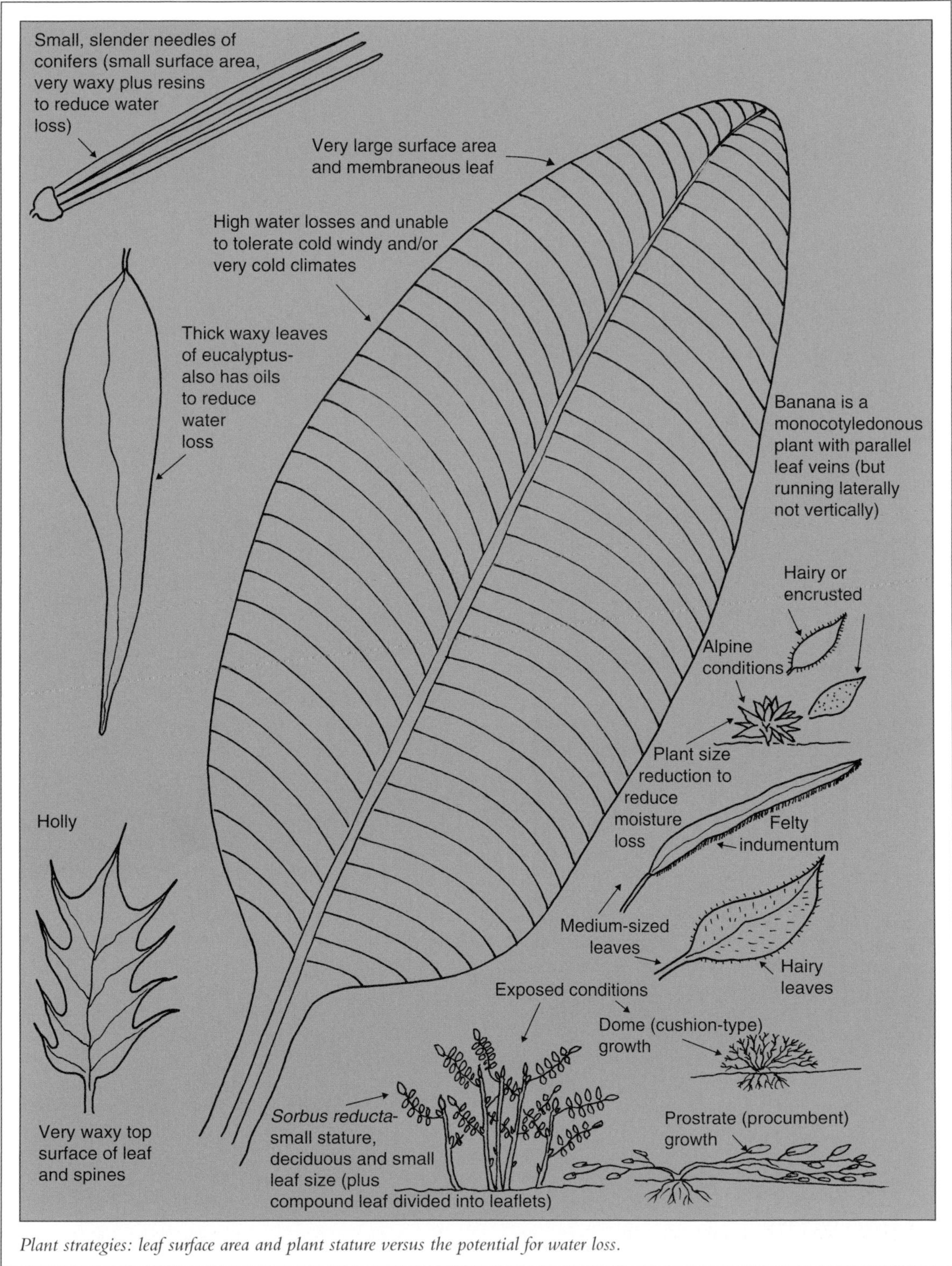

Plant strategies: leaf surface area and plant stature versus the potential for water loss.

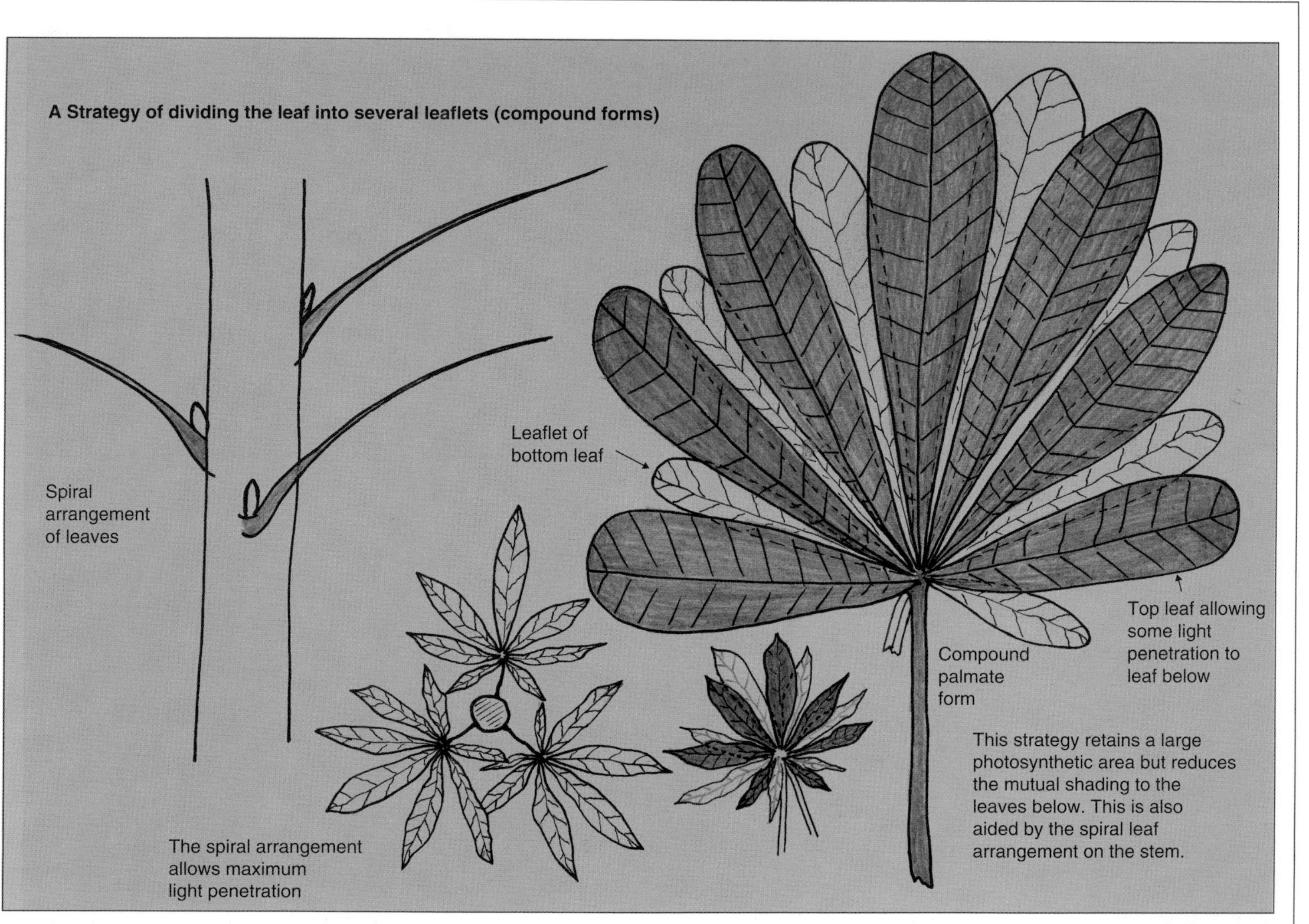

A Strategy of dividing the leaf into several leaflets (compound forms)
Spiral arrangement of leaves
Leaflet of bottom leaf
Top leaf allowing some light penetration to leaf below
Compound palmate form
This strategy retains a large photosynthetic area but reduces the mutual shading to the leaves below. This is also aided by the spiral leaf arrangement on the stem.
The spiral arrangement allows maximum light penetration

Light penetration.

are called tropisms (tropic movements), and are named according to the stimulus responsible for the movement. Attraction towards a stimulus is known as positive, and growing away from, or being aggravated by a stimulus, is termed negative. Geotropism is the movement of the growing points of plants, whose direction is stimulated by gravity. Phototropism is the movement of the growing points and leaves of plants, whose direction is stimulated by light.

Stems are termed negatively geotropic as they grow away from the pull of gravity, whereas roots are termed positively geotropic as they grow towards the stimulus of gravity. Green stems and leaves are termed positively phototropic because they position themselves towards a light source to gain as much radiant light as possible; root apices, on the other hand, are termed negatively phototropic because most are aggravated by light and prefer to grow into the darkness of the soil.

Positive phototropism (when green leaves and stems bend towards a natural or artificial light source) is probably the most commonly witnessed tropism, and is triggered by hormonal action (or non-action). Auxins (responsible for cell elongation and plant growth) are the main hormones involved. Auxins are destroyed by light, and one of the accepted theories of phototropism suggests that as they are transported through the vascular tissues, the auxins on the more strongly irradiated side of the stem are broken down by the light and lose their effectiveness. On the more poorly irradiated (shadier) side of the stem, however, the auxins retain their effectiveness and cause rapid growth, and it is the disparity of growth from one side of the stem to the other that causes the marked curvature in the stem that causes the foliage to face the light.

The attitude or orientation of plants is therefore affected by the interaction of positive geotropism and negative phototropism at the roots, negative geotropism and positive phototropism at green stems and leaves, the structural strength of the developing anatomy, and cell turgor (water pressure against the cell wall). Very young woody plants respond to tropisms like any other green plant. However, after year one, only the green growth is influenced, as once woody material is laid down it cannot alter orientation. The accumulative effect of the response to this stimulus leads to woody perennials having a definite 'face' or 'front' (because their woody stems 'retain' or 'fix' the orientation) – the direction that the main foliage density faces.

Flowers that follow the path of the sun during the day are said to be positively heliotropic and include sunflower (*Helianthus*), named because of it. Ironically, *Heliotropium*, whose generic name and common name (heliotrope) perfectly describe this phenomenon, has now been proved not to be affected by heliotropism.

Hydrotropism is more contentious, as it would appear to be only represented by the roots, and only by positive movements – towards water. There are no known examples of negative hydrotropism – aggravated by water. Excess water does, however, create anaerobic conditions and therefore root death, and as new root growth will occur from existing live tissue where there is sufficient oxygen, it gives the effect of roots growing away from the water, when actually they are growing into oxygen-rich (aerobic) soil. The true overriding attraction of roots towards water can be illustrated in crops such as tomato, when drip-feed watering systems are used. Drip nozzles create a 'cone' of moist soil with the point of the cone at soil level and the wide base of the cone in the lower profiles of the soil. Roots excavated from the soil are shown to have systems that fairly accurately follow the shape of the 'cone of wetting'.

If it is the case that roots are positively hydrotropic, then it accounts for the idea of roots 'foraging' for moisture (in aerobic conditions). If positive geotropism were the only effective or overriding stimulus, then all root systems would go straight down. In practice this is not the case, as along with the limiting factor of available oxygen, it is the interaction of positive geotropism, negative phototropism, positive chemotropism and positive hydrotropism that accounts for the near horizontal (rather than vertical) 'attitude/orientation' of roots.

Chemotropism describes plant tissues moving towards (positive), or away from (negative) a chemical stimulus. This is evident when 'attractant' or 'deterrent' substances are secreted by various parts of the plant. Examples include the secretions from a stigma that stimulates the germinating pollen tube to move down the style towards the ovary – a positive response. A negative example is given by the

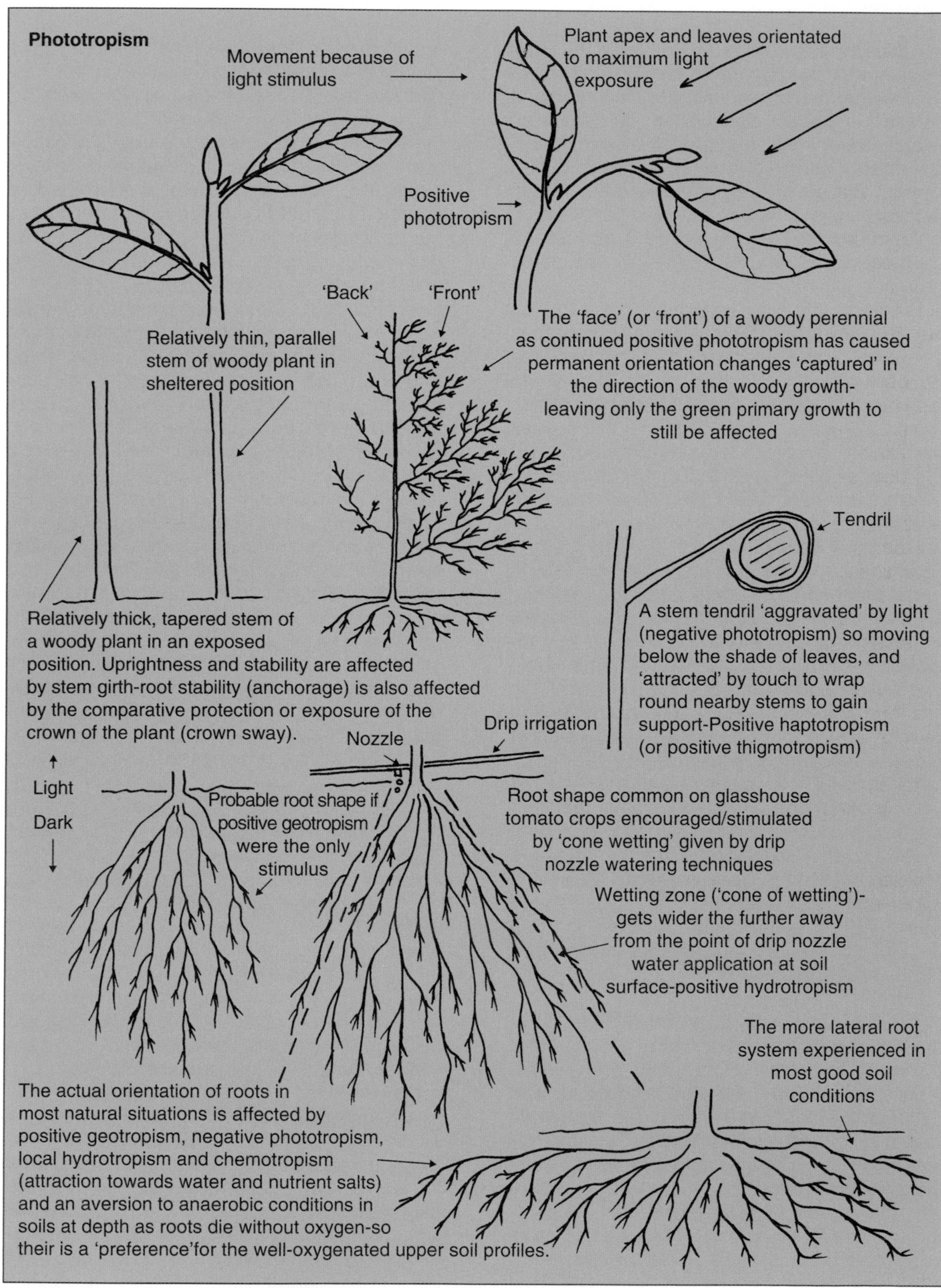

Plant strategies: leaf, stem and root orientation.

fact that some (not many) woody species do not easily invade the root space of others to compete, because of root secretions (inhibitors) in the soil from the other species. This phenomenon is known as allelopathy, and one common example is the reluctance of *Rhododendron* species and some *Sorbus* species to thrive in one another's company, and specific replant disease experienced when trying to establish plants in the Rosaceae in soils previously occupied by another rosaceous plant. Once you are alerted to this as a possibility, it is useful to look for symptoms that cannot be explained by other means. Perhaps the possibility of dissolved salts in the top section of the soil matrix, stimulating root movement towards them, is another stimulus to explain the more horizontal attitude adopted by roots?

Negative phototropism is experienced by the tendrils of some woody climbing plants with self-clinging mechanisms, as the tendrils seek the shade and are more likely to attach themselves to a wall, or other surface, for support. Thigmotropism (or haptotropism) is a plant movement caused by the stimulus of physical contact. The tendrils of some plants are positively haptotropic, and attach themselves once they make contact with the surface of any potential support. It is the interaction of positive haptotropism and negative phototropism that is responsible for some species being able to grow towards support and cling on with tendrils.

The flowers of peanut or groundnut (*Arachis hypogaea*) commence above soil level, but the fruits ultimately develop below the soil after pollination. The flower-bearing stalks (pedicels) are both negatively heliotropic (negatively phototropic?) and positively geotropic to facilitate this phenomenon.

Other hormonal responses are triggered by physical stimuli, including the stability (via their thickness) of stems, and the stability (anchorage) of roots, both being affected by the physical movement of the top of the plant. Plants in sheltered niches and protected (including mutually protected) areas do not increase in stem girth or root stability in the same way as they might in a more exposed environment. This phenomenon, which is particularly pronounced in woody plants, has led to the use of short stakes during the establishment phases of trees, as they allow a degree of crown sway and root movement to encourage stabilizing girth increment and root anchorage.

Where a movement is created by a stimulus whose direction is not relative towards or away from the stimulus, it is known as a nastic movement. Nastic movements include the movement of petals of daisy-like flowers – flowers 'opening' in the light and 'closing' at night ('day's-eye') known as photonasty. Other nastic movements include the response to touch (thigmonasty, sometimes known as seismonasty) – for example, sensitive plant (*Mimosa pudica*), where the leaves 'close' very quickly, and the closing of the two halves (valves) of the leaves of Venus flytrap (*Dionaea muscipula*) when an insect touches the trigger hairs on them. Thermonasty (temperature) is evident with the 'opening' of crocus and tulip flowers in warm weather, and their 'closing' in colder temperatures.

SECTION TWO

PLANT MANAGEMENT AND PRACTICE

CHAPTER 7

Plants: Their Use in the Landscape

Design Systems and Choices

The greater the range of both exotic and indigenous plants that can be used, the greater the diversity and potential aesthetic effect. The first criterion has to be the correct aerial and root (rhizosphere) environment for successful growth. Trying to grow shade-loving ferns in searing sunlight, for instance, is obviously doomed to failure, growing a species in the incorrect pH is futile, and plants that thrive in rich organic soils will not be successful in soils with low humus content. Hence, because we are dealing with living plants, the biology has to be right before anything else, and the desired visual/aesthetic effect will only be achieved by the correct choice of plants for the chosen positions – including their longevity. The chosen species must also tolerate the normal range of temperatures and exposure experienced on the site.

All of these things may be restrictive and reduce the final choice. If a client draws a sketch of their desired/perceived outcome, the skill of the designer is to decide on the plants that will attain that effect because they are successful in that position, and the effect will continue with the correct long-term management. Sometimes a desired effect cannot be achieved at all on the site available, and a trade-off is inevitable.

Horticulturists most commonly create mixtures of exotic plants – plants from other countries – to achieve their design objectives, and beds and borders feature heavily. Beds may be square, rectangular or circular (or any other geometric shape), or they may be more organically shaped in grass, forming what are known as island beds. Whatever their shape, the key factor is that they can be viewed from all, or most directions without restrictions. Borders may also be any shape, including rectangular or with undulating edges, but they will have viewing restrictions from at least one side (even up to three sides). The choice of bed or border will obviously make a difference in the choice of plant heights because they will be viewed from only one or from several directions.

Display of Hardy Annuals

Annuals by definition have to be resown every year, and with this comes a commitment to soil preparation and maintenance on an annual basis. The rewards are highly coloured summer displays that certainly can be very effective, but they have no permanent structure. Plans for the site are drawn to scale (1:50 is convenient) with the informal drifts superimposed over 1 m grid squares (at the same scale). The plan may then be easily transferred to the site by dividing it into 1 m squares and then using the grid lines as co-ordinates and as a guide to the proportions of the informal drift shapes chosen.

Hardy annuals are sown in the spring in the area where they are to be displayed – they are not produced at a remote site and transported in like bedding plants. They can be sown and displayed in regular or irregular beds and borders, but are nearly always sown in species-specific organic-shaped drifts within that bed or border, and usually in parallel drills within the irregular-outlined drift (in order to weed between them easily in the development stage). Furthermore, to retain the illusion of informality, the parallel drills of each drift are run at different angles to the drills in other drifts, and never parallel to the border edge. Drills are taken out with a sharp stick or the edge of a push hoe, and covered carefully using the hoe or the back of a rake. As the seedlings develop and increase in size the parallel lines are no longer visible and appear as a mass completely filling their irregular drift. Hardy annuals may be sown broadcast within their informal drifts and lightly

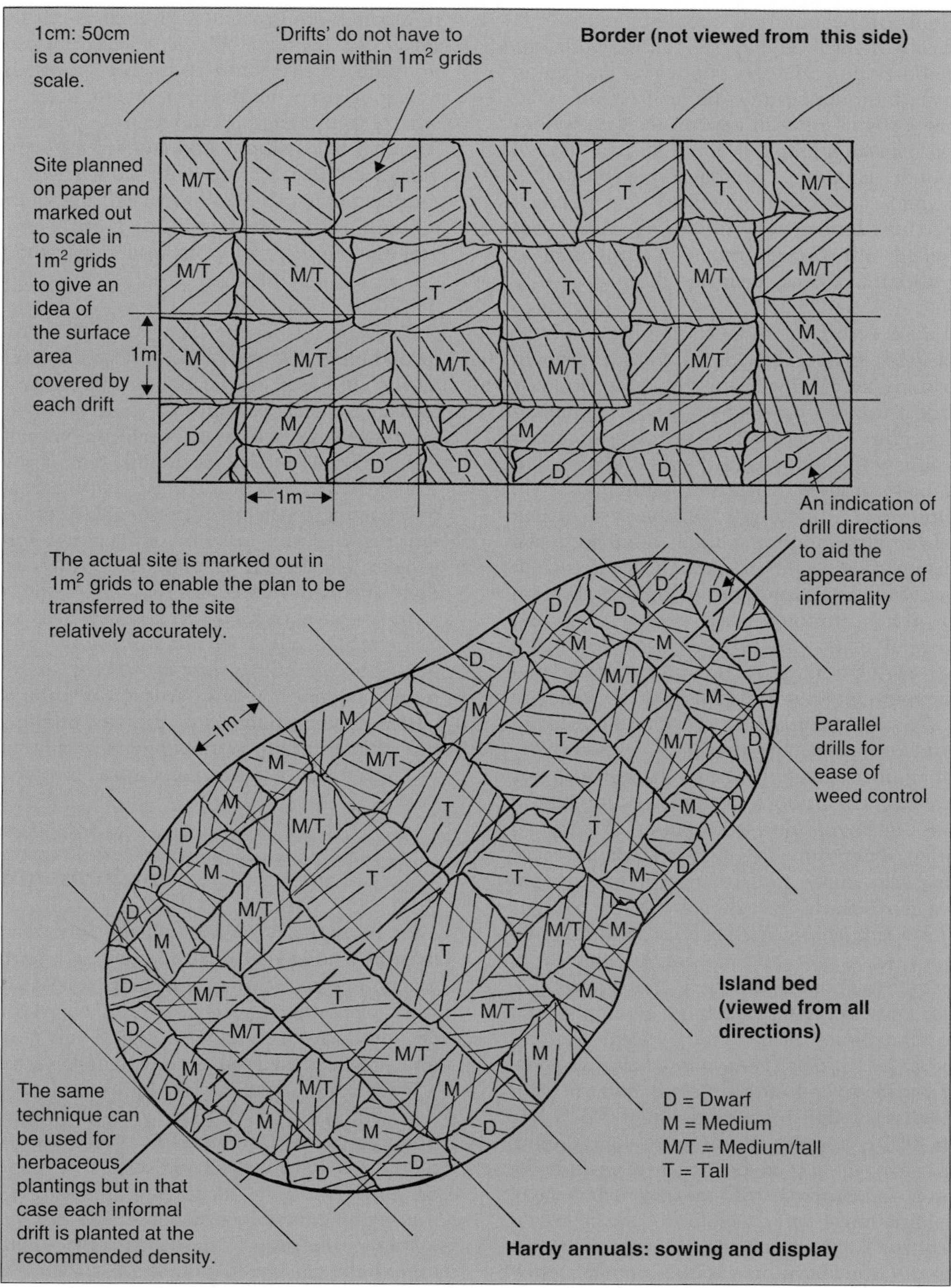
1cm: 50cm
is a convenient
scale.
'Drifts' do not have to
remain within 1m² grids
Border (not viewed from this side)
Site planned
on paper and
marked out
to scale in
1m² grids
to give an
idea of
the surface
area
covered by
each drift
1m
M/T
T
M
D
1m
An indication of
drill directions
to aid the
appearance of
informality
The actual site is marked out in
1m² grids to enable the plan to be
transferred to the site
relatively accurately.
1m
Parallel
drills for
ease of
weed control
Island bed
(viewed from all
directions)
The same
technique can
be used for
herbaceous
plantings but in that
case each informal
drift is planted at the
recommended density.
D = Dwarf
M = Medium
M/T = Medium/tall
T = Tall
Hardy annuals: sowing and display

Hardy annuals: sowing and display.

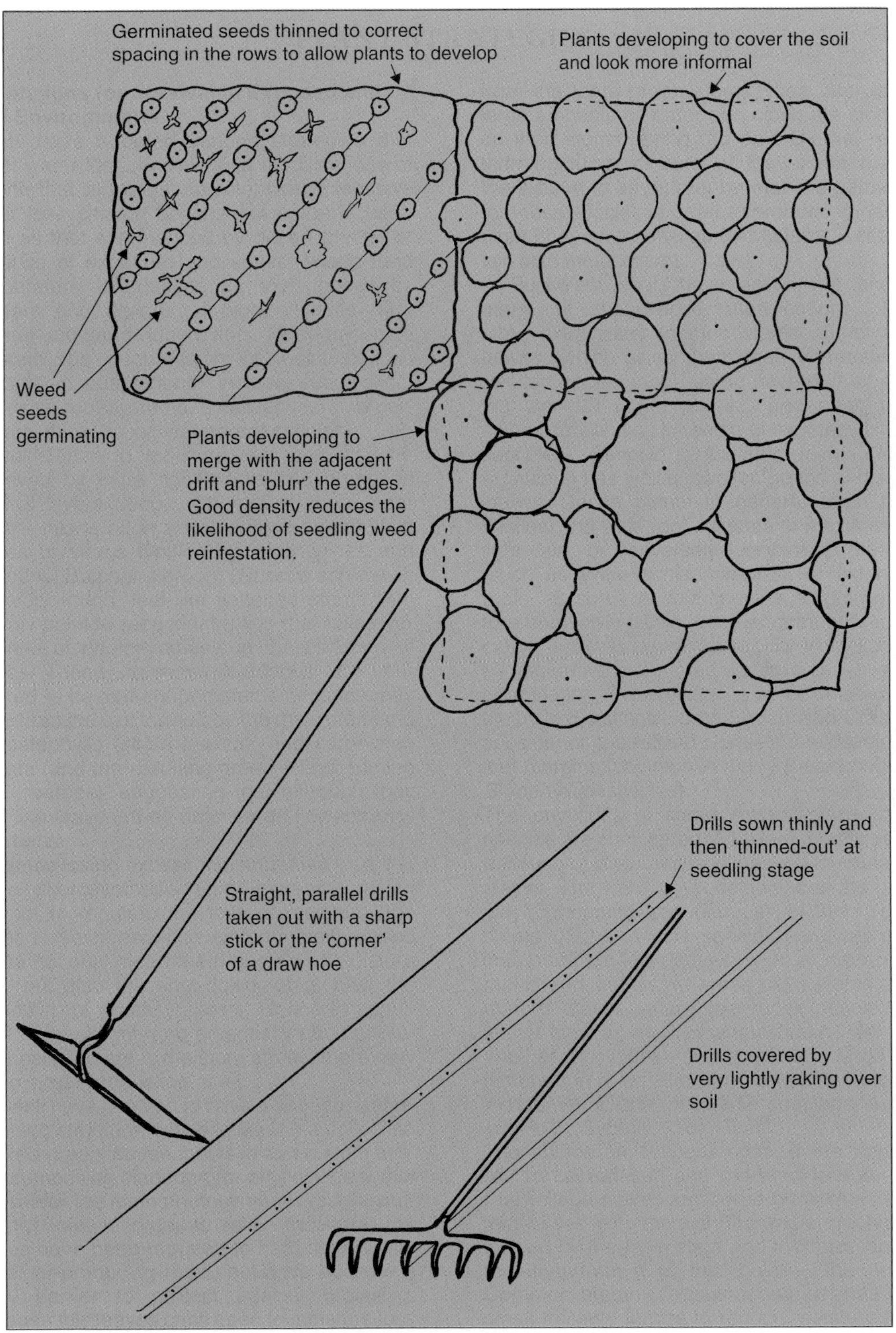

Hardy annuals: a closer look at individual 'drifts'.

Nasturtium (Tropaeolum majus).

A single-flowered cultivar of English marigold (Calendula officinalis).

raked or rolled into the soil surface after sowing, but broadcast sowing causes difficulties later for weed control.

Lower-growing types are chosen for the front, larger for the middle ground, and taller species in the centre or at the rear – features shared with nearly all display areas no matter what the plant types. What is decided as front or middle (or rear) will of course depend on whether a bed or border is being used for display.

EXAMPLES OF ANNUALS COMMONLY USED IN AESTHETIC DISPLAYS

These are plants used in annual beds/borders:

Edging/Front

California bluebell (*Phacelia campanularia*): blue, 15–20 cm

California poppy (*Eschscholzia californica*): orange, yellow, 30–45 cm

Poached egg plant (*Limnanthes douglasii*): yellow and white, 15–20 cm

Pot marigold (*Calendula officinale*): orange, 20–30 cm

Candytuft (*Iberis umbelata*): white, bushy, 15 cm–25 cm

Iberis amara: white, bushy, 30 cm

Alyssum maritimum: white, 15 cm

Salvia horminum (clary): purple, maroon, pink or red bracts, 20–30 cm

Nasturtium (*Tropaeolum majus*): orange, red, pink, yellow, 30 cm

Middle Ground and Taller Species

Moonwort (*Lunaria annua*): blue flowers and rounded fruits 50–75 cm

Gypsophila (*Gypsophila elegans*): pink, white, 30–45 cm

Poppy (*Papaver somniferum*): pink, 30–45 cm

Love-in-the-mist (*Nigella damascena*): pink, pale blue, white, 45 cm

Godetia/clarkia (*Clarkia amoenum*): salmon, pink, red, 45 cm

Cornflower (*Centaurea cyanus*): blue, 45 cm

Burning bush (*Kochia scoparia forma trichophylla*): green foliage turning red, 45–50 cm

Borage (*Borago officinalis*): blue, 45–50 cm

Bells of Ireland (*Moluccella laevis*): yellow/green, 60 cm

Annual tickseed (*Coreopsis tinctoria*): yellow, 60 cm

Annual mallow (*Lavatera trimestris*): pink, white, 60–75 cm

Cosmos: red, pink, 75 cm –1 m

Globe flower (*Helichrysum bracteatus*): orange, pink, 1 m

Dwarf sunflower cultivars (*Helianthus annuus*): yellow, 1–1.25 m

Love-lies-bleeding (*Amaranthus caudatus*): red, maroon, 1–1.25 m

Very Tall

Sunflower (*Helianthus annuus*)

Hardy annual species may be used to enhance flowery meadows by sowing some species in amongst an established mixed sward. However, because many of the species are naturally associated with the production cycle of field crops (notably cereal crops), they can also be used to emulate traditional cereal fields by sowing them into prepared soil at close spacing, or in amongst specially sown wheat or barley. Corncockle (*Agrostemma githago*), corn marigold (*Chrysanthemum segetum*) and field poppy (*Papaver rhoeas*) are examples.

Display of Spring Bedding Plants

Formal spring bedding schemes commonly include young plants of true biennials such as forget-me-not (*Myosotis*) and sweet william (*Dianthus barbatus*) planted out in formal rows. However, the short-lived herbaceous perennial *Bellis perennis* (common dog daisy), and the more persistent perennials polyanthus (and its cultivars) and wallflower (*Cheiranthus cheiri* cultivars), are also very commonly used. But because plants of these subjects do not flower in year one of their production from seed, they are treated in exactly the same way as true biennials such as sweet william. In spring bedding schemes, perennials are either moved or destroyed after their display of flower, as they will traditionally be in the way of the next planting of summer bedding.

Expensive and useful perennials such as polyanthus (*Primula vulgaris elatior* and cultivars) and primroses (*Primula vulgaris* and cultivars) may be retained in a nursery area after lifting from their display beds to be used elsewhere. Hardy primulas can be used successfully in herbaceous display areas or mixed beds and borders, so may also be used for these purposes after their initial use in formal spring bedding displays, and their production and high seed costs may be warranted because of this. Nevertheless, there is always an added cost of lifting them from the spring display area and planting them elsewhere, which militates in some ways against this practice. This operation is not considered worthwhile for wallflowers in spite of their perennial status, because they tend to get woody and 'leggy' at their base, to decline in aesthetics and to quickly exceed a suitable size for formal spring bedding display, and they do not prove particularly useful in herbaceous displays.

Species used for spring bedding are planted into their final positions (in prepared beds and borders) in the autumn to flower in the following spring, and because they usually follow summer bedding displays, this is as soon as the old summer bedding is removed. In some seasons, very difficult decisions have to be made regarding whether or not to lift summer bedding in really good condition for the sake of the following spring display. This will vary depending on the particular season and the longevity (with quality remaining) of the summer bedding plants, but is liable to be October/November. Early frost, or severe deterioration of the previous summer's bedding plant material, would encourage this for early October. Conversely, moist, mild seasons encouraging summer displays to continue well into October could stall the operation until November, if desired.

Traditional spring bedding displays often comprise a mix of wallflowers or sweet williams for the main ground planting, with either/or common daisy, forget-me-nots and primulas as edging plants. Although only a limited range of plants are suitable, more imaginative plantings are also possible (often involving the inclusion of bulbous herbaceous perennials such as tulips and daffodils). The range of stem lengths, flower types, cultivars and species of both *Narcissus* and *Tulipa* means that some may be used as main ground and some as edging types to good effect. *Narcissus* are mainly shades of yellow, although there are those with red and white coloration, some are dwarf types, and some have more than one flower head on a stem. Tulips also vary from dwarf to tall types and offer a wide range of colours from yellow, through orange to red, and even some very dark colours as well; but their flowers are relatively short-lived. Hyacinths offer a wide range of colours (but not heights), and *Chionodoxa* offers an alternative for edging, but both are early flowering. Snowdrop (*Galanthus nivalus*) and crocus are so much earlier to flower than the other types that they are generally considered unsuitable for formal bedding – but along with many *Narcissus*, very suitable for 'naturalizing'. Good quality spring bedding therefore enlists help from a range of plants from the herbaceous and biennial world.

Displaying Biennials

Some biennial species, such as foxglove, mullein and evening primrose, are much too large and flower too late to use *en masse* in formal bedding schemes. They can be included in more informal displays/designs specifically for biennial plants, used as architectural spot (or dot) plants to give height and extra interest, or used within herbaceous schemes, or in mixed beds/borders. Because of the ultimate height of some species they can be used informally grouped (to give a cottage garden effect) or as single specimens. Examples include the architecturally impressive Canary Island *Echium wildpretii* – a very tall, stately plant that grows to 2–3 m in the Boraginaceae family (essentially a massive viper's bugloss).

Also in this category are biennial *Verbascum* species such as the beautiful felted-leaved, yellow-flowered mullein (*Verbascum bombyciferum*) and its cultivars, and the striking yellow-flowered evening primrose (*Oenothera biennis*) that forms large, architecturally interesting spires of flower. Because they will not flower in their first year, and die after flowering, one-year-old plants (previously produced in a nursery area) are planted. These will flower in their first year within the display, and as they die, are replaced on a regular basis (unless they are in sufficient numbers and establishment to reseed naturally). Because of their prominence they are sorely missed if they are not replaced (or naturally self-sown) to retain the integrity of the design. But by good management of seedling regeneration (by seedling recognition and careful selection during weed control) they may perpetuate their benefits in the design and give a perennial effect, even though each individual is actually biennial.

Display of Summer Bedding

Bedding schemes are so called because plants are not sown *in situ* like hardy annuals, but are sown under protection and then young developing plants are 'bedded out' in dense blocks into their final display beds/borders. There is always a desire for structure in design systems, so bedding schemes are nearly always augmented with glasshouse perennials (half-hardy perennials), initially grown under protection and then bedded out for the summer months. The illusion of greater structural permanence is given by adding height and architectural benefits in the form of spot (or dot) plants

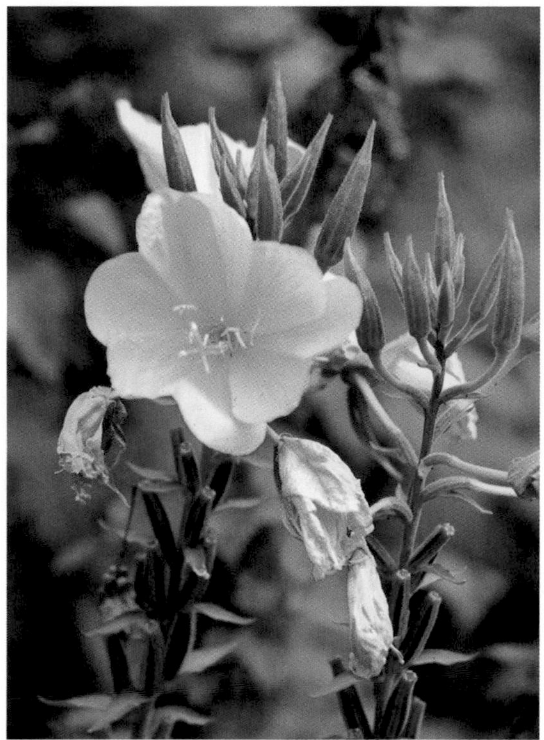

Evening primrose (Oenothera biennis), a biennial (as described in the scientific name); tall-growing.

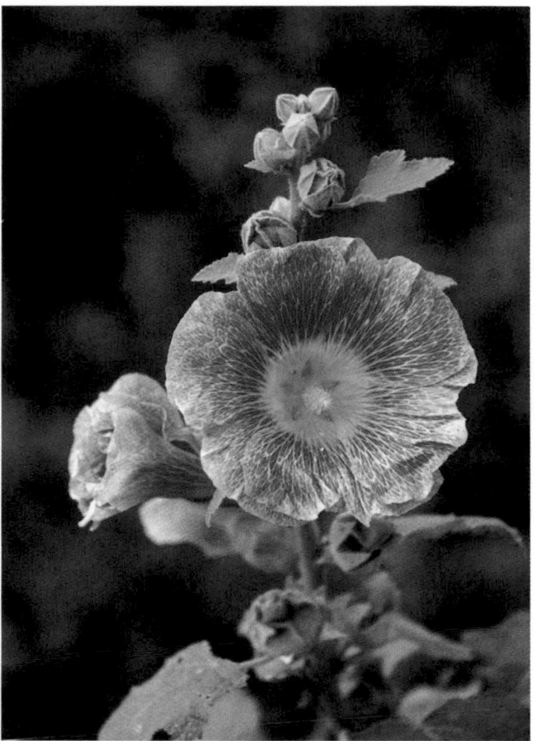

Hollyhock (Alcea rosea syn. althea rosea): a very tall, straggly perennial treated as a biennial.

in the design. Summer bedding displays are therefore achieved using a wide range of plants across several groups, including hardy annuals, half-hardy annuals, tender (glasshouse) perennials, hardy herbaceous perennials, and both tender and hardy woody perennials. The very wide choice of exotic plants and plant cultivars from across the range of major plant groups gives good colourful displays during the summer, and these are enhanced by the varying heights that can be achieved. Edging plants, foreground plants, middle ground plants, main ground plants, tall plants and spot (or dot) plants allow for such a pleasing variation.

Bedding schemes usually comprise half-hardy annuals such as *Dorotheanus bellidiformis* (syn. *Mesembryanthemum crinifolium*), *Lobelia erinus* cultivars, *Tagetes patulum* (including all dwarf French marigolds), *Nemesia strumosa, Ageratum houstonianum* cultivars, *Alyssum maritima* (a hardy annual treated as a half-hardy annual), and other low-growing types such as *Begonia semperflorens* (glasshouse perennial) as edging plants. They also include some hardy perennials such as *Antirrhinum majus* (and its many cultivars) that are treated as half-hardy annuals, and used as main ground plants along with half-hardy perennials such as *Lobelia cardinalis* and *Salvia farinacea*. Glasshouse perennials such as petunia, taller *Begonia semperflorens* cultivars, and geraniums (*Pelargonium zonale* and *Pelargonium regale*) can also be used as main ground feature plants. Foreground and middle ground plants include all the larger French marigold types, and the smaller African marigolds (*Tagetes erecta*), *Salvia splendens* cultivars, *Nicotiana alata* cultivars, annual *Dianthus* and *Celosia cristata* (all half-hardy annuals). Main ground planting choices also include *Fuchsia*, busy lizzies (*Impatiens*), dwarf bedding dahlias, and *Begonia* × *hybridatuberosa* cultivars (glasshouse perennials) – some forms of which can be used to give good height to the display.

Spot (or dot) plants can include half-hardy woody perennials such as *Datura* (*Brugmansia*), young *Eucalyptus globulus, Eucalyptus gunnii* and *Cordyline australis* (cabbage palm); hardy perennials such as *Phormium tenax* (New Zealand flax), and glasshouse perennials such as *Canna indica* (canna lily)

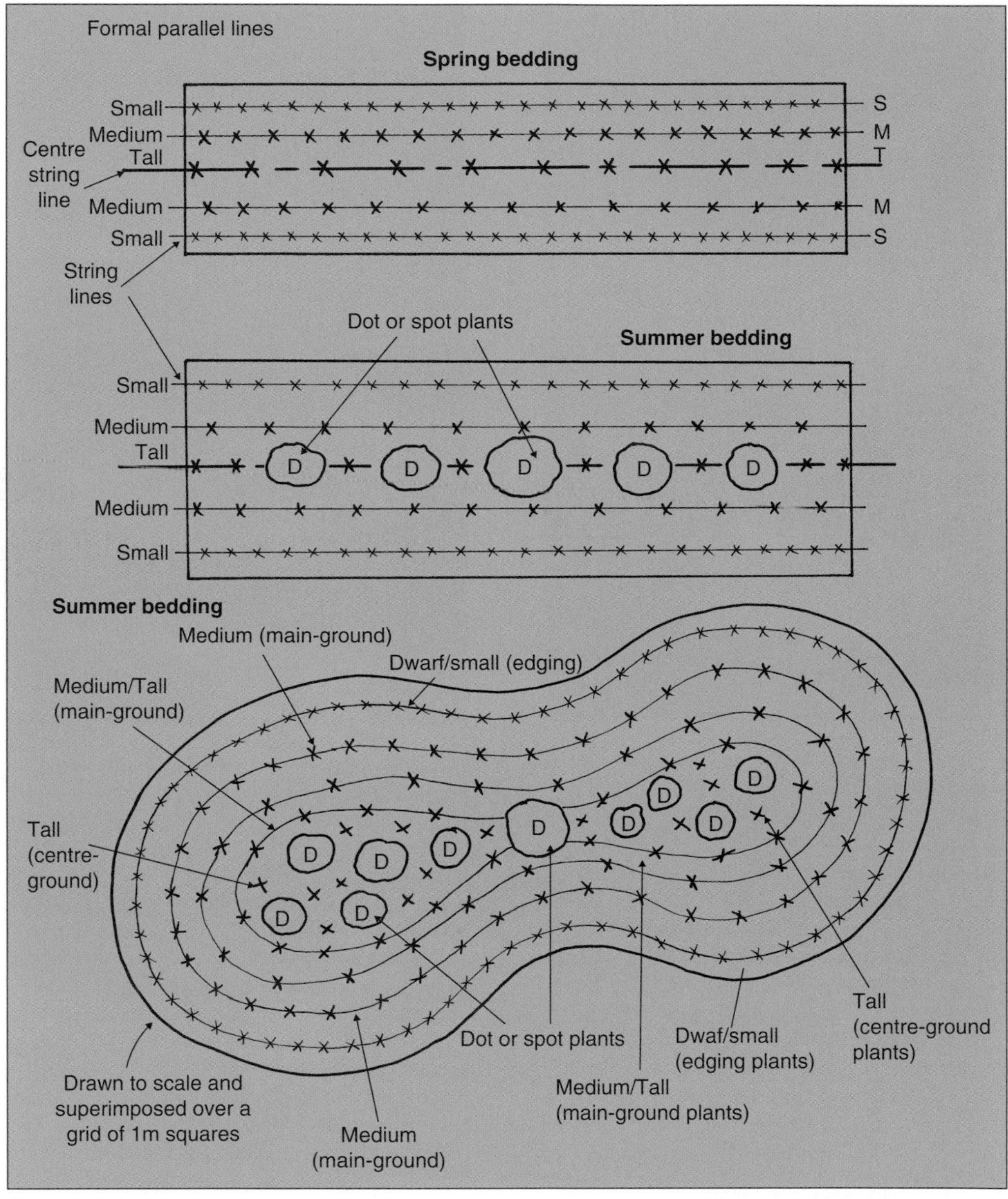

Spring and summer bedding schemes.

and *Strelitzia regina* (bird of paradise flower). Some species lend themselves to be trained as standards to give extra interest, for example *Fuchsia, Heliotrope, Plumbago capensis* and *Streptosolen jamesonii.* The production of plants to achieve good, colourful, and interesting designs therefore involves several methods of propagation, and several production environments.

Display of Herbaceous Plants

Herbaceous perennials (hardy perennials) include all leafy plant forms with underground storage organs, and they give wonderful summer displays with mixtures of height, texture and colour. Some forms do have an evergreen effect, even in temperate and cold climates, as they renew their

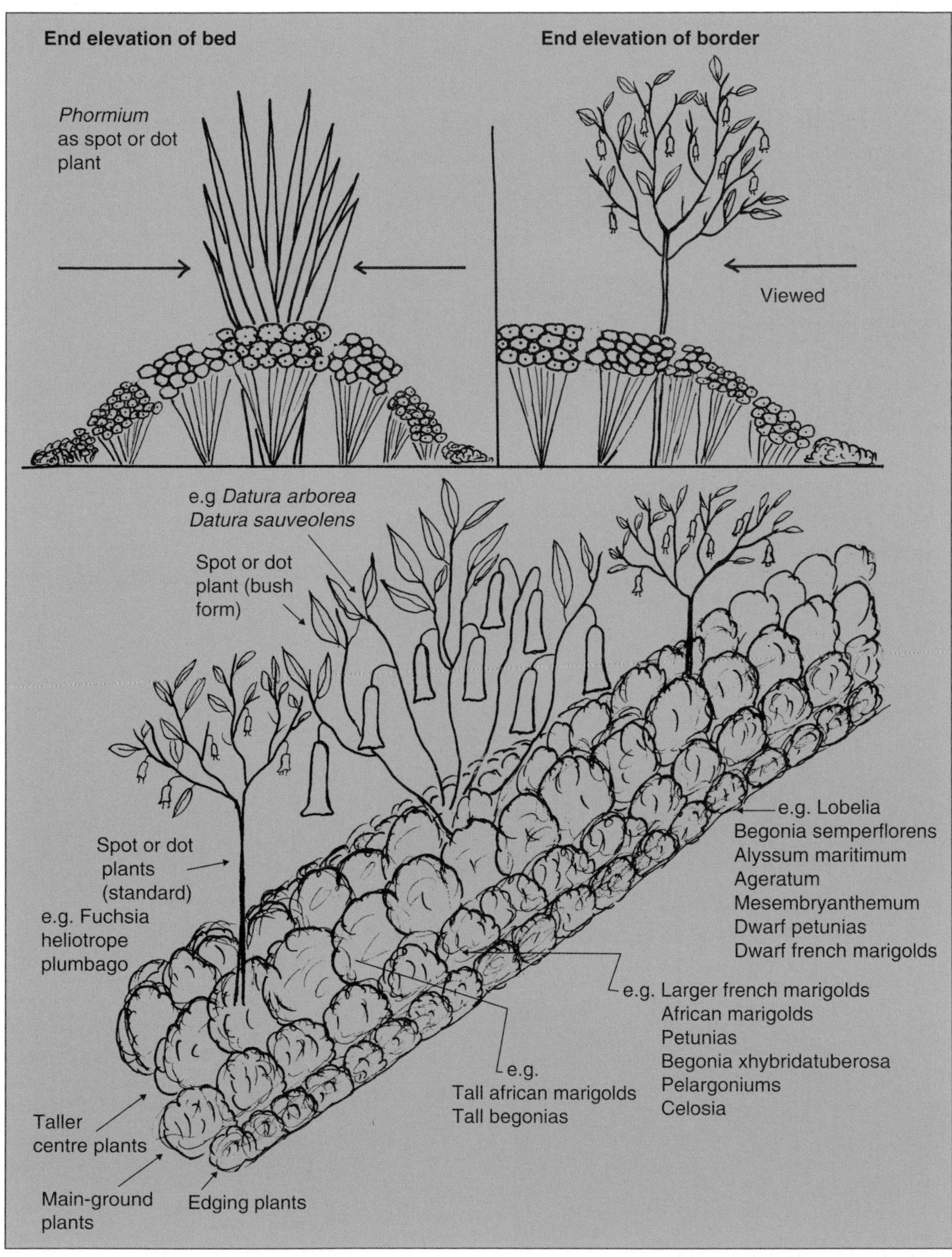
End elevation of bed
End elevation of border
Phormium
as spot or dot
plant
Viewed
e.g Datura arborea
Datura sauveolens
Spot or dot
plant (bush
form)
Spot or dot
plants
(standard)
e.g. Fuchsia
heliotrope
plumbago
e.g. Lobelia
Begonia semperflorens
Alyssum maritimum
Ageratum
Mesembryanthemum
Dwarf petunias
Dwarf french marigolds
e.g. Larger french marigolds
African marigolds
Petunias
Begonia xhybridatuberosa
Pelargoniums
Celosia
e.g.
Tall african marigolds
Tall begonias
Taller
centre plants
Main-ground
plants
Edging plants

Summer bedding display.

Innovative use of summer bedding at the entrance to Hayes Garden Centre, Ambleside, Cumbria.

Carpet bedding (using lots of small plants to form the patterns): Kitzbuhel, Austria.

An extreme form of bedding (using lots of small plants to form the patterns) at Mainau Island, Lake Constance, Germany.

Salvia farinacea, cultivar.

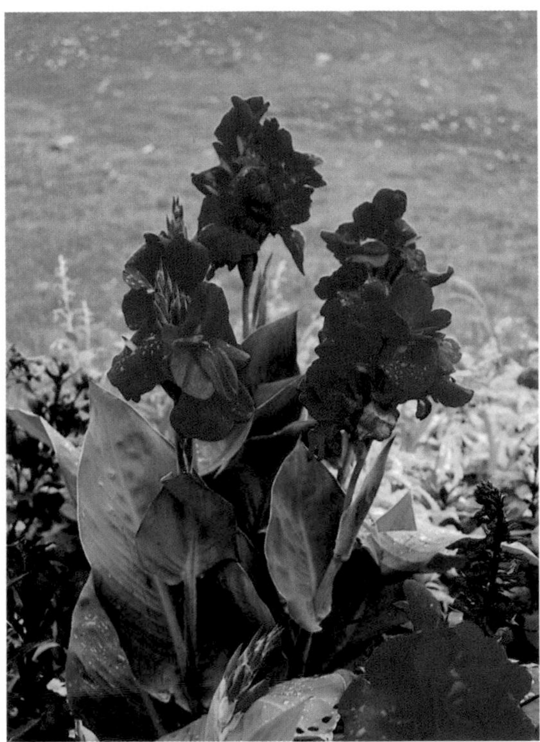

Canna lily (Canna indica), cultivar.

leaves sequentially and overwinter with leaves above soil level. Some evergreen species with sword-like leaves – for example *Phormium tenax* and some species of *Kniphofia* (red hot poker) – are noted for their architectural interest, and some species even have autumn colour that may persist over the winter period (such as *Bergenia cordifolia* – elephant's ears and its cultivars). Nevertheless, many herbaceous types have no growth above soil level in winter, as their main method of overcoming the low temperatures is for their foliage to supply sugars to the underground organ, and then to die back – so as regards design they are relatively short-lived. Many traditionally used herbaceous perennials are of this type: Michaelmas daisy (*Aster belgi-novi* and its many cultivars), ice plant (*Sedum spectabile* and its cultivars), *Delphinium* species and cultivars, and monkshood (*Aconitum napellus*).

Designs for herbaceous display take into account the same height considerations as for bedding schemes: low-growing plants as edging/low foreground plantings, taller species for foreground and middle ground, and heights ranked differently for a border than for a bed. Because beds are viewed from all directions, whether the bed is rectangular, square or irregular – like an island bed – the taller types are always planted in the centre, the medium-growing types surrounding them, and the lower-growing types form edging material all the way round the front of the bed to give a three-dimensional effect. Island beds (and beds in general) tend to use the more compact forms only, as very tall types need too much support to be successful. However, they still have the taller, compact types in the centre, whereas borders have the tallest types at the 'back' of the border (and are often sheltered because of this), and the shortest types at the front. So as far as shape and form are concerned, design is basically the same for herbaceous subjects as for most other plant display areas.

From a horticultural point of view, the key is to know which species and cultivars will fulfil the proposed heights, and have complimentary, harmonious or contrasting colours. Species with very bold leaves include *Ligularia przwalskia, Ligularia dentata* (with the cultivar 'Desdemona' being particularly effective), *Ligularia stenocephala, Rodgersia podophylla, Rodgersia aesculifolia, Rodgersia sambucifolia, Macleaya cordata, Rheum* species and *Gunnera manicata*. The massive scale of *Gunnera manicata*, and its liking for very boggy sites, makes it difficult to place with any other plants.

The next few pages comprise some basic forms and types of herbaceous perennials. They are not accurate shapes or specific to any one plant species, so are not intended as a means of botanic identification. They are instead offered as a matrix for design possibilities, with indications of the types and short lists of suggestions for species (with the flower colour and ultimate heights) that roughly adhere to these patterns and shapes. This information will be helpful if you need to fulfil a specific design objective. You are left to research yourself the times of flowering and the wide range on offer outside these suggestions – and once your plant repertoire is sufficient, you will find that suitable species can be brought to the front of your mind at will. These pages are not meant as a substitute for the superb identification books that are available. Furthermore, because the memory cannot always be relied upon, good reference material of the quality of the *RHS Encyclopedia of Perennials*, for example, is always a good idea, no matter what your level of knowledge.

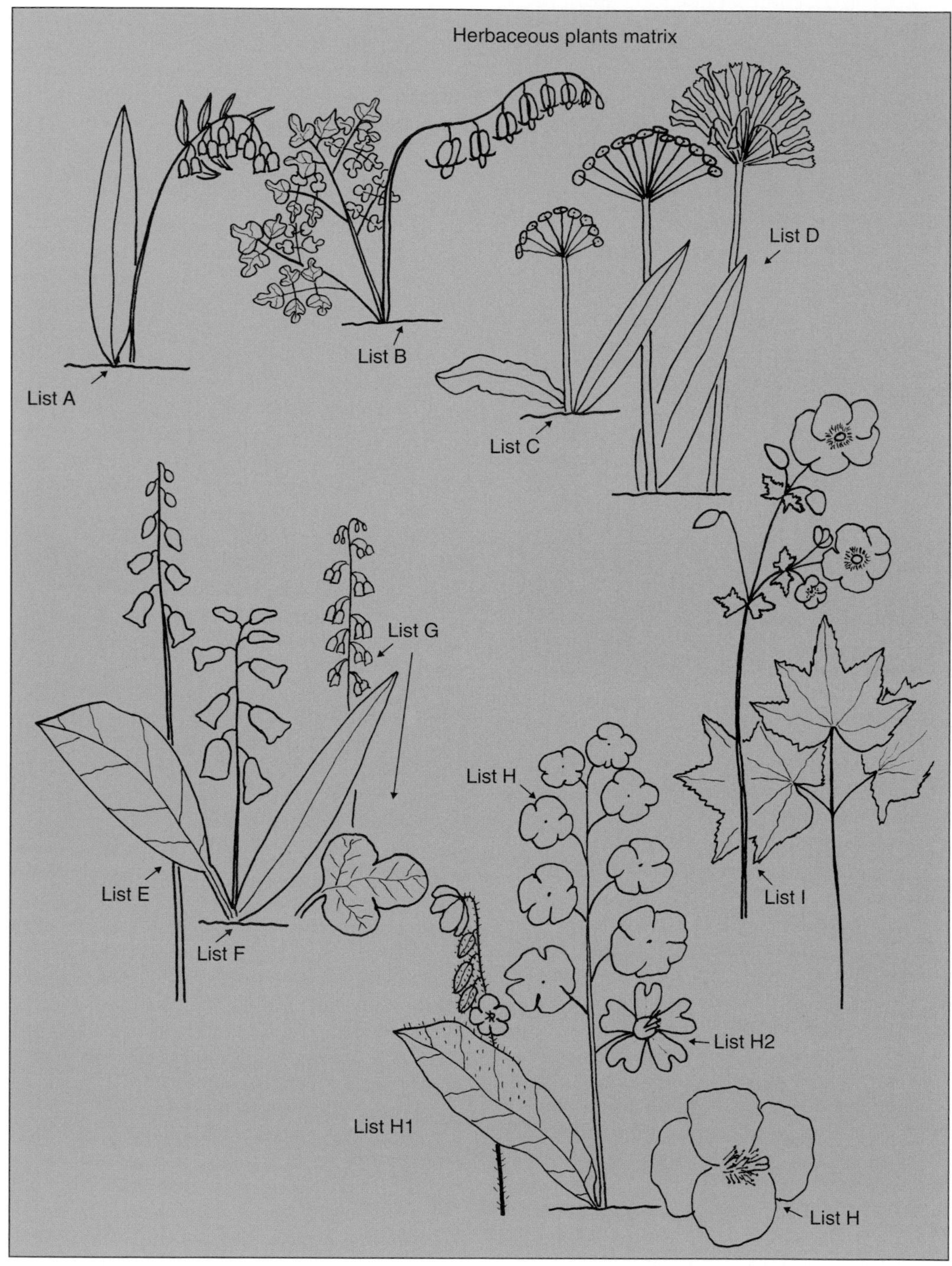

Herbaceous plants I.

Herbaceous Plants Matrix

A *Leucojum vernum (spring snowflake), white flowers, 10–15 cm*
Convallaria majalis (lily-of-the-valley), white flowers, 20–25 cm
Leucojum aestivum (summer snowflake), white flowers, 50–75 cm
Disporopsis fuscopicta, white/cream flowers and purple fruits, 50–75 cm
Polygonatum multiflorum (Solomon's seal), white flowers, 75 cm–1 m

B *Dicentra Canadensis, white, scented flowers with purple tinge, 25–35 cm*
Dicentra eximea, deep rose flowers, 45–60 cm
Dicentra spectabilis (bleeding heart), deep red/pink flowers, 75 cm–1.25 m
Dicentra spectabilis f. alba, white flowers, 60–80 cm

C *Primula denticulata (drumstick primula), pink/mauve flower heads, 25 cm*
Primula denticulata f. alba, white flowers, 25 cm

D *Allium schoenoprasum (chives), pale blue flowers, 20–25 cm*
Allium caeruleum, blue flowers, 45–75 cm
Allium aflatunense, blue flowers, 75 cm
Agapanthus praecox subsp. orientalis, pale blue flowers, 1 m
Allium giganteum, purple/blue flowers, 1.5–1.8 m

E *Digitalis purpurea (foxglove), biennial – range of colours white–purple, 1 –1.5 m*
Digitalis ferruginous, pale orange/brown flowers, 1.0–1.5 m

F *Hyacinthoides hispanicum (Spanish bluebell), blue flowers, 30 cm*
Hyacinthoides non-scriptus (common bluebell), blue flowers, 15–20 cm

G *Tiarella cordifolia (foamflower), white or pale pink flowers 20–30 cm*
Tiarella wherryi, white or pale pink flowers, 30–45 cm
Heuchera sanguinea (coral bells), pink-red flowers, 30–45 cm
Heuchera micrantha, green-pink flowers, 35–50 cm
Heuchera americana (alum root), small green flowers, 75–85 cm
X Heucherella (hybrid between Heuchera and Tiarella), white–pink, 30–45 cm

H *Romneya coulteri (Californian poppy), white, 1.5–1.8 m*
Oenothera tetragona, yellow flowers, 30–45 cm
Meconopsis betonicifolia (Himalayan blue poppy), clear blue flowers, 1–1.25 m
Meconopsis grandis, blue flowers, 1–1.25 m
Meconopsis nepaulensis, pink/red flowers, 1–1.25 m
Meconopsis cambrica, yellow or orange flowers, 30 cm

H1 *Borago pygmaea, pinkish-blue flowers, 25–60 cm*
Borago officinalis (annual), blue flowers, 30–45 cm

H2 *Phlox paniculata, pink-white flowers, 60 cm–1 m*
Phlox maculata, white/pink flowers, 75 cm–1 m
Lychnis coronaria, pink and white flowering forms, 1 m
Lychnis chalcedonica, scarlet flowers, 1 m
Lychnis x arkwrightii, red flowers, bronze foliage, 45 cm
Saponaria officinalis, pink/white flowers, 45–60 cm

I *Anemone japonica, pink-white flowers, 1–1.25 m*
Geum coccineum, red, orange and yellow flowers, 30–45 cm
Geum chiloense, red flowers, 30–45 cm
Geum x borisii, orange/red flowers, 30–45 cm

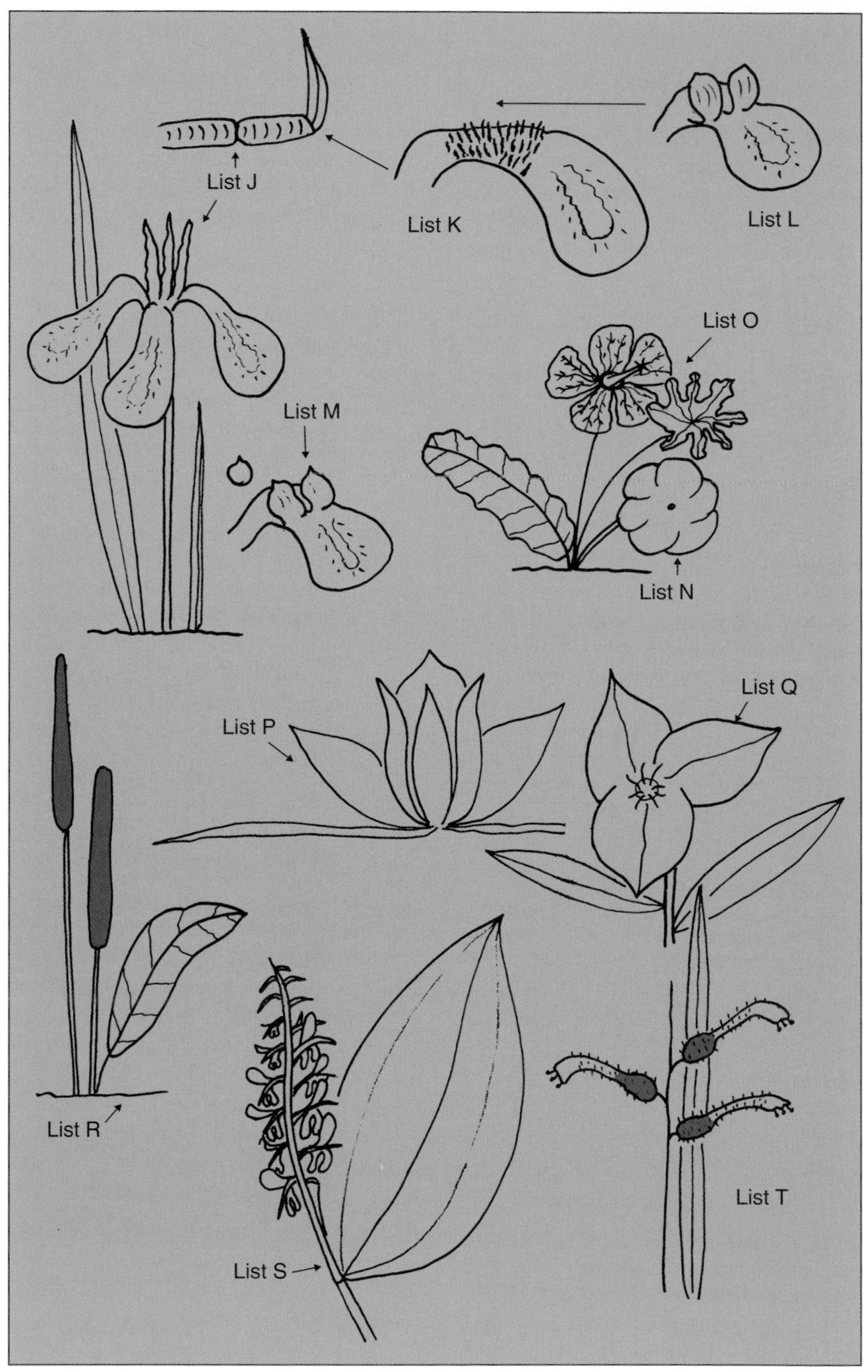

Herbaceous plants II.

J *Iris non-bearded (rhizomatous)*
Iris pseudoacorus (yellow flag), 1.5–2 m
Iris sibirica, blue, 50 cm–1 m
Iris ensata, blue-white, 1 m
Iris foetidissima, brown and purple, 45–60 cm

K *Iris bearded (rhizomatous)*
Iris germanica, blue, purple, violet, 75 cm–1 m
Iris acutiloba, purple flowers, 10–30 cm
Iris aphylla, purple-violet, 20–30 cm

L *Iris (rhizomatous) Pacific coast types*
Iris douglasiana, pale violet-purple, 45–75 cm.
Iris Missouriensis, blue-lilac-white, 60–75 cm

M *Iris (bulbous types)*
Iris bakeriana, blue, 10–15 cm
Iris aucheri, blue-white, 20–30 cm
Iris bucharica, yellow-white, 25–45 cm
Iris danfordiae, yellow, 10–15 cm
Iris reticulata, purple, 10–15 cm

N *Primula vulgaris (primrose), pale yellow, 20–30 cm*
Viola cornuta, lilac flowers, 15–20 cm
Viola odorata (sweet violet), white-pale blue flowers, 15–20 cm
Viola palustis (marsh violet), pale blue/lilac flowers, 15–20 cm
Viola elatior, blue flowers 25–45 cm
Anemone blanda , blue flowers, 20–45 cm
Anemone nemerosa (wood anemone), white-white/pink flowers, 15–25 cm
Meconopsis cambrica, yellow or orange flowers, 30–45 cm

O *Geranium psilostemon, pink/red-magenta flowers, 75 cm–1 m*
Geranium cinereum, white-pink flowers, 15–25 cm
Geranium pratense (meadow cranesbill), blue flowers, 30–75 cm
Geranium maderense, purple-red flowers, dark centres, 75 cm–1 m
Geranium sanguineum (bloody cranesbill), blue flowers, 25–45 cm

P *Trillium chloropetalum, pink-white sessile flowers, 25–30 cm*
Trillium sessile, maroon sessile flowers, 25–40 cm

Q *Trillium ovatum, white flowers, 25–40 cm*
Trillium grandiflorum, white flowers, 25–35 cm

R *Persicaria bistorta, pink flowers, 50–75 cm*
Persicaria affine, pink flowers, 10–25 cm
Persicaria amplexicaule, red flowers, 75 cm–1 m
Persicaria milletii, red flowers, 45–60 cm

S *Hedychium densiflorus, yellow/orange flowers, 1–1.8 m*

T *Anigozanthos falvidus (yellow kangaroo paw), yellow-green flowers with orange stamen, 50–60 cm*

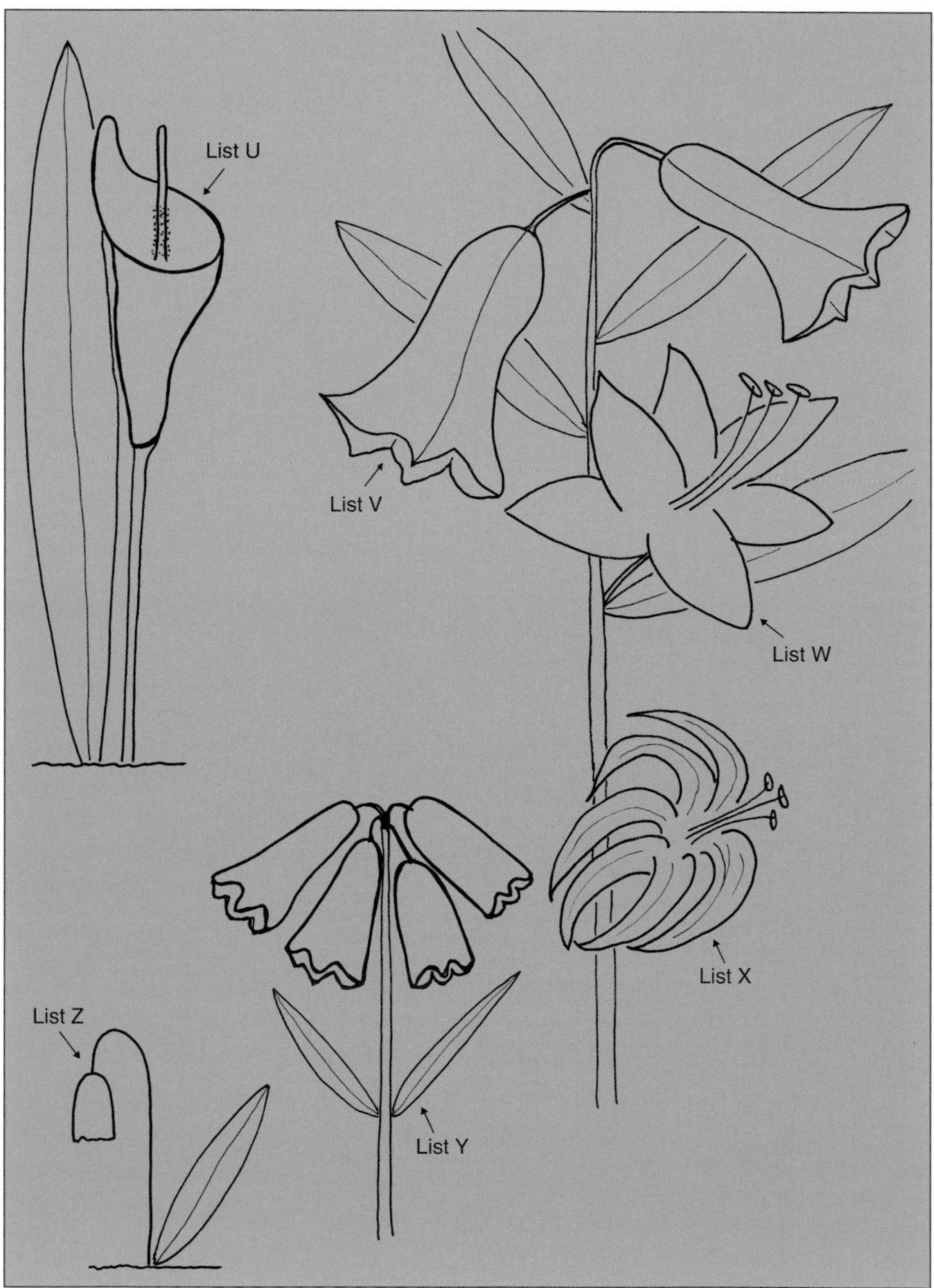

Herbaceous plants III.

U *Arums*
Zandtedeschia aethioipica, white spathes, 75 cm–1.25 m; many coloured cultivars also
Arisaema candidissimum, candy-striped spathes, 10–15 cm
Lysichiton americanus, large yellow spathes, very large leaves
Lysichiton camtschatcensis, large white spathes, very large leaves
Dracunculus vulgaris (dragon lily), deep maroon-blotched spathe, 75 cm–1 m
Arum creticum, yellow spathes, 40–60 cm
Arisarum proboscideum (mousetail plant), white spathe with dark streaks, 15–30 cm

V *Lily species with tube-like flowers – many colours, cultivars and a range of heights*
Lilium longiflorum, white flowers, 1 m
Lilium regale, white/pink, 1 m
Crinum moorei, pale pink flowers, 50–75 cm
Crinum x powellii, pale pink flowers, 75 cm–1 m
Cardiocrinum giganteum, cream/pink flowers, 2–3 m

W *Lily species with open flowers – many colours, cultivars and a range of heights*
Lilium bulbiferum, orange flowers, 1–1.5 m
Nerine bowdenii, pink flowers, 45–60 cm
Hemerocalis, many colours, cultivars and a range of heights

X *Lily species with reflex petals – many colours, cultivars and a range of heights*
Lilium pyrenaicum, yellow flowers, 1–1.5 m
Lilium lancifolium (tiger lily), orange flowers, 1–1.5 m
Lilium superbum, orange flowers, 1.5–2 m
Lilium speciosum, pink flowers, 1 m
Lilum hansonii, orange flowers, 1 m
Lilium chalcedonicum, red flowers, 1 m

Y *Fritillaria imperialis, orange-yellow, 50–75 cm*
Fritillaria pallidiflora, yellow/green flowers, 30–60 cm

Z *Fritillaria meleagris (snake's head fritillary), drooping heads of 30 cm*
Fritillaria pontica, green/yellow flowers, 20–45 cm
Fritillaria camschatcensis, purple-black flowers, 20–45 cm
Galanthus nivalis (snowdrop), very early white flowers, 10–15 cm
Erythronium americanum, yellow flowers, 20–30 cm

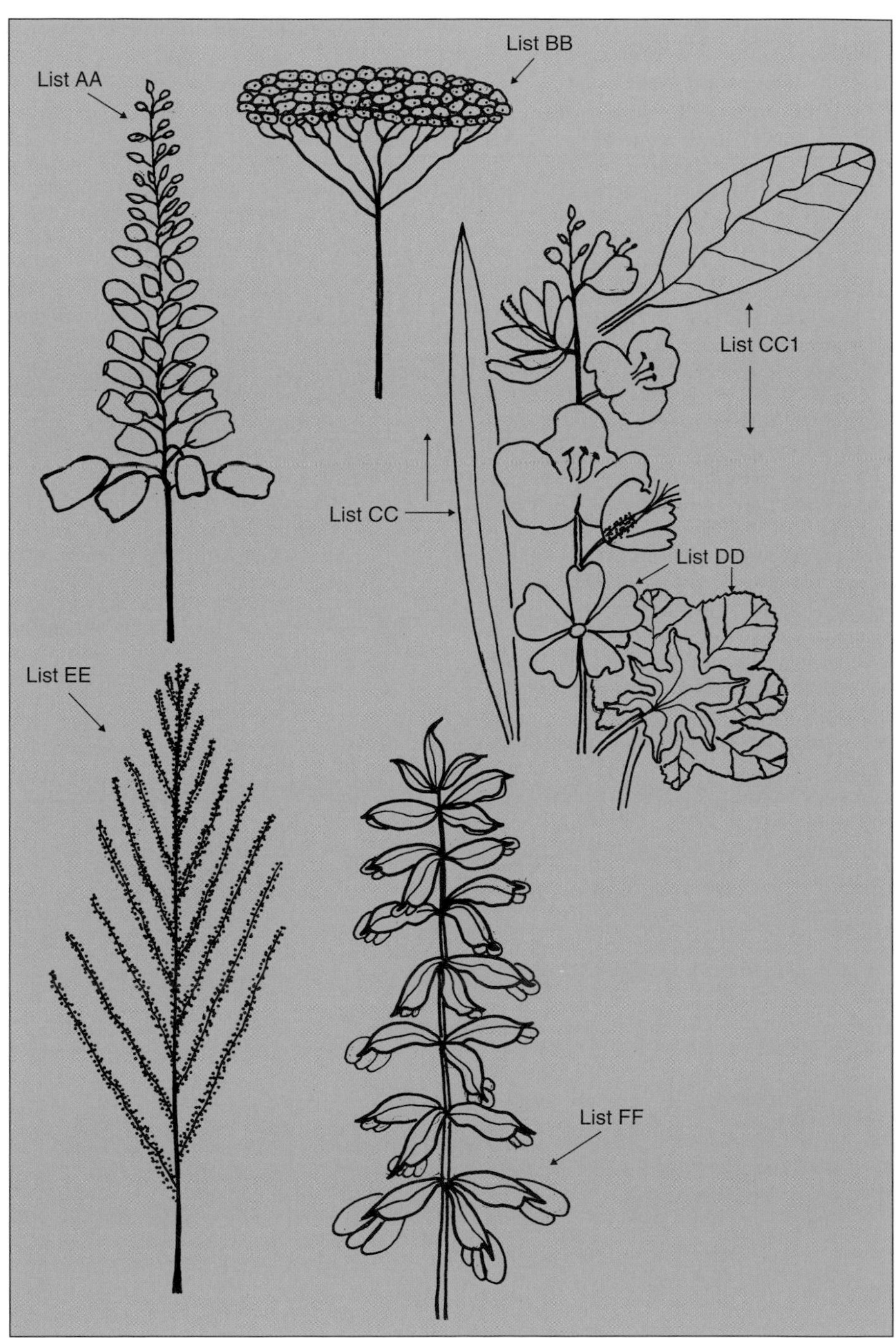

Herbaceous plants IV.

AA *Eremerus himalaicus, white flowers, 1.8–2.25 m*
Eremerus robusta, pink flowers, 1.8 m
Sanguisorba canadensis (white burnet), white flowers, 1–1.8 m
Delphinium, many colours, range of heights, 1–1.8 m
Epilobium angustifolium f. album (white rosebay willowherb), 1–1.5 m
Veronica virginica f. alba, white flowers, 1–1.25 m
Veronica gentionoides, pale blue flowers, 45–50 cm
Lupinus (lupin), many colours: pink, yellow, red, white, 75 cm–1.25 m
Asphodeline lutea (yellow asphodel), 75 cm–1 m
Asphodelus albus (white asphodel), 75 cm–1 m
Lobelia cardinalis, cardinal red flowers and dark foliage, 60 cm–1 m
Lythrum virgatum, pink flowers, 75 cm–1 m
Lythrum salicaria (purple loosestrife), purple flowers, 1–1.25 m
Lysimachia punctata, yellow flowers, 50–75 cm
Ornithogolum narbonense, white flowers, 25–45 cm
Camassia leichtlinii, white flowers, 1–1.25 m
Dictamnus albus, white flowers, 75 cm–1 m
Liriope muscari, deep lavender flowers, 30–45 cm
Eucomis pallidiflora (giant pineapple lily), greenish/yellow-white, 60–75 cm

BB *Achillea millefolium, many cultivars, yellow, pink, orange flowers, 60–75 cm*
Eupatorium purpureum, pink/purple flowers, 2 m
Verbena patagonica, pink/purple, 1–1.25 m
Sedum spectabile, pink/red, many cultivars, 30–45 cm
Ruta graveolens, glaucous foliage and yellow flowers, 30–45 cm
Peltiphyllum peltatum, its peltate leaf mentioned in both generic and specific names, white-pale pink, 1 m

CC *Gladiolus, many cultivars, red, pink, white, yellow, 1–1.2 m*
Schizostylus coccineus, red, pink, 30–40 cm

CC1 *Verbascum, several types, mostly biennial, yellow flowers, 1–1.8 m*

DD *Sidalcea, pink, mallow-like flowers, 1–1.25 m*
Lavatera thuringiaca (mallow), pink flowers, 1–2 m
Malva moschata (mallow), pink, 60 cm–1 m
Alcea rosea (hollyhock), straggly perennial treated as a biennial, many cultivars, 1–2 m

EE *Filipendula rubra, red flowers, 1.75–2.25 m*
Rheum palmatum, red flowers, 1.75–2.0 m
Macleaya cordata, cream-white flowers, 1.0–2.0 m
Aruncus doicus syn. Spiraea aruncus, white flowers, 1.5–1.8 m
Thalictrum lucidum, yellow flowers, 1–1.25 m
Rodgersia sambucifolia, white/cream, 1–1.25 m, very large leaves
Rodgersia aesculifolia, white/cream/pink flowers, 75 cm–1 m, very large leaves
Rodgersia podophylla, white/cream flowers, 75–1 m, very large leaves
Solidago, yellow flowers, 75 cm–1 m
Smilacena racemosa, white flowers, 60 cm –1 m
Astilbe, many cultivars, pink, red, white, 50 cm–1 m

FF *Acanthus mollis, flower spikes with white flowers in pink/purple bracts and very large leaves without spines, 1.25–1.5 m*
Acanthus spinnosus, spikes of white flowers with pink/purple bracts and large leaves with sharp spines, 1–1.25 m
Acanthus hirsutus, spikes of white flowers with pink/purple bracts and large, narrow, very prickly hairy leaves, 30–45 cm

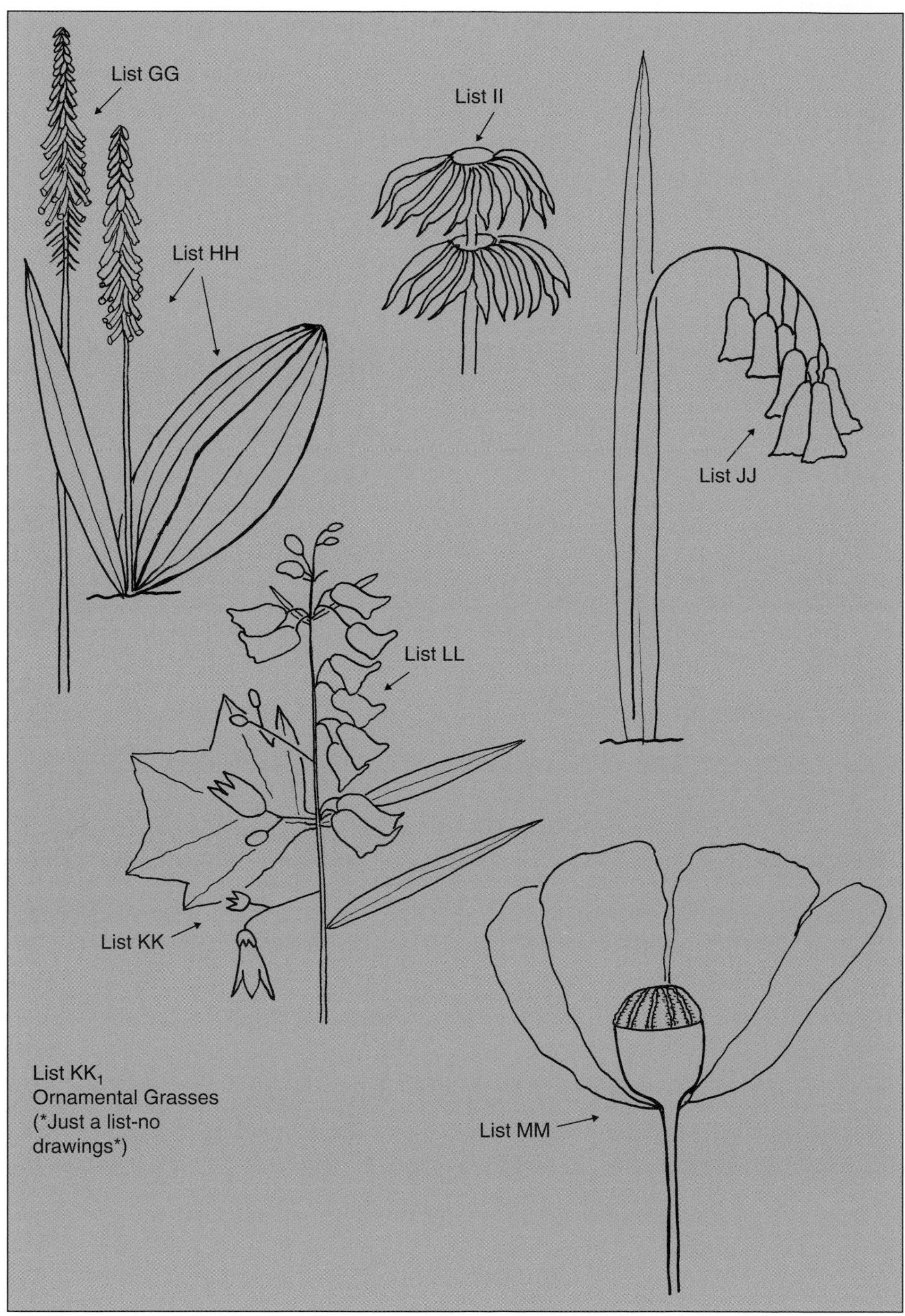

Herbaceous plants V.

GG *Red hot pokers – many cultivars – long, architectural leaves*
Kniphofia caulescens, red-cream flowers, 1–1.25 m
Kniphofia galpinii, red-yellow flowers, 50–75 cm
Kniphofia tringularis, red-orange flowers, 50–75 cm
Kniphofia northiae, yellow flowers, broad erect leaves
Kniphofia uvaria, red flowers, 1.5–1.8 m

HH *Hostas – mainly grown for their bold leaves, but have good, lily-like flowers as well in some years. Many cultivars, 25 cm–1 m.*
Hosta sieboldiana, very large, bold leaves, 75 cm–1 m

II *Monarda didyma (bergamot), red/maroon flowers, many cultivars, 75 cm–1 m*
Phlomis russeliana, yellow flowers, 75 cm–1 m
Phlomis cashmeriana, pink flowers, 1 m
Phlomis tuberosa, pink-purple flowers, 1–1.5 m
Leonotis leonurus, orange flowers, 2 m

JJ *Dierama pulcherrimum, arching stems with hanging pink flowers, 1–1.25 m*
Dierama pendulum, arching stems with hanging pink flowers, 1–1.25 m

KK *Kirengeshoma palmata, yellow flowers, 75 cm–1 m*

KK1 *Ornamental Grasses*
Cortaderia selloana (pampas), 1–1.5 m
Stipa giganteum (giant oats), 1.5–2.5 m
Miscanthus sinensis, 2–3 m
Pennisitum villosum, feathery tufts on the end of stems, 75 cm–1 m
Carex pendula (weeping sedge), flowers on pendulous stalks, 75 cm–1 m
Lagarus ovatus (hare's tail), fluffy tufts on short stems, 45 cm
Festuca ovina glauca (blue sheep's fescue), 25 cm

LL *Penstemon, many cultivars, red, pink, maroon, 30–75 cm*

MM *Papaver orientale, many cultivars, orange/red, red, white with and without blotches 75 cm–1.25 m*
Papaver bracteatum, similar to P. orientale but with dark blotches at the base of the red petals and distinct bracts immediately below the flower buds.
Papaver spicatum, red, 1 m
Many annuals including Papaver somniferum (peony-flowered), 60–90 cm
and Papaver rhoeas

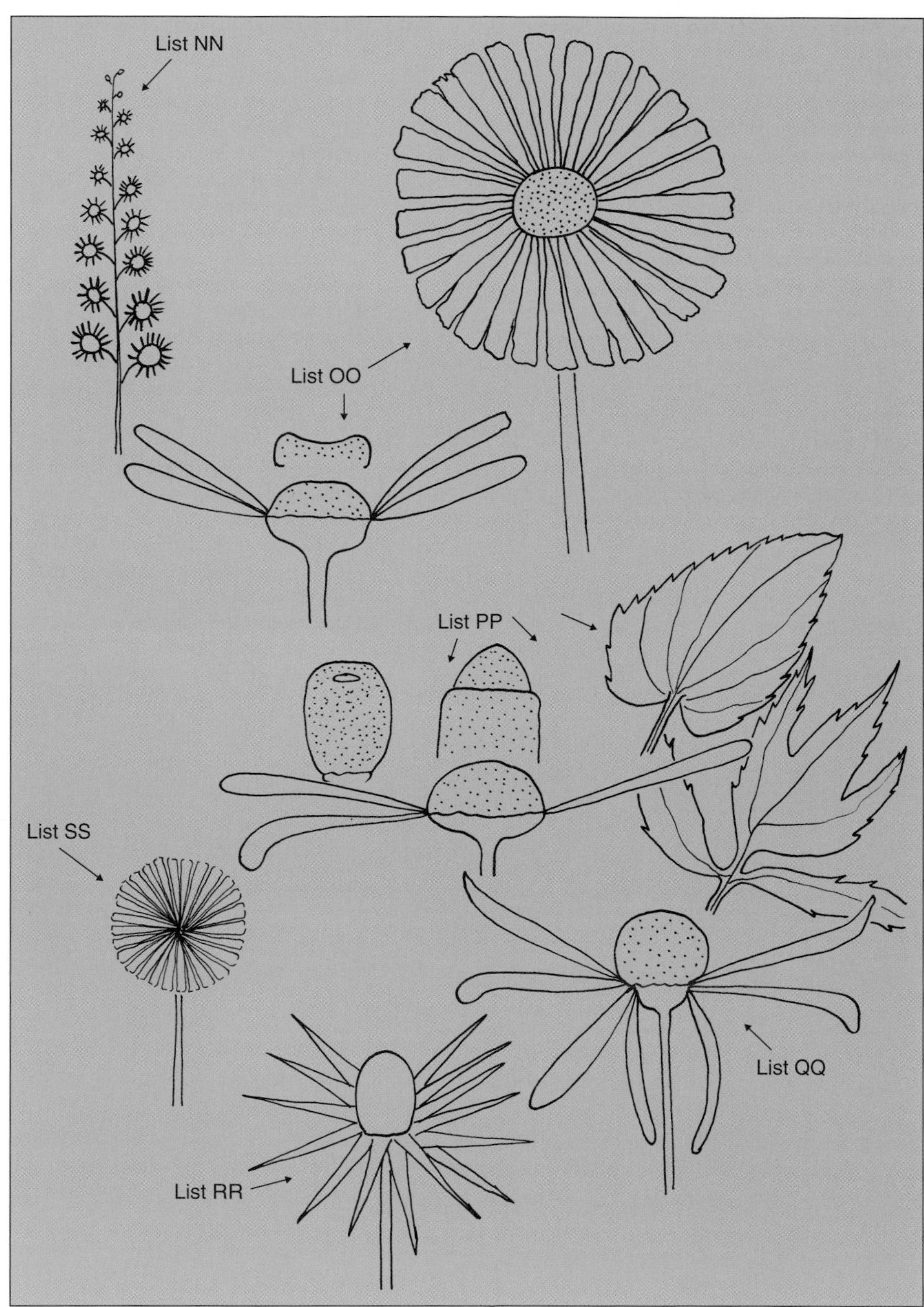

Herbaceous plants VI.

NN *Ligularia przewalskii, yellow flowers, 1–1.8 m*
Ligularia stenocephala, yellow flowers, 1–1.5 m
Chicorum intybus (chicory), short-lived perennial, pale blue flowers, 1–1.25 m
Senecio smithii, white/cream flowers, and large hairy leaves below

OO *Inula magnifica, yellow flowers, 1.5–1.8 m*
Helianthus × multiflorus, yellow flowers, 1.2–1.5 m
Chrysanthemum × superbum, white flowers, 80 cm–1 m
Anthemis tinctoria, lemon-yellow flowers, 75 cm–1 m
Coreopsis, many cultivars, yellow–yellow/orange, 45–75 cm
Osteospermum jucundum, pink flowers, 30–45 cm
Aster novae-anglica and A. novi-belgii (Michaelmas daisies), many cultivars, white, pink-mauve flowers
Dahlia, very many cultivars – wide range of colours and flower shapes, 20 cm–1.25 m
Erigeron alpinus, pink flowers, 20–30 cm, many cultivars

PP *Rudbeckia, many cultivars, yellow-orange flowers, 1–2 m*

QQ *Echinacea purpurea, 75 cm–1 m, and many cultivars*

RR *Eryngium alpinum, purple/blue flowers, 75–85 cm*
Eryngium tripartitum, blue flowers, 90 cm–1.25 m
Eryngium x oliverianum, blue flowers, 60 cm–1.2 m

SS *Disk florets form a globe-shaped head.*
Echinops sphaerocephalus, greenish/grey-white flower heads surrounded by sharp bracts, 1.8 m
Echinops bannaticus, blue flowers, 1–1.5 m
Echinops ritro, blue flowers, 1–1.5 m

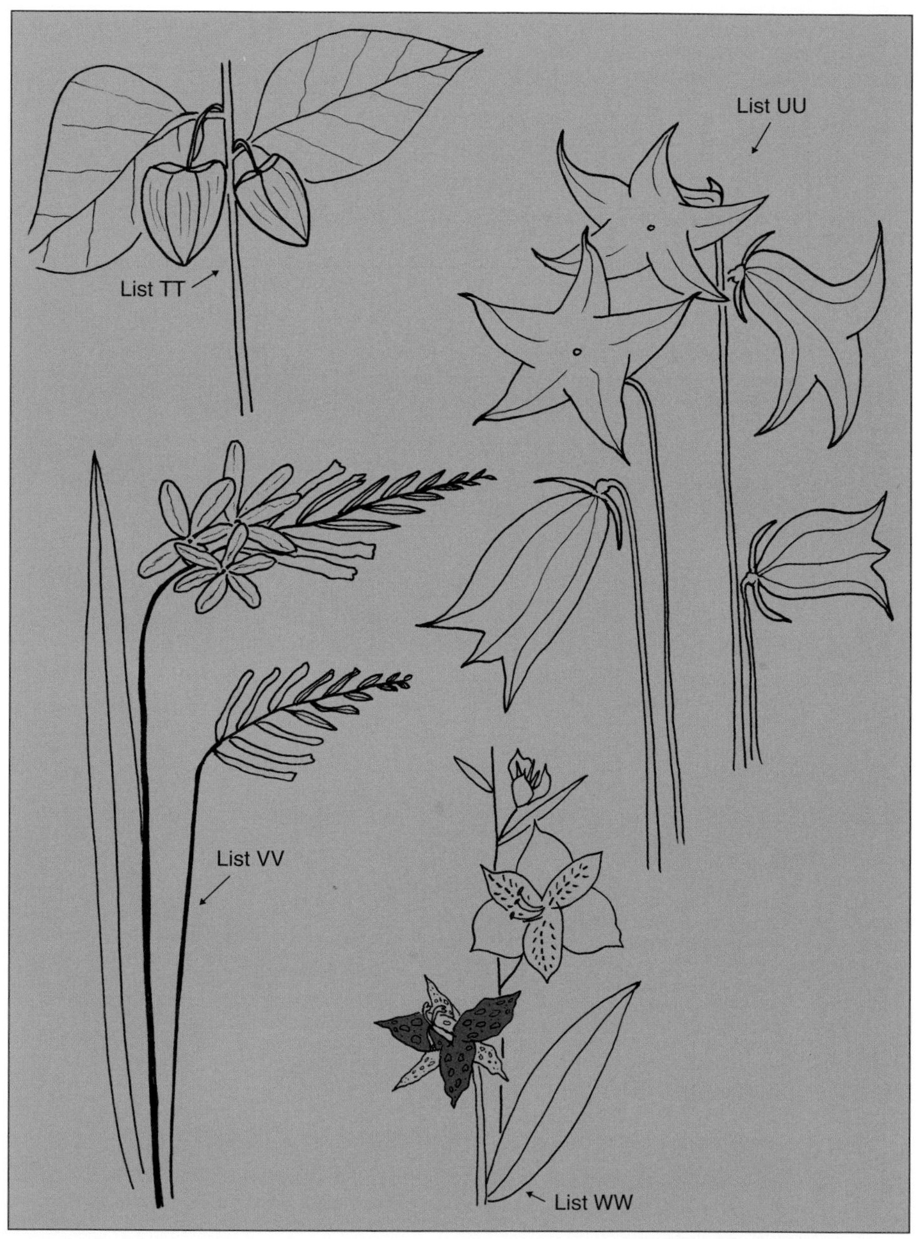

Herbaceous plants VII.

TT *Physalis alkekengi, red-orange lantern-shaped flowers, 50–75 cm*

UU *Campanulas and other bell-shaped flowers*
Campanula, many species and cultivars, white, pink-blue flowers
Platycodon grandiflorus, large blue flowers

VV *Crocosmia, many cultivars, red, orange and yellow flowers, 75 cm–1.25 m*

WW *Tricytris formosana, spotted flowers, 75 cm–1 m, some cultivars*
Alstroemeria, many cultivars, reds, oranges and yellows, 60 cm–1 m
Tigridia pavonia, white, yellow, orange to red spotted flowers, 30–45 cm

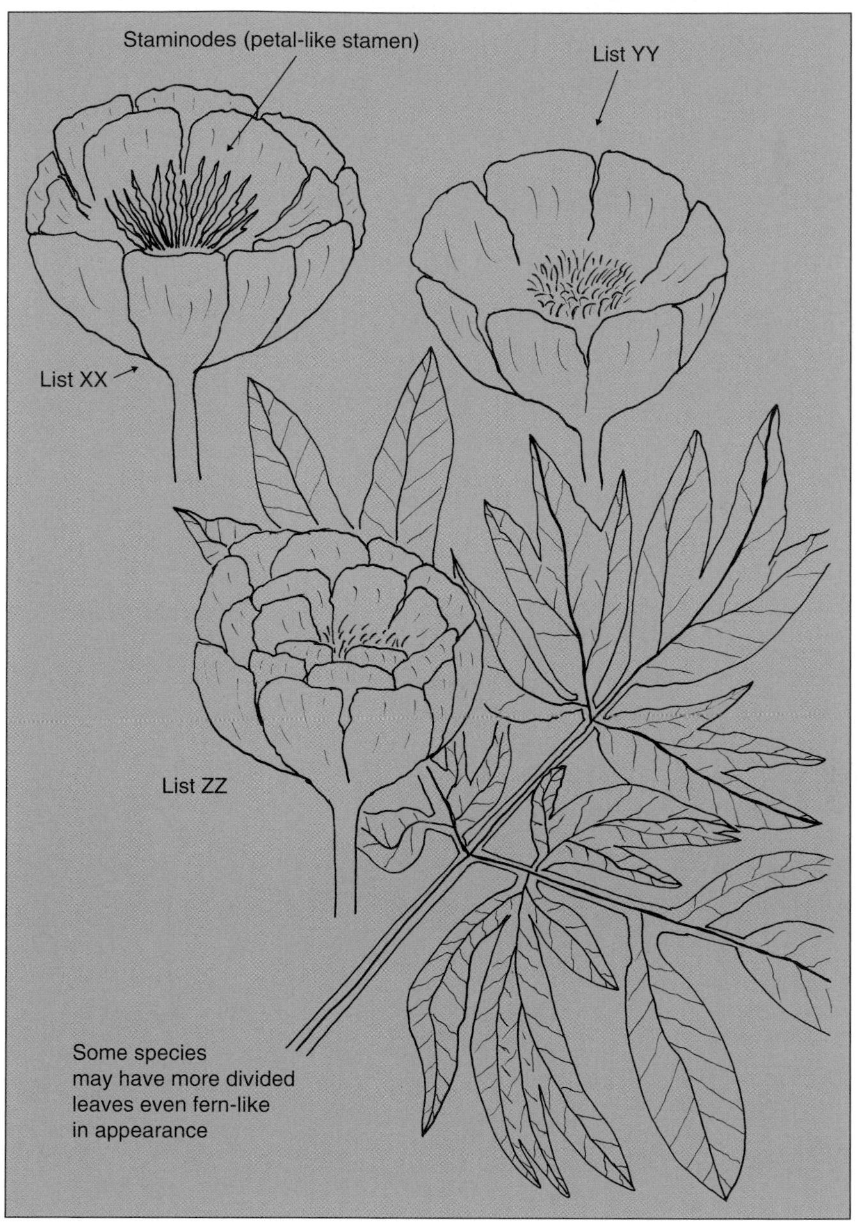

Herbaceous plants VIII.

XX *Peonies, anemone-flowered, Japanese or imperial types*
Paeonia lactiflora, white, pink or red; many cultivars, 75 cm–1 m
Paeonia tennuifolia, fern-like leaves, red flowers, 60–75 cm

YY *Peonies (single types); many cultivars, red, pink, white, 1 m*
Paeonia lutea var. ludlowii, sub-shrub with yellow flowers, 1–1.8 m
Paeonia anomola, red flowers, 50 cm
Paeonia emodi, white flowers, 60–75 cm
Paeonia officinalis, dark red flowers, 50–60 cm
Paeonia veitchii, pink-purple flowers, 50–60 cm

ZZ *Peonies (double types), many cultivars, 1–1.25 m*
Paeonia lactiflora, many cultivars, pink, red, white, 1–1.25 m

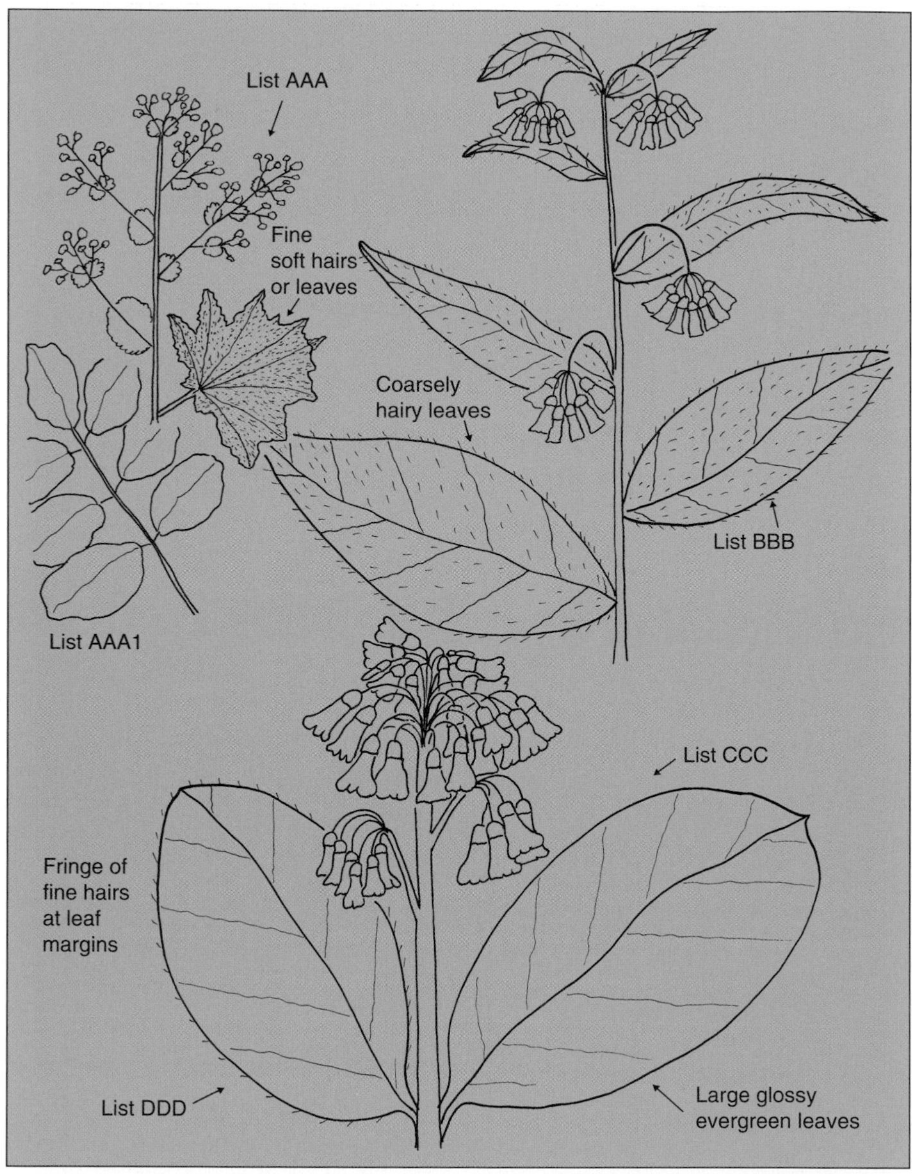

Herbaceous plants IX.

AAA *Alchemilla mollis, greenish-yellow flowers and soft hairy leaves*

AAA1 *Smyrnium perfoliata, yellow-green flowers*

BBB *Anchusa azurea, blue-purple flowers, 75 cm–1.25 m*
Pentaglottis sempervirens (alkanet), blue flowers, 45 cm–1 m
Pulmonaria angustifolia (lungwort), blue flowers, 30–45 cm
Pulmonaria longifolia (spotted lungwort), pink/blue flowers and white-spotted leaves, 25–45 cm
Symphytum caucasicum (comfrey), blue flowers, 60–75 cm
Symphytum x uplandicum (Russian comfrey), pink/blue, 45 cm–1 m

CCC *Bergenia cordifolia, red, pink flowers, large leaves, 45–75 cm*

DDD *Bergenia ciliata, white/pink flowers, 25–30 cm*

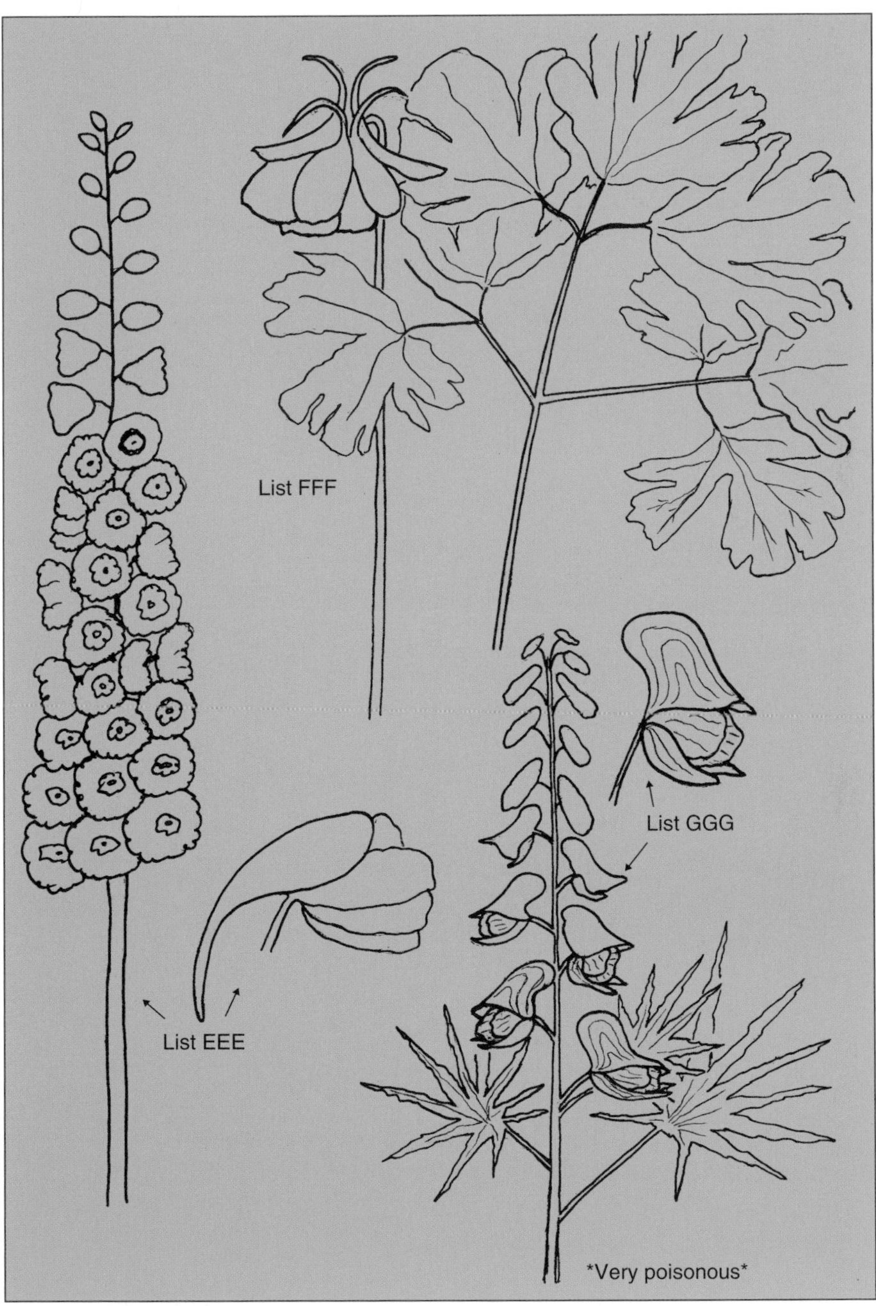

Herbaceous plants X.

EEE	*Delphiniums, many cultivars, white, yellow, pink, blue; tall, growing to 2 m*
FFF	*Aquilegia (columbines), many cultivars, white, blue, pink , red, 20–60cm*
	Aquilegia vulgaris, white, pale blue and pink flowers
GGG	*Aconitum napellus*
	Aconitum carmichaelii, dark blue flowers, 60cm–1.25 m

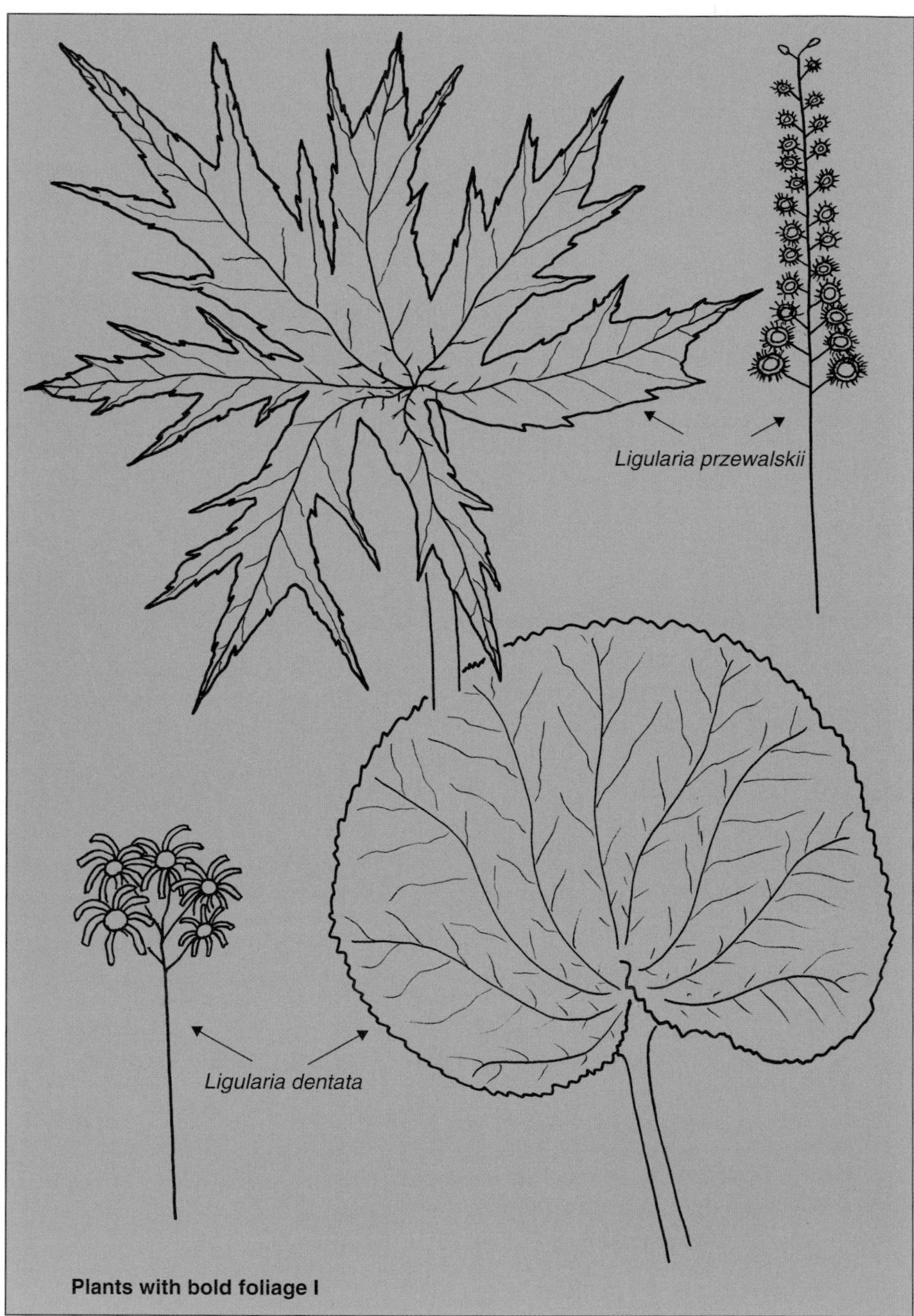

Herbaceous plants XI.

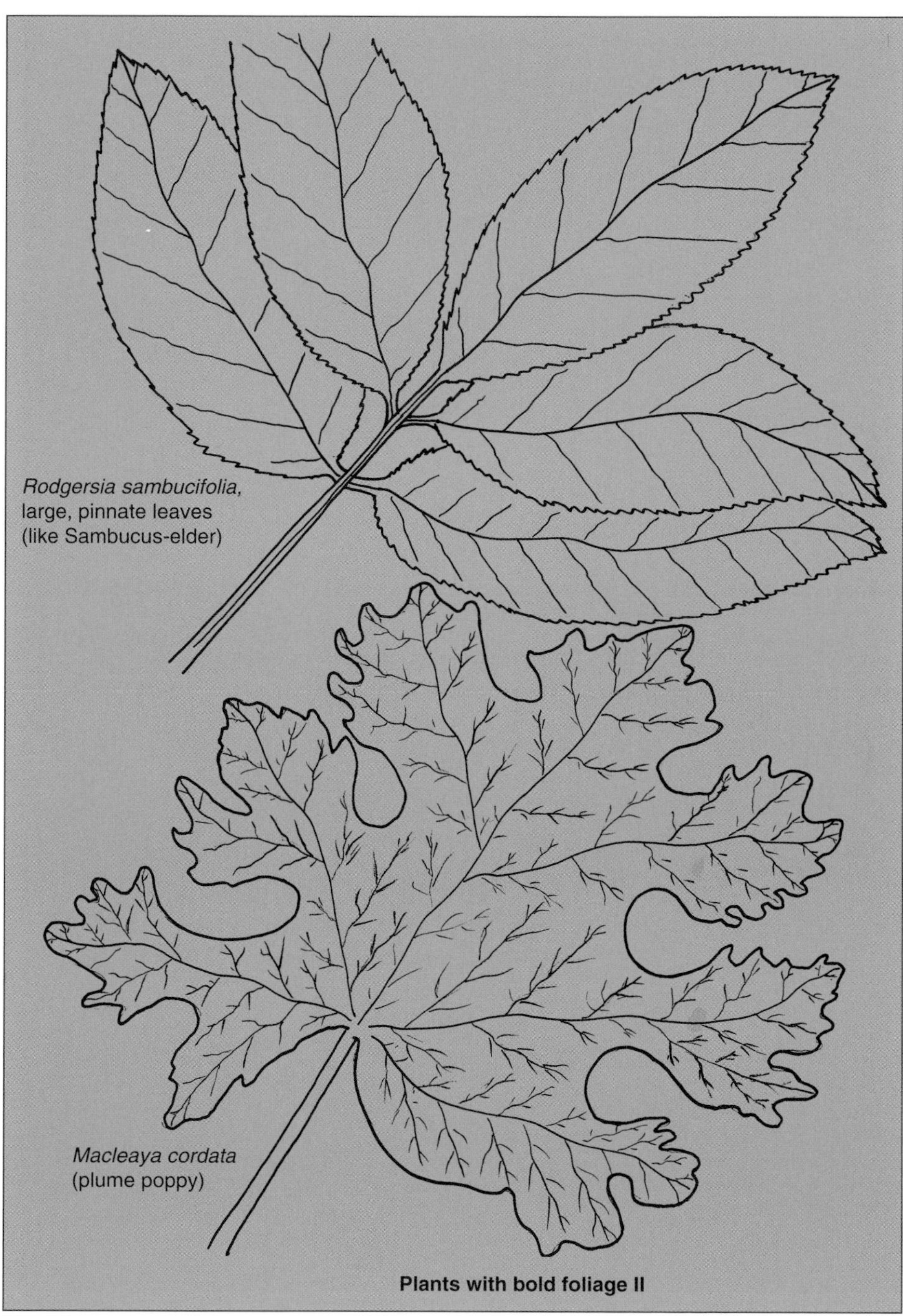

Herbaceous plants XII.

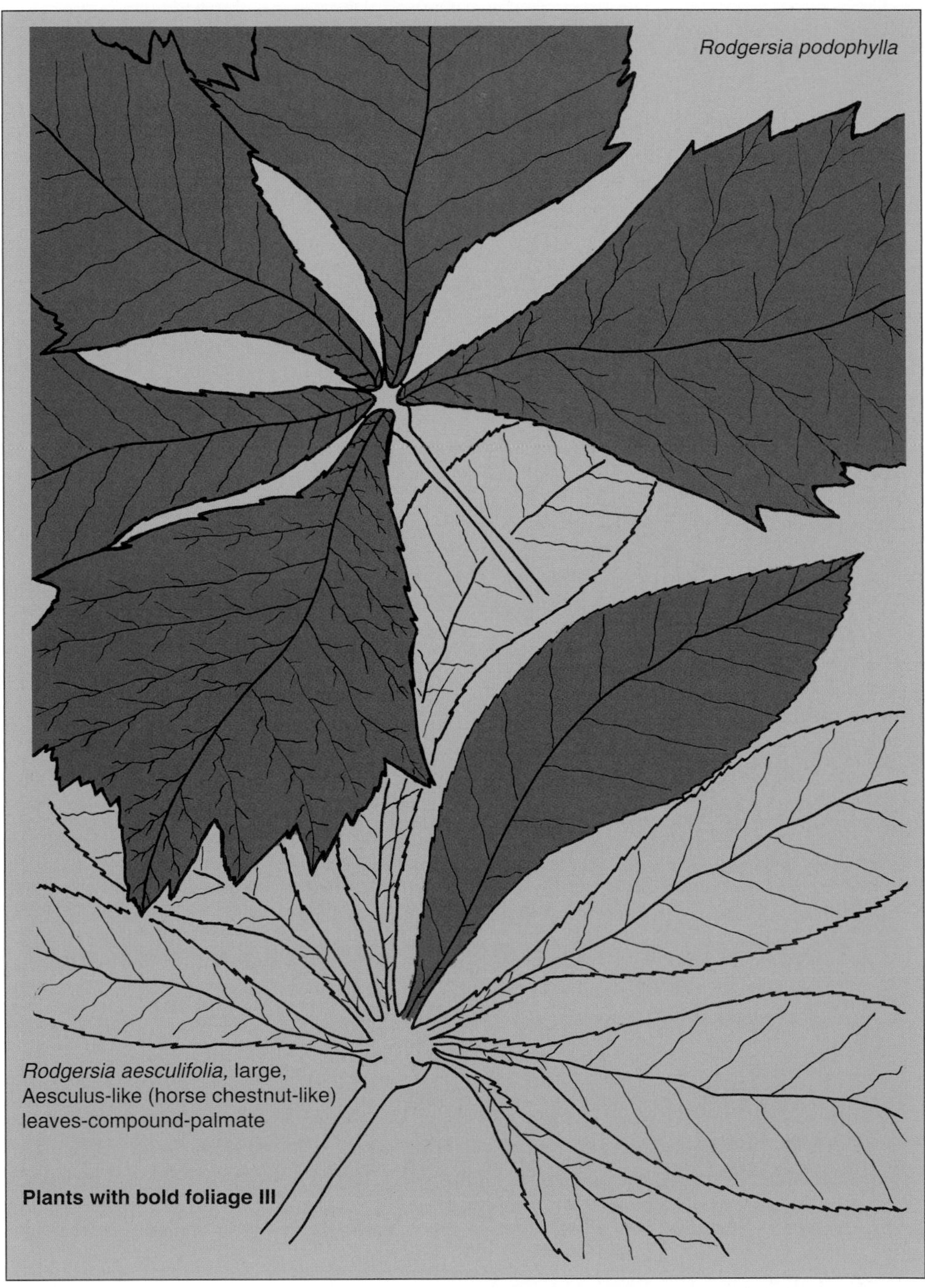

Herbaceous plants XIII.

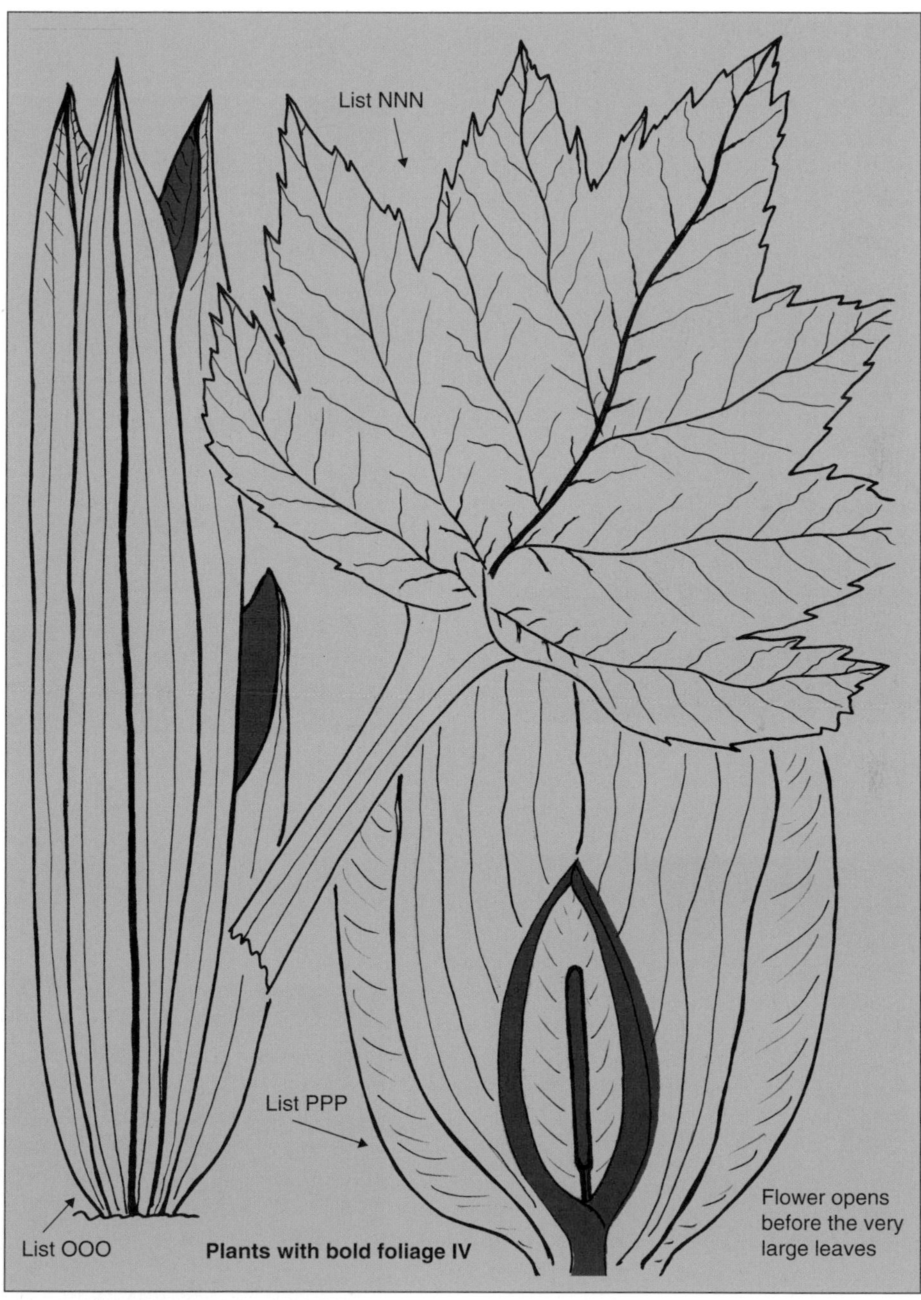

Herbaceous plants XIV.

NNN	*Rheum palmatum (ornamental rhubarb)*
OOO	*Phormium tenax*
	Phormium cookianum
	All kniphofias
PPP	*Lysichiton americanus*
	Lysichyton camtschatcensis
	Also with lily-like flowers, Hosta seiboldiana, 75cm–1 m

Herbaceous perennials

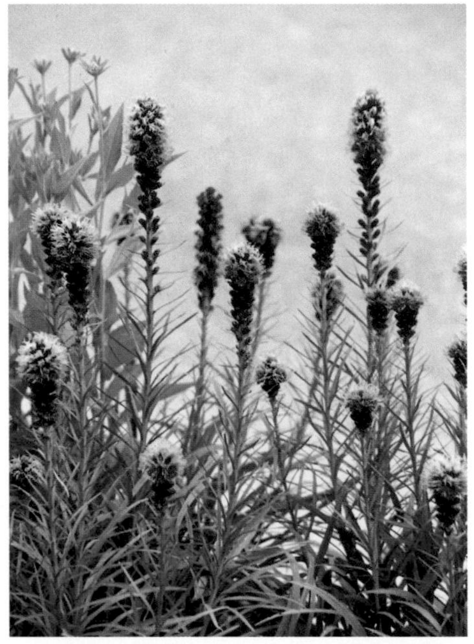

Liatris spicata.

Lobelia cardinalis cultivar.

Yellow loosestrife (Lysimachia punctata).

Hosta: to show the lily-like flowers.

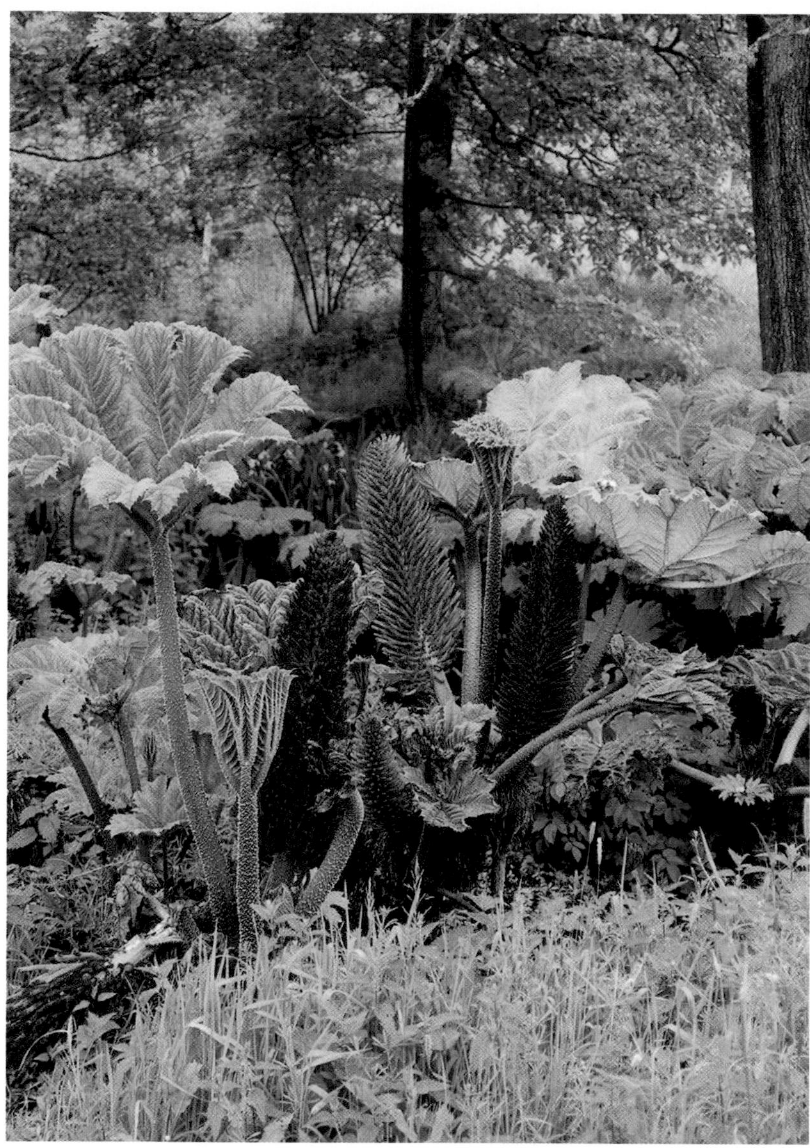

Gunnera manicata: the proportions of the plant are so large that it is difficult to place it near other plants – it is invasive, and needs a moist, protected site.

Groundcover Schemes

Groundcover schemes can be created using many plant types, and are not necessarily restricted to low-growing species as they extend to any plant genus, group or mixture, planted at sufficient densities to cover the soil; by definition they could include some forms of shrub border and mixed border.

Thoday *et al* identify four main design patterns for groundcover schemes: tapestry, patchwork, sheet with emergents, and tiered. Tapestry uses mainly low-growing species and emulates embroidery, with organically shaped drifts of colour that intertwine and have a close relationship with one another. A similar design pattern may also be used for herbaceous perennial display. Patchwork is created by more discrete block plantings ('drifts') of low-growing species forming the design – in fact a very similar pattern to that used when sowing hardy annuals, and the most common design pattern for herbaceous perennials. Sheet with emergents describes the use of low-growing species as the main groundcover (the 'sheet'), with isolated trees (the 'emergents') emerging from the 'sheet' of low-growing types. Tiered plantings are created by using woody plants of various sizes, from low hummocks at the front, to rounded shrubs in the foreground and taller trees at the rear, the scale of which can be varied depending on the site. The same design pattern is also very commonly used in shelterbelts for wind protection, although native and naturalized species are usually used in preference to exotics for this purpose.

Low-growing arboreals retain interest in the winter because of their shape and form, bark colour, tracery/outline and evergreen foliage – some even flower in the winter. Heathers in all their forms (such as *Calluna*, *Erica*, *Daboecia*) add colour throughout the year – callunas in particular, because of the many coloured foliage types available. Most herbaceous perennials are very seasonal, having excellent summer display but very little winter colour, and groundcover schemes need the inclusion of hardy perennials that retain some leaf over winter to increase interest. Elephant's ears (*Bergenia cordifolia*) and its cultivars give bold leaf patterns and superb reds and maroons in autumn/winter, and pampas grass (*Cortadieria*), dwarf bamboos, some *Sisyrinchium* species – including *Sisyrinchium striatum*, red hot pokers (*Knifophias*), and most *Iris* types with sword-like leaves – are used for height and architectural interest.

Lawns and Grass Areas

Lawns are the largest areas of groundcover systems that we have, and are usually formed from perennial grass species. The few common annual grasses that exist widely diverge in their use and popularity. Some species such as *Poa annua* (annual meadow grass) are considered 'weeds' of lawns because they die out to leave bare patches that are readily reinfested by other lawn weed species. Species such as *Briza maritima* (quaking grass), on the other hand, are considered ornamental and may be used in annual beds and borders, or in special ornamental grass feature displays amongst perennial types such as *Festuca* and *Miscanthus* species.

Lawns may also be produced by using other plant species: chamomile lawns are one option, mixed grass and wildflower swards another, and mosses yet another. In all these systems, species blanket the soil, retain the soil against erosion, but offer differing degrees of resistance to hard wear. Mosses are very useful because they happily follow the natural contours of the local topography and do not need mowing, so having to use the mower over inhospitable terrain is not an issue. Moss lawns make pleasing groundcover beneath trees where grasses do not flourish, and can look superb with their satin green effect; however, they offer very little resistance to foot-traffic damage, and the surface 'tears' very easily. The main management problem is ensuring there is sufficient water – usually by careful choice of the site initially, or irrigation – so sites that remain moist are best for moss lawns. However, moist sites –even firm and moist, rather than boggy – increase the probability of surface tear, relegating moss lawns to a 'look, don't touch (or use)' policy. All higher plants, including grasses, are the 'weeds' of moss lawns. Ironically, and as witnessed when prolific moss crops appear when using some herbicides on bare soil, moss lawns can be maintained weed free by the use of some residual herbicides (where moss resistance exists). Organic systems have to rely on hand weeding.

Grass lawns really need to be both ornamental and hard-wearing, which is not an easy balance to maintain. The most hard-wearing grass species are liable to form tussocks, and are not easy to maintain as level or 'true' surfaces; they are therefore limited in their use for small-ball sports or for producing a really fine surface. However, modern dwarf cultivars of hard-wearing ryegrass types have turned this on its head to some degree. Rhizomatous species help bind the soil together, and species that form stolons (overground runners) that root along their length help to bind the soil at or near the surface; for these reasons 'successful' grass areas usually comprise a mixture of grass types.

The main consideration for the choice and mix of grass species is therefore the use to which the lawn or grass area is to be put – whether the main properties required are considered to be wear resistance, the trueness of the surface after mowing, or ornamental features. Tough, tussock-forming species are regularly used in mixed sward, ornamental areas, giving a flowery meadow effect. Grass areas used for sports also require varying degrees of wear resistance and trueness of surface, and some comprise the finer grasses because of the tolerance to close mowing that these species show. The proposed mowing regime after grass establishment is a very important consideration when choosing the species mix.

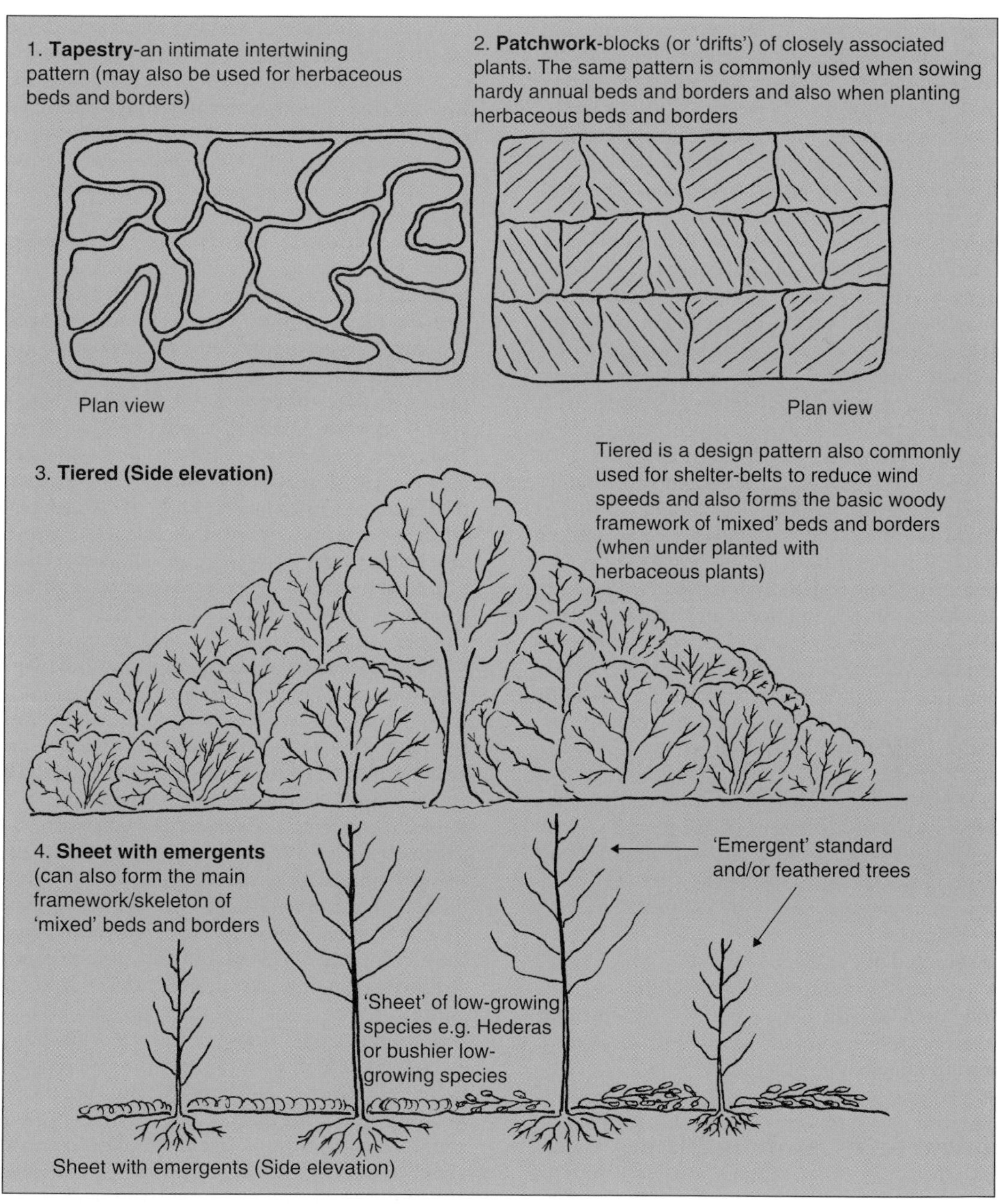

1. Tapestry-an intimate intertwining pattern (may also be used for herbaceous beds and borders)
2. Patchwork-blocks (or 'drifts') of closely associated plants. The same pattern is commonly used when sowing hardy annual beds and borders and also when planting herbaceous beds and borders
Plan view
Plan view
3. Tiered (Side elevation)
Tiered is a design pattern also commonly used for shelter-belts to reduce wind speeds and also forms the basic woody framework of 'mixed' beds and borders (when under planted with herbaceous plants)
4. Sheet with emergents (can also form the main framework/skeleton of 'mixed' beds and borders
'Emergent' standard and/or feathered trees
'Sheet' of low-growing species e.g. Hederas or bushier low-growing species
Sheet with emergents (Side elevation)

The four main design patterns used for groundcover schemes.

Height, Depth, Shade, Shape and Form

Using annual or tender perennial plants to create permanent designs is obviously useless, so although horticultural excellence can include all forms of showy yet temporary summer display, it is to woody perennials that we look for permanence and structure. It is this group of plants that are used to create the skeleton of most garden designs, whether on a domestic or a botanic garden scale. Permanent design structures are created by hard landscape features and the use of trees, shrubs and woody climbers arranged in groups or as single specimens. Mixed beds or borders comprise a mixture of woody and herbaceous plants, often (but not exclusively) of exotic origin. Single genera features include, for example, rose beds and borders dedicated to roses only, rhododendron collections, magnolia collections. etc.

Woody perennials are particularly good as a 'skeleton' or 'backbone' of landscape designs, both because of their permanence and height, and because they have such a large range of useful aesthetic features, including foliage, flowers, fruits, bark, shape and form. They also have sensory features such as smell or fragrance, texture and sound (leaf rustle). Woody plants of all types give a greater height, depth, shade and sense of enclosure than other plant groups, all useful tools for landscape and garden designers.

The habit of a plant is a natural growth pattern that recurs in a particular species, and may typify the potential shape or external outline of the plant. Hence habit, shape and form are often related. Habit also includes the particular branch patterns, shape and leaf forms. Trees may be multi-stemmed, standard (on a leg), upright (fastigiate), weeping (pendulous), with rounded crown, column- or pillar-like (columnar), pyramidal, narrowly or broadly conical. Shrubs may be (and usually are) multi-stemmed and bushy, may have lax, pendulous, arching stems (scandent or rambling), scrambling (mound-forming with arching stems), they may be semi-prostrate, prostrate (procumbent), very prostrate or creeping, or prostrate and rooting along their stems.

Grasses and alpine forms include mat-forming, tuft- or tussock-forming, rosette-forming and mound- or dome-forming types. Clump-forming types of many heights and sizes exist, including those with large, sword-like leaves.

Scandent and Climbing Plants

Scandent plants are those that produce very vigorous, long stems with no clinging ability, so support is gained by their arching stems or by draping over another plant. In the world of scandent woody plants, discerning between very vigorous lax-growing shrubs and non-clinging vigorous-stemmed climbers is not easy. Roses such as *Rosa moyesii* and *Rosa* 'Canary Bird' could be classed as either a large shrub or a non-clinging (scandent) climber. So also could lobster claw (*Clianthus puniceus)*, which, if your climate is tolerant of the species, should be grown not only for the aesthetic qualities of their flowers, but also for the length of flowering period. Other examples include hybrid climbing (and other very vigorous) roses, *Solanum crispum*, and some jasmine species (*Jasminum nudiflorum* and *Jasminum beesianum*).

True climbers have methods of self-support including twining stems, stem tendrils and leaf tendrils. Stem tendrils develop from axillary buds, and they either twine, or have sucker pads for support. Furthermore, the contraction and coiling of the tendrils pulls the main stems nearer to the host (whether this is another plant, a trellis, or a wall); examples include Virginia creeper (*Parthenocissus quinquefolia* – stem tendrils and sucker pads), and passion flower (*Passiflora caerulea* – stem tendrils only). The stem tendrils of white bryony (*Bryonia dioica*) are contrary coiled – the section nearest the main stem is coiled in a different direction to the section nearest the supporting host.

Many annual (and those treated as annual) climbers support themselves by twining stems – the stem gradually moves around the host for support, for example morning glory (*Ipomea*) and runner bean. However, sweet pea (*Lathyrus odoratus*) and cup-and-saucer plant (*Cobaea scandens*) have leaf tendrils, as does Chilean glory vine (*Ecremocarpus scaber*), which, because it is a fairly tender climber, is often treated as herbaceous. Leaf tendrils may arise as a single strand from the lamina tip, as found in glory lily (*Gloriosa rothschildiana*), or more commonly have three tendrils (a tripartite split, as in garden pea – *Pisum sativum*) at the tips of their compound leaves. Tendrils may also unusually arise from petiolate stipules, as found in *Smilax aspera*.

Climbing and scandent plants can be used against walls and/or scrambling over arches, trellises, and dead or alive trees, and because they invade the canopy of trees, add vertical and horizontal dimensions to plant display. Even soft, annual climbers such as canary creeper (*Tropaeolum canariensis)* can be effective ranging into the canopy of a small tree, and because of its growth type, with very little detriment to the host. The group of showy herbaceous perennial *Tropaeolum*, such as *T. speciosum* and *T. tuberosum*, thrive on the shelter gained from small evergreen trees, and may be more successful in this environment than when trained over a trellis.

Both deciduous and evergreen forms of clematis thrive when grown into trees with light, open

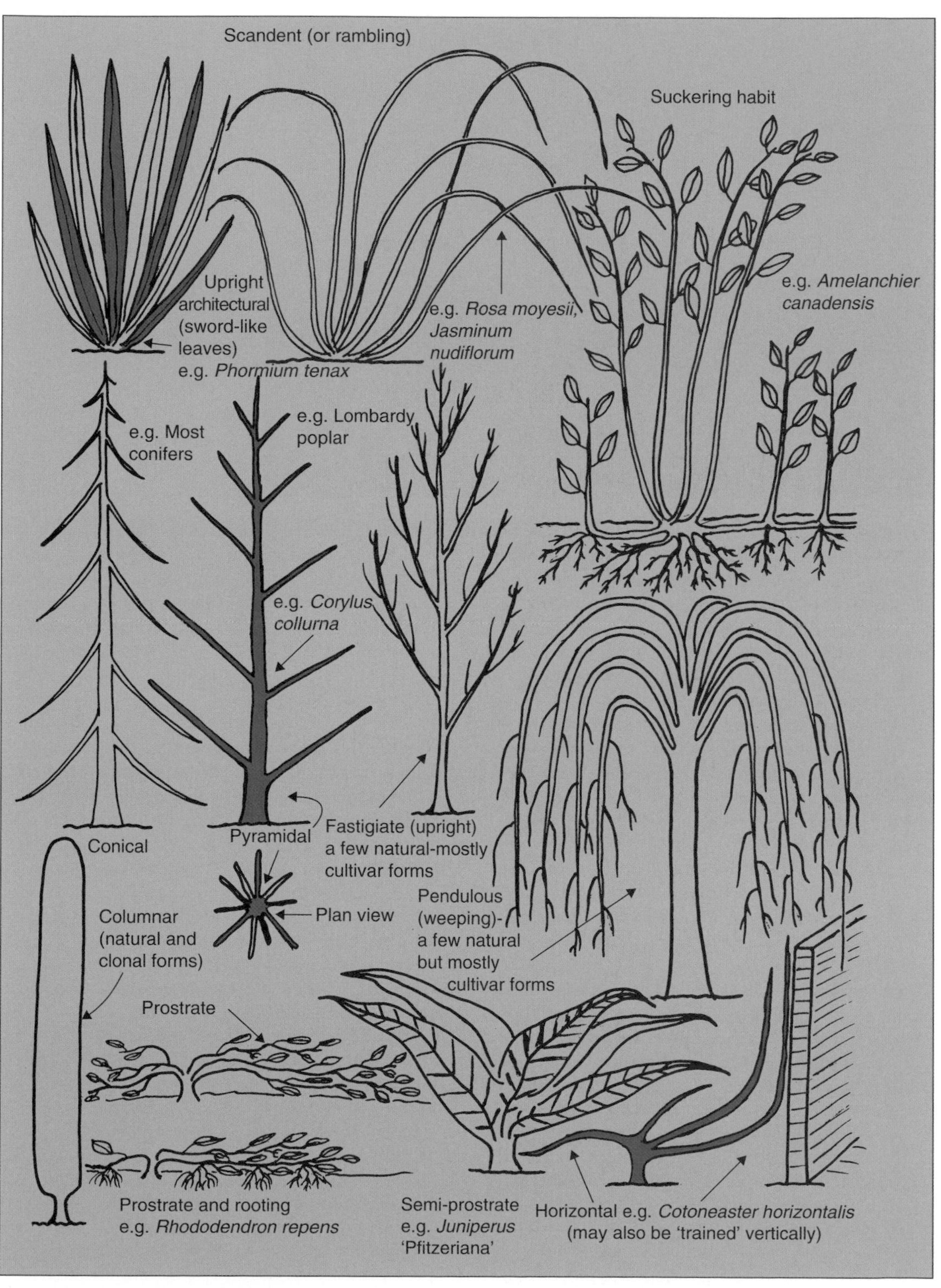
Scandent (or rambling)
Suckering habit
Upright architectural (sword-like leaves) e.g. Phormium tenax
e.g. Rosa moyesii, Jasminum nudiflorum
e.g. Amelanchier canadensis
e.g. Most conifers
e.g. Lombardy poplar
e.g. Corylus collurna
Conical
Pyramidal
Fastigiate (upright) a few natural-mostly cultivar forms
Plan view
Columnar (natural and clonal forms)
Pendulous (weeping)- a few natural but mostly cultivar forms
Prostrate
Prostrate and rooting e.g. Rhododendron repens
Semi-prostrate e.g. Juniperus 'Pfitzeriana'
Horizontal e.g. Cotoneaster horizontalis (may also be 'trained' vertically)

Shape and form (naturally occurring types).

Lobster claw (Clianthus puniceus) – also showing a small snail, and snail damage to the flowers.

canopies (particularly if their roots are initially established in good moist soil away from the dry base of the tree), and they can transform an otherwise quite dull tree species. The use of woody climbers such as Virginia creeper (*Parthenocisus quinquefolia*), Russian vine (*Polygonum bauldschuanicum*) and passion flower (*Passiflora caerulea*), with its coiled stem tendrils, are nearly always detrimental to host trees ultimately, as their very vigorous nature tends to completely blanket and smother the tree, reducing photosynthesis.

Many clematis have twining petioles for self-support and, along with some forms of bryony, have stems that twine together, giving a plaited rope effect to aid support, as does *Jasminum officinale*, even though its close relative *Jasminum nudiflorum* has scandent stems and is not self-supporting.

Both *Hydrangea anomala* subsp. *petiolaris* and *Hedera* (ivy) have fine adventitious roots for support, and ivy is probably the most overtly arboreal of all the climbers in nature, attaining very thick stem girths in favourable conditions. In some, albeit rare instances, host trees rot away to leave the very thick, woody-stemmed ivy to support the remains of the partially decayed hulk. Wisteria, which displays very large, showy flowers, has twining stems, and also ultimately becomes very woody. All climbing forms of honeysuckle (*Lonicera*) have effective aesthetic flowers, and the added bonus of fragrance. However, their twining method of self-support is very damaging to the host (strangulation).

Other examples of climbing plants include *Akebia quinata, Berberidopsis corallina, Campsis radicans, Lathyrus grandiflora* (everlasting pea), and the more tender *Mitraria coccinea,* and *Mutisia decurrens.*

Trained Shapes

'Trained' shapes may be created by the use of scandent/rambling or procumbent forms being budded or grafted on to an under-stock 'leg' to produce weeping standards, for example rambling roses on a leg to create weeping forms. *Cedrus atlantica forma glauca* 'Prostrata', a procumbent form, can be grafted on to an under-stock leg of *Cedrus atlantica* to form an artificially weeping form *Cedrus atlantica forma glauca* 'Pendula'. Naturally pendulous forms of shrubs and trees do exist, but the greater majority are produced by various propagation methods.

Cordons and oblique cordons may be created from rooted cuttings on their own roots (they are often in soft fruit, such as blackcurrant, redcurrant and gooseberry). However, apples and pears are grafted on to dwarfing (low vigor) under-stocks, which keeps them compact and encourages fruiting at a convenient height. Under-stocks often impart their relative vigour to the species or cultivar budded or grafted on to them, and intensive garden and commercial methods of apple and pear production are only possible because of this fact. Spill/spindle bush, cordon, espalier, pillar and

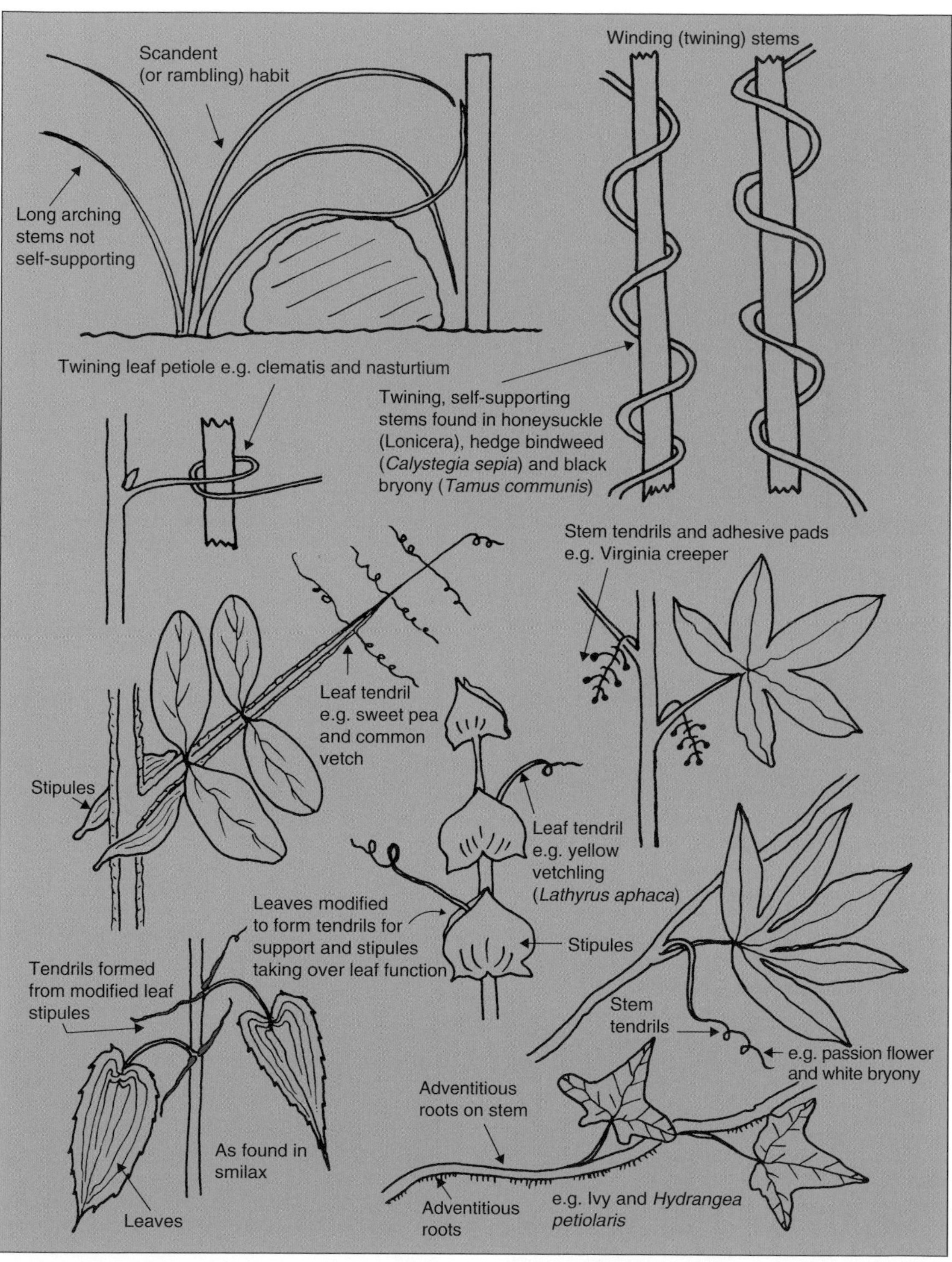
Scandent
(or rambling) habit
Long arching
stems not
self-supporting
Winding (twining) stems
Twining leaf petiole e.g. clematis and nasturtium
Twining, self-supporting
stems found in honeysuckle
(Lonicera), hedge bindweed
(*Calystegia sepia*) and black
bryony (*Tamus communis*)
Stem tendrils and adhesive pads
e.g. Virginia creeper
Leaf tendril
e.g. sweet pea
and common
vetch
Stipules
Leaf tendril
e.g. yellow
vetchling
(*Lathyrus aphaca*)
Leaves modified
to form tendrils for
support and stipules
taking over leaf function
Stipules
Tendrils formed
from modified leaf
stipules
Stem
tendrils
e.g. passion flower
and white bryony
Adventitious
roots on stem
As found in
smilax
Leaves
Adventitious
roots
e.g. Ivy and *Hydrangea*
petiolaris

Climbing plants and their methods of self-support.

Tropaeolum speciosum, contrasting beautifully with the conifer foliage. (*Cryptomeria japonica* 'Elegans')

Chilean glory vine (Eccremocarpus scaber).

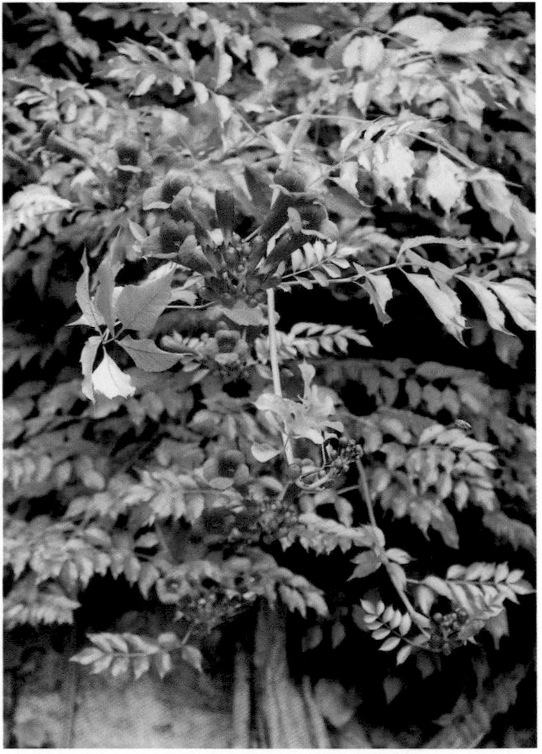

Trumpet vine (Campsis radicans).

Stem tendrils of Virginia creeper (Parthenocissus quinquefolia).

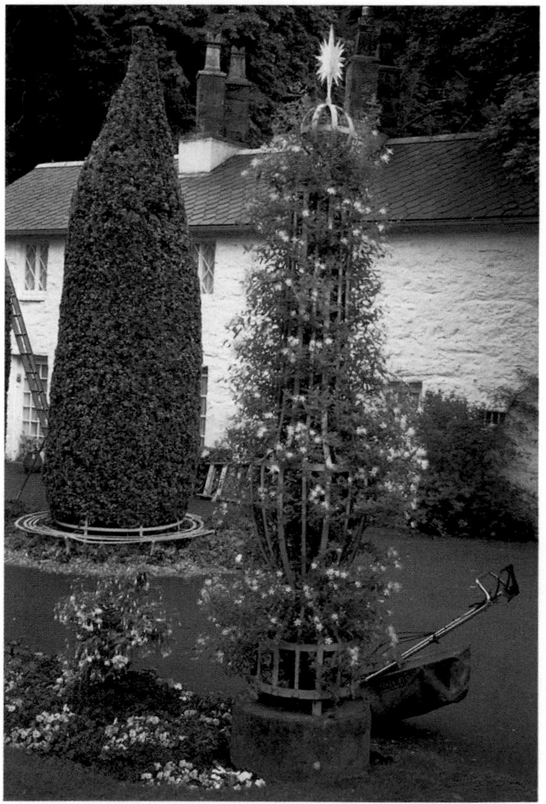

Canary creeper (Tropaeolum canariensis) at Port Merion, Wales.

dwarf bush all depend on this feature (and the fact that they are maintained by summer pruning, which removes leaves and reduces their energy budget). When ornamental cherry species and cultivars are grafted on to under-stocks of the very vigorous *Prunus avium* (wild cherry), it leads to dreadful disparities in the growth between cultivar and under-stock in top-worked trees (buds inserted into the top of an under-stock 'leg'). The 'Sheraton' cherry has a wild cherry under-stock, on to which is budded (or grafted) Tibetan cherry (*Prunus serrula*) to form a mahogany-coloured trunk, and on to the top of this is budded (using three buds) a flowering cherry cultivar (or even three different cherry cultivars, if desired).

Some conifers can have different forms created (accidentally or intentionally), depending from where the cutting or grafting material on the parent plant is harvested. Material harvested from apically dominant shoots will give a 'normal' apically dominant young plant. However, shoots harvested from lateral branches may give a different shaped progeny – a more procumbent form, in some instances. This phenomenon is known as topophysis, and sometimes leads to new cultivars being created (with varying degrees of 'success'). A prostrate form of Koster's blue spruce can be created artificially by this method (*Picea pungens* 'Kosterii Prostrata').

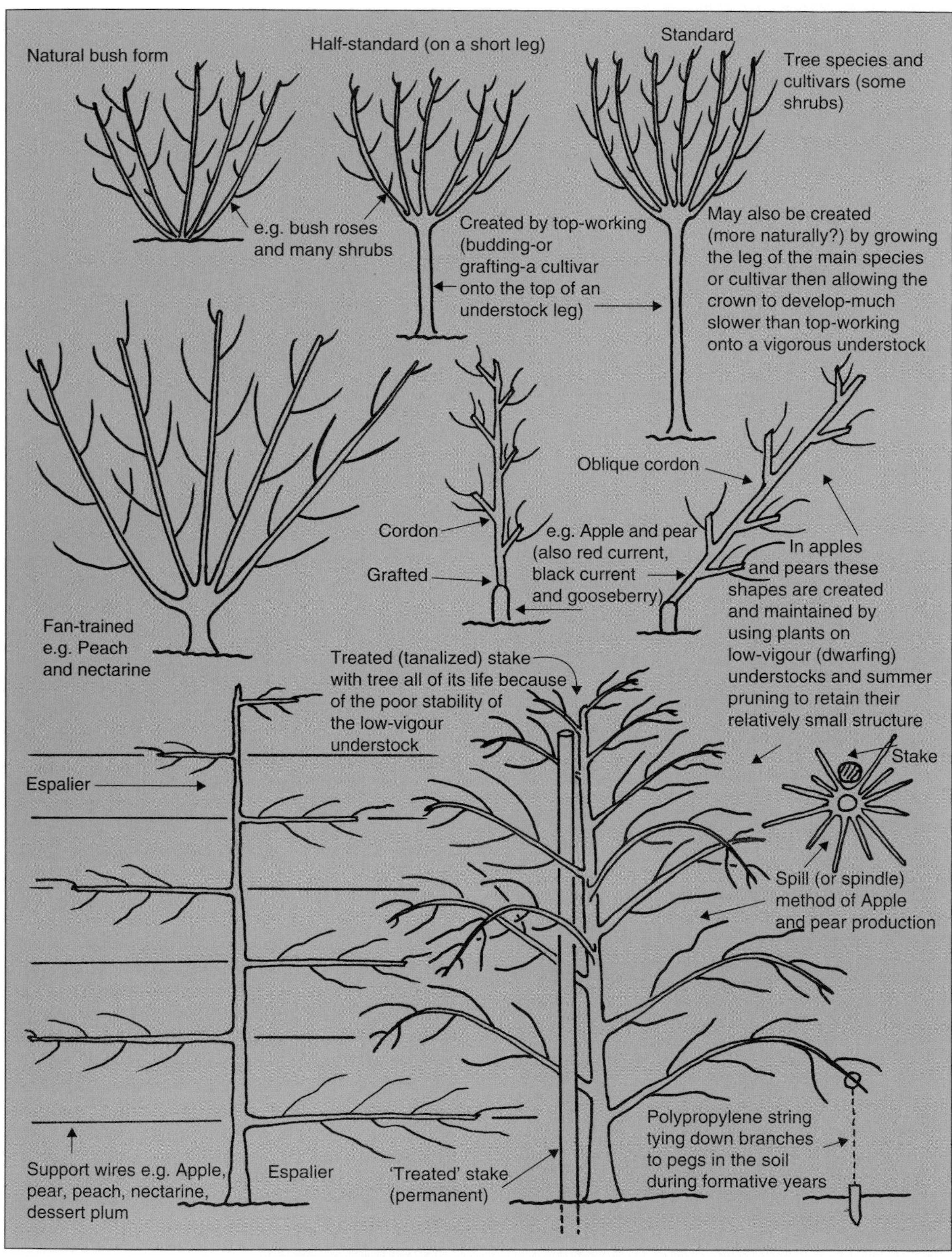
Natural bush form
Half-standard (on a short leg)
Standard
Tree species and cultivars (some shrubs)
e.g. bush roses and many shrubs
Created by top-working (budding-or grafting-a cultivar onto the top of an understock leg)
May also be created (more naturally?) by growing the leg of the main species or cultivar then allowing the crown to develop-much slower than top-working onto a vigorous understock
Oblique cordon
Cordon
Grafted
e.g. Apple and pear (also red current, black current and gooseberry)
In apples and pears these shapes are created and maintained by using plants on low-vigour (dwarfing) understocks and summer pruning to retain their relatively small structure
Fan-trained e.g. Peach and nectarine
Treated (tanalized) stake with tree all of its life because of the poor stability of the low-vigour understock
Espalier
Stake
Spill (or spindle) method of Apple and pear production
Polypropylene string tying down branches to pegs in the soil during formative years
Support wires e.g. Apple, pear, peach, nectarine, dessert plum
Espalier
'Treated' stake (permanent)

Shapes and forms 'trained', affected by their propagation and/or pruning method.

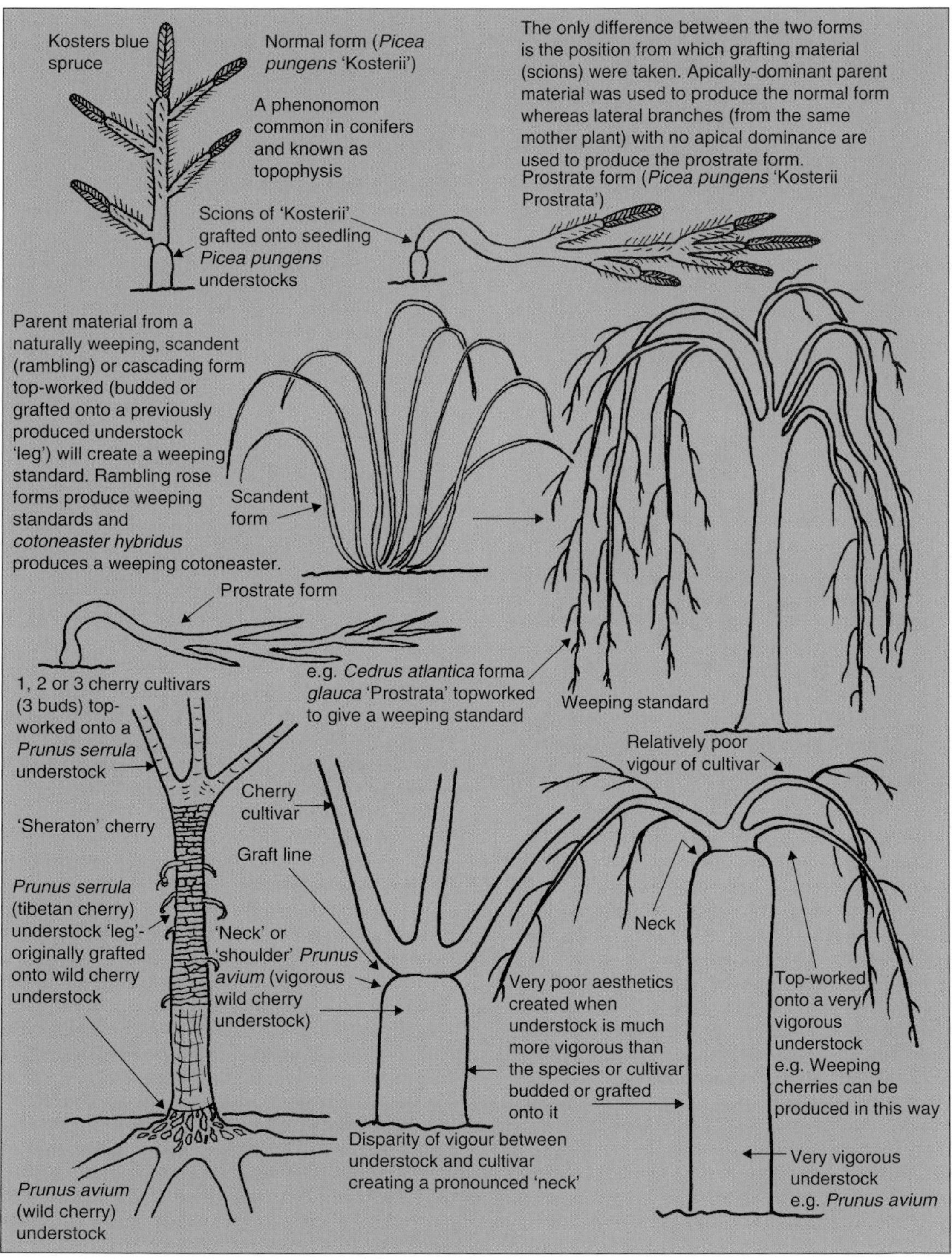
Kosters blue spruce
Normal form (*Picea pungens* 'Kosterii')
A phenonomon common in conifers and known as topophysis
The only difference between the two forms is the position from which grafting material (scions) were taken. Apically-dominant parent material was used to produce the normal form whereas lateral branches (from the same mother plant) with no apical dominance are used to produce the prostrate form.
Prostrate form (*Picea pungens* 'Kosterii Prostrata')
Scions of 'Kosterii' grafted onto seedling *Picea pungens* understocks
Parent material from a naturally weeping, scandent (rambling) or cascading form top-worked (budded or grafted onto a previously produced understock 'leg') will create a weeping standard. Rambling rose forms produce weeping standards and *cotoneaster hybridus* produces a weeping cotoneaster.
Scandent form
Prostrate form
e.g. *Cedrus atlantica* forma *glauca* 'Prostrata' topworked to give a weeping standard
Weeping standard
1, 2 or 3 cherry cultivars (3 buds) top-worked onto a *Prunus serrula* understock
'Sheraton' cherry
Cherry cultivar
Graft line
Prunus serrula (tibetan cherry) understock 'leg'- originally grafted onto wild cherry understock
'Neck' or 'shoulder' *Prunus avium* (vigorous wild cherry understock)
Prunus avium (wild cherry) understock
Disparity of vigour between understock and cultivar creating a pronounced 'neck'
Relatively poor vigour of cultivar
Neck
Very poor aesthetics created when understock is much more vigorous than the species or cultivar budded or grafted onto it
Top-worked onto a very vigorous understock e.g. Weeping cherries can be produced in this way
Very vigorous understock e.g. *Prunus avium*

Shape and form, as affected/created by the method of propagation.

'Roping' of clematis (in this case bound by twining petioles).

Shrubs

The wide range of shrubs available means they may be grown for their flowers, fruit, fragrance, bark, foliage type, foliage colour or autumn colour, or better still, a mixture of as many of these effects as possible.

Shrubs may be deciduous or evergreen, and sometimes within the same genus there may be deciduous and evergreen representatives. *Viburnum tinus* and *Viburnum rhytidohyllum* are both evergreen, yet *Viburnum fragrans* and *Viburnum plicatum* are both deciduous. *Berberis thunbergii* and *Berberis wilsonii* are deciduous, yet *Berberis candidula* and *Berberis verrucosa* are evergreen. Those grown for flower can have their effect at various times of year, and the flowering period can be prolonged by careful choice of subjects. There is a range of winter-flowering/early spring types, including the species *Lonicera fragrantissima* (an evergreen bush form of honeysuckle) – but the greatest choice of types is obviously in the main summer season, and includes common examples such as *Philadelphus*. Late-flowering types extend the flowering season and include *Rhododendron auriculatum* and *Eucryphia* × *nymensensis*.

Roses are amongst some of the most popular shrubs, and there are many types, depending on their natural forms, propagation system and training. Old roses include all the shrub roses, first-cross hybrids and species roses. Species roses include *Rosa moyesii, Rosa rugosa* and *Rosa glauca,* and subspecies, varieties and forms of species include *Rosa sericea forma pteracantha* with its superb winged thorns. First-cross species hybrids involving two parent species often retain many similarities to the original parents, and include *Rosa* 'Canary Bird'. Old-fashioned (or old garden) roses include many types that have been hybridized, involving many different parents over time; these include damask roses, gallica roses, tea roses (with their high fragrance) and moss roses (that have outgrowths on the stems, and calices that resemble mossy growth). Modern roses include large-flowered roses (hybrid tea) and cluster-flowered roses (floribunda roses). In recent decades, very compact mounded and trailing types of rose (called patio roses) have also been produced by rose breeders.

Many modern roses may be produced either as a bush or in standard form. Miniature bush roses are less than 12 in (30 cm) high, dwarf bush are less than 24 in (60 cm) high, and bush forms are more than 24 in (60 cm) high. Half standards are on a 24 in (60 cm) leg, and standards are on a 40 in (1 m) leg. Certain groups of roses lend themselves to being produced in particular shapes and forms. Large-flowered hybrids can be produced as bush, half-standard, standard (and climbing forms); cluster-flowered types as dwarf bush, bush, half-standard, standard (and climbing forms); and ramblers as scandent types, pillar or weeping standards. Miniature roses can be used in containers, densely planted in borders, or as groundcover, and patio roses may be used in large containers or as groundcover.

The vigorous scandent nature of rambling roses makes them ideal to train on wires, up trellises, round rustic arches and on a central stake as a pillar form. Most ramblers flower for about four weeks in the summer; however, perpetual flowering types extend over a longer period. Climbing roses fall into two main groups: climbing large-flowered hybrids (hybrid T), and climbing cluster-flowered hybrids (floribunda) resulting from natural 'sports' (genetic freaks) of their normal types with very vigorous shoots. There are also repeat and perpetual types derived from other material.

Rosa sericea forma pteracantha.

The Aesthetics of Individual Plants

Generally, small trees and large shrubs used in gardens with limited space need to have more than one main feature. Double-, triple- or multiple-featured species are needed in order to give maximum interest from the one (or few) specimens that can be used. So, good flowers and autumn colour, fruits and autumn colour, flowers and fruits, or excellent bark and autumn colour are useful. Furthermore, trees and shrubs with multiple features should be positioned in prominent positions to be easily enjoyed for their long period of interest – bark is a particularly good feature as it remains interesting throughout the year.

Bark

Bark is an essential physical feature of woody plants and can be an excellent feature at any time of year, but is often at its most notable during the autumn. This is because woody plants are at their largest girth for the current year at this time, which puts considerable strain on the bark, and ultimately manifests as stress lines, cracks and bark sloughing. Furthermore, deciduous subjects drop their leaves at this time of year, revealing the colour and texture of bark.

Arboreal plants can have smooth, rough, rhomboid (diamond-shaped) cracks, deeply fissured, platy or papery bark. Some coniferous subjects have very fibrous, resilient bark that may be aesthetic both for its texture and its strong red colour. Unfortunately, many of these species are unsuitable for small, enclosed spaces – for example, Wellingtonia or giant redwood (*Sequoiadendron giganteum*), coast redwood (*Sequoia sempervirens*) and Japanese red cedar (*Cryptomeria japonica*). Aesthetic effects depend on the chosen species, but include sloughing (to reveal new bark), and/or texture (silky, smooth, rough), and/or waxy, shiny, brightly coloured new growth (twigs), and/or striped, striated or snake bark.

There is a group of relatively small maples that give a striped bark effect and have wonderful shiny bark on new wood. Striped bark, striated bark or snake bark are synonymous generic terms for a number of species (approximately five to ten are readily available in commercial nurseries) that give this particular effect. The most commonly grown are Asiatic species: *Acer davidii* has green- and white-striped bark and waxy red new growth; *Acer rufinerve* has green- and white-striped bark and grey/blue buds; and *Acer capillipes* has green- and white-striped bark and waxy red new growth. *Acer forrestii* has white-striped bark and a very attractive leaf shape and *Acer hersii* has green- and white-striped bark. Moosewood (*Acer pensylvanicum*) is North American in origin, has large leaves, and also has green- and white-striped bark.

The peeling bark types all leave different textures and contrasting colours between the old sloughing bark and the newly formed bark below. They are divided into those that peel off in thin, papery plates, those that peel off radially into ribbons/ringlets, and those that peel off in slightly thicker organic-shaped plates, leaving several different colours to give a python skin effect. Examples include *Prunus serrula* (Tibetan cherry) that sloughs ribbons/ringlets of dark, mahogany-coloured bark, leaving silky smooth, sheraton-mahogany coloured new bark beneath; and *Prunus maackii* (Manchurian cherry) that sloughs ribbons/ringlets of rich, tawny/dark honey-coloured bark to reveal silky, light honey-coloured bark beneath.

Acer griseum (paper bark or peeling bark maple) sloughs bark in small, thin, mahogany-coloured, papery plates. *Arbutus menziesii, Arbutus andrachne* , *Arbutus* × *andrachnoides, Rhododendron thomsonii, Rhododendron barbatum* and *Myrtus luma* all slough with orange-red to red-brown papery bark. *Betula utilis jaquemontii* (an excellent varietal form of Himalayan birch) sloughs bark in papery sheets to leave a stark white new bark below, and shows excellent lenticel patterns, as does *Betula ermanii* (Erman's birch), which sloughs a papery, pale orange-white bark to leave a paler orange-white bark below. *Betula albosinensis septentrionale* (an excellent varietal form of Chinese birch) sloughs off orange-brown, papery bark to reveal orange-white bark beneath. *Eucalyptus pauciflora* subspi *niphophila* (snow gum) sloughs off thicker, organically shaped plates of grey bark to reveal green new bark below that gradually changes colour to olive, then brown, then grey – the so-called python skin effect. *Parrotia persica* (Persian ironwood), if grown as a large, multi-stemmed bush or as a small, single-stemmed tree, will slough bark in a python skin fashion on semi-mature or older plants.

Rubus biflorus has bark covered in a mealy white waxy 'bloom', *Cornus alba* has vivid red stems, *Cornus stolonifera* 'Flavaramea' has yellow-green stems, and *Salix alba* 'Vittalina' has yellow-orange bark on its stems.

The Aesthetics of Foliage

The most fundamental differences of foliage types in all plants are leaf shape, including simple, compound, large, bold and small leaves, and in woody plants deciduous versus evergreen. Leaf colour includes the summer colour of deciduous subjects with colourful leaves, colour all year round from colourful evergreen subjects, and species grown for specific autumn colours.

Autumn colour in its widest sense can include any main feature of a plant that occurs during the period September, October and November. It is not just the domain of deciduous woody perennials, as some hardy annuals such as *Kochia* (burning bush) can have autumn displays, and some herbaceous perennials such as *Sedum spectabile* 'Autumn Joy' for its late flowers, *Actaea rubra* for its red berries, and *Acanthus spinnosus* for its persistent bracts. *Bergenia cordifolia* (and its cultivars) has evergreen, leathery, persistent leaves that are noted for their red, yellow and orange hues during autumn. Some evergreen woody perennials also have autumn foliage colour – examples include *Mahonia japonica* and *Cryptomeria japonica* 'Elegans'. So, aesthetic features for autumn can include flowers, fruits/berries, bark and young twigs. However, the main understanding of the term 'autumn colour' is the change of leaf colour associated with deciduous woody subjects.

The display given by some deciduous subjects as they prepare themselves for winter is given by various pigments naturally found in leaves, including chlorophyll, which is green, and the caretenoides, which are divided into the carotenes – the reds, maroons and purples – and the xanthophylls (the yellows and golds). Because of its natural abundance, and the ease with which humans see the reflected green light, it is common for the green pigmentation of chlorophyll to mask the other more colourful pigments. For other pigments to show through they have to be in large amounts, and when they are in sufficient quantity to mask the green, the leaves present as a permanent (or summer) foliage colour. Indeed, in describing aesthetic features we distinguish between foliage colour and autumn colour. Deciduous subjects showing purples, reds and yellows during the summer months are described as having good foliage colour, whereas those having leaves that 'turn' in autumn are described as having good autumn colour.

The precursor to deciduous subjects losing their leaves in the autumn is a gradual preparation over approximately four to six weeks, during which time both day-length and temperature decrease. The hormonal response to this change in environment commences the production of a corky abscission layer that gradually blocks the vascular traces of the leaf where the leaf petiole meets the stem. This ensures that the tree or shrub does not haemorrhage precious fluids from its otherwise exposed vascular tissue ends, and cuts off the water supply to the leaf gradually.

Water is an essential ingredient of chlorophyll, and as the water supply diminishes, and ultimately ceases, the chlorophyll dies. The other pigments contained in the leaf, normally masked by the chlorophyll, are far less susceptible to the water loss, their demise is less immediate, and they

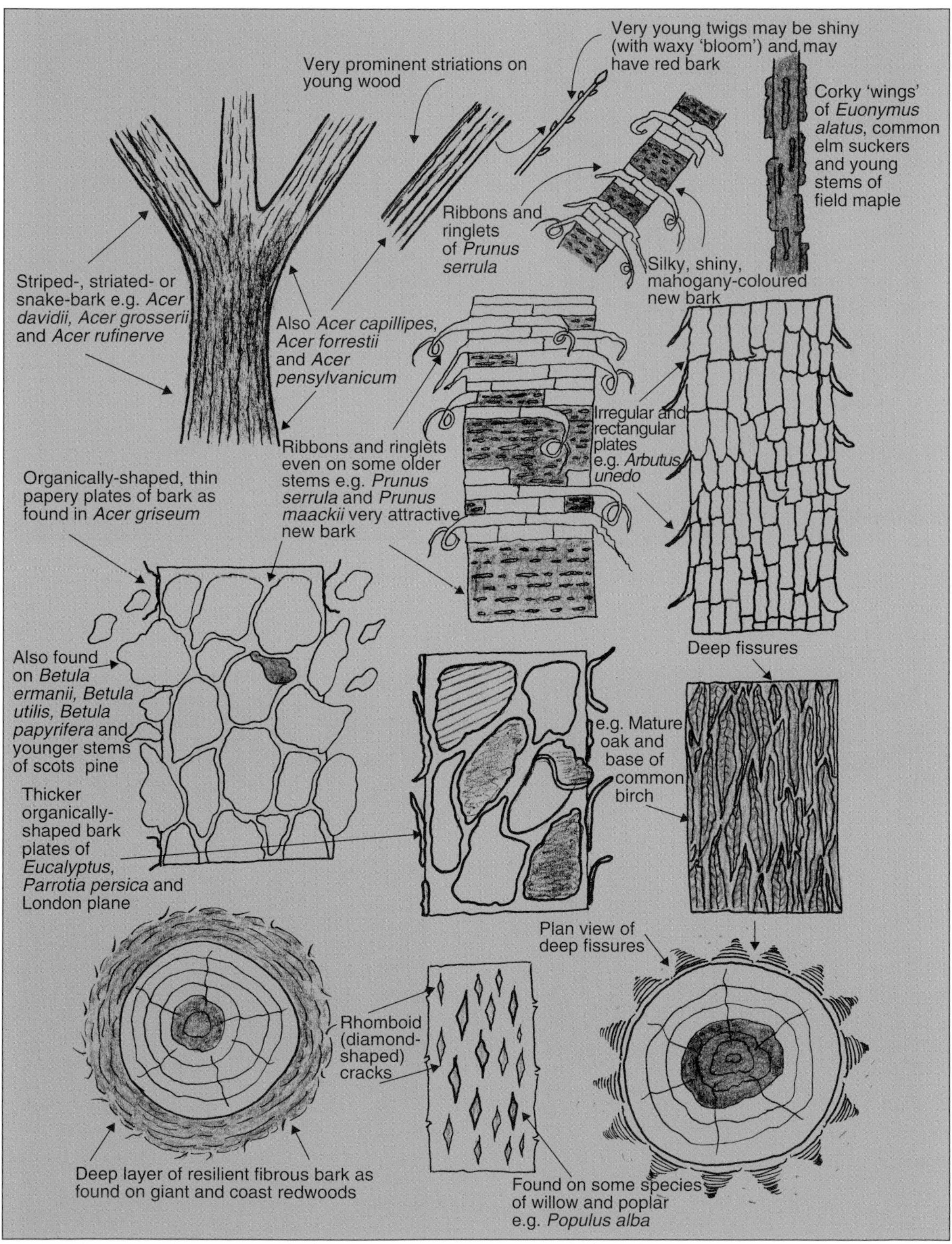

The structure, sloughing and aesthetics of bark.

The bark plates of the strawberry tree (Arbutus unedo).

Bark of Betula albo-sinensis.

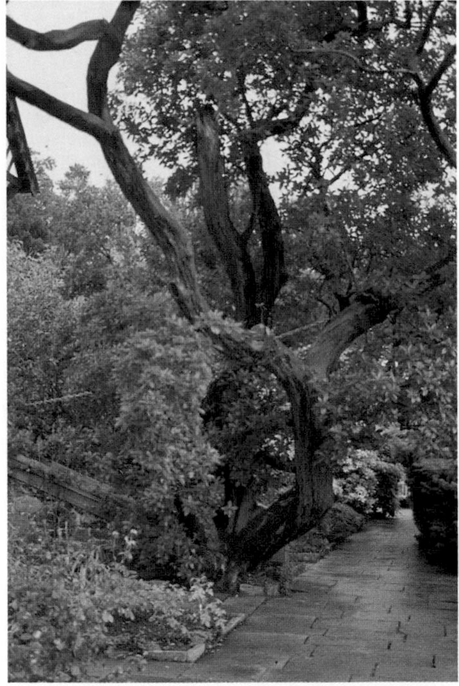

Peeling bark of Arbutus × andrachnoides.

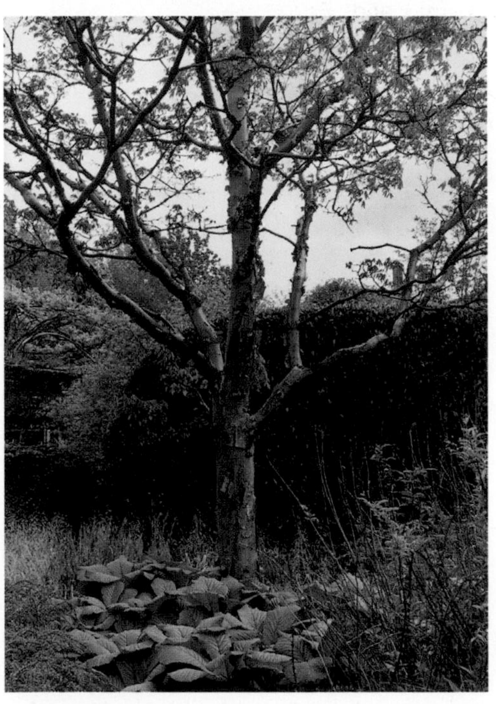

Bark of Acer griseum (paper-bark maple).

WOODY SPECIES WITH FEATURES INVOLVING FOLIAGE

Species with large simple bold leaves include *Rhododendron macabeanum, R. sinogrande, R. ficto-lacteum, R. praestans, R. hodgsonii, R. falconeri, R. orbiculare, Paulownia tomentosa, Fatsia japonica* and *Sorbus cuspidata* 'Mitchelli'.

Species for bold, compound leaves include *Gleditsia triacanthos* 'Sunburst', *Sorbus sargentii* (sargent's rowan), *Rhus typhina* 'Laciniata' (stag's horn sumach), *Dipteronia sinensis, Aralia chinensis, Decaisnea fargessi, Mahonia lomarifolia* and *Koelreutaria paniculata* (pride of India).

Species for coloured foliage include *Acer palmatum* 'Atropurpureum', *Gleditsia triacanthos* 'Sunburst', *Acer shirasawanum* 'Aureum', *Berberis thunbergii* 'Atropurpurea', *Philadelphus coronarius* 'Aureus', *Pieris forrestii* 'Wakehurst', *Photinia* × *fraseri* 'Red Robin', *Vaccineum glauco-album, Cotinus coggygria* 'Purpurea' and *Corylus maxima* 'Purpurea'.

Shrubs with good autumn colour include *Rhododendron luteum* – a very showy, fragrant, yellow, deciduous azalea with good autumn foliage. *Euonymus alatus* (winged spindle) is a bushy shrub with superb shrimp-pink tints (and also corky outgrowths on the stem), and *Euonymus europaeus* (spindle bush) has excellent shrimp-pink fruits with orange, protruding seeds and good autumn colour (when grown in open positions). *Euonymus planipes* and *Euonymus latifolius* have stunning, more showy fruits than *E. europaeus* and very good autumn colour. *Cotinus coggygria* (smoke bush), whose flowers give a fine smoky effect, has red and flame-tinted autumn colour. *Fothergilla major* has white flowers and excellent red/flame autumn colours, *Corylopsis pauciflora* has yellow catkin flowers and good purple-tinged autumn colour, and *Hamamelis mollis* (witchazel) has yellow, fragrant flowers in the winter and large felty leaves with excellent autumn colour. *Berberis thunbergii* has striking autumn colour, *Enkianthus campanulatus* is an excellent, acid-loving shrub with small, bell-shaped flowers, *Disanthus cercidifolius* has rounded leaves, and *Eucryphia glutinosa* is a very late, white-flowering bushy shrub – all these have good autumn colour.

Large shrubs/small trees with good autumn colour include *Cercidiphyllum japonicum* (katsura), which has rounded leaves and excellent pastel autumn colour; *Parrotia persica* (Persian ironwood), with excellent reds and/or yellow autumn colour; and *Stewartia pseudocamellia*, which has good white flowers and good autumn colour. *Aesculus parvaeflora* is a dense, suckering shrub with chocolate/maroon new growth, large white flowers and yellow autumn colour; *Rhus typhina* (stag's-horn sumach) has a large compound leaf, and excellent yellow and red autumn colour; and *Rhus typhina* 'Laciniata' (cut-leaved stag's-horn sumach) has large, dissected, compound leaves and excellent autumn colour. *Viburnum opulus* (the guelder rose) has striking, reddish, translucent fruits and good autumn colour; *Amelanchier canadensis* and *Amelanchier lamarkii* (snowy mespilus) are multi-stemmed, bushy trees with showy white flowers; and *Cornus kousa chinensis* is a small tree with excellent, showy white bracts that go pink: all these have very good autumn colour. *Acer shirasawanum* 'Aureum' has golden foliage colour and red-tinted autumn colour, *Acer japonicum* 'Aconitifolium' has wonderful dissected leaves and superb red autumn colour. *Acer palmatum* 'Osakazuki' has superb red and flame autumn tints, and is one of the best, as is *Acer circinatum* (vine maple).

Small to medium trees for autumn colour include *Acer capillipes; Acer davidii; Acer hersii; Acer forrestii; Acer rufinerve; Acer pensylvanicum. Acer cappodocicum* (Cappodocian maple) has excellent yellow and red tints, and *Koelreutaria paniculata* (pride of India) has very unusual compound leaves, bladder-like fruits and excellent yellow and red autumn colours.

continue to show as strong reds, flames, oranges, golds or yellows, depending on the concentration of carotenes and/or xanthophylls present.

There are known examples where, following or during the demise of the chlorophyll, synthesis of anthocyanins (commonly found as pigments in flowers) adds to the deeper pigmentation. However, in many cases much of the pigmentation (the carotenes and xanthophylls) is already present prior to leaf fall. Also, before abscission is complete, some (though not all) proteins and nutrients within the leaves travel back down the

The large leaves and colourful new shoots of Rhododendron macabeanum.

Autumn colour of the vine maple (Acer circinatum).

Autumn colour of Boston ivy (Parthenocissus tricuspidata).

phloem of the leaf trace to be stored in the stem. The rest of the nutrients are taken back to the soil at leaf fall – a superb method of recycling nutrients.

Substantial loss of water to the leaves because of drought conditions prior to, or even during, the early phases of abscission leads to a very different effect. Initially leaves commence an early colour change and premature leaf fall may result. However, if the stress is prolonged in the run-up to abscission, leaves will go brown, crinkle and desiccate, rather than remain entire, relatively engorged and presenting good colours until their fall. Because abscission must have its essential precursors of shortening days and lowering temperatures, desiccated leaves will not necessarily have a developed abscission layer and pull off easily, as it depends on the progress of abscission before the event.

The Aesthetics of Flowers

Flower colour, shape and unusualness all have aesthetic and interest value. Strong colours, strange outgrowths, weird shapes, dark spotting, stripes and heavy blotches all add to the unusual visual effects. The dark spotting/mottling on some *Tricytris* species, and the mottling on voodoo lily, all give stunning and remarkable effects.

Whether flowers are considered to be of aesthetic importance or not depends on individual opinion. Woody species grown for their flower are often lacking in other features, and species with good fruits or excellent autumn colour may have less than excellent flowers. Flowering is often most effective over a two- to three-week period, and then loses prominence. So to gain continuity, species are selected with various flowering times, and species that commence flowering very early or conversely very late (even into the winter) to extend the flowering feature, are at a premium. Obviously the more diverse the choice, the greater the number of individual plants needed, and in turn the larger surface area required for planting.

The Aesthetics of Fruits

Fruits, because of the time needed for their development, are liable to be features of mid-to-late summer and early autumn, but can persist over the winter. The family Rosaceae includes many of our most important and dependable ornamental and edible fruit-producing species. Cultivars of *Malus* known for their crab apples (pomes), *Pyrus* (pear), *Prunus* (cherry, plum), *Crataegus*, *Sorbus*, *Cotoneaster* and *Pyracantha* commonly produce masses of fruits. Apple-like fruits (pomes) of *Malus* species and cultivars have various hues of green, pink, cream and red, and some with dark maroon-coloured fruits have dark staining in the flesh. Both *Cotoneaster* and *Sorbus* provide a range of fruit colours,

SHRUB SPECIES WITH SHOWY FLOWERS

All roses are included in this category: *Aesculus parviflora, Caesalpina giliesii, Callistemon, Camellia, Crinodendron hookerianum, Clethra, Cytisus battendierii, Daphne, Desfontainea spinnosa, Deutsia, Enkianthus, Eucryphia, Forsythia* × *intermedia, Fothergilla, Fremontadendron, Hamamelis, Hibiscus, Kalmia, Laburnum, Magnolia, Mahonia, Menziesii, Nerium oleander, Osmanthus, Osmaria, Philadelphus, Pieris, Ribes, Stewartia, Styrax, Telopea truncata, Viburnum* and all *Rhododendron* including *R. cinnabarinum, R. decorum, R. racemosum, R. thomsonii, R. williamsianum, R. wardii, R. yunnanense.*

Early Flowering Shrubs

Rhododendron × *praecox* is a small evergreen shrub with very early purple flowers. *Stachyrus praecox* is a deciduous shrub with hanging catkins. *Lonicera fragrantissima* is a small semi-evergreen shrub that has very fragrant white flowers very early in the year, and all mahonias flower early in the year.

An unusually warm and sunny autumn can affect those species that would normally flower in the winter (or overwinter with developed flower buds and then normally flower very early in the following season). The mild weather speeds up the process, and some of the flower buds break to reveal flowers in the autumn instead ('autumn flush'), or those that naturally flower in the spring may flower over the winter period instead. The phenomenon has become increasingly more common in recent years (maybe the result of climate change?).

Furthermore, many species, instead of opening just a few of their flower buds earlier than normal, now come into full flower in the autumn. The result is a wonderful out-of-season display, though unfortunately, any flower buds that do break out of season cannot be 'magically' replaced in time to develop and overwinter for the next season. Hence the season that follows will show a paucity of flowers, or sometimes none at all. *Viburnum fragrans, Viburnum* × *bodnantense* 'Dawn', *Viburnum tinus* and *Mahonia* 'Charity' are all prone to this.

Later-Flowering Shrubs

This category includes *Rhododendron auriculatum, R.* 'Polar Bear', *Eucryphia* × *nymansensis, Eucryphia* × *nymansensis* 'Nymansay', *Eucryphia glutinosa, Clerodendrum trichotomum, Hibiscus syriacus,* hydrangeas.

Fragrant Plants

Foliage and stems can contain pungent, essential oils, and flowers can often be accompanied by a fragrance to attract insects, which may be persistent or may be more concentrated in the evening air, it depends on the species; *Petunia* species and cultivars and *Nicotiana* species and cultivars are renowned for their superb smell. Most, but not all, forms of honeysuckle have very fragrant flowers, including *Lonicera periclymenum* and its cultivars; other woody climbers with good fragrance include *Jasminum officinale* and the less hardy *Jasminum polyanthemum.*

Hammamelis mollis, Lonicera fragrantissima, Viburnum farreri, Viburnum × *bodnantense,* and *Viburnum* × *bodnantense* 'Dawn' are all good examples of hardy shrub species with strong fragrance. *Viburnum burkwoodii, Rhododendron decorum, Rhododendron discolor, Rhododendron luteum* (deciduous) and *Rhododendron auriculatum* all have excellent fragrance. Other examples include *Daphne, Magnolia stellata, M. grandiflora,* mock orange (*Philadelphus* species) and *Elaeagnu ebingii* (very small, insignificant, cream-coloured flowers, but with very strong perfume).

Cotoneaster rothchildianus has striking yellow berries, and *Cotoneaster affinis* has black berries. Some rose species are good for fruits (hips), most of which are shades of red or orange – for example, *Rosa moyesii* and *R. rugosa*. But no one should deny the aesthetics of common edible fruits such as raspberry, loganberry, strawberry, peach, plum and nectarine – only familiarity makes us treat their visual benefits with contempt.

The sub-shrub *Coriaria terminalis xanthocarpa* has superb yellow berries; the herbaceous *Actaea pachypoda* (syn. *Actaea alba*) has white berries; and *Actaea rubra* has red berries. The fruit capsules of poppies are very aesthetic in their shape and structure, and prolong their aesthetics for some time after flowering – likewise the orbicular siliculas of honesty (*Lunaria annua*) with their milky white/silvery membranes.

Crinodendron hookerianum.

Most deciduous euonymus species, all female and hermaphrodite forms of *Skimmia*, and *Viburnum opulus* have good fruits. *Piptanthus nepalensis* (Nepalese laburnum) has yellow flowers and very large green/yellow pods. Strange, ornate fruit structures and bladder-like capsules found in *Koelreutaria paniculata* (pride of India), *Colutea arborescens* (bladder senna), and *Staphylea,* all have aesthetic interest. Some very unusual fruits, such as the electric blue pods of *Decaisnea fargesii* and *Callicarpa bodinierii* var. *giraldii*, with profuse amounts of small, intensely purple/violet berries, can be stunning in their effect.

Plants and Clairvoyance!

The old wives' tale maintains that a particularly heavy show of berries in the late summer always heralds a cold winter, and often it proves to be correct. However, as complex as plants may be, they are not clairvoyant! Hence, a season of very unusual fruit abundance may well herald a cold winter, and make the old wives' tale that portends this come to pass. But the abundance of colourful fruit is purely a climatic/environmental response to the here-and-now (or what has already passed), but does not actually predict what is to come!

Good, warm, continually dry spring and summer seasons give ideal conditions for wood

Flower buds of Fremontadendron californicum.

The flowers of Fremontadendron californicum.

Telopea truncata.

to ripen, for flower buds to be formed, and for flowering to be prolific. In fact, a degree of stress created by very dry conditions temporarily creates the hormonal concentrations for very prolific flower production that under more excessive conditions would cause permanent damage, and even the plant's demise. If this period of dryness is followed by sufficient atmospheric humidity for good pollination and successful fertilization, and further environmental conditions are conducive to good fruit formation and fruit engorgement (moist conditions at soil level), then bumper fruit crops will result.

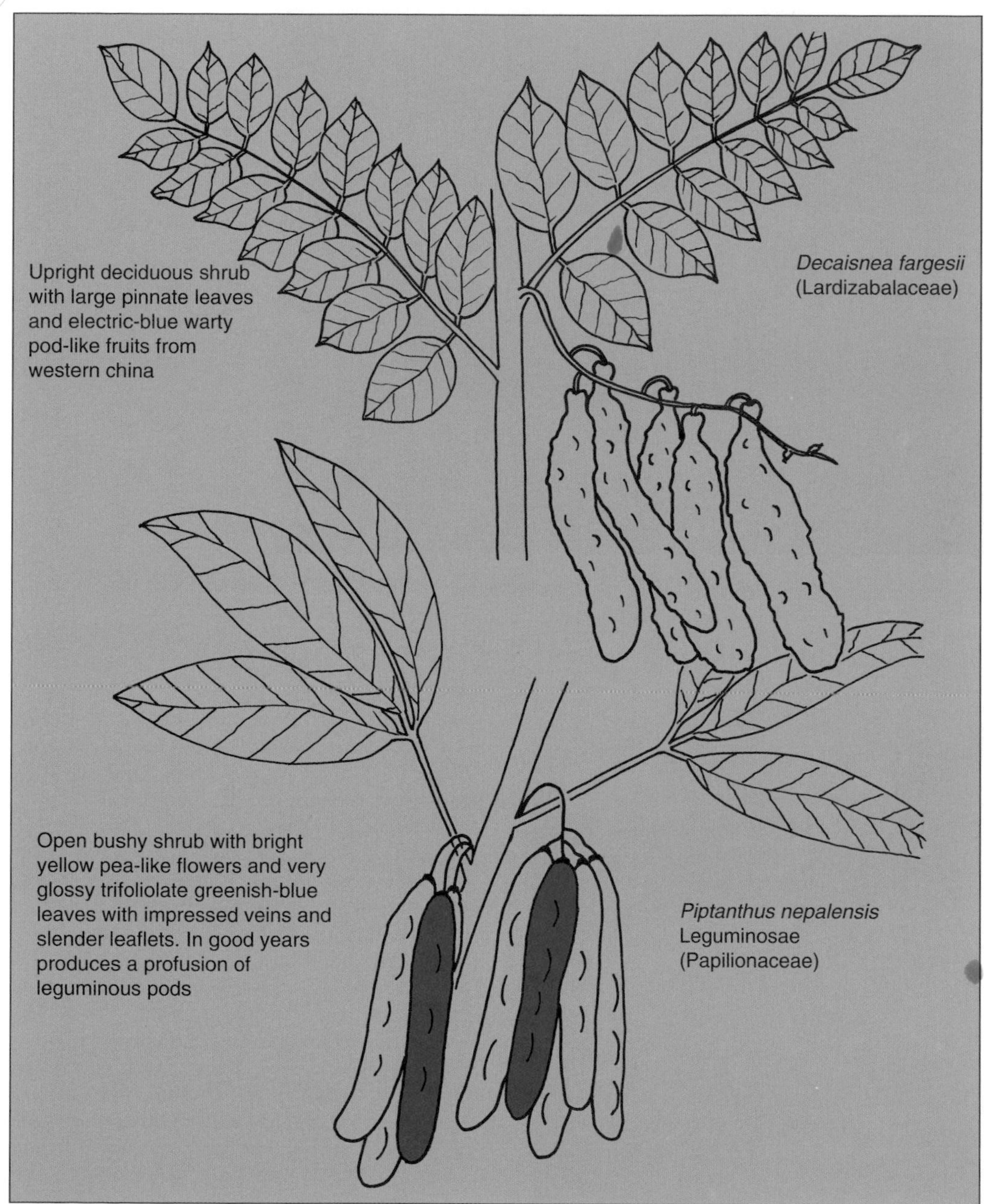

Upright deciduous shrub with large pinnate leaves and electric-blue warty pod-like fruits from western china
Decaisnea fargesii
(Lardizabalaceae)
Open bushy shrub with bright yellow pea-like flowers and very glossy trifoliolate greenish-blue leaves with impressed veins and slender leaflets. In good years produces a profusion of leguminous pods
Piptanthus nepalensis
Leguminosae
(Papilionaceae)

Showy fruits: Decaisnea fargesii and Piptanthus nepalensis.

Hedges

Hedges have many benefits: they mark boundaries, give privacy, have aesthetic value, create a foil or backdrop for other aesthetic features, provide a wind filter, create sheltered niches, create a habitat for wildlife, attenuate noise, or act as stock fencing. Informal mixed hedges may be used – and indeed, from an environmental and ecological aspect, mixed species are preferred.

Mixed hedges use a mixture of suitable species, rather than using only one species, and may, or

The fruits of Callicarpa bodinierii var. giraldii.

The bladder-like fruits of Staphylea pinnata.

may not, include a mixture of evergreen and deciduous woody subjects. They are very common in rural areas, and can include use for boundary demarcation and as a stock barrier. Informal, mixed hedges often comprise mixtures of native woody species, and do not usually suffer the problems associated with the susceptibility of monocrops to pests and diseases: the mixture of species effectively combats the build-up of pests and diseases.

They can also provide a good habitat for wildlife if managed properly. Very old hedges have built up a long-term relationship with native organisms, and they create excellent wildlife habitat, and 'corridors' for wildlife, allowing continuous, sinuous threads of habitat that interconnect with one another, and with other wider areas with habitat importance. The health and retention of ancient hedges is therefore very important. Furthermore, it is often beneficial to try to recreate a similar habitat by planting new hedges with a good number of native species. However, agriculturists commonly use single-species hedges of hawthorn (*Crataegus monogyna*), for its barrier thorns, and because it is low cost, and has a very quick establishment and growth.

Ornamental hedges comprise woody plants spaced close together in formal rows (sometimes

LARGE SHRUBS/SMALL/MEDIUM TREES WITH GOOD FRUITS

Arbutus unedo (strawberry tree) is an evergreen shrub with red, strawberry-like fruits. *Crataegus laciniata* (cut-leaved hawthorn) has cut, downy leaves and large red berries. All other *Crataegus* species and cultivars (hawthorns) have good fruits. *Cotoneaster* 'Cornubia' (tree cotoneaster) is an excellent evergreen with profuse red berries. *Malus* 'Golden Hornet' (golden crab apple) has a profusion of golden fruits, and most other *Malus* have good fruits, including *Malus* 'John Downie', which has a profusion of red-orange fruits. *Leycesteria formosa* (Himalayan honeysuckle) has insignificant white flowers, but showy maroon bracts and dark purple fruits. All *Ilex* (holly) species and cultivars (female and hermaphrodite forms) are noted for their excellent berries.

Small Trees with Coloured Fruits and Autumn Colour

Acer palmatum 'Osakazuki' has superb red schizocarps (the double-winged fruits typical of all maples), and also has excellent autumn colour. *Acer japonicum* 'Aconitifolium' also has red schizocarps, ornate leaves and very good autumn colour. *Sorbus vilmorinii* (vilmorin's rowan) has delicate, fern-like, compound leaves, coral-coloured berries turning through pink to white, and good autumn colour. *Sorbus* 'Joseph Rock' has yellow berries that are shown off to their best by the excellent wine-coloured autumn foliage. *Sorbus cashmeriana* (kashmir rowan) has large, succulent white berries and good autumn colour. *Sorbus hupehensis* (hupeh rowan) has small, white berries, is glaucous on the underside of the compound leaves, and has good autumn colour. *Sorbus sargentiana* (sargent's rowan) has very bold, large, compound leaves, large red berries and good autumn colour.

The fruits of Leycesteria formosa.

double rows), and clipped into a formal shape during an annual maintenance programme. Very different species may be needed where a mainly aesthetic feature is required, such as a flowering hedge or coloured foliage, rather than a hedge for stock fencing or as a barrier, and hedges grown for their flowers are nearly always represented by shrubs rather than larger trees. Both evergreen and deciduous plants can be used for formal hedges laid out in geometric patterns, traditionally including rectangles, squares, parallel lines and arcs, and other parts of circles including serpentine shapes. Single-species hedges (using many plants of only one species) are chosen for most formal hedges.

Although formal hedges may well act as a wind filter and protective barrier, they are mainly chosen for their aesthetic appeal. Nevertheless, it is still common practice to use formal hedges to create a sheltered niche, and a microclimate where a wider range of species may be grown within their protection. Hedges created by using only deciduous species such as beech (*Fagus sylvatica*) and hornbeam (*Carpinus betulus*) are effective

Sorbus 'Joseph Rock'.

functional windbreaks, but they can also be very aesthetic, particularly as these two species retain their rustling, brown, desiccated yet persistent leaves over the winter period.

The use of evergreen coniferous subjects for formal hedging, for their aesthetic effect and as a foil to other aesthetic displays, is highly recommended. However, less fortunate characteristics are their high moisture draw and poor wind-filter properties. The close-knit surface of a clipped evergreen hedge creates a lot of wind resistance and makes them a poor windbreak because they do not let wind penetrate easily, which would help to slow down wind-speeds, and the extra wind pressure this creates in fact often leads to windthrow. Furthermore, turbulence is set up when the wind hits the hedge, because it is forced over it, just as if it were hitting a solid brick wall, rather than filtering through it.

Species must be able to withstand regular cutting back, as trimming/clipping (pruning) is carried out regularly throughout the growing season. Hedges are often clipped to give a wider base than top, and the formal shapes used include tapered – the taper is physically a very strong shape – and loaf-shaped or rectangular. Where hedging includes topiary, all sorts of interesting, formal and organic shapes may be created, including designs involving the artistic pruning of individual plants into animal shapes, spirals, square pyramids and spheres. Regular maintenance is required to uphold their aesthetic effect.

Common yew (*Taxus baccata*) is very commonly chosen for topiary because it only needs clipping once or twice a year. Box (*Buxus sempervirens*), because of its relatively slow rate of growth, is another species used, and dwarf box is often chosen for dwarf hedging. Although the formal effect of hedges is enhanced by the uniformity of using one single species (or a low number of species) throughout the design pattern, there are some difficulties. With all forms of monocrop there is a tendency for pests and diseases to become entrenched, and box is having a difficult time at present combating a fungal disorder (*Vollutella*). In the last few decades box fell out of fashion; however, more recently people are wanting to reinstate old traditional plantings (parterres) that originally had box edging, and initiate new parterre-style plantings – although this has been hampered by the fungus problem.

Examples of other evergreen species used for hedging include X *Cupressocyparis leylandii*, which may be clipped as a formal evergreen hedge, or can be left to grow as a screen. *Leylandii* is, however, notorious for its growth rate, and, left completely unchecked or unmanaged, has massive stability and light preclusion problems in urban areas. Other evergreen examples include holly (*Ilex aquifolium*), Lawson's cypress (*Chamaecyparis lawsoniana*), laurel (*Prunus laurocerasus*) and the dwarf form *Prunus* 'Otto Lyken'.

Common yew is very often used for hedging because of its regenerative power after clipping,

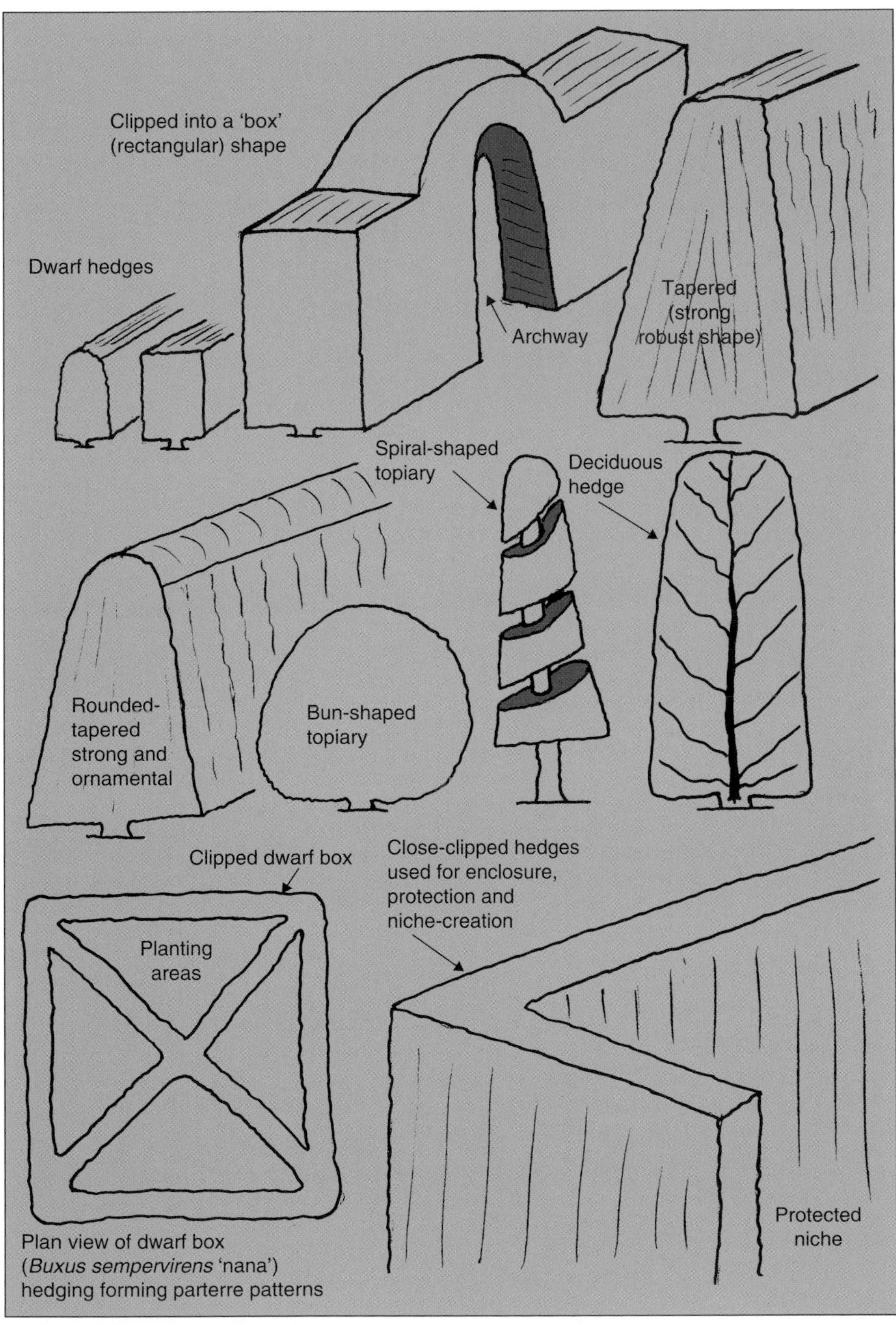
Clipped into a 'box'
(rectangular) shape
Dwarf hedges
Archway
Tapered
(strong
robust shape)
Spiral-shaped
topiary
Deciduous
hedge
Rounded-
tapered
strong and
ornamental
Bun-shaped
topiary
Clipped dwarf box
Close-clipped hedges
used for enclosure,
protection and
niche-creation
Planting
areas
Protected
niche
Plan view of dwarf box
(*Buxus sempervirens* 'nana')
hedging forming parterre patterns

Hedges: shapes and designs.

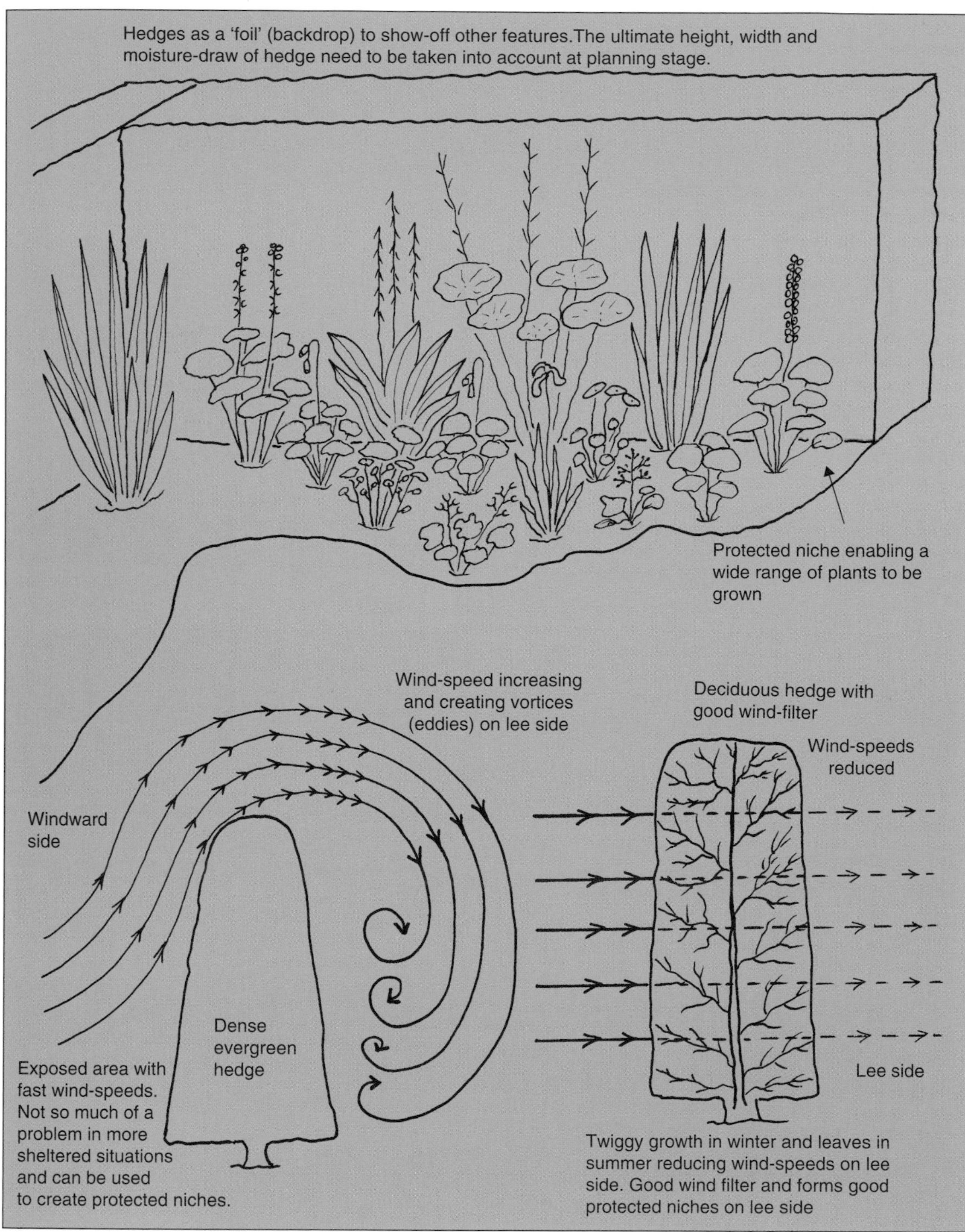

Hedges, wind resistance and protected niches.

and the good dense cover that is produced. With most species of evergreen constant clipping gives a very good effect; western hemlock (*Tsuga heterophylla*) in particular, because of the fine nature of the leaves, becomes particularly dense, making it extremely effective as a hedge both because of the leaf cover and aesthetically. But as always, it is a matter of personal choice, and if very colourful effects are wanted, they can be created by using evergreens with strongly coloured foliage.

Picea pungens forma glauca (a variation of blue Colorado spruce), *Picea pungens* 'Kosterii' (a more consistently blue cultivar of blue Colorado spruce), and *Photinia* × *fraserii* 'Red Robin' can give very strong effects as clipped hedges – although the cost of purchasing large numbers of plants may be prohibitive. Flowering hedges include low-growing types such as *Berberis verruculosa* and *Berberis buxifolia*, and taller forms such as *Berberis* × *stenophylla*, *Escallonia*, *Rosa multiflora*, *Berberis darwinii*, *Berberis julinae*, *Chaenomeles japonica* and *Rosa* 'Penzance Briar'.

Important Considerations

The most important criterion is to ensure that the site is suitable for the chosen species, including having adequate development room. Hedges have both height and width, and although there may be some control over their size (by clipping), this is limited. Also, hedges sometimes have a very heavy moisture draw, so the use of hedges to create microclimates in order to grow a wider range of plants within their protection can be ruined because plants come under drought pressure and never do well. Designs have to take all these factors into account.

Basically, the vigour and ultimate height of species, the ultimate aim of the hedge, whether it is deciduous or evergreen, and the soil pH requirements, can all influence species choice and are fundamental to success.*

Note that some trees are poisonous to animals, and care must be taken to avoid planting toxic species near fields where animals graze (as they may browse on them). Common yew (*Taxus baccata*) is a particularly toxic species.

*For more information on agricultural, mixed species and rural hedges, see *Resource Management – Hedges* by Murray MacLean. For more information on topiary, see *Topiary* by Chris Crowder and Michaeljon Ashworth. Both are published by The Crowood Press.

CHAPTER 8

Site Preparation, Sowing and Planting

Soil: The Environment of Plant Roots

The Composition and Properties of Soils

Soils comprise a mineral skeleton, organic matter (including humus), living organisms (microfauna, macrofauna and mesofauna – including earthworms), soil air (to support the respiration of root tissues), and soil water. In the natural world, the breakdown of rocks to form the mineral skeleton, and the constant addition of organic litter (partially decomposed to humus), forms the developing soil.

Soils are the environment of plant roots (the rhizosphere), and soils and other growing media, such as 'potting composts', need to supply anchorage, moisture, oxygen and nutrition to plants within the root zone. Soils comprise a matrix of mineral chips, including the clay fraction, and organic matter. The clay fraction is a rock flour comprising particulate sizes of 0.002 mm and below, and this very small size gives the clay fraction specific properties. Organic material within the matrix includes non-decayed organic litter, decaying and partially decayed (humified) organic material, and fully decayed organic matter that releases nutrient (mineral) salts. Humus is a black/brown, jelly-like, colloidal substance resulting from the partial decay of organic matter. Clays (minerals) and humus (organic) are both colloidal substances that have particular properties. Colloids comprise very fine particles that have cohesion (they stick to themselves), adhesion (they stick to other things), are very moisture-retentive (they hold water against gravity), and are very retentive to nutrient salts dissolved in water.

The adhesive and cohesive properties of both the humic and the clay fraction help create soil structure by forming aggregate lumps from the mineral particles of various sizes. Aggregate lumps bridge one another to form large pore spaces (macropores) and smaller pore spaces (mesopores) between them. Soil water drains freely, by gravity, from the macropores between aggregate lumps, whereas water is held against gravity within the smaller pores (mesopores) and the micropores created within each aggregate lump.

Because soils are a mixture of mineral and organic matter their properties will vary depending upon the proportions of each present in any specific soil. The mineral content is derived from naturally occurring rocks. Clays derive from eroded rocks and minerals, and it is the fine nature of the particles that gives clays their properties. Water encircles each fine clay particle, and the closeness of the molecules that this creates allows the very strong natural affinity of the water molecules to hold water against gravity. However, some of this water can be released by evaporation at the soil surface (energized by the sun), and some from the less fine pores (mesopores) by osmotic pressure at the foraging fine roots and root hairs of plants. Nutrient salts as electrolytic cations (positively charged ions) in the soil solution are held very strongly against gravity by the negatively charged (anion) clay particles.

The release of nitrogen into the soil from decomposing organic matter, and the subsequent taking up of soluble nitrogenous salts by plants, is cyclic. The plant takes up materials from the soil and releases them back into the soil at death. Organic matter of all types – for example dead animals, animal faeces, live, partially dead, and dead plant material – is initially broken down physically by chewing herbivorous or omnivorous mammals, and insects with biting mouthparts. The result is to produce sugar-rich mulches (detritus), and the action of breaking the material down into smaller particles renders it more easily attacked by fungi and other decomposing agents such as bacteria. The soft, chlorophytic material of leaves and young stems is easily broken down into its constituent parts; tougher, woody material is more difficult to break down. However, wood-rotting fungi secrete enzymes that digest the tissues of the wood, and they are very effective at gradually destroying the structure.

Soil Structure

Soil structure concerns the size and arrangement of the pore spaces between soil aggregate lumps (crumbs). Thus the size of the aggregate lumps will affect the relative size of the pore spaces created by bridging aggregate lumps. Larger pore spaces make for better structure, and anything that destroys the macropore pattern within the soil is said to damage or destroy structure. Poorly structured soils are easily compacted and can 'pan' on, near or below the soil surface. Human foot-traffic and heavy machinery all destroy soil structure and create an anaerobic rhizosphere (air at root level is excluded).

The large pore spaces formed by soil structure (macropores) cannot exist as a vacuum, and they will have either water or air in them. In non-impeded, well structured soils, air will always replace the outgoing water, creating a free-draining, aerobic (oxygen-containing) soil that is ideal for root penetration and growth. Well-structured soils therefore have good aeration from the air content in the large pore spaces, and good moisture retention by the minute pore spaces around the clay and humus particles (micropores). Hence, even though the soil is free draining and aerobic, it still holds large amounts of water against gravity in the finer pores. Plant roots penetrate the soil matrix, and as they forage in amongst the soil particles they find oxygen for respiration, and via osmotic pressure take up the water held against gravity in the mesopores (middle-sized pores); water is held against both gravity and osmotic pressure in the micropores created within each aggregate lump.

If the pore spaces fill up with water after rain and they do not drain, the soil will become waterlogged. However, if there is no impervious layer at lower levels, they are usually free draining, and gravity will pull the water out. 'Field capacity' is the condition when a soil holds the maximum amount of water possible against gravity – that is, the moisture content when the water has drained from the large pore spaces, but the maximum amount is held by the micropores within the aggregate lumps. Only water in excess of field capacity will move freely through the soil. If water is applied to soil that is not at field capacity it will be absorbed into the structure and will not percolate through.

Soil is very gradually wetted in layers known as 'fronts' – water must first wet the top surface (front one) thoroughly, before it can percolate to the next layer down (front two). A proportion of water applied to dry soils is absorbed, but the excess runs off (unless it is applied to an area that has already been wetted, which allows further percolation). Water will only percolate through soil in excess of field capacity in each layer – only as each layer is wetted will it percolate to the next one down. Vast quantities of water falling as rain (or irrigation) in relatively short periods of time will often be wasted as it runs off very easily into ditches, streams and rivers instead of penetrating the soil matrix.

Although the clay fraction (along with humus) is important for good soil structure, an excess of clay (which is a soft rock) in the soil profile can impede drainage and even be impervious to water, causing high water content – and correspondingly low oxygen content. The water table is the level to which surplus water rises, and in areas of free-draining topsoils the water table remains well below the soil surface. Low-lying, impervious barriers (such as tough bedrock or solid clay) causes the water table to rise, resulting in waterlogging, or even 'ponding' if it shows above soil level. Evidence of plants such as sedges, rushes, liverwort and mosses in relatively large numbers indicate poor drainage.

Within the soil matrix itself, evidence of poor drainage is shown by 'gleying' (mottling of the soil with shades of blue/grey running through the clay). Consistently high water tables fill all the soil pores with water, drive out the oxygen, and therefore create anaerobic conditions. Oxygen is needed for the respiration of plant roots, and to oxidize the iron compounds in the soil, creating the red and red/brown colours found in aerated soils. Low oxygen creates a reducing environment, not an oxidizing environment, and mottled blue/grey colours are the result.

Soils with consistently high or fluctuating water tables always cause problems for plants, and most plants will not tolerate these conditions unless the root system can find enough aerobic soil in the close vicinity to infest instead. Only the young, fine, fibrous roots on trees and shrubs can move, influenced by tropisms (stimuli); the old, cork-covered roots comprise lignified tissues that cannot grow away from the low oxygen condition and can therefore be 'drowned'. Even young roots can be overwhelmed by the anaerobic conditions, as waterlogging chases out soil air, prevents good

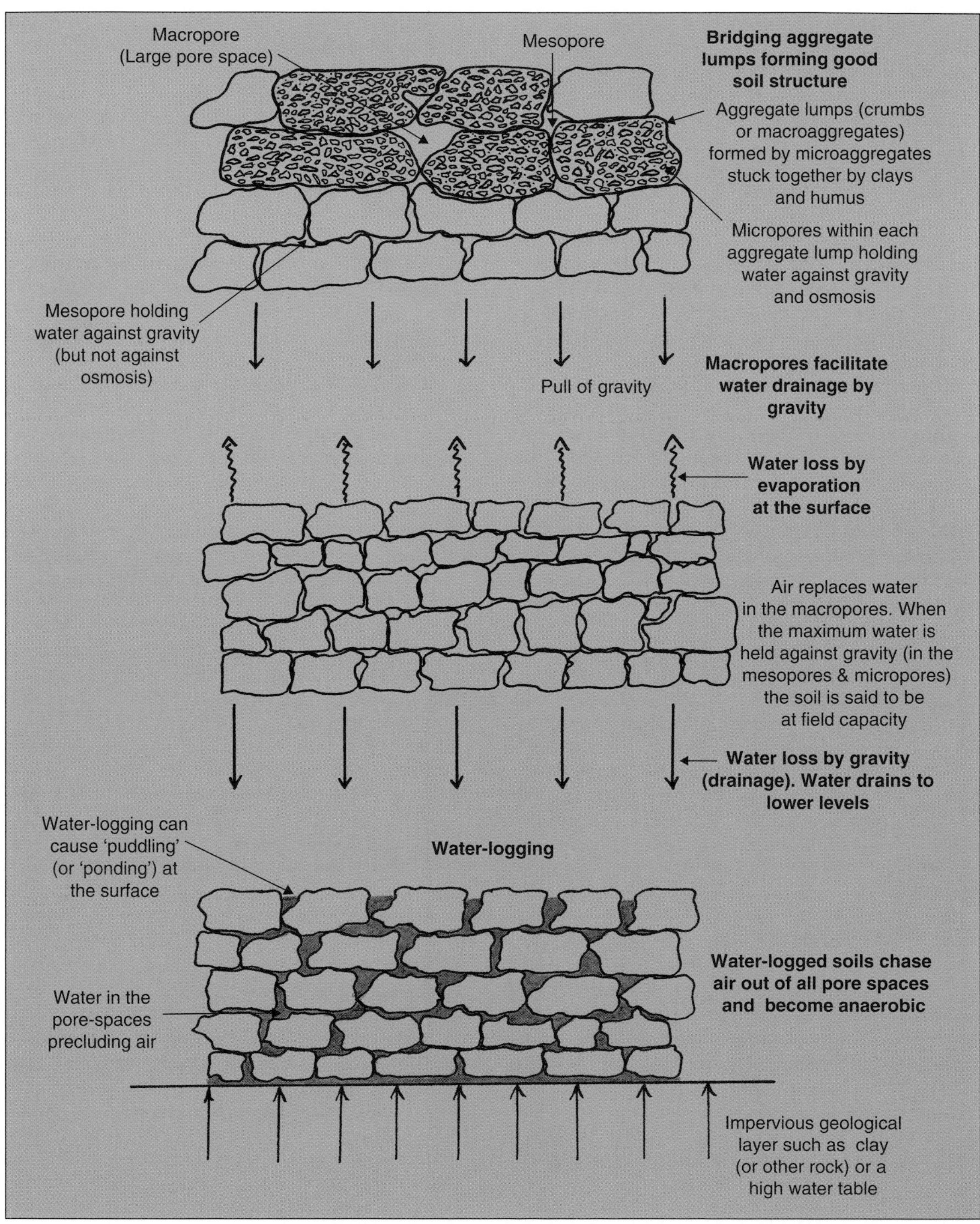

Macropore
(Large pore space)
Mesopore
Bridging aggregate lumps forming good soil structure
Aggregate lumps (crumbs or macroaggregates) formed by microaggregates stuck together by clays and humus
Micropores within each aggregate lump holding water against gravity and osmosis
Mesopore holding water against gravity (but not against osmosis)
Pull of gravity
Macropores facilitate water drainage by gravity
Water loss by evaporation at the surface
Air replaces water in the macropores. When the maximum water is held against gravity (in the mesopores & micropores) the soil is said to be at field capacity
Water loss by gravity (drainage). Water drains to lower levels
Water-logging can cause 'puddling' (or 'ponding') at the surface
Water-logging
Water-logged soils chase air out of all pore spaces and become anaerobic
Water in the pore-spaces precluding air
Impervious geological layer such as clay (or other rock) or a high water table

Soil structure, drainage, waterlogging and field capacity.

root penetration, and restricts root growth to the diminishing aerobic areas only. Conversely very low, or no moisture content in the soil, means there is insufficient moisture to retain the turgidity of cells and tissues, and the result is flaccidity (wilting). Soils that cannot support plant growth because of low water content (drought) are considered to be at permanent wilting point.

Soil Texture

Soil texture is the feel of the soil, and is also influenced by its clay content. Lower clay contents mean higher proportions of other soil fractions, such as silt or sand. Soil texture can be assessed by the tactile test – this is done by placing a sample of moist soil between the finger and thumb, and running the thumb over the sample, smearing it between them. If it feels gritty to the touch the soil is sandy; if it feels silky to the touch it is silty; if it feels very fine and smooth as it is smeared, and feels sticky, then it is a clay soil.

Another helpful assessment is carried out by forming a moulded shape out of a sample of soil by squeezing it in your hand. If the mould holds together very well and does not break up easily, the sample will have a high clay content (the clay content will give adhesion and cohesion). If it breaks into smaller fractions reluctantly, then it is probably a silty soil (having some clay fraction with the cohesion this brings). If it breaks to pieces easily, then it is probably a sandy soil (having little cohesion).

Clay soils are moisture and nutrient retentive, and are 'cold' (take longer to warm up). Sandy soils are very free draining, not very nutrient retentive, and warm up relatively quickly. Silty soils in general have properties intermediate between the two.

Site Preparation

Site preparation always commences with consideration of soil drainage, as this is fundamental to success. If the soil is naturally free draining then there is usually no need for a costly drainage system to be installed. However, there are instances where very efficient drainage is essential and expensive, and complex systems are installed no matter what the soil type. Bowling greens, football pitches and other sports pitches are the most common examples. On ornamental planting sites, if a drainage system has not been installed and there are doubts about the porosity of the soil, then sub-soiling or mole-draining may be used to aid drainage from lower levels. Very poorly structured soils may need expensive soil improvement involving substantial drainage systems and the use of ameliorating soil additives; but thankfully, most soils for ornamental plantings do not require this treatment.

The main methods of topsoil preparation prior to planting traditionally use soil inversion techniques. Ploughing has the greatest penetration depth of the mechanical methods of soil inversion, and is used for large-scale projects. Inverting the top vegetation means that it will die off, eventually break down, and become reconstituted with the soil to add to the humus content. Ploughing has severe limitations, including the fact that some deep-rooted, tap-rooted and slender rhizome-bearing weeds will survive the process to regenerate and reinfest the bare soil; and that heavy tractors will cause severe compaction and 'pans' – hard, impenetrable layers that build up in the soil when the ploughing process (as in agricultural practice) is annual. However, in preparation for landscape planting, if any existing pan is broken up by subsoiling, the problem will not normally reoccur. Ploughing cannot be carried out on small-scale areas, nor can it be carried out on poor and uneven terrain or on steep slopes. Greater depths of soil penetration are achieved by subsoilers (or multiple subsoilers linked together on the same frame, known as rippers); however, these merely tear into the soil to break up lower levels and improve drainage, but do not invert it.

Digging by hand is used for smaller-scale projects, and involves total inversion of the top spit (the length of a spade blade) of soil. Severe drainage problems are relieved by double digging (a system that involves digging the lower spit of soil, at subsoil level and inverting the top spit). The system is obviously costly in time and effort, but soil inversion techniques of this nature are the favoured option. However, it is not the only option: there is recurring interest in non-inversion systems that rely upon the constant and continual addition of organic matter to the soil as surface mulch.

There are major benefits to non-inversion systems, as the lack of disturbance prevents fresh seed lots coming to the soil surface from the seed bank, and therefore reinfestation of weeds will only occur from outside the system. Furthermore the nature/structure of the soil, because of the constant addition of materials that eventually break down to humus, allows any weed infestations to be removed very easily by hoeing or hand pulling.

Good soil structure, good moisture retention by the mulching effect, improved soil nutrition, an element of natural weed control from the blanket of mulch, and ease of weed control when it is necessary, make it a desirable system. Furthermore, it obviously sits well with the organic lobby if hand-weeding methods only (not chemical methods) are used. The down sides include the necessity to obtain large amounts of organic material, their availability, the large volumes involved, and the problems of storing, handling and applying these bulky organic materials frequently on to the site. Furthermore, although the system is accumulative and ultimately creates superb growing media, it cannot combat a site that commences with impervious or poorly draining subsoil without initial deep cultivation.

Producing a workable (friable) soil surface prior to seed-sowing or planting is important after soil inversion, and may be done by hand or using a rotary cultivator (rotavator). The aim is to reduce the soil crumb structure into a tilth (concerning relative particle sizes) that is good for the purpose – fine tilths are desired for seed sowing, and coarser tilths are usually used for planting into. By definition, creating fine, medium or coarser tilths destroys soil structure, as any operation that reduces the size of aggregate lumps, and therefore the size of associated pore spaces, destroys soil structure. There must therefore be a trade-off between creating a usable tilth and permanent soil damage.

Hand methods are unlikely to do severe structural damage, as a lot of effort is needed to break down the particle sizes with a rake, and cultivation is relatively near the soil surface. The use of rotary cultivators, on the other hand, although undoubtedly very useful and efficient in producing tilths prior to raking, and making the task much easier, can cause irreparable soil damage. Overuse/abuse of commercial-standard rotavators (particularly high-powered tractor-mounted versions) because of the constant smashing of the aggregate lumps, can be very detrimental to soil structure. Furthermore, there are major health and safety issues with such powerful, potentially dangerous machinery. So substantial safety footwear, good training prior to use, and healthy respect for the (often hidden) rotor blades, is essential.

The symptoms of over-rotavation manifest as poorly structured soils that become crusty with a surface pan when dry, and have the consistency of a pudding when wet (due to the closely packed particles resisting water percolation). Aeration of the surface becomes impeded, and poor plant growth results – particularly seedlings and small plants that do not penetrate to the lower and better oxygenated regions. Constant use of rotary cultivation can also lead to a poorly draining soil 'pan' at 100–150 mm depth, as their chopping action only ever penetrates to this depth as a maximum. Thus, deep cultivation should always be carried out before the use of rotary cultivators, with the only practical exception being to cultivate the very top of the soil surface to destroy annual and/or seedling perennial weeds.

Obviously, the non-inversionists do not need to tilth to any depth, and rotavation is not usually an option. Instead, shallow cultivation is carried out near the soil/mulch surface prior to seed sowing, and no preparation at all is necessary prior to planting as the soil is soft and penetrable because of the build-up of humus in the system.

Why a fine tilth for seed-sowing and coarser tilths for planting? Whatever the species, there is a direct correlation between seed size and sowing depth. Small seeds need to be sown at a shallower depth than larger seeds. Using the guide of sowing seeds at approximately twice their own depth ensures that germinating seedlings do not exhaust their self-contained and finite energy before they break the soil surface, at which time they can supplement their food supply by photosynthesis. Coarse tilths allow seeds to drop too deeply into the soil matrix, and there is no control over their final resting depth, whereas fine tilths allow fairly accurate drill depths to be used, that will not allow seeds to drop further down into the soil matrix. Established plants need to penetrate lower soil levels, so digging out planting pits through the prepared tilth at the top surface does not affect the outcome. In fact the idea of a tilth at all in planting areas is more for the aesthetics of the site than for any real practical benefits – and this holds true whether planting out vegetables or ornamentals.

Ploughing and digging are known as primary operations, and methods of tilthing and subsequent soil preparations are known as secondary operations, the final part of which includes consolidation and final surface raking. On a small scale consolidation is usually carried out by systematic 'treading', and on a larger scale using a roller. The aim is to prevent the soil from sinking in patches, forming uneven areas and indentations, but rather to be consolidated equally across the site to form a uniform surface. Final operations include raking in fertilizers prior to sowing or planting.

Sowing and Planting

Shallow drills are taken out for seed sowing, and this applies equally to vegetables, hardy annuals, biennials, or perennials – remember, vegetables are hardy annuals, biennials and perennials anyway. The only difference, therefore, is whether the sowing is to produce a food crop (in parallel lines to achieve uniform spacing and for ease of later weed management and crop harvesting), or in display beds for aesthetic reasons. Even those plants used for aesthetic purposes can be sown in parallel lines, but usually within organic-shaped drifts to give the appearance of informality. Grass seed and hardy trees and shrubs are nearly always sown broadcast over the soil surface. However, in some instances they may be sown equidistant in precision drills by hand or machine (space-sown).

The Production of Spring Bedding Plants

Wallflowers, sweet williams, forget-me-nots and *Bellis perennis* are usually sown outside. The soil on the seed-sowing site is prepared as for any other outdoor seed-sowing operation – digging, rotary cultivation, raking down by hand, light consolidation by treading, an application of phosphatic fertilizer, and then a final raking. Seed is sown thinly in straight drills marked out with a string line and drawn out in the soil by a sharp stick or, more commonly, the pointed edge of a draw hoe. Drills are produced in parallel lines for both ease of identification of the desired crop and subsequent weed control.

Sowing is carried out in May/June, and plants are lifted from the original seedbed in the early stages of development and lined out (transplanted) in July/August into prepared soil at wider spacing to continue their developmental process. Plants are therefore well developed and ready for transplanting into their final display areas by the oncoming autumn.

Primulas (including polyanthus, cowslips and primroses) do not germinate well in warm temperatures but can be sown under cold, but well ventilated, protection if desired, and if the facilities are available. Nevertheless, they are quite successful in outside beds, even if there may be a greater mortality rate than when they are sown in trays and grown in the relative protection of cold frames, cold polyethylene tunnels or cold glasshouses.

The Production of Summer Bedding

The production of mainstream summer bedding requires heated glass for their germination and initial development. Although *Alyssum maritima*, *Lobelia erinus*, *Nemesia*, *Petunia* and *Antirrhinum majus* cut across various plant groups, because all types are marketed in trays, for convenience, they tend to be produced in the same way – by seed under heated glass.

The intensity of the production system (and the associated high energy costs) necessitate practices that create uniform, dependable sowing surfaces, uniform seed sowing, uniform seed covering (when necessary), uniform temperatures, and therefore uniform germination. Seeds are sown broadcast over the surface of a good quality seed-sowing medium; the seed trays are prepared by overfilling them with the sowing medium, striking them off level with a wooden batten, and then compressing the medium with a purpose-built pressing board to form a uniform, level surface (below the level of the tray edges).

Seed-sowing densities and coverage very much depends on the species. Small and medium seeds are sown fairly densely, but leaving (albeit small amounts of) uniform space around each. Development is continued at wider spacing in new trays after 'pricking out'. Hence, densities may be relatively thick if pricking out into new trays is done efficiently and immediately as needed. Very fine seeds such as *Begonia semperflorens* (fibrous-rooted begonia) are sown thinly, then lightly pressed into the prepared surface with a clean pressing board; they are not covered with sowing medium afterwards. The moist environment maintained by stringent management will enable the seeds to imbibe water and germinate without the hindrance of a covering. Medium-sized seeds are covered with a thin, uniform layer of loose sowing medium after sowing – it is not compressed, as this may inhibit germination and/or seedling development. Larger seeds may be space sown – individually placed to ensure uniformity of spacing, by pressing each seed into the prepared surface. Furthermore, large seeds are also covered with a thin layer of loose compost (sowing medium), the insertion depth and the depth of covering together not to exceed twice the depth of the seeds.

Prepared trays may be watered overhead using a fine rose on a hosepipe or watering can. However, by far the best method, not least because it does not disturb the sowing medium surface, is to sub-irrigate by placing sown trays (either before, or preferably after sowing) into irrigation trays (drip trays). These are very efficient in ensuring the whole volume of the medium is thoroughly wetted. Irrigation trays are very shallow, and the water is carefully topped up so it is never deeper than the prepared surface of a seed tray, so will never flood the surface, with or without sown seeds.

F1 and F2 *Pelargonium zonale* types (geraniums of horticulture) and *Pelargonium peltatum* (trailing or ivy-leaved geraniums) are sown in January or February. Fine-seeded *Begonia semperflorens* (fibrous-rooted begonias), *Petunias* and *Lobelia erinus* types are sown in Januay/February. Medium-seeded types such as *Antirrhinum*, *Salvia splendens* and *Nemesia* are sown in February/March; African and French marigolds, and the really tough, quick-growing types such as *Alyssum maritimum* are sown the latest (March/April). Actual times depend on how quickly the species germinate and develop, and on the facilities available, but plants need to be at a marketable or planting size for May.

Sown, irrigated trays are covered with a sheet of glass to retain humidity in the top surface, and if large bench areas within suitably heated glasshouses are available, then the trays may be laid out, covered in a black polyethylene sheet and left to germinate. Optimum temperatures are expensive to maintain as most types require consistent temperatures of 55–60° F (17–20° C) for germination. Some large commercial producers (because of their throughput) do find it viable to heat the whole glasshouse, but other growers use various methods of reducing energy costs and rationalizing energy use in the first instance (when the higher temperatures are needed for successful germination); these include sectioning off specific glasshouse areas using polyethylene curtains and creating complete polyethylene tents (with heavily insulated roof areas) within the main glasshouse. Any well insulated system that reduces the volume of air that needs to be heated and increases heat efficiency is preferable to heating the whole glasshouse to high germination temperatures at this time of year.

If the luxury of masses of heated glasshouse bench space is not affordable, then sown trays may be stacked on top of one another within their insulated 'tents' (because in the initial stages of germination light is not usually important – apart from begonias). However, this can only be done in a regime of scrupulous and rigorous regular inspection to ensure trays that present germinating seedlings are removed from the stack and into the light as soon as possible to continue their development. A sheet of glass is only required on the top tray of each 'stack', and polyethylene sheets can be draped over each stack to reduce moisture loss and insulate available heat further. As soon as germination commences and radicles appear through the testa, the trays are spaced out into the main glasshouse area where temperatures can be slightly lower than those necessary for initial germination. However, any rapid drop in temperature can be detrimental during this phase.

Seedlings, as they develop, can be pricked out into fresh trays with new growing medium and at a wider spacing. Trays are prepared in exactly the same way as for seed sowing, but using a stronger compost (growing medium). Numbers per tray will vary according to the size and vigour of the types, but are usually in the range of twelve to forty-eight per tray. Species that produce large individual plants, such as pelargoniums and tuberous begonias, are pricked out at the lower numbers, whereas lobelia (which is pricked out using small clumps at each station, rather than singly) is usually spaced to produce higher densities in the trays. Seedlings should not be left for too long in the original sown trays as severe overcrowding leads to the onset of some fungal diseases and to poor seedling development. During the pricking-out process, care must be taken not to damage the roots of the seedlings when lifting them from the sown trays. Care must also be taken not to crush the hypocotyl of the seedlings when placing them in their new trays and lightly firming them into the prepared dibber hole with the side of the dibber. Multi-dibbers can be used to prepare a set number of equally spaced dibber holes in one operation, or modules with specific numbers of individual cells or 'plugs' may be used in preference to traditional seed trays.

Freshly pricked out trays are watered thoroughly overhead using a fine rose on a hose or watering can. Once the roots of the seedlings start to invade the new compost and plants develop in the trays, the necessity for high temperatures diminishes; thus the trays can be transferred to another, cooler part of the glasshouse, or to a separate glasshouse set at slightly lower temperatures. As the plants develop further and start to fill the growing space in the trays, they may have the temperature reduced yet again. Some types are tougher than others, and although optimum growth rates will not be achieved, they can be grown successfully at fairly cool temperatures as long as frost is kept off – alyssums, well developed petunias and antirrhinums are quite tolerant of lower temperatures. Even well developed, fibrous-rooted begonias are quite tough, and all these types can be successfully grown on in polyethylene tunnels or glasshouses that are on a low thermostat setting in case of frost. Conversely, the very frost-tender, temperature-sensitive types such as French and African marigolds have to be maintained at slightly higher temperatures.

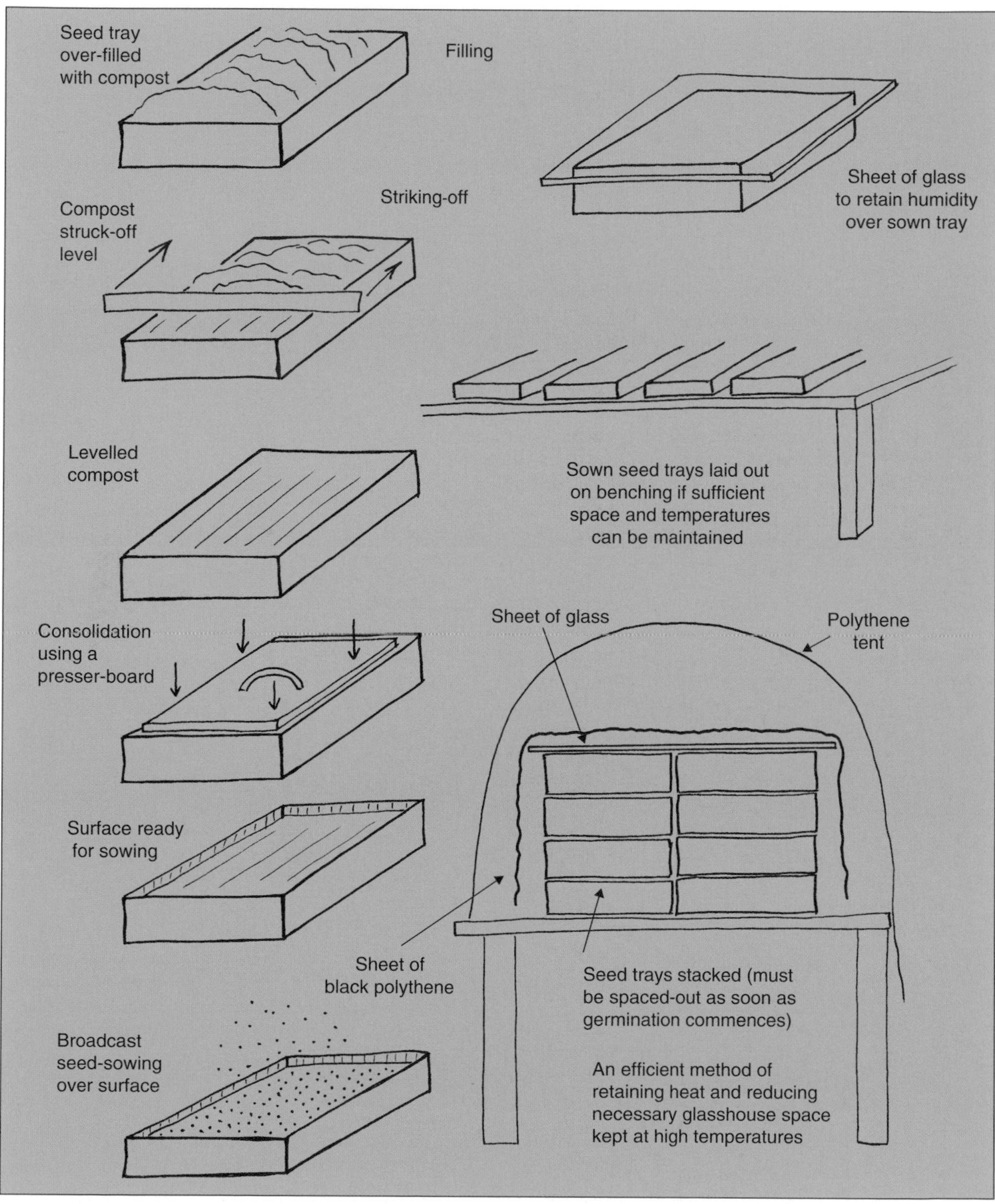

The production of summer bedding I.

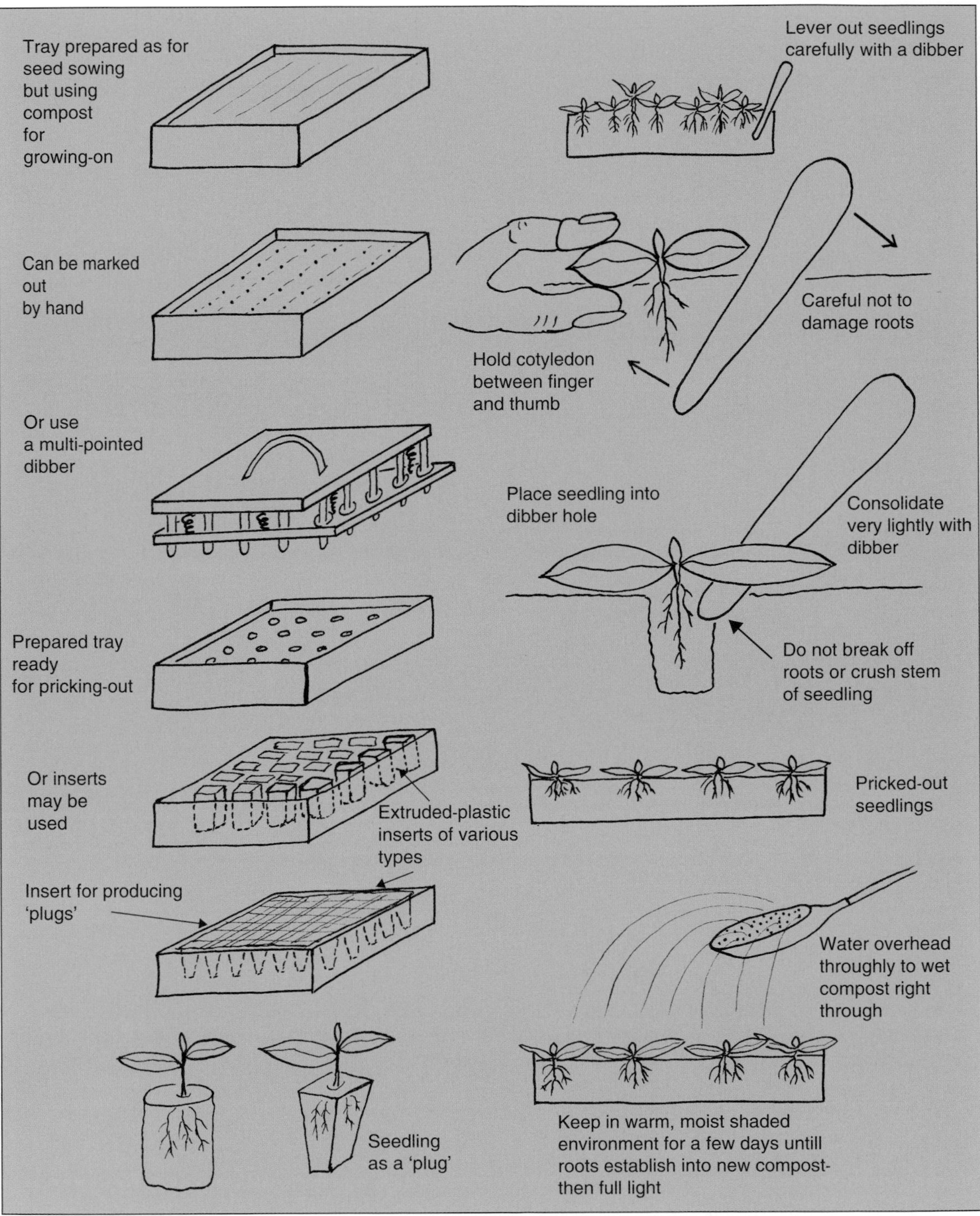
Tray prepared as for seed sowing but using compost for growing-on
Lever out seedlings carefully with a dibber
Can be marked out by hand
Careful not to damage roots
Hold cotyledon between finger and thumb
Or use a multi-pointed dibber
Place seedling into dibber hole
Consolidate very lightly with dibber
Prepared tray ready for pricking-out
Do not break off roots or crush stem of seedling
Or inserts may be used
Extruded-plastic inserts of various types
Pricked-out seedlings
Insert for producing 'plugs'
Water overhead throughly to wet compost right through
Seedling as a 'plug'
Keep in warm, moist shaded environment for a few days untill roots establish into new compost-then full light

The production of summer bedding II.

Summer bedding approximate spacing

List A *Very vigorous African marigolds, 35–40 cm apart*
Vigorous African marigolds, 25–30 cm apart
Antirrhinums (large types), 30 cm apart
Vigorous Afro-French marigolds, 25–30 cm apart
Begonia hybridatuberosa types, 30 cm apart
Pelargoniums/geraniums, 30 cm apart
Salvia farinacea types, 25–30 cm apart
French marigolds (taller types), 25 cm apart
Petunias (larger types), 25 cm apart
Salvias, 25 cm apart
Antirrhinums (compact forms), 25 cm apart
Alyssum maritimum 20 cm
French marigolds (dwarf types), 20 cm part
Begonia semperflorens (fibrous-rooted begonias), 20 cm apart
Ageratum, 20 cm apart
Lobelia erinus, 15–20 cm apart

Spring bedding approximate spacing

List B *Tall narcissus and tulips, 30–35 cm apart*
Tall wallflowers, 30 cm apart
Ornamental kales and cabbage, 30 cm apart
Sweet williams, 25–20 cm apart
Myosotis (forget-me-not), 25–30 cm apart
Dwarf wallflowers, 25 cm apart
Bellis perennis, 25 cm apart
Primulas/polyanthus, 25 cm apart
Dwarf narcissus and tulips, 25 cm apart

The process involves a gradual reduction of temperature so that the plants are hardy enough to plant out after the danger of frost is over – so lowering the temperature or shifting the trays into a cooler environment is necessary. By grouping the plant types according to their growing requirements, temperature regimes of the individual structures can be gradually adjusted accordingly. The final phase of the hardening-off (climatization) process may be under cold glass, or even by transferring them initially into heated frames, and then cold frames if desired. However, daily management tasks during the growing period are carried out much more easily in walk-through tunnels and glasshouses.

Seed-raised pelargoniums and tuberous begonia hybrids may be pricked out into individual pots for growing on, if desired. Pelargoniums may also be produced from softwood cuttings, and tuberous begonias from leaf cuttings. Furthermore, all forms of bedding can be purchased from specialist producers, either pricked out into trays or as 'plugs' in extruded plastic modules.

Planting Summer Bedding
Summer bedding plants are planted after the danger of frost is over (May/June, depending on the region), and usually displayed in borders or beds with a slightly raised soil surface. Preparation of the soil is as for any other seed-sowing or ornamental display area. Plants are 'bedded out' (planted from their trays into their display beds) in formal straight lines using a hand trowel. Beds are marked out with string lines, and well developed plants are laid out at the desired spacing (for short periods only, to prevent root desiccation). They are planted by taking out a planting hole of sufficient size with a hand trowel, and then dropping the trowel to firm them in thoroughly with your hands. Vigorous African marigolds and taller types should be spaced at 30–40 cm apart, medium growers 25–30 cm apart, and low-growing types 20–25 cm apart. Spot plants can be planted using a small spade.

Planting Hardy Biennials and Hardy Herbaceous Perennials
Soil preparation for herbaceous beds and borders is usually by hand because areas are rarely large enough to enlist the help of machinery. Because most herbaceous subjects prefer soils with good moisture content, organic manure is usually incorporated during the primary preparation processes in an attempt to increase the moisture-holding capacity of the soil. After initial preparations, the area is lightly consolidated by treading (or rolling on a large area), and then raked to a relatively fine tilth. After all primary and secondary cultivations are completed, including lightly raking a fertilizer into the top surface, the site is marked out ready for planting. Herbaceous subjects are usually planted in irregular, organic-shaped 'drifts' to give a feel of informality. The positions and sizes of individual drifts are marked on to the site either by scribing the shape on to the site with a sharp stick, or using light-coloured sand (or both for greater permanence if the site is not to be planted immediately).

Herbaceous subjects are usually planted at fairly close spacing to give a dense, matted effect. Small subjects are planted at densities of seven to nine plants of the same species per square metre, medium

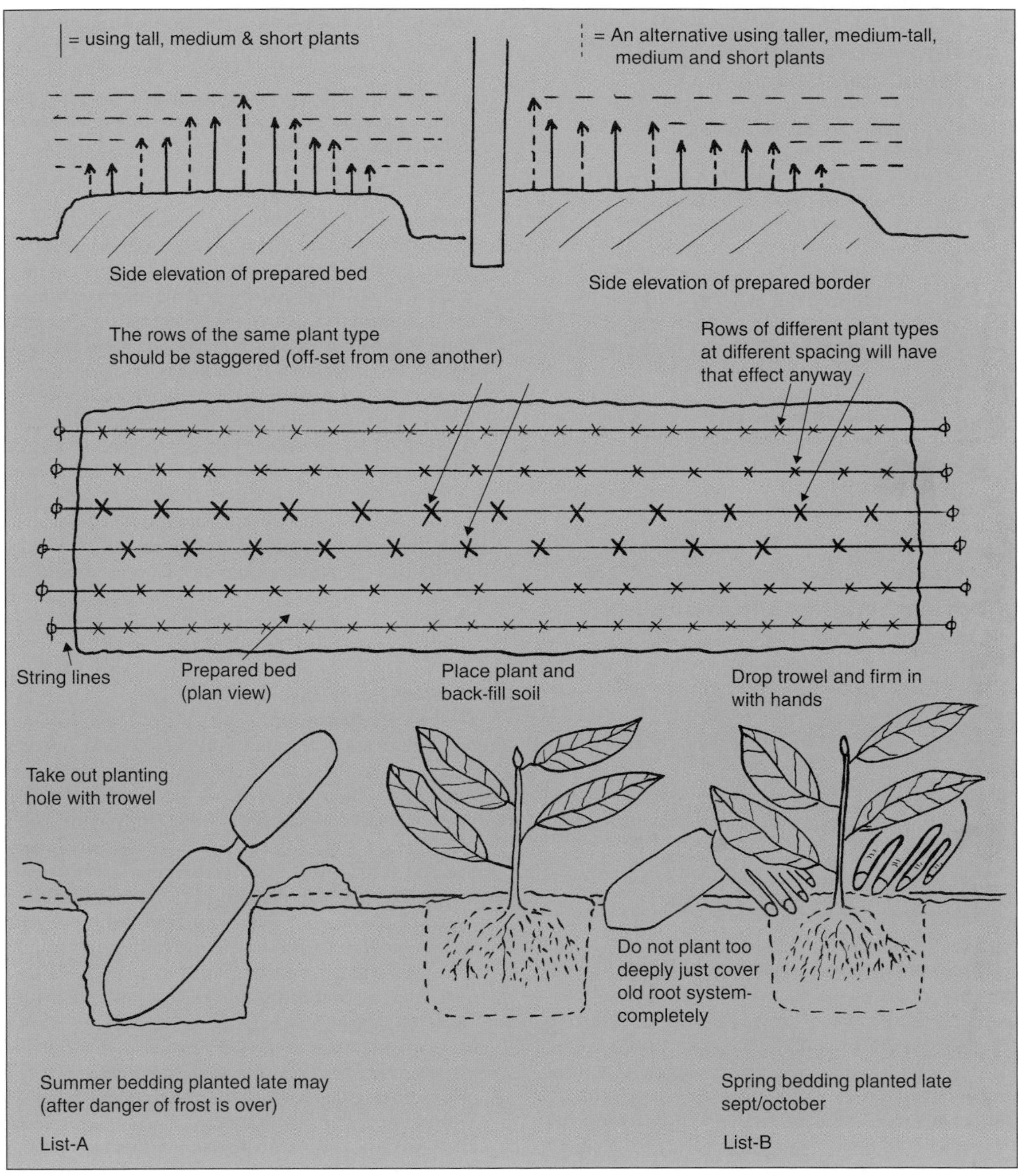

Planting spring and summer bedding.

subjects might be at five to seven plants per square metre, and larger species at lower densities of, say, three to five per square metre. The relatively high densities of herbaceous plantings inhibit weed infestation, but will not stop it completely. Furthermore, any infestation of perennial weeds is very difficult to control, as the weed plants have exactly the same subterranean organs for their survival as the aesthetic species chosen for display. Infestations of slender, rhizomatous species of weeds such as couch grass and ground elder are very difficult to eradicate, and herbicides cannot be used after planting, as they cannot distinguish between the desired and the unwanted plant species.

Planting through sheet mulches of black polyethylene to reduce or completely alleviate the need to weed afterwards is usually only considered for woody perennials, as herbaceous plants grow by increasing in size laterally through the soil. However, planting herbaceous plants through the very thin-gauge polyethylene sheets used to cover capillary beds in glasshouses (with large numbers of pin-prick holes in it) works very well to reduce weed infestation. The material is relatively cheap, and the minute holes allow water to percolate through but do not allow weeds to penetrate. If the hardy perennials are planted through a much larger-than-normal cross or even a 'star' cut into the polyethylene, then the plants are not too restricted in their lateral development – the slits can even be carefully extended as the plants progress. Furthermore, the maintenance programme of herbaceous plants involving lifting and dividing them every four years means the polyethylene can be replaced at that time (because it is so thin it starts to photo-degrade). Obviously the black sheet mulch needs to be covered with a mineral or organic mulch to make it more aesthetically acceptable.

Planting is best done in the autumn when soil conditions and moisture content are good. However, it is also possible in the spring if necessary (delphiniums and peonies prefer this), and it is even possible (yet not so desirable) during the winter months – though not in severe weather conditions (including frosty weather). Summer planting of container-grown plants is possible, but only if irrigation is available. Because of the relatively small size of the plants, the planting operation is carried out using a hand trowel, which is particularly useful as it can be dropped to the ground easily after excavating the planting hole, and plants 'firmed in' using the fingers of both hands. Larger plants can be planted with a small spade, and then lightly but firmly consolidated with the heel of the boot. Holes are excavated to ensure sufficient width and depth to accommodate the root system, and planting is carried out to leave a thin layer of soil covering the top layer of all fibrous-rooted and dense slender-rhizome systems. Fleshy tap-rooted species need deeper, more vertical planting holes than most other types, and fleshy rhizomes are planted to just cover the rhizome – even though they often work their way back to lie on the soil surface eventually.

In general, herbaceous subjects are relatively 'forgiving' in their planting depth. However, if bulbs, corms, tubers and fleshy rhizomes are planted too deep, the same problems will be encountered as with seed regarding their finite energy supply. So bulbs are planted at twice their own depth to leave the nose of the bulb one bulb depth below the surface. This is sufficient to keep the bulb moist and protected from severe weather and to trigger initial growth, but not so deep that energy reserves are completely depleted before the developing green leaves can break the surface and photosynthesize to top up the failing energy supply. Some scaly bulbs of *Lilium* species are planted at three times their own depth in soft organic soils.

Herbaceous species with dormant adventitious buds at the top of the crown should be planted so the buds are only just covered (very lightly) with soil – soil should only just 'invade' the top of the crown, and not cover it to any great depth. Use either your fingers as you plant each one, or a small rake afterwards, to disturb the surrounding soil: this leaves a 'professional finish' and makes the site look good before you leave it.

The site should now be mulched thoroughly and accurately to leave a level, undisturbed finish that does not seriously invade the crowns of the plants and undo all the good done by carrying out accurate planting depths.

Plant Support Systems

Pressing cut brushwood into the prepared soil after planting is exceptionally useful for supporting hardy annuals and herbaceous perennials that do not normally support themselves. Its informal look, low cost, relative unobtrusiveness and efficiency as a method of support makes brushwood a very popular option. However, more formal systems involving either canes or small-gauge stakes and twine, or purpose-built linking metal supports, can be used very successfully. However, canes pushed into the soil do have safety implications because they are not easily seen from 'end on' when you approach them; hence wide, purpose-made stop ends are recommended to prevent possible eye injury. Annual climbers such as sweet peas can be supported on cane 'wigwams' using 1.8–2.4m canes tied at the top. Specialized wooden, squarely conical pyramids can be purchased for the support of climbers of all types, including woody climbers; and trellis of various patterns is particularly useful

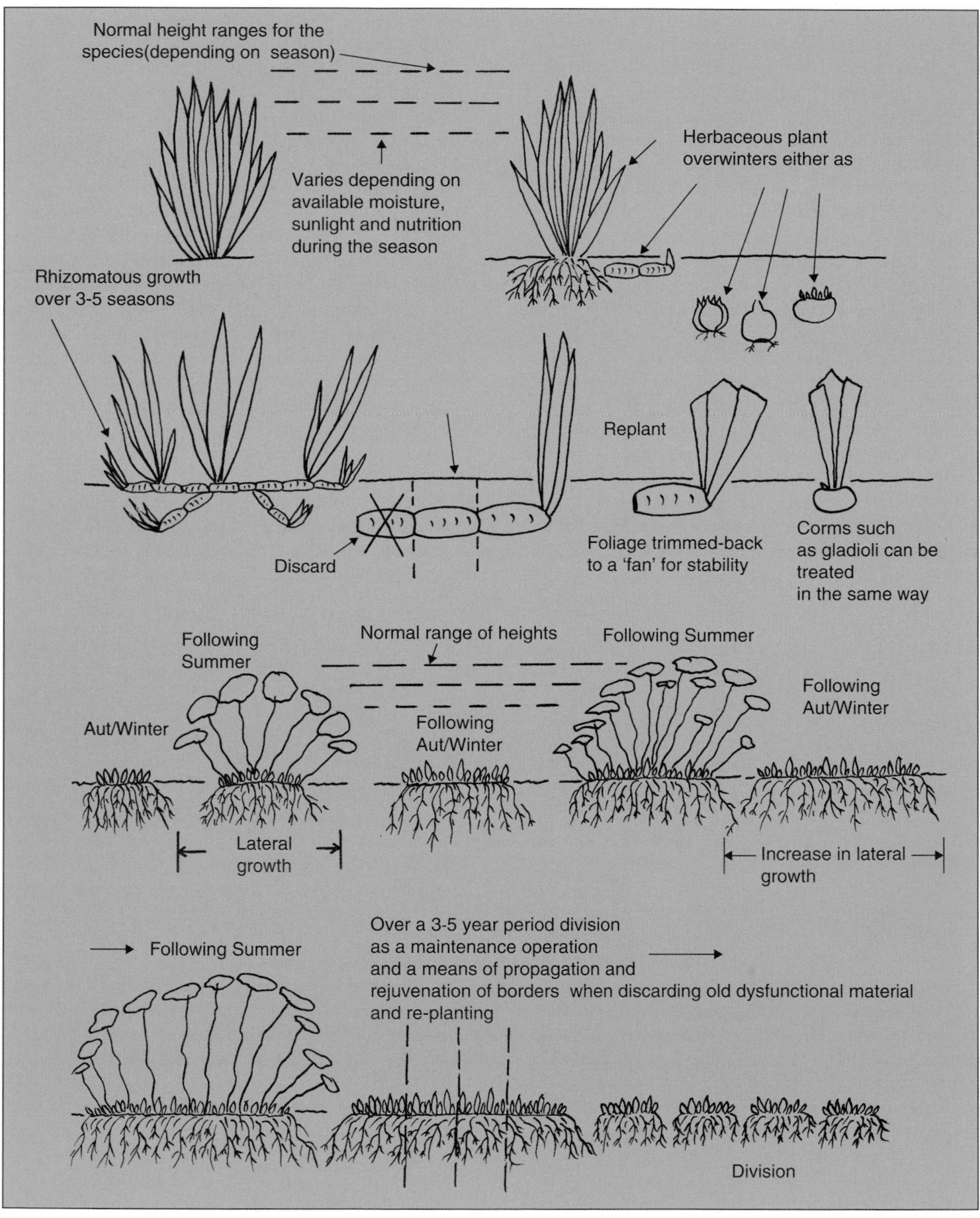

Dividing and planting herbaceous perennials I.

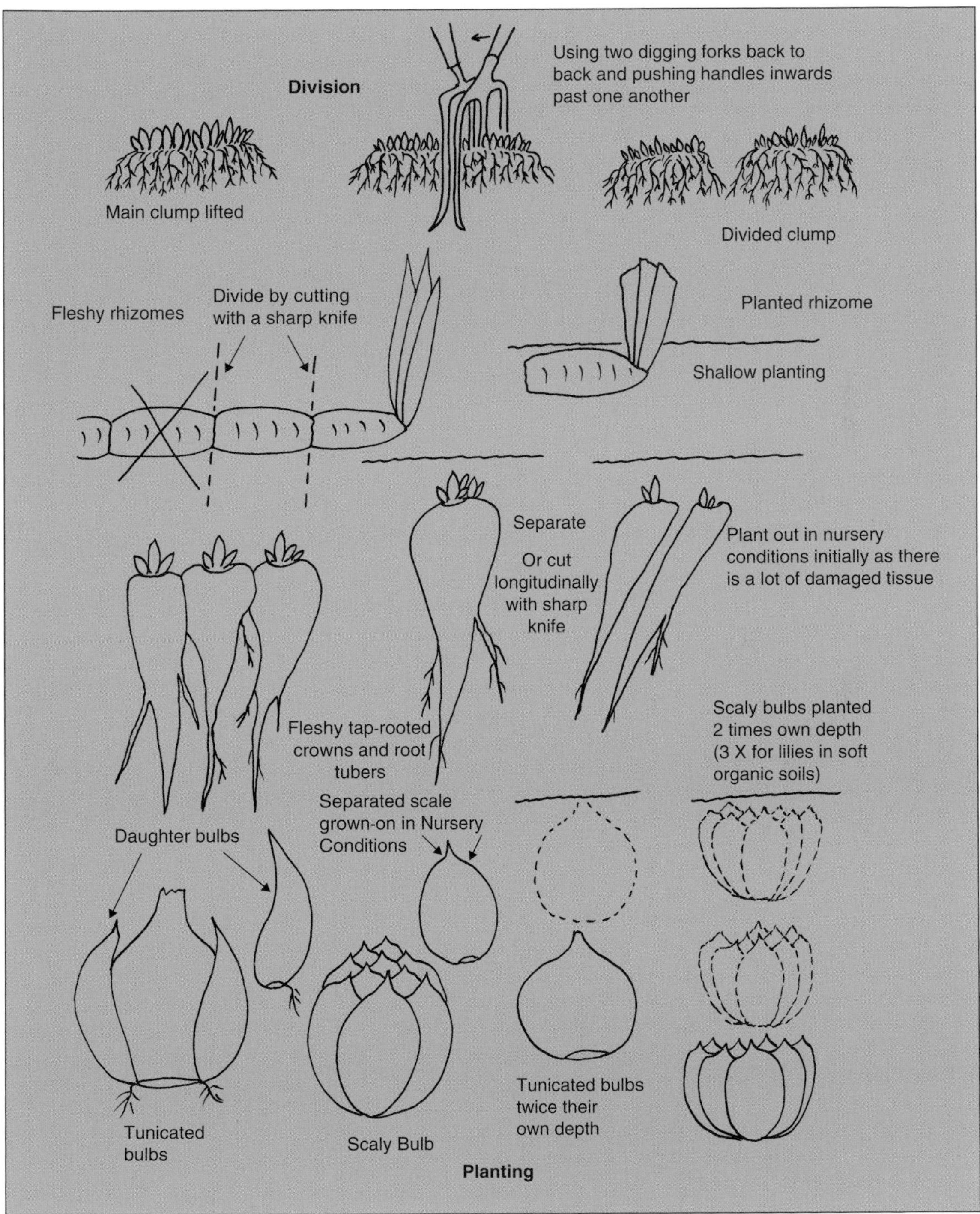
Division
Using two digging forks back to back and pushing handles inwards past one another
Main clump lifted
Divided clump
Fleshy rhizomes
Divide by cutting with a sharp knife
Planted rhizome
Shallow planting
Separate
Or cut longitudinally with sharp knife
Plant out in nursery conditions initially as there is a lot of damaged tissue
Fleshy tap-rooted crowns and root tubers
Scaly bulbs planted 2 times own depth (3 X for lilies in soft organic soils)
Separated scale grown-on in Nursery Conditions
Daughter bulbs
Tunicated bulbs
Scaly Bulb
Tunicated bulbs twice their own depth
Planting

Dividing and planting herbaceous perennials II.

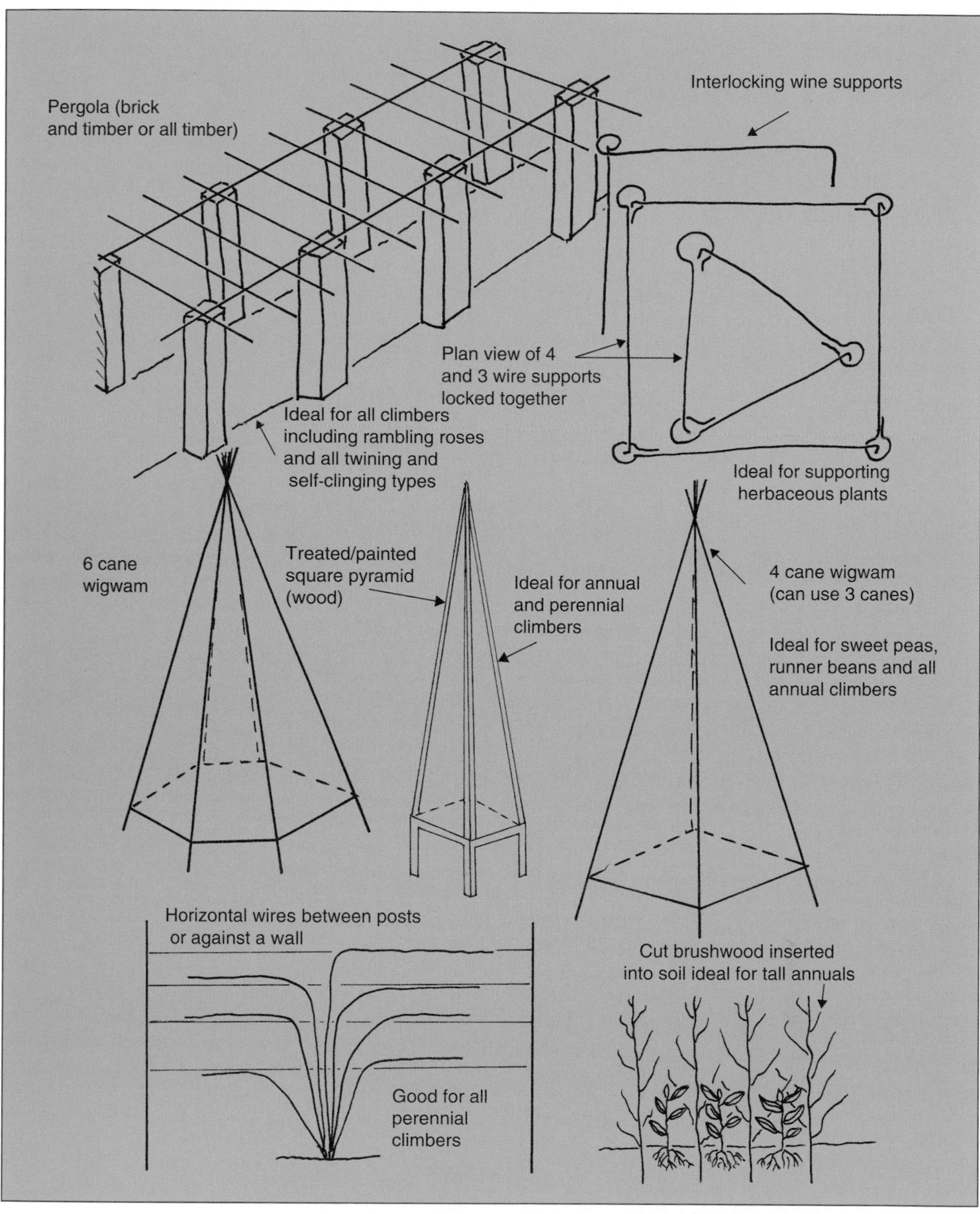
Pergola (brick
and timber or all timber)
Interlocking wine supports
Plan view of 4
and 3 wire supports
locked together
Ideal for all climbers
including rambling roses
and all twining and
self-clinging types
Ideal for supporting
herbaceous plants
6 cane
wigwam
Treated/painted
square pyramid
(wood)
Ideal for annual
and perennial
climbers
4 cane wigwam
(can use 3 canes)
Ideal for sweet peas,
runner beans and all
annual climbers
Horizontal wires between posts
or against a wall
Good for all
perennial
climbers
Cut brushwood inserted
into soil ideal for tall annuals

Plant support systems.

to this end – the more expensive types are often very aesthetic as well as functional.

Hardy Nursery Stock

Hardy nursery stock (HNS) comprises all types of hardy plants including biennials, herbaceous perennials (including alpine plants) and woody perennials. So HNS includes trees, shrubs of all types, wall shrubs and climbers, and hedging plants. Also included are hardy biennials, hardy herbaceous perennials and hardy rock garden plants.

The form of production (propagation) of nursery stock is relevant to cost, and even establishment. Forms include seedlings (young plants raised from seed), divisions (herbaceous species divided up into smaller segments), grafted plants, budded plants, layers (stems rooted whilst still attached to the parent plant), root cuttings and stem cuttings. A history of the growth environment and its subsequent treatment may be known (and is sometimes very important); terminology includes 'cell-grown' (produced in small, cell-like modules), 'transplant' (young plants lifted and transplanted), 'liner' (young plants to be lined out at a wider spacing), and 'container-grown' (spending its entire development in containers). 'Bare-root' plants are lifted with no rootball or soil around the roots, and 'rootballed' plants are lifted retaining a ball of roots and soil.

Seedling trees are young trees raised from seed and undisturbed since sowing. Seedlings vary in height depending on both the vigour of the species and the growth environment in which they have been raised, but are usually under 1 m high as one-year-old plants. Whips are the next size up from seedlings (therefore 1m and over in height), and are well developed, single-stemmed plants that may be seed-raised, cutting-produced, layered, grafted or budded plants. There is a cost difference between seedlings and whips because of the greater growing time, and also a greater difference in price between seedling whips and those produced by other methods, as production by seed is still the cheapest method and is reflected in the price. However, in some instances it is not possible to propagate trees that are 'true to type' by seed, so the more expensive production methods have to be employed.

Feathered whips are whip-size plants (1 m and over) but with well developed lateral branches ('feathers'). The cost differential applies again, depending on production time and method, with seed-raised being the lowest and grafting the highest.

Standard trees have been 'trained' by the removal of the lower side branches in the early stages of production. Their stem girth (thickness) may vary, but they are produced with a defined stem of 1.8 m clear of lateral branches. Standard trees are useful in situations where lower branches might create an obstruction – for example single-specimen trees planted in grass in public places such as parks and recreation areas, and also street trees.

Feathered trees are well developed young trees 1.8–3.5 m high that, unlike standard trees, feature retained lateral branches commencing near ground level. They may be propagated by any method (seed, cuttings, layering, budding or grafting), and are much larger than feathered whips.

Propagation of Hardy Trees and Shrubs by Seed

Propagation by seed is the only method resulting from a sexual process, and although having some limitations, nevertheless it has many benefits, including the relatively low costs involved in plant production, and the fact that plants produced from seed are usually vigorous, healthy progeny without inherent disease. Production by seed is very intensive, and many plants can be produced in a relatively small surface area. Difficulties include obtaining good quality seed of the desired species, the low viability experienced in the seeds of some species, and ensuring the progeny is 'true to type' and not influenced by the hybridization of similar and/or related species – cultivars cannot be produced from seed.

Individual trays or seedpans can be used for seed germination on a small scale, or for small seed lots on a large scale. There is a range of woody plants that have a higher temperature requirement and are sown under glass, but in general terms woody subjects are sown in outside seedbeds prepared approximately 1m wide, with a 'knocked-up' (raised) edge and a level top surface. The old 'Duneman' style of seedbed has a lot to offer because it incorporates forest litter into the sowing surface, thus inoculating it with mycorrhiza. However, it is difficult to find a legitimate source of forest litter on a large scale, which does not cause detrimental environmental effects when harvested.

Seeds are sown broadcast on to the prepared surface of the seedbed, and covered with approximately 5–15 mm of 3–5 mm grit (depending on seed size). The grit keeps the surface open and well aerated and prevents surface capping or the invasion of mosses and liverworts, and helps to retain moisture in the germination zone within the soil. It also reduces – although does not prevent – weed infestation, and as long as weeds are not allowed to

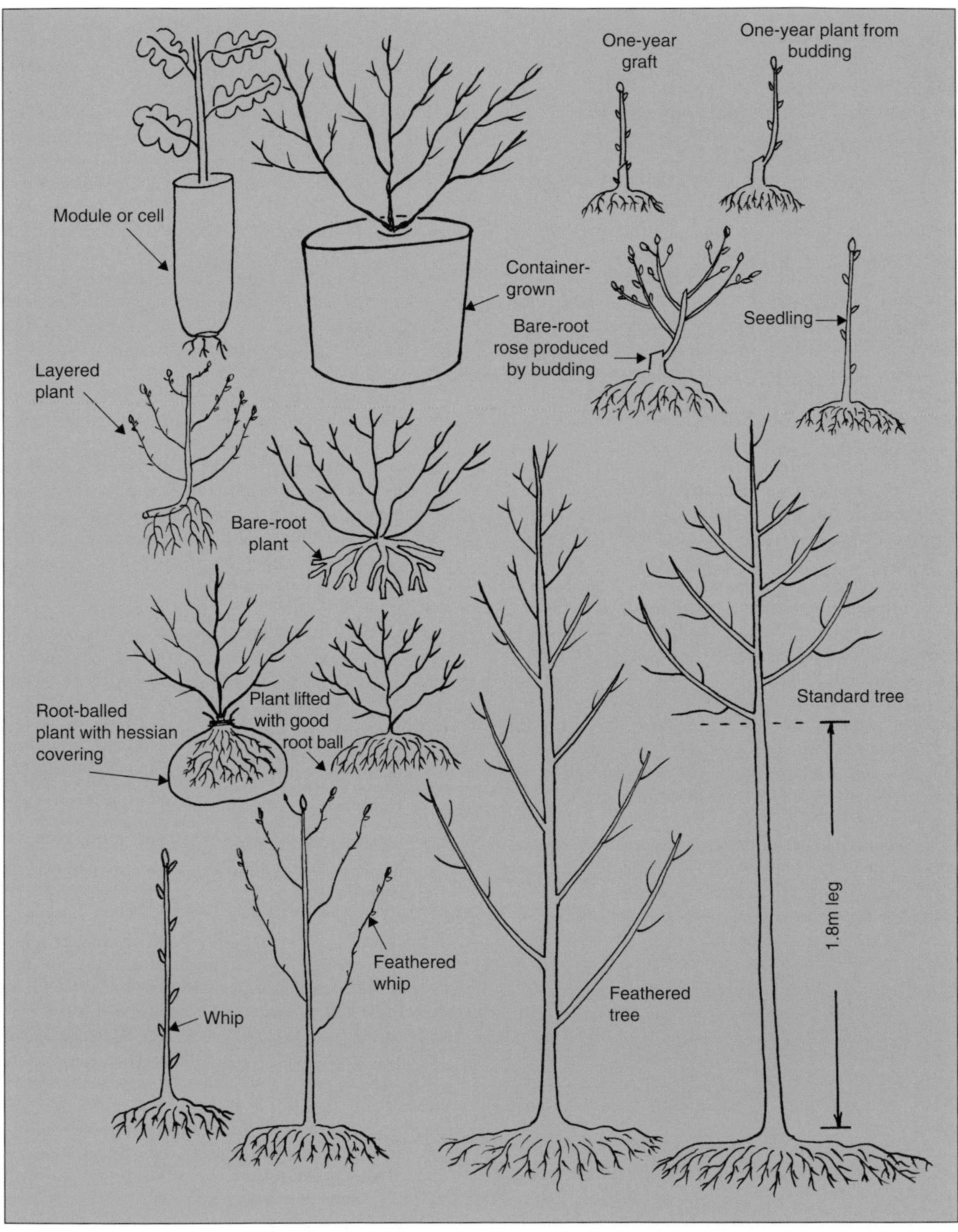

Stock sizes and types.

establish themselves too well, enables reinfecting weed seedlings to be pulled easily from the top surface.

Outdoor seedbeds are protected against birds and mammals by polyethylene netting draped over galvanized or polyethylene tubes tunnel-style. Some seedbeds feature an overhead spray line hung below the tunnel hoops that can be connected to a standpipe and turned on manually in dry periods. Sown seedpans are protected in cold frames, but without glass covers.

Methods of Breaking Dormancy

Unfortunately, good viability does not necessarily mean good germination. Often, plant species will have mechanisms that delay or prevent germination, resulting in inconsistent germination (or even non-germination) – in some woody species sporadic germination of untreated seed over one to three years is not uncommon. Breaking the dormancy of these species is particularly important in order that large outside seedbeds are not tied up for more than one season by poor uniformity of germination.

Production by seed is sometimes difficult just because the vast range of different species have different germination temperatures and conditions for success, and knowledge of these essential conditions may be sketchy in some instances. Sometimes growth inhibitors (hormones), such as abscisic acid, inhibit germination, and/or retard the growth and development of the hypocotyls, epicotyls and radicles of germinating seeds. Some species germinate best if harvested 'green' (the fruit has to be completely developed, but not allowed to go brown and desiccate) – however, from a practical standpoint this is sometimes difficult to arrange. The seeds are then sown as soon as possible after harvesting, as this reduces the build-up of natural inhibitors and prevents double dormancy (two-year dormancy) in susceptible subjects. Many *Meconopsis* species (Himalayan poppies) germinate more successfully if the fruit capsules are harvested 'green' and the seeds are sown immediately after harvesting: for example, *Meconopsis betonicifolia.* Also, *Acer capillipes*, *A. griseum*, *A. ginnala*, *A. japonicum* and *A. nikoense* all respond well to being sown green. *Camellia* species such as *Camellia sinensis* and *Camellia japonica* are sown directly after harvesting, but in a heated glasshouse rather than outside.

Various methods of breaking seed dormancy may be used, and are usually carried out in areas other than the seedbed, which prevents the seedbed from being tied up too long with non-germinating seeds.

Stratification is a pre-sowing treatment where alternate layers of seed (or harvested fruits) and sand are placed in a container (with drainage holes in the base). The container is then placed in an uncovered outside frame so that it is open to the elements, and after a prescribed period of time (usually six to eighteen months, depending on the species), the seed and sand mix is sown on to a seedbed where it will germinate quickly. The name 'stratification' accurately describes the strata-ed layers of seed and sand, but unfortunately does not describe any of the processes involved within the seed.

Stratification allows a resting period for the seed during which time the embryo can develop fully (if it has not already done so), and both food substances and growth inhibitors can break down in preparation for germination. A period of cold temperatures experienced in the outside conditions in winter will satisfy the low temperature requirements (vernalization) of some seeds. In the warmer conditions of summer, bacteria will start to break down the testa (seed coat) in preparation for both water uptake and ease of radicle penetration. The fruits of *Arbutus unedo* (strawberry tree) are harvested in the autumn and then stratified for six months prior to pulping, separating the seeds, and sowing in a heated glasshouse. The tough seeds of *Magnolia* species, such as *Magnolia sinensis* and *Magnolia wilsoni* are also sown in a heated glasshouse, but with an eighteen-month stratification period first. Stratification for six months is used on *Cotoneaster* species such as *Cotoneaster frigidus*, though after separating the seeds they are sown in outdoor seedbeds, and likewise for *Sorbus* species such as *Sorbus vilmoriniana*, and *Berberis* species such as *Berberis darwinii* and *Berberis wilsoni*. The fruits of holly species are stratified for eighteen months before the very hard seeds are separated and sown into an outside seedbed.

Chilling is a process whereby seeds are placed in moist peat, left to imbibe water for twenty-four hours, then placed in a domestic refrigerator (at about 3–4°C) for ten to twenty-one days (depending on species). Because layers (or really a mixture) of moist peat and seed are used, the process is sometimes described as a form of stratification. However, chilling only really fulfils the cold temperature (vernalization) requirement, and does not fulfil the other criteria associated with stratification. Many species of maple (*Acer*) and beech (*Fagus*) respond to chilling at 1–4°C for ten to fourteen days as a pre-germination requirement.

Chitting (not chilling) is a process whereby seeds are pre-germinated before sowing in order to give them the right conditions to commence

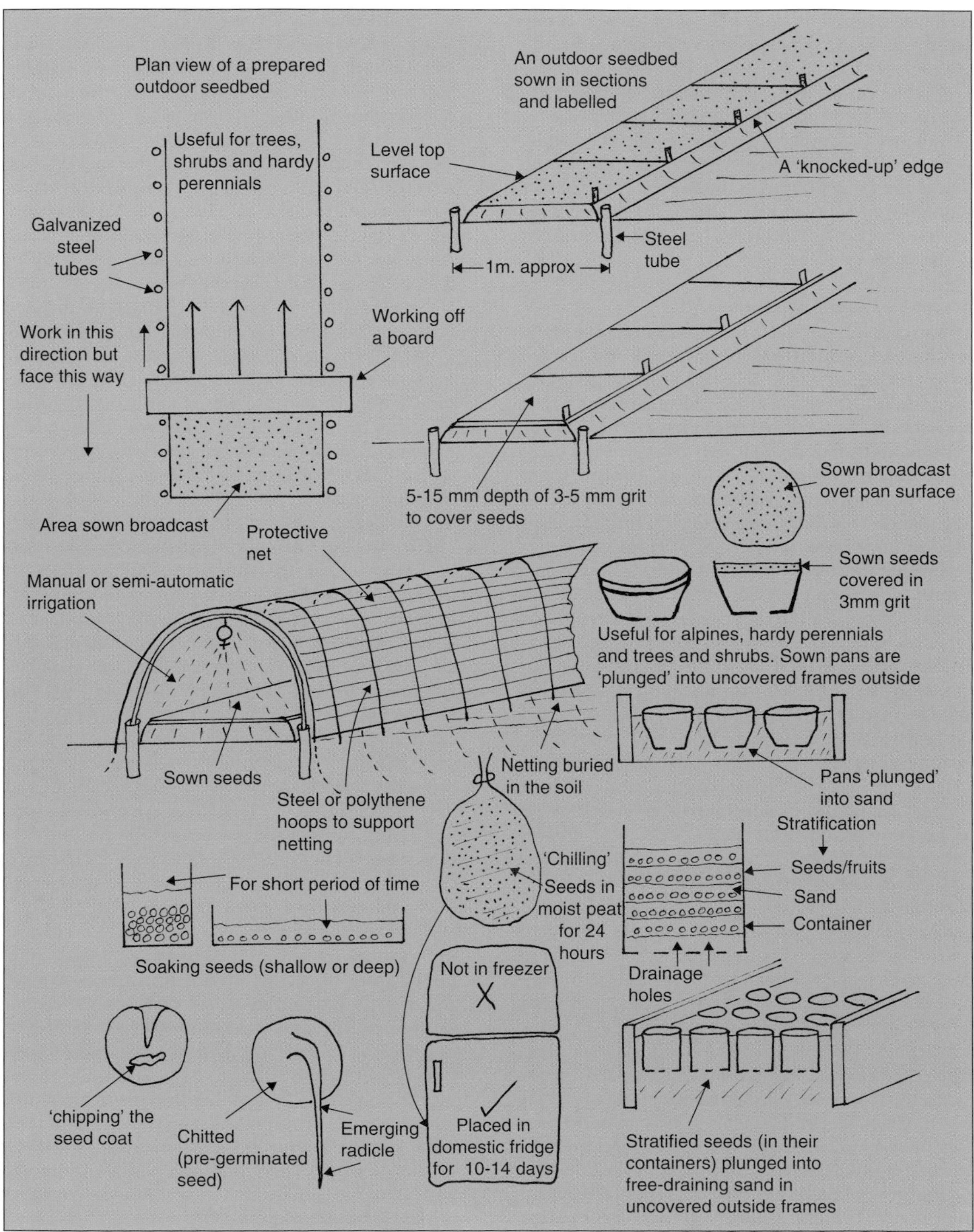

Preparing and sowing outdoor seedbeds and seedpans, stratification, soaking, chipping and chilling.

germination, and to give them a good start before they go to their sowing positions. Chitted seeds are sown with their radicles already emerged/ emerging from the testa, so must be kept moist directly after sowing. The process of chitting may be carried out in moist peat, or in inert sowing media such as vermiculite or perlite, on moistened cloths, or in specially prepared gels. The term is also used to describe the process of inducing shoots from 'seed potatoes' (these are not seeds, but selected stem tubers used to produce a new crop). In this instance the chosen potato tubers are stored in the dark to induce etiolated new shoots (long internodes and soft, non-pigmented tissues) in slatted crates to allow good air movement and reduce the risk of fungal diseases.

Soaking involves immersing seed into warm or cold water for two to twelve hours (depending on the species and toughness of the testa*); this ensures good water uptake through the micropyle of species with tough, waxy testas. The resulting engorged cells of the interior spongy layer of the testa then split the outer testa, thereby greatly increasing the rate of water intake further. Canna lily or Indian shot (*Canna indica*) and sweet peas (*Lathyrus odoratus*) are often treated in this way, with good results.

'Chipping' means to physically damage the seed coat, and can be carried out on a small scale on seeds that have very tough testas. A sharp knife is used to carefully expose internal tissue for better water uptake – though particular care must be taken not to damage the prominent radicle (or the rest of the embryo) during the process. On a large scale, 'scarification' is achieved by wearing away the testa of the seeds by the use of abrasive belts, or in a rotating drum with an abrasive powder added. For seeds with a very tough testa, such as canna lily, both chipping and soaking for a short period may be used. All forms of scarification leave the testa capable of better water and oxygen uptake, and improve the possibility of radicle penetration. However, there needs to be stringent and systematic control, with regular sampling to gauge the degree of degradation of the testa, or irreparable embryo damage could result.

Soil Preparation for Planting Trees and Shrubs

Shrubberies may be established by planting at high densities. High-density plantings are expensive, and often use 'fillers' (cheap shrubs) in the short term that are either culled or lifted (to use elsewhere) when thinning to the required ultimate density is required. Alternatively, top quality shrubs are planted out at their final spacing, allowing them room to develop *in situ*. It is traditional to plant new shrubs into areas prepared as beds (or borders), and trees into individual planting pits. We know that cultivated soil invites both weed infestation and regular reinfestation by seedling weeds, yet it is customary to prepare large beds/borders of soil to establish shrubberies: ironically it then becomes a struggle (and expensive) to keep such large amounts of cultivated soil free of weeds. Moreover we often over-prepare with regard to the surface area of disturbed soil. In fact shrubs – and any trees used within group plantings – could be pit-planted into cylinders of prepared soil cut into grass, just as you might for specimen standard or feathered trees.

The benefit of this is that the main surface area of the proposed bed can be maintained by mowing (using the versatility of an air-cushion or other rotary mower), which is relatively low in labour intensity, and only the circles of the pit-planted material need to be kept weed free. As the woody plants develop they will kill off the surrounding grass, leaving less and less to mow, and more and more soil that remains weed free within their increasing shade. The mulched circles are cut out wider as the plants grow, and can be kept free of weeds with a good depth of mulch. Furthermore, as the woody plants gradually shade out the areas around the initial circles, they also shade out the front edge of the site, leaving a natural-looking undulating edge. As the grass area diminishes, mowing becomes less necessary, and eventually stops altogether.

Ultimately the desired planting will establish with relatively little maintenance required, as the mulched area around the individual trees and shrubs can be expanded annually to retain the weed-free environment. Systems like this do away with the need to cultivate (and subsequently keep weed free) the very large areas of 'bare' soil associated with traditional preparation methods. Furthermore, the system lends itself to conversion to non-soil-inversion policies after planting. A very good alternative is to plant through a sheet mulch.

Factors Affecting the Successful Establishment of Trees and Shrubs

The key factors for the successful establishment of trees and shrubs are good quality, well grown plants with good root-to-shoot ratios, good handling conditions prior to planting (for instance, keeping rootballs moist), good planting technique, and essentially, good aftercare (particularly in the

*Soaking for excessive periods kills seeds.

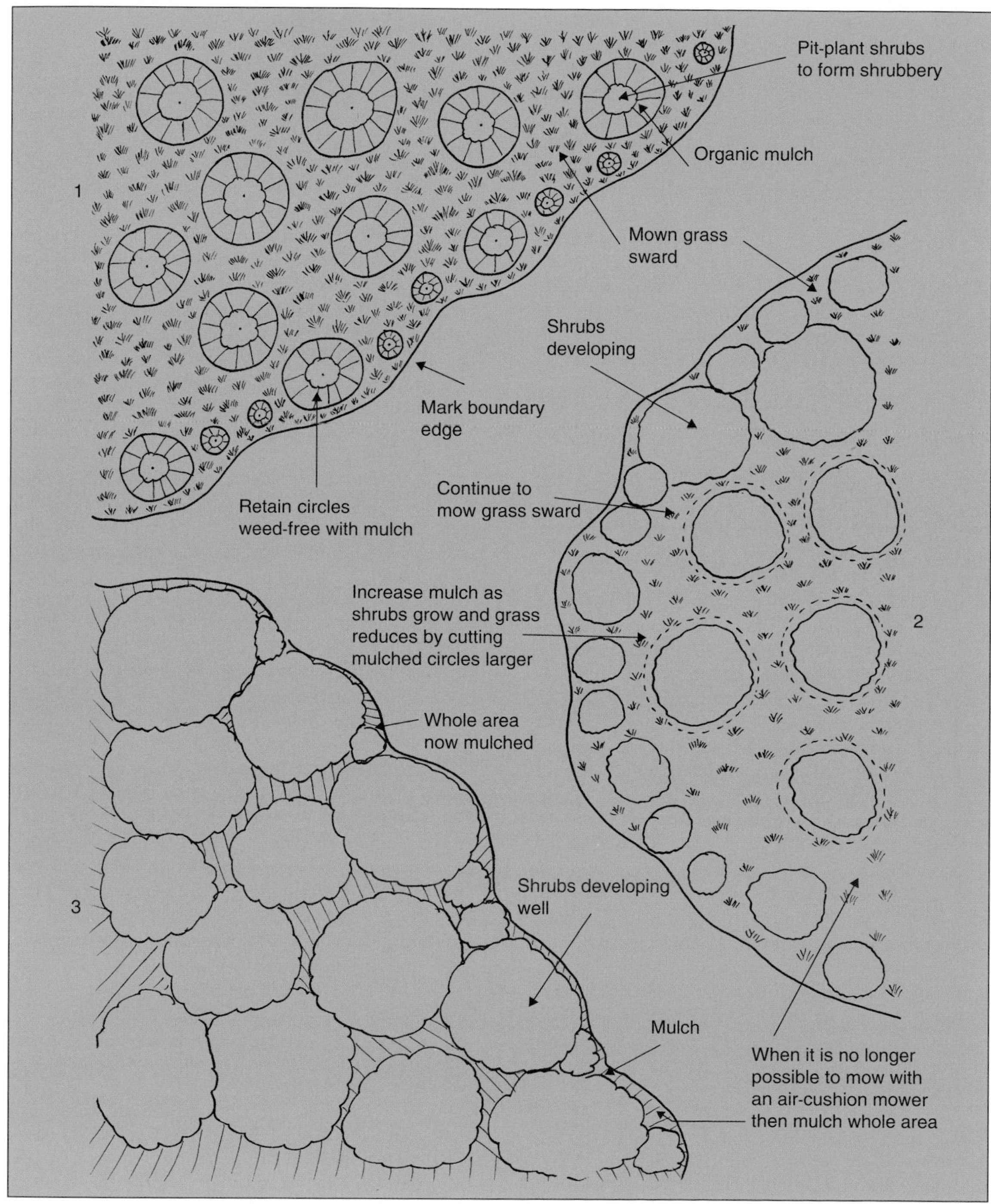

Pit-planting into a grass lawn as a possible way of establishing a shrubbery.

first three to four seasons after planting). However, the following all influence success: the correct choice of species for the site, the choice of stock sizes and types (whether bare-root, rootballed, cell-grown or container-grown), and the correct handling of stock prior to delivery, during delivery, and whilst on site. Furthermore planting in the correct season, good site conditions, the availability of irrigation, the essential use of mulches and/or chemical weed control, and adequate protection are also factors that will affect the plants' establishment into the new site and soil, their subsequent growth, and ultimate survival. Success in establishing trees and shrubs (purchased at any phase of their development) is very much dependent on good management after planting. However, nursery stock size will also have a bearing.

Root-to-Shoot Ratio

If all other factors of soil and aerial environment are equal, and the aftercare is correct, then smaller plants (young shrubs, seedling trees and whips) have a greater chance of survival than larger stock sizes, because their root-to-shoot ratio is so good by comparison. Small stock sizes have relatively little disturbance and damage to the root systems at 'lifting', and they retain good proportions of absorptive root. Larger stock sizes have much poorer root-to-shoot ratios, and although systems of 'preparation' in the nursery attempt to redress the balance, it is nevertheless difficult to do so.

The ratio becomes very important in the spring when the young plants increase their water demand due to high transpiration rates. In the first spring after planting, if the root system was damaged when it was lifted, and has not produced sufficient new lateral roots from the cut surfaces (with the fine absorptive roots and root hairs associated with this), then insufficient water will be transported to the breaking leaves at bud-burst. The result is a water deficit, and the new growth will wilt temporarily. If this continues for protracted periods, tissue desiccation occurs, which will lead to death in some young tissue, or even complete death of the tree or shrub.

'Lifting' Trees and Shrubs

Because of their extensive root systems, whenever shrubs or trees are 'lifted' from the nursery (or from other sites in the garden) they will necessarily lose some roots. It is impossible to lift woody plants from their original site to replant elsewhere without doing some root damage – the extent of which is obviously important when considering the root-to-shoot ratio and its influence on successful establishment.

The differences between fibrous-rooted species and species with thick, fleshy roots are fundamental. Although all the parameters already mentioned affect any plant's ultimate establishment, fibrous-rooted subjects – such as most rhododendrons, camellias, kalmias and many ornamental conifers – are the easiest to 'lift' from their original positions, and the easiest to re-establish afterwards. They are far more likely to retain a greater proportion of the important absorptive roots at lifting because they lift with a rootball with roots neatly clumped together in one concentrated area. Furthermore, there is also a greater likelihood that these species will have relatively shallow root systems when compared with types with thick, fleshy roots.

Far-reaching, thicker-rooted types will almost certainly leave a large proportion of their roots in the soil at lifting, as their root depth and structure means that the angle of the spade at that depth will cut off the roots, leaving large amounts still buried in the soil. Roots with relatively large diameters are severed, and re-establishment can only occur after new, more absorptive lateral roots have been produced from, or near, the cut surfaces. The diameter of severed roots on fibrous-rooted subjects is far less, and the relatively juvenile material that is damaged, repairs and forms lateral root branches very readily. For this reason alone it is critical to plant trees and shrubs at favourable times.

Planting in the Optimum Season

Planting in the correct season is very important because of the high moisture requirement for both evergreen and deciduous trees and shrubs in the spring. Slightly higher soil temperatures, compared with air temperatures, are found in the autumn, so root action continues in the autumn for up to two weeks after shoot growth has finished. A similar situation is true in the spring – root growth commences earlier than shoot growth. Winter (if soil conditions are conducive) and spring may or may not give sufficient time for severed roots to form new laterals for absorption, prior to the maximum water demand.

Autumn is therefore the optimum season, especially for bare-root deciduous species – usually late October to late November. The earlier in this time frame the better, as the double period of root growth and establishment, prior to the development of moisture-demanding leaves in the spring, gives a far greater chance of establishment. Hardy evergreens such as hardy conifers and holly can be

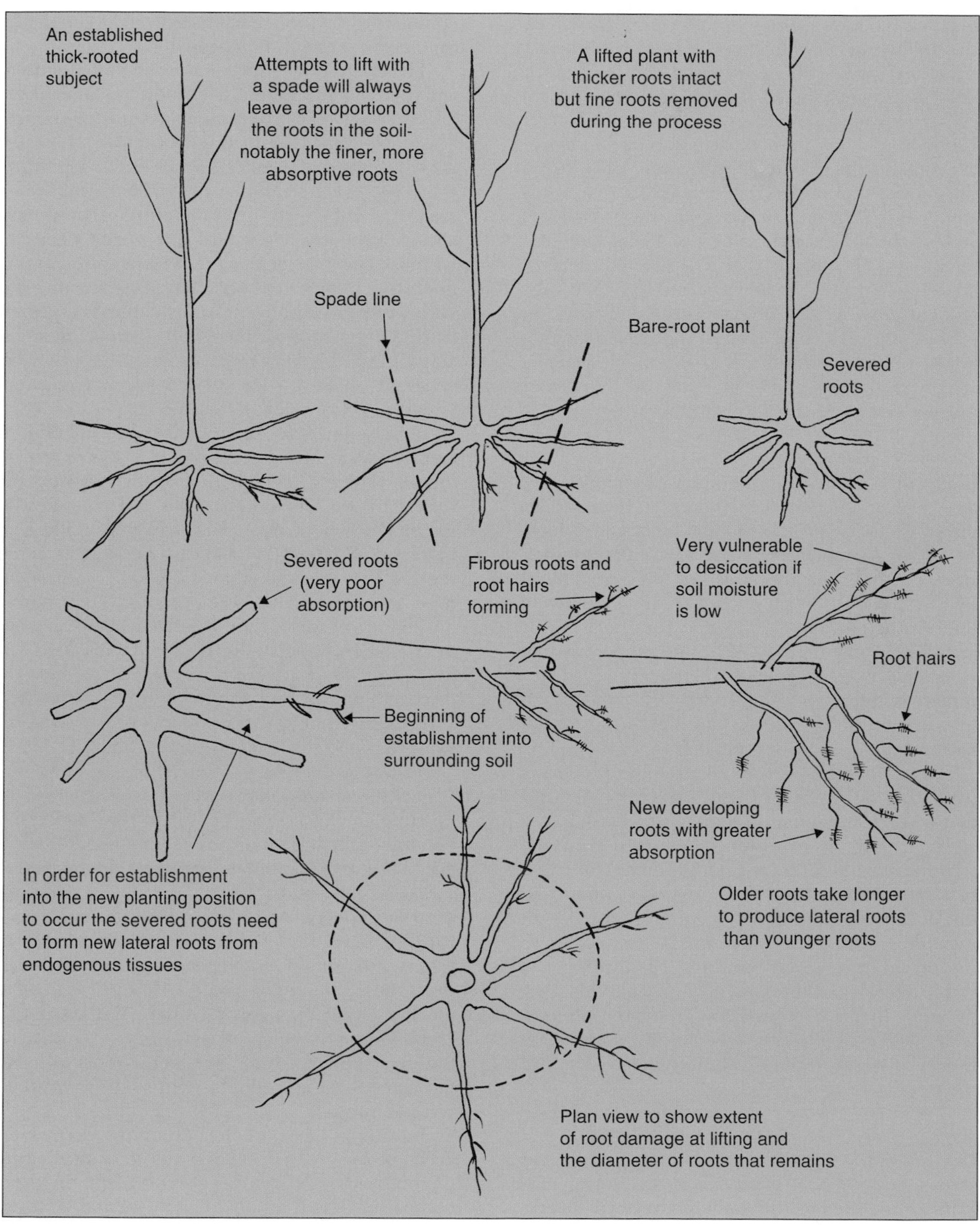
An established thick-rooted subject
Attempts to lift with a spade will always leave a proportion of the roots in the soil- notably the finer, more absorptive roots
A lifted plant with thicker roots intact but fine roots removed during the process
Spade line
Bare-root plant
Severed roots
Severed roots (very poor absorption)
Fibrous roots and root hairs forming
Very vulnerable to desiccation if soil moisture is low
Root hairs
Beginning of establishment into surrounding soil
New developing roots with greater absorption
In order for establishment into the new planting position to occur the severed roots need to form new lateral roots from endogenous tissues
Older roots take longer to produce lateral roots than younger roots
Plan view to show extent of root damage at lifting and the diameter of roots that remains

Cutting through tree and shrub roots at lifting.

planted at any time from October to March if conditions are suitable. However, evergreens from warmer climes (such as *Laurus nobilis* – bay) prefer an early spring planting so that they grow into a warming (rather than a cooling) soil – ideally February to March (however, more realistically this would be January to April). Times after this are too dry. Indeed all types, both deciduous and evergreen, can suffer serious damage and heavy losses in an unusually dry spring. For successful establishment, the root system of the newly planted stock needs to penetrate, infiltrate, and establish in the surrounding soil of the new site. Only then can the roots take up water efficiently and forage for moisture within the soil matrix before the foliage demands a lot of water via evaporation. Excessive moisture demands of the foliage from a poorly established root system causes stress, and if suffered for protracted periods, will cause death.

There are many modular systems of production for small stock sizes designed mainly for larger project use. Modularized or cell-grown plants aid the establishment of stock in the planting season, and also allow you to extend the planting season under some circumstances. However, even planting from modules out of season may require an irrigation system, no matter how basic, to ensure survival – especially in a dry spring.

The quality of container-grown plants is obtained from good potting media, good watering and feeding, good root establishment in the containers during the production process, and good root systems for onward planting. Container-grown trees and shrubs, if correctly produced, suffer very little disturbance to the root system at transplanting, so if planted at the correct time, and with correct aftercare, establish very successfully. They are also regularly planted out of the normal planting season, where even here their success rate is good. However, they can still be lost if irrigation is not carried out in dry years, or dry times of year, during their establishment phase. The soil at the planting station must be kept moist, and directly irrigated after planting and for some weeks afterwards; keeping the plants weed free and mulched after planting is an added insurance for success.

Ordering, Delivery, and Inspection for Quality of Stock

Plants must be ordered in plenty of time so that the delivery date precedes the correct/optimum planting time. Failure by suppliers to lift/prepare your order and deliver on time because of the volume of orders at this peak time of year is unfortunately common, and it accounts for the increasing number of enforced winter and spring plantings. On arrival, plants must be inspected to ensure they are in good health, have as good a root-to-shoot ratio as is possible, that the stems are not desiccating or dying, that buds are not breaking ('flushing'), and that during storage, transport and delivery the roots have remained alive. Loads should be delivered in a vehicle with covered sides, or be covered by a tarpaulin in an open-backed truck. Every effort must be made by the supplier to keep the roots of the lifted trees and shrubs moist, by covering them at all times from 'lifting' at the nursery to delivery on site – at which stage you take over the responsibility for their welfare.

Once delivered, the plants (in their packaging) should be kept out of direct sunlight so that the bags do not sweat and heat up. The plants may be removed from the delivery bags and be 'heeled in' to keep them moist until planting. 'Heeling in' bare-root and root-balled plants in trenches is a good method of temporary storage, and the trenches may be prepared in advance and protected from the weather in readiness for the arrival of large consignments. In heavy soils that are difficult to work, enclosed areas with raised edges of wood can be constructed and filled with peat, prepared soil, or spent hops to ease the heeling-in process. Plants arriving in frosty weather can cause problems, and it is essential that roots are covered with either peat, tarpaulins, plastic or hessian sheets/sacking to prevent them drying out, and they should then be stored in a frost-free shed or in a refrigerated cool store, or be heeled in to frost-free soil.

Protection of Roots on Site

It is essential to ensure that plant roots do not dry out whilst actually on the planting site. The roots of young stock (indeed the fine, fibrous, water-absorbing roots of any size stock) dry out very quickly (sometimes within twenty minutes of exposure). Trees and shrubs must be heeled in, or have their roots covered with wet sacks or polyethylene sheets (in the shade) to prevent desiccation.

Planting Shrubs, Standard and Feathered Trees

Pit Planting

Pit planting can be used for seedlings, whips, shrubs, feathered trees and standard trees, and rootballed and container-grown trees and shrubs.

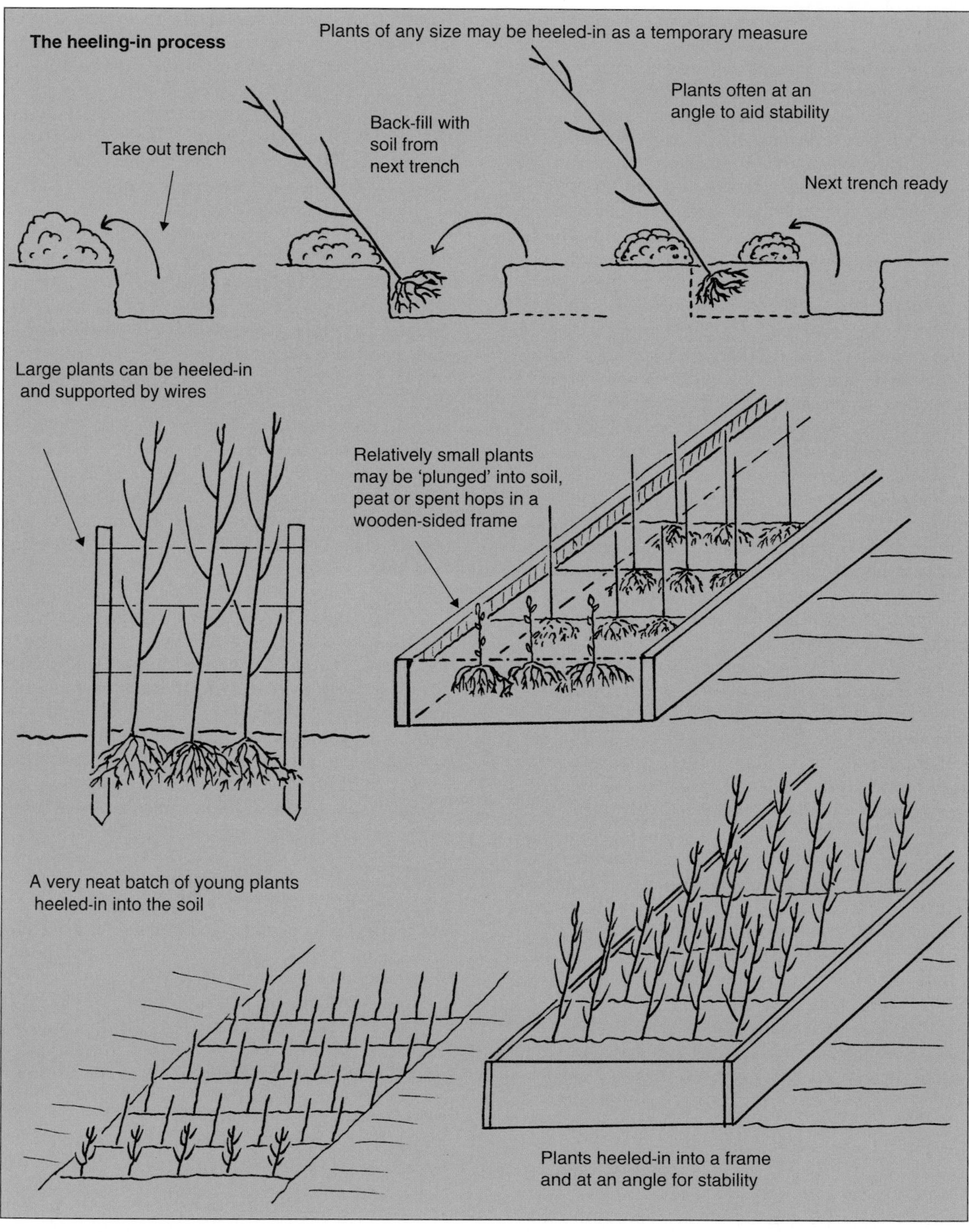
The heeling-in process
Plants of any size may be heeled-in as a temporary measure
Take out trench
Back-fill with soil from next trench
Plants often at an angle to aid stability
Next trench ready
Large plants can be heeled-in and supported by wires
Relatively small plants may be 'plunged' into soil, peat or spent hops in a wooden-sided frame
A very neat batch of young plants heeled-in into the soil
Plants heeled-in into a frame and at an angle for stability

Heeling in and temporary storage.

Traditionally, cylinder-shaped planting pits have been used. However, it is easier to keep the shape of a square-sided pit forming a cube when digging out, and this shape has also proved very successful for tree establishment. Large pits with lots of cultivated soil are preferable as they aid rapid establishment. But in general terms, prepared planting stations need to comfortably accommodate the root system of the tree or shrub to be planted, so inevitably the size of planting pits varies, and depends on the situation and the stock sizes used. For small feathered trees and light standard trees a complete cylinder (or cube) of soil is removed, with pits 1 m (even up to 1.5 m) wide and 0.5 m (even up to 0.75 m) deep. If drainage is in any way suspect the bottom of the pit should be forked over, and trees and shrubs should have at least 75 mm of good topsoil beneath their roots.

Normal horticultural practice, when planting individual specimen trees or large shrubs, is to strip off lawn turf (or other surface vegetation) and, after forking the bottom of the pit, to incorporate the turf into the lower part of the pit, chopping it up with the spade before partially backfilling the pit with topsoil. Because planting pits have a different compaction factor to the surrounding soil they can easily become water sinks for the surrounding area, so it is essential to ensure they drain freely and to consolidate the backfilled soil well at planting.

Planting fibrous-rooted subjects such as rhododendrons or camellias into woodland soils (with good amounts and depths of organic woodland litter) does not usually encounter excavation or drainage problems – and good natural infestations of mycorrhiza aid plant growth and establishment. Good garden soils that have been regularly cultivated also give good planting conditions with few problems. However, not all sites are like this, and in reality, soils that are heavily compacted and low in mycorrhizza are regularly experienced. Spoil comprising subsoil or inferior topsoil should be discarded, and levelled out on large rough sites, or in a more formal situation should be removed from the site completely. Furthermore, tree and shrub planting into lawn surfaces should be carried out using polyethylene sheets over the grass to prevent damage.

The nutrient additive traditionally associated with planting trees and shrubs has always been bonemeal, which releases its phosphates after bacterial action breaks it down. Because bone is such a tough material its breakdown is often a very slow process; even bone flour is slow to break down, but large coarse 'chunks' of bonemeal decrease the speed of breakdown even further. There is a good chance therefore, that bonemeal mixed with the planting spoil will not be available as soluble phosphates when the developing/establishing roots need them. Bonemeal placed at the bottom of the planting hole takes a long time to break down, and will almost certainly leach away from the root zone level when it does. The benefit of its inclusion, therefore, is doubtful. Perhaps fine bone flour, not coarse bonemeal, in the top third of the backfill may give the best results, or more soluble forms of phosphate spread on top of the planting station.

Fertilizer additions may be needed in cases of known low fertility, definite deficiencies, and for subsequent healthy plant growth after establishment. But unless there is evidence to the contrary, it is generally accepted that there is sufficient nutrition in the soil to support initial root establishment.

Backfilling the Planting Pit

Ideally, only good quality topsoil should be used for backfill, and in most circumstances the site will provide it. In rare cases soil may have to be imported on to site, or alternatively ameliorating materials can be added to improve the existing soil. However, although soil-structure ameliorants (soil improvers) by definition will 'improve' the soil, the addition of such materials has not been shown to improve initial plant establishment in trials.

When planting standard trees, prior to backfilling the excavated planting hole, a short stake is driven into the base of the prepared hole, just offset from the centre. For all trees and shrubs set the plant into the pit, spread out the roots, and place friable (easily worked) soil over the roots. During the process, shake the plant gently up and down to ensure that all air pockets and voids are filled, and that the roots are encased in moist soil. As the backfilling proceeds, consolidate the soil by carefully, yet firmly, 'heeling' around the plant, taking care not to damage the roots, stems and soil structure.

The planted subject must be at the correct planting depth, and final depths should ensure that soil covers the root system completely, but does not encroach a long way up the stem. This is achieved by ensuring that no more than 50 mm for most trees and shrubs, and 100 mm for very large trees and shrubs, covers the uppermost roots at the root to stem interface. To ensure the planting level remains correct, all surplus soil is removed, and not mounded up around the plant base. The finished soil level of the pit should be slightly above the existing ground level: 50 mm is adequate to allow for settling, and ultimately leaving a level finish to permit percolation of water to the roots.

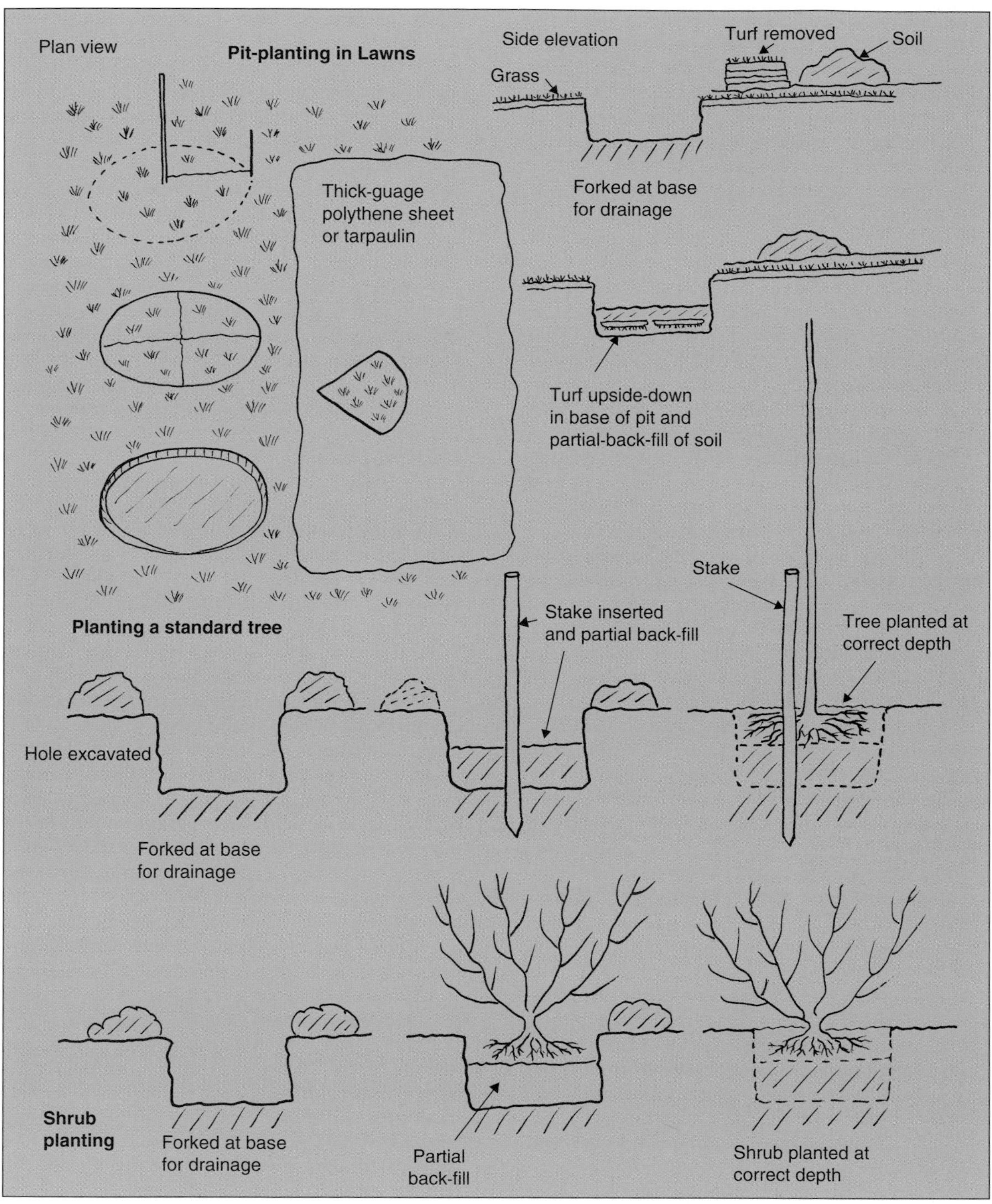
Plan view
Pit-planting in Lawns
Thick-guage polythene sheet or tarpaulin
Side elevation
Turf removed
Soil
Grass
Forked at base for drainage
Turf upside-down in base of pit and partial-back-fill of soil
Stake
Planting a standard tree
Stake inserted and partial back-fill
Tree planted at correct depth
Hole excavated
Forked at base for drainage
Shrub planting
Forked at base for drainage
Partial back-fill
Shrub planted at correct depth

Tree and shrub planting.

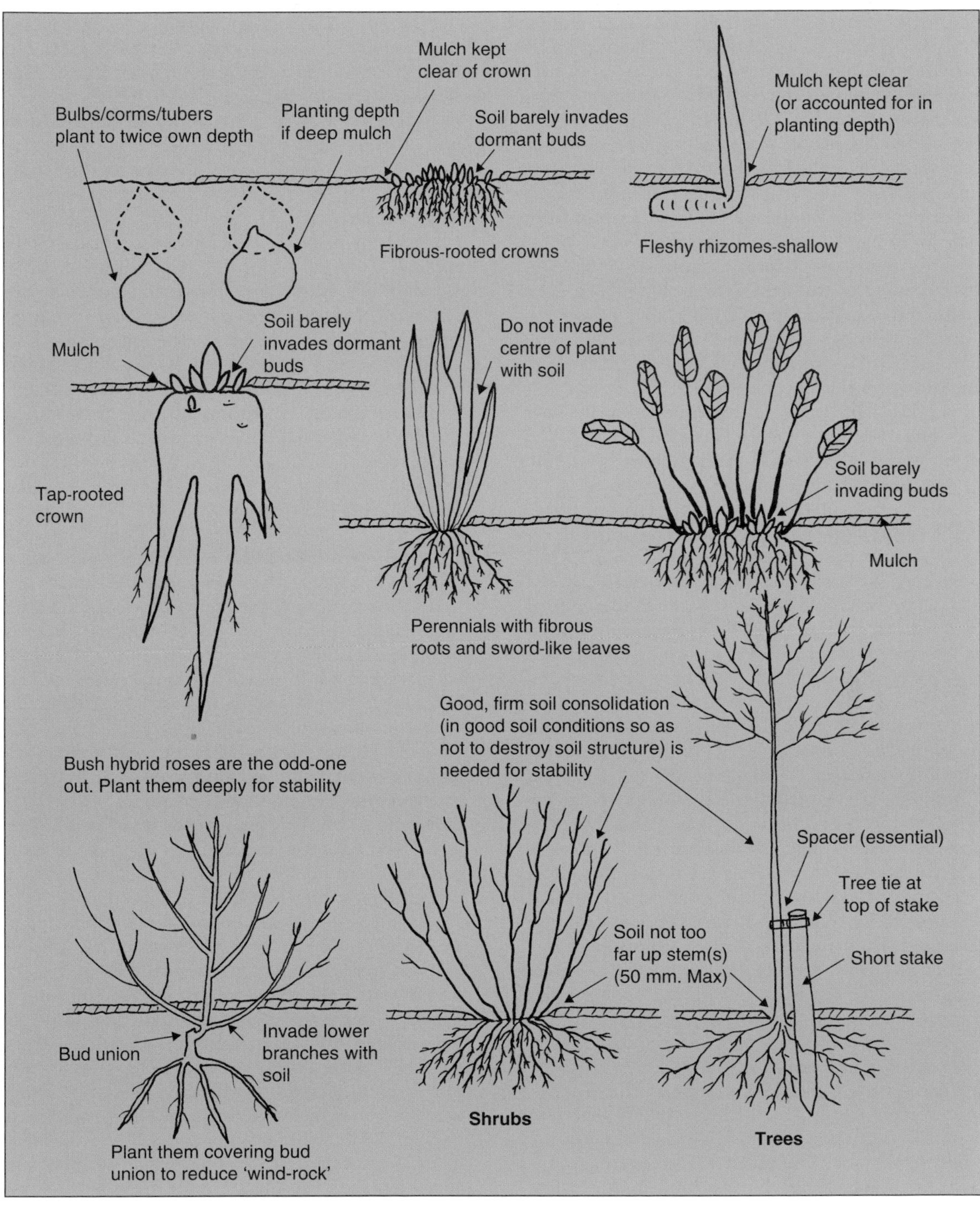

Comparing planting depths.

On very free-draining soils, and sites with a known drought history, 'finished' planting levels can be low and 'saucer-shaped' to aid moisture retention from rain and to facilitate efficient irrigation applications. Basic irrigation systems, comprising perforated plastic pipes to facilitate the swift movement of applied water down to the root system of the tree, can be installed at planting, or even permanent hand-operated (or semi-automatic) systems using products such as leaky hose.

The depth of planting is critical. If fibrous-rooted subjects are planted excessively deep they either die (which is unusual if all other factors are correct), or they establish and form a second rootball on top of the first. The effect of this may never be known if the tree or shrub is left *in situ* and never disturbed again. If, however, the plant is in the nursery for sale, or it is decided to move the plant within the garden (or landscape site), the operation is far more difficult than expected.

In the case of deciduous subjects with thick, fleshy roots, larger amounts of root are more likely to be left behind at lifting than fibrous-rooted types anyway. And those that have been planted too deep initially (in the nursery, or in the original place in the garden) often suffer because the stem tissues that are normally exposed to the atmosphere and are now under the soil become etiolated and soft. This is a precursor to forming adventitious roots to supplement the existing root system, and those that are not planted quite so deep may therefore be successful in this aim and survive well. Those that are planted much too deeply have too great an area of etiolated tissue, and water is absorbed over a long distance up the stem. The effect is known as 'flooding', as water coming in from the soft tissues puts pressure on the normal vascular water system and works against it. Often trees and shrubs will die in these circumstances – particularly those species that do not have a great propensity to produce adventitious roots from stem tissues. Those that do survive have a two-tier root system, the depth of which encourages an even greater proportion of the root system to be left behind when the plant is relifted. The depth of insertion of the spade necessary to lift the twin rootballs means that extensive damage will be done to the lower one. The situation is mitigated somewhat by the fact that the top (newer) rootball is probably the most efficient due to its relative juvenility and greater amount of good functioning tissue. Nevertheless, more damage to the rootball will be done than if it had been planted at the correct depth in the first place.

The key to successful 'lifting' is to cut all the way round the root system with the spade, and drive the spade in vertically and as deeply as possible before attempting to lever the plant from its original position. The depth of the rootball will obviously influence the success of this operation, and two-tiered rootballs exacerbate the difficulty. To lift specimens with a larger rootball, a vertical-sided trench is dug all the way round the root system prior to any upward leverage being applied.

After planting, any damaged branches are removed, and any necessary formative pruning, such as the removal of double leaders on trees, and crown balancing, is carried out. Research has shown the importance of mulches for tree establishment, and is extrapolated to shrubs, so an organic mulch should be applied at this stage. Shrubs do not normally require support (only in rare instances), nor do seedlings, whips or feathered whips, but standard and feathered trees do.

Protection from Mammals

In streets and pedestrian precincts, tree guards are desirable to protect the stems of trees, and tree grilles can be placed over the root area as a means of reducing soil compaction. In some rural areas and parkland, stock guards (cattle protectors) will be necessary to keep browsing animals at bay. Mammals such as rabbits, voles, mice, squirrels and deer can cause damage to plants. Rabbits eat soft food crops and ornamental and woody plants of many types, and voles undermine woody plants and cause root desiccation as well as gnawing at their bases. Mice often feed off seeds and can devastate seedbeds – particularly in cold or adverse weather conditions when other natural foods are in short supply.

Rabbit populations are most commonly controlled by shooting in rural situations. Animal – friendly systems suitable for urban areas include a ring of wire netting supported with a small square stake for individual plants, though this system is not suitable for dense plantings. Individual trees may be protected by spiral rabbit guards (soft plastic spirals used around the base of the tree stem) available in white and brown, 45 cm or 60 cm high, at a relatively low unit price. Tree shelters/ tree tubes come at a higher unit price but have the added benefit of accelerating the growth of young trees. They do not, however, allow lower branches to develop, so only the larger box types, and netted types, can be used on multi-stemmed trees and shrubs. Probably the most cost-effective solution for large-scale, high-density plantings during

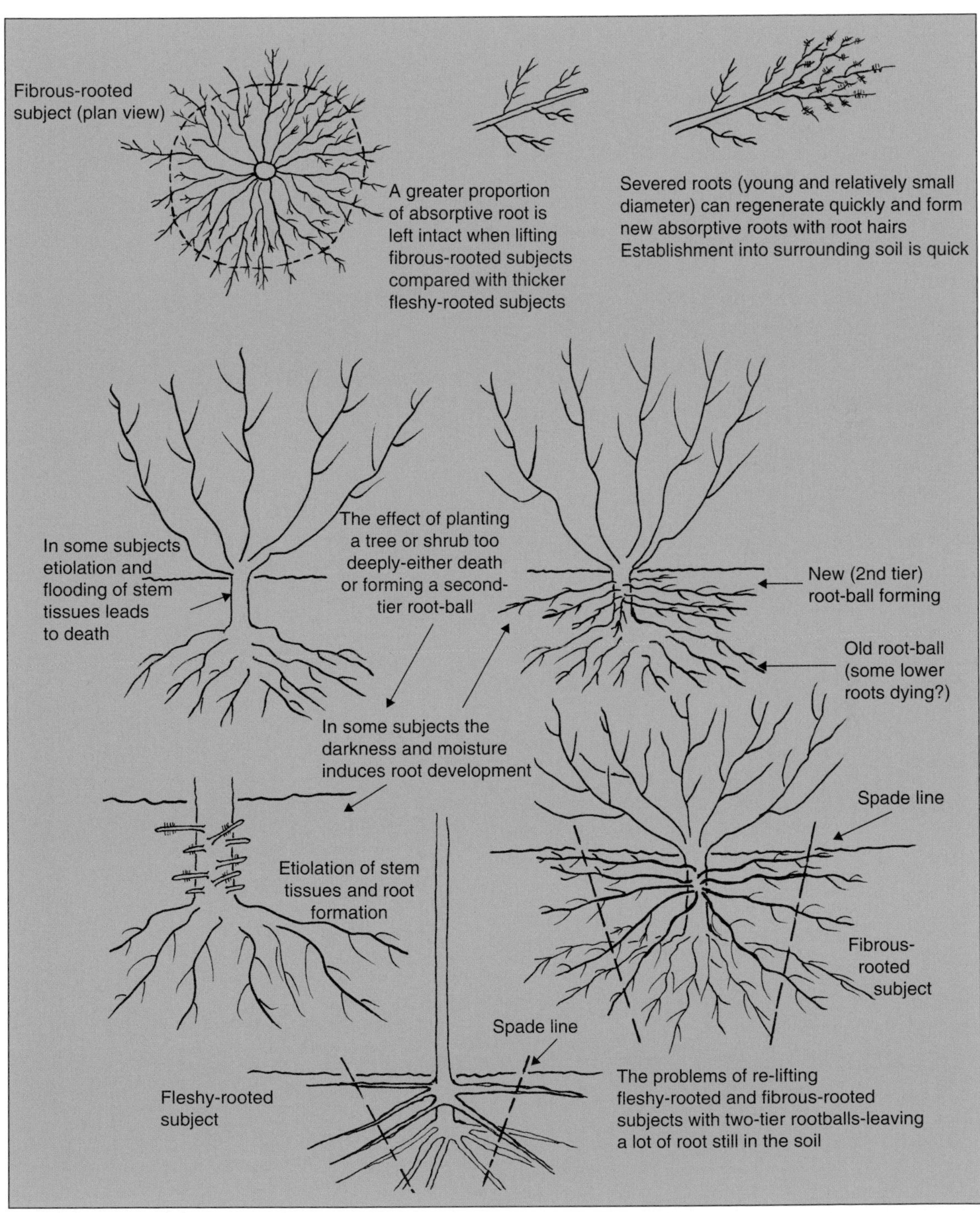
Fibrous-rooted subject (plan view)
A greater proportion of absorptive root is left intact when lifting fibrous-rooted subjects compared with thicker fleshy-rooted subjects
Severed roots (young and relatively small diameter) can regenerate quickly and form new absorptive roots with root hairs Establishment into surrounding soil is quick
In some subjects etiolation and flooding of stem tissues leads to death
The effect of planting a tree or shrub too deeply-either death or forming a second-tier root-ball
New (2nd tier) root-ball forming
Old root-ball (some lower roots dying?)
In some subjects the darkness and moisture induces root development
Etiolation of stem tissues and root formation
Spade line
Fibrous-rooted subject
Spade line
Fleshy-rooted subject
The problems of re-lifting fleshy-rooted and fibrous-rooted subjects with two-tier rootballs-leaving a lot of root still in the soil

Lifting shrubs and trees (including creating and lifting two-tiered rootballs).

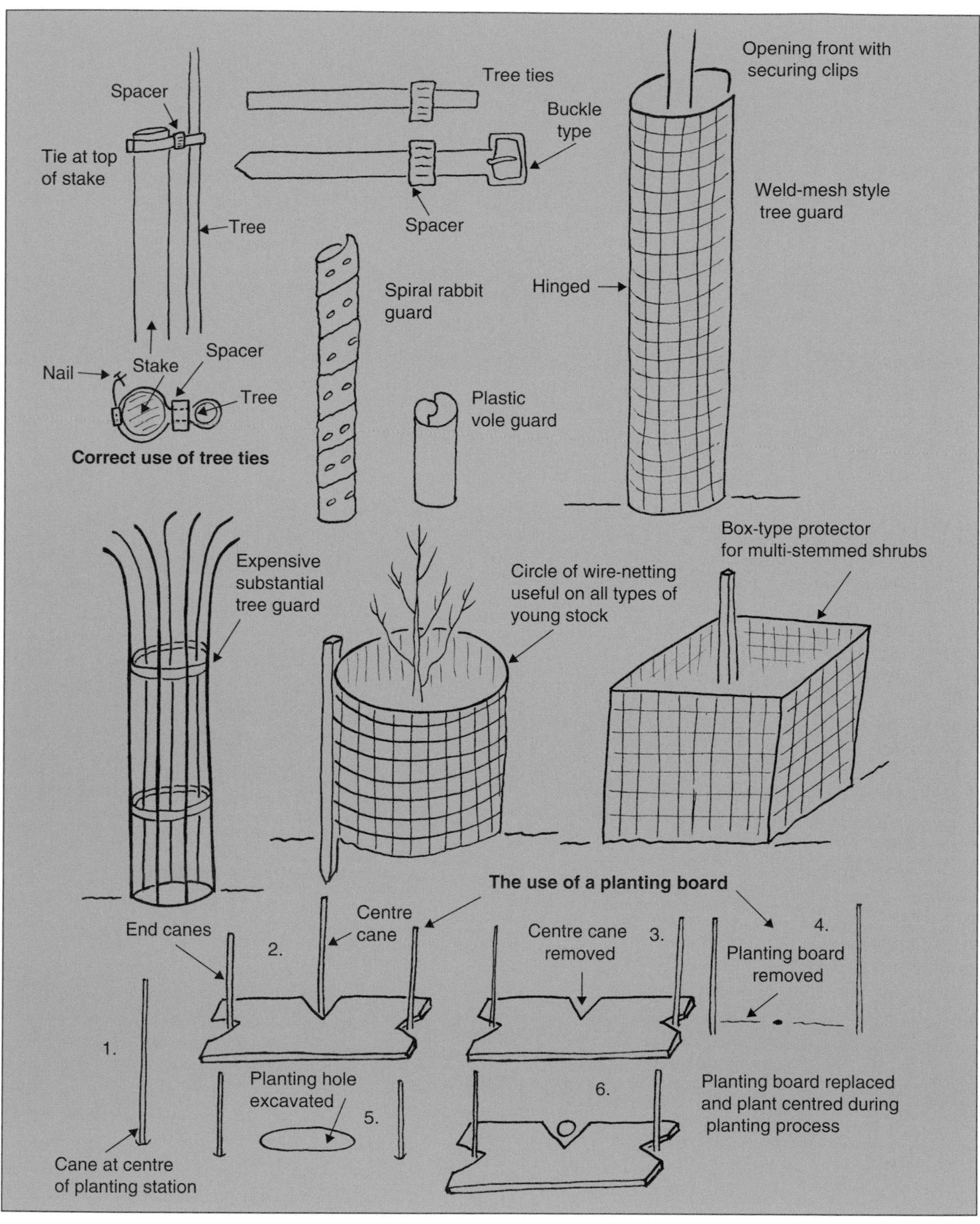

Protection and support during establishment.

their development phases is to fence off the area using rabbit netting around the entire site.

Field voles will characteristically de-bark trees and undermine a planting area, and can be a problem in large botanic gardens, arboreta, and areas that were previously pasture – any area that is (or was) predominantly grass. Their tunnelling habits expose young tree and shrub roots so that they desiccate, and in extreme cases the soil in the root zone may collapse completely. Some sheet mulches have the dual effect of both encouraging nesting (as the sheet acts as a shelter to keep the young dry) and preventing excessive root desiccation because of the undermining. Voles can be baited but are difficult to control, so vole guards, comprising a cylinder of very strong plastic placed around the base of the stem to prevent de-barking, are essential in areas of high vole populations.

Tree Support during Establishment

The most important aspect of the establishment of any plant is the progression of new roots into the surrounding soil of the original planting pit. True establishment into the new site is dependent upon good root growth. Short stakes on trees (approximately one third of the length of the clear 1.8m stem of standard trees) allow the necessary crown movement that is known to trigger good girth increment and root growth. Besides establishment benefits, there are financial benefits to using short stakes: short stakes are cheaper than long stakes, and sometimes two tree ties are necessary on long stakes, whereas only one is ever needed on short stakes.

Key concepts to ensuring the health of trees include always using effective 'spacers' on tree ties to keep the stake and the stem of the tree apart, and ensuring the tree tie is positioned at the very top of the stake (both of which will prevent the tree chafing on the stake). The early removal of tree ties in time to prevent damaging constrictions to the developing stem is essential. Many deaths of establishing trees are caused by tree ties girdling the main stem – far more than deaths from any form of intentional vandalism. Irrigation in dry weather and efficient weed control and mulching are also critical for success.

Establishing a Formal or Informal Hedge

Trees and shrubs require very similar conditions whether used in a hedge, in a plantation, or as individuals. However, because of the nature of hedges, there are obvious spacing and maintenance schedule differences. Selected plants for hedging ideally should be multi-stemmed, bushy plants furnished with branches right to ground level in order to create the density needed.

The optimum time to plant bare-root deciduous subjects is from October to November (realistically through to February if soil conditions are good), and evergreen subjects in February and March (realistically October to March). Only first quality plants, free of pests, diseases and serious physical damage should be chosen, and may be planted using basic techniques such as notch planting. However, because of the benefits to successful establishment, they are commonly planted using horticultural pit-planting or trench-planting techniques. Hedges may also be formed by pushing hardwood cuttings (setts) through a black polyethylene mulch over prepared soil.

Site Preparation

The site should be weed free (achieved either by herbicide treatment, or during hand digging, or both). Two lines 0.75 cm–1 m apart are put down over the soil as a guide, and ideally a trench taken out between the lines 60cm deep, and the bottom forked to assist drainage. Well rotted horse manure (or other organic material) may be included in the base of the trench for long-term benefits if desired (although this may encourage unnecessarily deep root penetration), and the soil should be replaced to completely backfill the trench. Throughout the process any obvious underground organs of perennial weeds should be removed, and the site should be consolidated and then cultivated to a reasonable tilth ready for planting. Ideally the site should be left to settle for ten to fourteen days before planting (often omitted commercially because of the extra cost of having to revisit the site).

Planting

A single planting line is run along the backfilled trench offset from the middle so that the plants are central after planting. Planting holes are made by taking out enough soil to receive the roots of individual plants without cramping them, and the plants are lightly shaken up and down at initial soil backfill to settle the soil around the roots. The plants must be firmed in well by consolidating the soil with the heel after ensuring the correct planting level initially, and at all times throughout the process. A greater density can be achieved in a shorter time by planting in two staggered rows. This increases the cost and is therefore usually reserved for establishment on exposed sites only. The planting site can be shaped to give semi-circles, curves or other formal geometric patterns if desired.

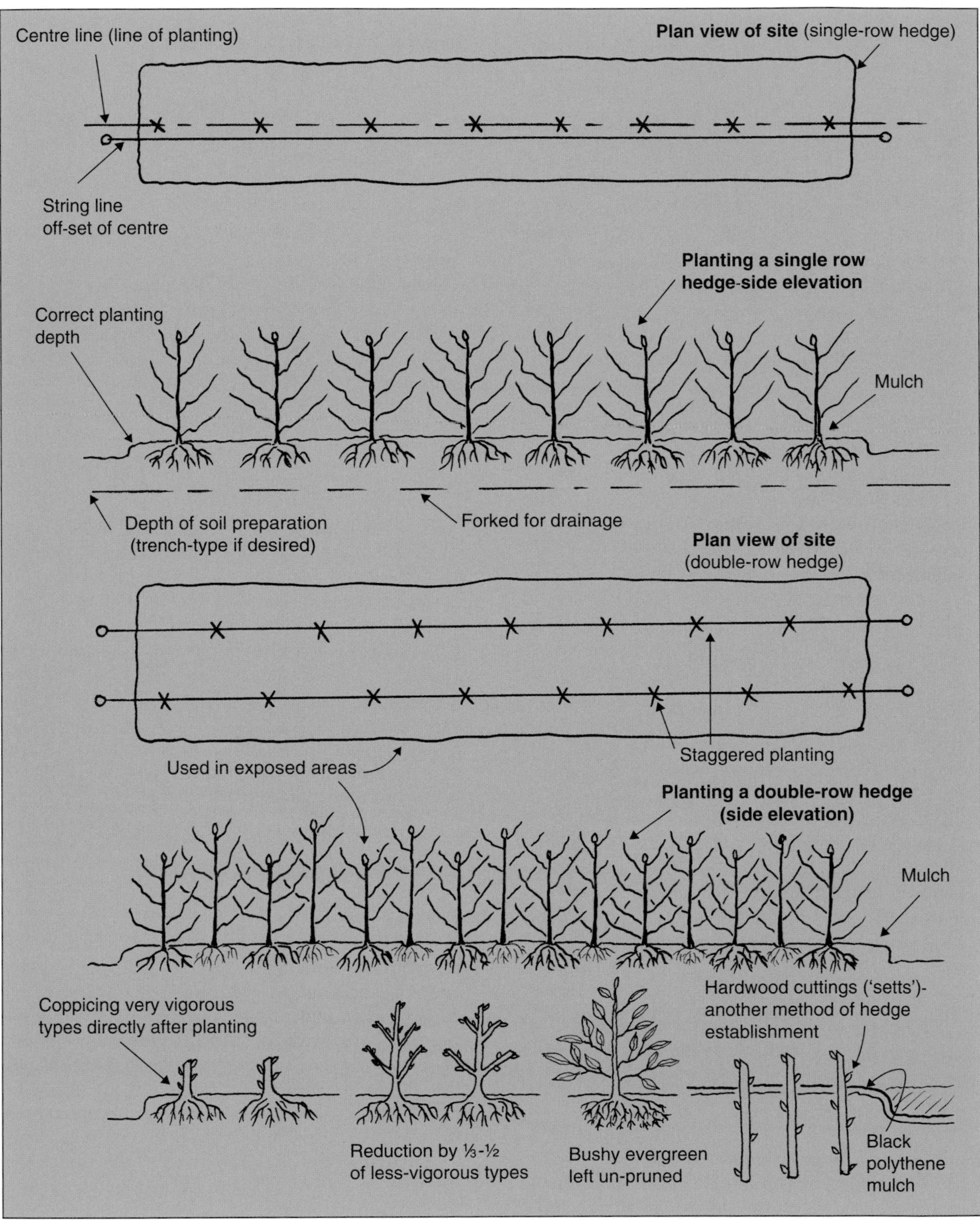
Centre line (line of planting)
Plan view of site (single-row hedge)
String line
off-set of centre
Planting a single row
hedge-side elevation
Correct planting
depth
Mulch
Depth of soil preparation
(trench-type if desired)
Forked for drainage
Plan view of site
(double-row hedge)
Staggered planting
Used in exposed areas
Planting a double-row hedge
(side elevation)
Mulch
Coppicing very vigorous
types directly after planting
Reduction by ⅓-½
of less-vigorous types
Bushy evergreen
left un-pruned
Hardwood cuttings ('setts')-
another method of hedge
establishment
Black
polythene
mulch

Preparation of the site and planting hedges.

SUGGESTED PLANT SPACING FOR HEDGES

Evergreen hedges	
Bush honeysuckle (*Lonicera nitida*)	45 cm apart
Lawson's cypress (*Chamaecyparis lawsoniana)*	60–75 cm apart
Common holly (*Ilex aquifolium*)	60 cm apart
Japanese euonymus (*Euonymus japonica*)	60 cm apart
Common yew (*Taxus baccata*)	60–75 cm apart
Western hemlock (*Tsuga heterophylla*)	60–75 cm apart
Spotted laurel (*Aucuba japonica*)	60 cm apart
Leyland (X *Cupressocyparis leylandii*)	100 cm apart
Photinia × *fraseri* 'Red Robin'	75 cm apart
Escallonia species and cultivars	60 cm apart
Dwarf hedging	
Prunus 'Otto Lyken'	30–40 cm apart
Berberis verruculosa	30–40 cm apart
Buxus sempervirens (Dwarf forms)	30 cm apart
Deciduous hedges	
Common hawthorn (*Crataegus monogyna*)	45–60 cm apart
Common beech (*Fagus sylvatica*)	45–60 cm apart
Common hornbeam (*Carpinus betulus)*	45–60 cm apart
Purple myrobalan (*Prunus cerasifera* 'Atropurpurea')	60 cm apart
Flowering hedges	
Berberis stenophylla (evergreen with yellow flowers)	60 cm apart
Berberis darwinii	60 cm apart
Berberis julianae	60 cm apart
Chaenomeles japonica (Japanese quince)	90 cm apart
Rosa 'Penzance Briar'	90 cm apart

Important Aftercare of Hedges in the Establishment Phase

Freshly planted material should be watered in if soil conditions are dry, and the site should be mulched and kept weed free at all times. As with all woody subjects, good weed control is essential for successful establishment, and planting through a sheet-mulch gives very good weed control and moist establishment conditions at the roots, so can be very successful. Any dead or diseased plants should be replaced immediately with first quality material.

Vigorous species such as blackthorn, hawthorn, plum and privet should be coppiced hard directly after planting (cut back hard at all main leaders, or vigorous shoots) to encourage bushiness and hedge density. Less vigorous species such as beech, hornbeam, *Lonicera nitida* and *Euonymus japonica* are trimmed more lightly by reducing the plant by up to half as a maximum. Evergreens such as Western hemlock, Lawson's cypress, holly, aucuba, bay, box, yew, *Photinia* × *frasrei* 'Red Robin', laurel and Portuguese laurel are best left entirely alone in their first season, and the same is true of deciduous species grown for their flowers (unless their branch structure is particularly poor).

Establishment of Grass Lawns and Amenity Turf

Establishment by Seed

Lawns and amenity grass areas can both be produced by sowing grass seed. Soil preparation is basically the same as for any other outdoor seed-sowing operation – deep cultivation by digging with a spade (or ploughing for large areas), followed by rotary cultivation (or breaking down by hand operations such as rough raking), followed by raking to a fine tilth. During preparation a level surface is always maintained, the soil is consolidated by treading (or rolling for large areas), and a high phosphate fertilizer may be applied prior to a final raking.

Grass seed is sold as known species mixtures with recommendations for specific usage, and both the species mix and the recommended rate of sowing will vary according to this. Seed is sown broadcast on to prepared soil, and may be sown using a calibrated sowing machine or by hand. Sown seed is lightly raked into the top surface. Germination is relatively quick, and any obvious omissions (bare patches) may be rectified by over-sowing immediately.

You may get very adept at uniform sowing, and if so, the operation can be carried out in one direction only with success. However, to ensure even coverage of seed by hand, the prepared soil surface is sown at half the rate in one direction and half the rate at right angles to it, which sounds fairly straightforward, but is probably best explained with examples. Dividing the plot into 1m wide strips is the best method for uniform sowing, and obviously therefore a square plot would entail using half of the measure of seed in one direction and half at right angles to it. For example, a plot 4 m × 4 m divided into 1 m wide strips would entail four 4 sq m plots in one direction, and the same at right angles to it. If the rate were 30 g per square metre, then each 1m wide strip would require 4 × 15 g (half the rate), which equals 60 g. This would mean 4 × 60 g of seed sown in one direction and 4 × 60 g sown at right angles to it – sown in 1m strips but adding up to 240 g in each direction, and making a total of 480 g over the site. This tallies with the straight arithmetic (4 m × 4 m = 16 sq m × 30 g = 480 g).

An example using a rectangle highlights the slight differences in the calculations. A plot 5 m × 3 m at the same sowing rate would total 15 sq m × 30 g = 450 g of seed. However, 1m wide strips in one direction would be 5 m long and equal 5 sq m in area each (there are three of these). 1 m wide strips at right angles to it would equal 3 sq m in area each, and there are five of these. Each strip of 5 sq m area needs 5 × 15 g (half the rate) each = 75 g, and as there are three of these the total seed for this area would be 3 × 75 g = 225 g. The strips of 3 sq m need 3 × 15 g each = 45 g, and as there are five of these, the total seed for the area sown at right angles to the first would be 5 × 45 g = 225 g, making the overall total for the area correct at 450 g (2 × 225 g). This means for each long strip, 75 g of seed is needed, and for each short strip 45g of seed is needed.

In all cases each strip has to be sown as uniformly as possible to ensure the correct sowing rate, and all weighed seed from any container must be completely used up over its designated area.

People will always regale you with stories of just how early, or in what very severe or arid conditions they have sown grass seed with success. Nevertheless, optimum times for sowing are those times with the greatest chance of moisture for the germination process. Hence, very late summer/early autumn is ideal in many ways, not least because in temperate climates this period follows a time with good soil preparation conditions, and is itself followed by a time of relatively mild temperatures and increasing rainfall. The conditions for site preparation should be moist but not soaking wet soil. Thus preparation and sowing at this time is relatively easy without creating excessive footprints or depressions, and the period of rainfall that is imminent supports developing seedlings. To add to this, the growing period that remains before the likelihood of very severe or cold weather allows germinating grasses to develop well before the onset of winter. However, it is equally true that spring-sown grasses have a long, continuous growing period in front of them. So as long as the spring weather is not too dry, this is another good time to sow. Each side of these times, and between these two main periods, are all possible, but bring with them greater elements of risk.

Sowings in the summer can be ideal because of good growing temperatures, but only if chance rain (or irrigation) intervenes. The midst of winter only ever works if the weather fortuitously remains very mild, and the normal periods of low temperatures do not occur. Nevertheless, taking all possibilities into account (from the chancy to the more ideal), there is a relatively long period in which success can be achieved. Practically, varying degrees and combinations of the parameters may exist, and informed decisions are made by individuals (and managers) based on these parameters as they seem at the time – the good or bad consequences of which have to be lived with.

Establishment by Turf

Soil preparation is as for establishment by seed, but once the site has been vacated after initial soil preparation, any return to the site involves working off boards for all operations to ensure the surface remains good for root-to-soil contact and without major depressions and compacted footmarks. Using boards for any operations on soil to prevent surface damage can be useful, and always aids operations on soil in less than optimum weather conditions, but is always essential for turf laying.

The critical factor for good turf establishment is good root-to-soil contact, and sufficient temperature and moisture for root growth. Root growth is still active during the autumn and early winter period, and also in the early spring. This fact, coupled with the relatively high moisture

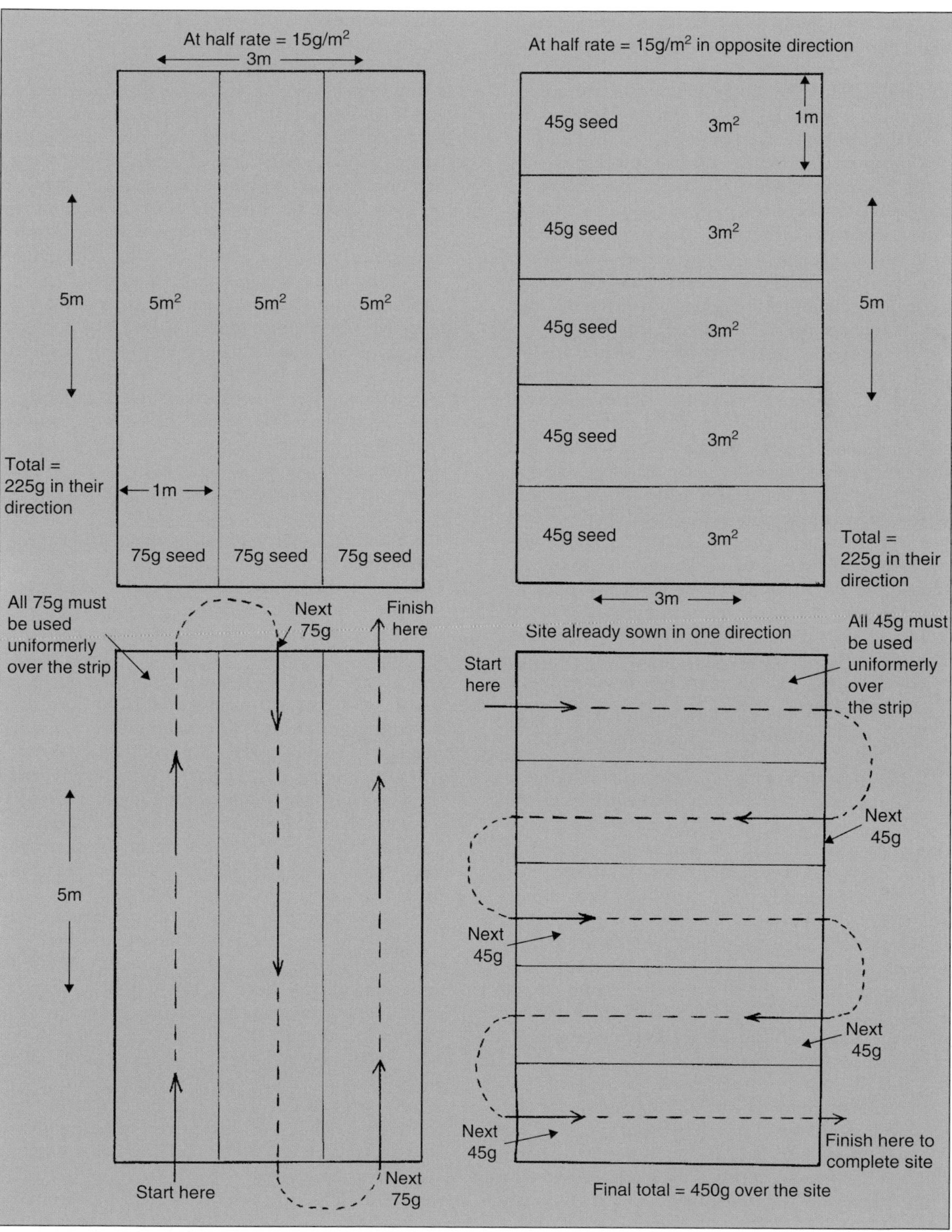

Sowing grass.

levels normally experienced at these times, makes them ideal times for laying turf. However, suitably mild (non-frosty) periods throughout the winter and spring also allow turf-laying to be carried out (the main criteria is usually whether the soil is workable, or too wet to cultivate).

Laying turf during the summer months is possible, but if conditions are very dry it may be impossible for the turf supplier to lift turf. Furthermore, an efficient irrigation system will be needed to ensure the turf knits together with the soil surface via root action and does not die. As long as these considerations are taken into account, turf can be successful at most times of the year.

The correct number of turves (plus a contingency for cuts and wastage – say, 10 to 15 per cent) should be ordered well in advance of the operation, and the site must be fully prepared prior to the pre-arranged delivery date. Furthermore, it is essential that someone is on site to receive the turf on arrival, both to ensure the quality and to oversee the unloading and stacking. Turf should be stacked in low piles conveniently placed around the site to reduce the distances that they need to be moved during the turf-laying operation (a policy of 'little and often').

Boards must be used wherever turves are moved from a stack to where they are to be laid (to prevent ruts and soil compaction, particularly if using a wheelbarrow). Boards must also be used at the current turf-laying position, and the process should never be contemplated without adequate boards for the job.

If the surrounds of the turf-laying site permit, the first line of turf can be laid working from outside the prepared site. If not, the first boards are put down on the prepared site a short distance (just over a turf width away) from the starting edge, and the first line of boards are walked on and used to lay the first line of turves. The turves are put down on to the prepared surface making sure that the ends are butted up well and are close together with no gaps.

The first line of turf is laid on the longest side of a regular-shaped plot and overshoots the right- or left-hand edge of the site (depending on your work direction), to be trimmed off at a later time. Once the first line of turf is in place, the boards are moved on top of them, and all operations for the second line of turves are carried out working off them. This helps to stabilize the turves by pressing them into the soil surface, which improves the butted joints and the root-to-soil contact, and prevents large footmarks, pits and hollows from forming in the laid turf. The second line is laid ensuring the long edges of the turf abut one another well, and commences at the site edge but overshoots the site by half a turf on the opposite side – this ensures that all turves are 'bonded' by half a turf length in a 'brickwork' fashion (for strength). The process continues by working off boards placed on the previously laid line of turf, and overlapping side edges and offsetting turves by half a length until the whole site is covered – any half-turf gaps are made up by the off-cuts from the turfs that initially overshoot the site. Irregular-shaped sites are tackled in the same way but commencing from the longest axis of the site – which may be the middle.

If the soil preparation (including consolidation) is carried out well, there is no need to roll in the turf, as your own human bodyweight over the working boards does this job for you. Efficient butting up, both at turf ends and at turf edges, makes it unnecessary to add bulky top dressings along the joints. The final operation before leaving the site is to trim off any unwanted turf with an edging iron and apply irrigation if necessary.

Maintaining Lawns and Grass Areas in the Establishment Phase

Freshly sown lawns can be over-sown on any bare patches. The grass needs to be lightly tipped as soon as it is practically possible, to encourage tillering (the development of lateral shoots), which increases the overall density (and therefore grass cover) of the sward. Improvements to grass density/soil cover in the early stages of development help guard against soil erosion and surface wear.

Late summer (or earlier) sowings may need their first mowing before winter, and if it is necessary, involves a very light 'tipping' with a sharp, well set mower to prevent damage at this critical phase. Obviously grass seedlings that do not have a fully developed root system, and a mower with very dull blades that pulls up the grass rather than cuts it (or a combination of both), are disastrous for the developing lawn. The first cut(s) should be light and not too close, as this would destroy the seat of new growth in grasses that develop low on the plant. The same reasoning exists for grasses that have safely overwintered before needing either their first (or second) cut ('tipping') in the spring. Ideally, grass surfaces need to be dry before any form of mowing, but this is even more important for cuts early in the development phase, to further reduce the chances of damage to young seedlings during the mowing operation. 'Arisings' from the mowing process (cut grass clippings) should always be removed at this phase either via the grass box attachment or by raking.

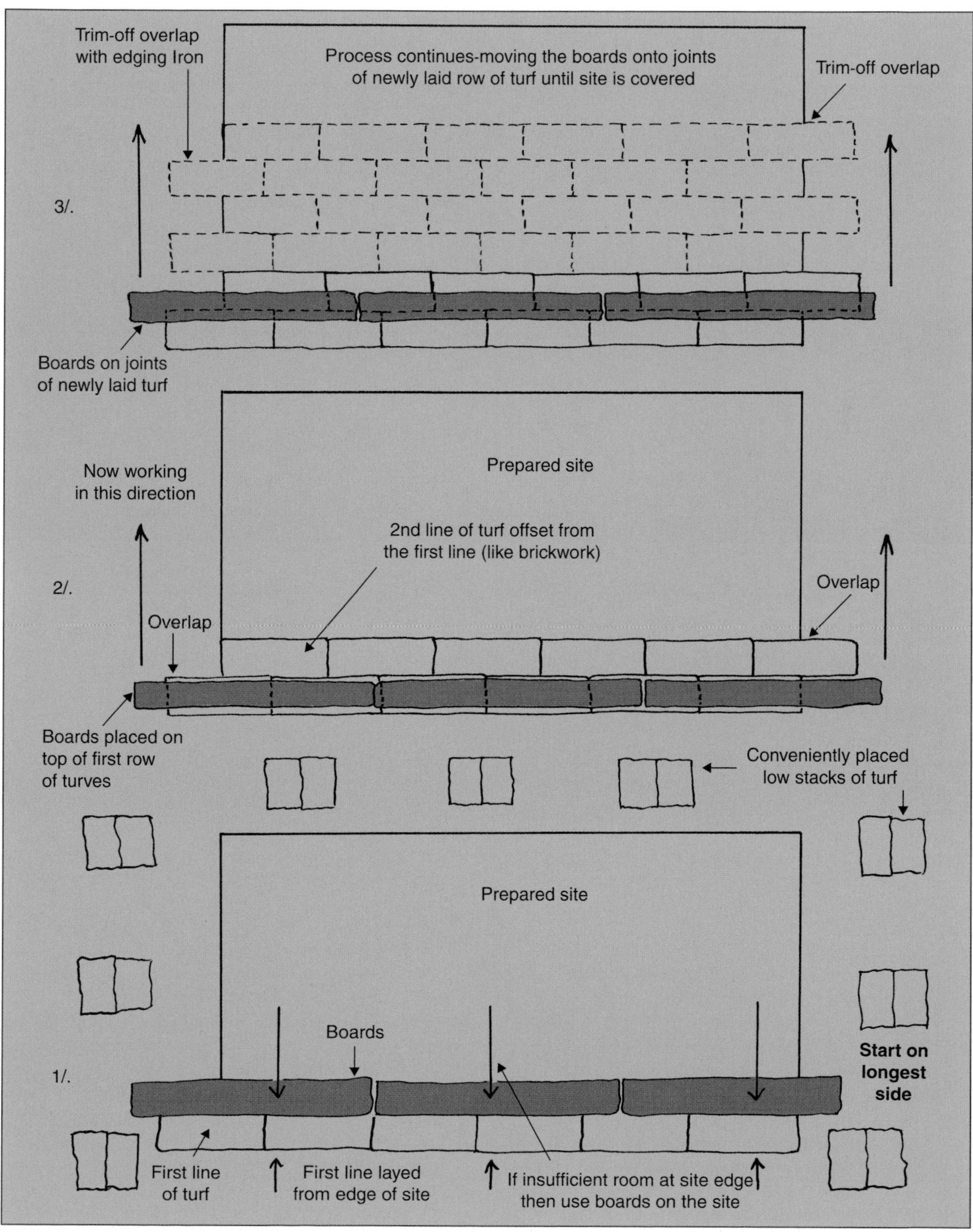
Trim-off overlap
with edging Iron
Process continues-moving the boards onto joints
of newly laid row of turf until site is covered
Trim-off overlap
3/.
Boards on joints
of newly laid turf
Now working
in this direction
Prepared site
2nd line of turf offset from
the first line (like brickwork)
2/.
Overlap
Overlap
Boards placed on
top of first row
of turves
Conveniently placed
low stacks of turf
Prepared site
Boards
1/.
Start on
longest
side
First line
of turf
First line layed
from edge of site
If insufficient room at site edge
then use boards on the site

Laying turf – rectangular and square sites.

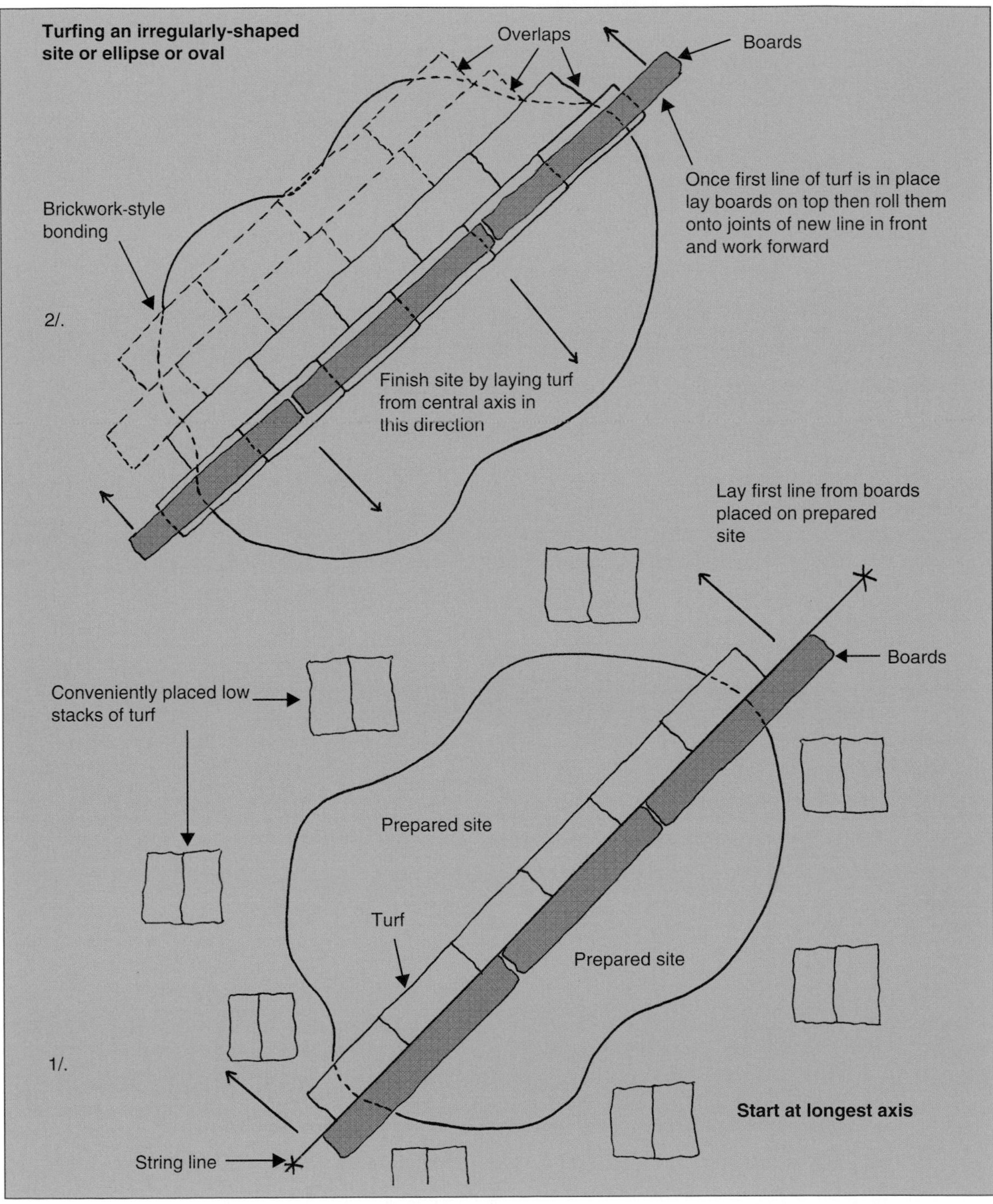

Laying turf – irregular-shaped sites.

In general terms, it is annual weed species that will infest freshly sown lawns, as they are typical of freshly disturbed soils. Nevertheless, some perennial weed species may germinate and create problems for the future. Usually, both weed types will be destroyed by the first mowing, as cutting off the main growth of annual species will destroy them, and many perennial species will not tolerate being mown off when they have not had time to develop underground organs. Regular mowing kills out most broad-leaved weeds because they do not produce shoots that are as low growing as grasses. However, some broad-leaved weed species are particularly well adapted to miss the main mowing height, and thus avoid being regularly cut off. Plantain is particularly adept at this and manages to survive in regularly mown lawns. Other broad-leaved weed species have some of their foliage cut off by mowing, but because they have deep underground organs they can regenerate relatively easily; such persistent species may need herbicide treatment.

Turf purchased from a reputable supplier should already be weed free, so problems during the development phases are less likely; they occur only after reinfestation at a later date. Only after the turves have knitted together and have commenced root initiation into the new site can they be walked on at all, so it will be several months after laying before they can be mown. First mowing must be carried out with care and with an efficient mower, and should not be necessary (or desirable) during the late autumn or winter months, as the soil is usually too wet for this to be carried out successfully: the mower would tear the surface, skid, smear and pull out young grass, rather than cut it efficiently under these conditions. Relatively low soil and ambient air temperatures would normally preclude the good growth that would make mowing necessary in the winter, however turf established in the early autumn could need tipping fairly soon after establishment.

---

*For more information on grass management, see *Sports Turf and Amenity Grassland Management* by Stewart Brown, published by The Crowood Press.

CHAPTER 9

Managing Planted Areas and Individual Plants

High, Medium, and Low Maintenance Strategies

Whether high, low or medium input maintenance strategies are adopted will obviously have an effect on the overall annual outlay for the management of a particular area. High input maintenance systems are usually considered synonymous with high quality – and of course this is not always true: there is no guarantee of a direct or complete correlation, because many other factors can influence the outcome – not least the definition of, and perceived concept of 'quality'. So it becomes very counter-productive if the high cost does not, in fact, result in the desired high quality. Furthermore, all forms of management have to be tempered with (even governed by) a financial constraint of some type. The degree of that constraint depends upon individual circumstances, and there is never a situation where an 'open purse' truly exists. Inevitably, therefore, some form of reasonable balance, compromise and common-sense approach prevails.

Although no one would advocate the use of high input (and therefore high cost) systems as the norm in all instances, specific high profile, prestigious, even specialized or noteworthy areas may require extra effort and expense to give the desired effect. It would seem that cost effectiveness rules, but it is difficult enough to measure cost effectiveness ordinarily, so what of gardens that are not open to the public and so do not attract an entry fee revenue? How are these assessed for their cost effectiveness when the maintenance of excellence can only be justified aesthetically, not financially? Areas of public access always bring with them the dilemma of low standards encouraging poor use, abuse and vandalism, and high quality areas helping to discourage vandalism and misuse. However, neither system can be guaranteed to be totally exempt of the other's benefits or failures. Furthermore, if damage does occur in high quality areas it is usually very expensive to repair, renew or replace the damaged features.

The low maintenance systems discussed are not necessarily recommendations, as they are often detrimental to aesthetics. However, it is sometimes necessary to reduce annual outlay in the management of gardens, and garden managers need to know the options open to them. Therefore the management of planted areas is discussed here, commencing from a series of low maintenance strategies and radiating outwards (upwards?) so that the options and prioritizing decisions, and the balance of expenditure over final outcome, is considered from the outset. Even domestic garden owners may need to adopt low maintenance/low input strategies – for various reasons, including lack of personal mobility or lack of time input. Low maintenance strategies include prioritizing the input on features that are to be retained, modifying existing features (or even the entire site) to create low maintenance areas, or commencing with a low maintenance design in the first place.

Starting from scratch obviously brings with it the benefit of getting it right from the beginning. If, however, you have inherited an existing garden or feature, then remodelling or modifying what already exists is the only option. It is perfectly possible for low maintenance designs and strategies to result in successful aesthetic features and garden areas. However, retaining the delicate balance of quality plantings and maintenance input is never easy, because up to a point, the quality of most garden features is directly proportional to the amount of time devoted to them.

Hard Landscape Features

Design options for scratch designs (or remodelling) include the creation of greater areas of hard landscape (including pergolas, patios, paths/paving, concrete, brick and stone walls and statuary), all relatively expensive to create, but very low in their maintenance needs if correctly installed. They offer a wide range of aesthetic effects, but care needs to be taken to ensure that the design and materials chosen blend in well with their surroundings and do not look incongruous. Furthermore, attention to detail is very important: it is pointless constructing something at high cost if it is not fit for the purpose. Many ostensibly good schemes are ruined because the details have not been checked over – for example, if the site is subject to flooding in poor weather conditions.

Soft Landscape Features

Soft landscape areas are those that involve any sort of planting (including plantings in containers), and include components such as soil, plants, mulches. They are also the most labour-intensive sectors of any design, because operations such as weed control, pruning and watering are the most time consuming. So any new designs or design changes planned should therefore look towards reducing the labour requirement in these areas, and if it can be done without obvious loss of quality and retaining the integrity of the original design, then it is truly successful.

Rock gardens are a mixture of hard and soft landscape, with weed control as their main expense. Containers are also a mixture, but their main maintenance problem is the ongoing cost and commitment involved in efficient and regular watering.

Using Greater Areas of Grass

Grass is an essential element of garden design as it unifies areas, gives a sense of space, and acts as a foil for other features. Planted beds and borders (with their area of bare soil) are very labour intensive by comparison to grass areas. Keeping grass tidy is easy and therefore very cost effective, and creating more grass areas, or grass areas from previously planted areas, is a policy that is commonly adopted by local authorities when budgets get tight: they tend to reduce horticultural excellence in favour of grass areas when 'rationalizing'. However, a green desert with no colourful features has little to offer, so a balance needs to be struck between the aesthetic, the cost-effective benefits of grass, and the overall design interest. Nevertheless, identifying some areas as suitable for extra lawn that can be incorporated into the main mowing regime, and substituting or moving other, more aesthetically interesting feature(s) elsewhere on the site, is a highly cost-effective way of reducing labour on a park or large garden scale. The main reason is that the site has to be visited to mow the existing grass anyway, and unless a greater capacity mower is now needed because of the increase in the grass area, then the capital outlay for the mower is usually a necessity anyway and has already been spent. However, there are alternatives to increasing grass areas – for instance, using easy-to-maintain systems such as geotextiles and mulching in beds and borders – without reducing the colourful features.

Irrigation

Increasing mulched areas and using manual and semi-automatic irrigation systems where possible greatly reduces plant maintenance time. Watering plants in containers, and especially hanging baskets, is a very time-consuming operation in the summer, and likewise, the successful establishment of new tree and shrub plantings often involves watering in a dry season. Basic, manually operated watering systems for hanging baskets are easy to install and save hours of time: these consist of capillary tubes coming off a main header pipe, attached to a single tap. Other simple irrigation systems to reduce the time involved in watering outdoor plantings are also inexpensive: 'Leaky Hose' is one example – a subterranean system that comprises a small-bore hose made of rubber chips and sealed off at one end. When the hose is pressurized by relatively low-pressure mains water, the individual chips of rubber are forced apart to slowly release drips of water into the soil. This inexpensive hose is installed at planting by threading it past the root systems of the trees and shrubs. Watering is therefore directly at root level and does not evaporate away during application; and if mulching is also practised, the rate of success of establishing trees and shrubs increases considerably.

Grouping Plants with Similar Maintenance Needs

Grouping together plants with similar maintenance requirements (pruning and/or chemical weed control) is always a useful way of reducing labour costs, and can be successful. However, it is difficult not to sacrifice the integrity, philosophy, ideals and aesthetics needed for successful design. Using more groundcover plantings of all types, and making more use of woody perennials as specimen trees, shrubberies and groups of low-growing arboreals where they are suitable, helps reduce maintenance. This can also extend to woodland gardens if room is available and scale permits, as they are ideal self-regulating systems.

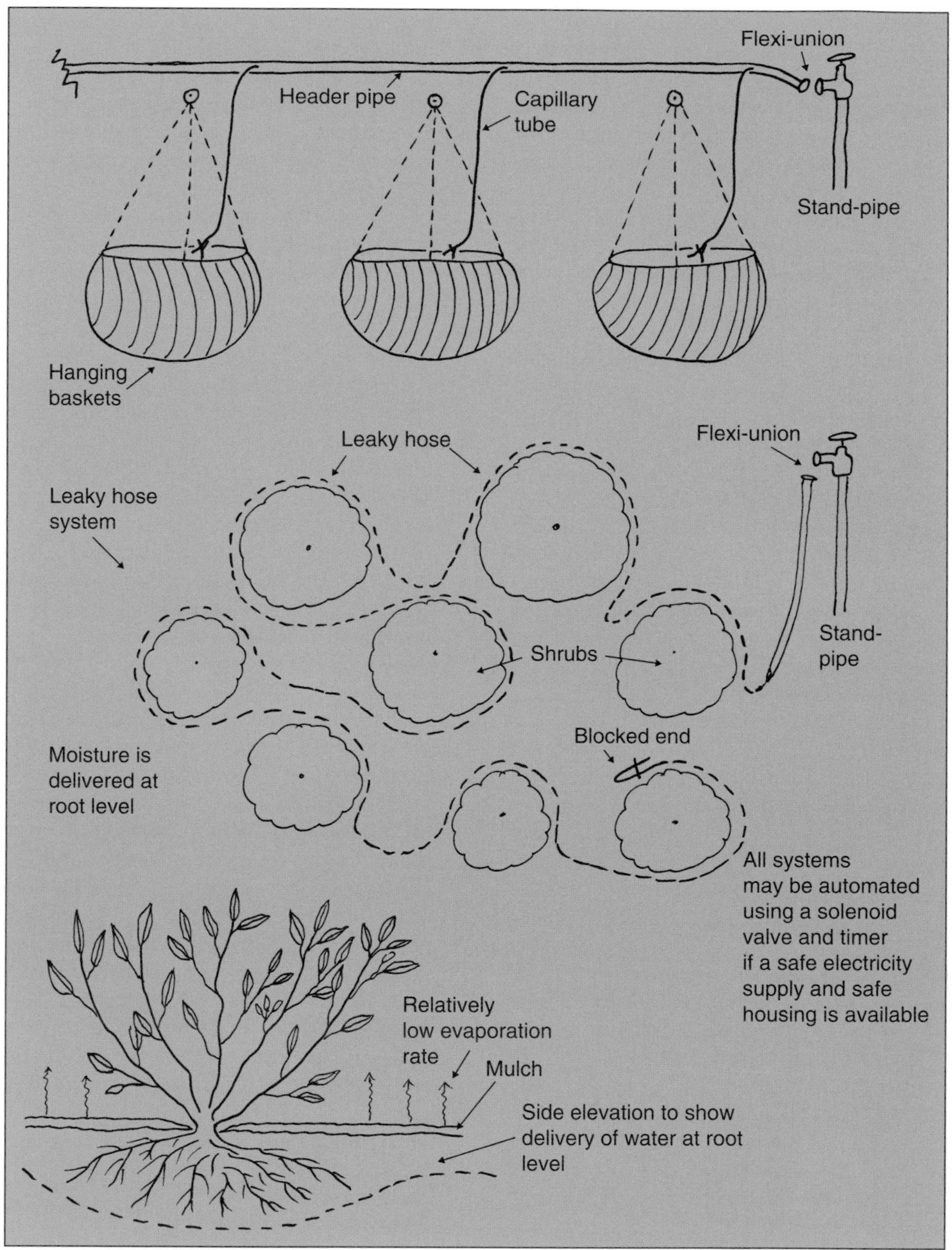

Simple irrigation systems.

Where more than one site is managed, woody species may be grouped according to the similarity of their pruning programme, so that the labour force can prune large volumes of shrubs with similar requirements in one main visit, leave the site, and then return for the next main group, and so on. The relative efficiency of not having to bring the labour force and equipment on site for many more shorter visits certainly helps to reduce costs. However, organizing the relevant species in this manner is certainly restrictive to imaginative design – for example, large areas of shrubs may be grouped together because they are all managed by coppicing in the spring, but in the early part of the year these same large areas are completely lacking in interest. Furthermore, plants of any one specific flowering period (or other features), instead of being mixed into a matrix to maintain interest throughout the year, become concentrated into blocks that display their feature and then become very dull. More importantly, the low number of site visits may also reduce effective weed control due to lack of monitoring.

Grouping woody plants together because of their resistance to specific herbicides will help the cost of annual herbicide application by increasing the efficiency of the operation. Rhododendrons (because of the range of flowering times that can be achieved by careful choice of types) are often grouped as single genera collections anyway. However, being shallow rooted (amongst other factors) they are not very resistant to the application of many types of herbicide. This, coupled with the fact that they do not tolerate hoeing or forking in amongst their roots (in the summer, in particular), means they are good candidates for mulching, and they generally look good and do well with this option. Perhaps now with mulching so strongly recommended, this should always be the preferred option – not just for its known benefits, but because it allows greater diversity of planting material with efficient weed control (and is moreover an organic option).

Tougher, woody species with deeper roots and a greater natural resistance to herbicide applications can be grouped elsewhere in the design; examples include bush willows, *Cornus alba, Cornus stolonifera* 'Flavaramea', *Ribes, Forsythia, Deutzia* and *Philadelphus*. In practice, grouping in this way does often look too utilitarian, but it can in fact be quite successful, because quite a high number of tougher, showy species of shrubs are tolerant to a wide range of available herbicides. Therefore good aesthetic areas can be maintained with normal herbicide management – but they will never make botanic garden status.

The herbaceous border is another example of grouping similar plants together, but they are renowned for being difficult to maintain, largely because they cannot be managed using herbicides, since most of these are designed to be toxic to perennial weeds that are themselves herbaceous perennials.

Groundcover schemes purport to be low maintenance, but in order to subdue weed growth they need to be at very high densities, and unfortunately they do need weed control and other maintenance in their early stages; neither annual nor perennial weeds are subdued in any way during the first one to four years of groundcover establishment. So be prepared for extra weed control problems in the early years of a scheme's development, before it achieves its full density. Weed control prior to planting groundcover schemes is essential, including using herbicides for tenacious perennial weeds.

Pruning

Pruning is the removal of plant material, and includes dead-heading annual, herbaceous and woody species. Pruning can be carried out regularly, as in hedge clipping, or it can be done annually, or periodically, or for reshaping or maintaining balance, for safety (removing obstructions), to reduce height (overall size), to retain good flower-to-wood ratios, or not at all. Every pruning operation does breach the system and therefore opens up the potential for introducing infection (and for transferring infection from one plant to another), so should always be carried out with this in mind.

Dead-heading

Dead-heading is the removal of old flowers (or flowering stems), and is a process that often encourages more flowers to be produced, and so increases the continuity of flowering. The early removal of dead flower heads prevents the production of seeds, and since this latter process uses valuable energy, dead-heading therefore leaves the plant with more energy for new flower and/or shoot production. The removal of dead flowers from tender subjects that are often used in bedding schemes – such as French and African marigolds, salvias and petunias – is always beneficial and, used alongside a good watering and feeding regime, will often prolong their flowering period considerably.

Dead heads can be removed using sharp scissors, secateurs or by pinching them off between finger and thumb.

Dead-heading herbaceous plants involves removing not just the dead flower, but the whole flower-bearing stem down to, or near, soil level; this should be done with sharp secateurs. Care must be taken not to cut so low as to damage any dormant adventitious buds at the very base of old flowering stems (or at the tops of the perennating organs), and when flowering stems are removed, not to damage or remove intact fully functioning leaves. The energy saved (and the potential energy-producing foliage saved) in this instance is used to encourage an increase in clump size laterally, including the production of new daughter bulbs, corms, rhizomes or other underground organs.

Rhododendrons tend to produce seed capsules fairly early, so as the petals are dying away, the seed heads are often already forming. Dead-heading rhododendrons involves the removal of seed capsules by grasping the whole seed head between finger and thumb, and removing it with a downwards and sideways pull. If this is carried out correctly the seed head breaks off neatly at the girdle scar created by the bud scales of the flower bud, without damage to the main stem or the developing axillary buds. Flowers of rhododendron species terminate growth, hence uninterrupted lateral extension growth from the axillary buds is very important, as it allows sufficient development and extension in the current season and ensures good wood that will ripen and set flower buds for the following year at its apex.

Dead-heading roses is usually done with sharp secateurs, cuts being made that remove all the dead flower heads just above the nearest axillary bud, leaving any dormant or developing axillary buds (that are often in close proximity) in good condition. If this is done very soon after petals begin to fade, then the minimum of available energy will be used for fruit and seed production. If it is beneficial to the shape and future appearance of the plant, the dead flowers may be removed using cuts much lower down the flowering stem, but leaving any lower dormant (or developing) buds intact. Dead-heading roses nearly always extends the flowering season to some degree, so is particularly useful for types that have large flushes of flowers in one go. It may also aid 'perpetual' flowering types, but less so.

Large-flowered bush (hybrid tea) roses may have one third to one half of their stem lengths reduced in November (known as 'heading back') because they are prone to 'wind-rock' in the winter months, which makes them unstable and can scour out the area of soil at the base of their stems.

Pruning Hedges

Hedge trimming is a form of light pruning. Hedges respond best if they are trimmed regularly to encourage branch density and the desired shape. Hawthorn, *Lonicera*, privet and plum should be clipped every six weeks from June to September, and the leaders lightly tipped back until the desired height is reached. Most conifers only regenerate from green tissues, so should not be pruned too severely. Holly, *Cupressus*, *Thuja*, yew, false cypress and *Euonymus japonica* should be clipped twice, once in July and once in September. *Aucuba*, laurel and *Elaeagnes* should be clipped once a year, in August (to ensure good wound healing before the onset of winter), and again the leaders tipped back until the desired height is reached. Beech and hornbeam are also clipped once a year in August and the leaders tipped back as above. Flowering subjects are generally pruned directly after flowering, although some species may be trimmed at other times. The top line of a hedge can be maintained using a string line as a guide, and many shapes can be maintained by eye. However, specially constructed wooden trammels may be used as a guide to retain specific shapes more accurately.

Small areas may be clipped by hand, using shears, but larger areas may require the use of mechanical hedge trimmers with either an electric or a petrol motor.* Note that health and safety issues concern the use of both hand and motorized cutting tools: all powered hedge trimmers have very sharp reciprocating blades, and care must be taken at all times to ensure both personal safety and the safety of others. Gloves and safety glasses (or goggles) must be worn, and special care taken not to cut through the cable on electric trimmers: circuit breakers must be fitted on these. The hedge should be inspected for hidden obstructions before starting the trimmer. It is essential to use stable scaffolding, tower scaffold or a hydraulic platform for any hedge above chest height, and the trimmer itself should never be used above chest height at any time. If steps are to be used they must be stable and fully extended to ensure a wide base. Low, sturdy

---

*Refer to guidance via the *Arboricultural Advisory Service/Forestry Commission Safety Leaflets* and any relevant legislation.

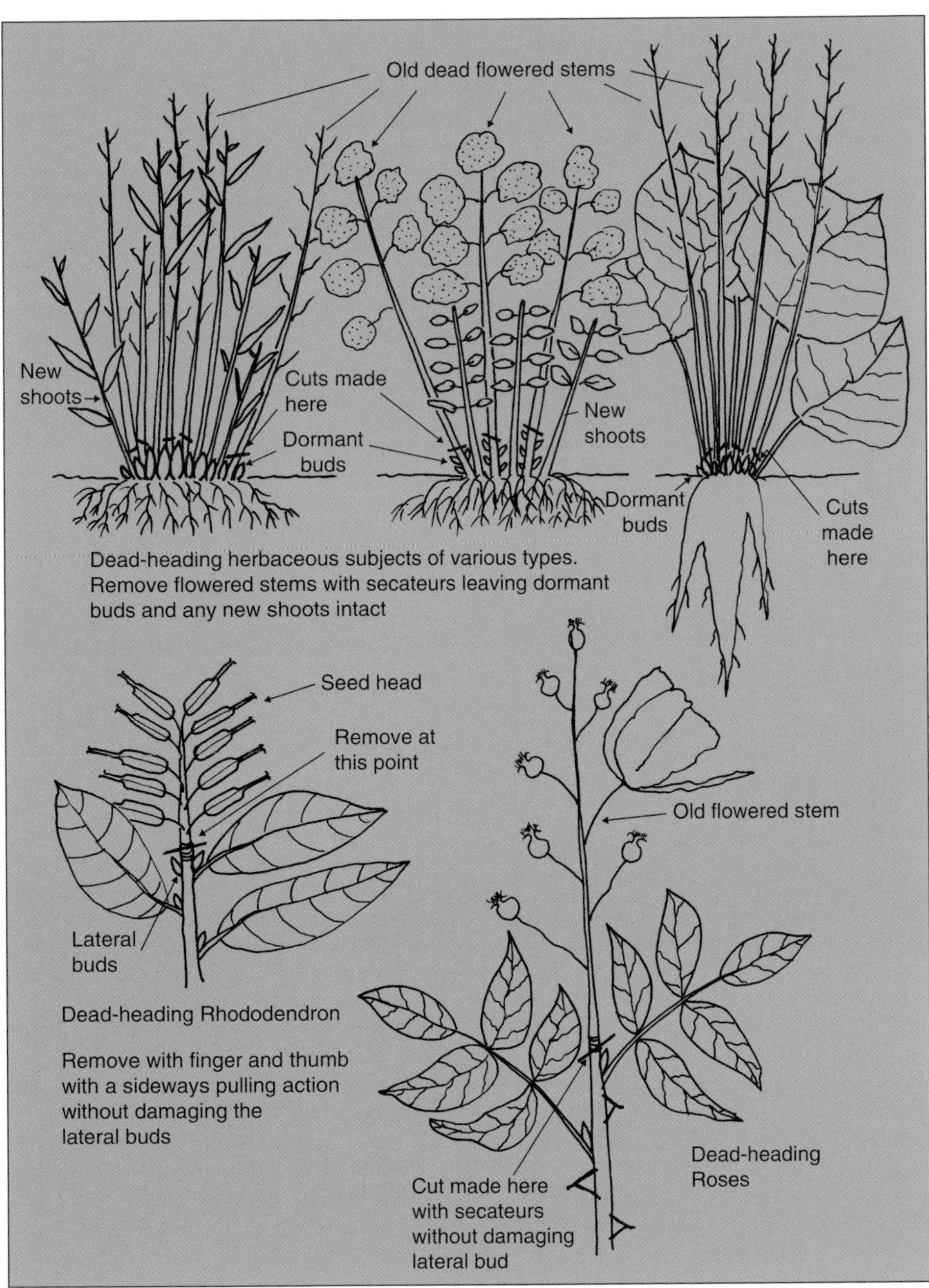

Old dead flowered stems
New shoots
Cuts made here
Dormant buds
New shoots
Dormant buds
Cuts made here
Dead-heading herbaceous subjects of various types. Remove flowered stems with secateurs leaving dormant buds and any new shoots intact
Seed head
Remove at this point
Lateral buds
Dead-heading Rhododendron
Remove with finger and thumb with a sideways pulling action without damaging the lateral buds
Old flowered stem
Dead-heading Roses
Cut made here with secateurs without damaging lateral bud

Pruning systems: dead-heading.

wooden boxes and a proper walkboard (not scaffold board) are preferable to steps for safe working.

Shrub Pruning

While some pruning operations must be carried out regularly – on hedges, on some woody fruit-bearing plants that require stringent pruning regimes to induce fruiting, and on plants that quickly get out of hand and/or bear a poor flower-to-wood ratio – most pruning is in fact a matter of choice, and is neither a necessary nor an essential annual requirement.

Key decisions on shrub pruning are based on whether the subject is evergreen or deciduous, and when it flowers (early or late in the season), which is related to the maturity of wood on which it flowers – whether it produces flowers on the current or previous season's wood. It is also important to recognize that on the same species, flower buds (producing flowers) and mixed buds (producing flowers and foliage) are fatter than the more slender vegetative buds (producing foliage and new extension growth only). Armed with this information (and knowledge of the correct pruning cuts), you can work out if, when and where to prune. Those that flower on the previous season's wood in the spring or early summer are pruned directly after flowering, and those that flower on the current season's wood (usually mid- to late summer) are pruned in the early spring.

Evergreen Subjects

Evergreen shrubs (apart from those used in formal hedging) do not normally need pruning at all; they may need pruning for remedial reasons, to retain their shape and size in a limited area, but certainly will not actually need a systematic annual pruning operation.

Rhododendrons, camellias, kalmias, evergreen berberis and cotoneaster species do not normally require annual pruning. However, should remedial pruning be necessary (including very severe coppicing techniques), then the important question is, on what wood does it flower? Taking rhododendron as an example, they flower on the previous season's growth, produce flower buds in the later part of the summer, and are carried as fat, well developed and prominent buds over the winter months (in common with camellias and magnolias). Potential flowering density therefore can be easily estimated in the previous season. Those buds that are successfully carried over the winter (usually all or most) then flower in the early spring, late spring or early summer (depending on the species), when the very final stages of their development is rapidly completed.

Obviously, therefore, any pruning operation carried out in the late summer/autumn, early winter, late winter or in the early spring would remove potential flowers for either the following or the current season, and would ruin or drastically reduce the aesthetic effect of the shrub. So, pruning at that time of year would only be carried out in an emergency. Mahonias do not technically need pruning, but some types do get rather upright without it. The answer is to place them correctly within the design, but they may be pruned to create a more compact habit, as witnessed by individuals that regularly have propagation material harvested from them.

If an individual specimen has become too large for its position, then it may need remedial pruning involving fairly severe reduction of the woody framework. The planned reduction of large rhododendrons (no matter how drastic) is ideally carried out directly after flowering, giving ample time for the regeneration of healthy new vegetative growth with the potential to produce developed flower buds in readiness for the following season. The development of flower buds is influenced by many factors, but mainly light intensity and the production of hormonal substances. It may well be that the removal of very large amounts of wood in a remedial process, even at the 'correct' time of year, will result in poor flower-bud production anyway, as energy is sacrificed and diverted to rapid vegetative growth instead. The production of many new (previously dormant) shoots, and rapid extension in length, are both favoured over flower buds in an attempt to redress the balance of the root-plate to shoot (branch-work) ratio. However, there may be sufficient energy and favourable conditions to produce some flower buds for the following season, and furthermore the flowers of the current season are not sacrificed.

Camellias, kalmias and evergreen magnolias would be treated in the same way, but remedial pruning involving drastic reductions of wood is rarer (but not unheard of) on these subjects. Other evergreen species such as *Griselinia littoralis, Prunus laurocerasus* (common laurel) and *Prunus lusitanica* (Portuguese laurel), whether grown as individual specimens or in thickets, can also be treated in the same way, but because they are not usually grown for their flowers (but mostly their evergreen effect), the timing is less critical. However, if pruning is necessary it should be carried out in the spring, or early or late summer to ensure both

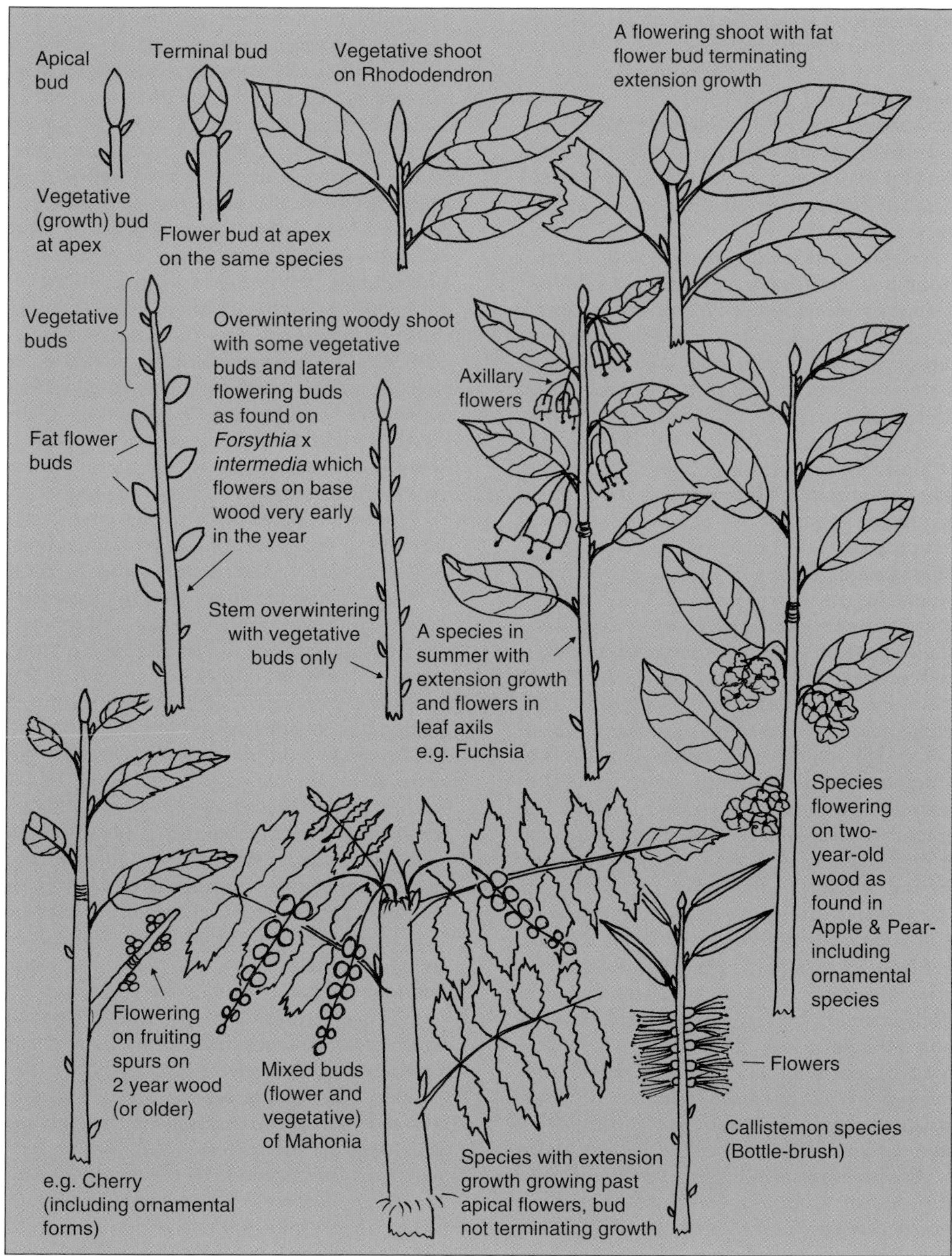

Flower buds and the maturity of wood.

the production of, and lignification of, new shoots prior to the onset of winter.

Deciduous subjects

Deciduous shrubs are divided into two main types: those that benefit (indeed are managed) by annual pruning; and those that are best left alone (just like evergreen shrubs) unless a remedial pruning is necessary.

Modern roses, including hybrid bush (large-flowered and cluster-flowered), climbing and rambling roses are maintained and managed by annual pruning techniques. The regular annual process maintains a healthy flower-to-wood ratio, and furthermore retains it at human level for best aesthetic effect. Species roses and first-cross species hybrid (old-fashioned) roses, on the other hand, are not pruned unless it becomes necessary to retain shape and/or avoid obstruction by their vigorous shoots.

Coppicing is a form of management involving the total removal of the top woody growth, and can be used on an annual basis on some species of shrubs to create rapid regeneration of young stems for aesthetic reasons – for instance, for winter bark colour. Bush willows and some dogwoods (*Cornus* species) grown for their showy bark are managed by annual coppicing because their flowers have no aesthetic merit, but the vigorous new stems covered with waxy new bark are induced to give the best effect. Coppicing encourages very rapid regrowth that lignifies at the end of the season and gives a very good bark effect in the autumn, winter and early spring; it is therefore usually carried out in the late spring to give the maximum period of time possible to benefit from the aesthetics of the bark. Species that are suitable include *Salix alba* 'Vittelina' (yellow bark), *Salix alba* 'Britzensis' (orange bark), *Salix daphnoides* (dark purple/violet bark), *Cornus alba* (red bark), *Cornus alba* 'Siberica' (red bark), and *Cornus stolonifera* 'Flavaramea' (yellow/green bark).

Coppicing is also a useful management tool for *Buddleia davidii* types, as these in particular tend to get out of hand if left unpruned, and aesthetically they are definitely improved by coppice management. If there is sufficient room, vigorous hardy fuchsias, such as *Fuchsia magellenica* 'Riccartonii', are best left alone to form a natural shape. However, they may be managed by coppicing if desired. The coppicing of these types is usually carried out in the early spring because they flower on the current season's growth and benefit from the maximum growth period of the regenerating new shoots before setting flower buds and flowering in the same summer. Later pruning would leave insufficient time for extension growth and bud development in the same season.

Some large-flowered bush roses (hybrid tea roses) respond well to very severe pruning, back to two to three buds from soil level (to outward-facing buds) in the early spring (in effect, coppicing). This system is commonly used, and results in long, vigorous shoots that terminate in flower buds during the same season (good for long-stemmed cut roses). The time needed for the vigorous extension growth to form flower buds does delay flowering until slightly later in the summer, and also it induces flowering over a relatively short period. But roses seldom disappoint regarding bud/flower formation, and the quality of blooms produced in this way is usually first class, so although display time is short, it is good.

The same types (hybrid tea roses) may be pruned less severely (moderate or medium pruning) where only half to two-thirds of the wood is removed. The result of this is to create medium-length wood, good flowers, and mid-season flowering. If light pruning is used – removing only one third or less (lightly tipping the stems) – it results in short-flowering stems, only moderate flower quality, but relatively early flowering.

Obviously you can use all three methods on one individual plant to gain an extended flowering period. However, traditionally, individual plants are pruned using the various degrees of harshness or lightness to give the desired effect. Pruning cluster-flowered bush roses (floribunda types), on the other hand, because they are notoriously uniform in their flowering and their effect is over very quickly, involves the use of all three degrees of pruning severity on each individual plant.

Climbing roses, like the bush forms, flower on the current season's growth, and their climbing (actually scandent) effect is created by their vigour. The management of climbing roses involves retaining a framework of old vigorous stems that are tied in to supports (for example, horizontal wires), and then pruning the shoots that arise from the lateral buds. The lateral shoots emanating from the older woody framework are spurred back to three to five buds long annually in the spring, and the new shoots encouraged by this process flower later in the same year. The main framework can be changed at will by removing one or more of the old main stems, and replacing them with conveniently placed, new vigorous shoots arising from or near the base of the plant.

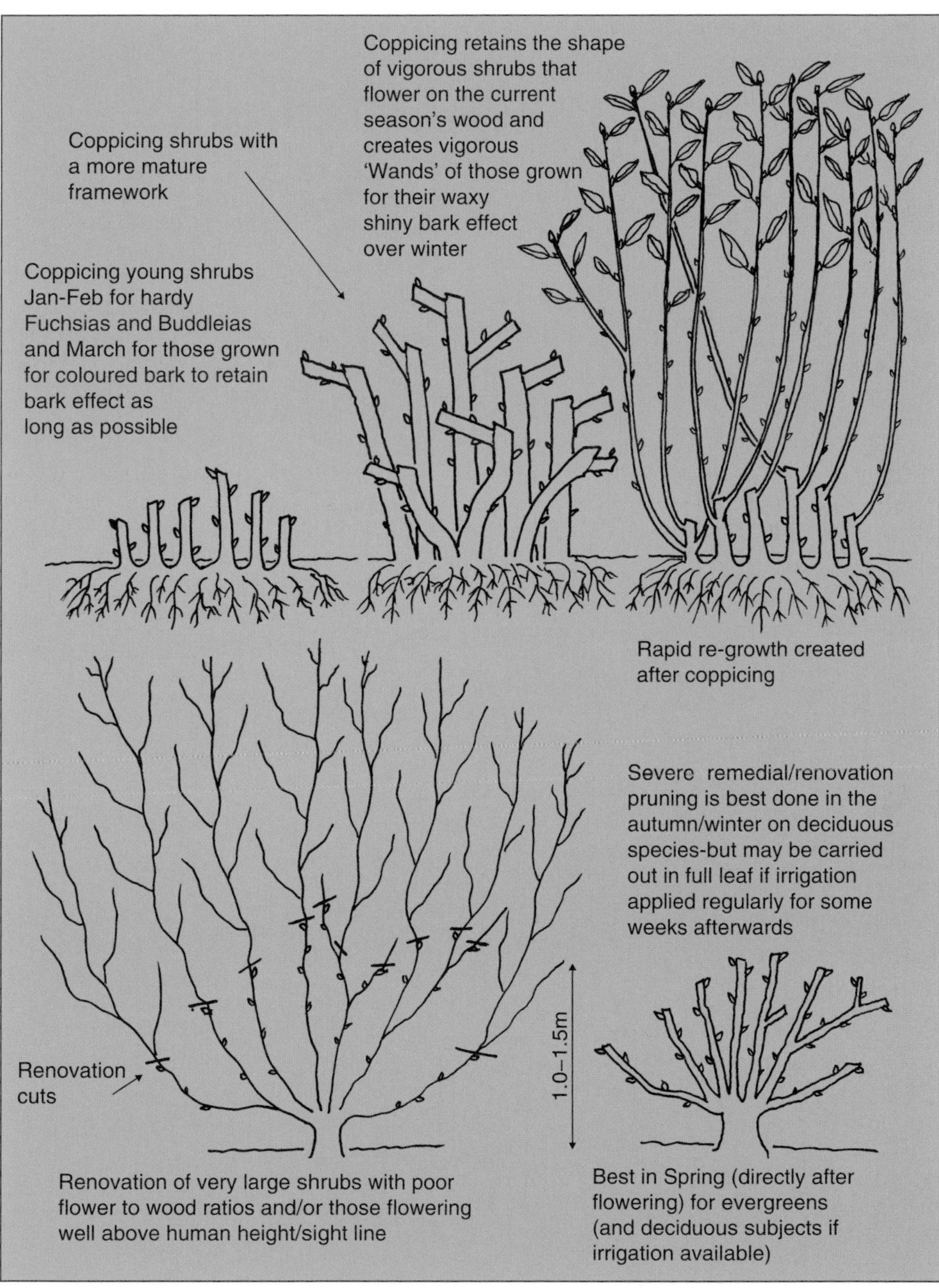
Coppicing retains the shape of vigorous shrubs that flower on the current season's wood and creates vigorous 'Wands' of those grown for their waxy shiny bark effect over winter
Coppicing shrubs with a more mature framework
Coppicing young shrubs Jan-Feb for hardy Fuchsias and Buddleias and March for those grown for coloured bark to retain bark effect as long as possible
Rapid re-growth created after coppicing
Severe remedial/renovation pruning is best done in the autumn/winter on deciduous species-but may be carried out in full leaf if irrigation applied regularly for some weeks afterwards
1.0–1.5m
Renovation cuts
Renovation of very large shrubs with poor flower to wood ratios and/or those flowering well above human height/sight line
Best in Spring (directly after flowering) for evergreens (and deciduous subjects if irrigation available)

Shrub pruning: coppicing and shrub renovation.

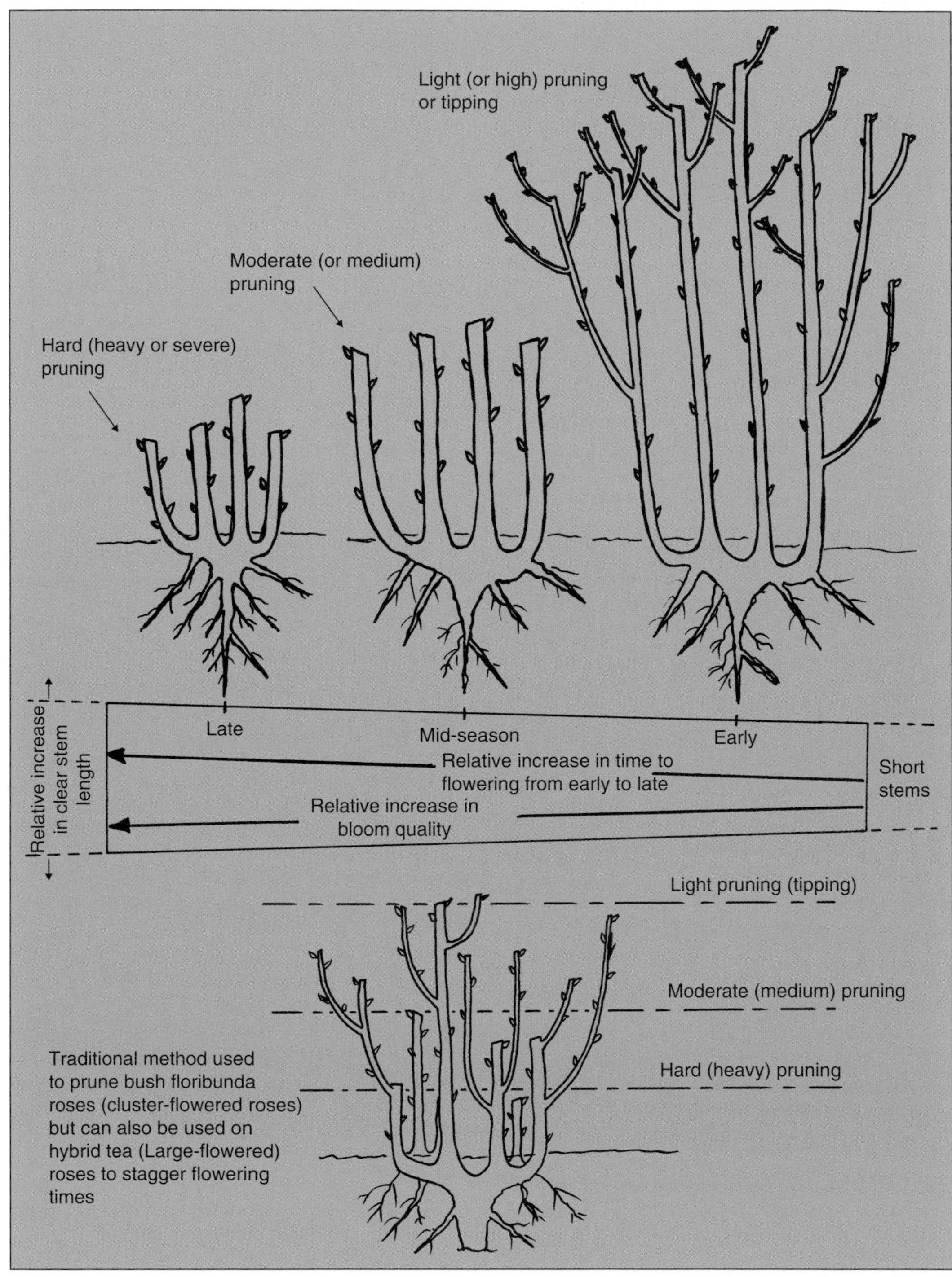

Pruning bush hybrid tea roses (large-flowered roses) and bush floribunda (cluster-flowered) roses.

PLANT STRATEGIES: REGROWTH AFTER PRUNING

Woody plants produce growth promotion hormones in their root systems that travel to the stems, and growth inhibiting hormones within the vegetative buds on the stems. There is a hierarchy within the bud system, as apical buds produce the most growth inhibitor, the next bud down produces slightly less, and so in turn down the stem, with the lowest buds producing least. The effect of one hormone on the other (promoter versus inhibitor) creates the growth rate 'norm' for any particular species. The removal of stems during pruning removes various sites of inhibitor production, but does not affect the quantities of growth promoter produced at the roots. This creates an imbalance of growth promoter over growth inhibitor, and the result is for dormant buds to 'break' and produce rapid extension growth. The rate of extension growth is dependent on how much material is removed at pruning.

Severe pruning removes nearly all the sites of production of growth inhibitor (apical and lesser buds), but does not reduce the amount of growth promoter produced at the roots that moves through the system; hence very rapid regrowth results. The removal of moderate amounts of material results in moderately vigorous regrowth, because only half the sites of production of growth inhibitor are removed (but it includes the most productive buds of the hierarchy). Light pruning removes only a few sites of production of growth inhibitor (albeit some of the most important sites at the apex and just below), so results in shorter regrowth. This effect has to be considered in all aspects of pruning, and is the reason for very rapid regrowth after coppicing.

The induced regrowth forms new buds that in turn produce growth inhibitor, and gradually restore the balance, bringing the growth rate back to the expected 'norm' (usually three to four years after severe coppicing). However, annual pruning processes do not allow this to happen, and very vigorous regrowth results every time the operation is carried out.

Root pruning has the opposite effect (growth promoter is reduced, but inhibitor remains at the same levels), and is used as a management tool to reduce aerial extension growth and to induce fruiting in vigorous subjects such as figs.

Vigorous new shoots with the potential to renew the framework should be tied in and 'protected' (not pruned back) in readiness for the renewal process at a later date. If climbing roses are not maintained annually they quickly become unmanageable, get out of hand, and have poorer quality flowers.

Rambling roses (unlike the other types) flower on the previous season's growth, so their pruning involves the removal of wood soon after flowering, and a continuous annual renewal process (the same technique as used for most raspberries). New shoots of rambling roses are very vigorous, and generally arise from ground level; their general readiness to do this, and the fact that they flower on the previous season's, rather than the current season's growth, distinguishes them from climbing roses. The pruning technique involves untying all stems, and removing all old flowered stems (cutting them back to ground level) directly after flowering (late summer/autumn), and tying in all new stems produced in the growing season so that they can successfully flower in the following year. As with climbing roses, failure to maintain them annually quickly results in major management problems and poor flower-to-wood ratios. Weeping standard roses (because they comprise rambling types on a standard leg) are pruned in the same way as rambling roses, by the removal of all old flowered stems after flowering.

It is important to realize that apart from the rose forms mentioned, and those shrubs responding to coppice techniques, the majority of deciduous shrubs do not require annual pruning. However, should these subjects need pruning for remedial reasons, it is critical to know their time of flowering if their aesthetic appearance is to be retained. Pruning common shrubs such as *Forsythia*, *Deutzia*, *Philadelphus* and *Ribes* in the autumn, for instance, would remove the potential flowers for the following spring, as flower buds develop in the summer and overwinter as developed buds ready to open in the following season. These shrubs are therefore pruned directly after flowering. Only those that flower on the current season's growth should be pruned in the spring.

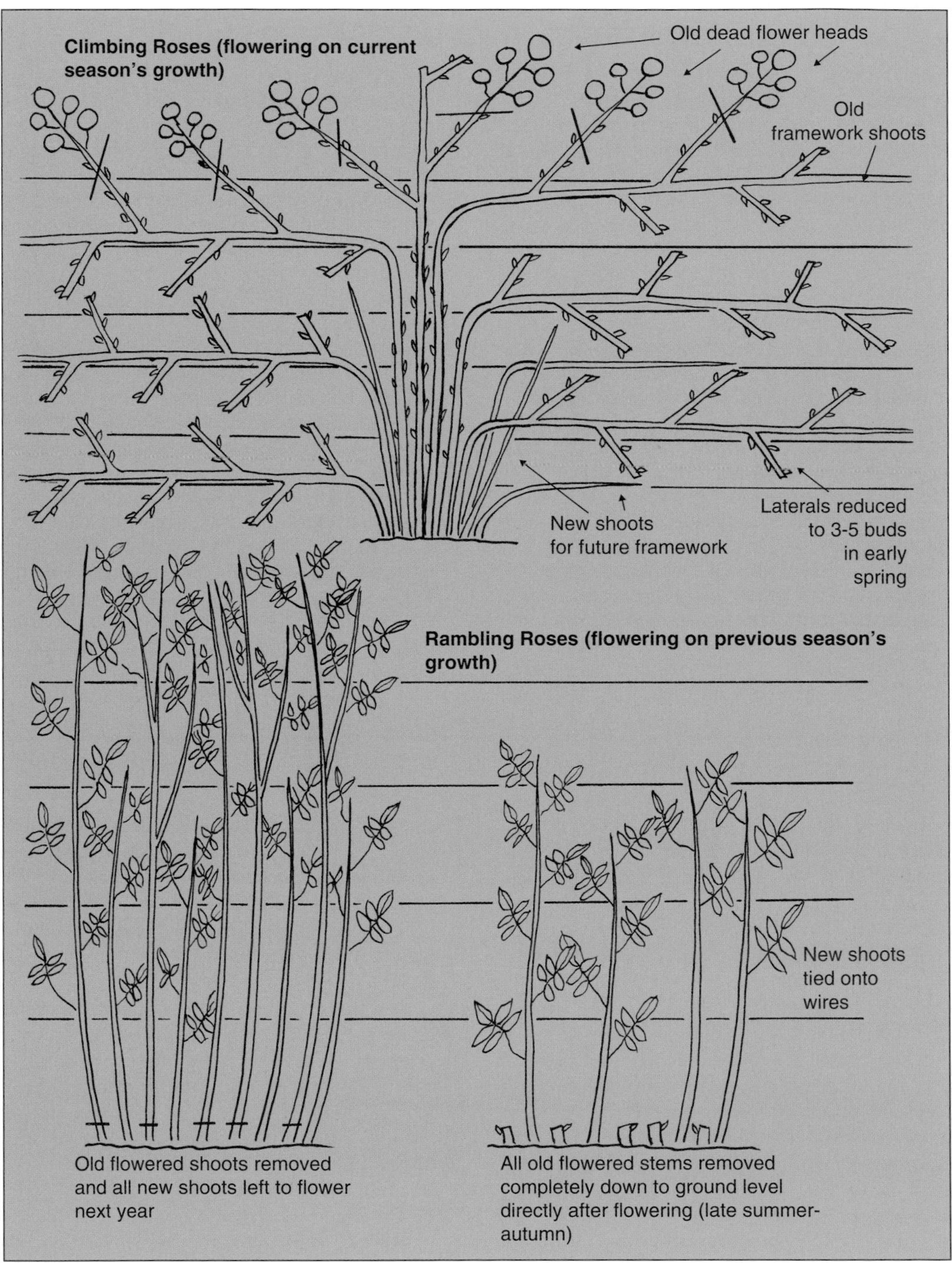
Climbing Roses (flowering on current season's growth)
Old dead flower heads
Old framework shoots
New shoots for future framework
Laterals reduced to 3-5 buds in early spring
Rambling Roses (flowering on previous season's growth)
New shoots tied onto wires
Old flowered shoots removed and all new shoots left to flower next year
All old flowered stems removed completely down to ground level directly after flowering (late summer-autumn)

Pruning climbing and rambling roses.

If a bush shrub does need pruning (and only if it does), proceed as follows:

1. Remove any branches that may be causing an obstruction to pathways or access points.
2. Remove dead, diseased, crossing or rubbing branches.
3. Open up the centre of the shrub for better air flow (and thereby also reduce the potential for fungal disorders).
4. Prune at the correct time of year for the species (spring if it flowers on the current season's wood, and directly after flowering if it flowers on the previous season's wood).
5. Prune back to either a dormant bud or a main lateral branch union, without leaving 'snags'. If you have to prune at anything other than the optimum time for the species, then try to balance the overall shape and the amount of budded (flowering) or potentially flowering wood for the coming season.

Renovation Techniques

Very large shrubs that have outgrown their position, and often have poor flower-to-wood ratios and flowers that are borne above normal human level, may require renovation/rejuvenation. This can be carried out by coppicing (cutting hard back to dormant buds) to leave stumps approximately 1m above soil level (or less in some instances). To ensure/encourage continuity of flowering of subjects that flower on the current season's growth, carry out the renovation pruning in the early spring; for those that flower on the previous season's growth, prune directly after flowering.

It is wise to water the roots of the cut-back shrubs in a dry spring, and particularly wise for subjects that are this harshly treated in the summer months. Most subjects are robust enough to tolerate this treatment without long-term detriment. However, further irrigation may be necessary to ensure success in prolonged drought conditions, because although there is no green tissue left to transpire and lose water, sufficient water is needed to encourage the dormant buds to break, and for new stem development during the summer. Furthermore, as the new growth develops, it produces fresh green transpiring material.

In any pruning operation, pruning cuts on small-diameter wood removing parts of a stem should be made just above a bud (leaving the bud intact). Cuts should be acute enough to aid water run-off, but not so acute as to leave an incomplete diameter of stem to support any developing buds. Pruning cuts that remove lateral branches from a main stem are best done using target pruning, a system that leaves the branch collar intact, with a cut that mirror images the branch bark ridge angle. (The branch bark ridge is an interference pattern in the bark tissues caused where the cylinder of bark of the main stem meets the cylinder of bark of a lateral branch, meeting at two different angles. The branch collar is a swelling on the main branch where the lateral branch meets it, and is created from main branch tissue as it overlaps the lateral branch tissue.)

The branch collar contains healthy, main-stem vascular cambium that is continuous with the higher tissues of the main stem, and it should never be breached when removing a lateral branch. Cuts are made standing off the branch collar, leaving it completely intact, and at an angle that creates a mirror image of (forms an isosceles triangle with) the branch bark ridge line. Pruning cuts using target pruning are considered particularly important when pruning trees, and often branch collars are more prominent around the large branches of trees. However, it is good practice to use the same technique on shrubs.

When a branch is removed from a shrub or tree, the vascular cambium (the tissue between the xylem and phloem responsible for new cell production) is exposed. The sudden lack of containment by the outer tissues, and the change in environment created by the sudden exposure of the internal tissues, triggers the vascular cambium into producing callus (disorganized, undifferentiated cells). Callus changes in nature relatively quickly, and ultimately the callus cells differentiate into woundwood (having all the tissues associated with a woody stem, including bark).

Suckers are juvenile shoots, but unlike epicormic shoots, they are initiated on root systems, not on stems, so are adventitious. Not all species have the ability to sucker, and certainly some species have a far greater propensity to do so than others. *Prunus avium* (common cherry), *Amelanchier Canadensis, Laurus nobilis* and *Rhus typhina* (sumach tree) all sucker readily.

The terminology 'sucker' is also used for shoots arising from an under-stock (rootstock) on a grafted plant. This holds true when the shoots arise from the root-plate of the under-stock. But technically the term should not be extended to include shoots arising from the base of the stem (or even part-way up the stem) of a grafted or budded

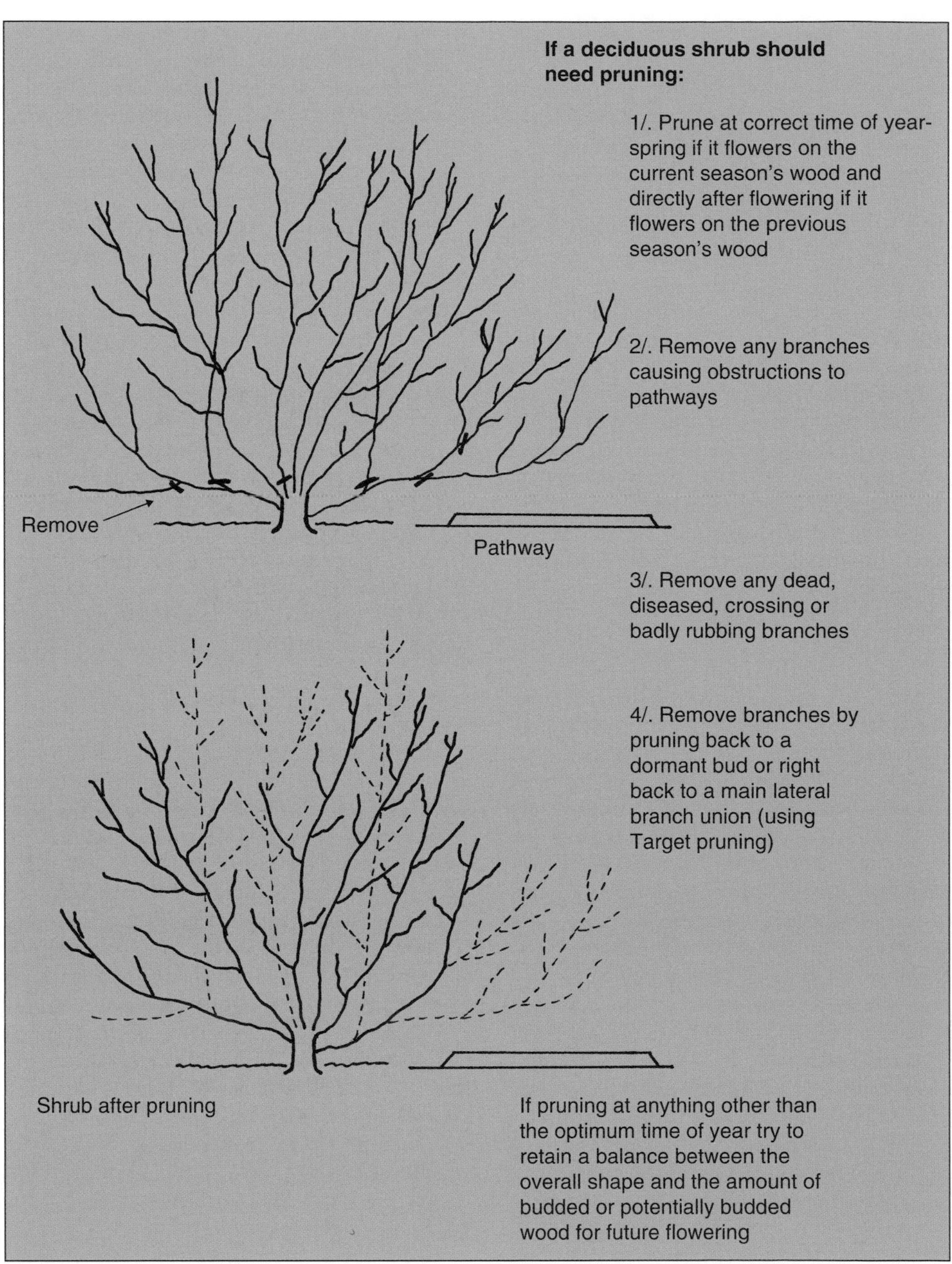

If a deciduous shrub should need pruning:
1/. Prune at correct time of year-spring if it flowers on the current season's wood and directly after flowering if it flowers on the previous season's wood
2/. Remove any branches causing obstructions to pathways
Remove
Pathway
3/. Remove any dead, diseased, crossing or badly rubbing branches
4/. Remove branches by pruning back to a dormant bud or right back to a main lateral branch union (using Target pruning)
Shrub after pruning
If pruning at anything other than the optimum time of year try to retain a balance between the overall shape and the amount of budded or potentially budded wood for future flowering

Pruning deciduous shrubs.

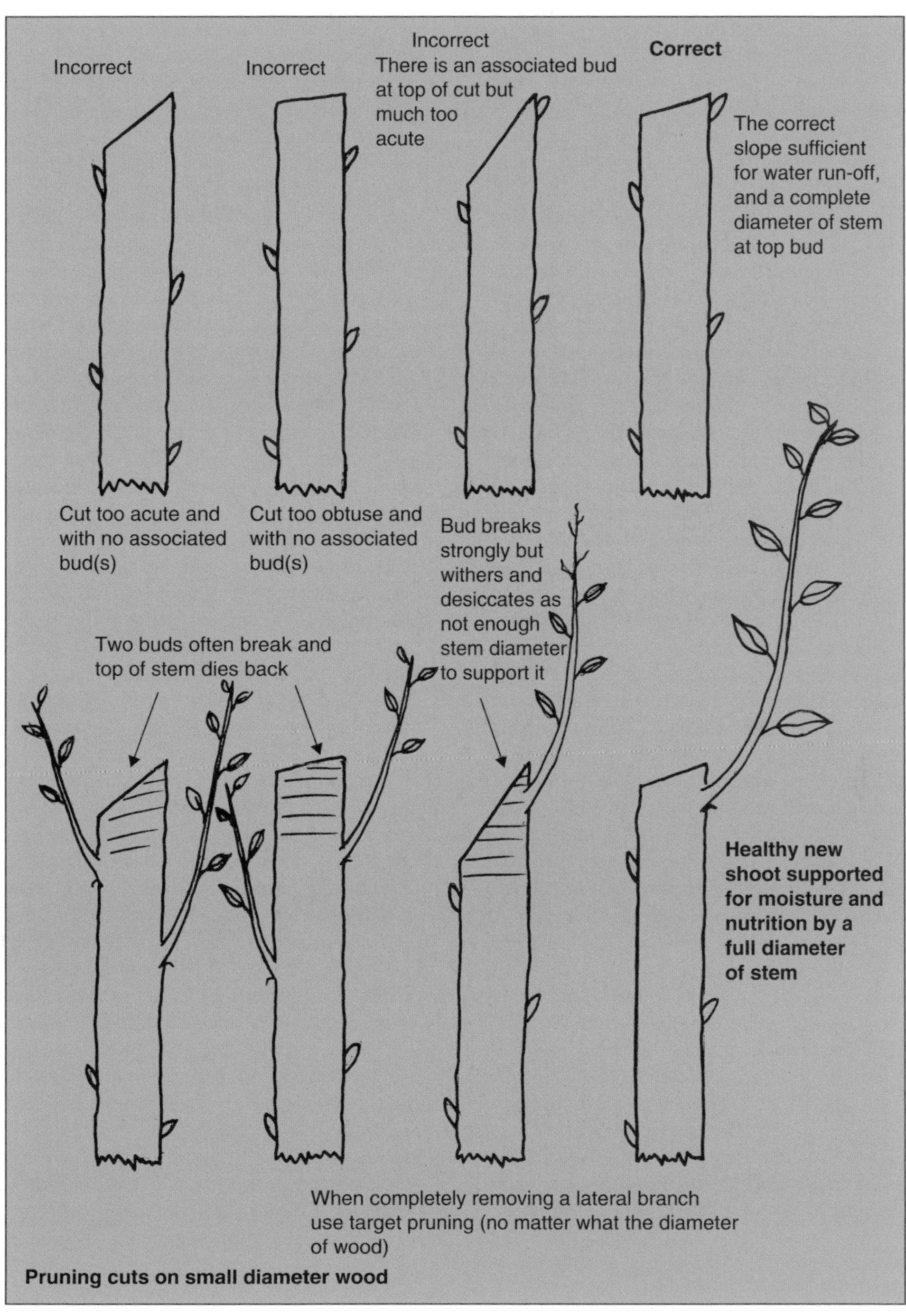
Incorrect
Incorrect
Incorrect
There is an associated bud at top of cut but much too acute
Correct
The correct slope sufficient for water run-off, and a complete diameter of stem at top bud
Cut too acute and with no associated bud(s)
Cut too obtuse and with no associated bud(s)
Bud breaks strongly but withers and desiccates as not enough stem diameter to support it
Two buds often break and top of stem dies back
Healthy new shoot supported for moisture and nutrition by a full diameter of stem
When completely removing a lateral branch use target pruning (no matter what the diameter of wood)
Pruning cuts on small diameter wood

Correct pruning cuts on small stems.

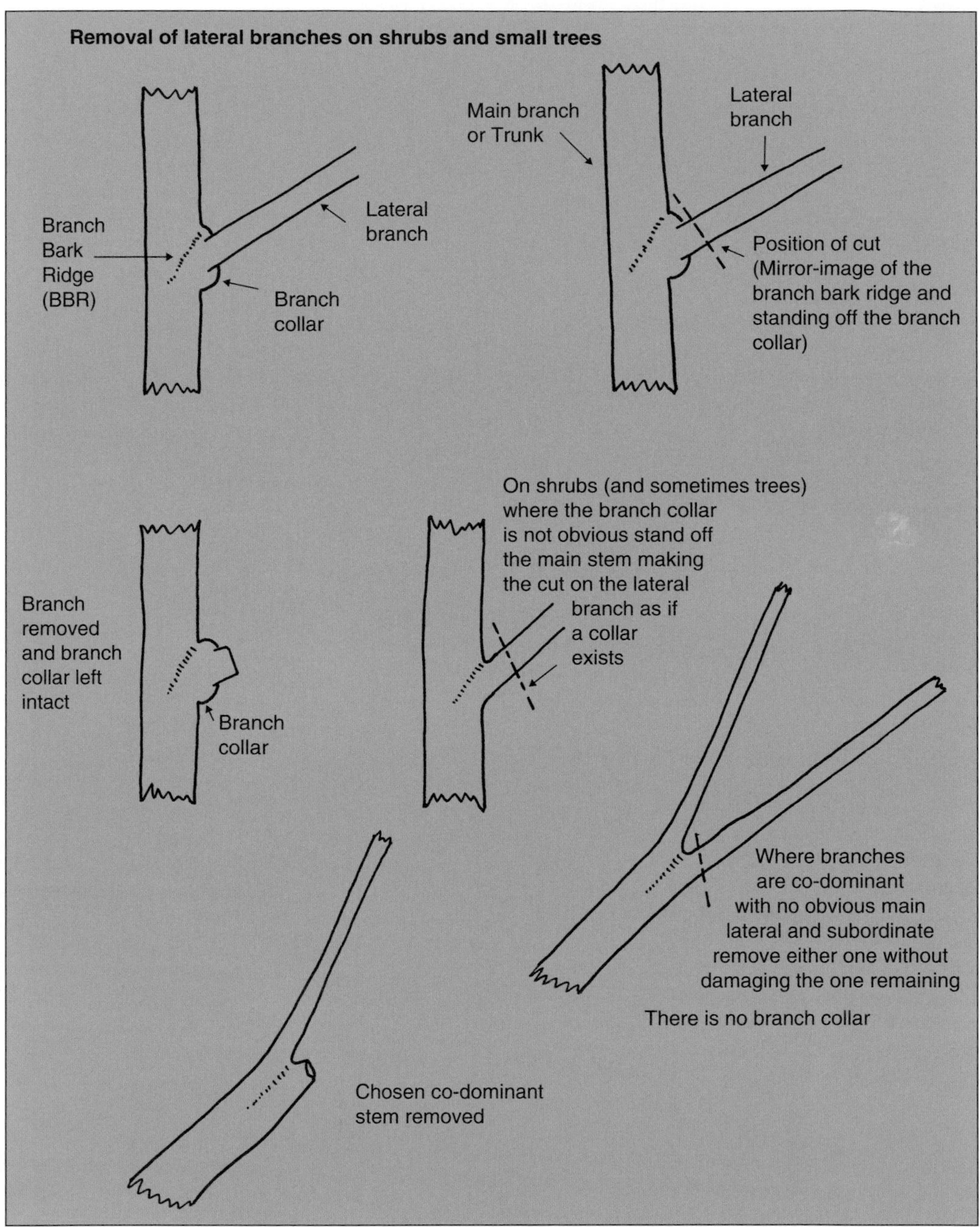

Target pruning and the correct removal of co-dominant stems.

plant, as these should be called epicormic shoots rather than suckers. However, as they are all similar in nature it probably doesn't matter. Stress in a grafted plant, caused either by compatibility problems between under-stock and the grafted cultivar, or by gradual crown demise, induces sucker production in the same way that stress induces excessive epicormic shoot production on the stems of some species.

Ornamental and fruiting cherry trees, produced by cultivars of cherry being grafted (or budded) low on to wild cherry under-stocks (rootstocks), will often show suckers of wild cherry arising from the roots at the base. Or, if the tree has been budded or grafted higher, epicormics can be produced anywhere on the under-stock stem, and suckers on the under-stock root-plate. Cultivars of *Rhododendron* will often present suckers and/or epicormic shoots of *Rhododendron ponticum* from their under-stock, and standard forms such as standard roses often sucker at the base, all of which, because of their vigour, will take over, and smother the main desired cultivar if not removed. Where standard forms are created from cuttings on their own roots, such as standard bay (*Laurus nobilis*), any suckers from the root-plate or epicormic shoots from anywhere up the stem are shoots of the true plant (bay), and not vigorous under-stock. Nevertheless they have to be removed to retain the desired standard shape.

Suckers are best removed by pulling them off (even with the relatively large amount of tissue damage this may create), because cutting them back with secateurs has the same effect as harsh pruning, and they will regenerate vigorously.

Removal of Reversion

Chimeras (comprising two genetically different tissue types) occur because of a genetic breakdown of one or more of the cell layers at the shoot apex. The continued periclinal divisions at the outer tissues of the shoot apex (the tunica) form tissues that are 'normal'/typical in shape but not in pigmentation. The genetically different cells are commonly unable to produce chlorophyll, and the variegation of leaves produced by this tissue is obviously caused by their poor chlorophyll content, shown by the low amount of green pigmentation.

Chimeras are often relatively stable and may continue to divide and produce a consistent variegation pattern because the genetic abnormality is faithfully reproduced by the periclinal divisions. However, the other (inner) tissues of the shoot apex (corpus) continue to divide both periclinally and anticlinally to retain 'normal' tissue patterns to the extension growth – thus leaving one or several shoots as variegated, and the rest of the growth as 'green'. Sometimes anticlinal divisions from lower tissues will 'contaminate' the area of chimera cells and produce green material at or near the variegated areas. This is termed 'reversion', meaning that the growth reverts back to its normal green coloration with full pigmentation.

Conversely, on rarer occasions, continued successful periclinal divisions in some areas of the plant may lead to many shoots (with associated variegated leaves) having no chlorophyll and appearing as completely white outgrowths. This is a relatively common occurrence in some forms of variegated holly (*Ilex*). Some variegated cultivars are more stable than others, but all have the potential to revert. The removal of reversion from variegated types, in order to retain the desired (arguably aesthetic) variegated effect, can be an ongoing maintenance operation with some unstable cultivars.

Maintenance programmes therefore include the careful and continuous removal of reverted branches from shrubs and trees back to wood that bears variegated material. All variegated forms can be prone to reversion, as can other genetic variants that create new cultivars, such as cut-leaved and purple-leaved versions of plants. Some cultivars, such as *Weigela florida* 'Variegata', *Cornus alba* 'Elegantissima' and *Cornus alba* 'Spathei', are more prone than others to reversion problems, as are many variegated hollies (*Ilex*).

Pruning Fruit Trees

Fruit tree pruning is an annual process, and pruning for fruit yield is in many ways no different to shrub or tree pruning for aesthetic effect. In both instances retaining the correct flowering wood is of paramount importance, as the main aim of pruning is to retain a good flower-/fruit-to-wood ratio. And again, knowledge of the maturity of wood on which they flower (and therefore fruit) is essential. Fruit trees flower on the previous season's growth. However, traditionally, bush apples and pears are pruned in the autumn or winter, so recognizing the difference between fat flowering or mixed buds and the more slender vegetative buds, and where they are liable to be formed, is very important for the process. This factor alone will prevent you from removing too many branches of flowering/fruiting potential.

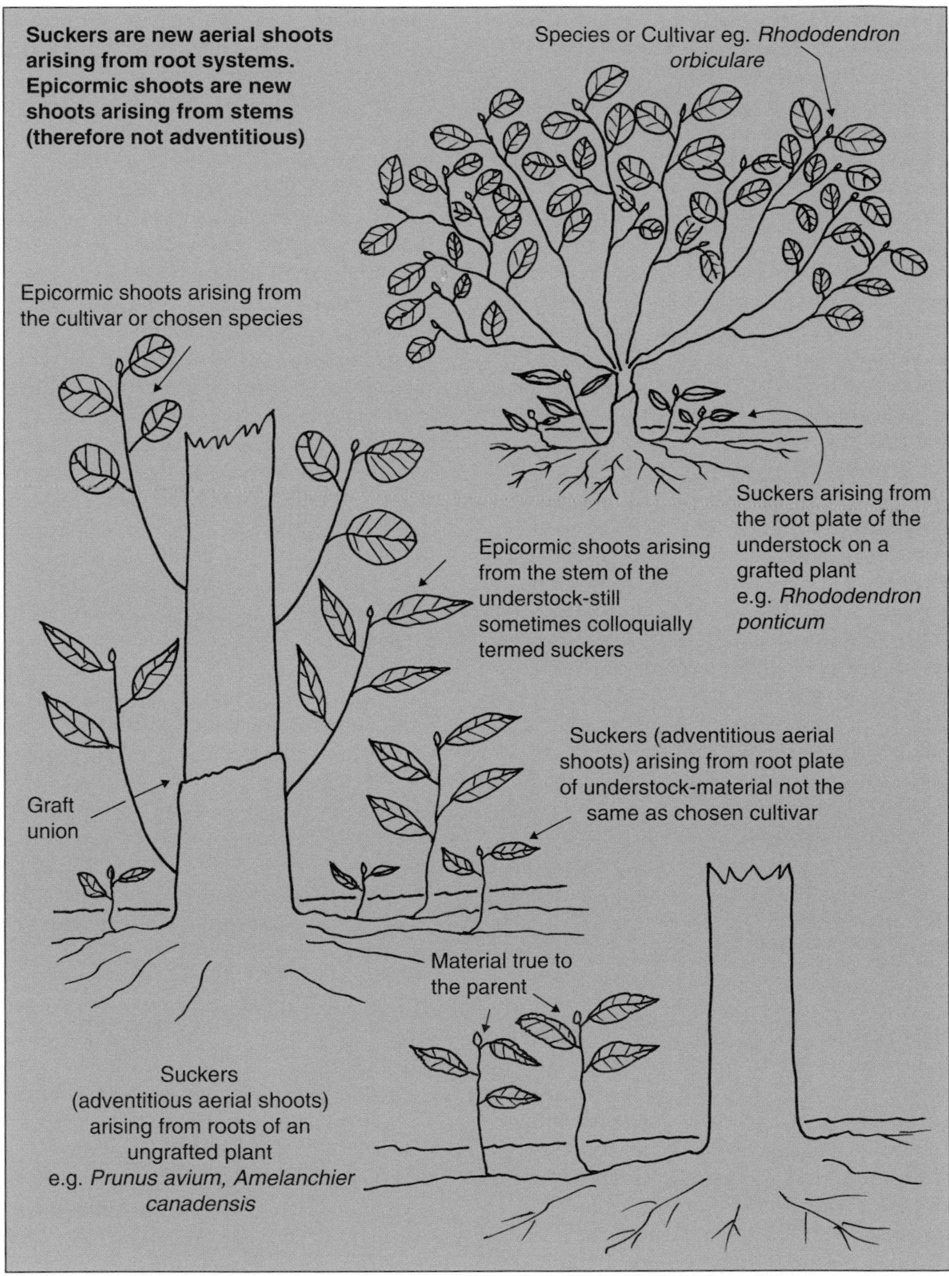
Suckers are new aerial shoots arising from root systems. Epicormic shoots are new shoots arising from stems (therefore not adventitious)
Species or Cultivar eg. *Rhododendron orbiculare*
Epicormic shoots arising from the cultivar or chosen species
Suckers arising from the root plate of the understock on a grafted plant e.g. *Rhododendron ponticum*
Epicormic shoots arising from the stem of the understock-still sometimes colloquially termed suckers
Suckers (adventitious aerial shoots) arising from root plate of understock-material not the same as chosen cultivar
Graft union
Material true to the parent
Suckers (adventitious aerial shoots) arising from roots of an ungrafted plant e.g. *Prunus avium, Amelanchier canadensis*

Suckers and epicormic shoots.

Different apple and pear cultivars vary in their fruit-bearing positions: some are tip-bearing and some are spur-bearing, but all apple and pear cultivars bear fruit on two-year-old wood at the earliest. Hence tip-bearers will flower (and bear fruit) near the tip of the shoot, not on the current season's growth, but on wood of the previous season and older. Spur-bearers produce specialized, short lateral growths (spurs) on two-year-old wood or older, and large older spurs are some distance from the shoot tips. Flower buds are therefore borne not on growth at the very tips of branches, but on wood behind the first girdle scar at least.

Removing large amounts of branch ends from tip-bearing types would obviously remove most of the fruiting potential. Pruning systems therefore rely upon the removal of selected branches only, leaving plenty of flowering wood, and keeping any tipping back of the remaining branches to the current season's growth only. The pruning of bush apple trees usually comprises the removal of any dead, diseased and poorly functioning/fruiting branches first, and then a few healthy branches radiating off the main framework to improve the shape and openness of the tree. Open-centred trees are desirable, as the relative ease of air movement through their crown reduces the risks of fungal infection. The pruning process therefore involves a balance between openness of crown and the retention of as much fruiting wood as possible. The same is also true of spur-bearing types, but these may have the ends of branches pruned back because the fruiting spurs will still be left intact.

Most systems of pruning involve the removal of some older wood and the retention of some non-fruiting wood that will ultimately improve the structural shape of the tree and bear fruit at a later date. Such systems of 'renewal' involve planning for the future structure and fruiting of the tree in the long term – for instance, that in five years' time branch 'A' can be removed and be replaced in dominance and effectiveness (fruiting and structurally) by branch 'B', which is not yet fully developed. Ultimately the aim is to retain good uniform branch-work (with good crown shape and branch spacing), coupled with continuity of fruiting.

Summer Pruning of Tree Fruit

Compact forms of apple and pear trees, including dwarf pyramid, cordon, espalier and spindle bush (spills) remain compact because of their under-stocks. The vigour of any rootstock (under-stock) on which they are grafted affects the ultimate vigour of the cultivar grafted on to it. In the case of apple trees, a lot of research work was carried out by East Malling Research Station in Kent. Their work categorized the various vigours of individual clones (genetically identical rootstocks) whose effect on grafted cultivars can be estimated with some accuracy. Dwarfing rootstocks keep the cultivar grafted on to them small and compact, whereas vigorous rootstocks have the opposite effect.

However, compact forms remain compact not only as a result of the chosen under-stock, but also by summer pruning. Although major reshaping and branch removal is carried out in the autumn or winter months, all compact forms involve summer pruning on an annual basis. The removal of relatively large amounts of leafy, sugar-producing growth by pruning back tip growth by four to five buds whilst in full leaf reduces the energy budget of the tree and therefore the rate and quantity of regrowth, to retain their compact habit. It ensures that the older, fruiting wood is not touched.

Plant Nutrition

Relative amounts of essential elements available to plants will depend on many factors, including the availability of specific elements from geological (rock) origin. The types and amounts of organic litter created by the local plants and organisms will also affect the soil type, as will the types and quantities of living organisms present in any particular soil capable of the diminution process, decomposition and the dispersal of detritus throughout the matrix. The soil type and structure affect nutrient availability, and the previous use of the site can also affect the fertility of the soil.

Plant foods come primarily from sugars and starches, but other essential elements are also required for their successful nutrition. Plants need nitrogen, phosphorous, potassium, calcium, sulphur and magnesium as essential elements in varying amounts, and deficiencies cause cell death. It may be that these nutrients are actually missing from the soil solution – often they have been leached away. In other instances nutrients may be 'locked up' in the soil matrix because of the pH of the soil, as the pH of a soil will alter the solubility, and therefore availability, of these essential elements. This may limit the choice of species for some sites. Alternatively, it may mean exploiting a particular site because of its pH value.

Water is an essential ingredient, not only for the process of photosynthesis, but also to act as a carrier of the soluble salts within the soil matrix, and to transport them from the soil matrix throughout the plant body. The essential elements include gases such as carbon dioxide, oxygen and nitrogen (all found in the atmosphere). Carbon dioxide is an important gaseous compound necessary for the process of photosynthesis, supplying the carbon for carbohydrates and releasing oxygen for respiration. Oxygen is essential for respiration, and nitrogen is essential for the chlorophyll molecule – but unfortunately atmospheric nitrogen cannot be taken up by plants, and the major part of plant uptake of nitrogen is by soluble nitrogenous salts.

The essential ingredients of the process of photosynthesis are light, carbon dioxide, water and chlorophyll. Sugars produced during photosynthesis may be used straightaway whilst in their soluble forms, or they may be stored as insoluble starches, to be used at a later date. Sugars, as well as being capable of conversion to sucrose and starch, can also be converted to other important plant foods such as fats, and can be used to form proteins. However, protein synthesis requires other chemical elements found as soluble forms in the soil – these include nitrogen, sulphur and phosphorous. Plant foods therefore include carbohydrates (sugars, glucose, sucrose and starches), fats, proteins, and nitrogenous compounds. So the conversion of non-living elements and compounds such as carbon dioxide (CO_2), water (H_2O), nitrates, sulphates and phosphates are used to produce the living tissue of the plant.

The importance of chlorophyll in cells is definitely to produce sugars, as cells without chlorophyll can still change glucose into starch and synthesize fats and proteins for assimilation. But the other essential ingredients necessary for plant food production and assimilation come from solutes in the soil matrix that makes up the rhizosphere (root environment). The essential nature of these soluble salts arises out of the fact that, not only are many used in protein synthesis, but some are actually essential in the chlorophyll molecule itself – examples being nitrogen, iron and magnesium. Thus a deficiency in any, or all of these soluble salts causes malfunctioning (or dead) chlorophyll, resulting in chlorosis (yellowing of the tissues) and, unless rectified, ultimate plant death.

Lime-Induced Chlorosis

Lack of solubility of nutrient elements affects their ultimate availability to plants, and the pH of the soil will influence nutrition, as mineral elements have differing solubility in different pH levels. Lime-loving plants are called calcicoles, and lime-hating, and therefore by definition acid-loving plants are called calcifuges. Calcifuges such as rhododendrons thrive in an acid (low pH) soil, and under these conditions, iron in particular is very soluble. Planting a calcifuge into an alkaline (high pH) soil means the plant cannot gain nutrition properly, because the iron is poorly soluble, and therefore not available to it at levels it is used to. The effect is known as 'lime-induced chlorosis', because due to the lack of iron the leaves go very yellow. It is a condition that cannot be maintained for long periods of time, as the malfunctioning chlorophyll does not produce sugars, and the energy levels of the plant diminish rapidly.

The effects of lime-induced chlorosis can be ameliorated by the use of fertilizers containing fritted trace elements (or chelates) that are soluble in the high alkaline conditions, or by the use of sequestered iron, which fulfils the need for iron directly to the roots as a soluble product. However, this may prove to be an expensive remedy, and getting the pH right in the first place, even limiting the range of species grown because of the pH, is always the best option for continued success.

The essential elements are divided into two groups: those used by the plant in relatively large amounts are called the macroelements, and those needed in very small amounts (yet nonetheless essential) are known as microelements (or trace elements). The macroelements are oxygen, carbon, nitrogen, phosphorous, sulphur, potassium, calcium and magnesium. The microelements are iron, manganese, zinc, copper, boron, molybdenum, chlorine, and cobalt. Excessive concentrations of any soluble nutrient salts are harmful to plants, but more than minor traces of microelements are very toxic (they are phytotoxic).

Nutrient salts usually exist in the soil as electrolytically charged ions, which influences their availability – not their scarcity, or otherwise, within the soil matrix, but their availability to the plant. The polarity of electrolytic ions will affect the affinity towards, or the repulsion from, soil particles. Clay particles are negatively charged and will, therefore, strongly attract positively charged ions, and will retain positively charged essential elements against gravity and other forces. Potassium, calcium and magnesium are examples of positively charged ions held tightly in soil crumbs via electrolytic forces, which affects their availability at any time, and means they resist leaching.

Phosphorous tends to be in sufficient quantities in temperate soils, but is often unavailable to plants due to its poor solubility across most pH levels. The difficulty is overcome by mycorrhizal colonization of roots processing phosphorous found in the soil matrix: mycorrhiza take sugars from the roots and in return convert phosphates from insoluble to soluble forms that can be used by the plant.

Nitrogen is relatively soluble, and for this reason is often in short supply, especially in high rainfall areas, because the soluble salts of nitrogen are easily removed from the soil by percolating rain and gravity-powered drainage (leaching). Poor nitrogen levels are typified by yellowing leaves (chlorosis), highlighting the essential nature of the presence of nitrogen in the chlorophyll molecule. It is often necessary, therefore, to add nitrogenous salts as a fertilizer. Soil fertility may be improved by the addition of chemical salts, decided upon after a soil nutrient test.

Plant nutrition may be aided by the broadcast application of artificial fertilizers to the soil surface (products comprising soluble inorganic chemical salts known to promote plant growth). Alternatively, organic materials can be added to the soil – materials that will be broken down by soil organisms to ultimately produce the very same inorganic salts needed for healthy plant growth.

An extra benefit of using organic fertilizers is the soil-improving qualities, which are properties not found when using applications of straight inorganic chemicals (artificial fertilizers). Mineral materials open up the soil structure, and the organic materials (following biodegradation) add humus to soils, thereby improving both the structure and the nutrient status of the soil. In the natural world, decomposition of organic litter, and erosion of rocks/minerals found in the soil, supply necessary mineral salts to plants. Organic materials in the final stages of their decomposition go through a process of mineralization to produce ammonium salts that supply nitrogen. The organic litter of both dead plants and animals ultimately biodegrades to form plant nutrients.

Growing Media and 'Composts'

Growing media used for container-grown plants need to have similar properties to soils – they must provide anchorage, moisture retention, drainage of free water, aeration, and suitable nutrients essential for healthy plant growth. Some mixtures used for the purpose actually include a specific soil type known as a 'loam' (loams comprise a balanced mixture of mineral and organic content) – these are known as loam-based composts, whereas loamless types are usually based on peat.

Traditionally, products used for seed-sowing, nursery potting and growing on in containers are known as composts. Their history holds the key to this terminology, which is still used today. Certainly in Victorian times (and probably before) both private gardens and commercial growers would use their own products for container-grown (pot-grown) plants. The chosen formulae may have included 'soil' (where weed infestation would have been a major problem), but was more likely to have been based on composted materials – hence potting 'compost'. The proportions of the various composted materials, soils and other materials in the mixtures varied considerably between growers, as did the results. Very successful formulations that gave particularly good results were kept as closely guarded secrets; so secrecy was the norm, and competition between growers was fierce. There was little atmosphere of co-operation between the different factions.

In an attempt to standardize growing media and create a standard quality, the John Innes Institute looked at various formulations for 'composts' (potting media). Their research led to a set formulation regarding the bulk ingredients, and this was modified for various uses by using differing amounts of an added nutrient. They are adjusted in 'strength' by adding more fertilizer according to the plant type, the phase of the plant growth and development, and the period of time a plant might be expected to remain in the container.

John Innes formulations have five main ingredients: a good loam (steam sterilized to kill harmful pathogens and weed seeds), sphagnum moss peat, coarse sand or grit, ground chalk to adjust the pH, and the level of nutrient considered suitable for the particular plant needs. Sphagnum moss peat is very low in nutrition but is a fairly uniform and inert material (with little or no weed infestation, and no bacteria or fungal spores), mainly used for its moisture-retaining properties. Washed silica sand is used to ensure that no pollutants or phytotoxic materials are present and that the pH is not too high. The amount of ground chalk is decided after a pH test of the bulk ingredients. Using standardized materials therefore cuts out the need to continually do pH tests.

The 'weakest' John Innes formula is used for seed sowing and has triple superphosphate (the most soluble form of phosphates) as the base fertilizer.

The inclusion of phosphates is designed to aid root development, commencing with the radicles of germinating seeds. The main potting 'composts' all have the same bulk ingredients – that is, the same proportions by volume (not weight) – and are numbered in ascending 'strength': one, two and three, depending on the amount of nutrient included. John Innes potting compost number one contains only a small amount of the base fertilizer (which comprises a mixture of organic and inorganic fertilizers), and is used for the initial potting of small plants. Number two contains more base fertilizer and is used for more developed plants, and number three has the most in the formulation and is used for larger plants and those liable to remain in the container for long periods of time. Special ericaceous mixes (which should be called calcifuge mixes) are formulated for acid-loving plants such as rhododendrons and heathers: these comprise the basic formula, but ensure the loam is acidic, using sphagnum moss (which is always acidic), and omitting the ground chalk.

John Innes formulations and those amended by the addition of extra peat are still used today. However, work done by UCLA (University of California, Los Angeles) has produced loamless formulae that have largely superseded the John Innes types. The research was probably borne out of the fact that loam is a naturally occurring soil, and is therefore unsustainable – its use in 'composts' diminishes supplies that can only be naturally reconstituted over very long periods of geological time.

UCLA formulations use only peat (75 per cent) and sand (or grit – 25 per cent), giving a relatively inert, weed-free medium without the need to use or sterilize loam. The peat content gives very good moisture uptake and retention against gravity, and the pore spaces (aided by the coarse sand, or grit) give good drainage. The main research work therefore involved finding out the correct nutrient formulations that would successfully hold against leaching as a result of the high rainfall in California, and that would give sufficient nutrition over a protracted period of time.

The extra bonus provided by the UCLA formulations (and all peat-based and peat-amended formulations) is their relatively low weight when compared with loam-based types. This, coupled with a move away from baked clay pots and the use instead of polyethylene or plastic containers, means that large numbers of full containers can be carried relatively easily in large carrying trays. The down side is that containers left to stand in outside frames or standing-out areas, when drying out tend to be unstable and fall over. The mineral content (sand or grit) of the UCLA formulations does mitigate against this a little, as does good management, in that irrigating will retain a constantly moist compost. In the same way that John Innes formulations have been modified by some, the UCLA formulations have been modified for cooler temperate climates and many formulations and proprietary brands of loamless growing media now exist.

Fertilizer mixes and the way that nutrition arrives at, and is meted out at the root zone, is critical for good plant health in container-grown plants. Technological advances in fertilizer manufacture have moved growing media on another stage further. Families of quick-release fertilizers comprising very soluble nutrients (straight chemical salts, usually in crystalline form) can quickly rectify specific nutrient deficiencies. Slow-release fertilizers comprise a mixture of nutrient salts compressed into globules, pellets or granules. The nutrients are released from their various layers by soil moisture slowly dissolving them. More recent technology introduced controlled-release fertilizers that comprise a mixture of soluble plant nutrients contained in resin-coated capsules. The resin coat has small cracks and splits in it that allows it to mete out the nutrition over a long period of time, controlled by available 'compost' moisture and the ambient temperature. Warm, moist periods (good growing conditions) fortuitously offer the best conditions for nutrient release. Furthermore, the introduction of controlled-release fertilizers has led to families of peat-only growing media that can be used not only for containers, but also for grow-bags usable for food and ornamental flowering crops.

Just as the use of loam in JI formulations is a major ecological problem, both JI and UCLA formulations (and most others) are unsustainable in their use of peat, which is also a naturally occurring material taking periods of 2,000 to 5,000 years to produce. Non-peat alternatives are being sought, with coconut fibre (coir) as one such alternative that has given promising but somewhat variable results so far.

Propagation Media

Most proprietary brands include formulations of seed-sowing media, and these have a relatively low nutrient content. Besides seed-sowing media, propagation media include formulations for rooting cuttings: these comprise mixtures of inert materials with an emphasis on moisture retention to increase moisture at the base of cuttings (the eventual root zone), yet with essential aeration. The moisture-retentive properties also help to release extra moisture into the local atmosphere to increase and sustain humidity levels. When we talk of propagation media, we are usually referring to the material used in mist units, fogging units, and propagation frames of various types. Many media for cuttings are based on peat and coarse sand/grit (50:50 per cent), peat and perlite mixes (60:40 per cent), or others including 100 per cent vermiculite and 100 per cent perlite, and mixes involving all types.

Media for rooting cuttings do not usually have added nutrition, and cuttings are mostly successful because they spend a relatively short time in the rooting environment, and are moved swiftly into potting media containing a higher nutrient status. However, some researchers have found that the inclusion of low levels of balanced nutrition in the rooting medium aids the rooting process and the subsequent early growth of the cuttings after rooting. Alternatives include the application of liquid nutrients over the top of the rooting cuttings to soak into the rooting medium, or even the use of proprietary foliar feeds (liquid nutrition absorbed into the leaves). This backs up the idea that the sugars produced by photosynthesis in softwood and semi-ripe cuttings also need the presence of nutrient salts for successful protein synthesis. Hardwood cuttings are the odd ones out, because the whole of their nutrition is derived from the foods stored in the stems of the prepared cuttings, as they bear no leaves (for photosynthesis) at the time of insertion.

Watering Techniques

Watering, like all other management operations, is a skill, and it should be treated as such: it should never be assumed to be easy, and that it can be carried out by untrained staff. Major damage can be caused to container-grown plants when watering from above by the incorrect use of a hose – with or without a rose: too weak a flow is probably insufficient to wet the soil or media, and if it is too rapid it can dislodge the plants or wash out the compost from the pots or containers, leaving the roots exposed above compost level.

The key to efficient watering is to make sure that all the layers of the growing medium right down to the root zone are wetted thoroughly. Soils (and other growing media) are wetted in 'fronts' (layers or levels commencing at the surface and working downwards); only when the topmost layer is at field capacity will water move down to the next layer, and so on to the layers below. So for water to percolate from the surface to the lower layers, the topmost layer must be thoroughly wetted first, then the next layer down, and so on. The depth of any wetting front is not actually quantified, but no lower 'front' will be wetted until the one above it is. It is therefore essential that all these 'wetting fronts' are penetrated in order for water to get right down to the root zone.

The effect of this can be seen after a shower of rain in previously very dry, drought conditions – the top surface of the soil is barely wetted, and only minor excavation will show the layer of soil just below the surface to be as dry as before. The effect on plant roots can be disastrous, as no water reaches the root zone. Furthermore, the very small quantities of rain delivered will evaporate from the surface very quickly in hot weather and be of no value, as its effect will not even last long enough to thoroughly wet the first 'front' and encourage further percolation if rain does continue. If, however, continual, steady rain falls for some time afterwards, water will slowly percolate layer after layer until it reaches the critical root zone level – and only then can plants take up water. Very heavy rain on dry soils just runs off the surface because it cannot percolate through the soil matrix quickly enough. Plant containers outside (even very large ones) will never gain sufficient watering from natural rainfall for their needs during the summer months, and even very heavy rain should only be considered a useful added extra, as it is very unlikely to have a significant effect on water levels at the root zone area.

For the same reasons, in protected environments, very fine mists delivered in propagation systems go no way to wetting the rooting medium. The application of spray/mists or fogs of water droplets keeps the surfaces of cuttings and local air very humid, but does not moisten the compost sufficiently to wet the lower levels. Added to this, the drying effect of the heating cables or pipes underneath creates very dry root zones. Hence it is a falsehood to consider that it is not necessary to water rooting media because of the continual application of fine mist overhead. Moisture levels throughout the whole volume of the medium should be routinely checked, and watering with a can and rose, or a hose or sub-irrigation system, is necessary to retain moisture at rooting levels.

Overhead watering of prepared seed trays can ruin prepared surfaces prior to sowing, and actually

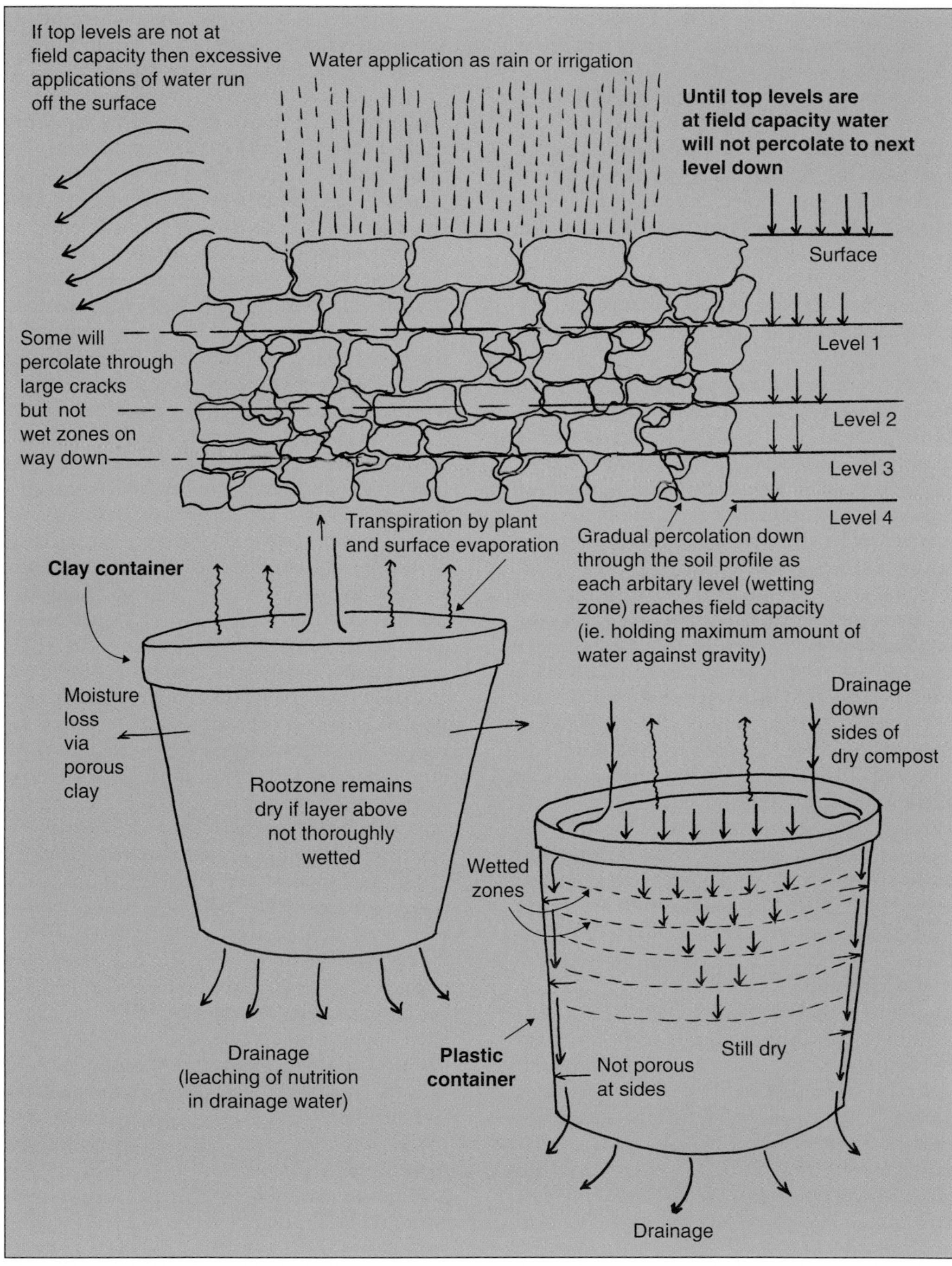
If top levels are not at field capacity then excessive applications of water run off the surface
Water application as rain or irrigation
Until top levels are at field capacity water will not percolate to next level down
Surface
Level 1
Level 2
Level 3
Level 4
Some will percolate through large cracks but not wet zones on way down
Transpiration by plant and surface evaporation
Gradual percolation down through the soil profile as each arbitary level (wetting zone) reaches field capacity (ie. holding maximum amount of water against gravity)
Clay container
Moisture loss via porous clay
Rootzone remains dry if layer above not thoroughly wetted
Drainage (leaching of nutrition in drainage water)
Drainage down sides of dry compost
Wetted zones
Plastic container
Still dry
Not porous at sides
Drainage

Wetting zones and watering.

displace seed on sown surfaces. Sub-irrigation systems are excellent for preventing moisture levels getting too low in containers, and particularly for rewetting containers with dry growing media. However, the system of using a drip tray (sub-irrigation tray) is not without its problems, as the plants/prepared seed trays have to be physically taken to the tray and replaced after sub-irrigation is complete – not left *in situ* as they are when using overhead systems. The depth of the drip tray, and more importantly, the depth of water within it, is critical. In the case of prepared or sown seed trays, sub-irrigation trays need to be topped up with water so that they give the maximum water level without completely submerging the finished level of the compost. Water should infiltrate to the top levels from below, and if the level in the drip tray is kept at the correct height, this will occur efficiently without damage to the prepared or sown surface.

Growing media in containers react in exactly the same way as soils, and containers that are not wetted thoroughly at each watering operation end up with very dry root zones. Furthermore, peat-based (and coir-based) composts have to be maintained constantly in a moist (not waterlogged) condition, as once they do dry out they are very difficult to rewet. Peat becomes difficult to wet when it is very dry because it floats and/or makes the water run off the surface and drain down the sides of the container without wetting the compost. Again, the rewetting has to be done by thoroughly soaking the container so that the wetting zones allow percolation of water to the roots. It is a common sight to see dead plants in containers, which now have moist compost, but have previously been allowed to dry out. Subsequent correct watering technique in an effort to revive the plants is unfortunately too late. Containerized/potted plants can be sub-irrigated in deep trays, and it is very important not to let them get so dry before sub-irrigating them that they float, become unstable and topple over. In all instances, leaving containers for excessively long periods in a sub-irrigation tray will cause anaerobic conditions at the root zone, and can cause irreparable tissue damage and even plant death.

To help with watering problems, peat-based growing media usually include 'wetters' – materials that aid initial water uptake and retain moisture levels in the peat. Wetters are detergent-based, and break down the surface tension of water and increase its ability to be absorbed quickly. Small amounts of old washing-up water can be used to water plants throughout the summer with the same effect as a proprietary wetter. There are no toxic effects if relatively low concentrations of detergent are used – if there is any doubt, then clean water applied immediately afterwards can be used to dilute the detergent concentration. Furthermore, subsequent water applications directly afterwards are notable in their speed of percolation down into the compost. This is because the wetting agent effect quickly wets each level of the compost and allows penetration to lower levels.

All forms of irrigation need to take into account this basic fundamental behaviour of water on soils and growing media. Insufficient volumes of water applied, or sufficient volumes applied too quickly, will all fall short of what is required, and it is as true of outside food crops, containers stood outside, and ornamentals planted outside, as it is of cricket wickets or lawns. Outdoor irrigation systems delivering water overhead need to have water-droplet sizes that are heavy enough not to be totally displaced by light winds (and therefore accidentally applied to the wrong area), yet not so coarse as to compact soils (or growing media in containers). The rate of water supply, the sizes of the nozzles used, and the droplet size that they administer, are all critical features that have to be considered when designing overhead irrigation systems.

Container-grown plants can be stood outside on sand beds, on capillary beds, in wooden-sided frames, or other areas covered with woven polyethylene products such as mypex. Woven polyethylene is used as a mulch because it allows water percolation and because of its weed-suppressing qualities, and for the same reasons it is an ideal material to cover bare soil used as a standing-out area for containerized stock – whether the area has constructed frames on it or not. Some products even have a coloured line woven into them to assist uniform and straight lining out. The stability of container-grown plants is also dependent on good natural shelter of the site (or shelter provided by hedges, lathes or polyethylene netting fences).

CHAPTER 10

Plant Protection

Weed Control

Weeds may be merely plants in the wrong place, but they cannot be truly classified as pernicious unless they are so dominant and invasive that they smother or otherwise affect the plants that are actually wanted in the design or crop. Weeds, therefore, tend to be those plants that infest areas and overwhelm them – unlike ecologically balanced situations where plants (except for a few tough examples) are unlikely to invade outside their 'comfort zone', and where conditions created by the cultivation practices for other plants are not likely to suit them. Common bluebell (*Hyacinthoides non-scriptus*) for example, because of the limited environmental conditions in which it is successful, should not be considered invasive. As soon as the 'invasion' takes fresh young plants away from the semi-shade and organic soils of its preferred woodland habitat, and produces isolated seedlings rather than plants in gregarious groups, they are not so successful. In fact, it would be better to try and save the young plants before they start their demise, by lifting the bulbs and re-establishing them back in their woodland environment where they may be successful once more and add to their *en masse* effect. Spanish bluebell is more invasive.

Vast drifts of native plant material in its natural niche and growing in balance with its surrounding species should never be considered to be weeds, and should never be 'controlled' for the sake of it. Weed control measures should never be abused to include ecological niches where they do no 'harm' to other plants (native or exotic). Tragically, herbage overkill in areas abutting the ornamental planting or crop is relatively common – just because there is some herbicide mix left over? It would be foolish to consider swathes of bear's garlic (*Allium ursinum*) or lesser celandine (*Ranunculus ficaria*), or self-sown crow's garlic (with its ready-made plantlets) to be pernicious or invasive weeds: these are far better viewed as wonderful native plants thriving in their particular niche.

Common dock, on the other hand, is 'successful' as an 'invasive' weed because of its deep underground storage organs and its tolerance of a wide range of soil types, in cultivated and uncultivated conditions, because it seeds profusely and because its vigour is such that it competes easily with vigorous grasses. Broad-leaved dock (*Rumex obtusifolius*)+ and curled dock (*Rumex crispus*)+ are in the same family as rhubarb (*Rheum*) – the Polygonaceae – and share rhubarb's fleshy taproot system of perennation. If rhubarb is left to flower, the inflorescence is remarkably similar to dock, as are the resultant seeds.

Also in the Polygonaceae, and sharing many of the features of dock (but in miniature), are the sorrels. Common sorrel (*Rumex acetosa*) and sheep's sorrel (*Rumex acetosella*) inhabit acidic soils and happily grow in amongst sparse grasses. Sorrels get their name from the fact that the leaves, if eaten, taste sour – very much like vinegar ('sorrel' means 'sour'). *Harvested leaves of sorrel actually make a wonderful accompaniment to salads and can be mixed with the young leaves of another indigenous 'weed' plant, dandelion (*Taraxacum officinale*). Dandelion is in the same family as lettuce (Asteraceae), and shares the white lactate (milky fluid) that exudes from severed vascular tissues, and tastes similar to it.

Even though it is often considered commercially essential, the desirability of using chemical weed control methods in amongst food crops is of course itself questionable. It might also be

---

*When harvesting native plants to eat, your plant recognition must be correct, and you must ensure that no toxic herbicides have been used on them.

+All come under the Weeds Act 1959.

questioned for control in ornamentals, because of the half-life in the soil of some products. It is definitely desirable to seek out the 'safest' types and reserve weed control to weeds within the Weed Act and amongst food crops (where the yield will be affected), or amongst ornamentals (where the aesthetic effect would be diminished and the desired design effect threatened).

Weeds compete with plants of all types for water and nutrients, surprisingly even including young establishing trees, as the young trees and the weeds both share the same root zone. Tall weeds can preclude light from small stock of all types, and in extreme cases blanket the desired plants completely. Weeds may also harbour pests and diseases, and some species dry out readily after death (or during the later part of their life cycle) to cause a potential fire hazard – examples include tall, straw-like grasses, heathers (*Erica* and *Calluna*) and bracken (*Pteridium perenne*). The important thing about weeds is that they are plants, and therefore will be ephemeral, annual, herbaceous or woody perennial, and within their particular plant groups they share all the features of the crops that they infest.

Essentially ephemeral and annual weeds are similar in effect. The most common examples include common groundsel (*Senecio vulgaris*) and thale cress (*Arabidopsis thaliana*), both very persistent annual/ephemeral weeds of disturbed soils. Thale cress is a particular problem of seedbeds, and its 'success' as a weed is due to its prolific seed production and because it has developed a resistance to many herbicides. Hairy bitter cress (*Cardamine hirsuta*) is a persistent weed of nursery beds, any disturbed soils, and particularly plant containers; fat hen (*Chenopodium album*) and common chickweed (*Stellaria media*) are common annual weeds of disturbed fertile soils. Lesser burdock (*Arctium minus*) and greater burdock (*Arctium lappa*) are biennial, as is hogweed (*Heracleum sphondylium*) and giant hogweed (*Heracleum mantegazzianum*)*. Giant hogweed, as well as being very invasive, has major health and safety implications as its sap is poisonous, caustic and irritant to humans, and may cause severe allergic reactions on the skin, and/or reddened wheals and blisters. It needs to be treated with particular care, and removal by hand is carried out by operatives wearing protective clothing that includes face masks/goggles and thick rubber gloves.

Methods of weed seed dispersal (and therefore sexual methods of weed infestation) vary between species. Ragwort (*Senecio jacobaea*)+ is a perennial weed in the Asteraceae family, with yellow, daisy-like flowers that produce prolific amounts of seeds; it is very poisonous to horses (and other animals). Rosebay willowherb (*Epilobium angustifolium*) has fine hairs on its seeds to help wind dispersal. The seeds of ragwort,+ creeping thistle (*Cirsium arvense*),+ common groundsel (*Senecio vulgaris*) and dandelion (*Taraxacum officinale*) are very efficiently dispersed by wind and air currents because of their parachute-style pappus. The seeds of hairy bitter cress (*Cardamine hirsuta*) and Himalayan balsam (*Impatiens balsamifera*) are dispersed by an explosive/propulsive mechanism brought about by differential drying of seedpod tissues. Himalayan balsam is also helped in its progress by stream and river water.

Many common species are dispersed because their seeds get into irrigation water from tanks, through pipes and open ditches/irrigation channels. Mammals (including humans) are responsible for the distribution of many weed seeds, including those with adhesive and barbed/hooked fruit types such as cleavers (*Galium aparine*) and lesser and greater burdock (*Arctium minus* and *Arctium lappa*). Humans, other mammals and birds are responsible for the dispersal of bramble (*Rubus*) and wild raspberry (*Rubus idaeus*) via their succulent fruits.

Germination of weed seeds is an on-going process in the correct conditions, because as soil is disturbed or cultivated, it will bring various levels of the seed bank (held in the soil) nearer to the surface at different times. Near the soil surface oxygen, moisture and favourable temperature (due to irradiation by the sun) will all be correct for germination.

The success of weed plants therefore depends upon the amount of viable seed produced, a successful/efficient method of dispersal, and the successful establishment of seedlings in the prevailing conditions. Hence, having a fast rate of growth and development aids their competitiveness, and having both seed and vegetative methods of reproduction also helps 'success'. Asexual (vegetative) methods of reproduction include the development of stolons (the best example is bramble – *Rubus fruticosus*) and suckers from root systems originating from adventitious buds on roots, for example wild raspberry (*Rubus ideaeus*), both of which reproduce freely by seed also.

+All come under the Weeds Act 1959.

*Comes under a Regulation of the Wildlife and Countryside Act 1981.

Common ragwort (Senecio jacobaea).[+]

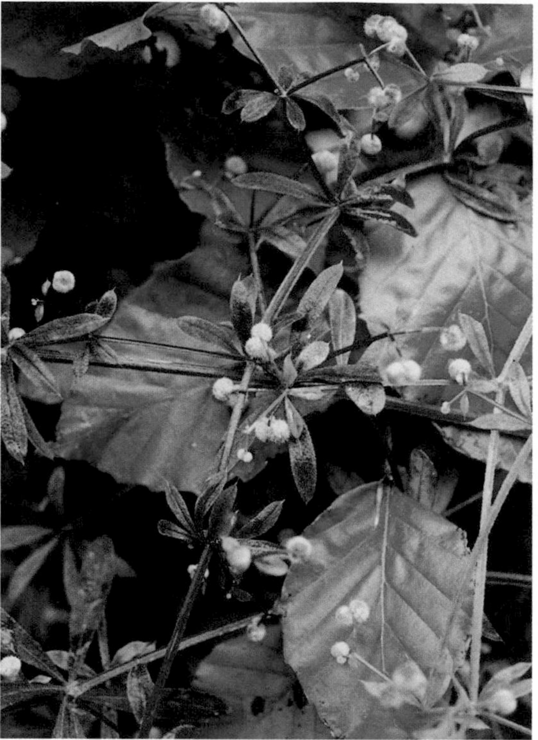

Cleavers (Galium aparine).

Herbaceous perennial weeds reproduce via their vegetative subterranean organs, as well as by seed. There are many weeds in this group that because of their perennation mechanisms, are difficult to control. Broad-leaved dock (*Rumex obtusifolius*)[+] has a fleshy taproot, procumbent pearlwort (*Sagina procumbens*) has a small taproot, and bracken (*Pteridium perenne*) has fleshy rhizomes. Deep, fleshy taproots often give rise to 'successful' weeds. However, slender rhizomes also facilitate successful weed infestation, and both hedge bindweed and field bindweed are problematic because of it. Both are in the same family (Convolvulaceae), as can be witnessed by the similarity of their flowers and their similar habit. However, the scale and size of the species is entirely different, and they are in different genera. Hedge bindweed (*Calystegia sepia*) is very vigorous and climbs fences (and into hedges) by a stem-winding action; field bindweed (*Convolvulus arvense*) has much smaller flowers and stems and tends to trail along the ground. Their depth of penetration and the extent of their slender rhizomes makes them difficult to control, even with chemical herbicides, as it is very difficult for systemic herbicides to reach their extremities. They commonly leave large volumes of their underground organs in the soil to regenerate, whatever the attempted method of control.

Sharing slender rhizomes as a perennation method, but not related in any way to the bindweeds or each other, are ground elder, creeping thistle, rorippa and couch grass. Ground elder (*Aegopodium podograria*) is a herbaceous weed only related to elder in its common name (due to its leaf shape) but by no other botanical connection. Its success as a very invasive weed species arises from its ability to produce mats of slender rhizomes. It establishes itself very quickly and invades the roots of other plant material easily, and is difficult to extricate – especially in amongst other herbaceous species. Creeping thistle (*Cirsium* arvense)[+] is in the same family as dandelion (Asteraceae), and is as successful at invading by seed (a parachute-style pappus) as it is because of its system of slender rhizomes. Likewise, the success of rosebay willowherb (*Epilobium angustifolium*) is because it has both slender rhizomes and wind-borne seeds.

[+]All come under the Weeds Act 1959.

Hedge bindweed (Calystegia sepia).

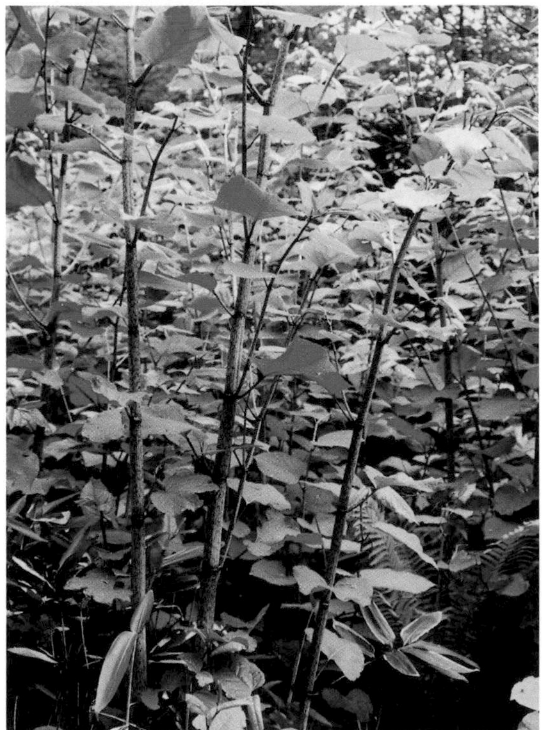

Japanese knotweed (Reynoutria japonica).

Creeping yellow cress (*Rorippa sylvestris*) is a very difficult weed to control. It is recognized by its distinctive cut leaf and yellow flowers that are very similar to cabbage, because it is in the brassica tribe of the cabbage, wallflower and cress family (Cruciferae).

*Japanese knotweed (*Reynoutria japonica*) is a very tall, vigorous herbaceous perennial with very large, fleshy rhizomes; it causes a great deal of damage to other plants because of its vigour and shading-out properties. Its resistance to control is related to its inordinate proportions, as its very thick rhizomes store massive amounts of sugars that give rise to its size, and it is very difficult to destroy chemically or by hand. The physical problem of a systemic herbicide absorbed at the leaves, actually reaching the rhizome at a concentration that may do damage, and the sheer size of the organs and the amounts of toxic chemicals that are needed, create the control problems.

The regeneration of herbaceous perennial weed species from their subterranean organs is a major problem for reinfestation. Regenerating after physical damage (via dormant adventitious buds on their taproots, bulbs, corms, tubers or rhizomes) is an effective mechanism for reproduction. Physical damage to the storage organs of weed plants encouraging regeneration may be either accidental or

*Comes under a regulation of the Wildlife and Countryside Act 1981.

intentional, or may be caused by incorrect herbicide dosage, or even correct herbicide dose but insufficient waiting time prior to ploughing or cultivating. Problems arise if a systemic/translocated herbicide is not given sufficient time to get round the system of the weed plant before subsequent ploughing or cultivating is commenced. The top part of the underground organ that has taken up a toxic amount of herbicide chemical is separated from the lower part (with minimal or no toxic content) by the plough or cultivator. The lower fragments of subterranean organs remaining in the soil can therefore regenerate and reinfest the area. Too quickly removing the top growth (for the sake of appearance) removes the leaves and surface tissue that still function to move liquids around the plant (even though the tissue may be desiccated – or dead).

The persistence/difficulty of controlling weed species is therefore related to their biology and structure, and/or to the method of their dispersal. Knowledge of their biology is therefore useful to aid decisions on their control. In general terms, annual weeds are easier to control than perennial weeds, so recognition of weeds from seedlings to fully developed plants is important in diagnosing how relatively easy or difficult they are to control.

It is essential to be able to recognize the difference between weed species and the desired sown plants just after germination, as well as when they are well established. Weed seedlings (whether of annual or perennial subjects) are relatively easy to control by chemical means, and mulches will also inhibit/control seedling weeds very successfully. However, established perennial weeds will not be suppressed sufficiently by organic mulches, or be controlled so easily by chemical methods.

Weed Control by Hand

Hand-pulling, hoeing and removal by fork are all perfectly feasible methods of weed control if the necessary labour is available. Hoeing is a particularly efficient method if the weed species are shallow rooted (and/or annual) – so that the individual plants die after removal of the top green material – and the soil condition is soft enough at the surface for the hoe to slide through it. There is no underground storage organ to regenerate. Hoeing using a Dutch push hoe is most common on areas that are regularly cultivated and therefore host regular flushes of germinating annual species of weeds. Swan-necked hoes need more effort to use, as they can be used with a chopping action, and allow slightly deeper soil penetration – even so, they are most effective on annual, rather than perennial weeds.

Hand-pulling and forking can be used for the removal of annual weeds, but are also a successful methods for the removal of deep-rooted biennial and herbaceous perennial weeds. However, successful and effective control by this method will not be achieved in anything other than good, moist soil conditions, or it will result in parts of the taproots and rhizomes being left in the soil to regenerate at a later date. Deep-rooted subjects such as dock, creeping thistle and bindweed are renowned for being difficult to remove completely. Very hard, dry, compacted soils will prevent hand-pulling and reduce the efficiency of forking.

Mechanical Methods of Control

Rotary cultivation is useless when used against established infestations of perennial weeds, as it merely chops off the top of the weeds and they are left to regenerate from their subterranean organs (fleshy taproots, rhizomes). Deep rotavation often propagates perennial weeds by chopping up the underground storage organs into small pieces, each one with the ability to regenerate and form a completely new plant. So it is very unlikely that rotary cultivation will control established perennial weeds, but it will help control both seedling and established annual weeds and seedling perennial weeds. Rotary cultivators are therefore only useful for the removal of surface weed during the growing season if there is sufficient room between the plants, used between row crops (including ornamental plants in nursery rows); but they have very few applications amongst ornamental plants grouped for display.

Ploughing inverts the soil so covers the weed growth that was once on the surface, and therefore has greater control of established weed species than rotary cultivation (and without the damage to the soil structure). Even so, ploughing can leave portions of particularly deep-rooted perennial weeds in the soil to regenerate. Furthermore, ploughing cannot be used in amongst a growing crop (either in rows or ornamentally grouped), and is therefore reserved as an operation to be carried out only in preparation for planting on a large scale.

Because hand-weeding is laborious and expensive, and most mechanical methods are not feasible in amongst groups of aesthetic plants (either because they would cause unacceptable surface root damage, or there is very little room between

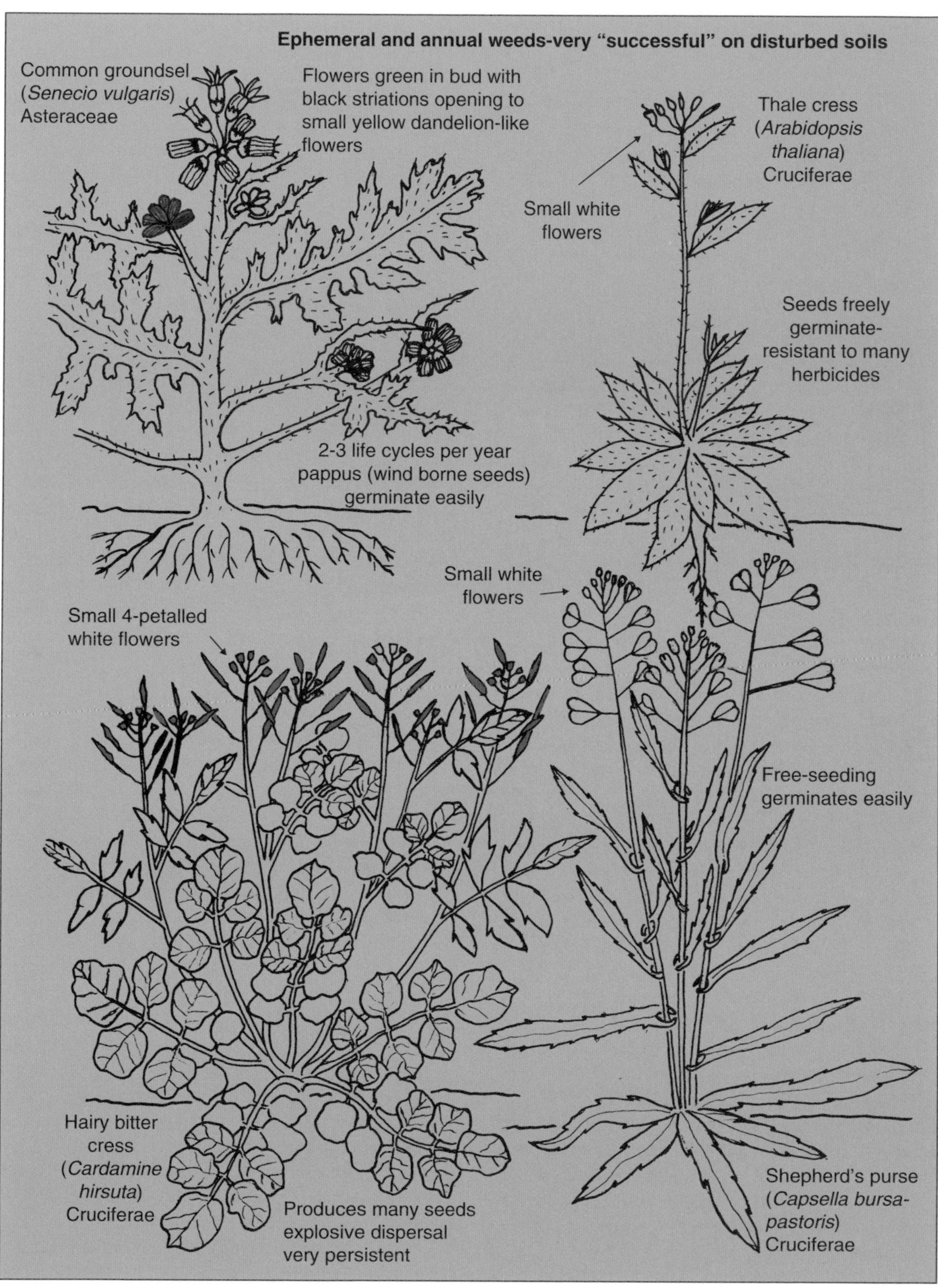
Ephemeral and annual weeds-very "successful" on disturbed soils
Common groundsel (Senecio vulgaris) Asteraceae
Flowers green in bud with black striations opening to small yellow dandelion-like flowers
2-3 life cycles per year pappus (wind borne seeds) germinate easily
Thale cress (Arabidopsis thaliana) Cruciferae
Small white flowers
Seeds freely germinate-resistant to many herbicides
Small white flowers
Small 4-petalled white flowers
Free-seeding germinates easily
Hairy bitter cress (Cardamine hirsuta) Cruciferae
Produces many seeds explosive dispersal very persistent
Shepherd's purse (Capsella bursa-pastoris) Cruciferae

Common weeds and their recognition.

the displayed plant material), mulching is a very viable and popular option.

Mulching

Mulching is one of the most cost-effective methods for weed control, and the benefits are many. Besides weed suppression, mulches also retain soil moisture and warm the soil, and if organic, add nutrients and improve soil structure.

Types of Mulches

Floating mulches are used to accelerate growth on food crops; they are placed over the top of a growing crop directly after sowing, and remain throughout the crop's early development. Floating mulches are the odd one out, as they do not suppress weeds – in fact the good growing environment that they create will accelerate weed growth along with the intended crop. They comprise thin, transparent polyethylene sheets with either holes or criss-cross lacerations throughout their surface area that allow oxygen exchange but also allow the crop to swell beneath the polyethylene sheet without damage, and are removed some time before the final maturity of the crop. Floating mulches retain moisture and create a warmer than normal microclimate, so they act rather like cloches to speed up the growth of vegetable and salad crops, such as lettuce and carrots, for early harvesting.

Sheet mulches can comprise black polyethylene, paper, and geotextiles such as polypropylene weaves (for example 'Mypex'), bitumastic felts and multidirectional fibre mats. Sheet mulches are laid on bare soil, and are planted through holes cut into the sheet mulch at soil level.

Organic Mulches comprise bulky organic materials, including rotted farmyard manure (FYM) and chicken litter, wood shavings, peat and bark chips, which because of their bulk need high-volume storage space. Unfortunately they are not as efficient at suppressing weeds as sheet mulches, and they do not suppress vigorous perennial weeds; however, they do have the benefit of adding nutrients to the soil, a feature not shared by sheet mulches.

If organic mulches are to be used as the main mechanism for labour reduction (or purely as an organic policy), then always design new beds, borders and vegetable gardens with very good access so that bulky organic materials can be transported easily to their destination. It will also be essential to have good rear access to the site to ferry bulky organic matter to their destination after delivery, and cunningly disguised areas for storing/stacking bulky organic materials may need to be created – disguised, that is, by plant features. Non-inversion systems of soil management involve the constant addition of surface mulches annually (or more regularly). Such systems rely upon the depth of mulch to suppress all weeds and increase nutrient status. The importance and effectiveness of mulching for weed control, moisture retention and good plant growth cannot be overstressed.

Mineral mulches include all types and sizes of mineral chips, gravels and grits. They are used to cap off alpines in order to suppress weeds and aid drainage in pots, and as coatings on rock garden plantings on terraces, planting pockets and screes – although they are only moderately effective in their weed suppression in these areas. They are also used to cover sheet mulches as an aesthetic finish, and are commonly used to good effect for this purpose, with the added benefit of acting as an anchor to prevent wind uplift of the sheet mulch below.

Chemical Control

Chemical control is implimented by the use of herbicides – materials applied to the weed plants that are toxic to them – plant poisons, in effect; the toxicity may be caused by caustic (burning) action, or it may interfere with the normal growth and physiology of the weed plant. Photosynthesis, cell division, cell development, respiration and rate of growth may all be affected by the application of certain materials. Herbicides may be absorbed and enter the plant via the leaves, or through the roots via the soil, or they may have both actions. Leaves will normally absorb applied liquids fairly readily – indeed, this is the basis of foliar feeds as well. However, the waxiness of the cuticle, the smoothness or hairiness of the leaves, their orientation and arrangement, and the local relative humidity, will all influence effectiveness and may reduce absorption rates and therefore plant intake of the material to toxic levels. The weather conditions during and after application can also influence eventual effect. Caustic (so-called contact) herbicides are usually absorbed quickly, and as long as it does not rain immediately after application, are fairly rain-fast. Systemic (or translocated) types (those that travel – are translocated – through the system) are not so quickly absorbed into the foliage, and they wash off easily if it rains fairly soon after application. Toxic amounts of the product are translocated only relatively small distances, but are designed to be absorbed by the leaves and moved to the underground organs of perennial weed plants. If by the time the product reaches the subterranean organs it is no longer at a concentration that is toxic to the weed, it is rendered ineffective, and will only kill the top foliage (the site of original absorption), rather than the whole weed plant. Perennial weeds will easily regenerate under these circumstances.

*Methods of application that help initial absorption and subsequent effectiveness include the use of a wetter mixed with the herbicide. Wetters are additives (adjuvents, usually detergent-based) that reduce the surface tension of the spray, and may even break down leaf-surface waxes. The size of the application droplets is also important for good absorption, hence machines that apply more uniform droplets (CDA: controlled droplet application) are the most efficient. Siphon and pump-based types such as knapsack sprayers are the most commonly used (possibly because of cost), but they deliver uneven droplet sizes and are therefore amongst the least efficient forms of application. Spray drift is difficult to guard against when using knapsack sprayers, even when special spray/wind guards are fitted. Spray drift is, of course, eliminated by the use of weed wands (methods of application whereby the herbicide is wiped directly on to individual weed plants).

Both seedling and established annual and perennial weeds can all be controlled by chemical application (herbicides). However, the formula used in each case is very different. Annual, and most seedling perennial weeds can be controlled by using contact herbicides (such as paraquat as 'Gramoxone') that merely burn off the soft green tissue responsible for photosynthesis. Once this tissue is killed, annual weed species (and seedling perennial weed species that have not yet developed their storage organs) have no means of continuing. Contact herbicides do not usually affect established bark-covered tissues. However, young woody stems with chlorophyll in their bark will be affected, as will any green foliage or green shoots that the spray may drift on to.

Established and more developed seedling perennial weeds require systemic (or translocated) herbicides to control them, for example glyphosate as 'Roundup'. These chemicals work when the weed plant absorbs a toxic amount of the product via the leaves, and this is moved (translocated) through the system by the normal physiological processes of the plant. For systemic herbicides to be effective they have to be applied at a dosage that will be toxic even when transported as far as the underground storage organs of the weed plant. So time for the toxic material to move to the storage organs is an important element of their effectiveness, and top herbage should not be removed until the requisite time has elapsed after application.

Hormonal herbicides are based on hormonal salts and also have a systemic action. Selective herbicides are designed to selectively control specific weed groups – for example, broad-leaved weeds (dicotyledons) in a grass or cereal (monocotyledonous) crop – and are usually hormonally based. Broadband herbicides are herbicide mixtures that often comprise more than one herbicide action, and are designed to control a relatively wide range of weeds.

Residual types (those applied to bare soil and which are retained for relatively long periods) are fairly rain-fast – indeed, their action depends upon moisture in the soil. Residual herbicides (for example, dichlobenil as 'Casaron') take up moisture and release toxic gases into the soil to kill seedling weeds and germinating weed seeds as they try to respire. Used around young trees and shrubs they work by releasing the gases in the very surface levels of the soil (seed germination zone) inhabited by the larger, corky roots. At the correct application rates and in the correct soil conditions they do not penetrate the fibrous and absorptive lower root zone where they would cause phytotoxic damage to the desired plants as well as the weed plants. For this reason they are not used around softer plant material or surface-rooting species such as *Rhododendron, Camellia* and so on. The susceptibility of host plants to damage should always be checked before use.

Incorrect and drastically uneven application rates can cause severe and irreversible crop/ornamental plant damage. Residual herbicides can be less effective than they might be (or even damaging) if the colloids of the soil do not hold the active ingredients against both gravity and leaching (as they percolate down to the fine absorptive roots

*Health and safety issues of chemical herbicide application revolve around protective clothing to prevent ingestion and absorption of toxic materials by the operative. Spray drift can be harmful to other non-weed plants, animals and humans. So correct, effective dosages, and efficient, safe applications, in good weather conditions, and after reading the directions on the label correctly, are essential for safe use.

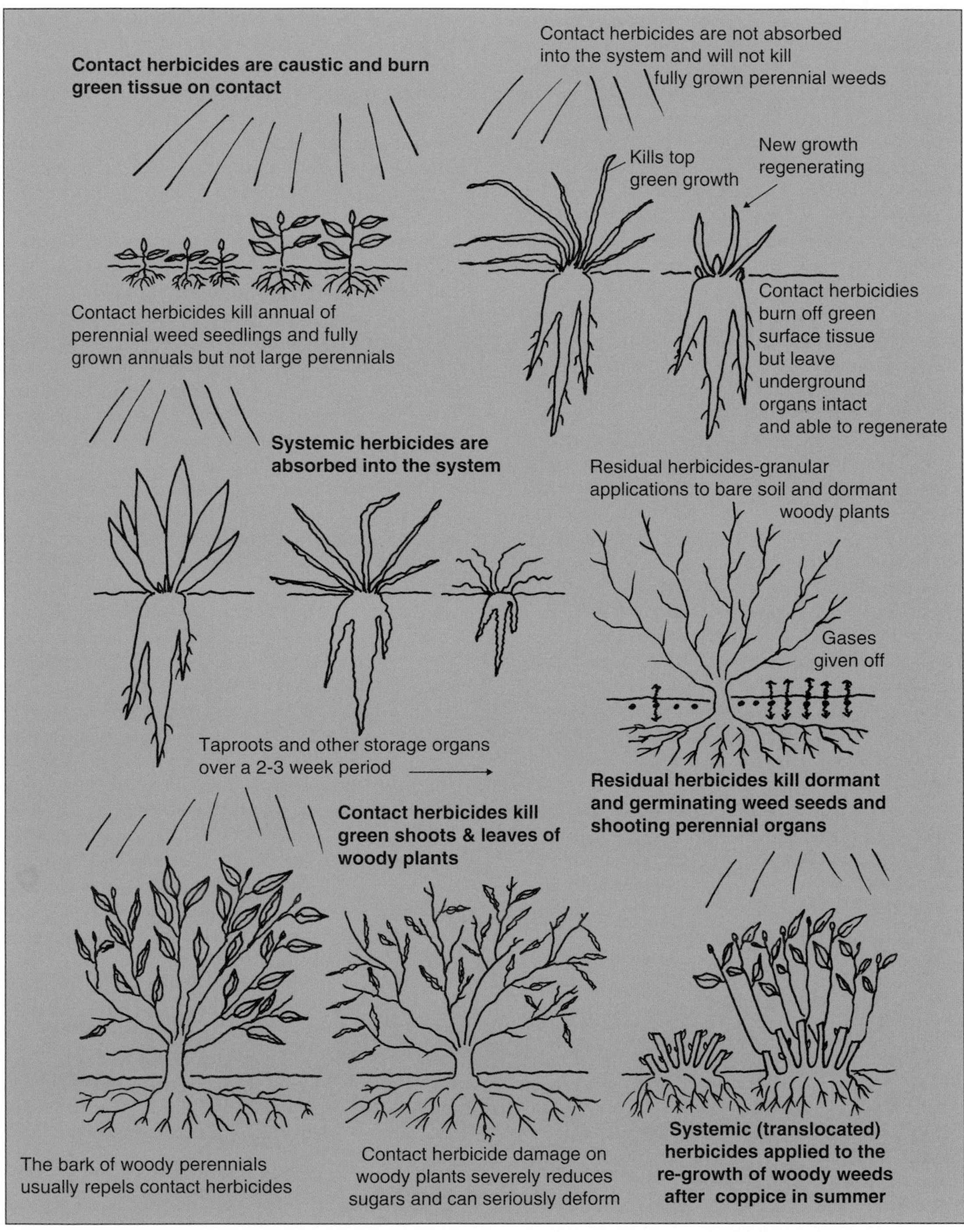

Herbicides and their action.

of the desired plants). Volatization (evaporation from the soil surface) is another hazard of incorrect application or poor soil types when using residual herbicides, as gases could damage aerial parts of the desired crop/plants as well as prevent weed growth/germination. Leaching, volatization and biodegradation all reduce the effective life of residual herbicides over time.

Freshly prepared soils, because of the soil inversion/mixing process, have the lower layers of dormant seed bank mixed into the top surface layers. These layers are well aerated, and in open, fully illuminated conditions the weed seed population germinates readily – and because there is no competition from the plant crop (or chosen aesthetic plants), they thrive. Annual and ephemeral weed species seed profusely and regularly and add to the seed bank all the time, for germination in both the short and the longer term. A method of weed control known as the 'stale seed-bed system' utilizes this fact. When the first crop of weed seedlings appears after soil preparation they are killed by herbicide spray, and care is then taken not to disturb the soil (no further cultivation) so the seed bank is not brought to the surface. This system leaves the soil relatively 'clean' in the top surface, and very little further weed germination occurs until reinfestation at a later date. This process depends upon the destruction of the first flush of seedling weeds, and then the separation in time of the chosen crop seedlings and any further weed seedling infestations.

Plant Pests

Plants may be damaged by both biotic and abiotic factors. Abiotic agencies are non-living and include frost, exposure, drought and nutrient deficiencies. Biotic agencies that affect plant growth include animal pests, bacterial and fungal diseases. Plant pests in the animal kingdom occur in three main phyla: Mollusca, Arthropoda and Mammalia.

Mollusc Damage on Plants

Slugs and snails are members of the class Gastropoda within the Mollusca phylum, and are typified by having a very visceral nature, a large muscular foot and a distinct head with two prominent antennae-style tentacles. Slugs (*Limax* species) have no shell and can survive in acid or alkaline conditions. Snails (*Helix* species), on the other hand, have chitinized shells that they secrete from their mantle, and they need calcium present in their environment in order to initially build and then maintain their shell. Snails therefore thrive in more alkaline conditions, as can be witnessed by their presence on old lime-mortar walls.

Both slugs and snails are herbivorous and have very strong mouth-parts and a rasping tongue. Their grazing can cause extensive damage to leaves, and they will eat any green tissue. Species of plants with green primary tissue (including green stems), and even woody subjects that retain a soft, chlorophyll-infested bark, can all suffer severely from snail and slug damage. Snails in particular are commonly found at height within the canopy of shrubs, or amongst the leaves of woody climbers.

Terrestrial gastropods (slugs and snails) initially make very large holes in the leaves of plants, often ultimately destroying the leaves and greatly reducing the sugar-producing (photosynthetic) area of individual perennial plants to a point where there is insufficient energy for the plant to survive. Annual plants, because they comprise only primary green growth, are easily destroyed by the action of slugs and snails – French and African marigolds are good examples of susceptible species. Herbaceous perennials are often severely damaged (*Hosta* species being a favourite), and may be killed – especially if the growing tips are eaten away; serious damage or even death to large shrubs (particularly soft sub-shrubs such as *Abutilon* and *Clianthus*) is not unheard of where large infestations exist.

Slugs and snails lay eggs that are easily recognized because they are round and white, or off-white, and laid in small clusters. There are many recommended controls, all of which go some way to reducing populations and minimizing damage; however, these pests remain a major problem. Beer traps are one method of control: a beer trap consists of a shallow container with vertical sides that is filled with beer (or some other lure) to attract the slugs and snails, which then drown when they fall into the liquid and cannot get out of the steep-sided container.

Fine gravels, grits and coarse sands also have a controlling effect, as molluscs are reluctant to move on to these materials – they seem to find it difficult to negotiate and move over them with their muscular foot. However, it would appear that they only find the surface irritating and uncomfortable, and although they may be reluctant, they certainly do not find it impossible to cross. If their main source of food is only accessible over such surfaces many do manage to achieve it – or they find a bridge across the coarse surface via other less desired plant material to gain access to their favoured plants. Snails can be incredibly resourceful, so it is sometimes necessary to cut back bridging vegetation in the area surrounding snail-favoured species. This also has the effect of reducing localized humidity, which may make the area less attractive to, and so less frequented by, slugs and snails.

Unfortunately, baited pellets (slug pellets) still remain one of the most efficient means of control. Thankfully, suppliers have worked hard to make modern versions more eco-friendly and reduce the risks to birds and mammals higher up the food chain. However, killing vast quantities of slugs and snails by any means (even with beer traps) cannot be viewed as ecologically desirable – but perhaps a necessity in certain circumstances in order to spare favoured plants. The option of only selecting plant species less favoured by slugs and snails greatly reduces the options for plant design systems. Those who really want to remain faithful to ecologically sensitive methods can remove the slugs and snails by hand and transport them (in a polyethylene bag with some plant material inside) to an area of wasteland where they can be carefully tipped out of the bag.

Insect Damage on Plants

Within the phyla Arthropoda are several classes, the most important of which, as regards plant damage, are the classes Insecta (those with a hard, chitinized exoskeleton and six jointed legs for mobility) and Arachnida (spiders and mites with eight legs). Many insects cause problems for plants, particularly members of the sub-class Pterygota (those with wings), and especially when they are in large groups or large infestations.

Insect pests can affect plants by leaf mining, sap sucking, eating portions of leaves and soft tissues, and total defoliation. They may also suck sap from bark-covered tissues, from shoots or buds, or may bore into wood, buds, seeds or bark. Other species may feed off root tissues, or eat fruits, or create galls. However, in general terms, trees and other plants exist alongside insects, and paradoxically insects, although they cause damage to food crops and ornamental plants, form an important part of the ecology of the natural world.

Depending on the life cycle and habit of the species, various forms of damage can be caused by insects. Some species have sucking mouth-parts only, some have biting mouth-parts as adults and in their larval stages, whilst others have biting mouth-parts only in their larval stages. Larvae, in particular, are voracious eaters and will destroy large amounts of tissue as they gorge themselves with food in order to build up sufficient energy reserves to carry out metamorphosis successfully. The main problems of insect infestations include the reduction of the photosynthetic area, with subsequent reduction of sugars. To what extent this causes a problem depends on the severity and longevity of any infestation: small infestations can cause minimal damage, while large infestations can cause serious problems, particularly when large groups of the same species of plants are grown near one another that encourage (and facilitate) the reproduction of vast quantities of the same insect species.

Healthy plants heal more easily than unhealthy plants when live tissue is damaged, and they are more resistant to insect attack because they produce sufficient amounts of the materials present in plant sap that act as an insect repellent. This is best illustrated by considering two rows of pine seedlings, in identical soil and nutrient conditions. Row 'A' is covered in black polythene for four daylight hours per day, which reduces photosynthesis, and therefore reduces the available energy budget. Row 'B' is left uncovered. If you were to sample some of the leaves/pine needles from row 'A' by biting into them, your reaction would be one of indifference: the taste would be mildly unpleasant, but acceptable. Chewing needles from row 'B' would be a different matter, however, and your reaction would be one of total distaste, and you would probably spit them out immediately. This is because the plants in row 'B' are healthy, and the strong and unpleasant taste is because they are producing enough terpenes and resins to repel insect predators. Plants from row 'A', on the other hand, are producing insufficient extractives because of their relatively poorer health, and the likelihood of row 'A' succumbing to an insect attack is therefore greater than row 'B'.

Unfortunately these materials are not foolproof, and although they may deter, they will never completely prevent insect attack. Even starting from the premise that intact, healthy, unbreached plants have very good resistance to pathogenic diseases, the combination of tissue puncturing, sap drawing, boring or tissue tearing, then injecting the tissues with a virulent disease, is a potent one. Insects acting in this way as vectors for damaging fungal, bacterial or viral diseases often cause the demise of previously very resistant individuals.

There are very good reasons why it is important to be able to recognize insect pests that attack plants, not least that it enables you to decide on how best to control them. You should also be able to assess the potential seriousness of the attack, the likely problems because of it, and the likelihood of long-term damage. Ironically, recognition may initially be decided by the symptoms of the damage caused, rather than actually recognizing the characteristics of individual insects. Damage symptoms do vary greatly from insect to insect, and are often good indicators of the species. Recognizing the type and pattern of frass (bore meal), entry holes, exit holes, egg-laying galleries, and which part of the leaves or stems are attacked, can be critical in insect identification. Likewise, knowing the position of the damage on the plant, whether damage is to soft leafy growth, soft stems, fruits (including nuts), or woody tissues, all aid diagnosis and give some indication as to the species responsible, its importance, and its potential severity. Pupae type, larvae type, adult insect type, and even excrement shape and size are all used to identify individual species.

The main insect pests of plants are in six orders of the class Insecta:

- Lepidoptera (butterflies and moths)
- Coleoptera (beetles and weevils)
- Hemiptera (aphids, woolly aphids, adelgids, scale insects and whitefly)
- Hymenoptera (wood wasps, gall wasps and sawflies)
- Thysanoptera (thrips)
- Diptera (two-winged flies, including leaf miners)

Lepidoptera Larvae Damage

Lepidoptera are those insects with scaly wings (lepidote = scaly, and ptera = wings), and the name refers to the rounded, overlapping, scale-like coverings found on the wings of moths and butterflies – the same derivation as that describing the scales on the underside of the leaves of lepidote rhododendrons. The adult phase of the Lepidoptera species (the imago) is harmless to plants, as it has no biting or puncturing mouth-parts. During this phase, which is often short-lived, feeding is by sucking nectar from flowers. Conversely, the larval stage of Lepidoptera (caterpillars) have biting mouth-parts, are voracious eaters, and gorge themselves on food until their bodies become greatly distended, and use the energy gained for their successful pupation and development into an adult. Many Lepidoptera larvae have stiff, bristly hairs that form a coarse protective layer against predation, and in some instances are very irritating, even causing a serious allergic reaction (urticaria) in humans. Brown-tail moth larvae can actually drop coarse hairs that are severe irritants, and serious infestations in trees in their preferred southern counties of the UK have caused a health risk to humans.

The adults of the large cabbage white butterfly (*Pieris brassicae*) and the small cabbage white butterfly (*Pieris rapae*) lay hundreds of eggs (two generations per year) on most leafy brassica crops and *Nasturtium* (also in the Cruciferae family). The many larvae that result have yellow-green bodies with long yellow stripes, and are voracious eaters.

The lackey moth (*Melacosoma neustria*) has larvae that are mainly blue with a reddish stripe on each side, below which is a reddish-orange stripe, and a central bluish-white stripe down the back. The larvae mainly feed on the juvenile branches of semi-mature trees or large deciduous shrubs, and they defoliate in large gregarious colonies (protected by a cocoon-like tent) during April to June.

Buff-tip moth (*Phalera bucephela*) larvae are hairy and yellow with black lines and a black head. They feed in gregarious groups from May to June, stripping the branches of deciduous subjects.

The brown-tail moth (*Euproctis chrysorrhoea – Liparis chrysorrhoea*) has larvae that are initially dark brown with two orange warts. Later the larvae become very hairy and have two lines of white hair tufts down their back. They feed as large gregarious groups on juvenile material and small trees in the Rosaceae, but will attack other species. Often associated with a silky, cocoon-like tent, they are very voracious feeders and can defoliate small subjects to death. The larvae shed hairs that can cause severe urticaria (a reaction to irritating hairs) on humans.

It is obviously important to know the feeding habits of insects when contemplating the possible damage of an infestation, and it is also useful to know what are likely insects to infest specific host plants. However, the social habits of some insects also affect their damage potential; thus insect species may be solitary, or they may colonize in small groups, or be very gregarious. Gregarious Lepidoptera larvae (caterpillars) can contain themselves in cocoon-like 'tents' made of off-white or grey membranous material. They form very large colonies and can cause massive defoliation – the larvae of the buff tip, brown-tailed and lackey moths are all good examples. However, gregarious colonies are not always necessarily Lepidoptera, as the larvae of some sawfly species in the Hymenoptera also congregate in this way.

Hymenoptera Larvae Damage

Members of the Hymenoptera are those with membranous wings (hymen = membrane), and include wasps, wood wasps and sawflies. Unlike the more haphazard feeding of caterpillars, sawfly larvae arrange themselves neatly along both margins of the leaves and eat progressively inwards towards, but seldom into, the midrib. At first glance they look like caterpillars, but closer inspection shows important disparities, namely the number of prolegs, and to which body segments these are attached: sawfly larvae have more than five pairs of prolegs (on some, or all, of segments two to seven, and the rear segment), whereas most Lepidoptera (except for loopers, who have less than five) have five pairs of prolegs, on segments three, four, five and six. Furthermore, some of the sawfly, including birch sawfly, when colonizing a leaf, 'arch' the tail end of their bodies upwards in unison if they are approached. Their gregarious nature already separates them from the solitary looper (geometrid), furthermore they do not leave the prolegs of the rear segments still attached in the way of a looper caterpillar, but leave the tail end free. The instantaneous uniform motion of the group is obviously a protective mechanism: it is disconcerting, and is designed to frighten off foraging birds.

Gooseberry sawfly (*Nematus ribesii*) larvae are green with black spots, and skeletonize leaves. Solomon's seal sawfly (*Phymatocera aterrima*) larvae can completely defoliate plants.

Birch sawfly (*Fenusa pusilla*) larvae feed by congregating around the leaf edges of birch species, causing kidney-shaped holes. Larvae appear from June to September and are light green in colour with a light, yellow-brown head. They are flattened and tapered at the rear.

Galls

Many galls are formed by the action of gall-forming wasps in the order Hymenoptera. The galls are formed from plant tissue as a reaction to the invading insect. The shape of the gall, created by the interaction of the insect and the plant, is specific to the particular insect and particular plant, and is consistent in shape and colour from the same partnership. Galls are seldom detrimental to any degree, but are of interest.

Mossy rose or robin's pin-cushion gall (*Diplolepis rosae*) is one of the most aesthetic, unusual and visually spectacular galls: it is caused by the gall wasp *Diplolepis rosae*. The gall is found on wild and cultivated roses, and comprises a large swelling with cord-like outgrowths (looking a bit like moss). The outgrowths commence green and later turn red.

Robin's pin-cushion gall.

Adult vine weevil damage on Rhododendron.

Damage by Coleoptera

Coleoptera are those with covered wings (coleo = covered), referring to the leathery elytra (wing covers) that are typical of beetles. Both weevils and beetles are in the order Coleoptera, and have biting mouth-parts in both their adult and larval stages, so are capable of extensive plant damage in both instances. Weevils are distinguished from beetles by their snout-like head (rostrum) and elbowed (crooked) antennae, whereas bark beetles have a smooth, rounded head and very distinct (often striated) wing cases (elytra); leaf beetles have highly coloured bodies and almost straight antennae.

Black vine weevil, or cyclamen weevil (*Otiorynchus sulcatus*), is endemic to Europe, and adult damage is typified by 'C'-shaped cuts in the leaf margins. The damage caused by the subterranean larvae is at the roots and root collar, where tissue is eaten and the roots die, causing the sudden collapse of young plants – particularly in the nursery phases. Vine weevil illustrates the resistance of some insects to toxins, because they are common on *Rhododendron*, the sap of which is very toxic to humans and other mammals. The adults are all parthenogenetic females (they do not need male fertilization), and they lay thousands of eggs. Adults are nocturnal feeders, are dark brownish-black, and have the characteristic 'snout-like'

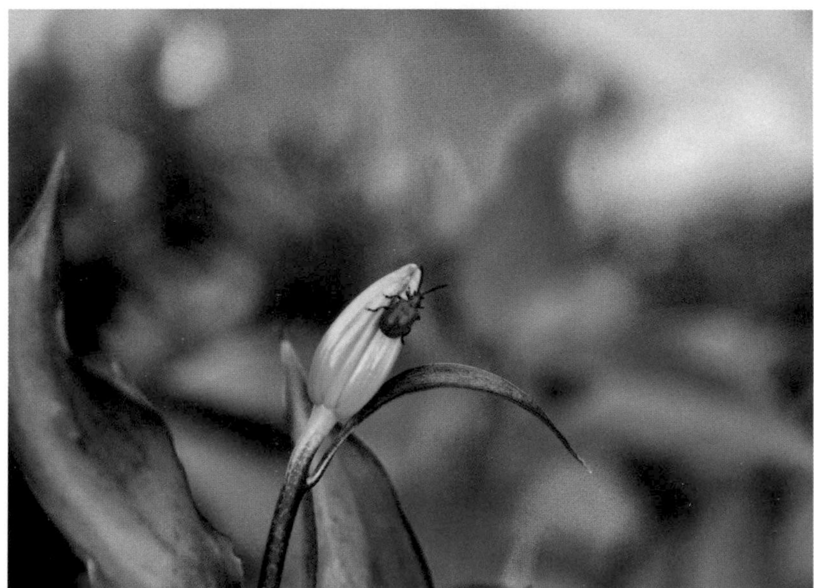

Lily beetle (Liliocerus lilii).

appendage of weevils (rostrum) and elbowed (crooked) antennae. They play 'possum' if approached, and will lie for hours without movement. One of the most effective controls is biological, using the nematode (parasitic eelworm) *Heterorhabditis megedis*.

The larvae of turnip gall weevil (*Ceutorhynchus pleurostigma*) also cause damage below soil level, creating large swellings (galls) in the roots of turnip. The damage caused is minimal, however, it can be mistaken for club root. The pea and bean weevil (*Sitona lineatus*) eats 'U'-shaped chunks from the leaf margins of peas and broad beans.

Cockshcafer larvae are very long (4 cm) and fat, and 'C'-shaped; like vine weevil larvae, they eat the roots of many plants.

The adults of the lily beetle (*Lilioceris lilii*), a leaf beetle, have bright red bodies and black legs, and feed on the leaves and flowers of lilies, *Fritillaria imperialis*, and related species.

The adult flea beetle (*Phyllotreta* species) is a small black beetle, often with a broad yellow stripe on each wing cover; they cause many small holes in the foliage of seedling and young developing brassicas, including cabbage, radish and wallflower.

There are three main elm bark-boring beetles: the large elm beetle (*Scolytus scolytus*), the small elm beetle (*Scolytus multistriatus*), and the Scandinavian elm beetle (*Scolytus laevis*). All these species bore into the bark of elm species and lay their eggs in the tissues below the bark; this causes physical damage, but the tree can usually 'accommodate' the damage with little detriment. Unfortunately the beetles carry with them a fungal disorder (a vascular wilt) called *Ophiostoma nova-ulmi* (Dutch elm disease), which is fatal to the tree.

Damage by Hemiptera

Hemiptera (hemi = half) are so called because some representatives have front wings with two different shapes; the front portion looks like half a wing and is hardened (sclerotinized), and is known as hemi-elytra. The order includes aphids, scale insects, leaf hoppers and adelges (adelgids), all having sucking mouth-parts. Aphids withdraw sap from the cells of living tissues, and by doing so kill them. The effect of this on soft chlorophytic material, such as leaves and green stems, is asymmetric growth patterns: because the dead tissues no longer grow, whereas the unaffected remaining live tissues continue to develop, distortions of various types are created, such as 'cupping' of leaves and twisted stems. Aphid damage is often associated with 'honeydew', which is cell sap leaked from the damaged tissues, and from the aphids themselves when they secrete cell sap in excess of what they can ingest and cope with. Honeydew acts as a perfect substrate for sooty mould (a pin mould), which covers the leaves in a black coating and is detrimental to plants because it blankets the photosynthetic tissues and reduces sugar production.

Also in this group are the carnivorous 'bugs' such as shield bugs, which unusually suck fluids from aphids. They are therefore a good biological control for the aphid group in general, as are the adults and larvae of lacewings in the Neuroptera, and ladybird larvae in the Coleoptera; however, their populations are unfortunately reduced by insecticides. Whitefly (including glasshouse whitefly) are also in this group (not in Diptera).

Cherry blackfly/black cherry aphid (*Myzus cerasi*) are very common on *Prunus* species of all types, but notably on *Prunus avium* (common wild cherry). Blackish-brown nymphs feed on the leaves and young stems of trees of any age. Copious honeydew causes a great deal of sooty mould. Garden blackfly (*Aphis fabae*) cause damage on many soft plants, and form dense black colonies on broad bean, French bean and runner bean in particular. Garden greenfly (*Acythosiphon pisum*) infest the soft growth of many common plants. Puncturing the underside of leaves (and young stems) causes a browning of the tissue (tissue death), which leads to twisting, curling and cupping of the leaves and distortion of the young shoots.

The nymphs of froghoppers (*Philaenus spumarius et al*) are hidden by a bubbly, frothy liquid that they secrete (known as cuckoo-spit). They suck sap from plants, but most species tend to do only very minor damage; they are not usually controlled.

Rose leafhopper (*Typhlocyba rosae*) sucks sap and causes damage to plants in the Rosaceae. Their name comes from the fact that they can jump quickly, but the small slender adults can, and do, fly. The nymphs are not covered in cuckoo-spit, like froghoppers. *Graphocephala fennahi* is a very colourful leafhopper with notable coloured stripes down the body; it is found on *Rhododendron*.

Capsid bugs (*Lyocoris pabulinus* and *Lygus rugulipennis*) cause damage to a wide range of plants in young leaves and flower buds. As damaged leaves expand, they appear tattered and full of enlarging holes, and flowers damaged in bud become distorted as they develop. Capsid bugs are found on woody perennials to pelargoniums.

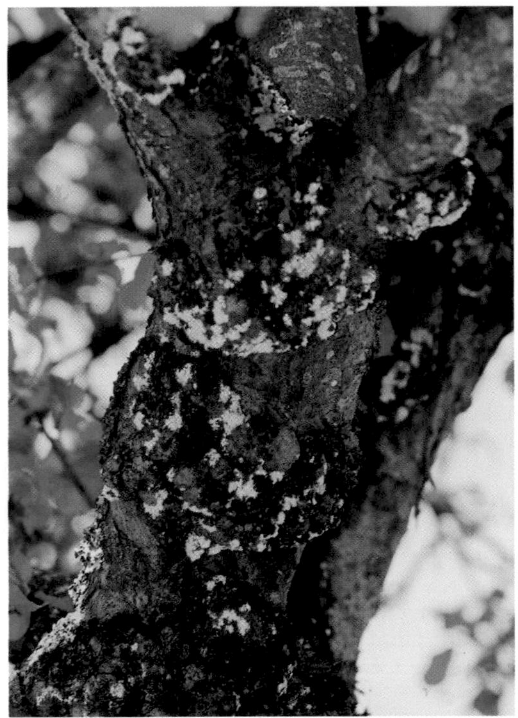

Woolly aphid on apple.

Horse chestnut scale on Lime (Tilia).

Apple woolly aphid (*Erisoma lanigerum*) is found on fruiting apple trees in particular, but also on ornamental apples, and other ornamentals in the Rosaceae. It appears as a white woolly mass, and the sap-sucking insects that are covered in the white wax can cause severe cankers and disfiguring stem galls.

Glasshouse whitefly (*Trialeurodes vaporariorum*) cause severe damage to tomato and other glasshouse crops, and ornamentals by adults and scales (on the underside of leaves) sucking sap. Outside, cabbage whitefly (*Aleyrodes proletella)* causes damage to brassica crops.

Mussel scale (*Lepidosaphes ulmi*) is shaped like mussel shells, and is common on apple. Brown scale (*Parthenolecanium corni*) is found on peach, fig and vines, *et al*. Soft scale (*Coccus hesperidia*) and hemispherical scale (*Saissetia coffeae*) are found on relatively soft-barked ornamental woody species grown under glass. All species suck sap and deplete the energy supply of the plant.

Horse chestnut scale (*Pulvinaria regalis*) is a flattish, pale brown or straw-coloured scale insect (as with all others, it is related to the aphids and adelges). However, the body of this scale is usually covered in a very woolly, white, waxy cuticle, which can be seen throughout the year. *Pulvinaria* is found on leaves and bark. Large infestations on bark cause a lot of damage to ornamental trees. It is not specific to horse chestnut but is also common on other *Aesculus*, *Tilia* and *Acer*.

Damage by Diptera

Members of the order Diptera have two wings (di = two); the original hind wings are now reduced to two stalked club-like appendages known as halteres. It includes some leaf-mining larvae and some that damage plant roots.

Chrysanthemum leaf miner (*Phytomyza syngenesiae*) and holly leaf miner (*Phytomyza ilicis*) both cause the typical white or silvery lines on leaves, as internal leaf tissue is destroyed, leaving only the transparent epidermi.

The larvae of carrot fly (*Psila rosae*) eat into the taproots of carrots and parsnips causing severe damage.

The larvae of crane fly (*Tipila* species) are large, greyish-brown maggots known as leatherjackets, and cause severe damage to plant roots, notably grasses.

Thysanoptera
Members of the Thysanoptera include thrips, a group of sap-sucking insects that can do severe damage to leaves.

Methods of Control
The use of substances that are toxic to insects will always create ecological and environmental problems unless they can be applied specifically, or are toxic to only a specific insect species, or a very small range of species. Killing large colonies of harmless native insects is not an option, even under the banner of commercial production.

Some insects are soft-bodied throughout, but most have tough, chitinized, shell-like exoskeletons in their adult phases. The chitin shell acts as a tough physical barrier, and because of this, and the thick waxy cuticle associated with many species, they not only shed water easily, but also shed, rather than absorb, any aqueous solution with additives aimed at their control. Hence, ingested stomach poisons (substances taken in through the mouth at feeding and applied as systemic compounds to the host plant) and biological parasitism appear to be the most successful options for control. Unfortunately, aphids are not controlled by systemic stomach poisons (ingested when feeding off plant material), so contact insecticides have to be used.

Arachnida
The class Arachnida includes those spiders that are carnivorous and do not cause problems for plants. However, mites are also included, some of which do damage to plants, with glasshouse red spider mite of major economic importance. Red spider mites are very small and difficult to see with the naked eye, so are best viewed with a × 10 hand lens. Glasshouse red spider mites (*Tetranychus urticae*) are straw coloured with two distinct dots on the body; they are gregarious (in very large groups), have sucking mouth-parts, and can cause severe plant damage by killing layers of cells. They only turn red later in their life cycle, and because of their size they can go largely undetected for long periods until the symptoms of the damage show externally. The mites mainly live on the underside of leaves, and produce a very fine webbing – so fine that it is not easily seen without definite and scrupulous inspection. As they puncture the green tissue, the cells gradually die, and eventually this shows as the most prominent symptom: a characteristic 'bronzing' effect showing through the top tissues of the leaves.

Red spider mites attack a large number of glasshouse crops and ornamentals, and in very warm seasons they also cause damage to ornamental bedding plants such as African marigold outside. Fruit tree red spider mite (*Panonchus ulmi*) infests apples, pears, plums, hawthorns and rowans, outdoors. Control, both inside and outside, is notoriously difficult, as they are very resistant to many substances applied to the crop – sometimes because insecticides, rather than arachnicides, are applied in error, but mainly because of their inherent resistance. Under glass, although the arachnids can reproduce readily and be very successful, at least under these controlled conditions some forms of biological control can be successful – an active predator *Phytoselius persimilis* – itself another mite – is used against glasshouse red spider mite, with some success.

Biological Control
Most biological controls are via parasitization of, or preying on, the plant-damaging insect. Specialized parasitic insect species with ovipositors (egg-laying tubes) can inject their eggs into a host insect: the host is killed when the eggs hatch into voracious larvae and eat their way out.

The use of *Encarsia formosa* against glasshouse whitefly (*Trialeurodes vaporariorum*) is now fairly common practice. The parasite can be readily purchased through specialist producers, and is imported into the glasshouse on parasitized whitefly scales held on pieces of tobacco leaf or on special cards. Adult *Encarsia* eat their way out of the whitefly scales and release themselves into the glasshouse space where they will reinfest whitefly scales by laying their eggs inside them. The process of the developing *Encarsia* eating their way out of the scale tissues kills the scales (and therefore the potential whitefly infestation). There always has to be at least a small amount of the whitefly population to support the parasite, so complete control is undesirable (and unlikely); but should this happen, it would also mean the demise of the parasite. The parasite population fluctuates anyway, and programmes of regular reintroduction are usually needed. Whitefly species also exist outdoors, including cabbage whitefly on brassicas.

The control of glasshouse mealybug is very difficult indeed. However, some success has been achieved using *Cryptolaemus montrouzeri* (a ladybird-like insect in the Coleoptera). *Cryptolaemus* are carnivorous predators, voracious eaters, and they actively hunt down mealybug (*Pseudococcus* species, a member of the aphid and scale group) – there are great similarities between this and ladybird larvae eating aphids outside in temperate climates. Ladybird larvae and some shield bugs (carnivores in the Hemiptera) are the best natural predators of problematic insects that we have.

Although *Cryptolaemus* can be very effective, they do require fairly well maintained glasshouse temperatures in order to be truly successful. Often, therefore, attempts to reduce heating costs (by reducing glasshouse temperatures in the winter) result in poor colonies of the predator, and correspondingly poor control of the pest.

Aqueous suspensions of fungal spores, bacteria and viruses can be applied to parasitize insects harmful to plants. The bacterium *Bacillus thuringiensis* is used successfully against a range of damaging Lepidoptera larvae, as are specific microviruses and so-called 'd' factors. Such measures are, of course, only necessary in intense commercial systems, where, because of the never-ending food supply, the colonies of damaging insects build up to epidemic proportions. Attacks on single individual plants can be devastating, but are seldom so.

In many instances common sense does not prevail, and even on a small scale, it is commonplace to see pesticides mixed and applied unnecessarily (with the associated cost, effort and application difficulties). Simply removing the infested tissue with secateurs and burning it would be so much easier, and often so much more effective. Aphids in particular can sometimes be controlled in this way, as we know they mainly infest leaves and green stems at young shoot tips.

Plant Diseases

Fungal Diseases

Fungal diseases are often secondary infections that infest after damage has been done by other agencies – including human activity such as vandalism and pruning. Other mammals, such as deer and voles, can also create areas for later fungal infection, as can insects. It is important to be able to identify plant diseases (and insects, for that matter), because without recognition you cannot seek advice for their control.

Moulds, Mildews and Rusts

Moulds and mildews cause damage at cellular level by invading individual cells and extracting the nutrition. Moulds, including grey mould (*Botrytis cinerea*), destroy the soft tissues of many plants and usually enter via tissue damage, but can also infect intact tissues. They thrive in damp conditions. *Botrytis cinerea* appears as grey, fluffy outgrowths and can be seen on any soft tissues, including fruits. *Botrytis* species can also cause leaf blotches such as chocolate spot on broad bean (*Botrytis fabae*) and tulip fire (*Botrytis tulipae*), typified by brown scorching on the leaf margins and brown spots on leaves. Downy mildews (*Peronospora* and *Bremia* species) also attack soft tissues and are somewhat similar in appearance as they are furry, but are more white or off-white, rather than dark grey, and are present mostly on the underside of leaves (which may twist to be revealed).

Powdery mildew on phlox.

Powdery mildews are typified by a white mealy layer on the leaf surface and are commonly found on soft leaves of a wide range of plants including begonia, phlox and Michaelmas daisy, and in general they appear during dry, hot seasons. In the woody plant world, powdery mildew is very prominent on young oak, sycamore, Norway maple and particularly on apple, where it can extensively disfigure new growth. Powdery mildew on apple is caused by *Podosphaera leucotricha* and on *Phlox paniculata* by *Erysiphe cichoracearum*.

Rusts are so called because of the groups of rust, or orange-red-coloured pustules that are associated with them. They often display their pustules on the underside of leaves. By killing leaf tissue they greatly reduce sugar production, and if infestations are high they will cause severe damage. Rusts are common on hollyhocks, fuchsias, pelargoniums, chrysanthemums, roses, pansies and antirrhinums; on vegetable crops, leek rust is caused by *Puccinia allii* and mint rust by *Puccinia menthae*.

Some microscopic fungi are carried in water droplets in the atmosphere or in the soil, and because of this are known as the 'water moulds'. *Phytophthera* species (along with some *Pithium* species) cause 'damping off' (and the subsequent total collapse of seedlings) in wet humid conditions, and attack the roots of many soft and arboreal subjects carried in soil moisture.

Phytophthora ramorum and P. kernoviae

Various species of *Phytophthora* have been identified on ornamental plants over a long period of time. Most cause severe die-back and root death. *Phytophthora ramorum* from North America was discovered in 2002 in the UK. The disease was not found on oak, which is its major host in California, where it is called sudden oak death (SOD), but on ornamentals, initially on *Pieris*, and then *Rhododendron* and *Camellia*. Unfortunately the disease is now present in many parts of Europe, where it affects established exotic ornamentals and young woody plants in the nursery, with *Rhododendron*, *Camellia*, *Viburnum*, *Arbutus* (strawberry tree), *Hamamelis* (witch hazel), *Pieris* (andromeda), *Syringa* (lilac), *Kalmia*, leucothoe and laurel as common hosts.

A separate and unfortunately more virulent strain of *Phytophthora* has now been isolated from dying and dead plants in some of the large UK collections in Cornwall. It has been responsible for the loss of some very large, important specimens of *Rhododendron* and *Camellia*, and has been named *Phytophthora kernoviae* (after Cornwall) because of it.

Entry of the disease is via any form of wound or breach in the system, and entry can occur via stomata and lenticels, both of which are involved in water movement. The disease seems to favour temperate climates and moist conditions, and movement of the disease from one site to another is probably in the moist soils of transported plants.

The symptoms on ornamentals are typified by severe leaf browning and die-back of shoots, and the plants very soon become aesthetically unacceptable and eventually die. On *Camellia*, laurel and lilac, the disease usually causes a leaf blight, which presents as brown, black, or brown/black necrotic patches on the leaf margins and leaf tips, and which gradually encroaches over the whole leaf; however, this often includes some shoot death. On *Viburnum* the disease may present as severe leaf blight (particularly on the evergreen species); it may also cause flower blight, but commonly kills stems, with necrosis starting at the base. On *Rhododendron* the disease sometimes causes a leaf blight alone, with blackening of the petioles, but often has shoot death with characteristic brown/black discoloration of the dying twigs that spreads towards the larger branch-work.

There are a few fungal diseases of trees that arboriculturists come across regularly, and because of their importance, horticulturists and garden managers of all types should also be aware of them (*see* below).

Meripulus giganteus (Giant Polypore)

The fruiting bodies of *Meripilus giganteus* are very large, comprising clustered layers of pale brown, tan or pinkish-orange-brown undulating frond-like growths, perhaps up to 80–100cm across in some instances, and bearing darker concentric rings. The undulating brackets are not only variable in colour, but may also have rounded or more acutely 'sharp' edges, depending on the individual. The brackets form right at the base of the trunk in the soil (at the root-to-trunk interface), or on large cork-covered roots radiating outwards from the central trunk. The fruiting bodies sometimes appear to be growing in soil remote from the main trunk, but closer inspection shows them to be attached to large conducting roots penetrating the soil.

Meripilus giganteus is commonly found on beech (*Fagus*), but is not specific to it because it also attacks sycamore, plane, lime and oak, including North American species. The fruiting bodies are not perennial, and they turn to a black, jelly-like mass after a few weeks, or immediately after frosts. *Meripilus giganteus* causes a root rot that quickly leads to low structural strength. Infections on trees in public areas create very hazardous situations, and for this reason many arborists consider the presence of fruiting bodies (particularly on beech, which is very prone) to be sufficient reason alone to fell the tree – and on North American red oak (*Quercus rubra*), considerable cause for concern.

Giant polypore (*Meripilus giganteus*) on beech.

Such a definitive decision – to fell a tree due to the presence of a particular species of fungus – is thankfully rare; generally the presence of most other fungal species may give cause for concern, but will lead to closer inspection of other factors before the fate of the tree is decided.

Armillaria mellea (Honey Fungus) and Related Species
Armillaria mellea (a form of honey fungus) is a pathogenic fungal disease that infects a wide range of tree and shrub species, causing a white root rot and butt rot. Nearly all other fungi types are readily spread by spores (although root-to-root contact and mycelial contact can also be involved), but *Armillaria* species have evolved vegetative mechanisms to aid infection. The species exists as sheets of white mycelium, but the mycelium can change in form, aggregate together to form 'rhizomorphs' (root-like structures – rhizo = root-like, and morph = form/shape) that are able to travel tens of metres through the soil to infect the next host. The depth of penetration of rhizomorphs is quite low, but is within the normal rhizosphere of trees (within the top 700–800 cm of soil), but most probably in the top sector of this range.

The white to white/brown rhizomorphs travel through the soil to a neighbouring host, where they enter the roots and travel up into the inner

Close-up of *Meripilus giganteus*.

bark (phloem tissues) to feed off the sugars of the host. The rhizomorphs become flattened by the pressure of pushing up under the bark, and also turn black, so they look like 'bootlaces' – hence the other common name 'bootlace' fungus.

Old stumps are undoubtedly the main infection sites, as they build up colonies of the fungus, and their rotting tissues supply energy for its successful production of fruiting bodies and rhizomorphs. The fruiting bodies are honey-coloured, mushroom-shaped, gilled-cap types that give rise to the common name, honey fungus. The mushrooms are relatively short-lived, and are found at the base of infected trees (or stumps), where they produce spores. However, because of the success of the species via rhizomorph infection, the spores have lost their importance and are rarely successful. Honey fungus is difficult to treat with chemical controls. There is no cure for infected trees, but because the fungus cannot live in the soil (without wood as a host), the most practical answer is to remove or grind up old infected stumps. If a barrier is to be used, then a neoprene or polythene barrier in a trench about 40–45 cm deep around the old infected tree, or dead stump, will help prevent cross contamination. Biological control can include the introduction of a competing fungus (*Trichoderma viride)* to cut stumps.

Death of small trees is usually within one to two years. Larger specimens take longer, but if infection is complete, then death will almost definitely ensue. If the disease causes damage that girdles the base of the tree, then demise will be relatively quick. Very large specimens may take several years to succumb, and the hazardous or non-hazardous state of trees in public places may need to be decided over a period of years.

Armillaria mellea can survive as a saprophyte, but is commonly found as a pernicious parasite. *Armillaria gallica*, another similar species of honey fungus with rhizomorphs, is more likely to appear as a secondary infection on trees that are already in a state of stress or demise.

Other Symptoms of Fungal Diseases

Vascular wilts cause severe damage to the vascular tissue (the veins of the tree – notably the xylem) e.g. Dutch elm disease. Verticillium wilts affect a diverse range of plants, and cause the foliage to wilt and the total collapse of the plant.

The term 'canker' is a generic term that does not tell you what the causal agent is, but merely describes the set of symptoms (that may be caused by various fungal, and sometimes bacterial, species). The protective nature of bark normally prevents canker infections. However, cracks, fissures, wounds and insect damage can allow entry.

Target cankers occur on main trunks, scaffold branches, and more rarely on lesser order branches. The typical symptom of a target canker shows as concentric rings of dead tissues. Target cankers are chronic conditions, not acute, and because so much superficial and deeper tissue damage occurs over an increasingly large area, the structural strength of the area lessens over long periods of time, and in extreme cases large trunks and major limbs can snap. The fungus *Nectria galligena*, found as a target canker on as wide a range of trees as apple, pear and oak, is a common example. Very severe, untreated infections will necessitate felling the tree either because of its weak state, or because aesthetically it has become so unattractive, or both. Some bacterial cankers gain entry by actively killing soft tissues (such as leaves) to do so.

Bacterial Diseases

Bacterial diseases involve the invasion of plant tissue by harmful microscopic, single-celled organisms (capable of being seen with a light microscope) that can cause canker symptoms and may also be responsible for shoot-tip die-back. Bacterial infections can lead to disfigured stems, large lesions/cankers, and ultimately death. Symptoms may also be associated with jelly-like, bacterial oozes (gumoses), and stained fluxes at certain times of the year. Because bacterial infections can lead to disfigured stems, they quickly create unacceptable aesthetics, and infected plants may need to be culled.

Fireblight

Fireblight is a bacterial disease that has economic important because of the large range of ornamental trees and shrubs in the Rosaceae family that it can affect. Hawthorns (*Crataegus*), plants in the genera *Sorbus* (particularly in the rowan group), *Cotoneaster* and *Pyracantha* are particularly susceptible, and pears (*Pyrus*) and apples (*Malus*) may also contract the disease (cherries are not susceptible for some reason). Fireblight is no longer a notifiable disease, but is nevertheless a very serious problem of both nursery and fully developed plants.

The leaves of infected plants turn brown, yet remain leathery and persistent and do not fall prematurely, which makes the plant look as though a bonfire has been lit underneath it and the leaf tissues have been burned. The disease enters via the flowers, causing them to die, and then moves through the vascular tissue. Brown, mummified flowers, turning black/brown and remaining persistently on the plant, are common, and brown, mummified fruits are very common. Entry of the disease into the main vascular tissues causes the die-back of tip growth, which is followed by the death of larger branches and the gradual demise of the tree or shrub – the death of even a well established plant can be within two to three years.

Bacterial Canker of Cherry

Bacterial canker of cherry is specific to cherry, and shares some of the symptoms of fireblight. However, it is less virulent, causing disfigurement and a slow demise over many years, rather than the more rapid death experienced with fireblight. Furthermore, bacterial canker of cherry enters the system via the leaves, not the flowers, and because of this the dead tissue (killed by contact with the bacterium) forms brown, circular patches on the leaves. Sometimes the flowers go brown, and occasionally there may be a few mummified fruits left on the tree. The symptom that really makes the disease appear like fireblight is that leaves go brown and papery and die early in the late summer and autumn – they can also be seen alongside live leaves in some instances. However, the defining factor is the dead patches in the leaves that are gradually scoured out by wind, giving the appearance of 'shot holes', because this is a symptom not associated with fireblight. Die-back of shoot tips occurs (along with a general reduction in crown vigour and density), and gumoses (large globules of oozing gum) appear from cracks and splits in the bark, with definite cankerous lesions appearing on both minor branches and main trunks.

Bleeding Canker of Horse Chestnut

Bleeding canker of horse chestnut (*Aesculus hippocastanum*) is a bacterial disease caused by the bacterium *Pseudomonas Syringae*. The symptoms commence as rusty brown, or darker, stains on the main trunk-even on semi-mature specimens. The staining gradually worsens and staining fluxes can run down the bark. In many cases the foliage in the crown yellows and gradually dies back. Severely infected young trees are felled and burnt as soon as it is obvious that the crown is in demise. Larger specimens are monitored for progression of the disease and may have to be felled eventually. Unfortunately, this bacterial disease is becoming increasingly prevalent.

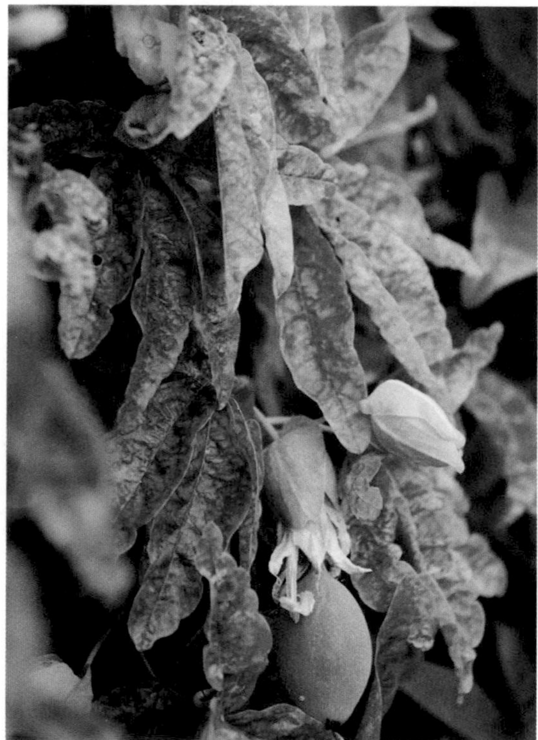

A mosaic virus on passion flower (Passiflora caerulea).

Virus Infections

Viral infections are caused by the invasion of viruses, which are sub-microscopic organisms (can only be seen with an electron microscope). Because they are so small, and may or may not show obvious external symptoms, they can be difficult to detect. In some species the symptoms lie latent for very long periods, and in other instances very minor external symptoms are shown. Some species do show distinct symptoms, usually mottled (or mosaic) patterns in the leaf tissue, caused when chlorophytic tissue is killed, and sometimes accompanied by a leaf roll. There are no cures for viral diseases whether they have latent or obvious symptoms.

Viral diseases are usually classified by both a description of their symptoms (if any), and the name of their host: for example, cherry leaf-roll virus, cucumber mosaic virus, gladiolus mosaic virus and tobacco mosaic virus. The host plant name used in the classification does not indicate that any virus is specific to that plant only – in fact the opposite is generally true, as one virus can have many hosts. Dahlia mosaic virus, for example, can be found on many plants other than dahlia. Classification is often reduced to an abbreviation comprising an acronym: for example, cherry leaf-roll virus is CLRV, and tobacco mosaic virus is TMV. Some ornamental plants such as *Abutilon pictum* 'Thompsonii' owe their variegation to viruses, and in order to retain their ornamental effect the virus has to be successfully transferred during propagation. This is not usually a problem, as lots of infected tissue is used during the operation.

Because of the difficulty of tracing latent viruses, helpful work on the subject is sparse. It could well be that viruses are far more prevalent in woody subjects than we realize, and that many species under-performing in aesthetics, quality and growth, besides coming under environmental stress or having genetic deficiencies, have viral infections.

Abiotic Agencies Affecting Plant Growth

Physiological Disorders

Any factors at aerial and sub-surface levels interfering with photosynthesis, transpiration and/or respiration of plants detrimentally are physiological disorders. Environmental factors – including drought (lack of rainfall), other climatic, geographic and topographic factors, exposure, root death, damage caused by atmospheric pollution, toxic materials in the soil, and nutrient deficiencies – can all cause physiological malfunctions of plants.

Subsurface (Root) Environment (the Rhizosphere)

Root-level environments often include thin or no topsoils (particularly in upland areas), or disturbed/moved soils, or compacted topsoils and/or subsoils. Soil cultivation machinery, itself brought in to improve soils, can also cause problems of compaction at topsoil and subsoil levels, as can excessive human foot traffic.

Lack of moisture (drought) may be caused by insufficient rainfall, excessively free-draining soil, or because the plant roots have been damaged and cannot, therefore, take up sufficient water. In any case the youngest tissue will become flaccid, and will come under stress. The effect will be to reduce sugar production (as photosynthesis will not go on when tissue is flaccid/wilting). If the stress carries on for a protracted period, a brown scorching appears on leaf margins, and all green tissues desiccate and eventually die. Any condition that causes root death (waterlogging, soil compaction, protracted drought, and toxins in the soil) can also result in the same symptoms, as root death prevents the efficient uptake of water (and soluble nutrient salts dissolved within it) through the plant. Hence, nutrient deficiency symptoms (chloroses) will show under these conditions because chlorophyll dies. Poor nutrition because of total lack of soil fertility is actually fairly unusual.

The most common physiological disorders are nutrient deficiencies. Soluble nutrient salts are essential for good health (many are directly involved in the chlorophyll molecule), and are normally present in the soil solution within the soil matrix. Poor nutrition, whether because it is not present within the soil, or because of lack of take-up by the plant, shows initially as a chlorosis, which is a leaf yellowing caused by the malfunction of chlorophyll. Any plant not functioning properly because of a physiological disorder will be weakened in general, will have low or no sugar reserves, a correspondingly low energy budget, and poor resistance to fungal and insect attack.

Nitrogen is relatively soluble, and for this reason is often in short supply, especially in high rainfall areas, because the soluble salts of nitrogen are easily removed from the soil by percolating rain and gravity-powered drainage. It is often necessary, therefore, to add nitrogenous salts as a fertilizer. Poor nitrogen levels are typified by yellowing leaves (chlorosis), highlighting the essential nature of the presence of nitrogen in the chlorophyll molecule.

However, a chlorosis is only a symptom of the problem, and the cause could be a number of things, from inadequate drainage, viral infection, incorrect soil pH, or inadequate nutrition. An application of nutrients may not be sufficient to 'cure' the problem, and a chlorosis is potentially fatal. Inadequate drainage causes anaerobic root conditions and ultimately root death. Loss of any volume of root can lead to poor transportation of available nutrition, and cause chlorosis. Some virus infections present as mosaics/variegations that include chloroses, so will affect the physiological processes locally. Some nutrients may be 'available' (as nutrient salts) within the soil, but due to local pH may not be soluble enough to be 'available' to the plant – conversely, there may not be sufficient nutrients in a soil.

Many of the most common nutrient deficiencies have some similar symptoms, and are not always easily distinguished one from another. Nitrogen deficiencies show as weakly growing plants that are pale green to yellow in colour instead of the normal healthy dark green, and that make weak, spindly growth with smaller-than-average leaves. Autumn-like tints as leaves become stressed are not unusual, and yields are reduced because fruits are small.

Magnesium deficiencies show as a marked chlorosis on the leaves between the veins, leaving the area around the veins starkly green in comparison; iron deficiencies look very similar (including lime-induced chlorosis). In magnesium deficiencies the chlorosis commences on the older leaves. Leaves often also become highly tinted with autumn-like oranges and flame reds as the chlorophyll suffers.

In phosphorous deficiencies, leaves tend to have a bluish or purplish hue to them, and other tints are unlikely. Some species show marked brown or purple blotching on the leaves. Unfortunately, potassium deficiencies also show as having bluish-green hues, and may also have brown scorching or spotting.

The Aerial Environment

Late frosts (occurring in the late spring) can cause severe damage to young, green (non-lignified) growth, because the freezing and thawing ruptures cells. Air frosts create masses of cold air that, being a cold fluid, moves downhill, and can collect in low-lying frost pockets where its effect is exacerbated.

Excessively cold temperatures slow down respiration – the energy-using, tissue-building functions of the plant. So long periods of exposure to unusually cold temperatures reduce growth and may kill tissue. The effect of high wind-speeds is to reduce growth rates and sometimes cause physical damage. Protected niches created by hedges and shelterbelts may be used to help reduce transpiration rates. Exposure due to estuarine or maritime conditions may be ameliorated by the careful use of shelterbelts using salt-tolerant/exposure-tolerant species, then planting other less tolerant species within the shelter of the hardier, more resistant shelterbelt species.

Excessive heat and higher-than-normal temperatures increase the transpiration rate of plants, which may cause temporary wilting initially, and if it continues for a protracted period, the outcome is tissue death. Because guard cells close down as a result of the stress caused by high transpiration rates, it lowers the rate of photosynthesis.

CHAPTER 11

Managing Plant Collections

Because plants from all over the world may be included in plant collections, their management can be very complex, and knowledge of the country of origin and, more importantly, their host environment is often essential for successful management. Sometimes emulating their natural environment is the only way to success, and knowledge of their root environment (rhizosphere) and specific pH needs is always essential. This chapter looks at the adaptations that plants have made in order to survive their specific environments, the methods of emulating these environments, the range of plants possible for plant collections, and their limitations.

Ecology is the interaction between living organisms and their environment. Environments can be quite complex, as ecosystems include not only the living organisms, but also the physical and chemical factors that form the habitat of those organisms. Physical factors include climate and the geology of the area. Climate obviously includes rain, cloud, temperature, wind and sun, all of which can be influenced by the relative shelter or exposure experienced by individual plants. Soils and underlying geology will also have a physical and chemical influence on plant communities, depending on their porosity, texture, cleavage, softness/hardness and so on. Furthermore, pH differences and availability of nutrient salts are affected not only by the decaying organic matter present, but also the inherent solubility of the local rock.

Ecosystems have to be self-sustaining in order to survive. Their success (or otherwise) depends upon the interaction of all representatives within the system. Plant communities are dynamic as regards reproduction, growth, competition, survival, ultimate death and replacement. When a plant dies, another will quickly fill the vacant niche; this creates an ever-changing patchwork of species in the short term, but a surprisingly stable pattern over time, if the ecology is well balanced. However, relatively small changes in the balance can cause massive, and mostly irreversible, damage. Small rises (or falls) in average temperature and/or rainfall will have disastrous effects on an ecosystem. Reductions in a local grazing population of rabbits can alter the balance of species within a community: because the more competitive plants have lost their predator, they become too vigorous and densely populated, and take over. Many upland areas would be densely wooded were it not for the pressures of sheep grazing.

The Importance of Plants in the Environment

Plants are the main agency for water regulation (via movement from the soil by plant roots, transpiration and evaporation at green surfaces); they are therefore very much involved in the hydrological cycle, which is affected by both the surface area that lower plants cover, and the height (and therefore volume) of the transpiring surfaces of arboreal plants.

Plants of all types give off oxygen as a by-product of photosynthesis, and the sugars (carbohydrates) produced from the process form the carbon store that is released and retrapped during the carbon cycle. Being able to produce their own nutrition from sunlight makes plants autotrophes (self-feeders). Green plants are therefore primary producers at the commencement of the feeding levels (trophic levels) of organisms. The energy for green plants to survive is taken from the non-living (abiotic) environment (heat and light), whereas all other organisms are reliant on the living (biotic) environment to feed – for example grazers, browsers, omnivores and carnivores. Humans and other animals are heterotrophes (needing several food sources to survive), and cannot gain nutrition by merely being exposed to sunlight, and can never be the primary source of the food chain because of their dependence on organisms that can feed themselves (plants). Viewed from this perspective, the importance of plants in our world is beyond any doubt, whether those plants are naturally occurring or produced specifically as a food crop.

Changes in energy occur at each trophic (nutrition) level. Ultimately the breakdown of organic matter, initially formed using heat/light energy, leads to the release of minerals (chemical energy). The breakdown of organic material (including plants/trees), as well as releasing carbon into the environment, also releases mineral nutrients back into the soil matrix (by the process of mineralization). The never-ending cyclical nature of these processes maintains our aerial atmosphere in a state that allows us (and plants and other animals) to respire and survive, produces condensation (via evaporation) to create rain for growth, and produces sugar-rich foods for our sustenance.

The matrix of plant species of naturally occurring systems is of course dependent on the species within it, and the balance and diversity of species that it creates very much influences the balance of interdependent fauna that it can support.

Obviously, tall-growing woody plants (trees) can be illuminated at their top and other exposed surfaces to produce sugars (the energy supply for further growth). This group of very large plants (after all, trees are just big botany) is critical (though no more critical than the other pieces of the jigsaw) for the success of our ecology. Light filters through the canopy and offers varying degrees of intensity to other lower-growing plant groups. Climbers facilitate the energy for rapid growth by having their roots in the soil matrix, and by covering (smothering?) their host tree to gain access to some of the precious light source – often precluding light from their host. Sharing, competition, death and regeneration produce a relatively 'balanced' (over time) species matrix known as 'climax vegetation'. Various species within the shrub layers utilize the filtered light and/or the light of natural open glades. Herbaceous forest-floor species manage to thrive (or be suppressed and survive in) the semi-shade on offer. Because of the biological need of green plants to be illuminated in order to photosynthesize, none will survive in exceptionally low, or no, light conditions.

Without the various open environments and the layers of forest, the varying degrees of light and shade, and the plants and animals that can take advantage of the niches on offer, we would not have the diversity that we enjoy. The dynamic matrix of plants involved in mixed woodland is something that designers have tried to emulate for centuries using exotic plants. Emulating the use of plants at all levels of the matrix can create very successful plantings, including fruiting and food plants, and in recent times, forms of forest gardening and perma-culture attempt to do this, and with some success. The various layers of the canopy are recognized as high canopy, low canopy, shrub layer, vertical layer (woody climbers), herbaceous layer, and forest floor. Even large-scale, yet simple, designs for shelterbelts, mixed plantings and groundcover schemes have successfully emulated part of the matrix, by using tiered layers of woody plants.

However, balanced, relatively stable vegetation matrices may not necessarily involve many or any woody plants at all. Open grasslands, estuaries, downlands and deserts all have a specific and relatively stable flora and associated fauna without much arboreal influence. Deserts support plants with adaptations to allow them to grow in their harsh environment (xerophytic adaptations), and they flower and seed very quickly when conditions are eventually correct.

Simulating Specific Environments in Botanic Collections

The range of soils and the temperate, yet relatively wide-ranging climates of North America and Europe, allow non-indigenous (exotic) plants from many countries to be grown successfully. This enables a wonderful range of botanically and aesthetically interesting plants to be introduced in landscape design. In many cases the successful management of plants necessitates understanding the environment from which the plants come, and the adaptations they have made in order to grow in different environments. Simple management mistakes, such as over-watering succulents in the winter, can be disastrous. Plant modifications and strategies, including xerophytic leaf adaptations such as indumentum, halophytic adaptations – salt resistance (halophytes) – and hydrophytic adaptations (hydrophytes) need to be considered. Epiphytic, parasitic and carnivorous adaptations are all relevant, as are both the plant and leaf size modifications found in some alpine plants.

Commonly, botanic gardens have hardy natives, hardy exotics and tender exotics as part of their collection, so a plant manager may need to simulate

PLANT STRATEGIES FOR SURVIVAL

Plants from all levels of their ecosystem adopt certain strategies for survival, and they may be successful in their particular niche until the environment in which they survive changes.

Plant Parasites

Some plant species survive without photosynthesis by being parasitic and gaining their nutrition directly from the vascular tissues of an illuminated sugar-producing host. They live 'off' their host plant, and actually penetrate the vascular tissue of their living host to derive nutrition from it. The most common example is mistletoe (*Viscum album* – the name is a reference to the white, sticky/viscous berries), which is actually a hemi-parasite as it has the ability to photosynthesize, but augments its nutrition by parasitizing apple trees, poplars and willows. Dodder, broomrape and toothwort are wholly parasitic (holoparasites), as they have no chlorophyll of their own. Common broomrape (*Orobanche minor*) and toothwort (*Lathraea squamaria*) are found feeding off the roots of trees and shrubs, and dodder (*Cuscuta epithymum*) and greater dodder (*Cuscuta europaea*) are commonly found feeding off heather.

Epiphytic Plants

Many species of plants have become epiphytic to survive. Epiphytes live 'on' but not 'off' other organisms. They photosynthesize and, unlike parasites, do not penetrate the vascular tissue of their host plant to gain nutrition. Epiphytes appear outwardly to be taking nutrition from their host because they are firmly attached to it, but true epiphytes only use their host for anchorage (epi = outside, phyte = plant), and gain their nutrition via photosynthesis from their own green parts. They share the aerial environment of their host, and therefore their success is dependent not only on the survival of their host, but also the atmosphere around the host. They function by gaining their own moisture and some nutrients from the surrounding atmospheric humidity, so only tend to be successful in areas of naturally high humidity.

Temperate zones have many epiphytes, with mosses, liverworts, lichens and ferns as common examples. Many of the same species can be found naturally growing terrestrially as well as epiphytically. Common polypody (*Polypodia vulgare*) is a temperate example, and large colonies are often present where the correct favourable host and a high local rainfall (and/or natural humidity via mists, low cloud or water spray from waterfalls) come together. Lichens are not solely plants, but comprise a symbiotic relationship between two organisms (an alga species – plant – and a fungus species). Many species of lichens in Europe and North America are epiphytic, and there are three main groups: crustate, foliose and fruticose. Crustate lichens lie closely attached to their host, and form thin layers of tissue with a crusty/crystalline effect; they are often grey, silver-grey, yellow or yellow-green – for example *Rhizocarpon geographicum.* Foliose types, as their name suggests, look like flattened leaves pressed against their host, are often highly chlorophytic, and are commonly green, orange or yellow. Examples include *Xanthoria parietina* and *Parmelia* species. Fruticose types look like small, bushy shrubs that hang from trees and are usually grey, grey/green or silver; examples include *Usnea florida* and *Evernia prunastri.* Not all lichens are epiphytes – there are many species of lichens that are largely terrestrial, for example, the fruticose reindeer lichen or reindeer moss (*Cladonia rangiferina*) or crustate on rocks.

Other mechanisms of epiphytes to gain nutrition include long aerial roots that drop to soil level from branches and take up moisture and dissolved nutrients. Some epiphytic orchids produce two main types of root, clinging roots for support, and absorbing roots (with velamen) to gain moisture and nutrients from surrounding decaying matter. Orchids such as *Dendrobium, Phaeleopsis* and *Vanilla* species (from whence the essence of vanilla is harvested) all have epiphytic species. Bromeliads (those in the family Bromeliaceae), such as edible pineapple (*Ananas comosus*), ornamental pineapples (*Bilbergia nutans, Aechmia fasciata* and *Vriesia splendens*), and ferns such as *Platycerium bifurcatum* (stag's horn fern) also survive by anchoring themselves to the bark of a host tree. Edible pineapple is found growing epiphytically in its native tropical habitat, but is cultivated for fruit in the tropics in terrestrial plantations – it is not an obligate epiphyte.

The commercial and practical benefits of being able to grow and harvest the crop at ground level are obvious. 'Vases' formed in the collective leaf bases of bromeliads have some similarities to insectivorous plants in that they form thick nutritious soups in their container-like structure (but have no motile parts). The 'soup' comprises water, dissolved plant organic matter, and some dissolved nutrient derived from drowned insects, which can be absorbed slowly into the leaf tissue and into the vascular tract to supplement photosynthates.

Insectivorous (Carnivorous) Plants

Some plants successfully growing in poor, boggy, nutrient-leached (and often highly acidic) soils cannot gain nutrition via their roots, so owe their success to specialized leaf modifications that allow them to ingest nutrition by absorbing nutrients derived from trapped insects.

Although insectivorous (carnivorous) plants have all evolved to survive very similar environmental conditions and with similar methods, they have not all evolved in exactly the same way (or in the same direction). Some plants in the Sarraceneaceae, such as *Sarracenia purpurea* and *Sarracenia flava* from North America, have evolved to form terrestrial plants with large pitcher-shaped leaves for trapping insects (pitfall types). The sundews (*Drosera* species in the Droseraceae) and butterworts (*Pinguicula* species) have evolved in similar conditions and for the same reasons, but have sticky glands to trap insects ('sticky flypaper' types). Both butterworts and bladderworts are carnivorous and in the same family (Lentibulaceae); however, bladderworts (*Urticularia spp.*) are found in water and enlist bladder-like leaves and a siphon effect to trap their prey, whereas butterworts (*Pinguicula spp.*) have partially rolled leaves and are terrestrial.

Species in the Nepenthaceae, such as *Nepenthes rafflesiana* from South-East Asia, are mainly epiphytic with large pitchers, so have adapted to be remote from the competition of the forest floor to fare better higher up in the canopy of trees. Others have adapted to have motile (moving) parts, such as the Venus flytrap (*Dionaea muscipula*) that closes on its prey (man-trap or bear-trap types). Trapped insects are gradually digested by enzymes, and the resulting nutritional material is absorbed into the leaf tissue so that it can be transported by the vascular tissue to be used or stored for use at a later date.

Pitfall types comprise leaves modified to pitcher-shaped, vase-shaped or funnel-shaped vessels, into which insects can slide – for example *Nepenthes* and *Sarracenia* species. Unsuspecting insects slip neatly into the vessel, at the base of which is a soup containing digestive enzymes. However, other additives within the soup are also known, including toxic materials and wetting agents that debilitate insects and increase their absorption to the enzymes. The pitcher-type leaves of *Sarracenia* species (terrestrial pitcher plants) have other mechanisms to aid their insect entrapment – some are there to entice insects initially, and others to help prevent any escape attempts. Fragrances/odours, nectaries and pigmentation that gives the modified leaves the appearance of flowers, are used to attract insects to their fate, very slippery surfaces (usually lubricated by the digestive enzyme liquids) aid capture, and downward-facing hairs inside the pitchers prevent trapped insects climbing back up the inside leaf surface.

Some have leaves evolved to form long, tube-like funnels with a partially open top (lobster-pot fashion). Because the leaves form a three-dimensional tube and the tissues are thin, it allows partial light penetration, resulting in a translucent light and an illuminated leaf venation (often pigmented) akin to a stained-glass window, which lures the unsuspecting insect into the neck of the pitcher. Once inside, the insect feels trapped and starts to panic. The top lid (even though it is not entire to allow entry in the first place) confuses and disorientates it, and the downward-facing hairs force it downwards into the enzyme soup at the base. The soluble parts of digested insects are absorbed by the soft tissues at the base of the pitcher (modified leaves), but the insoluble parts are left as remains.

In Venus flytrap, the leaves are modified to form hinged, semi-rounded flaps (valves) that, because of the thick hair-like appendages at their rims, appear rather like two eyelids with eyelashes. On the modified leaf lobe surface are three finer hairs that act as a trigger for messages to close the trap. It is the agitation by an insect that initially triggers the relatively rapid closure, and the continued insect movement in panic that finishes the process by inducing a tighter hold and the excretion of enzymes to digest its soft parts. The nutrients thus released are absorbed by the leaf tissues and taken into the vascular tissues. Cunningly, the trigger hairs have to be agitated more than once before the 'trap' is closed. Hence, insects alighting quickly only once on the trap, and dead inanimate insects, do not trigger the system.

The thick, hair-like/spine-like appendages at the rims of the 'traps' form a grid-like mesh that prevents the prey from escaping. However, because in the first instance the trap valves are held relatively loosely (and only after continued agitation do they close more tightly and exude digestive enzymes), small prey that would supply only low energy food, can escape. This mechanism prevents excessive energy being used by the plant for a negligible (or negative) return. Artificially triggering the system using a pencil to prod the hairs twice (or more) makes the plant use up valuable energy in closure (or at least in reopening), yet it gains nothing from it and so creates a negative energy budget. In hinged bear-trap types the non-nutritious chitinized shell (exoskeleton) of the insect is released, to be blown away by wind currents when it opens at a later date – it is a fairly long process, and in Venus flytrap (*Dionaea muscipula*) it might take several days.

Those with sticky glands (the so-called flypaper types) include the sundews (*Drosera* species). These glue the insects to the leaf surface by exuding mucus: the more the insect struggles, the more the mucus covers it. The glands that secrete the mucus are held proud of the leaf surface by stalk-like tentacles, and the tentacles themselves (with their mucus coating) wrap round the struggling insect. Furthermore, the leaves may roll inwards to help trap the insect, and digestive enzymes are exuded from the glands to break them down. Butterworts (*Pinguicula* species) are similar in their action, but the glands are not held on tentacles, they have evolved to have leaves split linearly into two halves (valves), each lined with hair-like glands. In this instance the two valves are slowly motile and can close (roll in) on the insect, placing the glands nearer the insect prey.

Bladderworts (*Urticularia* species) are insectivorous freshwater plants (carnivorous hydrophytes), and their many tiny bulbous bladders are used to create siphon-like suction to draw in their water-inhabiting prey.

Great Britain and Europe boast at least three genera and eight species of insectivorous plants that are commonly found. *Pinguicula vulgaris* (common butterwort) has small blue flowers, and *Pinguicula lusitanica* (pale butterwort) has pale blue ones. *Drosera rotundifolia* (round-leaved or common sundew) has small rounded leaves, *Drosera anglica* (great sundew) has larger, longer, more linear leaves, and *Drosera intermedia* (oblong-leaved sundew) is a species that, as its name suggests, has qualities intermediate between the other two. All species have small yet distinct red glands on their leaves – like red hairs with rounded ends. *Drosera* and *Pinguicula* species successfully inhabit upland acid heath, sphagnum blanket bog with a high water table, and sphagnum-covered areas of all types. Although *Drosera anglica* frequently inhabits the same conditions, it can also be found on lime (higher pH) areas.

There are three representatives of bladderwort: *Urticularia vulgaris* (greater bladderwort), *Urticularia minor* (lesser bladderwort) and *Urticularia intermedia* (intermediate bladderwort). Native insectivorous plants are of particular botanic interest, but they are all very small and not as obvious and aesthetically pleasing as their North and South American counterparts.

Adapting to High Water Levels (including Hydrophytes)

Some arboreal plants such as swamp cypress (*Taxodium distichum*) and mangrove species produce pneumatophores – specialized roots that come vertically out of the water (or the anaerobic mud and silt) and allow oxygen into the system. Each pneumatophore comprises a 'knee-like' outgrowth from the main bark-covered supportive root system. However, on the topmost part of the 'knee' the outer tissue is very thin, and soft parenchymatous cells very near the surface allow oxygen to enter and diffuse into the lower tissues. Plants with pneumatophores are considered to be semi-hydrophytes.

Many true hydrophytes are able to absorb water all over their tissue surfaces. Oxygen and carbon dioxide are taken out of solution from the water. Some, like common water lily (*Nymphaea alba*), have leaves that float on the surface of the water. Unlike land plants, because the water has a cooling effect, the

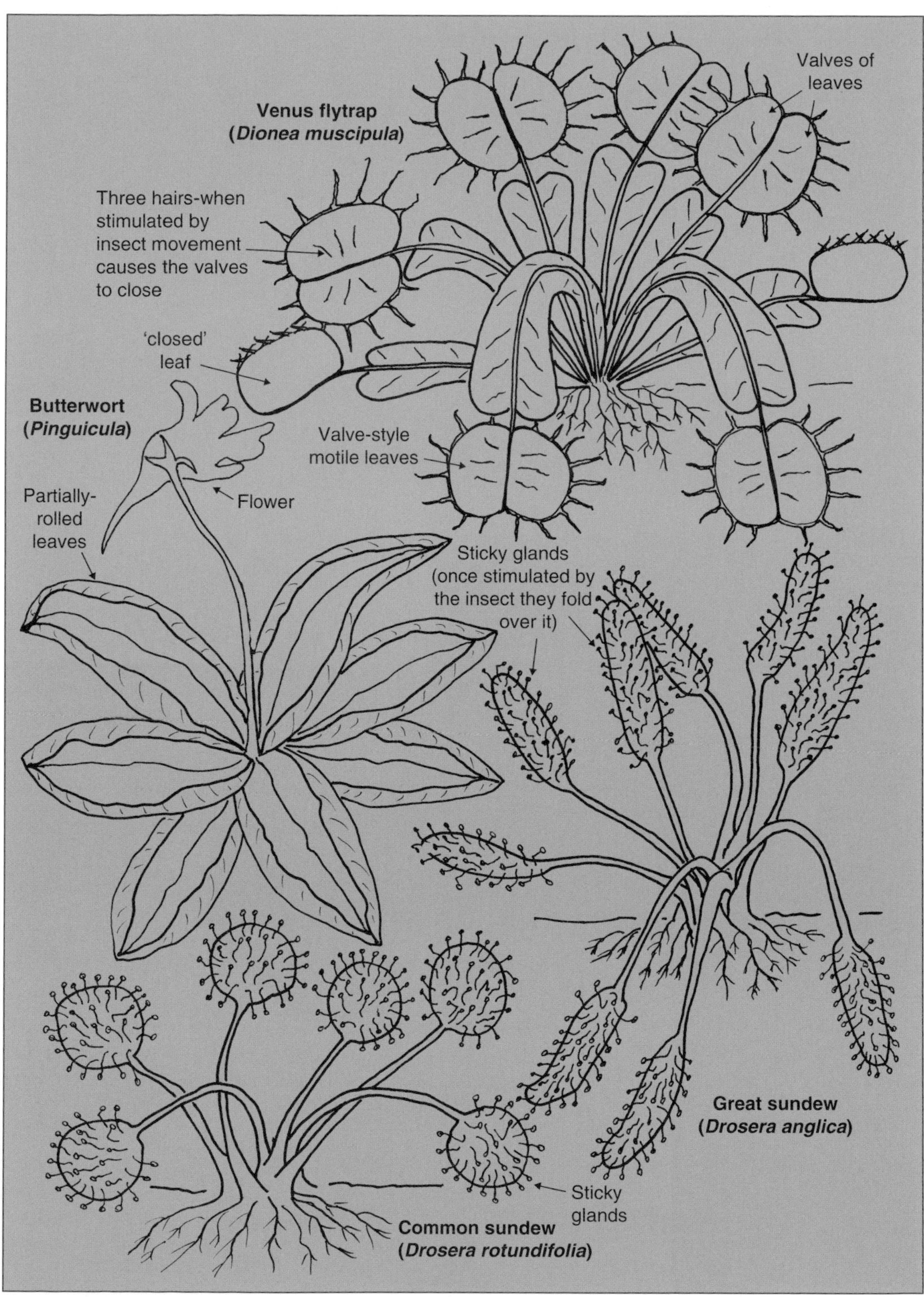

Insectivorous (carnivorous) plants (terrestrial forms).

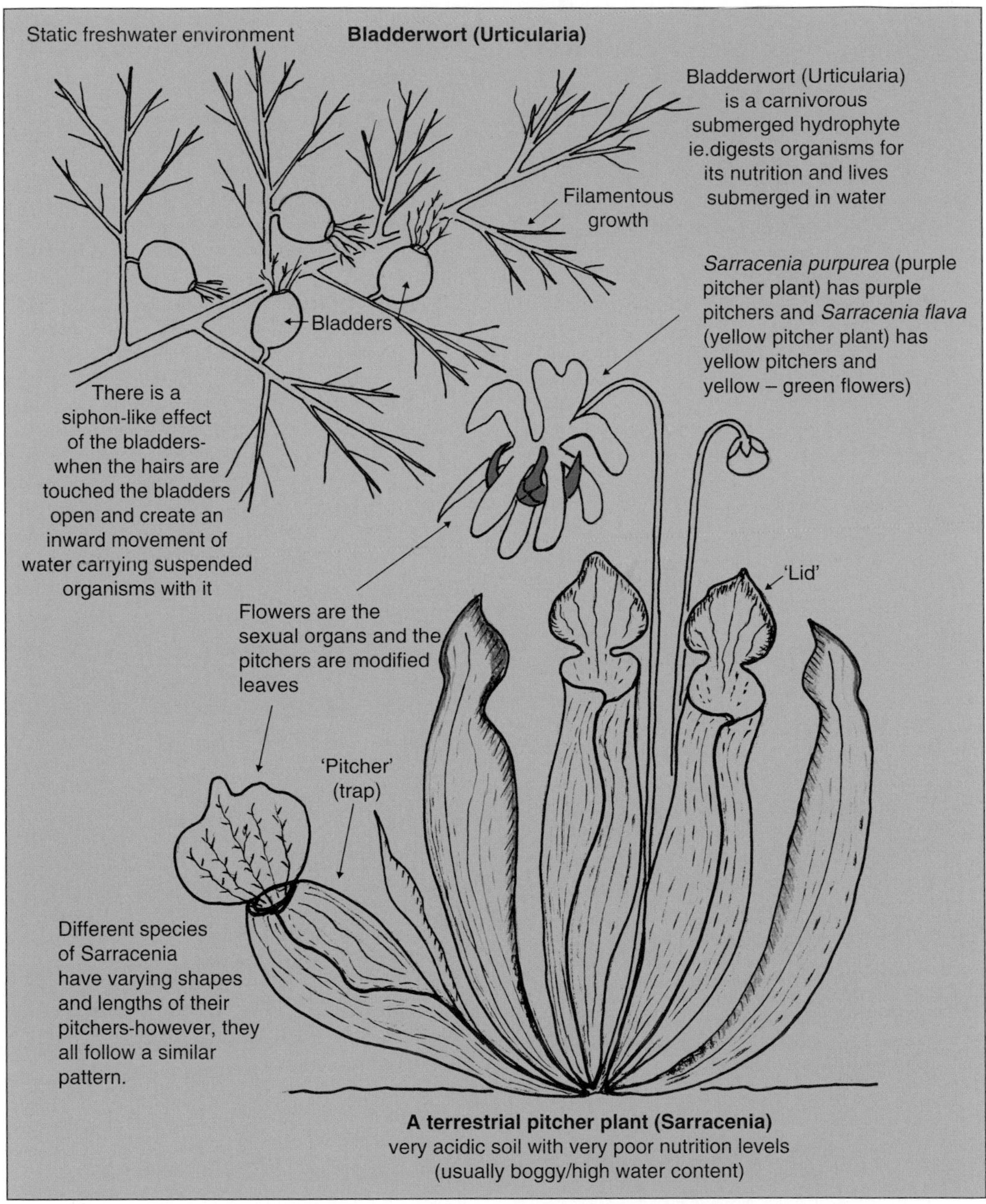
Static freshwater environment
Bladderwort (Urticularia)
Bladderwort (Urticularia) is a carnivorous submerged hydrophyte ie.digests organisms for its nutrition and lives submerged in water
Filamentous growth
Bladders
There is a siphon-like effect of the bladders- when the hairs are touched the bladders open and create an inward movement of water carrying suspended organisms with it
Sarracenia purpurea (purple pitcher plant) has purple pitchers and Sarracenia flava (yellow pitcher plant) has yellow pitchers and yellow – green flowers)
'Lid'
Flowers are the sexual organs and the pitchers are modified leaves
'Pitcher' (trap)
Different species of Sarracenia have varying shapes and lengths of their pitchers-however, they all follow a similar pattern.
A terrestrial pitcher plant (Sarracenia)
very acidic soil with very poor nutrition levels (usually boggy/high water content)

The pitchers (traps) of Sarracenia and the valves/bladders of Urticularia (bladderwort).

stomata are on the upper surface of the leaves, and they carry out normal gaseous exchange, alleviating the need to undergo gaseous exchange in water. In order to prevent the problems of the top leaf surface flooding during high water levels, the leaves and leaf petioles (and sometimes the roots and stems) contain specialized cells known as aerenchyma (aerenchymatous tissues) that comprise very large air spaces and act like bouyancy bags. Another strategy to prevent flooding is the production of either a very long or a corkscrew-shaped leaf petiole that gives a pontoon effect and always leaves the leaf surface level and floating on the surface.

Another modification is found in the upturned leaf margin of the massive floating leaves of Victoria lily (*Victoria regia*). The leaf margin is at right angles to the main leaf around the complete perimeter of the leaf lamina to prevent flooding on the top surface, so the stomata are always out of the water and carry on normal gaseous exchange.

Submerged hydrophytes tend to have modifications that help them remove oxygen and complete gaseous exchange in solution. They often have very finely sub-divided leaves that are therefore in close contact with a large amount of water. By this means they can take in and release gases into the surrounding water – the plant equivalent of gills. Some species have a mixture of large-lamina surface leaves with stoma, and finely divided submerged leaves with no stoma but the ability to diffuse gases – albeit slowly, for example water crowfoot (*Ranunculus aquatilis*). Bladderwort is a submerged, carnivorous hydrophyte.

Water margin plants are relatively normal in their structure as only the base of the plant is submerged in water, mud or silt with low oxygen content; their leaves are aerial and can carry on gaseous exchange via their stomata.

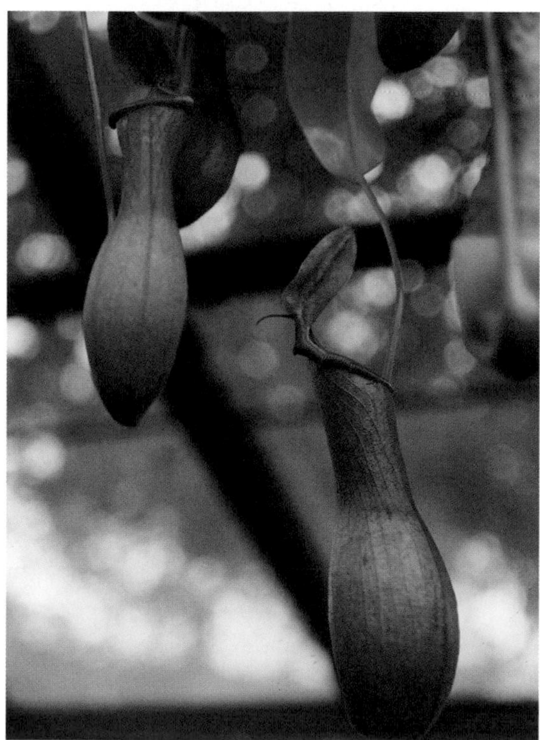

An epiphytic Nepenthes species.

a wide variety of climates for a large botanic collection. Ironically, some hardy *Cyclamen* such as *Cyclamen hederifolia* are very successful grown below the light canopy of trees, considering that their natural habitat comprises dry, baked ledges and scrub land at altitude. The secret of their success is that the tree canopy keeps the soil very dry in the summer, but allows some moisture percolation just prior to flowering in the autumn – thus emulating their necessary growing conditions, even though not their natural terrain or climate.

What is important to plant managers is how plants respond to the conditions in which they are now placed, and how to emulate/simulate the correct environment for their success. When all the parameters are added to the system, the possible combinations of requirements becomes quite complex. Examples include soil types (texture, consistency and moisture content – free-draining gravel gardens, moisture-retentive soils, water margins), plus pH considerations, and whether the natural local climate is suitable for the plant, or needs an artificial climate as provided by glasshouses or conservatories (heated or cold) and/or polyethylene tunnels/domes. The greater the range of plants grown, then the greater the range of environments needed for success; extreme examples of this might be a heated glasshouse and a rock garden.

There is an increasing trend to simulate specialized geographical conditions/climates in plant collections. Mediterranean gardens and Canary Island collections are popular in botanic gardens and add interest. Other collections can include medicinal plants, plants of economic/commercial importance, and ethno-botanical plants.

At present the best examples of housing plant collections with very diverse needs are probably the National Botanic Garden of Wales and the Eden Project, St Austell, Cornwall. At the Eden Project, economic factors (the Lottery and government aid), insight, innovation and leadership, good marketing and a keen client interest have all been brought together to create a very successful venture. This is all aided by good natural climate, special microclimates using the shelter and orientation of a derelict quarry, good natural light, and the clever use of height within the large domes. The Project site includes plants from various geographic locations and plant groups, including mediterranean, desert, cacti and succulents, temperate, rainforest and plants of economic importance, all displayed as an educational and scientific collection.

Biomes

It is the interaction of the soils, the underlying geology and flora and fauna therein, as well as the natural climate of the area, that all make up a specific ecosystem known as a biome (or biotic province). To recreate/simulate such areas is fraught with problems. Rock gardens, gravel gardens/dry gardens, mediterranean gardens, conservatories, water gardens, bog gardens, cacti and succulent collections, shade gardens, ferneries, insectivorous plant collections, herbaceous borders, meadow gardens and lawns, are all simulated systems used as botanic features and in horticultural design. None of these, however, is large enough to be at biome level, but are merely horticultural features that in order to succeed aesthetically and biologically, endeavour to simulate a specific ecology.

The Establishment and Management of Rock Gardens

Traditionally, exotic and indigenous hardy 'alpines' (plants naturally occurring in alpine/mountainous areas) are cultivated on outdoor rock gardens. Less hardy alpine species are grown under the protection of a glasshouse – but is it the lack of hardiness that forces them to be displayed under glass, or is it the fact that they will not tolerate the natural rainfall of their enforced/unnatural growing conditions, and under glass the amounts of water can be regulated?

Rock gardens at their best are exemplified by those at the Royal Botanic Gardens, Edinburgh, Scotland and the RHS Gardens, Wisley, Surrey, England, as both are superbly constructed attractive features. They set out to emulate the natural conditions encountered by various plants from higher altitudes: these include annuals, biennials, herbaceous and woody perennials. The broad term 'alpine plant' may say something about its origin (technically, only plants at elevations above the tree line qualify as alpines), but the term is actually used to encompass a wide range of plants from very different habitats. Alpine meadows grazed by cattle, and alpine meadows only grazed by indigenous mammals (at altitudes of, say, 1,500–3,000 m) are examples of the diversity of habitat. Dry southern slopes that are baked hard all summer (and wet and cold most of the winter), snow sheets and elevated moorland, bogs and heaths are further examples. Other examples include free-draining moving and fixed screes, rock terraces with rudimentary soils, porous rock/rock fissures, and rock crevices with little or no developed soil. Added to these are the climatic and pH requirements of the species, ranging from calcifuges – plants that will not grow in lime soils as they need a low pH – and calcicoles – plants that tolerate, or even do better, in alkaline soils because they need a higher pH: the list is far from exhaustive.

Alpine meadow plants require a grass sward for success, and rarely survive easily when isolated or in other environments – even though the grass sward acts as a competitor at aerial level (and as a supplier of synergism/symbiosis at root level?); some plants like the presence of others and fail without it, others will fail in the presence of some other species (alleleopathy). It can therefore be seen that both the physical influences (as in soil textures, drainage, permeability or otherwise, of underlying geology and climate) and the chemical (as in pH and the presence of soluble nutrients) constituents of the environment, plus the presence of an existing plant community, can have an effect on success.

Rock gardens attempt to simulate scree slopes, cliff faces, rock terraces and rock crevices, and involve rock construction work and the use of specialized growing media within the planting areas. Rock terraces are constructed and oriented in varying directions to gain as many aspects as possible in the space available, including sun-baked south-facing terraces and shady north-facing terraces, and are filled with growing media of specific pH, nutrient status and draining qualities. Free drainage and rock chips, for example, could be the main criteria required to simulate the conditions naturally found on scree slopes – the absolute opposite of bog, or water-perimeter plants that need impervious materials to create their conditions. The construction rock itself may be calcareous (with a high pH, such as limestone), or silicaceous (silica-based with a low pH, such as sandstone), both of which interact with rainwater quite readily, or it may be a more acidic, igneous rock that does not dissolve easily, such as granite. For permanence the rock should be frost resistant and should not shale away easily, one which if chosen at the outset needs very little or no maintenance – on the other hand it does lose some natural interaction.

During the construction phase, both the remaining soil and the backfill material (soil or other planting medium) of planting pockets, screes and terraces must be weed-free – especially free of perennial weeds, to guard against ongoing maintenance costs. Furthermore, the use of dense layers of mineral chips (of the correct pH for the plants used, and the best type for the design) to suppress annual weeds and to reduce the likelihood of slugs and snails visiting the site, helps considerably in reducing maintenance input. Even so, the planting pockets created within a rock garden can be very labour-intensive with regard to effective weed control, which is usually the main maintenance expense.

The chosen position (including orientation and altitude) of the rock garden is fixed. To re-create an accurate ecological niche with so many different parameters in the correct balance is not easy, particularly as one of the most important ingredients, climate – rainfall, exposure, sunshine hours, cloud and so on – cannot be simulated. The cost of any totally artificial environment is prohibitive, both to construct and to maintain; furthermore, the normal altitude for the chosen species can never be simulated. Very successful alpine gardens are produced in the western Alps, in Austria, for example, at the normal altitude for the plants on display, but can only be visited by those ambulant enough to do so. Regarding phenotype and ecotype, obviously you do not have to live on a mountain to grow mountain ash. However, the phenotypes of plants from higher altitudes can change quite drastically if grown in lowland conditions; for example, the phenotype of *Aquilegia alpina* can change in relatively few seed generations from 12–15 cm to 30–45 cm in height when grown on rock gardens at lower elevations.

Already there are many constituent parts that could mean success or failure to any plant communities that we introduce, yet traditionally we have been able to successfully cultivate a wide range of 'alpine' plants – even with all the problems. Fast drainage, poor drainage, rock fissure, rock crevice, porous rock substrate, high nutrient, low nutrient, high pH and low pH can all be created moderately easily, and we therefore tend to display features within each one of these parameters. However, if we are trying to recreate a natural ecosystem, then we have to use native species – but native to where? Furthermore, if we mix exotics into the collection, how can we use exotics from more than one country if they would not normally share the same biome or biotic province? This obviously highlights the difference between an aesthetic feature and a botanically or ecologically accurate biome.

We tend, therefore, to live with the trade-off of the best feature (not the best ecology) that can be sustained on the chosen site. The emphasis in these instances is therefore aesthetics, and in order to create successful aesthetically pleasing features, horticulturists have widened the group to include anything that looks reasonable on a rock garden. They include some dwarf conifer cultivars that might have the right proportions, but certainly not the right pedigree to be called 'alpines'. Aesthetics bring in visitors, who bring in finance, and can be the main ingredient to a successful (viable?) garden/botanic garden open to the public. Success, therefore, is judged on income, and not on scientific value. Even those who create the most successful rock gardens aesthetically have to console themselves with the fact that the success is within a limited environmental range, and there will be many desirable species that cannot be grown even though, ostensibly, they appear to like similar conditions to those on offer. The best simulated rock crevice created can only sustain rock crevice species that tolerate the pH generated by both the chosen soil and the degeneration of the rock used in construction.

Plants Suitable for Rock Gardens

Andromeda polifolia★#, *Androsace pyrenaica*★+", *Aquilegia alpina*★+, *Arcyostaphhylos uva-ursi*★#, *Armeria*★+" *Aster alpinus*★+", *Aurinia saxitilis*★+", *Cassiope lycopodioides*★#, *Cornus Canadensis*★#,

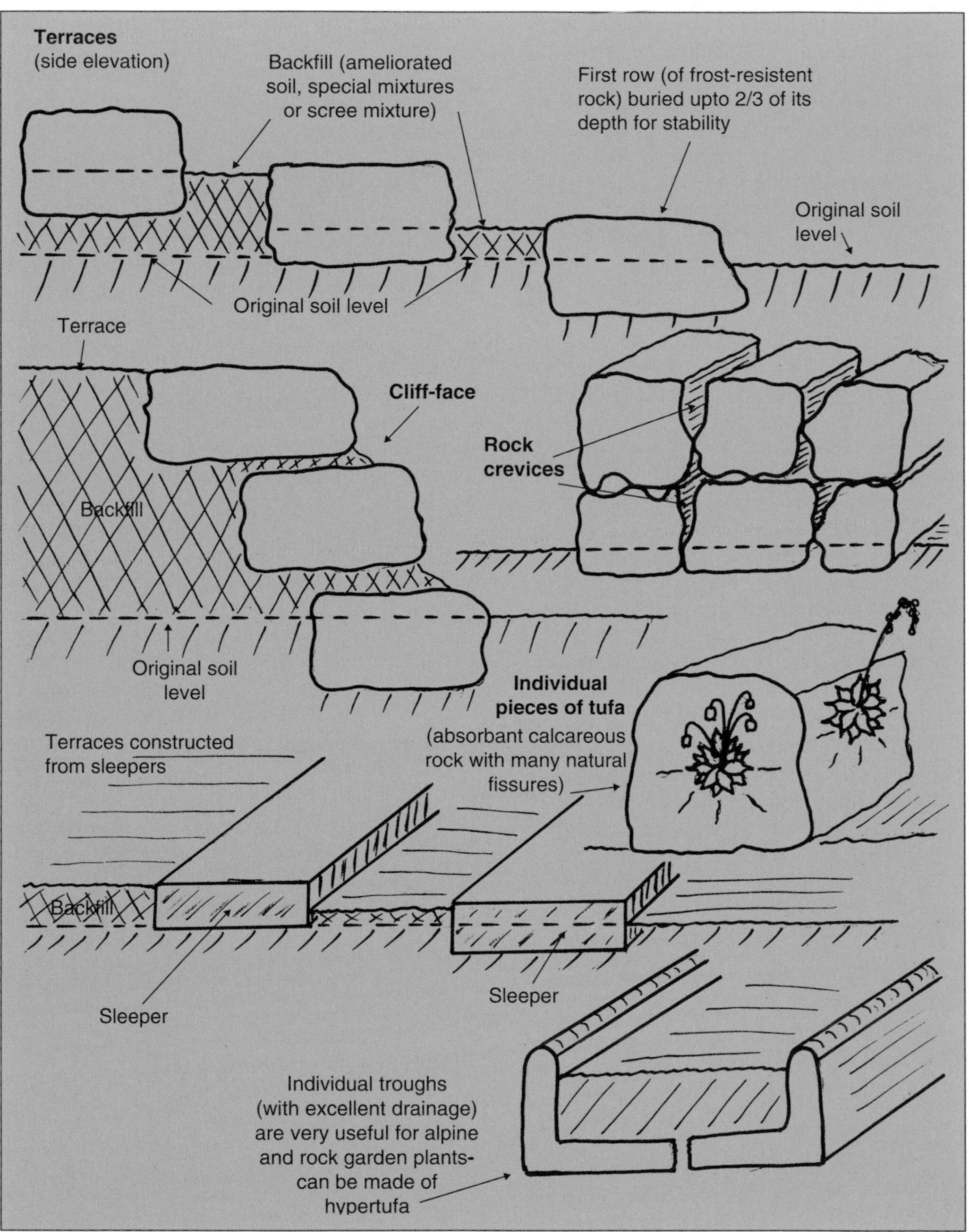

Rock garden construction: terraces, screes, cliff-faces and rock crevices.

*Corydalis lutea**", *Corydalis wilsonii**", *Cyathodes colensoi**#, *Dianthus**+, *Diascia rigescens**, *Draba**+^", *Dionysia**+^", *Erigeron alpinus**", *Erinus alpinus**+"^, *Gaultheria procumbens**#, *Gentiana**, *Geranium cinereum**, *Lewisia cotyledon**+", *Phlox subulata**+", *Phyllodoce caerulea**#, *Pulsatilla vernalis**+, *Pulsatilla vulgaris**+, *Rhododhypoxis baurii**+, *Rhodothamnus**+, *Saxifraga* *+^", *Salix lanata**#, *Salix reticulata**#, *Sedum spathulifolium**", *Sempervivum**+", *Shortia soldanelloides**#, *Soldanella alpina**#, *Silene schafta**+", *Sisyrinchium graminoides* (syn. *S. bermudiana*)*+, *Sorbus reducta**, *Vaccineum vitis-ideae**#, *Viola**#.

Key:
Terrace *
Scree +
Cliff-face and crevice"
Individual porous rocks ^
Acid litter and partial shade #

The other very useful feature of alpine/rock garden plants is their small stature and compact nature, and the vast range that it is possible to grow in relatively small areas. Comprehensive collections may be established in very small gardens using screes and terraces, and basically converting the whole available area into a rock garden. With the addition of large rocks with many crevices and fissures (for example, large pieces of tufa, an absorptive calcareous rock) planted up as individual features, drystone walls and hypertufa troughs, the various niches required for a large range of alpine plants can be accommodated.

Mediterranean Gardens

As our climate has changed, emulating mediterranean-style gardens in warm southern areas has increased in popularity. Many plants with xerophytic adaptations are used, including those with small leaves, hairy leaves, leaves encrusted with mealy materials, and those with large amounts of essential oils and usually pungent. The viscous oils in their sap, the small leaf-surface area, and the various coverings on their leaves and stems all help to reduce transpiration so they can tolerate hot, baking, dry summers. Plants such as lavender, French lavender (*Lavandula stoechas*), rosemary (*Rosmarinus officinalis*), *Perovskia*, *Caryopteris*, bay (*Laurus nobilis*), bottlebrush (*Callistemon citrinus*), catmint, oleander (*Nerium oleander*), *Helichrysum*, *Cistus* and *Helianthemum* are often used. Many sub-shrubs can be included in this list of suitable subjects, including *Phlomis fruticosa* and *Salvia fulgens*.

A little research into the origins of these plants very quickly shows that it is a mediterranean garden that is being emulated, and not a mediterranean biotic realm. The plants' origins are as diverse as South Africa, Great Britain, other parts of Europe and Australia – very few are native to the Mediterranean countries. Even in good years

Lewisia cotyledon.

the southern counties of the UK do not fully emulate the climate of the Mediterranean (which is itself a wide area with varying climates), and winter rainfall in the cooler temperate areas causes problems for some of the chosen xerophytes. Gravel gardens are often used to simulate mediterranean conditions; however, at best it is a very crude simulation, as neither the available sunlight nor the rainfall can be controlled – in fact the gravel garden attempts to ameliorate these adverse conditions by offering free drainage of excessive water (and some retention in films around individual mineral chips). But they cannot emulate the natural climate of these plants, and necessarily, therefore, the choice of species that would be successful even under these conditions is limited.

Anyone who has been to Tenerife (Canary Islands) will know of the marked change in climate (and associated flora) from the north to the south of the island. It appears like a distinct line drawn roughly across the centre, cutting the island in two, with arid conditions in the south, and moister, subtropical conditions in the north. The flora in the south is therefore based around leaf succulents, stem succulents and even some exotic cacti, whilst in the north, the mild but wetter climate supports a wide range of warm temperate, subtropical and even tropical species, including many arboreal and arborescent types. Moreover, a visit to Masca near Los Gigantes in the west of the island opens up an area comprising many different (yet mild) microclimates. The warming influence of the sea is apparent in the Masca gorge, and the physical shelter of the very high rocks, the influence of altitude on the central arete, the influence of various soil depths, and the effect of the orientation of the gorge (and particular orientations within it) are all marked and obvious within a relatively small area.

Areas with very thin topsoil over hard rock, even with the slightly higher rainfall of the west, create free-draining, arid conditions that suit many leaf- and stem-succulent species in the Crassulaceae and Euphorbiaceae families – Aeonium species such as *Aeonium arborescens*. Alongside these species grow canary palms (*Phoenix canariensis*) that both thrive and regenerate readily in deeper, moister, more organic soils. Amongst the date palms are other cultivated and naturally regenerated fruiting crops, including banana, loquat, oranges and lemons, all of which fruit wonderfully well – particularly if they are lucky enough to be in a favoured sunny aspect and sheltered from strong winds. Walking down into the shady parts of the gorge, into the wind-sheltered warm areas, and then back out into full sunlight, illustrates the benefits of each environment (from a plant growth and fruit ripening perspective) very well. Because of the superb conditions found in this type of environment, the natural regeneration of indigenous, introduced, naturalized, cultivated native and cultivated exotic plants is very common. The secret of such a wide variety of plants (regarding heights, species, woodiness, succulence and so on) growing next to one another so successfully is the mix of soil depths, soil types, shelter, open areas, exposure and warmth from the sea; also the warmth radiating from rocks that have warmed up during the day and give off heat at night.

Other areas of Europe (and North America) do not all have such favoured growing conditions, but many cool temperate climates can gain immensely from specific locations, and specific positions and orientations within those locations. Land masses forming a peninsular and jutting out into the sea get the radiator-style warming effect of the sea on both sides, which greatly influences the local climate of the relatively narrow strip of land, and supplies wonderful growing conditions for a wide range of both native and exotic plants. In the UK, Cornwall is a good example of this phenomenon, with the Falmouth area in particular supporting a very wide range of plants, including agaves growing (and flowering) outside. To a certain extent the coastal regions of both Devon and Cornwall have the benefit of this milder climate, and are notorious for their diversity of exotic plant material. In Devon, Salcombe, Dartmouth and Kingswear are notable in the range of more tender plants that they can support successfully, and both Greenway Gardens and Colyton Fishacre are in this area. Dartmouth and Kingswear benefit not only from their southerly location within the country, but also their proximity to both the River Dart estuary and the sea, and a good range of subtropical plants are possible outside in the sheltered, sunnier aspects of these areas, including banana, loquat and true palms. The same is true of similar aspects at the mouth of the River Tamar on the Devon-Cornwall border, and gardens at Trebah, near Falmouth, Cornwall are spectacular because of the range that they can grow. They give us a wonderful lesson in location, location, location, position, position, position and orientation, orientation, orientation! The Isles of Scilly off Cornwall can support even more exotics because of their frost-free climate, and the gardens at Tresco exemplify this. However, establishing woody plants (trees in particular) does have its problems because of the exposure to strong, salt-laden winds.

What may be more surprising is the range of plants that may be grown much farther north in the UK. For the same reasons already outlined, the Logan peninsular in Scotland (not that far from Glasgow) has a good climate and has Logan Botanic Gardens with its superb range of exotic plants not normally found growing outside all year round in the UK. Even though Logan Botanic Gardens is on the west, its position tends to fend off excessive rainfall, and many of the plants grown outside on this site are succulents. The peninsular effect is enhanced even further on an island with the sea on all sides, the proof of this provided by the botanic gardens at Ventnor on the Isle of Wight; for the same reasons, the west coast of Ireland also boasts superb plant collections. However, the effect is also surprisingly evident further north at Brodick Gardens on the Isle of Arran off the west coast of Scotland, where the relatively high rainfall makes it a haven for large-leaved rhododendrons and New Zealand tree ferns.

Even more exotic wonders exist at Inverewe Gardens in the far north-west of Scotland, where success is due to the proximity of the gardens to both a sea loch and the sea. Furthermore, just like Logan Botanics (and the Isle of Arran that bit further south), the Gulf Stream (North Atlantic Drift) is a massive influence, bringing warm currents very near these sites, and coupled with sea breezes usually keeps them frost free in winter. For the best success in these areas, and in order to grow the widest range of plants possible, plants must be protected in some way against both salt-laden spray and the damaging south-west winds. This is usually achieved by planting salt-resistant species (such as pine) as a windbreak/shelterbelt, and then planting the exotic collection in the lee of it.

There are many examples elsewhere in the world where the already favourable local climate can be enhanced by good choice of position, location and orientation; peninsular, isthma and islands within these areas are of paramount importance in this.

Plants Grown Outside in favoured Conditions

The following are examples of plants grown outside in sheltered niches in areas of North America, Eire, Great Britain and mainland Europe with a favourable climate:

Magnolia campbelli; *Telopea truncata*; *Grevillea rosemarinifolia*; *Phormium tenax*; *Phormium cookianum*; *Cordyline australis*; *Phlomis fruticosa*; *Ilicium floridanum*; *Peaeonia luteum*; *Rhododendron ficto-lacteum*; *Rhododendron falconeri*; *Lobelia cardinalis*; *Cardiocrinum giganteum*; *Romneya coulteri*; *Calistemon citrinus*; *Myrtus luma*; *Passiflora caerulea*; *Actinidia sinensis*; *Eccremocarpus scaber*; *Eucalyptus viminalis*; *Paulownia tomentosa*; *Paulownia fargessi*; *Catalpa bignonoides*; *Fatsia japonica*; *Cestrum newelli*; *Abutilon vitaefolia*; *Trachycarpus fortunei*; *Fuchsia magellenica*; *Dicksonia antarctica*; *Echium wildpretii*; *Cytisus battendierii*; *Tropaeolum tuberosum*; *Mutisia decurrens*; *Albizia distachya*; *Feijoa sellowiana*; *Sparmannia Africana*; *Agapanthus africanus*; *Punica granatum*; *Brugmansia suaveolens*; *Nerium oleander*; *Amygdalus communis*; *Morus nigra*; *Eriobotrya japonica*; *Melianthus major.*

The following are examples of plants grown outside in Madeira, the Canaries, southern Spain and the Mediterranean:

All the plants in the list above, plus *Araucaria heterophylla; Erythrina crista-galli; Euphorbia pulcherrima; Hibiscus rosa-sinensis; Cycas revoluta; Ficus benjamina; Ficus elastica; Ficus lyrata; Macademia integrifolia; Grevillea banksii; Tecoma alata; Syzygium jambos (Eugenia jambos); Psidium guajava; Coffea arabica; Coccoloba uvifera; Cassia didymobotria;*

Melianthus major.

The Masca Gorge, Masca, Tenerife.

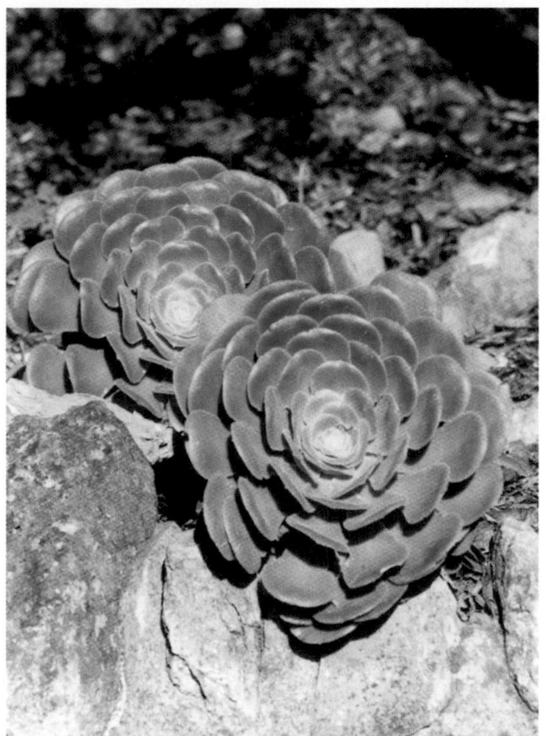

Aeonium species growing outside in the Canary Islands.

Carica papaya; Caesalpina pulcherrima; Citrus sinensis; Passiflora ligularis; Metrosideros excelsa; Aloe vera; Ricinus communis; Opuntia ficus-indica; Ananas comosus; Tecoma stans; Stephanotis floribunda; Thevetia peruviana; Allamanda cathartica; Agave sisalana; Pyrostegia venusta, Solandra maxima; Plumeria rubra var. *acutifolia* (frangipani).

Also *Eriobotrya japonica* (loquat), *Punica granatum* (pomegranate), *Annona reticulata* (custard apple), banana (*Musa*) and avocado (*Persea americana*) are all used as fruiting orchard plants in these countries. *Myrtus ugni* (Chilean guava), the strange elephantoide *Phytolacca dioica*, a wide range of true palms such as Canary date palm (*Phoenix canariensis*) and *Chamaerops humilis*, and arborescent monocots such as the screw pine (*Pandanus utilis*) can all be grown. *Jacaranda mimosifolia, Grevillia robusta* (silk oak), *Chorisia speciosa* (kapok tree or floss silk tree), *Schinus terebinthifolius* (Brazilian pepper tree), *Brachychiton acerifolia* (flame tree), *Brachychiton discolor* (scrub bottle tree), *Bauhinia variegata* and *Spathodea campanulata* (red tulip tree) are commonly used as street trees in some of these areas.

Protected Environments

Because of their day length, and their moisture and temperature differences, it is just as difficult for

Red Tulip Tree (Spathodea campanulata) used as a street tree in Los Christianos, Tenerife.

Close-up of the flowers of Spathodea campanulata commonly used as a street tree in Madeira and the Canary Islands.

Close-up of the flower of Albizia.

The stem-hugging fruits of papaya (Carica papaya) growing outside in Los Gigantes, Tenerife.

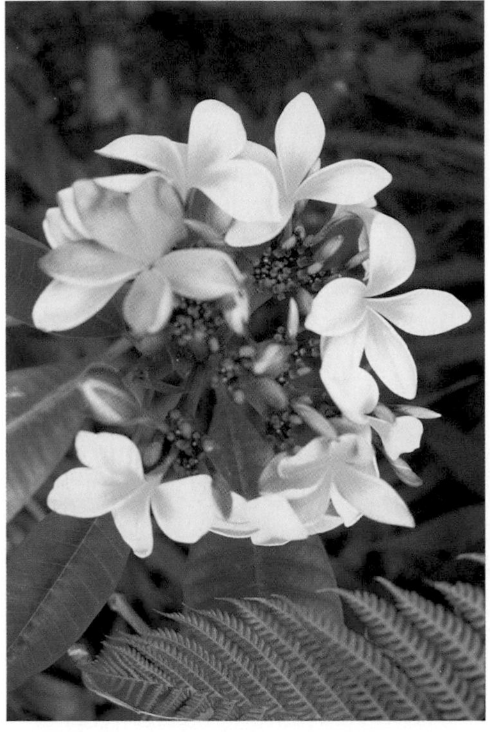

Frangipani (Plumeria rubra var. acutifolia).

Banana (Musa).

Kapok tree (Chorisia speciosa).

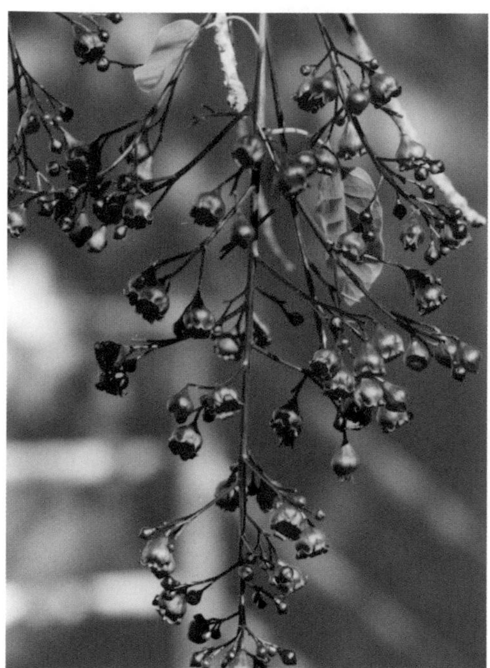

Close-up of the flowers of Brachychiton acerifolia (flame tree) used as a street tree in the Canary Islands.

The fruits of Brachychiton acerifolia.

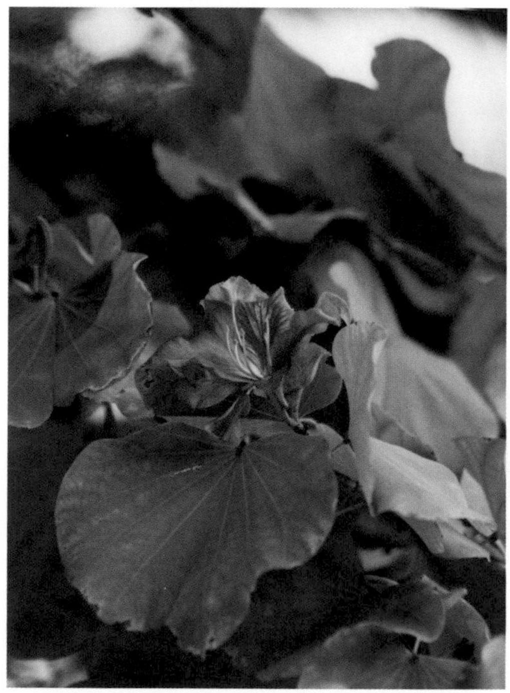

The flower of Bauhinia variegata (orchid-flower tree).

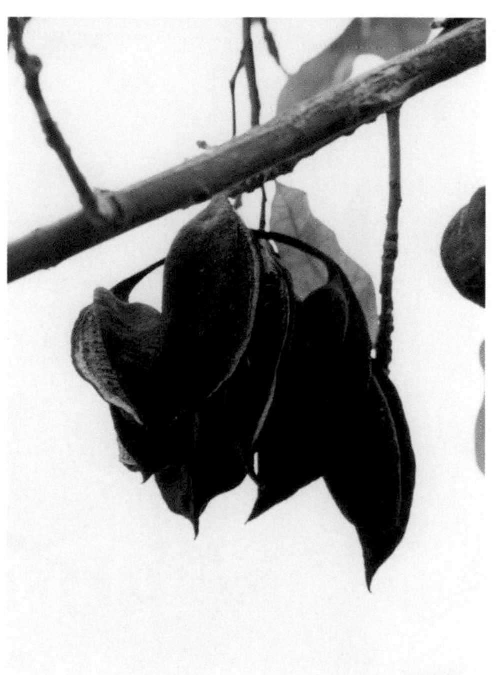

Close-up of the dehiscing fruits of Brachychiton acerifolia.

Erythrina crista-galli.

warm countries to display plants from cool temperate climates as it is for temperate countries to display tender plants from warm climates. Constantly keeping large arboreals cool and moist in hot climates has just as many logistical problems as keeping plants dry and warm in temperate climates. Growing plants in the lee of protective hedges and windbreaks — or even within the protection of solid walls, if the orientation is such to prevent accelerating eddying, funnelling or venturi effects — can increase the possible species grown. Heat leakage from buildings and good orientation may give up to a 2° temperature lift in the winter in some instances; this makes some urban courtyards and areas surrounded by heated offices prime sites for less hardy plants.

In temperate climates, botanic gardens have some of their features under cover (in display houses, palm houses, conservatories). In the protected environment of the glasshouse where the aerial temperature and amount of water added to the system can be controlled, a wider range of plants can be grown than is normally possible outside. However, the available light (both intensity and day length) cannot be controlled (without exceptional and unacceptable expense), and the requirements of plants are so diverse that the problems are huge. What of plants that originate

*Cassia didymobotria. (*syn. *Senna didymobotria).*

A close-up of the spines of kapok tree (Chorisia speciosa).

from areas that have hot dry summers and wet winters, compared with plants from areas that have hot dry summers and mild (or cold) and dry winters? How could they be successfully kept under the same roof? It may not be adequate scientifically, but the most successful designs aesthetically are those that are relegated to a fairly limited range of species that are truly successful in the most affordable environment on offer.

The cost of putting up these protective structures is prohibitive, and likewise the running costs, because the energy and staff needed to maintain this sort of environment is always expensive. Local authorities and parks departments sometimes create and manage large displays as amenity glasshouses for public use (often called 'winter gardens'), and university-based botanic gardens may also do the same. In these instances the cost is met by the public through taxation, and their viability regularly comes under debate because of their running costs. Many universities with a botany background have had to 'rationalize' their collections, usually by scaling down or by removing or finding other uses for their most energy-intensive systems. The cost of private/domestic glasshouse displays has to be met by the private owner, and if he/she thinks it is worth it for the benefits and pleasure that it provides, then the cost is accepted.

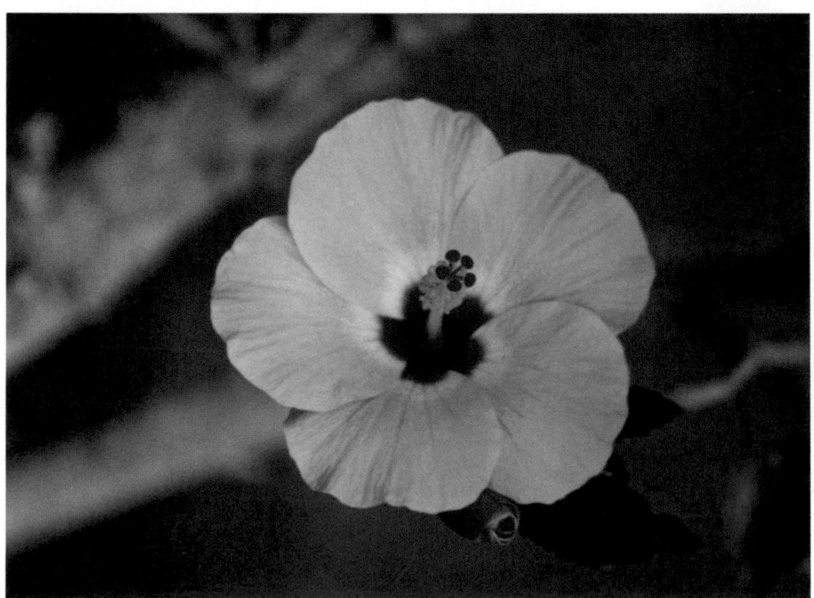

A striking cultivar of Chinese hibiscus (Hibiscus rosa-sinensis) growing outside in Tenerife, the Canary Islands.

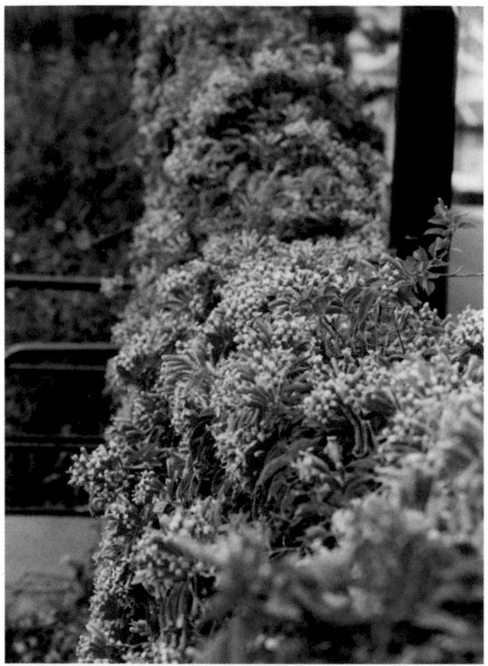

Pyrostegia venusta growing along a fence line in Los Gigantes, Tenerife.

Close-up of the flowers of Pyrostegia venusta.

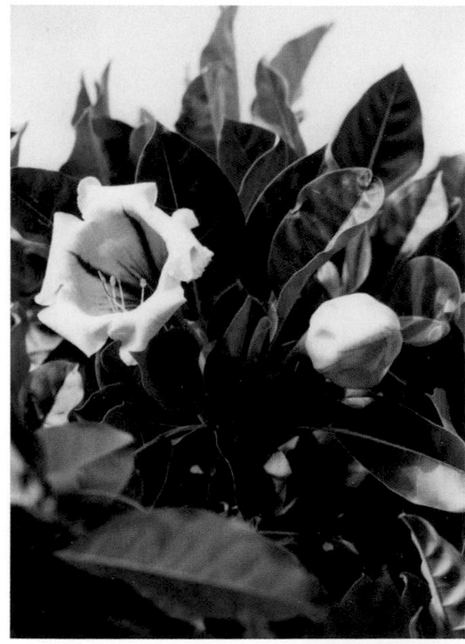

The vigorous tropical climber Solandra maxima (golden chalice vine) growing outside without protection in Tenerife.

Because the natural soil left after the construction of display houses is usually poorly structured, it has to be improved in some way, or a system of pathways and retained edges constructed – most commonly both. The retained planting areas are filled with either ameliorated soil or a specially formulated growing medium.

The basic garden conservatory at whatever scale contains a mish-mash of species with varying needs. Often the diversity of sub-surface (root-zone) needs can be accommodated by using individual containers, as the planting media used can be produced or purchased, and texture, pH and nutrient status can be tailored to the needs of the individual plants. Furthermore, additives of various types can be incorporated as necessary. In practice, even professionally managed conservatories often have sub-standard displays with hard, muddy, poorly textured growing media, and insufficient attention to repotting and compost management.

Some tender perennial and annual climbers are treated as annuals outside in temperate climates for their summer effect; for example, the leaf-petiole climber *Rhodochiton atrosanguineum*, *Ipomea* (morning

Eriobotrya japonica (loquat).

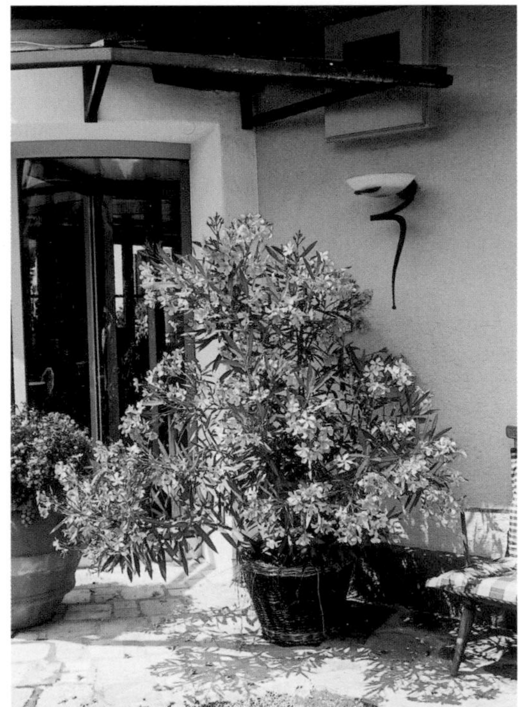

Oleander (Nerium oleander), being used just for the summer months outside the Edelweiss Hotel, Kitzbuhel, Austria.

glory), *Cobaea scandans* (cup and saucer plant) and *Thunbergia alata* (black-eyed Susan). Perennial forms may, of course, be grown in glasshouse displays continually, and semi-hardy types such as *Jasminum polyanthum* (for its very fragrant flowers) and *Passiflora cauerulea* (for its orange fruits) are common. Tender perennial climbers commonly retained in glasshouse collections include *Lapageria rosea*, *Allemanda cathartica*, *Passiflora quadrangularis*, *Canarina canariensis*, *Bougainvillea glabra*, *Stephanotis floribunda*, and the very vigorous South American climber with very large yellow flowers, *Solandra maxima* (golden chalice vine). Shrubs of various types may also be included in glasshouse and conservatory display, and they range from the relatively hardy *Nerium oleander, Acacia dealbata and Eriobotrya japonica* to more tender species such as *Tibouchina urvilleana* (syn. *T. semidecandra*) and *Streptosolen jamesonii*. Any of the previously listed examples grown outside in Madeira, Canaries, southern Spain and the Mediterranean may also be used if development room is sufficient.

The choice of cladding for protected environments has increased in recent decades; glass, polyethylene, double-glazed polyethylene bubbles and polycarbon are all examples. The choice of material for the main structure is also varied, wood, steel, aluminium and carbon-fibre framework

The foliage and fruits of Ficus lyrata.

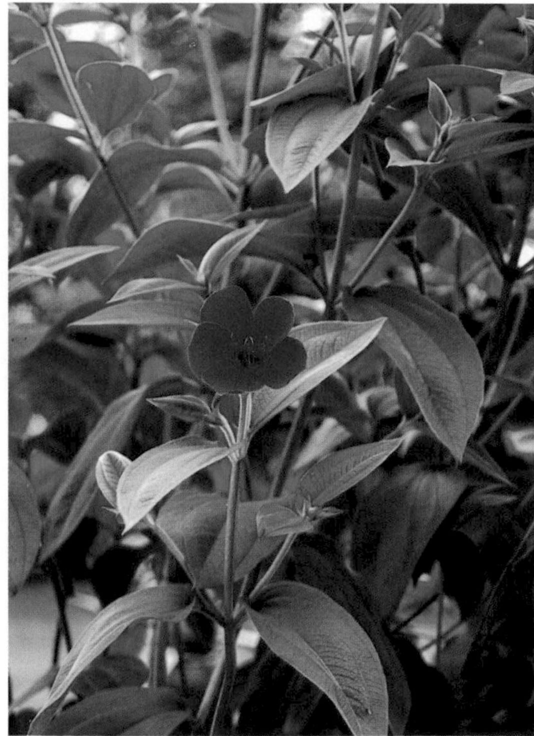

Tibouchina urvilleana (syn. T. semi-decandra).

being examples. These may be used as an internal skeleton, portal framing, or exoskeleton (to leave a clear area inside). The hexagonal units of the domes of the biomes in the Eden Project are a great leap forwards in the technology in this field. The biome 'rainforest', when kept under cover in botanic collections in naturally temperate climates, is usually achieved by ensuring adequate temperature in a heated glasshouse (or polyethylene dome), and high humidity. Collections of tropical epiphytes can be successfully included in simulated rainforest.

Humidity can be increased in protected environments by the addition of areas with partially retained water levels for marginal and bog plants, and areas of complete water retention for a water feature. However, even though evaporation from the water surface increases on hot days (thus increasing the humidity) and decreases on cool days (which is probably beneficial), there is no actual control over this. With or without a water feature, ancillary irrigation is essential under cover – the main benefit being that you have total control over how much, or how little, water is applied. An installation may be as basic as a hose system, or it may include sprinklers, mist sprays and fogging units. Sprinklers generally create relatively large droplets that fall to the soil straightaway – hence water delivery at root/soil level is good, but heavy

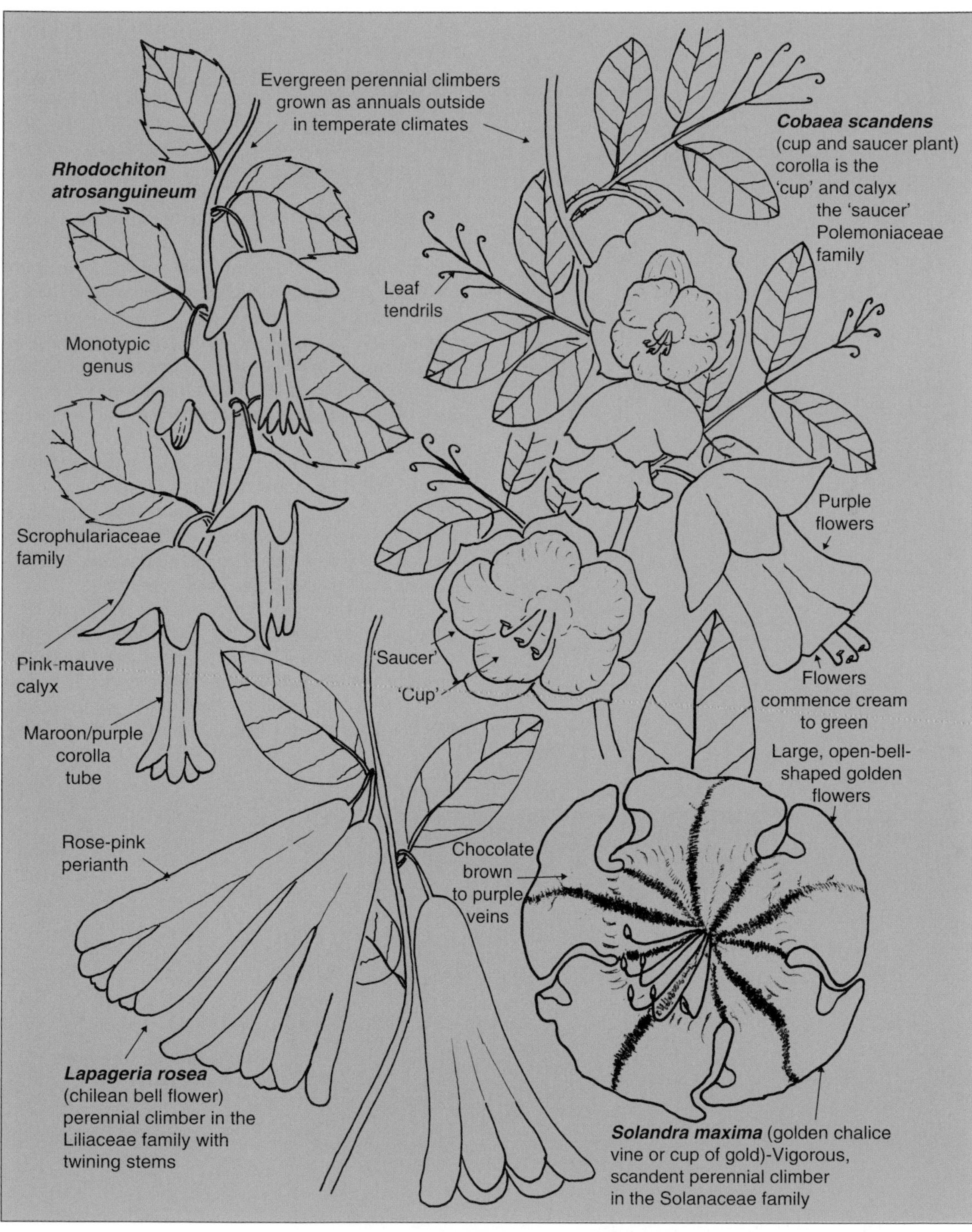
Evergreen perennial climbers grown as annuals outside in temperate climates
Rhodochiton atrosanguineum
Monotypic genus
Scrophulariaceae family
Pink-mauve calyx
Maroon/purple corolla tube
Cobaea scandens (cup and saucer plant) corolla is the 'cup' and calyx the 'saucer' Polemoniaceae family
Leaf tendrils
Purple flowers
'Saucer'
'Cup'
Flowers commence cream to green
Large, open-bell-shaped golden flowers
Rose-pink perianth
Chocolate brown to purple veins
Lapageria rosea (chilean bell flower) perennial climber in the Liliaceae family with twining stems
Solandra maxima (golden chalice vine or cup of gold)-Vigorous, scandent perennial climber in the Solanaceae family

Climbers for the conservatory.

Cup and saucer plant (Cobaea scandans) growing in Ventnor Botanic Gardens, Ventnor, Isle of Wight.

and can cause compaction, but humidity is increased by evaporation from the soil.

Misting units, comprising fine spray created by water being forced through a nozzle and hitting a baffle plate, create a varying range of droplet sizes. They range from very small droplets that remain suspended, to larger droplets that fall to ground easily – some can wet the soil, albeit slowly, others will remain suspended in the atmosphere, and in very hot temperatures these may evaporate again to aid humidity. Fogging units (water from a fine jet being hit by a stream of compressed air) create very fine droplets of relatively uniform size that remain suspended in the air for much longer periods of time and are more efficient in keeping relative humidity high. All these irrigation systems can be manually, semi-automatically or automatically operated.

Simulated rainforest can be energy intensive because high humidity and high temperatures need to be maintained even in winter, and an application of water immediately reduces temperature. Hence the two systems work against one another, and in fact, in naturally cool temperate climates, because of the poor light and low ambient temperature, the amounts of irrigation are usually reduced during the winter months. A reduction in humidity levels can only be effected by lowering the water application rate and/or increasing the ventilation, and because both of these lower

Bougainvillea glabra growing outside in Tenerife.

temperature, sophisticated automatic systems are used on a large scale because they can integrate temperature, ventilation and irrigation control.

The Eden Project, St Austell, Cornwall, England has several biomes and biotic provinces represented, each one having good 'display' space. One particularly successful system is achievable because of the size of, and the interconnectedness of the 'domes'. Very clever use has been made of height, because there are different biomes represented at varying heights (and therefore varying temperatures and humidity): desert (and mediterranean) and warm temperate at the lowest levels, sub-tropical at the next level, and the moist tropical rainforest biome at the highest level. This makes use of the fact that hot air rises – so those plants needing the highest maintained temperatures are at the highest elevations within the domes.

This progression from cacti and succulents, through sub-tropical plants to tropical rainforest, can all be achieved with no ancillary heating in the summer and minimal heating in the winter months. How far it goes as a sustainable system is debatable as it does need an energy input, but good orientation, good protection/shelter, and because of the topography of the site, good natural sunlight – all because of the geographic location – besides new cladding technology, all help energy conservation. A superb mix of initial funding, technological know-how, an inviting geographic location, the range of plants grown, good marketing, funding to expand, and funding for sufficient and experienced staff, all contribute to the success of the project.

Epiphyte Collections

Bromeliads (plants in the pineapple group) can engender interest and be very pleasing to the eye when presented in a well crafted display. Most bromeliads are displayed epiphytically, and *Tillandsia*, *Vriesia*, *Billbergia*, *Neoregelia*, *Aechmia* and *Ananas* species can all be used successfully. Epiphytic bromeliads are displayed in glasshouses (or outside in warm climates) on old rotting tree trunks or on 'artificial trees' constructed from various materials (including metal or wood) covered in large pieces of bark from cork oak (*Quercus suber*)*. They are often displayed accompanied by epiphytes in other plant groups that thrive in the same conditions (such as epiphytic orchids *Dendrobium* and *Coelogyne*), and even epiphytic cacti and ferns can be added. Some individuals of all types may also be grown in hanging baskets, and some may be planted terrestrially at the base of the 'trees' in specially prepared beds or containers.

The key to the successful culture of most epiphytic species is to ensure high aerial humidity, particularly when they are first introduced to their artificial home – but also continually throughout their life. Most species can be wired on to the 'tree' initially with fine, soft wire that may be removed at a later date when the plant becomes stable over time via aerial roots. Alternatively, they may be planted into specialized planting media (usually based on high levels of cork chips) in natural planting pockets, or formed from wire netting (or other net-like materials). The planting pockets are wired on to the 'tree' and the epiphytes are planted freely within them – usually without support. However, the need for extra support by wiring is common when larger specimens are used in the establishment phases.

Although orchids, bromeliads and ferns lend themselves to being managed in a general epiphyte collection, individual species may have different cultural needs, so a trade-off may be necessary regarding the optimum conditions for any one species. Relative success is nevertheless usually experienced even when growing a very diverse range of epiphytic types. Epiphytic cacti species such as Christmas cactus (*Schlumbergera* × *buckleyi*) and Easter cactus (*Rhipsalidopsis gaertneri*) can be successfully established and displayed in adjacent areas. Their watering regime is not unlike that of epiphytic orchids, but is less consistent with the humidity and includes periods of very little water, or none at all, in some months.

Insectivorous (Carnivorous) Plant Collections

Hardy insectivorous (carnivorous) plants may be displayed outside in natural (or specially constructed) areas with moist acidic soils. However, both hardy types and the more tender types are often displayed in cold glasshouses so that their aerial and root-zone water levels can be monitored more closely and controlled more easily. Epiphytic *Nepenthes* species can be grown on upright logs, in hanging baskets or on constructed 'trees'. Their ornate pitchers always engender interest, as do

*Safety is paramount when erecting parts of trees or constructing 'artificial trees' for epiphyte display. Structures must be sound and safe.

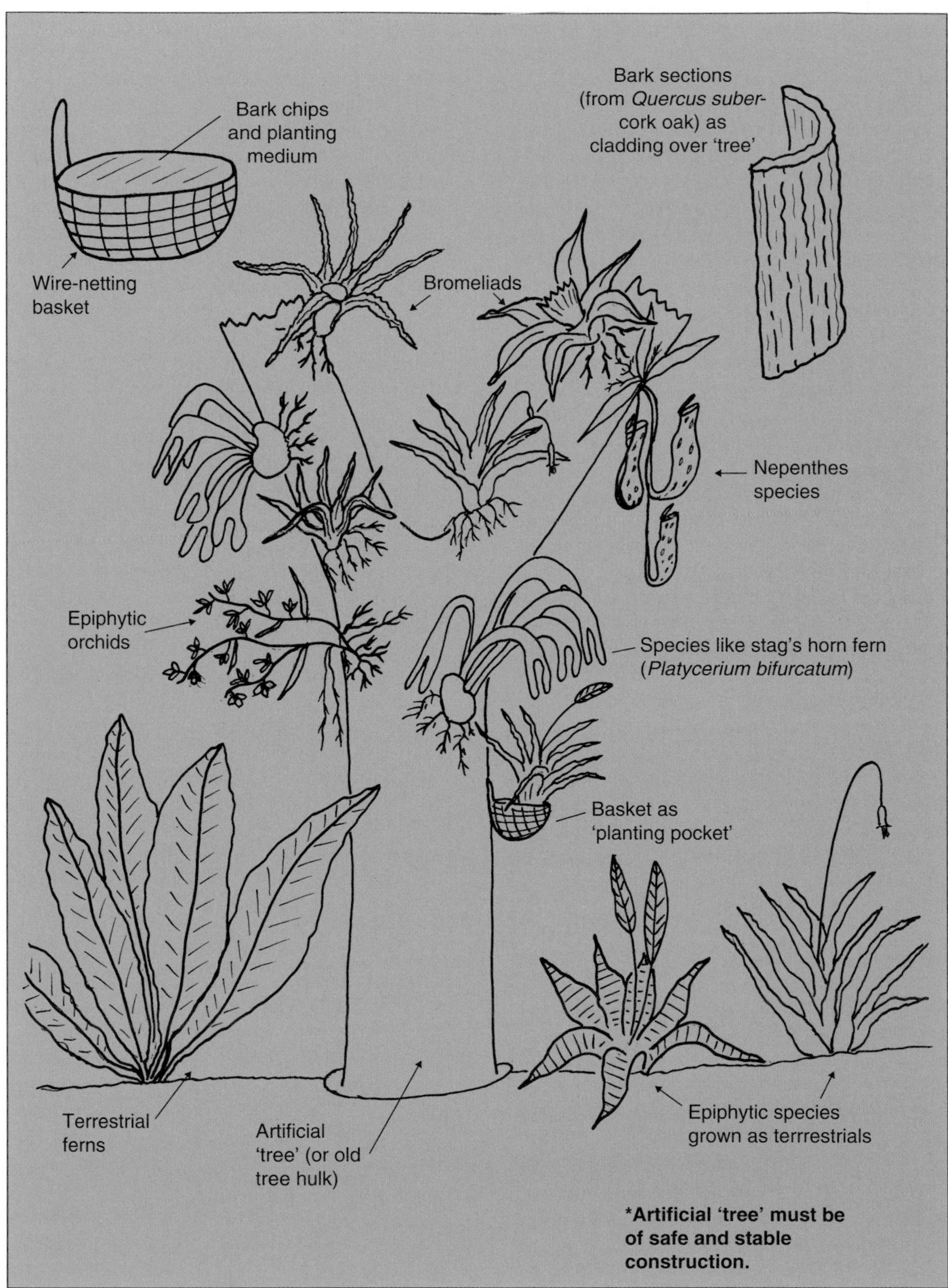

Display of epiphytes in protected environments.

An epiphyte collection.

terrestrial species such as Venus flytrap (*Dionaea muscipula*) and the sundews (*Drosera species*). However, the Sarracenias with upright pitchers are probably the most impressive of the terrestrial types – for example, purple sarracenia (*Sarracenia purpurea*) and yellow sarracenia (*Sarracenia flava*).

Most terrestrial species are herbaceous, and may be grown in specially created 'bogs' with retaining edges and waterproof bases; alternatively on a smaller scale (or for convenience) individual plants may be kept in waterproof durable plastic trays. There is some merit in having even a wide-ranging collection in individual trays, as soil-borne root diseases (which are rare) and physiological problems caused by the growing medium can be isolated (and rectified) easily. However, individual containers may be 'hidden' by growing media to create a more 'natural' and 'landscaped' appearance. This reintroduces the possibilities of infection from tray to tray, but does not prevent each tray being lifted and isolated if necessary. Growing media for terrestrial forms are based on mixtures of sphagnum moss and sphagnum moss peat, as sphagnum moss survives in (and helps maintain) correct pH levels (to ensure acidity). Other inert materials (usually minerals such as vermiculite, or those of volcanic origin, such as perlite) may also be included to aid moisture retention in the long term. No nutrition is included in the growing medium, as a low nutrient environment causes/triggers the carnivorous adaptations in the first place, and thus management techniques have to include maintaining a low nutrient status in the growing medium. Nutrition must be via trapped insects only, as the addition of applied nutrients rapidly kills the plants.

Watering regimes are critical, and terrestrial species fare best if they have low levels of free water at their bases to ensure the growing medium remains very moist at all times, but not so as to drown the roots.

CHAPTER 12

The Management of Gardens Open to the Public

Botanic gardens and other gardens open to the public have collections that include soft herbaceous plants and woody subjects. Arboreta contain arboreal (tree and shrub) collections, and pineta have coniferous collections as their main emphasis, even though all may include the display of other complimentary plant groups and features as well. Gardens are dynamic and cannot stand still: they must either develop and go forward, or fall into entropy and degenerate. Ironically, however, they also benefit from periods of stability and continuity.

The management of gardens open to the public begins with an assessment of the existing collection for its botanic and aesthetic standards. Integral to this is a site survey to recognize, classify, log the location of, and catalogue existing plant species on the site, and identify local landform (topography), soil type, texture and pH. Current themes, generic groups, mixed groups, notable specialist plant groups, and specific collections (including national collections) are also recorded. Any survey will also include the infrastructure of the site, noting existing buildings, their location (and existing and/or potential use), and any pathways and roads. Visitor numbers, extra visitor attraction potential, vehicle management, people flow, milling spaces, methods of people regulation/control, expenditure and income will also be documented, and as a priority the survey will highlight health and safety implications, including potentially hazardous trees or structures.

A Management Plan

The need for close direction is very important indeed – but 'close' does not mean stifling or uncomfortably overbearing. Certainly there is a need for close scrutiny, motivation, direction and good organization to produce successful features, and staff need to know who is driving the ideas through, and need to be appraised of any new ideas in order to respond to them – but they also need to feel involved. It may be the head gardener, the botanic director, the garden manager or the owner (or a mixture of all) who drives the successful ideas forwards, but a good working relationship with their staff is essential.

From the initial information, a management plan is drawn up that considers the present and future aims and objectives, maintenance work schedules, the means of implementing successful work schedules, future direction and development, and the budget necessary. Senior staff must be skilled and include trained horticulturists, although some skills can be imported using contractors when necessary, as this is often – though perhaps questionably – perceived as being more cost effective.

To successfully implement the management plan and new ideas, the manager/owner needs to be specific, clear and accurate in conveying the information to the work team. However, the motivation of staff is as important as their skills, and this can be helped by not only giving credit to those who deserve it, but also allowing them, even encouraging them, to have some input into management decisions. Giving people a vehicle to show their knowledge, express their opinions and feel an integral part of the team, can only be good. Sometimes advice from someone with a practical background may prevent unnecessary mistakes in implementing policies.

Management decisions specific to the site necessarily have to instruct on the best, and particularly the safest practices that will be used to fulfil the stated aims and objectives most efficiently. Instructions must be clearly relayed to the work team to ensure correct understanding and implementation, and any changes in objectives, and subsequent policy direction, must be relayed to those concerned quickly, as this will affect day-to-day operations and may affect morale.

Aesthetic Plant Display in Botanic Gardens

Most botanic collections have a conservation role to ensure the survival of rare species, and the exchange of propagation material by botanic institutions is essential to this end, as is the sourcing of seed from the wild. Included in the aims of botanic gardens are education and dissemination of information, and scientific and academic roles (including conservation and ecological research). So sometimes their aesthetics can be almost a by-product (albeit a very pleasing by-product) of their other main roles.

New acquisitions for development need to be obtained from legitimate sources with known provenance in order to uphold the integrity of the collection. How far this purist attitude is taken in private collections depends on the direction and decisions of the owners/managers of the garden, and how dependent it is on visitor numbers or outside sources for finance. Obviously botanic considerations may influence management decisions to a greater extent, however, healthy and increasing visitor numbers is nearly always an important consideration, if not a limiting factor financially. For this reason the introduction of many species, sometimes of lower scientific/botanic merit (if this can be measured) but known for their aesthetic qualities, and extending the seasonal appeal, is common – science/botany versus aesthetics.

Displays of bedding plants or other plant groups designed specifically with bright colours as the main priority are relatively rare in botanic gardens. Conversely, they are a common feature in domestic gardens, local authority and public parks of all types. Such displays incorporate plants of horticultural interest from a wide variety of plant groups and Taxa, but with no consideration of their origin or botany. They are short-lived in their effect, but are colourful and effective, and may be useful but only as an added extra to encourage visitors. Even important collections of woody species benefit from the aesthetics of free-flowering herbaceous perennials displayed near them.

The appeal of individual plants as well as their group effect is important, and aesthetic interest includes their layout and design, and sometimes the use of art (such as sculpture or statuary) within the designs. Bold, beautiful sculpture can evoke very strong feelings and add another dimension to this effect. Other art forms such as paintings can also be included, and music and poetry (written and performed) can be an effective addition to a botanic project. Labyrinths and mazes, rills, lakes, streams and other water features can also be useful aesthetic additions, and plants with architectural interest such as bamboos, *Cordyline*, *Phormium* and *Yucca* always add to the aesthetic effect.

Displays of wonderfully productive edible fruiting species (even some vegetables) and their attractive effect are not necessarily mutually exclusive. Apple, plum and cherry trees laden with their fruit in reds, yellows, greens, purples and maroons, all hold their own aesthetically. Mixed plantings that include a range of edible plants of all types merely emulate the systems found in the natural world, which in turn have been reproduced as 'mixed borders' by horticulturists. The stark red stems and very bold leaves of rhubarb (*Rheum palmatum*) are

Statuary and sculpture – 'Die Pieta' by Anna Chromy in Salzburg would make a wonderful addition to any botanic collection. Unfortunately, not all sculpture is quite so powerful or can evoke so much emotion.

very attractive – after all, rhubarb is another herbaceous perennial, and being edible is just a useful added extra. The superb dark feathery foliage of purple fennel (*Foeniculum vulgaris* 'Purpurea') also proves the point. The use of mixtures of woody, herbaceous, biennial and annual plants for both showy appearance and food has been developed in forms of permaculture and sustainable gardening (notably forest gardening).

Flowers are always relevant because of their particular emphasis on colour, intricate shapes and fragrance; and leaves are important, their colour, texture, shape, size, summer colour, autumn colour, evergreen-ness and deciduousness all aesthetically interesting features – and other, more sensory aspects can be included in their appeal, such as leaf rustle, shine/matt, hairiness/pubescence and pungence, for example.

Bark can be very attractive, and sections devoted to bark as a main aesthetic display give interest all year and can be very effective as a 'crowd puller'; they may even be collected together and 'marketed' as a 'bark park' to engender interest.

Plants of Economic Importance

Many botanic collections have a section devoted to plants of economic importance, and often these include plants from the tropics and subtropics such as coffee, tea, sisal, hemp and rubber. However, to display such plants in cooler temperate climes, an infrastructure is necessary that is costly to set up and maintain; displays of this type must therefore be popular and well visited to warrant the cost. The educational value of plants of economic importance is often sidelined because of the cost-to-interest ratio. It is notoriously difficult to make a display of this nature attractive enough to appeal to all but the very interested and already 'converted': even if you take crops that are of economic importance to, and grown in, your own country, it is difficult to maintain public interest in these short-term food crops set up for educational purposes – small plots of wheat, barley and sugar beet have limited appeal. The Eden Project has been very brave in this area and they do display everyday crops as well as the more exotic (in the true sense of the word).

Ethno-Botanical Uses of Plants

Ethno-botanical sections within botanic collections aim to show how native peoples have utilized plants for medicinal and ceremonial reasons, and more importantly, clothing, sustenance and other functional purposes. There is obviously an overlap with 'plants of economic importance' and 'medicinal plants'. However, ethno-botanical displays are often dedicated to one race, ethnic group, or regional peoples' activities. They often illustrate the chosen group's use of plants for food, textiles and house building – in fact, used anywhere within their culture: thus fibres are used in bowstrings, plant poisons to kill prey (and enemies), and plants to give hallucinatory/mind-altering effects. Or, simple olfactory stimulants used in medicine, religious ceremonies or psychologically based rituals can be included – this could include the use of incense in some religions of our own culture, for example.

Plants and plant products may be used for shelter, housing (such as wooden and bamboo houses, and houses on stilts), clothing (weaving) and food (fruits and so on). Foods from plants include banana, pawpaw, papaya, guava, mango, date palm and tomato grown outside in the warmer regions of the world, including the Mediterranean, the subtropics and the tropics. Potato, parsnip, beet, cabbage, kale, cauliflower, turnip and carrots are all grown outside in temperate climates, and tomato, cucumber and peppers grown under glass.

Trees are used for fuel, timber for construction, fencing, pallets and cork, and are also used to produce fruit crops. Apples, pears, cherries, apricot, peaches, nectarine, almond, plums, walnut and sweet chestnut in temperate regions, and peaches, nectarines, apricots, oranges, lemons, limes, avocado, custard apple, sweet chestnut, walnut, almond, fig, olive and loquat in Californian and mediterranean climates. Other woody plants may also have ethno-botanical uses, with the North American canoe-bark or paper-bark birch (*Betula papyrifera*) as a common example. Tea (*Camellia sinensis*, varieties and cultivars), coffee (*Coffea arabica*, varieties and cultivars) and rubber (*Hevea brazilliensis*, varieties and cultivars) are all grown commercially in sub-tropical and tropical climates.

Banana, of course, is used for its fruit, but what may not be quite so obvious are the uses of the leaves. In its native habitat, banana leaves may be used as plates for food, as cladding on basic housing, and when torn into strips, woven into baskets and other utensils. Bamboo is famously known as the staple food of the panda – and food for humans as bamboo shoots. Bamboo is also used as a house construction material for basic housing, and is used as scaffolding to help in the construction of more substantial housing.

Cycads such as *Cycas revoluta* are related to the conifers, not the palms. However, *Cycas revoluta* has a common name (albeit misleading) of sago palm, and indigenous peoples use the 'heart' of the growing tip as a food source. The down side of this nutritious food is that the plant is often completely killed when the growing tip is harvested; hence tree-size specimens are quickly destroyed, and native stocks easily decline to unsustainable numbers. Large-scale commercial production of palms for palm oil, displacing/destroying native plant diversity, can have a similar effect.

Palms of all types can have portions of their compound frond-like leaves woven into baskets and utensils. Specific species of palms also supply food, with dates and coconut being the most common examples. Coconut also gives usable fibres (copra) harvested from the outer layers of the fruit (husk).

Fibre production (including for use in string and ropes) is a common feature of a range of monocotyledonous plants including New Zealand flax (*Phormium tenax*), and the bow string plant or mother-in-law's tongue (*Sansevieria trifasciata*). Some arborescent monocotyledonous species are also used for the commercial production of string and ropes – these include sisal (*Agave sisalana*) and yucca. Flax (*Linum perenne*) and some annual forms of flax are dicotyledonous plants grown for their fibres (flax) and linseed oil production.

Collections of Poisonous Plants**

Digitalis purpurea (foxglove) and *Atropa bella-donna* (deadly nightshade – the source of atropine) can both be used in the control of heart problems at the correct dosage. Castor oil (*Ricinus communis*) is used as a laxative, and *Papaver somniferum* (opium poppy) and *Cannabis sativa* (hemp/cannabis) are used medically as narcotic drugs for pain relief. However, they can all be fatal if used 'raw' (prior to treatment), or if taken at anything but the correct dosage.

Most of the nightshades, including deadly nightshade (*Atropa bella-donna*), black nightshade (*Solanum nigrum subspecies nigrum*) and bittersweet (*Solanum dulcamera*) are very poisonous, as are apple of peru (*Nicandra physoloides*) and thorn apple (*Datura stramonium*), often found as adventitives. The Apiaceae family have many poisonous examples, including hemlock (*Conium maculatum*), hemlock water dropwort (*Oenanthe crocata*), hogweed (*Heracleum sphondylium*) and giant hogweed (*Heracleum mantegazzianum*) – the sap of which is both poisonous and very caustic.

The leaves of both rhubarb (*Rheum palmatum*) and wood sorrel (*Oxalis acetosella*) produce oxalic acid, which is extremely toxic. Monkshood (*Aconitum napellus*), cuckoo pint or wild arum (*Arum maculatum*), exotic arums *Zantedeschia*, *Arisaema*, *Lysichiton* and *Dracunculus,* and two very common ornamentals, lily-of-the-valley (*Convollaria majalis*) and meadow crocus (*Colchicum autumnale*), are also very poisonous. Other very common ornamentals that are poisonous include *Iris, Lupinus, Dicentra, Delphinium, Ranunculus, Lobelia* and *Physalis alkekengi. Euphorbia* and *Helleborus* are poisonous, and many *Primula* species irritate the eyes and skin.

Examples of poisonous species in the woody world include *Laburnum alpinum, Laburnum anagroides, Nerium oleander, Rhododendron species, Buxus sempervirens* (common box), *Brugmansia* (syn. *Datura*) and *Taxus baccata* (common yew). Ironically, extracts from common yew (*Taxus baccata*) are being actively researched as a cure for some forms of cancer.

Medicinal plants are often featured in botanic collections, and in some instances, such as the Chelsea Physic Garden, collections are specifically dedicated to this plant group; but they are difficult to display without good quality interpretative information accompanying them. The interpretative material can illustrate how attractive chosen plants can be, when that particular stage of growth is not in feature. Furthermore, a pictorial story alongside the otherwise 'dull' plant can increase interest, and certainly aid education.

Other groups that are notoriously difficult to cultivate, display, manage and make interesting, are fungi and the cryptogams (Thallophyta, Bryophyta and Pteridophyta): the mosses, lichens, liverworts and algae. They are no more or less sporadic in their main features coming to prominence than other plants, but arguably do not share the aesthetic appeal of higher organisms such as the angiosperms – certainly according to popular

**Individual poisonous plants appear throughout aesthetic plantings, and should be clearly labelled as such to prevent accidental ingestion, even if to do so is difficult. Furthermore, collections of poisonous plants are best located in an enclosed area, labelled with a warning sign, and viewed from the boundary of the enclosure.

opinion, as gauged by visitors' reactions. A great deal of interpretative work is needed to make this group of plants (and fungi, which are not plants) interesting and popular – that's not to imply unimportance, merely their short-term effect and lack of 'pulling power' to a non-discerning public. Some of the highly coloured, capped fungi generate interest, but they are difficult to manage in a collection, and have absolutely nothing to offer once their annual fruiting bodies have died down.

Botanic collections, whilst often needing to attract visitors to help with the finance for their upkeep, obviously place more emphasis on a plant's botanic features, country of origin, provenance and taxonomic relationships than may be the case in other gardens. Furthermore, they must uphold educational, scientific and research ideals. Botanic gardens therefore often have interpretative centres to help visitors' education and understanding.

Horticultural features may be used as a vehicle to display important botanic information; however, they are seldom entirely suited to the task. Both botanic gardens and horticultural-based parks can offer themed and specialized features such as Japanese gardens, alpine gardens/rock gardens, rhododendron gardens, Asian gardens, Himalayan gorges, bamboo gardens and bonsai gardens. A more scientific approach is to display plants in taxonomic groups or by geographic location.

Display in Taxonomic Groups

Taxonomic groupings can include plants from specific orders, families or specific genera, groups of different species within a genus, or plants with any other taxonomic relationships. There is no reason why taxonomic groups for study and comparison cannot be set up in specific and interesting ornamental designs. Indeed, informally (or even aesthetically) placed generic groups are relatively common in garden and botanic display – rhododendrons and magnolias, for example, often feature as specialized plant collections on their own, or as part of larger collections. However, taxonomic groups are often set up in formal beds, even though it is difficult to place representative family and/or generic collections with the wide range of plant types/groups, wide variations in requirements, hardiness differences and other factors (including pH) in one allocated area for comparison. Protection, or even artificial and heated environments, may be needed in order to represent the true range – and with heated structures come high construction costs and ongoing energy costs.

Collections According to Family

Using the family Solanaceae as an example is of particular interest, because it has within it both some very poisonous and some very edible species. The green (later turning red) succulent fruits of tomato are very flavourful and nutritious, as are the fruits of *Physallis edulis* (Cape gooseberry), peppers (*Capsicums*) and aubergines – yet many other representatives of the family are poisonous (stems, leaves and fruits in some instances). Tomato (*Lycopersicon esculentum*) is treated as a half-hardy annual or a glasshouse perennial in temperate climates. However, it will thrive outside during the summer months, and is very happy outside in mediterranean climates. Potato (*Solanum tuberosum* – a reference to its stem tubers) is a hardy perennial with poisonous green (later turning brown) fruits, but with edible, swollen, subterranean stem tubers that form part of the staple diet of millions of people throughout Europe and North America. Deadly nightshade (*Atropa bella-donna*), as its common name suggests, is a plant in which all parts are very poisonous.

Brugmansia (*Datura*) is a half-hardy sub-shrub in temperate climates, and can only survive outside in protected areas in our warmest regions. It bears massive, fragrant flowers that resemble the flowers of tobacco but on a far larger scale. The sap from all parts of the plant is very poisonous and can be fatal if ingested or if sufficient amounts are absorbed via the skin. Tobacco (*Nicotiana*) has very large leaves and is poisonous because of its nicotine content. All share a common flower structure that places them in the Solanaceae. Tomato may have yellow flowers, black nightshade (*Solanum nigrum subsp. nigrum*) white flowers, and bittersweet (*Solanum dulcamera*) lilac flowers – but otherwise the flowers look remarkably similar. However, they do not share size, hardiness, vigour, or common requirements for cultivation. Furthermore, tomato and potato have compound leaves, and *Brugmansia* and tobacco have large, simple leaves.

Geographic location follows a theme only so far – many come from South America, but others from Europe; and as illustrated above, there is no common thread regarding toxicity or edibility across the family. The cultural requirements, aesthetic value (compare *Brugmansia* with black nightshade), economic importance (compare potato with *Brugmansia*), and hazard rating (compare *Atropa*

Fruits of aubergine. (Solanaceae)

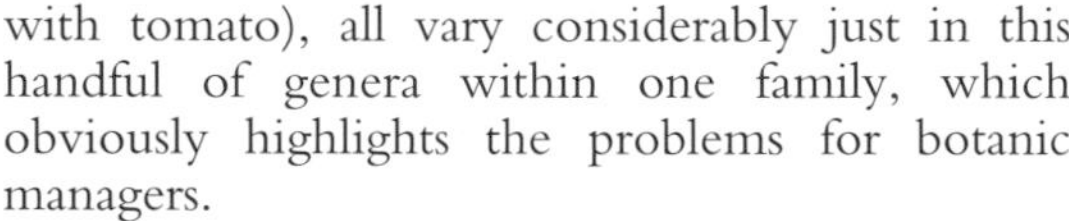

Fruits of the ornamental Solanum lacianatum – compare these with the tomato and the nightshades.

with tomato), all vary considerably just in this handful of genera within one family, which obviously highlights the problems for botanic managers.

The same exercise carried out for the Papilionaceae (Leguminosae) reveals similar problems. The defining features that place plants in the Papilionaceae family are the pea-like flowers and the resultant pod-like fruits. However, *Robinia pseudoacacia* (false acacia) is a very tall-growing, thorn-bearing, deciduous woody perennial (arboreal) with white flowers. Sweet pea (*Lathyrus odoratus*) is a vigorous, self-clinging, annual climber; and everlasting pea (*Lathyrus latifolius*) is a climbing, evergreen perennial, both having very colourful, fragrant flowers. Edible pea (*Pisum sativum*) and 'mange-tout' are both hardy annuals, having edible fruits and edible fruit cases respectively. All the peas have leaf tendrils for support.

Common laburnum (*Laburnum anagroides*) and Scotch laburnum (*Laburnum alpinum*) are small to medium-sized deciduous trees with showy yellow flowers and very poisonous seeds. Common gorse (*Ulex europeaeus*) is a spiny stemmed, woody shrub; common broom (*Sarothamnus scoparius* syn. *Cytisus scoparius*) is a smooth-stemmed, woody shrub; and Spanish broom (*Spartium junceum*) is a woody shrub whose stems look like rushes (*Juncus*) – all carry yellow flowers. *Wisteria sinensis* is a woody climber from China with pale mauve flowers. Nepalese laburnum (*Piptanthus nepalensis*) is an evergreen or semi-evergreen shrub with pale yellow, pea-like flowers and comes from Nepal; and *Cytisus battandieri* is a scandent woody shrub with fragrant yellow flowers (smelling of pineapple). Red and white clover, the trefoils, vetches and rest harrow (*Ononis repens*) are all herbaceous.

Tackling a family as large as the Rosaceae is very daunting, as it includes the genera *Rosa*, *Cotoneaster*, *Sorbus*, *Crataegus*, *Prunus*, *Malus*, *Pyracantha*, *Pyrus*, *Potentilla* and so on. A taxonomic display designed to contain such a collection would require a lot of room and have to cater for many differing needs. Some of its members are herbaceous perennials, whilst the majority are woody perennials. Herbaceous types include strawberry (*Fragaria*),

Geum (some of which have coloured flowers, rather than white flowers, but are very similar to strawberry), and herbaceous *Potentilla* that resembles *Fragaria* in many ways, and within the genus *Potentilla* there are also shrubby forms (*Potentilla fruticosa*). Meadowsweet (*Filipendula ulmaria*) requires totally different environmental conditions to most of the other genera within the family, as it thrives in soil with a high water table.

Display of Plants by Geographic Origin

Displaying plants from diverse families, genera and species in groups relating to their natural geographic origin can be very educational. The 'Himalayan Gorge' is represented very well at Wakehurst Place, West Sussex, and has proved to be a very successful botanic, geographic and horticultural feature. Part of its success arises from the fact that the range of plants includes some very prominent, highly coloured, large-flowered rhododendrons, and related plants with 'pulling power'.

Japanese gardens fill a useful geographic display niche, and are very popular with visitors. The secret of their success and popularity is probably an interest in a combination of factors, including plants, culture and architecture – and perhaps the Japanese garden is unique in this?

Plants of North America, South America, Asia, Australia, New Zealand and Africa can be successful as visitor attractions and tools of scientific study, but they actually represent species from a very wide collection area, and therefore with enormous diversity in terrain, climate and environment. The effect of this vast range of plant species is twofold: on the one hand, it makes it difficult to truly reflect the diversity and range in a limited display space; but on the other, such a wide range to choose from offers a greater chance of being able to use botanically interesting species that have excellent aesthetic features as well.*

The diverse regional climates within the temperate countries of Europe (and the temperate zones of North America) allow a very wide range of plants from all over the world to be grown. Many plant displays geographically based in cool temperate climates use plants brought from the higher elevations of their native country, so they might be hardy enough to be grown in their new climate without severe damage. The provenance of a selected species can sometimes be the key to success or failure in the adopted climate.

Often botanically intriguing peculiarities have little aesthetic or intrinsic interest to the majority of visitors. *Welwitschia mirabilis*, for example, exists as two strap-like leaves attached to a large taproot all its life. Its 'claim to fame' is its resistance to drought, so although its structure is of definite botanic interest, it has very little aesthetic appeal. A taxonomic display of the family Proteaceae, on the other hand, is not only of botanic interest, and is represented in Australasia, South Africa and South America, but also creates a superb aesthetic spectacle, judged by anyone's standards. However, if aesthetic appeal were the only criterion for plant collections, then many plants would not be grown or represented in collections, and might even be lost to cultivation (and their native habitat).

Using Indigenous Plants in Botanic Display

Indigenous plants are much under-used, and arguably underestimated in botanic gardens and collections. They have long been branded by many as 'weeds', and it seems that it is difficult to change the mind-set of those who consider all exotics to be 'exciting' and all indigenous plants to be bland, dull and uninteresting. The 'green' revolution sees more interest in gardening than ever before, but usually involves only 'token areas' of native plants.

It is perhaps not so surprising that indigenous plants are viewed as weeds when we struggle to rid our lawns of the native daisy, dandelion, hawkweed, cat's ear and so on – all of which are colourful, aesthetically interesting plants in their own right – for the sake of lawn uniformity. Large areas of such colourful plants anywhere else in the world would be considered worth cherishing. Complacency must be a large part of the formula for ignoring – even vehemently destroying – many native plants.

Lesser celandine (*Ranunculus ficaria*) has attractive yellow flowers, borne early in the year, and glossy green leaves, which die back to the tuber relatively early in the season. It is therefore not competitive to surrounding plants because it is separated from them in both time and space. So it is perplexing to see people spending a lot of effort and expense to rid herbaceous borders of lesser

*Plant World, near Newton Abbot, Devon has plant collections laid out in groups representing the five continents.

celandine (which may be invasive but is not pernicious), only to replace it with a poorer quality, relatively expensive, exotic ground-covering species. It is difficult to see how this sort of thinking can lead to successful use of indigenous plants in designs.

Our native plants include all plant groups, from ephemeral to woody perennial. They can be displayed with wildlife habitat in mind, and if this is given priority, the aesthetics may have to take a back seat – for example, leaving nettles for butterflies. However, the use of cuckoo flower (*Cardamine pratense*) as a host for orange-tip butterflies goes some way to balancing aesthetics and wildlife, and more examples of this nature would help change entrenched ideas. Unfortunately, *Epilobium angustifolium* is not usually held in much esteem because of its invasive nature, and the elephant hawkmoth is an exotic species not indigeneous to the UK – but the use of rosebay willowherb to attract elephant hawk moths could surely be seen as a plus?

Most native annuals are only successful when cultivated in disturbed soils – *Papaver rhoeas* (common field poppy) and *Agrostemma githago* (corn cockle) are examples of plants that need agricultural-style disturbed soils to succeed. *Linum usitassimum* (common flax), *Malva sylvestris* (common mallow), *Malva moschata* (musk mallow) and *Centaurea cyanus* (cornflower) can be displayed by sowing annually in rectangular, field-like areas, or irregular, organically shaped drifts, on a small or larger scale. Some can be sown in amongst small areas of cereal crops (sown at low densities) to be truly authentic! The effect is not an easy one to maintain, however, and can look very untidy at the end of the season as they die back and the soil shows through. Hence, many people revert to using grass sward with mostly perennial natives injected into it, to attain a more permanent flowery meadow effect: this involves less maintenance (only infrequent mowing), and avoids the problem of soil showing through, or the need to recultivate (and resow) every year. Nevertheless, in either case, late summer 'scruffiness' is a problem, and management of some sort is required.

With perennial (grass sward-based) systems the injected natives compete with, and eventually create a harmonious balance with, the existing species. Annual systems do not compete easily with vigorous weeds, and weed control in the development phase is absolutely critical for success. It is very important therefore to decide exactly what effect is required, as it will have implications on preparation, outlay and future management. Is an agricultural, arable crop-style flowering field required, or a permanent grass sward with mostly competing perennial species that give an ancient flowery meadow effect?

Mixed Sward Systems

Examples of native herbaceous perennials that can be considered weeds in some contexts (for example in formal lawns), but are considered useful aesthetic plants in other contexts (especially *en masse*) include dandelion (*Taraxacum officinale*) and orange hawkweed (*Hieraceum aurantiacum subspecies aurantiacum*). Other examples include red campion (*Silene dioica*), meadow crane's-bill (*Geranium pratense*) and meadow buttercup (*Ranunculus acris*). They are most successful and effective when displayed as mixed sward systems attempting to emulate ancient flowery meadows, and are created by simulating traditional cultural routines of mowing/grazing.

Very effective displays of indigenous plants can be created by mowing in 'zones' – that is, mowing different areas at different heights, and at different times. Mowing at three different heights – close, medium and high – encourages different species to establish. Areas managed by relatively close-mowing regimes will encourage the establishment of large communities of common lawn plants; these include germander speedwell (*Veronica chamaedrys*), common daisy (*Bellis perennis*), common field speedwell (*Veronica persica*), dandelion (*Taraxacum officinale*), common cat's ear (*Hypochaeris radicata*), self-heal (*Prunella vulgaris*), common primrose (*Primula vulgaris*), common cowslip (*Primula veris*), creeping Jenny (*Lysimachia nummularia*) and ground ivy (*Glechoma hederacea*). Optimum effect can be induced by mowing just above their development height early on in the season, and then leaving them to flower and set seed. Low mowing from then on, and particularly again at the very tail end of the season, leaves the sward already short for good flowering effect in the early part of the following season. This system 'misses' very low growers when mowing, and others by mowing at the right time; it probably emulates 'downland' as selectively grazed by rabbits and other herbivorous mammals.

Areas that are managed by leaving grasses to get longer before mowing – that is, by reducing mowing frequency to perhaps three times per year – and mowing at a higher level (using a rotary mower set high, for instance) induces other common species to establish. These might include *Geranium pratense* (meadow crane's-bill), *Geranium dissectum* (cut-leaved crane's-bill), *Hieraceum aurantiacum* (orange hawkweed), *Hieraceum maculatum* (spotted hawkweed), *Leucanthemum vulgare* (ox-eye daisy) and *Sanguisorba minor* (salad burnet). They will be left to flower and set seed before mowing is continued.

The third management option is to leave a mixed sward area throughout the growing season, and only mow once per year (twice as a maximum), using a reciprocating, flail (or other high capacity) mower after most species have flowered and set seed in the late summer. This method encourages many taller-growing perennials. Species might include all those listed above, as well as *Silene dioica* (red campion), *Hypericum perforatum* (perforate St John's wort), *Heracleum sphondylium* (hogweed), *Conium maculatum* (hemlock), *Centaurea scabiosa* (greater knapweed), *Ballota nigra* (black horehound), *Verbascum nigrum* (dark mullein), *Vicia cracca* (tufted vetch), *Vicia sativa* (common vetch), *Filipendula ulmaria* (meadowsweet) and *Sanguisorba officinalis* (great burnet).

Species cannot be induced to colonize if the seed bank is not naturally available, hence introducing the desired species into the existing sward is often the only option. 'Plugs' – container-grown or bare-root young plants of suitable indigenous species – can be introduced into any of the areas, whatever the 'zone' chosen; alternatively, small areas may be sown with native plant seeds. Preparation involves clearing small 'patches' throughout the existing sward by hand (or perhaps by chemical herbicide), and then planting, or sowing, into these areas. *Leucanthemum vulgare* (ox-eye daisy), *Geranium pratense* (meadow geranium), *Primula veris* (cowslip), *Knautia arvensis* (field scabious) and *Primula vulgaris* (primrose) are commonly introduced into an existing sward in large blocks for extra aesthetic affect. Other effects can be obtained by sowing (or planting) tall-growing biennials in larger, irregular-shaped drifts (in designated areas marked off formally or informally to avoid unwanted mowing); the following can all be used: *Oenothera biennis* (evening primrose), *Dipsacus fullonum* (common teasel), *Verbascum nigrum* (dark mullein), *Verbascum thapsus* (great mullein), *Verbascum blattaria* (moth mullein) and *Verbascum lychnitis* (white mullein), *Chicorium intybus* (chicory – a short-lived perennial) and *Tragopogon pratensis* (goatsbeard – an annual/perennial).

Some species require very specific regimes for success, which are not easily achieved; often these are kept to small, compact, easily managed areas. Examples include *Fritillaria meleagris* (snake's head fritillary), *Dactylorhiza majalis* (southern marsh orchid), *Epipactus palustris* (marsh helleborine), *Ornithogalum pyrenaicum* (star of Bethlehem), *Narcissus pseudonarcissus* (wild daffodil), *Colchicum autumnale* (meadow saffron), and *Pulsatilla vulgaris* (pasque-flower).

Very specific, acidic, moist regimes are needed to grow the following species successfully: *Drosera rotundifolia* (round-leaved sundew) and *Drosera anglica* (great sundew), *Narthecium ossifragum* (bog asphodel), cotton grass (*Eriophorum angustifolium*), cross-leaved heather (*Erica tetralix*) and *Pinguicula vulgaris* (common butterwort). Moist, boggy conditions are also needed for *Caltha palustris* (kingcup or marsh marigold), *Lychnis flos-cuculi* (ragged robin), *Myosotis scorpioides* (water forget-me-not), *Alisma plantago-aquatica* (common water plantain), *Filipendula ulmaria* (meadowsweet), *Iris pseudoacorus* (yellow flag) and *Lythrum salicaria* (purple loosestrife).

The Establishment and Management of Naturalized Bulbs

Some native and exotic bulbous subjects can be successfully naturalized into a grass sward: 'naturalizing' means planting bulbs into an existing grass sward and leaving them largely to their own devices with minimal management and aftercare. Narcissus, Spanish bluebell, snowdrops, star of Bethlehem (*Ornithogalum*), spring snowflake (*Leucojum aestivale*) and crocus are all suitable.

Bulbs are usually planted in small groups in organic-shaped drifts. Planting is carried out using either a special bulb planter, or a mattock (even though this might seem excessive). Bulb planters remove a core of soil, the bulb(s) is/are then placed in the hole and the core replaced over the bulb(s). However, a mattock is very useful for large areas of bulb planting, where large groups are to be planted at each station, and particularly useful for planting bulbs in informal drifts. The adze-shaped end of the mattock is used to flap back divots of turf, and the fairly large planting pockets created with each mattock stroke allows several bulbs to be planted at each station. Bulbs are clumped close together, and each clump adds to the informal pattern that appears to be (but actually isn't) random, to give the most natural effect. The rule of thumb for bulb planting also holds true in this situation: plant at approximately twice the depth of the bulb. The 'flap' of turf is then replaced over the planted bulbs, and lightly firmed with the foot. Most bulbous subjects are happy to compete with grasses in the spring, particularly as their subterranean food storage organs initiate early growth and give them a head start on the grasses and other vegetation within the sward.

The dilemma is that the most colourful effects are induced by mowing grass regularly; however, naturalized areas must not be mown directly after flowering: mowing must wait until the green foliage of bulbous subjects naturally dies down to ground level. This ensures that the leaves are left to photosynthesize for as long as possible to produce sugars that are returned to the bulb via the phloem and stored for later use. If this process is not efficient the bulbs will use more energy than they have stored, and the energy deficit will cause degeneration of the bulbs, poor flowering or none at all, and gradual demise. Naturalized bulbs therefore do well in mixed swards, where they are effective (perhaps not as showy as in regularly mown grass areas – but where the leaves are mown off too early) because the other species within the mixed sward are left to flower and set seed before mowing commences. Even the practice of tying the leaves of *Narcissus* and suchlike into large knots to tidy up areas in mown grass drastically reduces the photosynthesizing leaf surface area, and therefore reduces the potential for flowering in future years.

Maintenance of Existing Features

It is essential to have sufficient staff numbers to keep maintenance quality high; however, the use of organic mulches such as bark or mineral mulches will help in this latter objective. Some areas, sometimes even high profile areas, can be covered with geotextiles/sheet mulches to control weeds, as long as the organic or mineral material used to cover the sheet mulch is in keeping with the area, and is maintained intact so that it always looks aesthetically pleasing. Buildings must be maintained in good order both for their appearance and for public safety.

Grass swards need to be kept mown and in good order, not only for their good appearance, but also for the practical consideration of maintaining a good surface for visitors to walk on without them getting excessively wet feet in poor conditions. Good quality, clearly designated, and well maintained pathways near these areas will encourage their use and reduce the potential wear on grass surfaces in wet weather.

Management of Grass Surfaces

Grass areas are unfortunately prone to tearing and scarring, and this can lead to a very slippery surface in wet weather. It is easy to forget that a lawn's unified green surface does in fact comprise green plant material, with all the biological factors that this entails – so it is a mistake to think that growing grass under anything but good conditions is easy. Lawns and other grass surfaces are not necessarily that 'precious', but they have to be treated with respect and consideration, and be maintained correctly in order for them to be successful. Besides the health of individual grasses, their structure is such that they will not resist the ravages of physical wear in the same way as a hard surface. Lawns are open to surface erosion, and once this commences, grass surfaces can often maintain only a very tenuous hold.

Erosion as a result of foot traffic is a major problem for historic and botanic gardens. Grass surfaces subjected to excessive foot traffic will suffer the loss of binding at, near, or just below the soil surface, and this will lead to surface tear, and surface and green tissue erosion, and compacted soil will show through the very thin sward, which will quickly deteriorate. Areas of repeated erosion of this type may need to have reinforcing materials included within the turf, or even be covered in hard landscape materials as a permanent solution.

All maintenance operations are designed to reduce, alleviate or mollify these problems, and the starting point is therefore always with the correct choice and mix of species for the desired site use. The good binding effect of rhizomatous and stoloniferous species, the wear tolerance of clump-forming (tussock-forming) species, and the percentage of each type in the mix, are all very important factors. However, the grasses (and other plants) that make up lawns are dynamic, and any species mix, no matter how wisely chosen, will change over time as the more vigorous species start to predominate. The way the changes happen, and the degree of those changes, will be affected by the ultimate use (or misuse) of the area and the adopted management regimes.

Management regimes involve the addition of nutrients (fertilizers), adding organic materials, relieving surface compaction, and mowing to encourage tillering (lateral growth) and therefore shoot density (surface cover) and to reduce grass height both for tidiness/aesthetics and ease of access for visitors. Periodic, regular (even cyclic) work schedules are based around mowing systems in the growing season, and top dressing, seed sowing, surface scarification and hollow-tine coring in the autumn months. One regular maintenance operation that is unique to ornamental lawns is lawn edging, which establishes and maintains a vertical cut surface at plant border boundaries and around hard landscape features. Lawn edging may be carried out and maintained by hand using an edging iron, or mechanically using machines with a rotating blade once the initial edge has been formed with an edging iron.

Areas of lawn that 'thin' due to wear (literally the grinding down of soft green tissues by foot traffic or other agencies) can be resown ('oversown') with a grass seed mixture to bring the density back up to standard. However, there are a few important parameters that need to be considered. First, the grass seed must be suitable for the area, it must be stored correctly so that it remains viable, and it must be sown in suitable conditions for successful germination. Furthermore, even successfully germinated grass seedlings can still be prevented from attaining maturity and doing their 'patching-up' job if they are not cordoned off to prevent visitor access during their development phases, or if the first mowing operation after sowing rips the developing roots of the seedlings from their sown positions. Hence the suitability, sharpness and efficiency of the mower, and when the surface is mown relative to the development of the seedlings, all have an effect on the success or otherwise of their establishment.

If surface and immediate sub-surface water can be moved away quickly from lawns by drainage, then the surface becomes more cohesive, has greater traction, and is less liable to suffer from damage by surface tear. Surface water and waterlogging create a very slippery surface as soil particles displace easily because of their water-lubricated state, and ultimately the soil will become compacted. Long periods of anaerobic conditions experienced during waterlogging, and poor drainage generally, cause permanent problems to grass plants, including root death and fungal disorders. Bare patches caused by excessive wear can be oversown at any time, and if carried out in anything but ideal moisture conditions, irrigation can be used. If little time will elapse before the surface will be once more in use, turf can be let into the damaged surface, and pinned into place with soft pegs (for safety). However, it is a pointless exercise unless the area is 'protected' from foot traffic during development, and if the drainage problem that caused the wear is not alleviated first.

The relief of surface compaction, improving surface aeration and drainage, regular mowing, and the removal of 'thatch' (where it is present) feature in maintenance regimes for all grass surfaces, including ornamental lawns. Thatch is created by a build-up of dead and/or live lignified tissues on the surface. As grass species get older, some in particular produce older stems that form a mat at ground level. This material is removed by scarification, which can be carried out by hand using a spring-tined rake, or by high-capacity specialist machines (scarifiers). Slit-tining will also go some way to reducing thatch by cutting into the older, more lignified stems vertically from above. This procedure can be carried out on all grass surfaces during the autumn, in a non-frosty winter, and in the early spring months to relieve surface compaction. It can also be carried out during moist periods in the summer, as the tines are solid and knife-like and do not remove soil but displace it instead, so it does not disturb the soil surface too much during the process. Compaction relief is considered to be effective, but relatively short-lived. It is precisely for these reasons that slitting (rather than coring/hollow-tining) is used regularly on cricket wickets during the playing season.

Aeration using hollow tines actually removes cores of soil and relieves compaction for longer periods of time than slit tines. The cores of hollow-tining can be collected up and removed, or they may be broken down over the soil surface to disperse them evenly, and incorporated with a very open, resilient bulky organic top-dressing used to fill the open core holes and help ameliorate the soil condition. Such top-dressings are usually based on coarse sand and peat and are intended to open up the soil and improve drainage, thereby improving soil structure and its moisture-retaining capability for the summer months. The moist autumn months are perfect for the task. However, mild winter and spring months are also possibilities on areas where there is no winter use by visitors, or where they can be given alternative routes. One of the major problems is keeping grass areas in a constant state of readiness for use, whilst still carrying out routine repairs and maintenance. Sometimes the frequency of use of an area can even preclude major renovations being carried out at all.

Weed control may be by hand on a small scale, but professional turf and large grass areas unfortunately need herbicides for their management.

These usually comprise chemical constituents that are in toxic amounts for broad-leaved weeds but are non-toxic to grass species – the so-called hormone-based selective herbicides. Grasses are particularly nitrogen 'hungry', and need good and regular nutrition to make up for the nutrients that are taken away in the tissues every time they are mown if they are to stay dark green and healthy. Because of its relative solubility, nitrogen in particular readily leaches from temperate soils and is often in short supply. Hence, most fertilizers for grassed areas, whilst still supplying other nutrients, are strongly biased in favour of nitrogen. Proprietary brands of 'weed and feed' that comprise both herbicide and fertilizer granules, and allow both functions to be carried out in one 'pass', are readily available for both small and large scale areas.

The term 'top dressing' is sometimes used to refer to the application of nutrients (fertilizers), but is probably best reserved for the application of bulky organic (and mineral) materials over the soil surface. Bulky top dressings are used to level out uneven areas prior to oversowing with seed, as blanket coverings over entire areas (as a soil ameliorant), and to fill and level grass areas after hollow-tining (adding soil ameliorants at the same time). The greatest volumes of bulky top dressings are applied after hollow-tining, as a large proportion 'disappears' down the core holes. Fairly obviously it would be no good applying a bulky top dressing over the top of unmown or unkempt grass. Furthermore, top dressings should never be applied so thickly as to blanket the photosynthetic tissues of grass, thereby causing low energy budgets and foliage death because they preclude the light.

Mowing

Mowing is a frequent necessity throughout the growing season, and is still required, although less so, in the early and late autumn (depending on the season) – and sometimes even in early winter and early spring. Main ornamental areas can be cut to 8–12 mm high, as any lower would require the use of fine grass mixtures to tolerate the regime, but these in turn would not tolerate the severe wear they are liable to encounter. Rough grass is mown with either reciprocating or rotary mowers, the size and capacity of which depends on the area of the site. Utility lawns may be mown with rotary mowers or with a cylinder mower with a relatively low number of blades (four to five).

All grass areas that are close-mown need a mower with seven to nine blades (or more) on their helical cylinder to ensure the quality of cut. Quality of cut is a function of several factors, including the number of blades on the cylinder, the speed of rotation of the cylinder, and the forward speed of the mower. A high number of blades, high blade revolutions, correctly set blades, and a slow forward speed give the best quality cut. Other parameters include how level the site is, the prevailing condition of the sward surface and how well the blades are kept sharp.

The capacity of the mower is a product of the blade width, forward speed and manoeuvrability/versatility of the machine, with wide blades, a high forward speed and good flexible access ability giving the highest capacities – that is, the highest amount of material removed in the shortest space of time. A high capacity performance with a good quality cut comes from cylinders arranged in groups (including tractor-drawn gang mowers and purpose-built, self-propelled sit-upon mowers). Self-propelled units often have slewing clutches or hydrostatic drives to aid their access versatility. Very high capacity performances (relative to their mowing width) come from powerful flail mowers (so-called 'jungle busters'). However, these are only suitable/desirable for rough grass areas, although they do reduce grass mowings to a mulch-like material that readily decomposes and returns some nutrients to the soil.

Mowings from rotary mowers can be left on rough grass sites, but are best collected in a grass-catcher and removed from utility lawns (or better quality areas) because they create unsightly 'drift' left on the surface. Apart from those connected together in large 'gangs', cylinder mowers always come with grass collecting boxes because they are used on a relatively level surface, and drift is unacceptable on quality ornamental lawns (and would interfere with play on any sports turf). Grass collecting boxes are of varying capacities, but always need emptying into a bulk container. Some 'integrated' grass management machines include a system that will spread fertilizers and bulky top dressings, mow grass, and also move the cut grass into a large capacity, grass collecting trailer (or bulk container).

'Drift' left on site also encourages a high worm count because of the constant addition of organic matter. This is the big dilemma of grass management, because although larger worm numbers improve surface aeration and soil mixing, they also leave worm casts that are considered unsightly, and which when flattened, create ideal spots for weed seed germination. Obviously, very accurate level sports surfaces may find worms a major problem, but the destruction of worms would seem unwarranted on both ecological and soil improvement grounds. The situation can be alleviated by the use of brushes or a switch to distribute worm-cast soil over the surface.

People Flow

Ornamental lawns are usually multi-use/multi-purpose, and have to cope with varying degrees of foot traffic and general abuse. The most wear occurs where lawns reduce in width and effectively 'funnel' pedestrians into a space with a relatively small surface area – hence practical and thoughtful spatial design of lawn areas is as important as the choice of species. The very worst wear appears at 'desire lines': informal pathways that avoid natural obstructions and other features, created by constant use (abuse). Good design, with a sufficient network of convenient and usable, specially constructed pathways (including hard surface pathways), avoids the creation of desire lines. Conversely, a philosophical approach to the problem includes not constructing pathways at all in the first instance, waiting for natural desire lines to be produced, and eventually making these the permanent pathways by formalizing their existence.

The management of people flow is a major aspect of gardens open to the public, and good pathways with the correct surfaces are necessary for successful visitor experience. Although a grass surface makes an ideal, natural foil for aesthetic plantings, unfortunately it tears easily in wet weather and becomes hazardous and slippery. Botanic gardens (including arboreta) often have large visitor numbers during the autumn (for autumn colour) when rainfall is high. Soil structure is destroyed by foot-traffic compaction – and furthermore, compaction around trees and shrubs ultimately causes root death. Hard-wearing surfaces may need to be included in some areas where foot traffic is particularly high, though unfortunately at great cost. Conversely, it is sometimes beneficial to use resilient materials to cut down compaction, such as bark chips, where pathways come near the root plates of trees and shrubs. The available surface area of visitors' main routes through the garden will also influence potential wear. Areas visited by large numbers of people need to be large and have spacious 'milling areas' to help soak up and reduce the foot-traffic damage/erosion.

Improving Aesthetics and Visitor Appeal

Visitor numbers are usually (but not always) proportional to the amount of funding available for the maintenance and development of the botanic collection. However, increasing visitor numbers brings with it the problems of both extra foot-traffic erosion and the need for extra car parking. Methods of increasing visitor numbers include new developmental plantings, extending each end of the flowering season (or other feature), a good tea room and clean toilets (sufficient to cope with the visitor numbers), quality maintenance (including correct grass cutting), efficient signage, good footpaths/main footways, convenient car parking and efficient litter management. Good staff (of the right quality) are needed for organized theme days, guided walks, plant sales, cultural events, school trips, and so on.

Good, clear signage is important for the direction, education and regulation of visitors, and meaningful information provided on good quality, interpretative signs is essential for successful visitor management. Pre-arrival signs (including brown tourist signs on main roads) are essential to direct visitors towards the botanic collection, as are threshold signs at the main entrance and reception areas. Within the botanic garden, orientation information signs (so people know where they are, and what they may see) and direction signs (signage and maps to show the routes and flow through the garden) are both essential. The garden must have information signs showing opening times, where to find specific areas, and topical/featured plants. There must also be regulatory signs that state any rules and regulations, particularly those that encourage safe use of the area. Mobile arrows can be used to move visitors away from troublesome, worn or hazardous areas, and when maintenance work is being carried out.

Plant identification can be given on signs indicating plants within groups, but is best provided by labelling individual plants. Efficient, clear, effective, informative, durable and cost-effective individual labelling has always been a major difficulty in botanic gardens of all types. Getting the correct, durable materials engraved at a reasonable cost is always difficult, and there can be high replacement costs associated with their theft or damage.

Interpretation centres are essential to give educational and botanical information, information on conservation aims, relationships with other gardens and arboreta, and anything else that will help visitors relate to the work and philosophy of the garden. Permanent educational displays are always good. However, if the necessary skilled staff exist to carry it out, living exhibits and freshly cut specimens of topical plants with temporary labelling of a good standard, always add interest. Good quality tea rooms and site shops (including plant sales) can be a very successful way of increasing income.

Visitor appeal is always increased by the use of quality and colourful plantings, and visitor numbers will often be successfully increased by extending the length of attraction at each end of the season by the careful choice of plants for early and late flowering. Adding other genera to both specialist single genus and mixed genera displays will help to extend the season. Rhododendron gardens are renowned for having greatest interest from April/May until June, and the careful use of very early flowering types such as *Rhododendron* × *praecox* and late types such as *R. auriculatum* (and its hybrids) can help extend the flowering period at each end of the season. Late-flowering plant examples include *Eucryphia glutinosa*, *Eucryphia* × *nymansensis*, *Eucryphia* × *nymansensis* 'Nymansay', *Eucryphia cordifolia* and *Hydrangea aspera sargentiana*, *Hydrangea aspera villosa* and *Hydrangea paniculata* 'Grandiflora'. *Holodiscus discolor* can be included for its late flowers, and species noted for bold foliage, coloured foliage, autumn colour, bark and berries also extend interest.

Although we celebrate the diversity of our native and introduced (exotic) botanic heritage, their very diversity creates difficulties for their sourcing, successful management, cultivation, maintenance and display. Furthermore, as all botanic managers know, it does not stop at difficulties arising out of plant management, but includes successful visitor and financial management.

Gardens Associated with Historic Houses

Decisions on the management of historic gardens have many aspects in common with general garden management problems. Under all circumstances the smooth day-to-day running of the site depends on a clear line of authority to ensure that decisions are implemented as discussed at planning level. And of course there must be sufficient budget for development plans to be implemented, and for the necessary staff. However, they also have their own difficulties revolving around how to cost-effectively develop and maintain features to a high standard whilst retaining the ethos, philosophy and integrity of the garden, and keeping it in 'period'.

Historic garden management and development decisions centre round conservation/preservation versus restoration, development versus maintenance of existing features, contemporary developments versus traditional-looking designs, and botanic collection versus aesthetic appearance. Necessarily there is nearly always a blend of all these issues, and a trade-off that satisfies as many aspects as possible. It is notoriously difficult to stick to dogmatic policies that insist on purist ideals throughout, because all the issues cause tensions with one another.

Preservation means to preserve/save what already exists, and conservation means to protect from decay, harm, or loss – and both are difficult to maintain because of the transient nature of plants (albeit over different time scales, depending on the species involved). The same is also true of buildings and other man-made structures (many over a shorter time-scale than the most persistent plants, namely trees). Unfortunately, all require maintenance (and expense) of some sort, and where they are inherited in states of severe disrepair, will need restoration.

In many instances there are not sufficient proportions of existing period features on site to preserve and form an exciting, crowd-pulling feature. So often we have to look towards restoration, complete renewal, or brand new features and development projects – even perhaps moving from the traditional to the more contemporary. Restoration means to 'restore to its former condition', and it usually involves the use of historical records and old plans of the particular garden. Future development depends on whether there are any hidden features that can be restored/ resurrected to their former glory, or whether more contemporary designs can be safely incorporated – or both. There may be a need for a garden historian and/or garden archeologist, or some other form of consultant to help with development of the site within the chosen parameters. Often even the best examples of period features need expensive restoration of some sort. However, where no accurate records of a specific garden exist, it may be necessary to attempt to provide an 'accurate impression' of the supposed previous features – that is, to recreate as accurately as possible what it is thought would, or could, have been there.

Developing the Gardens

Improving the aesthetic appearance is obviously inextricably linked to garden development, and the introduction of new garden features needs to include exciting and colourful areas if they are to attract visitor interest. However, more colour alone is unlikely to suffice, particularly if it is out of keeping with the existing ethos, features and themes of the garden. Recognition and consistency of style and philosophy are very important and give a unifying effect to the garden. However, modern additions, even of a totally different style and design to existing features, can be a big crowd draw if they are of good quality – the water displays at Alnwick Castle are a good example. You can have good quality 'additions' that add interest to the garden without detracting from it. These are usually most successful when they form a new stand-alone addition, away from, or detached from, the main period garden.

Developing the gardens and taking them forwards within the desired framework is not an easy task. Decisions have to be made on how many new features should be included (the number will naturally be contained by budget and available space), and whether new developments should be completely new, or added to existing, well tried features. Should all new features be low maintenance designs, or should they be more complex (and more interesting because of it?) and given the necessary resources in order to be successful? If new features are not given some form of priority over resources they can fall into disrepair easily. Should the garden build on existing, specialist collections, or create new, specialized areas? Specialist collections can include generic groups such as rhododendrons, magnolias, camellias and sorbi, but most gardens (even specialist gardens) comprise large groups of single genera and/or mixes of genera in their collections, including ericaceous plants of all types, conifers, herbaceous plants, topiary, woodland gardens, tender plant collections, and so on.

Should all new and existing features be linked in some way to the history of the site, or can they be novel and innovative in other areas? How far can purist ideals be taken when it comes to botanic integrity for the period – would this be interesting only to botanists and garden historians, and not to visitors? Would visitors rather see new, exciting features that are not necessarily linked to previous history? After all, the Victorians considered our native evergreens to be boring, and were only too glad to introduce exciting new plants to their botanic collections, and gardens would not develop at all without innovation and new additions. If there is an Elizabethan, Georgian and Victorian influence to the gardens, to which influence/period should the new development be linked?

Good quality hard landscape features may help attract visitors, as might artistic works, sculpture, and/or statuary, but new features need to be lasting/durable, of a suitable scale for the site, and need to fire the imagination of the public. What about a maze or a labyrinth? Unfortunately, large capital outlay is usually associated with such projects.

Many public gardens and parks have changed their practices under the banner of 'rationalization', and aesthetic features have been reduced drastically in favour of grass. At present there is a mind-set that moves against the addition of new features because of associated costs – historic gardens and other privately owned gardens open to the public cannot afford to do this, as visitor interest will wane.

When developments are decided upon, perhaps a consultant garden historian would help with decisions on the plants used, and the ornamentation or associated features of the period. For this route, however, there does have to be access to original, or later, garden plans, in order for the project to be successful and without excessive supposition. Enlisting the help of a garden historian could have a high cost attached, but it could also fulfil the need for objective scrutiny from an outsider, the opinion of someone who does not have direct daily contact with the garden. Is the high cost associated with a 'consultant' of whatever type warranted or necessary? A garden archaeologist could be enlisted to find previously hidden features that could be restored in keeping with their period, as the restoration/renovation of existing features of merit is always newsworthy, and creates local, sometimes national interest. Reworked features can be as impressive as brand new ones if quality and aesthetic appeal are improved.

Woodland Gardens

New developments could include a woodland garden, as these can be used as areas of botanic, aesthetic, amenity and wildlife interest. They may be developed from existing woodland or exotic woody plant collections, or created from

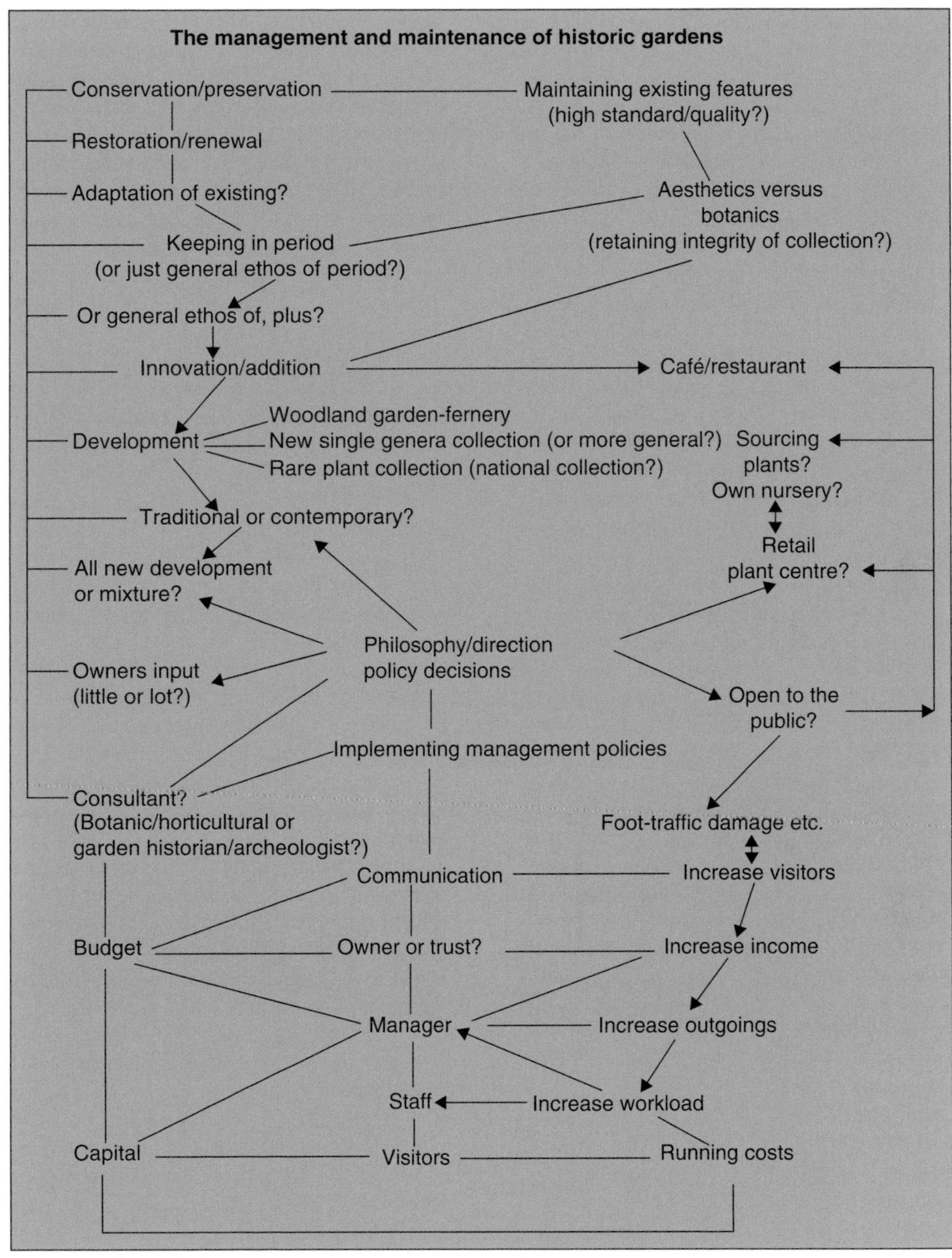

Managing historic gardens: the major decisions.

scratch using woody plants grown as groups. Woodland gardens most commonly comprise a skeleton of indigenous trees, or an indigenous/exotic mix, with exotics injected into the matrix and informal pathways meandering through. There is also a place for a range of interesting herbaceous subjects, including ferns, within a woodland garden.

One of the most difficult cultural problems, yet central to the whole idea of the woodland garden, is the varying light factor created by the different canopy densities, as the light factors can vary from full light to dense shade, dappled shade and filtered light. Managing these light factors, and/or carefully selecting the correct species for existing light factors, plays a great part in the success or otherwise of woodland garden features. A sound knowledge of matrix species, and knowing the light requirements of the chosen exotics that are to be included, is essential for success.

Woodland gardens take advantage of the shelter created by the growing habit of their matrix, of the differing heights, widths and wind-reducing properties; and the niches and microclimates created by mutual protection, shelter and wind filter in woodlands can be exploited to great effect. Thinning or coppicing woodland areas, felling areas as 'planting bays', and creating central clearings, all give different environments, enabling the introduction of exotic plants into the system. Selective thinning alters both the light factor and the potential competition for nutrition in an area. Thinned (or naturally low density) canopy that leaves good gaps between individual trees and gives dappled shade can support the growth of both herbaceous perennials and exotic arboreals planted ('injected') throughout the matrix wherever it is beneficial.

Advantage may also be taken of woodland shelter by creating planting bays with different orientations, creating varying shelters and varying light factors to explore, including shady and protected, partially shaded and protected, and areas of full sun and protection. Best results will only come with knowledge of the most favoured requirements of the chosen species. In many instances plants that would not grow successfully as free-standing specimens in exposed positions will be very successful in woodland conditions in our climate. The balance of the three 'S's – namely shade, shelter and sunlight – has to be maintained, and obviously, shade goes up with increasing shelter, and conversely sunlight intensity goes down. Exposure on the other hand goes up with increasing light factors; but the degree of exposure is also very much dependent on the orientation of the planting 'bay'. Facing south-west leaves plants open to the mild yet physically damaging strong winds typical of this direction, but facing north-east is more detrimental because of the low temperatures and cold desiccating winds encountered.

Gauging the range of plants suitable to face any particular direction is not easy, and there is nearly always congestion in the bays facing favoured directions. There must therefore be a trade-off somewhere along the line, with the most 'important' species in the most favoured positions, and other species having to make do with second or third preferences. It also depends on whether or not a planting bay in a particular orientation is in the lee of other protection, or is fully exposed to the elements in that direction. Moreover the existence or establishment of surrounding woodland distant from the planting bays can increase the shelter effect in the bays, even on those with poor orientation. Furthermore, the size of the gap between the shelterbelt block of woodland giving the lee-side protection and the planting bay will also have an effect. Even relatively large distances that allow good light penetration to the planting bays do not totally nullify the shelter effect – so these are very beneficial in increasing the range of plants that can be grown. For example a wide, sweeping, north-east-facing planting bay is the most difficult to find suitable candidates for. However, the same bay in the lee of another distant shelterbelt-style woodland planting opens up far more possibilities. In the same way, winds can be reduced to less damaging speeds for south-west-facing bays, and the same situation for south-east-facing bays opens up far greater possibilities for a more diverse planting because of the good light factor and the protection from cold easterly winds.

Only truly shade-loving species will fail to thrive in sunny, sheltered positions. Camellias tolerate dappled shade, yet can thrive in sheltered, full sun positions. However, best success with camellias comes from a sheltered, full light, north-facing aspect, as more southerly aspects can cause premature flower-bud drop. Camellia flower buds develop during the summer months, and by late summer/early autumn they are fully developed and overwinter as fat, rounded buds. Hence the potential for floral display in the following spring is obvious by the autumn. However, it is only 'potential', as bud mortality may be uncharacteristically high in certain circumstances. Premature bud drop is caused when buds freeze in adverse weather conditions during the winter and very early spring. The frozen buds that thaw slowly in dappled shade or fairly sheltered northerly aspects usually come to no harm, but plants in more sunny, south-facing orientations drop some of their buds because the previously frozen buds thaw too quickly on warm, sunny winter days and the delicate cells rupture. Camellias planted in containers

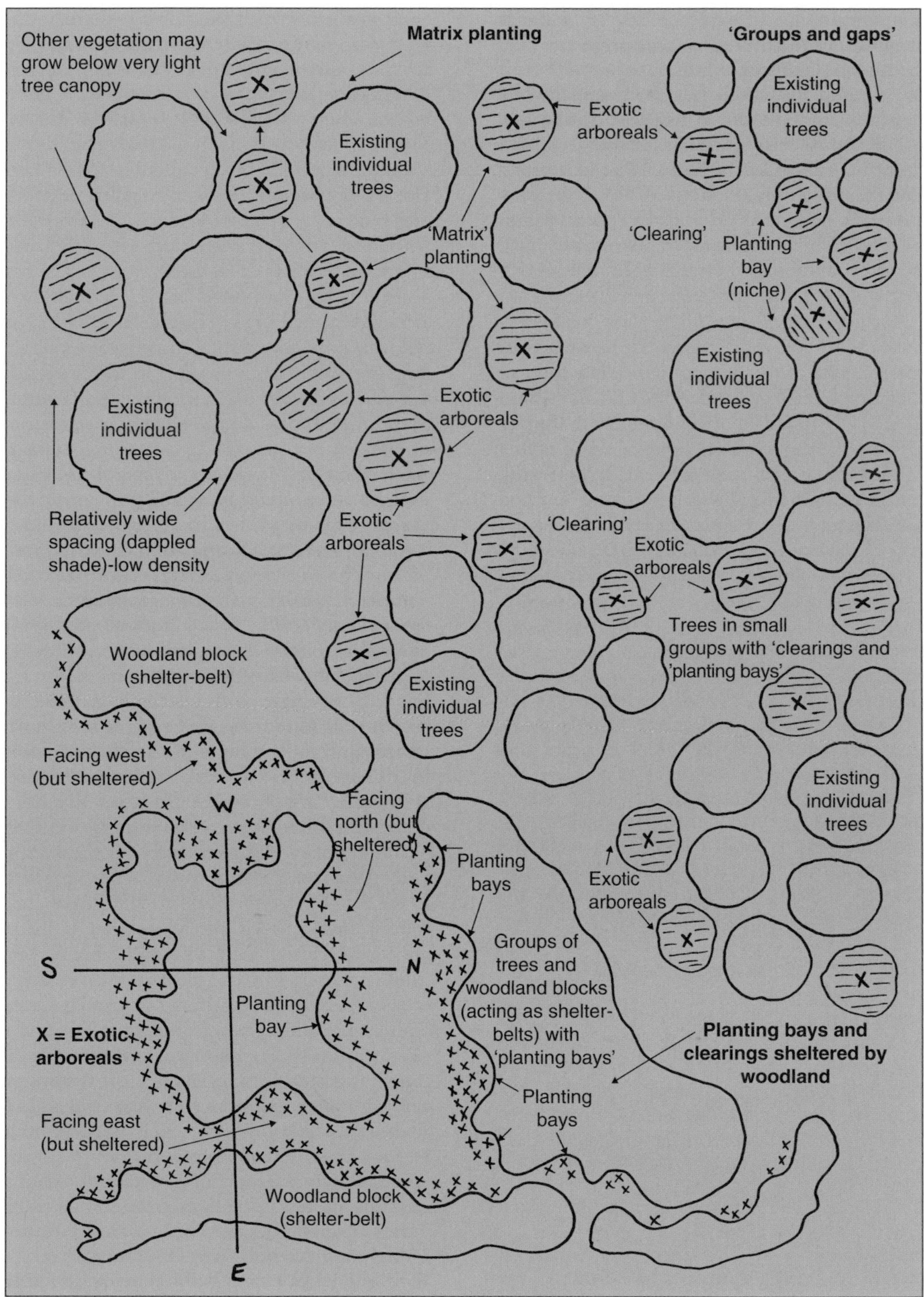

Cultivation, orientation and protection of exotic arboreals in woodland.

Indigenous Woody Species used in Woodland Gardens

Shrubs needing good light to fruit well: *Cornus sanguinea* (common dogwood); *Euonymus europaeus* (common spindle); *Viburnum opulus* (guelder rose); *Sambucus nigra* (common elder); *Sambucus racemosa* (red-berried elder); *Viburnum lantana* (wayfaring tree); *Ligustrum vulgare* (wild privet); *Daphne laureola* (spurge laurel); *Rosa canina* (dog rose); *Rosa arvensis* (field rose); *Rubus fruticosus* (bramble). Woody climbers: *Hedera helix* (common ivy); *Lonicera periclymenun* (common honeysuckle).

Indigenous Herbaceous Species used in Woodland Gardens

Hyacinthoides non-scriptus syn. *Endymion non-scriptus* (common bluebell); *Narcissus pseudo-narcissus* (common daffodil); *Iris foetidissima* (stinking iris); *Helleborus foetidus* (stinking hellebore); *Dactylorhiza fuchsii* (common spotted orchid); *Galanthus nivalis* (snowdrop); *Allium ursinum* (ramsom's/bear's garlic); *Lamiastrum galeobdolon* (yellow archangel); *Pentaglottis sempervirens* (green alkanet); *Cyclamen hederifolium* (sowbread); *Lysimachia nummularia* (creeping Jenny); *Viola odorata* (sweet violet); *Viola riviniana* (common dog violet); *Oxalis acetosella* (wood sorrel); *Meconopsis cambrica* (Welsh poppy); *Anemone nemerosa* (wood anemone).

Moist, boggy conditions, or near water: *Scrophularia nodosa* (common figwort); *Iris pseudoacorus* (yellow flag); *Dactylorhiza incarnata* (early marsh orchid); *Dactylorhiza praetermissa* (southern marsh orchid); *Filipendula ulmaria* (meadowsweet); *Osmunda regalis* (royal fern).

Exotic Woody Species Successful in Woodland Gardens

In dappled shade or full light with protection, as found at woodland edges or within woodland glades: *Acer palmatum*; *Acer japonicum*; *Staphylea colchica*; *Staphylea pinnata*; *Eucryphia* x *nymansensis*; *Eucryphia intermedia*; *Eucryphia cordifolia*; *Eucryphia glutinosa*; *Embothrium coccinea*; *Corylopsis pauciflora*; *Corylopsis willmottianum*; *Clethra arborea*; *Clethra delavayii*; *Magnolia tripetala*; *Magnolia campbellii*; *Magnolia macrophylla*; *Magnolia kobus*; *Magnolia wilsonii*; *Magnolia sinensis*; *Paeonia delavayii.*

Many *Rhododendron* species including *Rhododendron lutescens*; *Rhododendron augustinii*; *Rhododendron auriculatum, Rhododendron sino-grande, Rhododendron calophytum, Rhododendron fictolacteum, Rhododendron macabeanum, Rhododendron falconeri*; *Rhododendron* (azalea types) for example *Azalea lutea*; *Crinodendron hookerianum*; *Illicium floridanum*; *Styrax japonica*; *Styrax obassia, Styrax wilsonii*; *Kalmia latifolia*; *Halesia monticola*; *Stewartia pseudocamellia*; *Oxydendrum arboreum*; *Fothergilla monticola*; *Drimys winteri*; *Decaisnea fargesii*; *Cercidiphyllum japonicum*; *Hydrangea quercifolia*; *Cornus nuttallii*; *Cornus florida*; *Cornus kousa*; *Cornus kousa chinensis*; *Acer davidii*; *Acer capillipes*; *Acer carpinifolia*; *Acer griseum*; *Pieris japonica*; *Enkianthus campanulatus*; *Cladrastis lutea*; *Parrotia persica*; *Telopea truncata*; *Myrtus luma*; *Hamamelis mollis*; *Euonymus latifolius*; *Stachyrus praecox*; *Mahonia lomarifolia*; *Camellia* species and hybrids.

Exotic, Herbaceous (and very low Arboreal) Species used in Full Light or Dappled Shade

Moist or boggy conditions or alongside water: *Primula florindae* (Himalayan cowslip), *Primula denticulata* (drumstick primula), *Primula denticulata forma alba* (white drumstick primula), candelabra primulas (including *Primula aurantiaca, Primula beesiana, Primula belodoxa, Primula japonica, Primula pulverulenta* and *Primula bulleyana*), *Lysichitum americanus* (yellow skunk cabbage), *Matteuccia struthiopteris* (ostrich fern).

In acidic leaf-litter/organic soils: *Meconopsis betonicifolia* (Himalayan blue poppy), *Meconopsis grandis, Meconopsis integrifolia, Meconopsis nepaulensis, Meconopsis paniculata, Meconopsis cambrica, Cornus canadensis, Vaccineum vitis-idea, Pernettya mucronata, Cornus seucica, Linnaea borealis.*

and placed on south-facing patios close to the house for maximum floral effect can also have premature bud drop for the same reasons. Furthermore, the warm aspect causes the potting medium to dry quickly in the summer months, drastically reducing the chances of the plant remaining in good health and producing flower buds in the first place, and certainly causing premature bud drop if they have formed previously. Whether any orientation is with or without shelter, the 'right place for the right plant' is the only real principle.

The Establishment of a Fernery

Victorian and Edwardian interest in ferns was very high, and a good level of interest continues today. Ironically, the inordinate interest in ferns shown by previous generations, and their subsequent quest for live specimens, led to the extinction of many species by over-collection.

Ferns (Pteridophytes) have well developed vascular tissues, and are far easier to display than the other lower order plants. Allowing for light factor requirements, specific pH ranges/soil types, and moisture levels, ferneries can be very successful botanic and aesthetic features, and probably engender the most interest out of all of the lower order organisms.

For shade-loving and woodland species, ferneries can be constructed in dappled (or denser) shade, and made of rotten (or potentially rotten) logs infiltrated by organic soils or leaf mould. Naturally shady sites (including level beds) that preclude the successful cultivation of other plants can be used to advantage by planting ferns (that will thrive). Ferneries may also be created in amongst a woodland matrix, either as a specific feature or as a significant part of a woodland garden, by capitalizing on the various light intensities. Rocks, in a stable and natural-looking construction, infiltrated with leaf litter-infested soils (not unlike a woodland rock garden) also make a successful fernery if orientated correctly. Furthermore, this type of construction gives the chance for some aspects to be in full sun, and even for pH adjustments by the use of alkaline rocks for species such as hart's tongue fern; however, acidic rocks must be used for the bulk of the species.

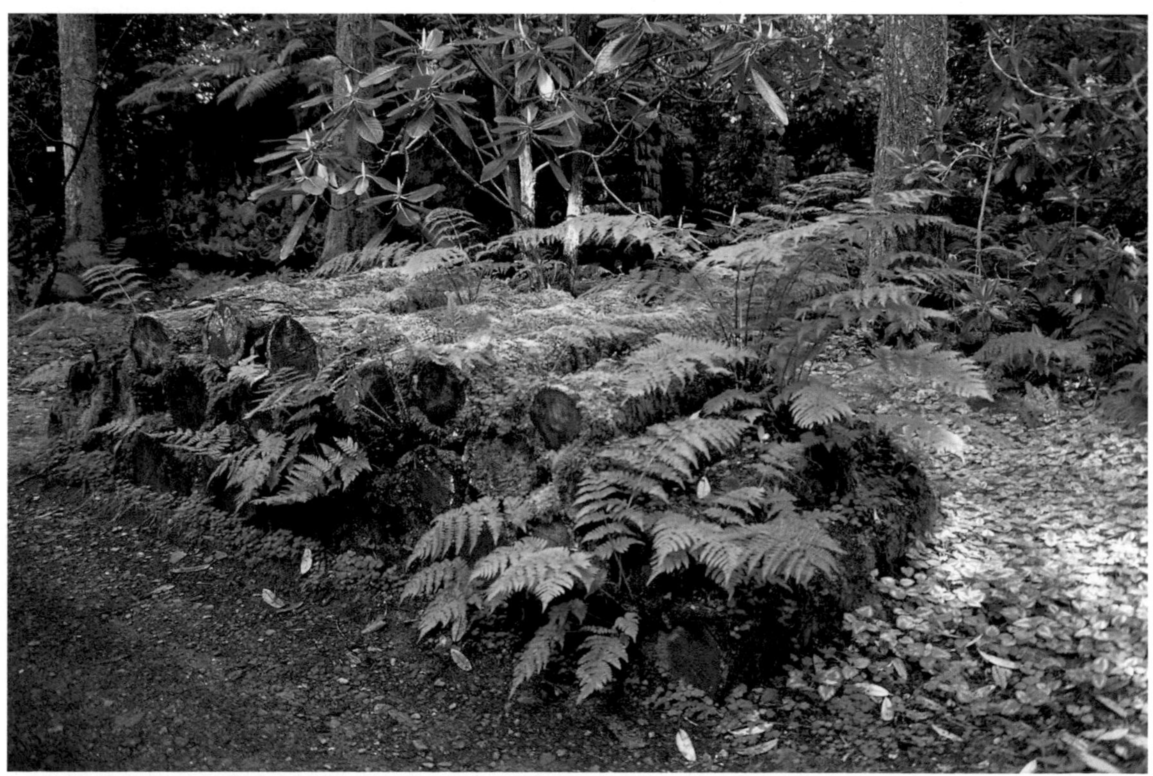

Ferns growing on rotten logs.

Forest Gardens

The forest garden is another development possibility, and based on the same basic tenets as the woodland garden, so may sit happily alongside and blend seamlessly with it. They both use the various woodland layers to protect and nurture the chosen species planted, though again, choice of planting positions (shelter versus available light) is critical.

The emulation of the naturally occurring tiers/layers of plants found in woodland includes the use of both woodland tree species and tree fruits (such as apples, pears and cherries) to form some of the basic structure (medium to high canopy). Soft fruits (bush fruits) such as gooseberry, blackcurrant, redcurrant and whitecurrant are used amongst other native and/or exotic species to form the shrub layer. Walnut and hazel (or cob nut) can also be used within their respective layer/level, as can cane fruits such as raspberry and loganberry, and woody climbers such as blackberry. Common crab apple (*Malus sylvestris*) is naturally a tree of light woodlands (*sylvatica* = of the woods).

Woody species, softer herbaceous types and leafy crops all need conditions with sufficient light to initiate, produce and develop their fruit in order to be successful. Even natural woodland species such as hazel (*Corylus avellana*) need some degree of sunlight both for photosynthesis and to generate warmth in order to bear good fruit. The North American blueberry (*Vaccineum corymbosum*), which attains heights of 1–1.8 m needs very good sunlight for successful cropping, and is equally at home in open cultivated plantations as it is amongst other woody subjects in a forest garden, as long as the pH is correct (acid soil). Its smaller counterpart, the bilberry (*Vaccineum myrtillus*), is less at home if it is located out of the dappled shade of open woodland and woodland edges. Loganberries, raspberries and blackberries may all grow in partial shade, but they will not fruit well without moisture, nutrition and sunlight at critical times – and certainly will not ripen well without them.

Aesthetic appearances have some importance in forest garden systems, but the main aim is to produce edible food crops in a sustainable, continuous system, so incorrectly positioned plants that cannot function in this way, defeat the object of the exercise. Some crops serve both food and aesthetic appeal well: rhubarb, for instance, is both food and ornament; although Swiss chard purports to serve both, it is actually weighed heavily in favour of aesthetic appearance over food value. Apples, pears, plums and cherries definitely fulfil both aims, and are actually vastly underestimated for both their attractive looks and their cropping value (including their storage and wine-making properties). Apple, pear and cherry can have relatively large crops and can be very successful if their canopies are in positions of good light.

The fruits of herbaceous plants such as strawberry, and other perennial, biennial and annual food crops such as spinach, celery, carrot and parsley, all need good light, so can be used in woodland clearings and at woodland edges. Chicory and angelica, which are both biennials, happily grow in amongst other light herbage at clearing edges. However, being able to grow certain species, and actually having a use or need for them, is another thing.

Technically, growing any form of herbaceous plants in the forest floor should be possible, but they should at least blend into the forest floor level in order not to look incongruous. Growing brassicas for instance, not only takes plants such as cabbage away from its coastline heritage, but they also look incongruous because of their large leaf-surface area, which is not really emulated by any naturally inhabiting temperate forest plant. Growing cabbage, cauliflower and Brussels sprouts in forest gardens brings with it other attendant problems, such as suitable pH requirements in a naturally acid or neutral soil condition. People are as reluctant to alter the natural pH to suit one particular crop type (because of the effect it might have on the local ecology), as they are to make their inclusion look like an allotment or raised flowerbed, which would be totally unacceptable in a semi-natural situation.

Hence, realistically, production is best left to woody perennial species, hedgerow species, forest floor species and forest-related species in the main. The best philosophy is probably one of diversity of crops with relatively small output from each, rather than going for quantity – bulk production is not easily attainable. Obviously sloes (*Prunus spinosa*), elderberry (*Sambucus nigra*), damson (*Prunus* domestica) and myrobalan or cherry plum (*Prunus cerasifera*), although having limited use (mainly preserve and wine-making), could all be used successfully because of their woodland pedigree. The more organically shaped the design and the more ecologically considered (wise use of the various layers), the better the results will be both from the aesthetic point of view, and as regards cropping.

The Management of Individual Trees

Garden managers are seldom experts in arboricultural practice, yet they usually inherit trees as part of the plant collection. Even managers of arboreta may not be practising arborists. Arboriculture is a specialized area, and tree surgery should be left to those with the necessary skills and safe practices: it therefore often involves the use of arboricultural contractors. However, because of the presence of trees in nearly all horticultural/botanical collections, garden managers should at least know how to recognize potential and actual problems. Knowledge of the most common signs of structural and biological breakdown can only be good (refer to *Trees: Their Use, Management, Cultivation and Biology* by the same author).

Bibliography and Recommended Further Reading

Abercrombie, M., Hickman, C.J. and Johnson, M.L. (1984) *Dictionary of Biology* Penguin

Allaby, Michael (ed.) (1992) *Concise Oxford Dictionary of Botany* Oxford University Press

Beales, Amanda *Roses, a Colour Guide* Crowood Press

Blamey, M., Fitter, R., Fitter, A. (2003) *Wild Flowers of Britain and Ireland* A & C Black

Brickell, C. (ed.) (2006) *Encyclopedia of Plants and Flowers* Revised and updated Royal Horticultural Society; Dorling Kindersley

Brown, Stewart *Sports Turf and Amenity Grassland Management* Crowood Press

Cox, Kenneth *Rhododendrons and Azaleas* Crowood Press

Crowder, Chris and Ashworth, Michaeljon *Topiary* Crowood Press

Davies, Paul and Gibbons, Bob (1993) *Field Guide to Wild Flowers of Southern Europe* Crowood Press

Davis, Bryan, Eagle, David and Finney, Brian *Resource Management – Soil* Crowood Press

Gibbons, Bob *Field Guide to Insects of Britain and Northern Europe* Crowood Press

Hickey, M. and King, C. (2000) *Cambridge Illustrated Glossary of Botanical Terms* Cambridge University Press

Hillier (1991) *The Hillier Manual of Trees and Shrubs* Hillier

Maclean, Murray *Resource Management – Hedges* Crowood Press

Maclean, Murray *Resource Management – Hedges and Hedgelaying* Crowood Press

Marshall, Keith *Orchids, a Guide to Cultivation* Crowood Press

McGavin, George C. (1992) *Insects of the Northern Hemisphere* Dragon's World (RSNC)

McMillan-Browse, P.D.A. (1985) *Hardy Woody Plants from Seed* Grower Books

Payne, Graham *Garden Plants for Mediterranean Climates* Crowood Press

Phillips, Roger and Rix, Martyn (1989) *Shrubs* Pan

Phillips Roger and Rix Martyn (2002) The Botanical Garden Vol. 1 Trees and Shrubs. MacMillan

Philips Roger and Rix Martyn (2002) The Botanical Garden Vol. 2 Perennials and Annuals. MacMillan

Rice, G. (ed.) (2006) *Royal Horticultural Society Encyclopedia of Perennials* Dorling Kindersley

Simon, E.W., Dormer, K.J. and Harstshorne, J.N. (1973) *Lowson's Botany* University Tutorial Press

Stearn, William T. (1992) *Stearn's Dictionary of Plant Names for Gardeners* Cassell

Watson, Bob (2006) *Trees: Their Use, Management, Cultivation and Biology* Crowood Press

Weir, E.T., Stocking, R.C. and Barbour, M.G. (1974) *Botany: An Introduction to Plant Biology* John Wiley and Sons

Welch, Charles *Breeding New Plants and Flowers* Crowood Press

Glossary of Terms and Concepts

Abscission The precursor to leaf fall, triggered by decreasing temperatures and decreasing day length. Cork is laid down in between the leaf petiole (or rachis) and the stem that cuts off the water supply to the leaf. The cork seals the vascular tracts and prevents the plant 'bleeding' to death (haemorrhaging cell sap and other liquids).

Absorptive roots The fine, fibrous roots at the periphery of the root-plate.

Achene The botanic name for nuts, i.e. fruits with a hard lignified shell on the outside such as acorns, and the hard central part of succulent fruits such as cherry, peach, apricot and plum. The hard nuts attached to wings as found singly in ash and elm species, and found fused together in pairs in maples, are also achenes.

Actinomorphic Used to describe regular flowers, i.e. flowers that may be divided into two equal halves when a line is drawn anywhere through their centre.

Adventitious Appearing where it would not normally be expected. Examples include the adventitious roots on the stems of common ivy, and suckers (adventitious shoots) that arise from the root-plate of some tree and shrub species, i.e. roots on shoots and shoots on roots, instead of the norm of shoots on shoots and roots on roots.

Adventitives Plants found in places where they would not normally be expected, e.g. cannabis (*Cannabis sativum*) found on waste ground, out of its normal indigenous range.

Aggregate fruit (etaerio) A fruit typified by raspberry and blackberry, which comprises a collection of drupes held in an aggregate mass. A drupe is a pulpy, succulent fruit whose flesh surrounds a single-seeded nut, or a collection of dry nut-like fruits (achenes) as found in clematis.

Aggregate lumps (or crumbs) Microaggregates (i.e. small aggregate lumps) bind into larger units (macroaggregates) formed when mineral chips are stuck together by clays and humus. The large pore spaces (macropores) that are created when aggregate lumps bridge one another, gives rise to soil structure.

Allelopathy The phenomenon whereby some species exude materials from their roots that are toxic to some other species, and prevent or inhibit the establishment of the same (or other) species near them.

Anaerobic Without oxygen. Anaerobic conditions prevent respiration and cause tissue death. Soil compaction causes anaerobic conditions and root death.

Anemophilous 'Wind liking' – a reference to wind-pollinated flowers.

Androecium (andrecium) A collective term for the male organs of angiosperm flowers, i.e. the stamen.

Angiosperm One of the main plant groups (Angiospermae) comprising the flowering plants with enclosed seeds (enclosed in ovaries). The Angiospermae is further divided into the Monocotyledonae (having one seed leaf), and Dicotyledonae (having two seed leaves). See also Gymnosperms.

Anther The pollen-bearing organ of the stamen attached to the stalk-like filament on an angiosperm flower.

Anthocyanins A group of pigments (glycosides) occurring in cell sap that present in flowers as pinks, reds or blues, depending on the pH.

Anticlinal Describing cells with cell walls being laid down at right angles to the root: shoot axis.

Apical dominance Having a definite, distinct 'leader' (leading shoot), i.e. the apical shoot has dominance over others.

Apical meristem The meristematic tissues (responsible for rapid cell division) found at shoot apices – *see* **Meristems**.

Apomictic (not mixing) The propensity for some species to self-fertilize prior to the corolla opening and therefore avoiding hybridization.

Apoplast (apoplastic system) The interconnection of dead cells and cellular spaces.

Arborescent plants Referring to non-woody plants that attain tree-like proportions. They include a few monocotyledons including true palms, some palm-like plants such as cordylines, and some fern species (Pteridophytes), and even some gymnosperms called cycads (distant relatives of the conifers). They are not technically arboreal as they do not share all of the features of other woody plants, and they gain height only very slowly (a set of leaf bases per year) facilitated by a central growth bud. Palms, ferns and cycads do not increase in stem thickness by secondary tissues, and are not covered in bark – considered one of the discerning features of woody plants. Cordylines and dracaenas often have a bark-covered exterior, but they increase in girth annually only very slowly.

Arboriculture The study, culture and management of woody (arboreal) plants (trees and shrubs), including the structural assessment, disease diagnosis and maintenance of amenity trees and woody plants in rural and urban areas.

Armillaria mellea A species of fungus with honey-coloured, mushroom-type fruiting bodies responsible for a root and butt rot on trees. It is the honey-coloured fruiting bodies that give rise to the common name 'honey fungus'. The species also has rhizomorphs (aggregated mycelia in a rod-like shape) that travel below the soil like a root and can infect the next host tree. The rhizomorphs commence creamy white or brown in the soil, but turn black on entering the tree, and can be seen below the bark on infected trees as black, shoe string-like structures that give rise to their other common name of 'bootlace fungus'.

Asexual (vegetative) Methods of propagation: other than by seed, using pieces of plant material, e.g. cuttings, budding, grafting, layering and micropropagation.

Asteraceae (formerly Compositae) A large family of plants that have composite flowers, i.e. flower heads comprising a collection of very small flowers known as florets, some of which are petal-like in nature (ray florets) and some that are cone-shaped with no petal-like structure (disc florets). Examples of composite flowers include the following: daisies, dandelions, dahlias, chrysanthemums.

Autumn colour The aesthetic effect given by many deciduous woody species in the autumn. During the summer healthy chlorophyll (the green pigment responsible for photosynthesis) masks the other pigments that are naturally found in the membranous leaves of deciduous subjects, i.e. the yellows and oranges (carotenes and xanthophylls), and the reds and purples (anthocyanins or xanthonins). When the abscission layer of cork is laid down in the autumn, between the petiole and the stem, to plug the vascular tracts, the water supply is cut off to the leaf. The result is the relatively quick death of chlorophyll, but not the more persistent yellow and red pigments.

Auxins A group of hormones found naturally in the plant responsible for plant growth and root promotion. They may also be manufactured artificially and used to aid the rooting of cuttings.

Available water The water held between field capacity and permanent wilting point.

Axil The angle formed by the leaf petiole (leaf stalk) and the stem.

Axillary buds Buds in the angle between the stem and the leaf petiole (the axil).

Bark The outer covering of woody plants is known as the periderm (as opposed to the green epidermis of the primary growth). The periderm from the outside tissues inwards comprises phellum (cork), phellogen (cork cambium) and phelloderm (cork cortex). Bark comprises four tissues, as it includes old phloem tissues as well. The exterior cork cells are covered with suberin (a water-proofing wax). The outer bark consists of dead cork cells (phellum), which will slough off at some time, but may be retained for a time in plates.

Bark sloughing Bark sloughing is the natural process of losing bark, carried out by healthy trees. Sloughing can occur regularly (annually, or over longer periods of time). Species that illustrate it well include peeling-bark cherry (*Prunus serrula*) and Erman's birch (*Betula ermanii*). Some species adopt a strategy of retaining large persistent plates of bark with wide gaps in between them. In all instances a new ring of bark is produced by the phellogen (cork cambium) forming a periderm underneath the old bark.

Berries A type of fruit comprising a pulpy, succulent flesh enclosing several seeds. *See also* **Drupe**.

Bifurcation The process of dividing into two (as in roots, leaves or stems dividing into two) – the epiphytic fern *Platycerium bifurcatum* leaves split into two distinct lobes.

Binomial system The system of nomenclature (naming) used on an everyday basis by foresters, arborists and horticulturists that involves only two names out of the main classification system, i.e. the genus and the species (generic and specific names), e.g. *Clianthus puniceus*.

Biological control A method of insect, fungus or other pathogen control by the use of another organism (natural or introduced) to parasitize or eat the damaging organism.

Bolting A plant's premature production of flowers and seeds.

Bracts A form of cataphyll. Leaf-like appendages that can take on the role of petals as in poinsettia (*Euphorbia pulcherrima*), *Cornus florida*, *Cornus kousa*, and *Davidia involucrata* (handkerchief tree).

Branch bark ridge An interference pattern in the bark tissues caused where the cylinder of bark of the main stem meets the cylinder of bark of a lateral branch, as they meet at two different angles.

Branch collar A swelling on the main stem where a lateral branch meets it. The branch collar contains healthy main stem vascular cambium that is continuous with the higher tissues, and it should never be breached when removing a lateral branch of any diameter.

Bud blast A fungal disease of *Rhododendron* species, typified by black rods protruding from the flower buds.

Bud break The term used to describe the process when leaves first start to force bud scales open – the original growth and development of leaves when they break through the bud scales in spring.

Buds Embryo shoots and/or developing flowers. **Vegetative buds** produce new stems and associated leaves, **flower buds** produce flowers, **mixed buds** comprise both vegetative shoots and flower buds. **Apical buds** are situated at the apex of a shoot. If the apical bud contains flowers and stops the extension growth of the shoot it is known as a **terminal bud**. **Axillary buds** are situated in the leaf axil (i.e. the angle between the leaf petiole and the stem). Vegetative buds (and mixed buds) are protected by exterior, wax-covered **bud scales**, whereas individual flower buds are protected by an external cover created by the calyx (a collection of tough, leathery, persistent sepals). Buds can be arranged in pairs (opposite), or singly (alternate), or in multiple numbers encircling the stem (whorled).

'Bagging' The process of placing a muslin bag over a flower to prevent the ingress of alien pollen.

Bud scales (scale leaves) A form of cataphyll. Scale-like, modified leaves, usually covered in waxes (e.g. suberin) that form a protective layer covering vegetative and mixed buds.

Callus Tissue comprising soft, undifferentiated, cellulose-walled cells without orientation. Callus can differentiate into new roots, shoots, or wound wood, depending on the environment in which it is produced. Callus is produced by the vascular cambium at wounds, including prepared cuttings.

Calyx The protective outer tissues of a flower, comprising tough, persistent sepals.

CAM plants Those plants relying on crassulaic acid metabolism because their stoma shut down during the day in their natural hot desert environment.

Cambium (plural cambia) A specialized meristematic area (area of rapid cell division), comprising sheets of brick-like cells that are able to divide in several directions, e.g. the vascular cambium, which eventually forms a cylinder of meristematic tissue in woody stems.

Canker Symptoms of fungal and bacterial pathogens usually expressed as swellings and lesions on the stem. Target cankers have a distinct form – they comprise concentric rings of dead tissues that form a shape like a target.

Carotenoides Plant pigments including the carotenes and the xanthophylls responsible for yellow, orange and red colours and found in chloroplasts and plastids.

Carpel Each carpel comprises a stigma (the receptive end), style (the stalk-like attachment), and ovary (the compartment with ovules inside). The gynaecium (female part of the flower) comprises the pistil with one single, or several carpels.

Casparian strip A group of cells in the root that are heavily waterproofed with suberin, and prevent incoming water from the roots invading phloem tissues as it tracks laterally across the cortex.

Cataphylls Specialized non-foliage leaves including bracts, scale leaves and stipules.

Cell maturation The process of cell maturation commences after initial cell division, and has three main phases: elongation (extension in length), vacuolation (forming a vacuole full of cell sap), and differentiation (changes in the cell to facilitate specialized functions).

Chelated *See* **Fritted trace elements**.

Chloroplasts Small units of chlorophyll.

Chlorophyll The green pigment found in plant leaves (and green stems) responsible for photosynthesis (sugar production).

Chlorosis Yellowing of the plant tissues because of poorly functioning and/or dying chlorophyll caused by nutrient deficiencies.

Cladode (phylloclade) A stem that looks like a leaf. Flattened forms are found in *Ruscus aculeatus*, and swollen, water-storing forms are found in some cacti.

Classification The placing of plants into their position in the main taxonomic hierarchy.

Clone A group of plants that have identical genetic make-up. Propagating plants derived from the same mother plant creates clones.

Coleoptera The group of insects with covered wings, i.e. having leathery wing cases (elytra), e.g. beetles and weevils. Coleo = covered and ptera = wings.

Collenchyma Tissue comprising cells with an extra cellulose layer in the cell walls, but no lignin.

Complete flowers Flowers with all four whorls of modified leaves (calyx, corolla, androecium and gynoecium) present.

Compositae *See* **Asteraceae**

Compound leaves Leaves having a rachis (an extension of the petiole) with finger-like leaflets arranged in the shape of a palm (compound palmate), or with leaflets positioned on each side of a central rachis (pinnate), or, if this pattern is repeated again, doubly pinnate (bipinnate).

Container-grown Plants that have been grown in containers throughout their development.

Containerized Plants that have been 'lifted' from the soil and placed into a container, not grown in containers throughout their development.

Coppice The process of cutting a woody stem hard back to produce a stump that regenerates via the vascular cambium with vigorous coppice shoots (induced epicormic shoots).

Core : skin hypothesis A hypothesis describing the growth of woody plants postulating that the 'core' of internal tissues is overlain by a 'skin' of external tissues annually.

Cork (phellum) The strong, resilient material, waterproofed by suberin, that makes up external bark tissues.

Cork cambium *See* **Phellogen**

Corolla The collective name for the structure on a flower made up of individual petals or comprising a tube made up of fused petals.

Corona The trumpet-like tube that forms part of the flowers of *Narcissus*.

Corona filaments The brightly coloured hair-like appendages found on the flowers of passion flower and some clematis.

Corpus The tissue at a stem apex responsible for the production of the main body of the plant. *See also* **Tunica**.

Cortex Packing tissue found in various parts of the plant and sometimes bearing a specific name because of its position or function. Mesophyll is a specific type of cortex forming the middle tissues of certain leaves.

Cotyledon Seed leaf found in seed embryos.

Crossing over The process of gene interchange that allows for variation.

Crumbs/ soil Crumbs (or aggregate lumps) *See* **Aggregate lumps.**

Cuticle The outer layer of wax (cutin) found on green stems and leaves.

Cutin One of the three important waxes: cutin, found as an external coating on green stems and leaves; lignin, found lining the walls of some specialized cells; and suberin, found in and on bark, and on bud scales.

Cymose arrangements A flat-topped or almost pyramidal inflorescence with the youngest flowers at the base and the oldest flowers at the tip.

Deciduous Deciduous trees and shrubs lose their leaves in the autumn. They are typified by possessing thin, membranous leaves that drop (after abscission) in the autumn – often accompanied by an autumn colour. Provision for extension growth for the following season is given by dormant buds with thick, leathery bud scales.

Deciduous conifers Most conifers are evergreen. However, there are six genera of deciduous conifers: *Larix*, with two species, *Larix decidua* and *Larix kaempferi*; all the others are monotypic, i.e. represented by one species only within the genus – *Pseudolarix amabilis* (false or golden larch), *Ginkgo biloba* (maidenhair tree), *Metasequoia glyptosroboides* (dawn redwood), *Taxodium distichum* (swamp cypress), and *Glyptostrobus pensilis*.

Decomposition The breakdown of organic material into its mineral constituents (mineralization); or partial decomposition, the formation of humus.

Detritivores Organisms that live off dead organic material.

Dichogamy When the two sexual organs mature at two different times: protandry is when the stamen matures first, and in protogyny it is the stigma first.

Dicotyledonous plants Plants that have two seed leaves in their embryo. Plants with only one seed in the embryo are termed 'monocotyledonous'.

Differentiation The process of cells changing from soft parenchymatous to cells with specific functions. Differentiation is the final phase of cell maturation that involves elongation, vacuolation and then differentiation.

Diffusion Molecular movement of water driven by a concentration gradient responsible for the transport of water between cells; very slowly and over short distances.

Dimorphism Having two different flower forms in the same species of plant – relevant to pollination because the sexual organs are at different levels within the different types.

Dioecious Dioecious arrangements have mono-sexed (single-sexed) flowers, with male flowers on one individual plant, and single-sexed female flowers on another individual plant, e.g. *Viburnum davidii*, sea buckthorn (*Hippophae rhamnoides*), and common holly (*Ilex aquifolium*).

Diptera The group of insects with two wings, i.e. the two-winged flies (di = two, and ptera = wings).

Dorsi-ventral Leaves with a distinct upper (dorsal) and lower (ventral) surface. *See also* **Iso-bilateral**.

Droseraceae The family containing the group of insectivorous plants known as the sundews, e.g. *Drosera rotundifolia*

Drupe A type of fruit resembling a berry with a pulpy, succulent flesh surrounding a one-seeded nut (achene).

Dysfunctional tissues Tissues that no longer function because of age, and/or breakage by compression. A common example is the build-up of old xylem tissue in the centre of an old woody stem.

Ecology The study of plant and animal communities and their interactions.

Elongation The first process of cell maturation when cells grow in length.

Emasculate The process of removing the stamen of a flower either to retrieve pollen, or to remove the possibility of pollution by unwanted pollen in the plant breeding process.

Endemic Naturally found in an area. **Non-endemic** Not naturally found in the area – unusual or not normal in the area.

Endodermis A layer of cells surrounding the stele of the root. The endodermis comprises some cells that are suberized in strips or bands (casparian strips or bands) to waterproof them, and some that remain unaltered by waxes (passage cells), thereby channelling water (tracking laterally across the cortex) to ensure it reaches the xylem correctly.

Entomophilous Insect-liking, a reference to insect-pollinated flowers

Epicormic shoots Young, vigorous shoots arising from the external tissues of a stem (epi = outside of, cormic = stem). Epicormic shoots are usually induced if a branch is removed or is broken off and the light factor changes (sprouts), or if a woody plant is coppiced or pollarded.

Epidermal hairs Protective hairs comprising one single, specialized, elongated epidermal cell found on the epidermis of stems and leaves of some plants. Absorptive root hairs are also epidermal hairs.

Epidermis The outer skin-like tissues found on stems, roots and leaves.

Epigeal A form of germination where the seed leaves appear above soil level (epi = outside, and geal = germination).

Epiphytes Plants that attach themselves to the outside tissues of a host tree and grow using the external tissues of the host as anchorage only, i.e. without penetrating the tissues of the host and without taking nutrition from it (epi = outside, phyte = plant).

Etaerio An aggregate fruit such as the collection of dry achenes in clematis, and the collection of succulent drupelets found in blackberry and raspberry.

Etiolation The effect on a plant when poor light conditions cause long internodes, low or no chlorophyll content, and small leaves.

Ethnobotany Describing the study of the various uses of plant materials and products by ethnic/indigenous peoples, e.g. for food, medicine, construction.

Evergreen Evergreen species retain the current seasons' leaves, and the leaves of the previous season as a minimum. The third whorl of leaves may be retained in good environmental conditions, and in extreme cases even a fourth whorl of leaves may be retained. The norm is years one, two, and perhaps three. Older leaves are usually aborted, as they become shaded, dysfunctional, or less efficient.

Exoskeleton The outside chitinized casing of insects, i.e. their external skeleton (exo = external to).

False fruit *See* Pseudocarp.

Family A large taxonomic division of plants comprising a collection of genera having many similarities, yet some dissimilarities, e.g. Rosaceae (comprising the genera *Rosa*, *Sorbus*, *Prunus*, *Cotoneaster* and so on) and Papilionaceae – formerly Leguminosae (comprising the genera *Wisteria*, *Laburnum*, *Lathyrus*, *Cytisus*, *Piptanthus* and so on).

Fertilization The fusing together of male gametes (found in pollen) and female gametes (found in the egg nucleus) to form a zygote (the precursor to an embryo), comprising both male and female genetic information.

Fibres Long, tough, strong, fully lignified tissues that give flexibility and strength.

Fibrous roots The finely divided, water-absorbing roots found at the periphery of the root-plate.

Filament The stalk-like part of the stamen (male sexual organ of a flower) that holds the pollen-bearing organs (anthers).

Field capacity The point where soil holds the maximum amount of water against gravity, i.e. when all free water has drained from the macropores, and the maximum amount of water is held against gravity by the micropores and mesopores.

Fireblight Fireblight (*Erwinia amylovora*) is a bacterial disease that will kill many plants in the Rosaceae. *Crataegus* (hawthorn), *Pyracantha* (firethorn), *Malus* (apple), *Pyrus* (pear), *Cotoneaster* and *Sorbus* are all susceptible. However, cherry and rose are not susceptible. The main symptoms are shoot die-back, brown, leathery, persistent leaves, and mummified flower trusses. Entry of the bacteria is via the flowers (unlike bacterial canker of cherry, where entry is via the leaves).

Floret Small flowers found in congregations of two to three in grasses. Also used by some to describe individual flowers in the composite flower heads of plants in the Asteraceae.

Flower The organ of a plant responsible for sexual reproduction.

Frass (bore meal) The material excavated by boring insects when they enter wood or other tough tissues and create a bore hole (the pattern of frass can be a distinguishing feature of some species of insect).

Friability The ease or otherwise that a soil will 'work'.

Fritted trace elements (chelates) Forms of fertilizer added to the soil that allow nutrients to become soluble (and therefore available to the plant) even in highly alkaline conditions.

Fruits Fruits are the product of fertilization of ovules within a carpel. The main types are dry, nut-like fruits such as achenes (nuts) or acorns and hazel nuts; dry-winged fruits (a nut with a samara, or wing), e.g. ash keys and maple schizocarps; legumes (simple fruits comprising one carpel in the shape of a pod), e.g. laburnum, sweet pea and gorse. Succulent fruits such as berries (succulent, fleshy fruits with several seeds enclosed), e.g. rowan berries; drupes (succulent fruits with one nut inside), e.g. cherry and plum; aggregate fruits such as raspberries and blackberries (collections of drupes held on a swollen receptacle); and false fruits (pseudocarps) such as apple, with a swollen receptacle enclosing the carpel.

Fruticose Describing a shrubby habit.

Fungi A group of organisms, originally included in the plant kingdom, but in a division noted for having no chlorophyll, and now, because they neither flower or have chlorophyll, in a kingdom of their own.

Geniculate Knee-like bends in the stems of some grasses, leaving the posterior part of the stem resting on the soil and the shorter anterior part erect.

Genus (plural genera) A taxonomic group of similar and related plants held in families.

Germ tube (pollen tube) The long tube that penetrates the stigma and style of a flower emanating from a germinating pollen grain.

Girdle scars The crescent-shaped scars found on woody stems formed from the tightly held bud scales encircling the stem. The distance between girdle scars, once laid down, is constant, and except in a few exceptional cases, represents one year's growth in length.

Girth increments Annual increases in the stem diameter of woody plants created by the addition of annual rings of secondary xylem tissues.

Gley (Glei) Water-logged and anaerobic soils, often with grey, or blue-grey mottling created by reduced iron compounds (ferrous) Fe 2+.

Grex The progeny derived from the result of hybridization. Each individual seedling resulting from hybridization is known as a grex.

Gymnosperms A group of related plants with 'naked seeds', i.e. seeds born on scales, as found in the cones of conifers and the strobili of cycads.

Gynecium (gynaecium or gynoecium) The collective term for the female organs of the angiosperm flower. Gynaecia comprise pistils, and each pistil may comprise only one, or several carpels, that can be free-standing or fused together.

Heliotropism Parts of a plant (usually the flowers) either (positively) attracted towards or (negatively) away from the sun.

Hemiparasite (semi-parasite) A plant that appears to be parasitic on a tree host because it punctures tissues and takes moisture and some nutrition, but has its own photosynthetic tissues which can fulfil all or part of its energy needs, e.g. *Viscum album* (mistletoe).

Hemiptera Order of insects including aphids and scales.

Hermaphrodite Bearing both sexes. Hermaphrodite (perfect) flowers have both male and female organs on the same flower. Examples include *Iris*, apple (*Malus*) and *Camellia*.

Holoparasite A species that is wholly parasitic on its host as it has no chlorophyll, e.g. toothwort.

Homogamy Where both male and female organs of a hermaphrodite flower mature at the same time, therefore an essential precursor to natural self-pollination.

Honeydew A sugary exudate from aphids produced as they feed on cell sap. Honeydew is commonly covered by a black fungus (a pin-mould called sooty mould).

Honey guides A series of dark lines or dots on the petals that guide insects to the nectaries.

Humification The process of forming humic material (humus) from the partial decomposition of organic matter. Full decomposition leads to mineralization, i.e. breaking organic matter down into its constituent parts.

Humus A black/brown, jelly-like material with adhesive and cohesive properties, formed from the partial breakdown of organic matter, and along with clays, forms the binding materials of soil aggregate lumps. Humus binds mineral particles together and helps increase aggregate lump size.

Hymenoptera The group of insects with very membranous wings, e.g. wasps and sawflies (hymen = membranous, and ptera = wings).

Hypogeal A form of germination where the seed leaves remain below the soil level (hypo = below).

Indumentum Congregations of thick, often white, cream, pale or dark brown felty hairs on the underside of leaves that reduce water loss.

Inflorescence The arrangement of flowers upon a stem. *See* **Racemose** and **Cymose**.

Imperfect flowers Mono-sexed flowers with only male or female organs, not both.

Incomplete flowers Flowers with one or more of the four main whorls (calyx, corolla, androecium or gynoecium) missing.

Inferior ovary The ovary of a flower positioned behind the point of attachment of the other floral parts.

Interfasicular cambium Tissue destined to be vascular cambium, and positioned between the existing vascular bundles, which will ultimately produce new bundles.

Intrafasicular cambium The vascular cambium, within a vascular bundle, situated between the xylem and phloem.

Invertebrates All animals without backbones, including insects that have instead a strong chitin exoskeleton.

***In vitro* culture** Systems of aseptic micropropagation that are traditionally carried out using glass containers (*in-vitro* = in glass), although mostly using plastic containers now. *See also* **Meristem culture**.

Isobilateral Leaves with two approximately equal sides, illuminated on both sides and with no obvious upper and lower surface.

Larva (plural: Larvae) Active, voracious eating phase of insects. Larvae include caterpillars (Lepidoptera – butterflies and moths), grubs (Coleoptera – beetles and weevils) and maggots (Diptera – flies).

Leaching The process of nutrient loss from the soil when soluble nutrient salts are moved downwards in the drainage water by gravity, and carried away from the root zone of the plant.

Leaf break (bud-burst or leafing out) Terms used to describe the original growth and development of leaves when they break through the bud scales in spring.

Leaf scars The scar left by the leaf petiole (or leaf rachis) at abscission (leaf fall). The scar is the same shape as the base of the petiole/rachis, and contains corky dots where the vascular traces have been sealed with cork.

Leaves The photosynthetic organs of plants. Mesophytic leaves have an upper and lower epidermis that sandwich other tissues (the mesophyll), and are either dorsiventral (having a definite top and bottom) and are mainly illuminated on their top side (as in deciduous trees, broad-leaved weeds), or they are isobilateral (being relatively equal on both sides) as in grasses, cordylines, sisal. Isobilateral leaves are held vertically or nearly vertically, and are illuminated on both sides. Leaves may be simple – having an entire leaf lamina (blade) attached to the stem by a leaf stalk (leaf petiole); or compound – having a central rachis (an extension of the petiole) with leaflets positioned on each side.

Legume The simple pod produced by plants in the Papilionaceae family (formerly Leguminosae), e.g. Judas tree (*Cercis siliquastrum*), laburnum, robinea, sweet pea (*Lathyrus odorata*), everlasting pea (*Lathyrus latifolius*), broad bean (*Vicia fabia*).

Lentibulariaceae A family of plants that includes the insectivorous butterworts and bladderworts, e.g. common butterwort (*Pinguicula vulgaris*) and greater bladderwort (*Urticularia vulgaris*).

Lenticels Apertures in the bark of woody stems containing soft cells that allow the diffusion of gases. Lenticels are usually pale brown or white, and are loosely associated with parenchyma rays (medullary rays) that facilitate the lateral movement of gases across the tissues. Lenticels differ from stomata, not only because they are in bark (not a green epidermis), but also because they are not regulatory – the aperture remains the same once laid down, whereas stomata are regulated in their aperture size by guard cells.

Lepidoptera The group of insects with scales on their wings, e.g. moths and butterflies. Lepidote = bearing rounded (spot-like) scales.

Lichens Living organisms comprising a symbiotic relationship between a fungus species and an alga species. The two species do not live separately but only as the symbiotic organism known as a lichen. There are many species, but only three or four main types, the most common being crustose (or crustate) lichens with crystalline crusty form; foliose lichens where the alga look leaf-like; and fruticose lichens resembling small shrubs. Lichens appear as epiphytes on trees, doing no harm.

Lignification The process of lignin wax being laid down on the inside of cell walls, resulting in death, but not necessarily dysfunction, of the cells. Also, for the same reason, used as a generic term to describe the processes when woody plant tissues harden up prior to the onset of winter. Technically, lignification is only part of the process, and it is suberized cork on the outside of the tissues that is more obviously apparent externally during this process.

Lined out Terminology used to describe plants that are lifted from their place of propagation and planted in the field at wider spacing.

Longevity A term describing the life span of a seed or plant, i.e. how long it stays alive. Relevant to how long a seed may stay alive (viable) in storage, or prior to sowing.

Macrofauna The larger animals associated with soil formation.

Macropores The large pore spaces in the soil matrix, created by bridging aggregate lumps, which drain water easily by gravity.

Meripilus giganteus A serious, wood-rotting fungus, found at the base of trees or protruding through soil when attacking large, cork-covered roots. The fungus comprises masses of overlapping, pale brown/fawn, frond-like growths. Commonly found on beech, but can attack other species including American oaks and, more unusually, English oaks. *Meripilus giganteus* causes a root rot in the main corky roots, which renders the tree very unstable relatively quickly.

Meristems (meristematic areas) Areas of rapid cell division. Examples include apical meristems (at shoot tips and root tips), vascular cambia, phellogen. Apical meristems comprise an amorphous mass of dividing tissues that differentiate to form all the tissue patterns associated with roots and stems. The vascular cambium comprises sheets of brick-like cells that differentiate to produce new xylem on the inside and new phloem on the outside. Vascular cambium cells, released when a branch is removed, can differentiate into new stems (epicormic shoots), or new roots (if environmental conditions dictate), but they usually differentiate into wound wood. Cells produced at the cork cambium are not brick-like in shape, which leads some authorities to call this tissue the phellogen instead, as these differentiate to produce new cork (phellum) on the outside and new cork cortex (phelloderm) on the inside.

Meristem culture One of the many forms of aseptic micropropagation, but in this case specifically using excised meristematic tissue from the apical meristem area of a young shoot of the plant.

Mesofauna The medium-sized animals/organisms associated with soil formation that can be seen with the naked eye and include earthworms.

Mesophyll The middle layers of tissues found in most leaves (meso = middle and phyll = leaf). See Leaves.

Mesophytic leaves Leaves with an upper and lower epidermis that sandwiches the other tissues (the mesophyll). *See* **Leaves**.

Mesopores The medium-sized pores in the soil matrix that can release water to plant roots via osmosis.

Microfauna The very small and microscopic organisms associated with soil formation, including bacteria.

Micropores The very fine pore spaces within the aggregate lumps of soil that hold water against both gravity and osmosis.

Micropropagation A generic term for systems of propagation using small pieces of tissue (or even cells) in aseptic conditions.

Micropyle Small pore in the seed coat that imbibes water.

Mineralization The breakdown of organic matter into its constituent parts, and the release of mineral salts into the soil. The bacteria Nitrosomonas and Nitrobacter are responsible for mineralization.

Monocarpic Species that are normally perennial, but only flower once (when environmental conditions are correct), and then die, e.g. many species of bamboo and *agave americana*.

Monocotyledonous Plants with only one seed leaf in the embryo – as opposed to dicotyledonous plants that have two seed leaves in the embryo.

Monoecious A monoecious arrangement comprises mono-sexed (single-sexed) flowers, and having single-sexed male flowers and single-sexed female flowers on the same plant, e.g. most conifers, common hazel (*Corylus avellana*), and common alder (*Alnus glutinosa*). See also Dioecious.

Monopodial Describing growth patterns mainly in one direction, as found in rhizomes and the juvenile (excurrent) phases of trees.

Monotypic Referring to a genus with only one representative species within it. The genus *Ginkgo* has only the one species (*biloba*) within it – hence *Ginkgo* is a monotypic genus.

Multi-stems (multi-stemmed) A description of trees with three or more branches arising from near ground level.

Mycorrhiza The generic term for a group of soil-borne fungi that colonize roots. The infection creates a symbiotic (mutually beneficial) relationship. The fungi take sugars from the roots, and in return they increase the absorptive area of the roots, and process insoluble phosphatic compounds, converting them into soluble phosphates (and therefore making them available to the plant).

Nastic movements Movements of a plant that are not attracted towards or aggravated by a stimulus.

Nectar A sugary solution produced by many plants that attracts insects.

Nectaries Small reservoirs formed in the petals that hold nectar.

Nectar guides *See* **Honey guides**.

Node (nodal region) The swollen area where the leaf petiole (or rachis) meets the stem.

Nomenclature The system of naming plants and animals. On a day-to-day basis the binomial system is used, i.e. two names – the generic name and the specific epithet forming the specific name, e.g. *Acer pseudoplatanus* (common sycamore).

Nuts A form of dry fruit with only one seed inside and having a hard or leathery protective outer layer; includes acorns, hazel nuts etc. *See* **Achenes**.

Nymph An immature phase that some (not all) insects go through, which unlike the pupal stage, is very active.

Organic matter Organic matter is usually added to the top horizon by biological action. However, it may accumulate on the surface if communities of living organisms are not large enough to mix it. Hence there may be distinct organic litter on the surface, or the top layer may contain intimate humus mixed within it.

Osmosis The movement of water from a weak solution towards a more concentrated solution through a semi-permeable membrane.

Oxidation The reaction of minerals with oxygen. It is the concentration of oxidized iron in soils (ferric iron) that gives the reddish coloration to soil profiles.

Palisade mesophyll Middle tissues (just below the upper epidermis of mesophytic leaves). It is orientated vertically and contains lots of chlorophyll, thus is perfectly adapted for irradiation and photosynthesis.

Papilionaceae A family of plants (including trees) all with pod-like fruits and pea-like flowers, e.g. Judas tree (*Cercis siliquastrum*), laburnum, robinea, sweet pea (*Lathyrus odorata*), everlasting pea (*Lathyrus latifolius*) and broad bean (*Vicia fabia*).

Pappus The fruits of some species in the Asteraceae forming a parachute-like head of hair-like appendages with an achene attached dispersed by wind, e.g. dandelion and goatsbeard.

Parasites Organisms that penetrate and feed off a host plant. Parasitic species of plant, i.e. taking nutrition from the host plant, include toothwort, dodder and broomrape.

Parenchyma cells Live cells with soft, malleable cell walls, cytoplasm and cell inclusions including nuclei still capable of division.

Parenchyma rays (medullary rays) Radial sheets of live, soft, parenchyma cells that punctuate the tough xylem tissues and form part of the symplastic system. Their function is to allow lateral transport of liquids and gases (including water vapour, carbon dioxide and oxygen from the lenticels).

Parthenogenic (Parthenogenetic) Virgin birth, e.g. insects that have races of all female adults that can lay fertile eggs without the necessity for male fertilization, such as vine weevil.

Parthenocarpic fruits Fruits produced without fertilization.

Parthenocarpy The process of unfertilized fruit production.

Partial decomposition The formation of humus. Decomposition – the breakdown of organic material into its mineral constituents (mineralization).

Pedicels Lesser, small flower stems (sometimes radiating from the peduncle) and actually attached directly to the flower.

Peduncle A flower stem branching off the main flower stem (rachis).

Perfect flowers Flowers with both male and female organs (hermaphrodite).

Perianth A collective term for the calyx and corolla together, i.e. the sepals and petals. Some species have their sepals and petals fused into a tissue that is indistinguishable from petals, each one called a perianth segment (tepal), and collectively called a fused perianth.

Periclinal Describing cells with cell walls being laid down in line with (parallel with) the root:shoot axis.

Pericycle A single layer of parenchymatous cells situated just below the endodermis of roots that is able to become meristematic and produce new root initials.

Permanent wilting point The soil is at permanent wilting point when it can no longer provide sufficient water to support plant life.

Permeability The ease with which water passes through the soil.

Petals Leaf-like structures that are the individual parts of the corolla and usually form the highly coloured parts of a flower for insect attraction.

Petiole The leaf stalk of a simple leaf. The petiole attaches the leaf lamina to the stem and scribes an angle with the stem known as the axil.

pH The acidity or alkalinity of a soil measured on a scale of 1 to 14 with 7 as neutral, above 7 as alkaline (with graduations of alkalinity up to 14, the most alkaline), and below 7 as acidic (the most acidic is therefore 1).

Phelloderm Sometimes known as the cork cortex – produced by the internal tissues of the phellogen within the bark-producing periderm.

Phellogen Sometimes known as the cork cambium. A sheet of meristematic tissue responsible for producing new cork cortex (phelloderm) on the inside, and new cork (phellum) on the outside – that along with phloem forms the periderm.

Phloem Tissue comprising open, elongate, dead cells with sieve plates (partially deteriorated end cell walls), associated live companion cells, and sometimes bundles of fibres. The fibres are dead cells and give structural strength and flexibility, and although the phloem vessels are also dead, their tube-like nature is responsible for the downward movement of dissolved sugars from the leaves. The function of the companion cells is unclear.

Photoperiodism A plant's response to day length – the number of hours of daylight.

Photosynthesis The process of sugar production fuelled by the sun in the presence of chlorophyll as a catalyst, and carbon dioxide.

Phylloclade *See* **Cladode**

Phyllode Modified petiole or rachis that is flattened to look like a leaf blade.

Piliferous layer (piliferous area) The area of root hairs on the epidermis of a root.

Pinnate A form of compound leaf with leaflets radiating each side of a central stalk – an extended petiole called a rachis.

Pistil The main female sexual part making up the gynoecium. Each pistil may comprise a single carpel, or more than one carpel that may be free-standing or fused together.

Placentation The arrangement of ovules within a carpel.

Pollen The material exuded by the male organs and responsible for fertilizing ovules.

Pollen donor The plant whose male organs of the flower are used in plant breeding to supply the pollen.

Pollen recipient The female (or mother) plant (egg donor) that receives pollen in plant breeding.

Pollen sacs The tissue sacs forming the anther of a stigma and which contain pollen.

Pollen tube The result of a pollen grain germinating. The pollen tube moves down the style of the carpel towards the ovary.

Pollination The transfer of pollen from an anther to a stigma.

Polycotyledonous Trees in the Gymnospermae (conifers) that have several seed leaves in their embryo.

Pome A form of false fruit (pseudocarp), where the receptacle rather than the carpel swells to form a succulent fruit; typified by apple and pear.

Primary growth Growth responsible for the increase in length of stems and roots. Primary growth features soft, parenchymatous tissues, produced at the apical meristem in stems, and the root-tip meristem in roots. Primary growth in stems is chlorophytic – not so in roots.

Procambium The initial undeveloped tissue found at all stem and root apices responsible for developing into the cambium and ultimately laying down all tissue patterns of the stem and root.

Promiscuity The propensity for many species to hybridize readily.

Protandry Where the male parts of a flower (anthers of the stamen) ripen before the female parts. *See also* **Protogyny**.

Proteaceae The family of plants that includes *Grevillea*, *Banksia*, *Embothrium* and *Telopea*.

Protogyny Where the female parts of a flower (the stigma of the carpel) ripen before the male parts. *See also* **Protandry**.

Pseudocarp A false fruit formed by any part other than the carpel, e.g. apples (pomes) a swollen receptacle enclosing the true fruit (the carpel forming the core).

Pteridophytes Non seed-bearing plants in the Pteridophyta, i.e. ferns (having fronds that look like wings): ptera = wings.

Quiescence (quiescent zones) Periods or areas of 'quietness' (non-activity), e.g. quiescence in seeds when they are inactive because all of the necessary conditions of germination have not been met; and the small area of low growth in a young root, i.e. the quiescent zone.

Racemose arrangements Racemes are non-branching, and in racemose arrangements the oldest flowers are at the base of the inflorescence.

Rachis The central leaf stalk (an extended petiole) of compound leaves and also technically the main non-branching flowering stem from which peduncles (branching stem) and pedicels (actually attached to the flower) arise.

Receptacle Swollen, fleshy end of a flower stem to which all the other floral parts are attached.

Respiration The energy-using process that requires oxygen and breaks down sugar substrates (produced by photosynthesis) to release their locked-up energy to the plant. Carbon dioxide is released during the process.

Reversion The process of a variegated plant reverting back to green.

Rhizobium These are the bacteria found in the root nodules of leguminous plants (e.g. clover, laburnum, alfalfa and cercis) that 'fix' elemental nitrogen. Rhizobium have a symbiotic (mutually beneficial) relationship with the roots of leguminous species taking sugars from the roots, and in return they process elemental nitrogen and change it into soluble forms that the plant can take up and use.

Rhizome A specialized storage organ comprising a condensed underground stem, as found in some ferns including bracken, couch grass, bindweed, bamboo and banana.

Rhizomorphs Cream to brown, rod-like growths formed from aggregations of mycelia that travel beneath the soil to infect the next host. They are typical of honey fungus (*Armillaria mellia* – and related species). Once the rhizomorphs enter the tree below the bark they form black 'bootlace' structures that give the fungal species its other common name (bootlace fungus).

Ring barking Damage to trees (including by insects to young trees) where the damage girdles the entire circumference of the tree.

Root architecture A description of the shape of root-plates/root crowns.

Root cap The renewable tissue that protects the very end of the root and is sacrificed to prevent damage to the root tip (which contains the very important root-tip meristem).

Root crown (root-plate) The extensive root system of a tree, from the thick, corky, non-absorbing roots meeting the stem, to the fibrous absorptive roots at the perimeter.

Root hairs Specialized epidermal cells that are responsible for water uptake from the soil matrix. The zone of root hairs on the root is above the area of elongation, (i.e. in the zone of differentiation), and is called the piliferous layer or piliferous area.

Root tip The growing tip of the root that is situated just behind the root cap and carries the root-tip meristem responsible for root growth.

Root-tip meristem *See* **Meristems**.

Samara Wing-like outgrowths found on some fruits and seeds, e.g. the double samara (two wings) of maple fruits that aid their dispersal.

Scale leaves Scale-like, modified leaves (cataphylls), usually covered in waxes (e.g. suberin), that form a protective layer covering vegetative and mixed buds (bud scales) in deciduous subjects. Also forming a papery outer covering to bulbs and corms (papery scale leaves) that protect the concentric rings of scale leaves inside tunicated bulbs (fleshy scale leaves).

Sclerenchyma (sclerenchymatous tissue) Tissues that are heavily lignified and include nut shells, fibres, conifer tracheids and xylem vessels.

Secondary growth (secondary tissues) Tissues responsible for an increase in girth in the stems and roots of trees and shrubs. Primary growth is responsible for an increase in stem length, secondary growth does not affect stem length, only girth. The addition of secondary xylem accounts for the main increase in stem and root girth.

Secondary infection An infection that occurs after the main pathogen has weakened the plant. A good example is *Armillaria gallica*, only infecting a tree when the tree is already in a state of demise.

Seed An embryo plant developed from a fertilized ovule.

Sessile Without stalks. Can refer to leaf stalks (petioles) or flower stalks.

Sexual cells Cells from the sexual organs of the flower and carrying either male or female genetic information. Pollen cells produced in the anthers of the stamen and egg cells produced in the ovules.

Shrub A woody perennial with many persistent woody stems arising from, or near, ground level.

Simple leaves Leaves may be simple – having an entire leaf lamina (blade), either sessile or attached to the stem by a leaf stalk (leaf petiole); or compound – having a central rachis (an extension of the petiole) with leaflets.

Snake bark (striated bark or striped bark) A generic term for a number of maple (*Acer*) species that have a striped bark effect – particularly noticeable on younger wood. Most are Asiatic species, but one notable exception is North American (moosewood – *Acer pensylvanicum*). The Asian species include *Acer capillipes*, *Acer davidii*, *Acer forrestii*, *Acer rufinerve*.

Soil moisture deficit The water precipitation minus water evaporation and plant water loss to the atmosphere.

Soil saturation Soil is saturated when all the pores (macropores, mesopores and micropores) are filled with water – hence oxygen levels will be low.

Soil structure Concerns the aggregate lumps (macroaggregates) and the large pore spaces (macropores) formed by the bridging aggregate lumps. Soil water can move out of large pore spaces by gravity (air takes its place), whereas soil held by the fine pore spaces (micropores) within aggregate lumps will not move out by gravity; water held by mesopores can be removed by osmotic pressure from plant roots.

Soil texture The feel of a soil, which varies depending on the various mixtures of particles present (particularly clay).

Somatic cells Cells produced asexually by mitotic division, that make up the tissues of the main body of the plant. *See also* **Sexual cells**.

Spadix The rod-shaped structure containing the sexual organs found in the centre of the spathe in members of the Araceae.

Spathe A leaf-like, inverted cone-shaped structure that forms the outside tissues of the flower of plants in the arum lily family (Araceae).

Species A group of similar plants classified within a genus. Being a species denotes that it can be found in the wild somewhere in the world.

Specific epithet The name that accompanies the generic name to positively classify a plant. The two together (generic and specific – binomial system) are known as the specific name.

Spermatophytes A group of plants in the main phylum Spermatophyta – the seed bearers. Other phyla (plant divisions) include the non-flowering (non-seed bearing) Bryophyta (mosses and liverworts), and the Pteridophyta (ferns).

Spike A sessile raceme – the upright stem of flowers maturing at the bottom first and with no pedicels (short stalks).

Spongy mesophyll Middle tissues found in mesophytic leaves that are large open tissues specially adapted for gaseous exchange, and associated with the stomata in the lower epidermis.

Standard tree A common form of trained tree with a clear 1.8m leg.

Stamen The male organ of an angiosperm flower comprising the anther and filament. The anther comprises long pollen sacs and is attached to the stalk-like appendage called the filament.

Stele The central 'core' of a root comprising the vascular tissues.

Stigma The receptive part of the female organ of an angiosperm flower. If conditions are correct, the receptive stigma will receive pollen prior to fertilization.

Stipules A form of cataphyll comprising small, leaf-like outgrowths on the stem that lie as flat flaps near the node or near condensed nodes at buds, e.g. *Pelargonium zonale*.

Stoma (plural Stomata), also Stomatal pore**:** The pore-like orifices found on green stems and leaves, regulated by guard cells, and responsible for gaseous exchange.

Stomatal cavity Open areas within the spongy mesophyll associated with stomata, and that facilitate gaseous exchange at each stoma.

Stratification Placing seeds (or complete fruits) inside a container in alternate layers of sand and leaving in an open frame to weather to break dormancy.

Striated Bark (Striped bark/snake bark) Generic terms for a number of maples (*acers*). *See* **Snake Bark.**

Style A hollow, filamentous tube that is part of the carpel and terminates in the stigma at one end and the ovary at the other.

Suberin A waterproofing wax found within the cork cells of bark, covering the bud scales of some species, and in the casparian strip of roots. The process of laying down suberin is suberization. *See also* **Lignin** and **Cutin**.

Sucker A form of aerial shoot arising from a root-plate or root system. Also used for aerial shoots arising from the root system (or even the base of the stem) of the under-stock of a grafted or budded plant.

Succession The process of successive groups of plants thriving in their environment, and in so doing, preparing the environment for the next successive group of species. Herbaceous marginal plants creating deeper aerated silt, that ultimately colonizes with shrubby species, which in turn colonizes with higher arboreals (trees), is an example. The final colonization is by climax vegetation.

Succulents Group of plants from arid environments with water-storage systems. Cacti have no leaves as they have evolved to spines and the swollen, green, leaf-like stems store water and photosynthesize. Leaf succulents store water in their swollen leaves.

Superior ovary An ovary of a flower situated above (in front of) the point of attachment of the other floral parts. *See also* **Inferior ovary.**

Symbiotic relationships Relationships between two organisms that are mutually beneficial, (e.g. mycorrhiza on roots, and algae and fungi forming lichens).

Symplast (symplastic system) The system of connecting live tissues in plants that allows chemical messages to be transported. The symplast in trees comprises the interconnection of cork cambium, phloem parenchyma, vascular cambium, xylem parenchyma, and parenchyma rays. *See also* **Apoplast**.

Sympodial A description of growth patterns that occur in more than one direction and form complicated systems of branching.

Tactile test The test for soil texture by smearing a soil sample between finger and thumb and testing to see how easily it moulds into a shape and then breaks back down again. Clay soils feel smooth and can 'polish', silty soils feel 'silky', and sandy soils feel 'gritty'. The adhesive and cohesive properties of clays make clay soils easy to mould into a shape, but difficult to break down again. Silty soils have some cohesion, but moulded shapes can be broken down relatively easily. Sandy soils are not usually cohesive enough to be moulded into a shape. Loams (comprising a balanced mixture of clay, silt, sand and humus) are cohesive enough to form a moulded shape, but also break down to their component aggregate lumps easily.

Taproot The principal root that develops from the emerging radicle of a seed, which may or may not persist to later stages of development. They form a method of perennation in some biennials and some herbaceous perennials.

Target pruning Removal of a branch using a final cut to remove the lateral branch that stands off the branch collar (leaving it intact), and is made at an angle that mirror images the angle of the branch bark ridge.

Taxonomy The systematic study of plants involving identification (recognition), classification (the placing of plants into their position in the main taxonomic hierarchy) and nomenclature (the system of naming plants).

Tilth Relating to the size of the aggregate lumps (the crumb structure) of soils – coarse, medium or fine.

Tomentosum Congregations of relatively coarse hairs found on the topside of some leaves and on some stems that aid the reduction of water loss.

Totipotency The phenomenon that all individual parenchymatous plant/tree cells, no matter which part of the plant/tree they come from, have the ability to divide and differentiate to form any set of tissues, any organ, or complete set of organs that make up the plant/tree. The genetic information is in the DNA in the nucleus of the cell, and the process is triggered by environmental and chemical (hormonal) factors that decide the final outcome.

Tracheid A unit of the xylem found in plants, particularly conifers. Tracheids facilitate upward water transport, but differ from xylem vessels in their smaller diameter, and the fact that they retain some of their cell end walls, meeting at a slope. Their difference in make-up to the larger diameter xylem vessels with no end walls (found in herbaceous and deciduous subjects) is thought to help prevent embolisms (air bubbles) in the xylem in the cold exposed environments common for conifers.

Transpiration The relatively rapid transport of water through the vascular tissues (xylem) of the plant, powered by evaporation at leaf surfaces.

Transplants Seedling plants that have been lifted from the seedbed, transplanted at wider spacing, and grown on before sale.

Tree A woody perennial with a distinct trunk or trunks.

Tropisms Responses by the plant towards (positive) or away from (negative) a stimulus (such as light).

Tunica The tissues outside the corpus at a stem apex responsible for production of the epidermis and outer tissues of the stem. *See also* **Corpus**.

Tunicated A form of bulb bearing a papery outside scale leaf (the tunic), usually brown, e.g. *Narcissus*, or may be coloured, e.g. *Hyacinthus*.

Vacuolation The middle phase of cell maturation where the vacuole forms and the cytoplasm becomes a thin membrane plastered to the cell wall. The vacuole fills with cell sap.

Vascular bundle A compact bundle of veins found in the primary growth of stems. Secondary growth sees the once individual bundles coalesce and become indistinguishable from one another, ultimately forming a complete ring of tissue.

Vascular cambium The ring (later cylinder) of tissues comprising brick-like meristematic cells situated between the xylem and phloem, and ultimately responsible for secondary growth and girth increase.

Vascular tissues The fluid-transporting, vein-like tissues of the plant often arranged in bundles and comprising xylem, phloem, vascular cambium and associated non-transporting fibres.

Vector A carrier, such as wind or an insect that carries a disease (viral, bacterial, or fungal) from plant to plant.

Vegetative Referring to growth that is purely vegetative (not flowering), e.g. vegetative buds responsible only for extension in stem length. Vegetative is also used to describe methods of asexual propagation that use pieces of vegetative (living/growing) parent material, rather than seed.

Vegetative buds Buds responsible for new stems only, i.e. not flowers, as these require flower buds or mixed buds.

Vegetative growth Stem tissue bearing leaves, buds etc.

Viability Referring to the life of seeds, i.e. how long (the longevity) a seed will remain viable (live) and therefore still able to germinate.

Water loss from the soil The three main ways water is lost from the soil are by drainage (via gravity from the macropores), via evaporation from the soil surface, and by transpiration of plants (including trees) via the transpiration stream.

Whips Seedling trees grown on to a larger size (1m approx). Feathered whips have side growths ('feathers').

Xanthophylls Plant pigments in the carotenoide group responsible for yellow and gold coloration.

Xerophytic adaptations Adaptations of plants to reduce water loss caused by arid or exposed conditions. Examples include indumentum, cacti having water – storing stems and no leaves and Conifers bearing needles (instead of large moisture-losing leaves), resin ducts, sunken stoma, and heavily waxed needle surfaces.

Xylem Tissue comprising groups of open, tube-like cells (xylem vessels), responsible for the upward movement of water and dissolved nutrient salts through the plant. Xylem vessels may have several types of cell-wall thickening (lignification), with specific names for the various lignin patterns, e.g. annular, spiral, reticulate and pitted.

Zygomorphic The name used to describe irregular flowers, i.e. those flowers that can only be divided into two equal halves by one particular line drawn through the centre.

Index of Generic and Common Plant Names

A plant being listed does not necessarily represent any discussion of its features as it may only be mentioned because of its place in the taxonomic hierarchy, or as an example, and it may not be discussed in detail. Page numbers in bold indicate that the plant (or part of the plant) is illustrated in some way – either as a photograph or drawing.

Index